CONSERVATION BIOLOGY

Foundations, Concepts, Applications

FRED VAN DYKE
Wheaton College

KRISTA L. CLEMENTS
Editorial Consultant
University of Arkansas

Boston   Burr Ridge, IL   Dubuque, IA   Madison, WI   New York   San Francisco   St. Louis
Bangkok   Bogotá   Caracas   Kuala Lumpur   Lisbon   London   Madrid   Mexico City
Milan   Montreal   New Delhi   Santiago   Seoul   Singapore   Sydney   Taipei   Toronto

McGraw-Hill Higher Education

*A Division of The **McGraw-Hill** Companies*

CONSERVATION BIOLOGY: FOUNDATIONS, CONCEPTS, APPLICATIONS

Published by McGraw-Hill, a business unit of The McGraw-Hill Companies, Inc., 1221 Avenue of the Americas, New York, NY 10020. Copyright © 2003 by The McGraw-Hill Companies, Inc. All rights reserved. No part of this publication may be reproduced or distributed in any form or by any means, or stored in a database or retrieval system, without the prior written consent of The McGraw-Hill Companies, Inc., including, but not limited to, in any network or other electronic storage or transmission, or broadcast for distance learning.

Some ancillaries, including electronic and print components, may not be available to customers outside the United States.

 This book is printed on recycled, acid-free paper containing 10% postconsumer waste.

International   1 2 3 4 5 6 7 8 9 0 QPD/QPD 0 9 8 7 6 5 4 3 2
Domestic   1 2 3 4 5 6 7 8 9 0 QPD/QPD 0 9 8 7 6 5 4 3 2

ISBN 0–07–239770–5
ISBN 0–07–119906–3 (ISE)

Publisher: *Margaret J. Kemp*
Senior developmental editor: *Kathleen R. Loewenberg*
Marketing manager: *Heather K. Wagner*
Senior project manager: *Mary E. Powers*
Senior production supervisor: *Sandy Ludovissy*
Senior media project manager: *Tammy Juran*
Lead media technology producer: *Judi David*
Designer: *K. Wayne Harms*
Cover/interior designer: *Kay Fulton*
Cover image: *Scott MacButch/Ecotone*
Senior photo research coordinator: *Lori Hancock*
Photo research: *Billie Porter*
Supplement producer: *Brenda A. Ernzen*
Compositor: *GTS Graphics, Inc.*
Typeface: *9.5/11.5 Times Roman*
Printer: *Quebecor World Dubuque, IA*

The credits section for this book begins on page 399 and is considered an extension of the copyright page.

Library of Congress Cataloging-in-Publication Data

Van Dyke, Fred, 1954–
   Conservation biology : foundations, concepts, applications / Fred Van Dyke. — 1st ed.
      p.   cm.
   Includes index.
   ISBN 0–07–239770–5 — ISBN 0–07–119906–3 (ISE)
   1. Conservation biology. I. Title.

QH75 .V37   2003
333.95′16—dc21

2002070880
CIP

INTERNATIONAL EDITION ISBN 0–07–119906–3
Copyright © 2003. Exclusive rights by The McGraw-Hill Companies, Inc., for manufacture and export. This book cannot be re-exported from the country to which it is sold by McGraw-Hill. The International Edition is not available in North America.

www.mhhe.com

DEDICATION

To Linda,
who has been with me
in trial and triumph.

CONTENTS

FOREWORD

When I teach biology at Oxford University, I ask my students to write an essay that answers the question, "Why should I give a damn about biodiversity?" To answer that question, the students must ask themselves whether we need all the species on the planet, and if so, why, and if not, why not? Their essays always contain arguments along the lines that we should conserve biodiversity because it is useful to us. However, that is a difficult argument to win since all species are not useful to us. The real reason that conservation of biodiversity has value is because we say it has value. This statement is an amalgamation of Michael Soulé's *normative postulates of conservation* (something you will learn about in chapter 1).

By now you are probably asking yourself, "What's all this stuff about values? I thought this was a biology course, not a philosophy course." Well, that's what makes conservation biology special as a discipline; it is *value laden*. It is also what makes the term *conservation biology* somewhat of a misnomer, and the subject rather difficult to teach. Conserving biodiversity is much more than the science of maintaining viable populations of all species. Conservation is also moral philosophy, social justice, economics, and politics. Finding a lecturer in a biology department who is well versed in all these areas is a rare thing indeed. Perhaps as a consequence, biology lecturers have tended to stick to the biological issues that concern conservation: management of populations, communities, and ecosystems; habitat restoration; causes of extinction and declining populations. This is certainly what is reflected in the last generation of conservation textbooks. But this is a narrow, unsatisfying, and perhaps misrepresentative view of the discipline. When taught in this way, conservation seems intellectually depauperate—the slightly awkward offspring of ecology. This is a great disservice to both the discipline and to the students who come to learn it. When viewed more broadly and more critically, conservation is one of the most intellectually stimulating subjects you can study!

As a lecturer who has struggled with many of these issues, and been frustrated by the textbooks previously available for undergraduate courses, I am extremely excited about this new text. It is the first to deal critically with the underlying philosophical issues, and to integrate policy and economic perspectives with what is otherwise a refreshing, up-to-date treatment of the fundamental biological principles. I am even more exited by Dr. Van Dyke's efforts to explain the intellectual antecedents of the core ideas in conservation, and to place these in their historical context. Historical perspective is extremely important for understanding the intellectual development of a discipline and its role in the world.

You, the student, will find this text unique because it is practical—not in the sense that a cookbook is practical, but in the sense of envisioning yourself as a contributor to the great public debates on the future of our environment, regardless of your chosen profession. If you hope that your future profession will be that of a conservationist, then there is even more great advice and intellectual stimulation to be found in these pages. Chapter 13 in particular is a spirited treatise on the requirements for students to become effective actors in conservation rather than just concerned citizens.

It is rare that a lecturer is excited about a textbook, but I am excited about this one. In the end, we often choose a course text because it is the least dissatisfying of the lot, and we really wish we had the time and energy to write a textbook that better suits the course as we envision it. In this respect, this book is a rare gem. I can't imagine doing a better job than Dr. Van Dyke has done with this book, and I am genuinely looking forward to using it in my course. I'm sure that you will find it as intellectually stimulating as I did.

Jonathan Newman
University of Oxford

PREFACE

Few things are as exciting as seeing a new insight in its original form, with all the ambiguity and uncertainty that accompanies the first articulation of an idea. For that reason, I have always loved studying biology through its primary literature, and I've always hated biology textbooks. Most of the textbooks I read as a student and have used as a teacher resembled well-ordered intellectual cemeteries, not gardens of ideas in bloom. These texts presented their concepts as a kind of catechism in which students were taught to recite the "correct" responses to formalized questions whose answers were known long ago. The body of ideas that had once held the life of the discipline was dismembered into chapters that bore little apparent relation to one another.

WHY I WROTE THIS TEXTBOOK

I believe that conservation biology should be taught as a vital unity of thought and practice expressed through a coherent foundation of concepts, theories, facts, and values, not as a loose assemblage of impressive, but dismembered disciplinary expertise. A unified textbook of conservation biology does not attempt to present every subject that conservation biologists have studied, but instead defines the context and relationships of controlling ideas, problems, and applications of the discipline. Critical facts and case histories are important, but are meaningless without context. We do not remember facts that we memorized years, or even days, ago simply for the purpose of passing a test. Rather, we remember information that skilled teachers imbued with meaning that inspired and enlightened us, and that led to an understanding of our own discipline, and our place and purpose in it.

HOW THIS TEXT IS ORGANIZED

Guided by these convictions, I have organized this textbook around three fundamental questions about conservation biology that give meaning to diverse research and management efforts. Readers should proceed with these questions in mind. First, how did conservation biology become a distinct discipline, and what keeps conservation biology from being absorbed into related disciplines? Second, what are the fundamental intellectual, conceptual, and practical problems that conservation biologists must address and solve? Third, what is the role of conservation biology in achieving "success" in conservation in ways that affect all dimensions of the human experience?

The question of conservation biology's origin and distinctions is examined in Part One, Foundations. Here, readers explore the recent history of conservation and conservation biology (chapter 1); the legal foundations and policies that empower conservation (chapter 2); the sources of ethics and values that give meaning, depth, and direction to conservation biology's goals (chapter 3); the emphasis conservation biology places on the concept, measurement, and value of biodiversity (chapter 4); and the intellectual paradigms that have formed conservation biology's theoretical foundation and directed its research agenda (chapter 5). Part Two, Concepts, explores the more detailed conceptual, theoretical, and analytical framework of basic problems that conservation biology attempts to solve. These chapters examine ways to conserve genetic diversity (chapter 6), conserve populations and species (chapter 7), protect and manage terrestrial and aquatic habitat (chapters 8 and 9), and manage ecosystems (chapter 10). The final section (Part Three, Applications) examines, in three chapters, some ways to apply the fruits of research and management at various scales. First, I explore direct applications of conservation science to the problems of restoring species, habitats, and ecosystems (chapter 11). Second, I consider ways to achieve broader, socially based applications of conservation to problems of economics and sustainable development (chapter 12). Finally, I describe practical ways for students to make career commitments to conservation, to function effectively within a community of conservation professionals, and to prepare for emerging trends in the field (chapter 13).

A PEDAGOGY THAT ENCOURAGES REFLECTION

In a classroom setting, education has two dimensions—presentation and process. Textbooks and teachers have traditionally emphasized the former and neglected the latter.

Presentation is important, and I have done my best to organize ideas carefully and make every sentence count. However, presentation of information can occur with incredible speed (especially if the speaker is armed with good presentation software and a laptop) without any learning taking place. Process, not presentation, is learning's limiting step. Thus, I interrupt each chapter at certain points with a question or related group of questions, boxed in inserts labeled **Points of Engagement.** These questions raise issues and problems about material that has been presented, and provide students an opportunity to work with, reflect upon, or discuss concepts and their implications before proceeding further. I am sure that many instructors can improve upon my questions or replace them altogether with substitutes of their own. But I do recommend the technique to engage students regularly in the flow of ideas.

For similar reasons, and because of my enthusiasm for primary literature, at the end of each chapter one will find four groups of questions arranged to facilitate **Directed Discussions** about assigned readings from peer-reviewed journals. These differ from the usual Questions for Discussion that appear at the ends of chapters in most textbooks. In the format you will find here, each question leads logically to the next, and students must spend time reading scientific literature to see how ideas were initially presented. I admit this approach requires more effort from instructors and students, but unprepared, speculative, and unfocused discussions have been of little value in my own teaching and preparation of students for a professional environment.

In addition to the two important features above, each chapter opens with a thoughtful **Quote** and a list of **Learning Objectives,** written in such a way as to encourage active learning and critical evaluation. Important **Key Terms** are highlighted to signify definitions and help students review chapter material. These key terms are also defined in the end-of-book **Glossary.** Finally, a comprehensive list of **Literature Cited** concludes each chapter and offers the student a bibliography to aid research or further reading.

WEB-ENHANCED

The World Wide Web is a valuable resource for students and instructors, and offers vast amounts of information to supplement what you will read in this textbook. At the end of each chapter there is a list of topics that relate to and enrich the material just covered. This list of **Web Resources** is repeated as hot links on the book's accompanying website. The website also offers test questions, images to use in classroom presentations, links to professional organizations, global issues and case studies, and animations.

UNIQUE APPROACH

This text strives to be genuinely interdisciplinary in its approach. Although an understanding of biological facts and concepts is essential and given pride of place, the text takes seriously the contributions of law, political science, economics, ethics, sociology, and other disciplines to the modern conservation effort. Additionally, my book does not conceal issues of ambiguity and uncertainty in conservation science, or issues of controversy in conservation ethics and policy. I do not believe that we should shield students, at any level, from the inherently controversial, often contentious, nature of the scientific effort, and we should not try to protect them from the messy uncertainties that inevitably arise when we attempt to translate research results into management decisions and policy directives. Many years of working with state and federal conservation agencies, combined with the shared experiences of other conservation professionals whom I respect, have convinced me that such uncertainty is best acknowledged quickly and forthrightly, lest it lead to recommendations that are more precise than accurate, and produce professionals who are more arrogant than useful. Thus, this text not only explains, but also critiques, the foundational practices, techniques, and concepts of conservation biology. My purpose in taking this approach is not to create a spirit of negativism or confusion, but to provoke current and future conservation biologists to examine their foundational premises carefully and make continued efforts to improve all aspects of conservation practice.

This text also takes a stand on some issues. My intent in doing so is not to indoctrinate students or to eliminate the possibility of alternative positions. Indeed, I do my best to present fairly all reasonable alternatives in serious controversies, but cases will arise in which action becomes impossible without commitment. Different commitments lead to different actions, and different actions lead to different outcomes, but such commitments are unavoidable and essential to the practice of conservation biology.

ACKNOWLEDGMENTS

I am most grateful for the reviewers who noted faults in earlier drafts of this text. These excellent scientists and teachers supplied many helpful suggestions to improve the book, and I have done my best to follow their advice. Errors that remain are entirely my responsibility. Kathy Loewenberg, Mary Powers, Marge Kemp, Dianne Berning, and others at McGraw Hill persevered with me to make the book more attractive and functional for every reader. Elizabeth Truesdell, Brian Darby, Nathan De Jager, Krista Anderson, and David Hoekman contributed greatly to the preparation of figures and tables, and suffered long and without complaint under the many demands I placed on their time and talent. Teresa Cerchio assisted me with many final adjustments of tables and figures, often on short notice, arranged innumerable express mailings of manuscripts, and was an unfailing source of cheer and encouragement under the most stressful conditions. My greatest debt is to Krista Clements, who served from start to finish as my editorial advisor and consultant. Her superb editorial skills transformed many pages of hopelessly muddled expression into readable and professional prose and corrected innumerable errors in the process. Her ruthless and relentless editor's eye, however, was bound to a kind and encouraging heart that helped me persevere throughout the many discouragements of this effort. Finally, I am most grateful to my wife Linda for her patience and support, especially through the final months of work when I was rarely seen at home. I hope to make amends for this neglect in the coming year.

Fred Van Dyke

Reviewers

Vickie L. Backus
  Middlebury College
Lawrence S. Barden
  *University of North
  Carolina, Charlotte*
David Ehrenfeld
  *Cook College, Rutgers
  University*
Laura Jackson
  *The University of
  Northern Iowa*
Frances C. James
  Florida State University
Susan R. Kephart
  Willamette University
Mark E. Knauss
  Shorter College
Jonathan A. Newman
  University of Oxford
Richard A. Niesenbaum
  Muhlenberg College

Kristian S. Omland
  Union College
Andrew G. Peterson
  Columbia University
William F. Porter
  *State University of New
  York, College of
  Environmental Science
  and Forestry*
Eleanor J. Sterling
  *American Museum of
  Natural History
  Center for Biodiversity
  and Conservation*
Stephen C. Trombulak
  Middlebury College
Robert W. Yost
  *Indiana
  University–Purdue
  University, Indianapolis*

ABOUT THE AUTHOR

Fred Van Dyke is a professor of biology at Wheaton College (Illinois). He has previously served on the faculties of Northwestern College (Iowa) and the Au Sable Institute for Environmental Studies, as a wildlife biologist for the Montana Department of Fish, Wildlife and Parks, and as a scientific and professional consultant to the U.S. National Park Service, the U.S. Forest Service, and the Pew Charitable Trust. He is the author of numerous publications on animal home range and habitat use, management and conservation of animal populations, management of successional processes to conserve habitat, and conservation values and ethics.

PART ONE

Foundations

One of the definitions of *foundation* is "the basis upon which something stands or is supported." The first five chapters of this book are an invitation to explore the basis of that support in conservation biology. Foundational perspective is essential, because only by knowing the past, can we place ourselves in the present, and only in that perspective can we act with foresight toward the future. Examining these foundations help us to appreciate the historical, social, and scientific origins of a discipline in order that we might better understand why it arose, why it persists, and how it may develop.

Each chapter in this section examines a different foundational element of conservation biology. Chapter 1 explores the historical events that created an environment in which conservation biology was able to emerge as a distinctive discipline, and explains why its distinctive traits are, in part, a reflection of its history. Chapter 2 examines modern conservation law and policy. This chapter is an intentional reminder that conservation biology did not grow in influence and prestige only, or even primarily, through scientific achievement. Rather, the scientific achievements of conservation biology became significant to non-scientists largely because a growing body of law and policy elevated these achievements to a position of influence in government regulation and political process. Conservation biologists who do not understand this interaction do not understand their discipline. Without such understanding, one cannot be effective in transforming research findings into management policies and legal directives. Chapter 3 examines the values and ethics upon which conservation biology is based. Some hold the view that the role of science, and scientists, is merely to provide decision makers with value free information, upon which they will then determine management actions. Whatever else might be said of this view, it fails to accurately reflect conservation biology's historical understanding of itself, just as it also fails to apprehend the historical and social forces that gave conservation biology engagement with law and public policy. Such a view even fails to understand what management actions in conservation are. Every management action is an expression of a value judgment, and every such judgment must be explained and defended to a skeptical, non-scientific public, who will not be content to simply hear *what* action was taken, but will want to know *why* it was taken. Examination of values and ethics in conservation biology will help you to acquire the theoretical background and develop the intellectual skills necessary to evaluate a variety of ethical positions that articulate the values that motivate conservation biologists to support specific management policies and long-term research efforts.

Chapter 4 examines in detail what is arguably the most pervasive and normative concept of conservation biology; the preservation of biodiversity. Biodiversity is a term the public often misunderstands and misperceives as a faddish label for attempts to preserve absolutely everything everywhere. Chapter 4 engages this often misunderstood concept by carefully and critically exploring what biodiversity is (definition), how it is assessed with scientific method and precision (measurement), meaningful reasons why it should be preserved (value), and where we are likely to find it (distribution). Finally, chapter 5 concludes this section with an examination of conservation biology's most important scientific paradigms and their intellectual development. Here we examine not a history of events, as in chapter 1, but a history of ideas, ideas that came to form the theoretical basis for the research and management agenda that characterizes what conservation biology is today.

CHAPTER 1

The History and Distinctions of Conservation Biology

Every great movement must experience three stages; ridicule, discussion, adoption.
—John Stuart Mill

In this chapter, you will learn about:

1 the origins and history of conservation and conservation biology.

2 the conceptual distinctions and distinguishing scientific paradigms of conservation biology.

3 the kinds of problems that conservation biologists investigate and attempt to solve.

FUNDAMENTAL PERSPECTIVES AND QUESTIONS FOR AN INQUIRY INTO CONSERVATION BIOLOGY

In 1978, at a banquet gathering of academic scientists, zookeepers, and wildlife conservationists at the San Diego Wild Animal Park, biologist Michael Soulé made an impassioned plea to his colleagues: with the world experiencing an epidemic of species extinction, it was time for academics and conservationists to join forces to save threatened and endangered species. Soulé's words sparked both controversy and criticism, but few were left unmoved. That meeting, now ambitiously called the First International Conference on Conservation Biology, led to new beginnings. A new scientific organization, the Society for Conservation Biology, and a new discipline, conservation biology, were born.

By 1992, the Society for Conservation Biology had more than 5,000 members. Sixteen graduate programs in conservation biology were operating at major universities, and more than $2.4 million in competitive funding in conservation biology was being made available from the National Science Foundation (Gibbons 1992). By 1996, conservation biology boasted a number of undergraduate textbooks (Cox 1997; Primack 1993; Hunter 1996; Meffe and Carroll 1997) and dozens of titles on more advanced dimensions of the discipline. Its growth was, in the words of founding member Stanley A. Temple, wildlife ecologist at the University of Wisconsin–Madison, "incredible" (Gibbons 1992).

Conservation biology is defined by some as a "crisis discipline" whose goal is to provide principles and tools for preserving biodiversity (Soulé 1985). It is to biology and ecology what surgery is to physiology or war to political science in the sense that it often must respond to emergency situations with incomplete information. But even though crisis oriented, conservation biology always has a long-term goal in view: the viability and persistence of functioning ecosystems (Soulé 1985). Conservation biology is synthetic, eclectic, and multidisciplinary in structure; it does not draw all its theories and models from, or make all of its applications to, biology. Thus, it has the effect of bridging the dichotomy between pure and applied science. In the eyes of its founders, conservation biology intentionally stresses *intrinsic values* (values associated with the inherent worth of an object or quality) of biodiversity and entire communities, not simply economic or utilitarian values (Soulé 1985). The first step toward understanding what conservation biology *is* begins with an understanding of the field's unique conceptual and intellectual attributes.

Although conservation biology is arguably a new and distinctive discipline, few, if any, of the major scientific paradigms employed in conservation biology are original, nor are the problems it attempts to solve. Conservationists, biologists, and conservation dilemmas existed long before there was conservation biology, and there were many individuals who practiced conservation as biologists. The fundamental questions we will answer in this chapter are: (1) What are the historical, scientific, and academic origins of conservation biology? (2) What characteristics distinguish conservation biology from its closest scientific relatives, such as ecology, wildlife management, and environmental science? And finally, (3) what are the chief criticisms and potential deficiencies of conservation biology as a discipline, and how can they be addressed in the long run?

ETHICAL ROOTS AND CONCEPTUAL FOUNDATIONS

Conservation biology draws much of its ethical support from three primary ideas. The first is that *all living creatures possess intrinsic value.* By logical extension, assigning intrinsic value to nonhuman creatures can lead to assigning to them certain rights, or at least treating them as if they did have rights. Ethicist Roderick Nash has argued that concepts of the rights of nature grew out of the older concept of "natural rights," the idea that certain segments of society, individual people, or even animals or inanimate objects, possess rights to continuing existence by virtue of the fact that they exist already (Nash 1989). The concept of natural rights, which can be traced ultimately to both Greek and Roman law and Judeo-Christian traditions, forms a foundation for the belief that natural objects and nonhuman creatures have intrinsic value, not merely commodity or utilitarian value (value based on usefulness for humans). The concept is elegantly, if pithily, summarized by that master of the English epigram, Alexander Pope. In his *Essay on Man,* Pope remarked,

> *Has God, thou fool!, work'd solely for thy good,*
> *Thy job, thy pastime, attire, thy food? . . .*
> *Know, nature's children all divide her care;*
> *The fur that warms a monarch, warm'd a bear.* (Pope 1733, cited in Nash 1989).

With "nature's children" clearly alluding to everything alive, it is but a small step intellectually to expand the concept of rights, including legal protections, to nonhuman creatures and nonliving landscapes. Although not all conservation biologists would agree that natural objects have rights, the historical foundations of conservation biology are imbued with this assumption. Aldo Leopold, the father of the "land ethic", believed that the land was more than just a collection of commodities. In his argument for a land ethic, Leopold asserted that it was the land's intrinsic value that led to its "rights." Speaking against those who saw the land only as a repository of "natural resources," Leopold wrote, "A land ethic, of course, cannot prevent the alteration, management, and use of these 'resources,' but it does affirm their right to continued existence, and, at least in spots, their continued existence in a natural state" (Leopold 1966:240). Many of the contemporary laws that give conservation biology its influence and legal power assume, without stating so explicitly, that natural areas and species possess rights. For example, one of the cornerstones of conservation law, the Endangered Species Act, assumes that endangered species have the right to life and liberty, apart from any value or service they provide to humans (Petulla 1988).

Not all of the ethical foundations and values of conservation biology are rooted in the idea of intrinsic value of nonhuman creatures and natural objects. Conservation biology also affirms utilitarian values of biodiversity and ecosystems. A second fundamental idea is that *the physical environment and the living organisms in natural ecosystems perform vital services and produce goods essential to the continuance of human civilizations. Further, the species within the ecosystems are vital for maintaining their function and structure.* This thesis, most clearly articulated by the nineteenth-century conservationist George Perkins Marsh in his classic work, *Man and Nature; Or, Physical Geography as Modified by Human Action* (originally published in 1864) is well supported by both ancient history and contemporary science (Marsh 1965). With abundant historical data, Marsh demonstrated that major declines in human prosperity and in the vitality of traditionally great civilizations were often rooted in environmental degradation. Marsh expressed his ideas in a difficult-to-read style of paragraph-length sentences. But his points eventually emerged. "It appears then," wrote Marsh, "that the fairest and fruitfulest provinces of the Roman Empire . . . was endowed with the greatest superiority of soil, climate, and position, which had been carried to the highest pitch of physical improvement, and which thus combined the natural and artificial conditions best fitting it for the habitation and enjoyment of a dense and highly refined and cultivated population, is now completely exhausted of its fertility, or so diminished in productiveness, as . . . to be no longer capable of affording sustenance to civilized man. If to this realm of desolation we add the now wasted and solitary soils of Persia and the remoter East, that once fed their millions with milk and honey, we shall see that a territory larger than all Europe . . . has been entirely withdrawn from human use, or, at best, is thinly inhabited by tribes too few in numbers, too poor in superfluous products, and too little advanced in culture and the social arts, to contribute anything to the general moral or material interests of the great commonwealth of man" (Marsh 1965:10). Marsh's fundamental concept, that human welfare, prosperity, and even survival are inextricably linked to the natural world, is at the root of the urgency of conservation biology's agenda.

A third foundational concept of conservation biology is that *the physical environment and its creatures add value, knowledge, and meaning to the experience of being human, and to the appreciation of higher values and virtues.* In secular traditions, this perception of value may take the form of claims that human virtue or human civilization is dependent upon contact with the natural environment, such as Thoreau's famous dictum, "In Wildness is the preservation of the world" (Thoreau 1995:672) or Aldo Leopold's claim that "Wilderness is the raw material out of which man has hammered the artifact called civilization" (Leopold 1966:264). Many religious traditions, such as Judaism and Christianity, state the concept more directly and more authoritatively by identifying the physical world as one part of God's self-revelation to humans ("The heavens are telling the glory of God . . .": Psalm 19:1). In eastern religious traditions, the interrelatedness of all things, including and especially human beings and their environment, is repeatedly stressed. In Buddhism, for example, it is articulated as the Law of Dependent Co-Arising that all events and beings are interdependent and interrelated. In such a view the universe is understood as a mutually causal web of relationship, with every action and every individual contributing to the nature and outcome of many others (Kalupahana 1987). When conservation biologists express appreciation for the aesthetic values of the physical world and nonhuman creatures, they draw from and are supported by multiple intellectual and religious traditions. Thus, it is critical to study such sources as carefully and analytically as possible, both for their own sake, and in order to truly understand their contributions to conservation biology. We will review some of this history in brief now, and in detail later in chapter 3, Values and Ethics in Conservation.

THE PROBLEM OF PERSPECTIVE: HOW HAS CONSERVATION DEVELOPED IN HISTORICAL CONTEXT?

Definition and Understanding

To understand the history of conservation, one must begin by defining conservation by its goals, not by its effects. A group of people living in equilibrium with their environment is not necessarily an example of conservation, and such people are not necessarily conservationists (Alvard 1993). They may be very wasteful, but cause no damage because their environment produces more than they need. Or they may be very inefficient, desirous of taking more from the world, but ineffectual in doing so. If their efficiency improves, they will take more, and may do increasing harm to their environment. In this case their "conservation" will be inversely related to their technology. Either way, this kind of conservation will disappear very quickly if the environment becomes less productive or if the people become more efficient at exploiting it.

Genuine and enduring conservation can occur only when humans knowingly use resources at less than maximum sustainable rates or forgo the use of some resources altogether. This kind of conservation is motivated by appreciating an intrinsic value of the resource itself or providing a long-term supply of the resources for others, including others still to come in future generations. These motivations are not mutually exclusive, and both are ethical in nature. Throughout history, human beings have shown themselves to be capable of both of these motives. They also have shown the capacity for intensely selfish motives to get as much of a resource as they can for their own needs and pleasures. The history of conservation is a history of this ethical tension. Conservation has benefits for humans, but it requires restraint and incurs costs.

Problems associated with human use of and relationship to the land and nonhuman creatures are not new, and they are not limited to technological societies. In ancient Greece, Plato compared present land conditions to the past in one of his dialogues through the character Critias. Critias, in a conversation with Socrates, laments of the land that

> . . . of old its yield was most copious as well as excellent
> . . . . By comparison with the original territory, what is
> left now is, so to say, the skeleton of a body wasted by

disease; the rich, soft soil has been carried off and only the bare framework of the district left. . . . There were also many other lofty cultivated trees which provided unlimited fodder for beasts. Besides, the soil got the benefit of the yearly "water from Zeus," which was not lost, as it is today, by running off a barren ground to the sea; a plentiful supply of it was received into the soil and stored up in the layers of nonporous potter's clay. Thus the moisture absorbed in the higher regions percolated to the hollows, and so all quarters were lavishly provided with springs and rivers. Even to this day the sanctuaries at their former sources survive to prove the truth of our present account of the country (Hamilton and Cairns 1961:1216–17)

Before Plato, Judaism extended the religious principle and command of Sabbath to include the land. The nation of Israel recorded God's command, through his prophet Moses, that the land must receive a rest from cultivation every seventh year: ". . . in the seventh year the land is to have a Sabbath rest, a Sabbath to the Lord. Do not sow your fields or prune your vineyards. Do not reap what grows of itself or harvest the grapes of your untended vines. The land is to have a year of rest" (Leviticus 25:4–5). The ancient Israelites, like the ancient Greeks, abused their land and did not keep this commandment, but it was not forgotten. Israel's prophets state that failure to observe the land Sabbath was one of the reasons for Israel's eventual exile from Palestine. Describing the defeat of Judah and the destruction of Jerusalem by Nebuchadnezzar in 586 B.C., the writer of the Second Book of Chronicles records that Nebuchadnezzar

> carried into exile into Babylon the remnant, who escaped
> from the sword, and they became servants to him and his
> sons until the kingdom of Persia came to power. The land
> enjoyed its Sabbath rests; all the days of its desolation it
> rested, until the seventy years were completed in
> fulfillment of the word of the Lord spoken by Jeremiah
> (II Chronicles 36:20–21).

These historical cases from Greece and Palestine are but two of many examples that demonstrate that, whether given a command from God or a rebuke from a scholar, people throughout history have often neglected and degraded the world around them. Neither Plato's remembrances of a better landscape nor religious commands that revealed land as an object of God's care were sufficient to keep people from selfish behavior that degraded their environment.

Many early efforts in conservation were not achieved by ethics, but enforced by punishment and coercion. In many cultures, conservation began with prohibitions against the use of some or all resources on privately owned land by any person except the landowner and his family. For example, European royalty and, later, other wealthy individuals, set aside land as hunting and forest preserves, forbidding "common people" to kill game animals, or even to gather sticks, within preserve boundaries. Trespassers could be imprisoned or even killed for their offenses. Although such prohibitions were, to some extent, only expressions of privilege, they represented an early recognition of the limits of resource use, namely, that natural systems, such as forests, could not produce sufficient resources for exploitation by everyone. Unless such exploitation was limited, the system would degrade.

In Western societies, cultural expressions of conservation advanced from preservation by prohibition to active manipulation and management of natural resources. In Europe, wealthy individuals employed gamekeepers, whose function was to ensure an abundance of favored species for hunting. Gamekeepers accomplished this, in part, by keeping out vagrants and poachers, but also through such activities as killing predators, introducing and translocating game animals to increase their densities, and even manipulating habitat by cutting trees or planting desirable species of vegetation.

Such activities, although sometimes beneficial in limited ways, did not achieve a comprehensive or consistent approach to conservation because they operated within the limits of established social boundaries of class, rank, and economic status characteristic of feudal societies and aristocracies. Indeed they were not designed for, nor did they attempt to achieve, this result. The development of a democratic political system led to a number of innovations in a systematic and populist-based approach to resource conservation.

Origins of Conservation Institution and Policy in the United States

Today the cause of conservation is a worldwide effort, with original and significant contributions arising from a diversity of nations and peoples. However, many of conservation's most foundational institutions and concepts, such as national parks, forest preserves, wildlife refuges, wilderness areas, endangered species protection, and the legal frameworks that systematically support and preserve such institutions, promote purity of water and air, and evaluate the environmental consequences of government actions, are fundamentally and essentially U.S. inventions. Thus, a brief survey of the history of conservation in the United States is an appropriate starting point for understanding modern conservation efforts worldwide.

The Washburn Expedition: The American Vision of the National Park

In the northwestern corner of what would one day become the state of Wyoming, John Colter, who left the Lewis and Clark expedition to become an independent trapper and explorer, found, in 1807, a region of incredible scenic beauty and unbelievable wonders, including geysers, "pots" of boiling mud, petrified forests, and hot springs. His reports of the area were not believed by most, and led the area to be disparagingly and disbelievingly referred to as "Colter's Hell." A mountain man and military scout, Jim Bridger, explored the area 23 years later. He returned with even more incredible stories, including tales of a river fed by geysers and thermal springs that "got hot on the bottom" (what is today called the Firehole River), cliffs of black glass (obsidian), and springs belching sulfurous steam. But his reports, sometimes embellished (Bridger spoke facetiously of "petrified birds that sing petrified songs in petrified trees"), were also greeted with skepticism, and often referred to as "Jim Bridger's lies" (Frome 1987).

Despite public skepticism, persistent rumors of the region's incredible natural wonders eventually led to a number of expeditions to the area, building its reputation and making Colter and Bridger's amazing claims more credible. Finally, in August of 1870, an expedition led by the surveyor general of Montana, Henry D. Washburn, Nathaniel Pitt Langford, and Lieutenant Gustavus C. Doane commanding a military escort left Fort Ellis, Montana, and proceeded into "Colter's Hell," the area surrounding the upper reaches of the Yellowstone River. Each night they posted a watch in order to, in the words of one member, "keep the Indians from breaking the eighth commandment" (Trumbell 1871). The members of the Washburn expedition found the area even more amazing than the tall tales they had heard and meticulously described and sketched what they encountered. One member, Judge Cornelius Hedges, described some thermal springs they observed near one camp in a meadow:

> *The westernmost spring had an oval shaped basin, twenty by forty feet in diameter. Its greenish-yellow water was hot, and bubbles of steam or gas were constantly rising from various parts of its surface. This spring, with two others, was situated in about an east and west line, and at the upper side of the basin, which opened south, toward the creek. The central one of these three was the largest of all, and was in constant, violent agitation, like a seething caldron over a fiery furnace. The water was often thrown higher than our heads, and fearful volumes of stifling, sulphureous vapors were constantly escaping.* (Trumbell 1871).

At rest in their camp on a September evening at the junction of the Firehole and Gibbon Rivers, their exploration nearly complete, the leaders of the expedition discussed by firelight the remarkable wonders of the area and what ought to be done about them. Judge Hedges, suggested that neither they nor others should exploit the area for commercial development. Rather, Hedges argued that the entire region should be set aside in perpetuity as a "national park" for the enjoyment of its citizens, as well as visitors from around the world (Frome 1987). It was an idea never before proposed, that a government should preserve the best part of its natural heritage in a manner that would make such heritage open and accessible to every person, a uniquely democratic vision of conservation. What has been called "the greatest idea in American history" rose from the sparks of the evening's campfire to become an ideal emulated worldwide—the concept of the national park.

Convinced of the rightness of their vision, the leading members of the expedition returned to the East to initiate a campaign for Yellowstone's preservation as a national park. A scientific expedition the following year commissioned by the U.S. Geological Service confirmed their reports, adding photographs, paintings, and credibility to their campaign (fig. 1.1). On March 1, 1872, Congress approved a bill creating Yellowstone National Park "as a public or pleasuring ground for the enjoyment of the people" (Petulla 1977). Although the area was initially protected by the U.S. Army, Congress would create the National Park Service in 1916, directing it to "conserve the scenery and the natural and historic objects and the wildlife therein, and to provide for the enjoyment of the same in such

Figure 1.1
The Grand Canyon of the Yellowstone by Thomas Moran, official artist and illustrator of the 1871 U.S. Geological Survey expedition to Yellowstone National Park.

manner and by such means as will leave them unimpaired for the enjoyment of future generations." Today, much of the worldwide effort in conservation biology, as well as important dimensions of its theory and application, rests on the legitimacy of designing and establishing conservation reserves, and on the assumption that such reserves are an important part of national and cultural heritage. Such a foundational premise of conservation is now so familiar that it is taken for granted. In truth, the idea of conservation reserves, which had its beginnings in the American idea of a national park, is a radical and relatively recent concept. The establishment of Yellowstone National Park was a watershed event in the development of the conservation movement. It created an enduring paradigm in U.S. conservation, one that would be spread throughout the world.

POINTS OF ENGAGEMENT—QUESTION 1

How might the history of conservation in the United States, and perhaps the world, have been different if the members of the Yellowstone expedition had seen the Yellowstone region as an opportunity for personal financial gain and private entrepreneurship?

The Romantic-Transcendentalist Ethic: Conservation Begins as a Moral Movement Led by Private Citizens

The public support to preserve Yellowstone in the 1870s was not generated simply by asserting its natural wonders or scientific value. Both the public and their elected representatives in government were persuaded that it was not merely *possible* to preserve Yellowstone, but that it was morally *right* to do so. The establishment of Yellowstone as a national park is but one example of an important principle needed to understand the development of conservation in America. Conservation in the United States was first framed in the form of a moral argument, not a scientific paradigm.

Prior to the Washburn expedition, writers like Ralph Waldo Emerson and Henry David Thoreau were among notable American authors and intellectuals of the nineteenth century who contributed to the formation of the romantic-transcendentalist ethic, although Thoreau's work was not widely read or appreciated until many years after his death. This ethic argued that the highest and best use of nature was not the extraction of its resources as commodities, but the appreciation of its intrinsic values and aesthetic qualities through which the human spirit was renewed and reformed. Emerson referred to nature as a temple where one could draw near and commune with God (Emerson 1836; Callicott 1990). Thoreau argued that civilization degraded and weakened the human spirit (Thoreau 1863). He insisted that human character was invigorated and purified through contact with nature in pristine form. "For I believe," wrote Thoreau,

> *"that climate does thus react on man—as there is something in the mountain air that feeds the spirit and inspires. Will not man grow to greater perfection intellectually as well as physically under these influences? Or is it unimportant how many foggy days there are in his life? I trust that we shall be more imaginative, that our thoughts will be clearer, fresher, and more ethereal, as our sky—our understanding more comprehensive and broader, like our plains—our intellect generally on a grander scale, like our thunder and lightning, over rivers and mountains and forests—and our hearts shall even correspond in breadth and depth and grandeur to our inland seas"* (Thoreau 1975:671).

The American conservationist, John Muir (fig. 1.2), concerned with the more practical problem of attempting to save the Sierra Nevada Mountains of California from logging, mining, hydrological development, and other forms of commercial exploitation, adapted the romantic-transcendentalist ethic to make it the foundation of a national campaign to appreciate and preserve nature. Muir invented and manufactured mechanical equipment, but nearly lost his eyesight in an Indianapolis carriage factory when a file he was using slipped and went through one eye. Eventually recovering his sight but shaken by this near life-altering experience, he retired to the wilderness in an attitude of repentance and reexamination of his life. After traveling on foot for over a thousand miles through parts of Florida, Georgia, Kentucky, and Tennessee, he eventually reached the Sierras. There, Muir had an intense religious experience, gaining a sense of profound fulfillment and exaltation, as well as an intense oneness with the land around him (Petulla 1977). Seeing destruction and degradation of the Sierras and other natural areas everywhere, Muir at once began a public campaign to save wilderness in the United States.

Muir was a persuasive writer, and his articles in major newspapers and national magazines urged Americans to (temporarily) leave the cities and enjoy the wilderness. Because his experience had been religious, his writing used religious language even more than earlier transcendentalists like Emerson or Thoreau, as exemplified by the title of one of his early and influential articles in the Sacramento (California) *Record-Union:* "God's First Temples: How Shall We Preserve Our Forests?" (Petulla 1977). Speaking of the Grand Canyon in Arizona, Muir wrote, "Instead of being filled with air, the vast space between the walls is crowded with Nature's grandest buildings, a sublime city of them, painted in every color, and adorned with richly fretted cornice and battlement spire and tower in endless variety of style and architecture. Every architectural invention of man has been anticipated, and far more, in this grandest of God's terrestrial cities" (Muir 1997:742–43). Like Emerson and Thoreau, Muir believed that contact with wilderness was essential to the formation of human character and virtue. "The mountains are fountains of men as well of rivers, of glaciers, of fertile soil," he wrote. "The great poets, philosophers, prophets, able men whose thoughts and deeds have moved the world, have come down from the mountains—mountain dwellers who

Figure 1.2

John Muir, nineteenth-century preservationist and founder of the Sierra Club.

have grown strong there with the forest trees in Nature's workshops" (Wolfe 1946:vi).

Although Muir drew portions of his philosophical and religious inspirations from the transcendentalists, he also extended them to more practical applications. Specifically, Muir came to believe that the best parts of nature should be left in an undisturbed state, and he began to work tirelessly to achieve that end. More than any other U.S. conservationist, it was Muir who defined a new school of thought and activism in U.S. conservation—preservationism. Muir and other preservationists condemned the destruction of nature to satisfy what he considered the greedy appetite of materialism. He made it clear in his writings that people who used nature as a place for religious worship, aesthetic contemplation, inner healing, rest, and relaxation were making a "better" (that is, morally superior) use of nature than those who cut trees, dammed rivers, mined minerals, or plowed the soil. Nature, Muir believed, was to be preserved in an undisturbed state so that these higher uses and values could be appreciated and enjoyed. Thus, Muir was instrumental in framing the debate in American conservation around the essential question: What is the best use of nature and natural resources? For over a century, right up to the present time, that question has been at the core of the U.S. environmental debate and remains a key to understanding the development of conservation biology as both a scientific discipline and a cultural force.

John Muir not only shaped many of the future ideals of conservation, but also created the organizational structures and practices that would become the model of the nongovernmental conservation organization, or NGO. In his campaign to save the Sierras, and especially Yosemite Valley, Muir began to see the necessity of public support, political will, and permanent legislation. Muir was gradually becoming convinced that only federal ownership could protect wilderness from the evils of private exploitation, but he also saw the necessity of public opinion as the leverage that would move the government to act.

To arouse public sentiment, Muir began to write extensively of the wonders and beauties of Yosemite, publishing much of his work in California newspapers. Muir's writings attracted the attention of Robert Underwood Johnson, associate editor of *Century* magazine, the nation's leading literary monthly. Johnson arranged a trip with Muir into the wilderness areas above Yosemite Valley. Inspired by the beauties of Yosemite's glaciers, mountains, and valleys, and by one another, Muir wrote articles about Yosemite for the *Century* that attracted wide readership, while Johnson began a congressional lobbying effort to have Yosemite, then a state preserve, declared a national park (Nash 1967). It was a successful collaboration. In 1890, Secretary of Interior John W. Noble, inspired by Muir's writings, convinced President Benjamin Harrison to support a Yosemite park bill that Johnson had helped promote. The Yosemite bill was signed into law on October 1, 1890, and a cavalry patrol was dispatched to guard the area (Wolfe 1946:251).

With his political success in preserving Yosemite, Muir began to appreciate the need for a more permanent organizational structure to advance the goals of conservation. Ever since 1889 Johnson had been urging Muir to form an association to "preserve California's monuments and natural wonders—or at least Yosemite" (Wolfe 1946:254). By 1892, in the face of growing environmental destruction from logging, grazing, and mining in his beloved Sierras, Muir acted. Meeting with a small group of carefully chosen associates in the office of attorney Warren Olney of San Francisco on May 28, 1892, Muir incorporated an organization called the Sierra Club, an NGO that has become synonymous throughout the world with effective conservation activism, political influence, legislation, and litigation. Its charted purpose was "To explore, enjoy and render accessible the mountain regions of the Pacific Coast; to publish authentic information concerning them; to enlist the support and cooperation of the people and the government in preserving the forests and other natural features of the Sierra Nevada Mountains" (Wolfe 1946:254).

It is perhaps prophetic that an organization that would come to use law and litigation so effectively was birthed in a lawyer's office. But the Sierra Club had little time to celebrate its own beginnings. Timber and grazing interests introduced a bill in the U.S. Congress in the same year to log nearly half of Yosemite National Park. The bill had already passed the House of Representatives by a large vote, but the Sierra Club acted before it reached the Senate. Members of the club wrote and personally lobbied senators to defeat the bill, while Muir gave extensive interviews to newspapers in which he demanded full protection for Yosemite. Muir also sent personal telegrams to many influential government leaders. The bill was tabled. It remained in committee for 3 years before finally expiring, but Muir and the Sierra Club ultimately prevailed.

In the Sierra Club's first meeting in Warren Olney's law office, John Muir was elected president. He held that office until his death in 1914. The organization that he formed and led established a model of focused mission, political involvement, public disclosure, and legislative advocacy that have become a pattern for NGOs like The Nature Conservancy, The World Wildlife Fund, The Wilderness Society, and others throughout the world. Thus Muir was instrumental not only in framing the debate about the goals of conservation, but also in framing the

means of public involvement through which such goals could be attained.

The Federal Government in Conservation

Even as Muir was shaping the key questions of the great environmental debate in the United States, the end of the western frontier (unsettled or sparsely settled land available for private ownership in the western United States) also began to effect a profound change in American environmental attitudes. Settlers had been formerly faced with a hostile, often life-threatening environment on the frontier that could be made livable only by strenuous individual effort and significant environmental alteration. Yet, for all its hardship, westward expansion gave a continuing impression that there would always be new lands to settle and more natural resources to find and use. As the frontier came to an end and the density of human populations in the United States began to grow even in remote areas, U.S. citizens now found themselves living in a nation of increasingly well-defined physical and environmental limits. At the same time, the use and exploitation of natural resources began to shift from individual effort to corporate effort, and from simple exertions of human and animal labor to increasingly sophisticated applications of advanced technologies. Freed from many of the former limitations, corporate interests in mining, lumber, fishing, and grazing began to exploit and alter resources over large areas. The state of Michigan, for example, which was the nation's largest producer of timber as recently as 1870, had removed most of its commercially valuable timber by 1920, and effectively cut itself out of the lumber business, especially because of such technological developments as the narrow-gauge railroad and improved techniques of nonwinter logging. In states further west, similar trends were in motion.

The establishment of Yellowstone as an American national park in 1872 had set a legal and political precedent for the federal government to take a more active role in conservation. In 1873, the U.S. Congress, in an effort to aid western homesteaders, passed the Timber Culture Act, which permitted the clearing of up to 160 acres of timber if the owner planted trees on 40 acres. Although intended to help individual families, the law was used most effectively by timber companies to clear large tracts of western forests with minimal reforestation. In 1891, at the urging of a number of scientific and professional societies, President Harrison, aided by Secretary of Interior Noble, succeeded in persuading Congress to pass a bill repealing the Timber Culture Act and granting the president (in a relatively unnoticed rider on the bill) authority to set aside forest reservations (Petulla 1977). The Forest Preservation Act, as it came to be called, was designed primarily to protect U.S. forests in western states from suffering the same fate as their counterparts in the states of Michigan, Minnesota, and Wisconsin.

As many private citizens were becoming increasingly vocal in their opposition to the consumptive use of natural resources, particularly on public (federal) lands, the power of the federal government, as well as its involvement in conservation, was beginning to grow. The assassination of President William McKinley in 1901 brought his young vice-president, Theodore Roosevelt, to the Oval Office (fig. 1.3). In his youth, Roosevelt was trained as a biologist at Harvard, but frustrated

Figure 1.3
Theodore Roosevelt, the twenty-sixth president of the United States
(1901–1909), greatly supported the role of the federal government
in conservation.

Figure 1.4
Gifford Pinchot, early head of the U.S. Forest Service and father of
the resource conservation ethic. From an original staff of only 123
in 1898, Pinchot built the Forest Service to an organization of
1,500 people administering 150 million acres of public land within
10 years.

by the emphasis on micro- rather than macrobiology, Roosevelt,
an active outdoorsman, began to take an interest in western
lands, particularly in the use of forest and wildlife resources. He
saw firsthand the loss of wildlife resulting from hunting. Convinced that such losses were not the result of hunting per se but
of unethical behavior and a lack of personal virtue in hunters, he
worked with many of the founding members of the Boone and
Crockett Club to establish an ethical code for hunters.

Alerted and persuaded by Muir and others of growing
environmental degradation in the western United States,
Roosevelt increasingly viewed the large corporations that were
profiting from western lands with distrust. Known as "The
Trustbuster" before the end of his first administration for his
zeal in breaking up industrial monopolies in the East, Roosevelt saw the corporate practices of logging and mining in the
West with similar hostility. He came to see such practices as
not only wasteful but undemocratic. To Roosevelt, it was clear
that a handful of individuals and their companies were reaping
most of the profits from natural resources that rightfully be-
longed to all citizens. Roosevelt increasingly saw federal regulation as both appropriate and necessary because the federal
government was, in his view, the only institution powerful
enough to oppose the corporate timber and mining interests
that threatened to destroy the landscape. An international
strategist, Roosevelt also recognized the value of the forests for
strategic purposes and as raw material for national defense.

On this basis, Roosevelt used the provisions of the Forest
Preservation Act aggressively and often, setting aside large
forested areas in the western states. The new forest reserves
were placed under the jurisdiction of the Division of Forestry
within the U.S. Department of Agriculture. There they became
the centerpiece of a comprehensive conservation program
developed by Gifford Pinchot (fig. 1.4), who had headed the
division since 1898. Pinchot was a forester trained in European
traditions of scientific forest management. He was not a preservationist like Muir and did not subscribe to the romantic-
transcendentalist ethic of nature preservation for the sake of
moral values. He saw the timber in the new reserves as an
exploitable resource, to be used with careful application of scientific management. Pinchot can be called the father of a new
ethic, the so-called resource conservation ethic—a view distinct from romantic-transcendentalist views. Pinchot crystal-
lized the philosophy of his movement in a simple, memorable
slogan (which he credited to a contemporary, W. J. McGee):
"the greatest good for the greatest number for the longest time"
(Callicott 1990).

The resource conservation ethic rested on two key intellectual pillars. The first was equity—resources should be justly
and fairly distributed among present and future generations.

The second was efficiency—resources should not be used wastefully. Pinchot and others who advocated a resource conservation ethic were not especially concerned about the "best" use of nature (which they called "natural resources") in the sense of moral superiority. They were concerned about the "fair" (i.e., democratic) and "sustainable" use of resources. Thus, they believed that all interests in resource use, both consumptive and nonconsumptive, should be considered and, when possible, satisfied. Time and again Pinchot made clear that, in his view, conservation did not mean protecting or preserving nature. Rather, it meant wise and efficient use of natural resources, with a goal of controlling nature to meet human needs over the long term (Nash 1989).

A superbly gifted administrator, Pinchot built the Division of Forestry from an initial staff of 123 (including 60 student assistants) in 1898 to an organization of 1,500 employees in charge of over 150 million acres of forests within 10 years (Petulla 1988). Eventually the Division of Forestry was reorganized under Roosevelt's administration and Pinchot's direction to become the U.S. Forest Service, and Roosevelt consulted Pinchot on all matters relating to conservation, not forest resources alone.

Roosevelt did not confine his own conservation efforts exclusively to marketable resources. Beginning with the establishment of Pelican Island (Florida) as a wildlife sanctuary in 1902, Roosevelt began an ambitious program to create federally protected wildlife sanctuaries throughout the United States. In all, Roosevelt established 52 such sanctuaries during his administration, including 17 in a single day. These lands were the beginning of what would grow to become the National Wildlife Refuge System of the U.S. Fish and Wildlife Service (USFWS), although the USFWS itself would not exist as a federal agency until many years later. Roosevelt established such reserves as a way of making a public and intentional statement of the inherent value of wildlife and forest resources.

Although Roosevelt appreciated inherent and intrinsic values of wildlife and natural areas, he was, like Pinchot, pragmatic and utilitarian in his comprehensive national strategies for managing natural resources. He understood that organic resources were inherently renewable, and thus, with Pinchot, contributed to formulating the ideas of *sustainable use* and *maximum sustained yield,* conceptual pillars of modern U.S. Forest Service management. Roosevelt's utilitarian but sustainable vision of conservation and resource use is perhaps nowhere more evident than in his fourth annual (1904) message to the U.S. Senate and House of Representatives. Speaking of U.S. national forests, Roosevelt stated, "It is the cardinal principle of the forest-reserve policy of this Administration that the reserves are for use. Whatever interferes with the use of their resources is to be avoided by every possible means. But these resources must be used in such a way as to make them permanent" (Roosevelt 1926:235).

Ultimately, the first quarter of the twentieth century saw the resource conservation ethic prevail over the romantic-transcendentalist ethic. The resource conservation ethic was more acceptable to the increasingly secular culture of U.S. government. The resource conservation ethic emphasized utilitarianism, which affirmed the democratic principle of individual choice, and it was perceived to address the needs of citizens more practically and democratically than romantic-transcendentalist views. However, the debate was far from over. A new paradigm would emerge, synthesizing both views, yet distinct from them, and would arise from a professional trained within the ranks of the Forest Service itself.

Academia and Conservation

Ecology, unknown even as a word until the late nineteenth century, began to flourish as an academic discipline at universities in the United States by the early 1900s. The early work of ecologists like Henry Cowles at the University of Chicago brought the systematic study of plant succession to new levels of sophistication and professional respect. Other prominent ecologists such as Frederic Clements and Victor Shelford advanced the concept of ecological communities as objects of research and played key roles in the organization of the Ecological Society of America. Their high view of the importance of ecological communities and biomes is best expressed in their own words. "The biome or plant-animal formation," they wrote, "is the basic community unit. . . . The concept of the biome is a logical outcome of the treatment of the plant community as a complex organism, or superorganism, with characteristic development and structure. As such a social organism, it was considered to possess characteristics, powers, and potentialities not belonging to any of its constituents or parts" (Clements and Shelford 1939:20). Although later studies would force significant reassessment of this view of biomes, the Ecological Society of America and the ecosystem studies it promoted under the leadership of Clements, Shelford, and others, laid a foundation of systematic and scientific information that increasingly became the basis for the professional management of resources and ecosystems by federal agencies.

Even so, the connections between scientific study and conservation practice remained tenuous. Such connections would become more explicit only when an agency scientist of the U.S. Forest Service began to develop a new paradigm for connecting science and conservation, and then returned to academia to teach the new discipline he had created. The discipline was game management, and the individual was Aldo Leopold (fig. 1.5). Both were to have profound effects on the development of the U.S. conservation effort.

Aldo Leopold, born and raised in Iowa in the 1880s when even this state still had many characteristics of a wilderness, was educated as a forester at Yale University, studying in the tradition of Gifford Pinchot and the concept of sustained yield. Leopold eventually joined the U.S. Forest Service, where his keen interest in wildlife populations was not lost on his superiors. Increasingly permitted to conduct investigations of game populations in national forests, Leopold became convinced that the practice of game management, although still an art, could nevertheless be informed by and practiced as a science. His experiences and reflections culminated in the classic textbook *Game Management,* published in 1932, which in turn led to the University of Wisconsin at Madison to offer him a faculty position in the field he had created. Leopold accepted the offer and helped organize one of the first academic departments in wildlife management and wildlife ecology at a state university in the United States.

Figure 1.5
Aldo Leopold, early twentieth-century conservationist and father of
the modern land ethic.

The synthesis of Leopold's experience in the Forest
Service and his studies in academia left him increasingly convinced that the resource conservation ethic was inadequate,
principally because it was untrue. The land was not, as the
resource conservation ethic portrayed it, simply a collection of
separate *commodities* that could each be harvested independently. The land was a system of interdependent *processes,* and
the outcome of those processes, when they functioned properly, was sustained production of the commodities, such as
soil, water, timber, wildlife, and forage for wild and domestic
animals. Leopold also recognized that some systems were best
"managed" by almost complete protection (i.e., preservation),
but he was not, as he has sometimes been portrayed, a true
preservationist in the tradition of John Muir. He believed in
management, but enlightened management that focused on
managing ecological processes, not individual populations of
plants and animals. During his years in the Forest Service, his
work and vision for setting aside large areas that were still
roadless and relatively unexploited ("wilderness") within national forests contributed, after his death, to the formation of
the National Wilderness System, The Wilderness Society, and,
ultimately, the passage of the National Wilderness Preservation
Act in 1964.

Although Leopold's views were informed by analysis of
science and the experience of management, he also realized
that an ethical transformation had to take place in natural
resource managers and the public. Like the transcendentalists
before him, Leopold became convinced that there was not only
an efficient and democratic way to manage resources, but there
was a "best" way to manage them. Such management recognized that the system—in Leopold's words, "the land"—had
intrinsic value, and included a concept of ethics that recognized this value and worked for the land's health and continued
productivity. Without these values, human selfishness and
consumptivism would increasingly thwart the most informed

science and the most enlightened management. Further,
Leopold argued, like Emerson, Thoreau, and Muir, that a recognition of the value of the land demanded changes in personal
attitudes and behavior—that government conservation policy
required the support of personal conservation virtue. Leopold
came to believe that the "conservation" of his day, based on
nothing but the enlightened self-interest of the resource-
conservation ethic, was only a counterfeit that masked the need
for the genuine ethical transformation required to change relations between humans and their environment. "I had a bird dog
named Gus," Leopold wrote. "When Gus couldn't find pheasants he worked up an enthusiasm for Sora rails and mead-
owlarks. This whipped up zeal for unsatisfactory substitutes
masked his failure to find the real thing. It assuaged his inner
frustration. We conservationists are like that" (Leopold
1966:200). It was this motivation to expose what he perceived
as the shallowness of the resource-conservation ethic that drove
Leopold to craft and contribute a radical alternative.

Leopold was neither original nor unique in his view that
ethics ought to include land and nonhuman creatures, or that
conservation requires private virtue as well as public law. Such
concepts are rooted in multiple ethical and religious traditions
in a variety of cultures (chapter 3). But Leopold's original contribution was to combine this ethical conservation with practical
experience in resource management, and then to inform both
with scientific expertise. Leopold wrote, "It is inconceivable to
me that an ethical relation to land can exist without love, respect, and admiration for land, and a high regard for its value.
By value, I of course mean something far broader than mere
economic value; I mean value in the philosophical sense"
(Leopold 1966:261). Leopold's writings on the ethical aspects
of land management were published posthumously as a collection of essays titled *A Sand County Almanac.* Leopold's "land
ethic" made the ethical treatment of land and resources a central
issue in conservation, and a point of serious discussion in academia. Leopold's work returned attention to the fundamental
question with which John Muir had initiated the great American
debate in conservation: What is the best use of nature and
natural resources?

Leopold's coupling of values (the land ethic) to applied
science (game management) in an academic setting began to
change fundamental assumptions not only about the best use of
natural resources, but also about the nature and purpose of all
ecological studies. These changes opened the door for the future
development of a value-laden, mission-driven approach to science and conservation, without which the field of conservation
biology could not have emerged.

The Emergence of Public Environmental Awareness

Leopold's writings, although of major significance, were not
immediately appreciated or widely read during the following
decade of the 1950s. Resource management was viewed primarily from a utilitarian perspective, and problems associated
with pollution of air and water were assumed to be local, directly related to particular sources such as factories, power
plants, and smelters, and, although inconvenient, nonthreaten-
ing. It remained for another biologist, Rachel Carson (fig. 1.6),
to alter this conventional wisdom.

Figure 1.6
Rachel Carson, U.S. Fish and Wildlife Service biologist and author of *Silent Spring* (1962), a seminal book in the modern environmental movement.

Carson, a U.S. Fish and Wildlife Service biologist, was also a gifted popular writer who had the ability to take the subjects of natural history, such as a marine ecosystem or the migration of an eel, and turn them into fascinatingly readable accounts of nature, or even novels in which animals were the main characters. Employed as a scientist with the U.S. Fish and Wildlife Service from 1935 to 1952, Carson's first book, *Under the Sea Wind*, published in 1941, enjoyed little commercial success. However, her second effort, *The Sea Around Us,* was so successful that it allowed Carson to leave the Fish and Wildlife Service and devote herself entirely to writing. It was in this second career as an author that Carson would change the culture of conservation, and indeed initiate the American environmental movement, with her fourth book published in 1962, *Silent Spring.*

From both experience and reading, Carson became convinced that the increasing use of agricultural and industrial pesticides was causing unforeseen and unintended effects on ecosystems. One of the first popular writers to appreciate the concept of biological amplification, Carson perceived that low levels of certain pesticides in plants increased exponentially in the insects that fed upon such plants. In turn, pesticide concentrations again increased several orders of magnitude in birds that ate the insects, leading to direct mortality or reproductive

dysfunction. Able to be carried by both air and water, pesticides applied in one location spread to others, even to relatively uninhabited areas otherwise little affected by humans. "The central problem of our age," wrote Carson, "has therefore become the contamination of man's total environment with such substances of incredible potential for harm—substances that accumulate in the tissues of plants and animals and even penetrate the germ cells to shatter or alter the very material of heredity upon which the shape of the future depends" (Carson 1962:8).

Carefully and compellingly describing the effects of pesticides, Carson opened her text with a quote from the English poet John Keats: "the sedge is wither'd from the lake, and no birds sing." Building from such imaginative and appropriate imagery, Carson foretold that the effects of pesticides, if left unregulated, would bring to the world the onset of a "silent spring," in which no sounds of birds or other living creatures would be heard.

Carson's book was not without its detractors, especially chemical and pesticide companies who alleged that her claims were unsubstantiated by sufficient scientific research. But her combination of eloquence, rationality, and scientific professionalism aroused both public sentiment and political will. Yet Carson had done more than sound a warning on the single issue of pesticides. Her work laid the foundation for the public understanding of three key principles. First, pollution and environmental degradation affect creatures at many different levels of ecosystem function through the cycling of matter. Second, environmental degradation and pollution were not merely local annoyances, but regional problems with global connections. Third, such degradation and pollution were not merely irritating, but life threatening. Years later, former Secretary of the Interior Stewart Udall called Carson's work "a masterstroke. . . . It shifted the debate over pesticides into a context where ecological, not economic, values would predominate" (Udall 1988). *Silent Spring* proved to be the seminal work that would inspire a new generation of public environmental activism and, perhaps more important, a new body of environmental law that created a climate in which conservation biology would grow. A detailed examination of these laws, and the policies that grew from them, is the subject of chapter 2 (The Legal Foundations of Conservation Biology). At the same time that public activism was increasing and environmental law was being formed, changes in traditional academic disciplines within the applied sciences of resource management were creating an academic environment for conservation biology to emerge as a distinct intellectual discipline.

CONSERVATION BIOLOGY EMERGES AS A PROFESSIONAL DISCIPLINE: WHY FRUSTRATION LED TO INNOVATION

Through the work of Leopold and others, applied sciences in resource management gained academic respectability in state universities after the 1930s. The most pervasive and influential of these disciplines were forestry, silviculture, fisheries management, (outdoor) recreation management, range management, and wildlife ecology (the modern version of game management). In addition, the traditionally "pure" discipline of ecology increasingly featured studies of species or ecosystems with clear implications for conservation. The inaugural issue of *The Journal of Wildlife Management*, the official journal of The Wildlife Society, defined wildlife management as "the practical ecology of all vertebrates and their plant and animal associates" (Bennett et al. 1937). Early issues of *The Journal of Wildlife Management* in the 1940s showed promise of embracing this definition, featuring a number of multiple-species studies and nongame studies (Bunnell and Dupuis 1995). However, in the years that followed, *The Journal of Wildlife Management* became increasingly dominated by studies of game mammals and birds. Although studies of nongame species began to increase in the 1970s, they still made up less than one-third of contributed papers in *The Journal of Wildlife Management* in the early 1990s.

The developing emphasis on game species within The Wildlife Society was arguably a reflection of public interest in these species, an obvious need to address their management, and the apparent ability of wildlife scientists to solve the practical problems associated with such management. Hunters were growing more numerous and affluent. They were willing to make direct payments (for example, to buy duck stamps) or to pay special taxes (such as those legislated by the Pittman-Robertson Act) for game species, payments that the general public was not willing to make for all species in general. The need to restore populations of larger species, especially larger mammals and birds, decimated by earlier, unregulated hunting was increasingly apparent in the years following The Wildlife Society's establishment. With such populations in jeopardy, management filled an obvious need. The apparent ability to meet that need grew as wildlife scientists mastered the techniques of trapping and transplanting individual animals, making more precise assessments of habitat, and more successfully regulating hunter harvest.

Wildlife management and other applied sciences such as forestry, fisheries management, range management, and others also were hindered from embracing studies of biodiversity and multiple species in other ways besides their emphasis on economically valuable species and commodity uses of resources. Their paradigm of conservation science rested primarily on two intellectual pillars. One was the emphasis on studies of individual species (autecology) of interest to conservation. The other was the study of individual types of habitats. Under this approach, conservation was essentially a case-by-case effort. Effective conservation was based on knowing everything possible about the natural history of the species of interest and then preserving as much of its habitat as possible. Thus, refuge design for wildlife was typically based on preservation of habitat for a particular species, whether it was the Kirtland's warbler (*Dendroica kirtlandii*) on specially purchased land in Michigan, the snow goose (*Chen caerulescens*) at DeSoto National Wildlife Refuge in Iowa, or the Joshua tree (*Yucca brevifolia*) at Joshua Tree National Monument in California (fig. 1.7). Understanding the natural history of a species and identifying its patterns of habitat use were considered the first steps in any conservation plan (Simberloff 1988). Such effort was often effective in accomplishing its objectives, but it did not determine or establish general principles that could be applied to all species or to produce unifying theories of refuge design.

The applied resource management sciences in general and wildlife ecology in particular prospered in the climate of a growing environmental and conservation movement of the 1960s and 1970s. Legally supported by laws like the Endangered Species Act and aided by associated funding, studies in wildlife ecology and management became increasingly important as sources of scientific information for management and recovery plans for threatened animal populations. Wildlife ecology expanded to include specialties such as "nongame wildlife management" and "urban wildlife management." Wildlife ecologists and managers also increased their focus on conservation issues, and wildlife biologists wrote textbooks on "biological conservation" and started a journal with the same name.

Although the applied sciences were becoming more inclusive in their definition and conception of "wildlife," tensions between the applied sciences and the conservation movement were increasing. With prophetic insight, Aldo Leopold himself had written of these tensions decades earlier and predicted the outcome. In an essay titled, "Land Health and the A-B Cleavage" (Leopold 1966), Leopold wrote that there was a single (A-B) split common to many of the academic specialties in resource management. "Group A," wrote Leopold, "regards the land as soil, and its function as commodity production; another group (B) regards the land as biota, and its function as something broader" (Leopold 1966:258–59). Resource management fields such as wildlife management, forestry, and range management (the A group) did not fully embrace either the values espoused by Leopold and others or the growing emphasis on nongame species. Although many biologists in wildlife management and other sciences respected Leopold's ethical position that valued all species in the context of their communities, the bulk of money and effort consistently went toward enhancing populations of species with commercial or recreational value for humans (Soulé 1985).

The A-B cleavage was not confined to a single field, such as wildlife management, but was pervasive throughout the applied sciences. The 1960s and 1970s had seen the development of major new ideas in population biology and community ecology, such as the theory of island biogeography. Many scientists began testing the predictions of the new paradigms in the problems of conservation. But the results of their experiments were not always appreciated in traditional applied sciences such as wildlife ecology, fisheries management, forestry, and range management. These disciplines had become departmentalized in major universities and isolated from one another. Their isolation led to alienation and even hostility. The seemingly natural exchange of ideas and infusion to applications in conservation problems was inhibited (Soulé 1986).

Individuals in the applied sciences did not always understand or effectively respond to the growing chorus of voices in the developing field of environmental ethics that claimed that all species, not merely game animals and fish, livestock, or plants of commodity value, possessed intrinsic values in addition to utilitarian values. Conversely, active conservationists (Leopold's B group) were failing to infuse their land ethic into resource management and the academy. Stress on Leopold's A-B cleavage was increasing, and the pressure could only be relieved by a split. But even as this tension was preparing both academia and the conservation movement for the emergence

(a)

(b)

(c)

Figure 1.7

The (*a*) Kirtland's warbler, (*b*) snow goose, and (*c*) Joshua tree are examples of species whose threatened status or popular appeal led to the establishment of conservation reserves set aside primarily for their benefit.

(*a*) *Photo by Richard Baetson (USFWS); (b) photo by Dave Menke; (c) courtesy of Don Klosterman.*

of conservation biology, the foundational discipline of both—ecology—was undergoing significant changes (chapter 5) that would help create the need for a science with a fundamentally different perspective and an array of conceptually distinctive traits.

THE PROBLEM OF IDENTITY: WHAT ARE THE CONCEPTUALLY DISTINCTIVE CHARACTERISTICS OF CONSERVATION BIOLOGY?

To survive and grow, a discipline requires a unique conceptual framework and a set of identifiable intellectual distinctions that can be shared by a professional community with a common mission. Michael Soulé, one of the founders of conservation biology, said, "Disciplines are not logical constructs; they are social crystallizations which occur when a group of people agree that association and discourse serve their interests. Conservation biology began when a critical mass of people agreed that they were conservation biologists" (Soulé 1986:3).

Conservation biology has been described as the science of scarcity and abundance, and more precisely defined as "application of biology to the care and protection of plants and animals to prevent their loss or waste" (Meffe and Carroll 1997). This statement is a reflection of one of the most important qualities of conservation biology. Born out of the crisis of worldwide extinctions and loss of species, the focus of conservation biology is on the preservation of biodiversity (the entire range of all species), not merely the management of individual species. Core disciplines that inform conservation biology's attempts to achieve this goal are, according to Soulé, ecology, systematics, genetics, and behavior. Related disciplines in the applied sciences such as wildlife ecology, fisheries management, forestry, and range management also draw much of their source data from similar backgrounds. However, the latter fields have traditionally selected subjects for research on the basis of either common characteristics or common management applications. Conservation biology, in contrast, focuses on the study and preservation of the diversity of life itself.

Conservation biology's second distinctive trait as a scientific discipline is that it is both *value laden* and *mission driven*. Ethical norms are characteristic of mission- or crisis-driven disciplines, and conservation biology is not an exception to this pattern. Integral and distinctive to conservation biology's identity as a discipline is its explicit recognition of what Michael Soulé called "normative postulates" (Soulé 1985). Four such normative postulates are offered by Soulé. The first is that d*iversity of organisms is good,* and its negative corollary is that *the untimely extinction of populations and species is bad.* Second, *ecological complexity is good.* The second postulate assumes the first postulate, but explicitly adds value to the preservation of habitat and ecosystem diversity. The third postulate is that *evolution is good,* or more precisely, that it is desirable to maintain the genetic potential of populations that permits adaptation and innovation in a changing environment, as well as ongoing speciation. Soulé's final postulate is that *biotic diversity has intrinsic value, regardless of its utilitarian value.* Stating this as a normative value makes it explicit that conservation biology is committed to the study and understanding of all species and their relationships, and is defined not merely by its *interest* in biodiversity but by its commitment to the *value* of biodiversity.

A third characteristic of conservation biology, strongly related the second, is that it is *mission oriented* and *advocacy oriented.* In fact, conservation biology has been explicitly defined as *a mission-oriented discipline comprising both pure and applied science* (Soulé and Wilcox 1980:1). Stanley A. Temple, a founding member of the Society for Conservation Biology, has stated as conservation biology's mission, "to develop new guiding principles and new technologies to allow society to preserve biodiversity" (Gibbons 1992). Given the intrinsic value of species, conservation biology perceives that the best and highest application of scientific knowledge about species is to ensure their preservation. But conservation biology does not confine this effort to research and management. Rather, many conservation biologists also assert the importance of communication by scientists of their spontaneous inner experience and appreciation for the creatures they investigate (Naess 1986), claiming that no one has more expertise, or right, to express a love for nature than those who have given their lives to its study (Soulé 1986). Indeed, environmental philosopher Arne Naess, in his keynote address at the Second International Conference on Conservation Biology in 1985, told his audience of conservation biologists that they had "obligations to *announce* what has intrinsic value" (Naess 1986, emphasis his).

If they are to make such announcements, conservation biologists cannot confine themselves to being only scientists, managers, naturalists, and teachers. They also must act as advocates. Some older and more established scientific disciplines, even in the applied sciences like forestry or wildlife management, have traditionally equated advocacy with a loss of objectivity. Some conservation biologists also share this view. Advocacy carries risks. Speaking out is dangerous because complexities of scientific issues in conservation, although they may be understood by scientists, can be easily misunderstood by the public, and such misunderstandings can do more harm than good. But, as a discipline, conservation biology's emphasis on action to save species and habitats and its declared intention to announce the values of nature encourage and indeed often demand that its practitioners act as what Daniel Rohlf called *focused advocates* (Rohlf 1995a). Rohlf defines a focused advocate as a person or group reporting data concerning an area in which he or she has expertise as well as deeply held convictions, and who works to ensure that the information presented is correctly interpreted and rightly applied. Many conservation biologists believe that such focused advocacy, specifically the development and implementation of government regulations and policies that conserve biodiversity, is an inherent responsibility of a conservation biologist (Noss 1989; Thomas and Salwasser 1989; Dudley 1995; Rohlf 1995b).

A fourth characteristic, closely related to the third, is that conservation biology originated as a *crisis-oriented* discipline. Its mission, the preservation of biodiversity, is perceived by its practitioners as not merely important, but urgent. Conservation biology, then, emphasizes the need for *immediate application* of scientific information for the benefit of threatened species, and the discipline's attention and choice of subjects for study is often driven by how immediately the subject is threatened with destruction or extinction. Conservation biology's need for cross-disciplinary integration and rapid investigation and response does not always provide it with the luxury of long reflection and multiple replications of studies before action is taken. Historically, most scientific disciplines, including biology, have tended to view unfavorable or premature application of scientific results as worse than no action, and thus have emphasized the importance of minimizing risk and maximizing reliability. In conservation biology, however, failure to act when a population is declining or a habitat is being degraded may ensure the extinction of the species or the loss of its environment. The late population ecologist Graeme Caughley and his colleague Anne Gunn put it well: "In conservation biology, avoiding saying there is not an effect when there is may be more important than ensuring the warranted rejection of the null hypothesis" (Caughley and Gunn 1996:6). Thus, conservation biologists are more willing to tolerate the risk of inappropriate action than the irreparable losses that may be associated with no action.

Conservation biology is more likely than its related academic disciplines to cross disciplinary lines. This is conservation biology's fifth distinctive trait, its *integrative and multidisciplinary nature*. Although rooted in the core discipline of biology, studies in conservation biology routinely cross disciplinary boundaries among major taxa, such as plants and animals, vertebrates and invertebrates, and between biological and physical processes. Further, conservation biology, because it is value laden and mission oriented, routinely investigates and addresses issues of ethics, human behavior and culture, law, politics, and sociology. Thus, the biological information generated from studies in conservation biology is inextricably linked to and dependent upon an intimate partnership with the social sciences if it is to achieve the actual purposes of conservation.

POINTS OF ENGAGEMENT—QUESTION 2

For most of the twentieth century, Western science has presented and described itself as value neutral. Is value neutrality an expression of objectivity or apathy in conservation? Does an idea become less true if one feels personally connected to and involved in the idea's implications? In the future, do you foresee that conservation biology will "mature" into a "value-free" discipline, or will its "normative postulates" continue to be important elements that inspire its work and attract new practitioners?

A sixth characteristic of conservation biology is that it is *a science concerned with evolutionary time*. In their emphasis on the preservation of biodiversity, conservation biologists seek not merely the preservation of present *types* of organisms, but the preservation of their genetic heritage (representing their evolutionary history and potential) and the preservation of ecosystem processes that promote adaptation, innovation, and speciation to maintain and enhance future biodiversity.

A seventh characteristic of conservation biology is that it is an *adaptive science*, unapologetically imperfect and, at times, imprecise. Although the recent strategy of *adaptive management* is not exclusive or original to conservation biology, it is a concept uniquely at home in this discipline. Compared with more traditional, management-oriented disciplines in the life sciences that have tended to see management actions and their responses in a cause-and-effect relationship, conservation biology is characterized by its tendency to treat management actions themselves as experiments. Thus, it tends to both expect and accept a higher degree of uncertainty from the response of a system or a population. The response is then treated more as an experimental result than a perfectly predictable outcome, and the management strategy itself may be revised in light of the results obtained.

The eighth and final distinctive trait of conservation biology is that it is *a legally empowered science*. In contemporary culture, conservation biology is unusual among scientific disciplines because it has received unique legal, political, and cultural incentives and reinforcements to act. The body of environmental legislation passed in the last three decades has done three things for the field of conservation biology. First, it has given legal incentives and approval to much of conservation biology's agenda to preserve biodiversity. Second, the same laws have shaped human cultural values that affirm many of the values characteristic of conservation biology and reinforce the value-laden nature of the discipline. Third, such legislation has attracted a strong body of scientific support because many in the scientific community approve of this legislation, are interested in answering questions that the legislation makes pertinent, and benefit from the funding the legislation authorizes. Although the discipline of conservation biology might still exist without such environmental legislation, it would be substantially less influential.

None of these traits is the exclusive property of conservation biology. For example, cancer researchers are also engaged in a mission-oriented, value-laden approach to their science. And it is a rare, and poor, scientist who does not love the subject that she or he studies, whether it be eukaryotic bacteria, marine algae, or venomous snakes, and who is not prepared to act and speak as an advocate of its value. However, taken together, this constellation of qualities does provide a defining picture of what conservation biology is, why it emerged as a distinctive discipline at a unique historical moment, and why it was not absorbed into the older, antecedent disciplines that provided its theoretical and practical foundations.

CRITICISMS OF CONSERVATION BIOLOGY AND ITS OVERLAP WITH OTHER DISCIPLINES

Conservation biology's integrity as an academic discipline rests largely on its claim to provide distinctive approaches and contributions to biology. That claim has not gone unchallenged. Both contemporary and historical perspectives have alleged that conservation biology has significant overlap with other disciplines, even to the point of identity.

Not surprisingly, the field that has been the source of some of the most serious criticisms of conservation biology has originated from the discipline that is intellectually and conceptually closest to it, wildlife ecology. From its inception, The Wildlife Society, the largest and most influential professional organization of wildlife ecologists, has embraced the concept of "conservation biology." In an article in the inaugural issue of *The Journal of Wildlife Management,* Paul Errington and Frances Hamerstrom even refer to the new field as "conservation biology" (Errington and Hamerstrom 1937). Many in the wildlife profession have questioned whether conservation biology really offered anything new or different from the fields of wildlife management and wildlife ecology. One example of such a challenge was made by wildlife ecologist James Teer, who stated that "what the society [for Conservation Biology] proposes to be, the profession of wildlife ecology and management has been for all its history" (Teer 1988:572). An unpublished but well-known joke among traditionalists in the applied sciences has been that studies in conservation biology consist of "data-free analysis" (Gibbons 1992). The new field has been accused of sacrificing depth for breadth, mixing up disciplines in a cavalier fashion, and being overly theoretical (Gibbons 1992).

Some professionals have attempted to assess the differences and distinctions associated with conservation biology in more rigorous and systematic ways. For example, Jensen and Krausman (1993), who compared articles from *The Journal of Wildlife Management,* the *Wildlife Society Bulletin* (both official publications of The Wildlife Society), and *Conservation Biology,* found the articles from *Conservation Biology* to be less restricted to North America, less restricted to game species, and more likely to be written by authors with an academic affiliation. They concluded that the discipline of conservation biology was not simply repeating or imitating the studies associated with wildlife management and would enhance and complement human understanding of biological resources.

Other scientists, while noting important similarities and valuable complementary differences, did not hesitate to criticize literature in conservation biology as generally less rigorous than literature in wildlife management, often lacking in reproducible methods and overly vulnerable to the volatility of surrounding society (Bunnell and Dupuis 1995). Such criticisms provoked spirited responses from conservation biologists who argued that the literature in conservation biology was unfairly evaluated (Noss 1995). Conservation biologists also have claimed that their field is, in fact, not more value laden than that of wildlife ecology or wildlife management, only that conservation biologists are more forthright about their values. Further, conservation biologists assert that such values do not bias the science of conservation biology, but help define the problems chosen for study (Noss 1995).

Most, if not all, of the early animosities between conservation biologists and wildlife ecologists have abated, thanks in part to a number of joint research ventures and conferences. The early controversies actually proved helpful in clarifying many of the unique contributions and goals of conservation biology and defining its place in both science and conservation.

THE PROBLEMS ADDRESSED BY CONSERVATION BIOLOGISTS

Although the worldwide crisis of species extinction might at first make the problem of conservation biology appear relatively simple to define ("save species"), practicing biologists know that the work of preserving biodiversity is a complex effort requiring solutions to a myriad of related problems. Demonstrating the truth of the old ecological adage that "everything is connected to everything else," conservation biologists have discovered that there are no dimensions of the biodiversity problem that can be attacked in isolation. This realization and reality are the primary forces driving conservation biology's multidisciplinary character.

Saying that problems are interrelated may sound profound, but it is no substitute for critically explaining and evaluating *how* they are related. The principal problems and their important relationships must be identified and explained. Without such an explanation, any student making an initial attempt to study this field would be quickly overwhelmed by an avalanche of information without any compass or context. When the important problems and controlling ideas of conservation biology are identified and explained, the specific management efforts and case histories of conservation can begin to make sense in a broader context of what problems the discipline is trying to solve. This is particularly appropriate for conservation biology because of the mission-oriented, problem-solution approach that the discipline takes to scientific inquiry.

Using this approach, we will attempt to give meaningful organization and context to the enormity of data associated with studies in conservation biology by focusing on five major problems that the discipline attempts to address. These problems are closely and inextricably integrated, and major studies and scientific paradigms in conservation biology usually have implications for more than one major problem. In that sense, this approach is imperfect and artificial. But it remains valuable because it provides one way to see all of the efforts of the discipline in meaningful relation to one another.

Problem One: The conservation of genetic diversity.

The attempt to preserve biodiversity is not merely an attempt to save the existing cohort of the world's species, like a biological version of getting all the baseball cards in a particular set. Understanding biodiversity begins with recognizing that species are not only morphologically unique, but also genetically unique. Their genetic identities and variation represent their evolutionary histories of adaptation, not merely the source of their physical appearance. Thus, if such genetic diversity is lost, a species loses its ability to adapt to a changing world.

The study of genetic problems in small populations was an important stimulus leading to the emergence of conservation biology as a distinct scientific discipline. The larger problem of preserving genetic diversity within individual populations can be solved only by dealing with four, more specific problems that disproportionately threaten small populations. These threats are inbreeding depression, genetic drift, the fixation of harmful alleles (sometimes referred to as "mutational meltdown"), and hybridization (Simberloff 1997). These problems will receive detailed examination in chapter 6, but require a brief introduction here.

One of the earliest threats to be identified in studies of genetic considerations of rare species was the problem of **inbreeding depression** (Moore 1962; Hooper 1971). Inbreeding depression can occur in small populations of normally outbreeding species when matings begin to occur more frequently between close relatives. Heterozygosity is reduced, and recessive alleles, occurring in a homozygous condition, are more often expressed. In some cases, development may not proceed normally after birth if there are too many alleles in a homozygous condition. Some effects of inbreeding depression include reduced number of offspring, lower immune response, physical deformities and abnormalities, and reduced longevity. The smaller the size of the population, the more likely this is to occur (Simberloff 1997).

A second, related problem in conserving genetic diversity is that of **genetic drift,** the random fluctuations in gene frequencies that occur as a result of nonrepresentative zygotes created during breeding. The proportion of such nonrepresentative matings tends to increase in small populations. The smaller the population, the greater the probability that the sample (random matings of individuals) may represent neither the average nor the range of characteristics found in the population. Variability lost via genetic drift can be counteracted by mutations, but mutations are rare events. With less genetic variation in the population, the power of selection to create adaptation is reduced. Thus, a small population experiencing a sudden change in its environment may lack the time and genetic variation necessary to adapt.

A third problem associated with the genetics of small populations is the **fixation of harmful alleles** (harmful forms of a given gene) (Simberloff 1997). Most mutations are harmful. In a small population, random mutations and genetic recombinations can lead to a disproportionate occurrence of harmful alleles, even to the extreme that all individuals in the population possess only the harmful allele and no other. Such a condition (fixation) can cause the population to decline further, which in turn can lead to the fixation of additional harmful alleles. This buildup of harmful alleles, called "mutational meltdown," can eventually lead to the extinction of the population.

The fourth problem is that of **hybridization** and **introgression.** Members of small populations may interbreed with members of closely related species that are more numerous, especially nonindigenous or nonnative species. Or they may breed with other indigenous species when a habitat barrier breaks down. This is especially true when a rare, wild species exists in an area with a closely related, but more common, domestic species. The consequence is the loss of the adapted gene complexes of the rare species.

But loss of genetic diversity also affects populations negatively in other ways. Every population has what could be called a minimum threshold of genetic diversity. When the level of genetic diversity is at or above this threshold, adaptive forces of natural selection prevail as the primary determinants of the population's density and distribution. But below such a threshold, random, nonadaptive forces are the determining factors. Below the critical level of genetic diversity, only a few successive instances of "bad luck" may be required to lead a population through a random sequence of events that lead to a permanent loss of genetic diversity and, possibly, extinction.

The conservation of genetic diversity is also critical to the preservation of biodiversity. Natural selection, acting on a variety of genotypes, generates novel gene combinations that can lead to the production of new species. If this diversity is reduced, so is the world's capacity to maintain biodiversity. This preservation of genetic diversity also affects the design of nature reserves. A reserve design informed by good science does not simply protect a "spot" from disturbance. An ideally designed reserve is one that, in some way, actually conserves and fosters the continuing process of speciation (Frankel 1974).

Studies of conservation genetics, especially of the problems associated with inbreeding depression and genetic drift, were fundamental to the earliest attempts to determine a *minimum viable population,* (Franklin 1980; Soulé 1980; Frankel and Soulé 1981), a concept (discussed in more detail in the following section) that was foundational to conservation biology's initial efforts at preserving genetic diversity. Thus, studies of genetic diversity and its conservation are critical to conserving populations because genetics is a primary consideration in establishing the size of a minimum viable population that can persist into the foreseeable future.

Problem Two: The conservation of species. Conservation biology has been called, from its inception, the "science of scarcity and abundance" because of its historic and continuing focus on endangered species and its understanding that the "mission" of the discipline is to save them. But saving species is not as easy as it sounds. A common misconception among the general public is that, once an endangered species is identified, "saving" that species consists of affording it complete legal protection from hunting, trapping, or trade and preserving its current habitat. In this view, the preservation of species would require only that they be protected from any disturbance or change in current conditions.

Although some species can benefit from this kind of ship-in-a-bottle approach, the road to recovery for most species is rarely so simple. Understanding the conservation status of a species and predicting that species' future levels of density and abundance are achieved primarily through the techniques of *population demography,* the quantitative analysis of the processes (primarily birth, death, immigration, and emigration) that determine population size and growth. Such analysis can contribute, when combined with an understanding of environmental and genetic factors, to an understanding of the causes of the population's decline. Additionally, the analyses associated with population demography not only must determine rates of change in a population, but also must determine the *minimum population size* required for that population to persist at a specified level of probability, for a specified length of time (standard parameters: 95% probability of persistence for several centuries without loss of fitness). This number is commonly referred to as the **minimum viable population (MVP).** Just as reduced numbers of individuals threaten the genetic diversity of a population, so reduced numbers of individuals make the population vulnerable to events that can lead to its extinction, and the two problems are inextricably related. When the number of individuals in a population falls below some critical minimum, subsequent fluctuations tend to be controlled by random, nonadaptive forces. At this point, a simple and short series of adverse environmental conditions or events can lead the population randomly to the point of no return: extinction. A minimum viable population is not simply one that can persist under average conditions. It must be large enough and diverse enough to suffer the "slings and arrows of outrageous fortune" that will express themselves in the form of environmental variation, demographic variation, genetic variation, and natural catastrophe.

The quantitative determination of MVP is part of a process referred to as **population viability analysis (PVA).** PVA entails the evaluation of data and models that predict a population's persistence to some arbitrary future time. The goal of such an analysis is to provide managers with insight into how to change existing conditions and parameters in order to reduce the population's probability of extinction.

PVA rests on the principle that the larger the population, the greater the probability of its persistence. Confounding variables that modify this basic relationship include environmental uncertainty, natural catastrophe, demographic uncertainty, and genetic uncertainty (Soulé and Kohm 1989). Small populations are more likely than larger ones to go extinct because of problems like inbreeding depression, genetic drift, or chance effects of birth and death (Boyce 1992), all of which will be examined in

later chapters. For now, it is sufficient to say that a comprehensive PVA must model genetic, demographic, and ecological processes simultaneously. To do this, a PVA makes use of experimental manipulation, biogeographic patterns, theoretical models, simulation models, and genetics (Shaffer 1981), but most models do not include information on all these dimensions of analysis for one species. Experimental manipulation is particularly difficult to achieve because the time frames needed are too long compared to the time frame for decision making and because many populations targeted for such analysis are already dangerously small and threatened. Despite the complexities of the problems associated with PVA, its development represented a major breakthrough toward solutions in conserving species and populations because it provided the means to develop a quantitative and predictive relationship between population size and population persistence (i.e., its probability of extinction) (Shaffer 1981).

Emphasis on the preservation of species, particularly the preservation of endangered species, also has generated a collection of concepts and ideas that influence the research and applications of conservation biology. The first of these, commonly referred to as the **small-population paradigm,** is concerned mainly with the effects of low numbers on a population's persistence, and with managing threats to small populations, especially stochastic (random) threats. The seriousness of the threat depends more on the population's size than on its life history attributes. From the study of case histories of small populations, conservation biologists have attempted to develop a general theory of the causes and effects of threats to small populations and how they can be mitigated. Questions within this paradigm are restricted to stochastic influences on the population's current dynamics: demographic and environmental stochasticity, inbreeding depression, and genetic drift (Shaffer 1981). How or why the population became small is not the issue.

PVA is a technique that forms an important part of the foundation of the small-population paradigm. It focuses on how to ensure that small populations persist, but does not identify why a population became small or what must be done to make it bigger. Thus, the key to conserving populations and species is to achieve a correct diagnosis of the problem, not simply to analyze how to maintain existing conditions (Caughley and Gunn 1996).

This need for a more comprehensive long-term solution has led naturally to development of the so-called **declining-population paradigm.** In contrast to the small-population paradigm, the declining-population paradigm is concerned with mitigating or reversing threats to population persistence. The declining-population paradigm focuses on the deterministic processes that caused the population to decline in the first place, not on the random processes that threaten any population when it becomes small. In the area of management and conservation, the declining-population paradigm is primarily concerned with what might be done to reverse the population's decline. Thus, processes of importance in the declining-population paradigm differ from one species to the next, and life history characteristics affect the analysis of a population in relation to its history of decline. This makes it more difficult to arrive at broadly applicable principles of population decline associated with the declining-population paradigm than to determine general factors that can threaten any small population, regardless of its life history. Although some have minimized the importance of distinguishing or differentiating between these paradigms, the differences are important and must be explicitly recognized in the analysis and conservation of populations. Michael Soulé (1986) noted that "the extinction problem has little to do with the death rattle of the final actor. The curtain in the last act is but a punctuation mark—it is not interesting in itself. What biologists want to know about is the process of decline in range and numbers." Or, as Graeme Caughley said, "The mark of a species in trouble is not so much rarity as such but the rate of its decline" (Caughley and Gunn 1996). Although the small-population paradigm provides approaches for dealing with the immediate problems of managing populations that have already reached critically low levels, research on the declining-population paradigm can lead to understanding and managing the fundamental processes of population decline, thus resulting in a more proactive approach to prevent populations from reaching endangered status.

The last major paradigm worthy of mention, at this point, is what is commonly called **metapopulation theory.** The concept of metapopulations is considerably older than conservation biology itself. In 1954 Andrewartha and Birch noted, "A natural population occupying any considerable area will be made up of a number of . . . local populations or colonies" (p. 657), and represented the metapopulation concept clearly with elegant illustrations remarkably similar to those used in metapopulation literature today (fig. 1.8). However it was not until the late 1960s and early 1970s that the idea of metapopulations became an explicit model in population biology (den Boer 1968; Levins 1968, 1969, 1970). Levins offered perhaps the first definition of a metapopulation as "any real population [that] is a population of local populations which are established by colonists, survive for a while, send out migrants, and eventually disappear" (Levins 1970).

Today, metapopulations are defined more generally as "groups of small populations connected by occasional movements of individuals between them" (Simberloff 1998). A variety of different conceptual models have now been developed (fig. 1.9), but are united by common theories of metapopulation dynamics that emphasize asynchronous dynamics of local populations and dispersal among them, combined with density-dependent processes as the primary factors affecting the size and rate of dispersal of the subpopulations (Hanski 1991). Today metapopulation theory is arguably the most widely used paradigm in conservation biology (Simberloff 1997).

Metapopulations are collections of subpopulations of a species in a given area, each occupying a suitable patch of habitat in a landscape of otherwise unsuitable habitat. The expected lifetime of a population is proportional to two variables. One is the lifetime of subpopulations in individual habitat patches. The other is the fraction of the habitat patches colonized according to random probabilities over time, which are functions of average dispersal rates and average dispersal distances of individuals in the subpopulations. In other words, the persistence of a metapopulation is a balance between extinctions and colonizations of subpopulations in individual habitats (Hanski 1997). In chapter 5 (The Historic and Foundational Paradigms of Conservation Biology) and chapter 7 (The Conservation of Populations), you will see how these assumptions and expectations can be clarified more rigorously. For now, we can generalize by

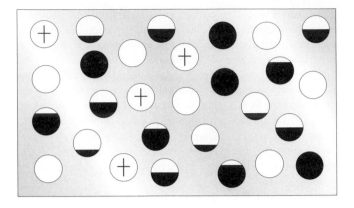

Figure 1.8

Diagrammatic representation of an arrangement of local populations ("metapopulation") based on Andrewartha and Birch (1954). Empty circles represent favorable habitats that individuals do not occupy. Partially or completely filled circles represent favorable habitats and relative densities of individuals in them as a proportion of the habitat's maximum capacity. Crosses indicate habitats in which local populations recently became extinct.

saying that, in metapopulations, the persistence of individual subpopulations is a balance between dispersal and extinction, and that balance is influenced by four factors: (1) demographic uncertainty, (2) environmental uncertainty, (3) natural catastrophes, and (4) genetic uncertainty.

Although subpopulations may suffer extinctions, colonization from other subpopulations can reestablish populations in those patches before all the subpopulations become extinct. There is thus a balance between local extinctions and recolonizations, with the consequence that the persistence of the regional metapopulation is greater than that of a similar population that is not spatially subdivided (Wiens 1997).

Metapopulation theory has been an attractive paradigm for conservation biology for several reasons. First, many species have patchy distributions, which fits the metapopulation paradigm. Second, habitat fragmentation, a major concern in conservation biology, tends to increase patchiness in distributions of populations, often creating metapopulations from populations that formerly were contiguous. Third, metapopulation theory is attractive because it reinforces the hope of many conservation biologists that disappearance of populations from local patches or fragments need not inevitably lead to extinction of populations over broader scales (Wiens 1997). Fourth, a metapopulation normally has more genetic variation than a single population, and thus the risk of extinction is lessened because all populations are unlikely to be exposed to the same demographic events simultaneously (Caughley and Gunn 1996). Finally, metapopulation theory has important implications for the design of reserves. Metapopulation dynamics predicts that species exist in fragmented, but demographically connected populations, and that the risk of extinction is inversely proportional to the number of fragments occupied. Thus metapopulation theory stresses the number of reserves as well as connections and dispersal between reserves (Caughley and Gunn 1996).

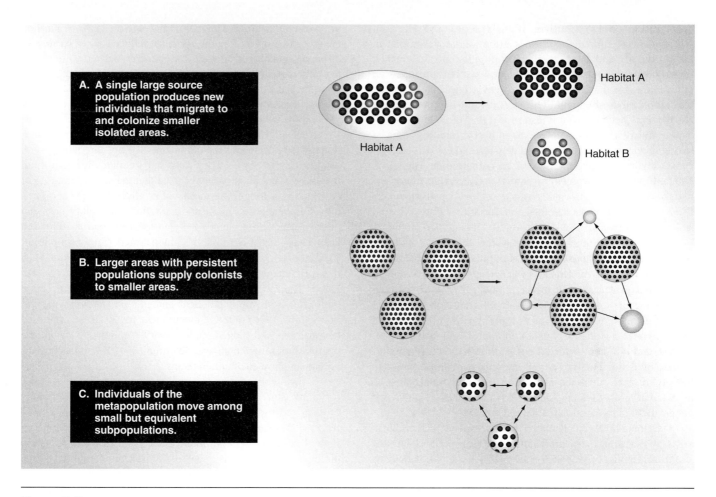

Figure 1.9

Three variations of the metapopulation concept. Although different in detail, all represent metapopulations as spatially distinct groups (subpopulations) that disperse to or among physically separated habitats.

The metapopulation paradigm is a conceptual model that, like all models, simplifies the reality of actual populations in space and time. Some metapopulation models assume, among other things, an infinite number of patches, an equal probability of extinction in each patch, extinctions that are independent of one another in different patches, and equal probabilities of dispersal and recolonization from one patch to another (Caughley and Gunn 1996). These assumptions are necessary to make the models manageable, but they are not always true. This is not a problem as long as the limitations of the metapopulation paradigm are recognized, and as long as the paradigm is used to generate hypotheses for testing how real populations actually behave. But it is a problem if the hypotheses that the metapopulation paradigm generates are assumed to be the same as facts about real populations.

Several different models of metapopulations exist, and each will be addressed in detail when we reach a point demanding more rigorous analysis of the problem of conserving endangered populations and species (chapter 7). Because metapopulation theory demands such close integration of models of populations and their habitats, it provides us with the natural bridge to the third major problem addressed in the field of conservation biology.

Problem Three: The conservation of habitat. Although much of PVA focuses on variations in genetics and population demography, the ultimate causes of most extinctions are ecological in nature. Loss or degradation of habitat will be the most significant factor in future species extinctions (Wilcox and Murphy 1985; Pimm and Gilpin 1989). Many key ideas about population conservation are, in fact, habitat dependent. For example, metapopulation theory proposes that species exist in a dynamic balance between extinction and recolonization of local populations in patches of habitat. Thus, it is impossible to conserve individual populations separately from their regional context (Hoopes and Harrison 1998). In understanding the importance of habitat conservation, metapopulation theory is complemented by the *source-sink theory* of habitat use and availability. The source-sink view of habitats asserts that populations exist in heterogeneous habitats that include areas in which population surpluses are produced (sources) and areas in which the population cannot replace itself without immigration (sinks) (Pulliam 1988; Hoopes and Harrison 1998). Source-sink theory draws attention to variations in habitat quality and their effects on a population's survival. It reveals that a population's persistence may be dependent on only a few key habitats, and not always those in which the species is most abundant. This theory and the models and experiments derived from it also reveal that the relative abundance and distribution of source and sink habitats may be strong determinants of population density, and even of the outcome of species interactions, such as competition (Danielson 1991). Finally, the problem of habitat conservation in conservation biology is informed by the *theory of disturbance dynamics,* which proposes that communities exist in spatial and temporal mosaics, and that diversity depends on maintaining the appropriate level of disturbance.

It is impossible to conserve populations without simultaneously conserving their habitats. Likewise, it is impossible to understand processes that affect population growth without considering the effects of differences in habitat on population demography. In addition, habitats can, and usually do, affect rates of genetic exchange, fixation of alleles, and drift in populations, and are therefore vital elements in the conservation of genetic diversity. Finally, habitats are objects of conservation in their own right. Habitats, in relation to populations, must be assessed in terms of both quantity and quality, because a lack of either can lead to the loss of a population.

Closely related to the loss of biodiversity is the worldwide crisis of **habitat loss, habitat fragmentation,** and **habitat isolation.** Habitat fragmentation refers to a disruption in patterns or processes associated with habitats. Some consider that habitat fragmentation is not simply one among many threats to species extinction, but the primary cause of it (Wilcox and Murphy 1985). Throughout the world, large blocks of habitat are increasingly subdivided into smaller blocks or shredded into tattered remnants of the original. Fragmentation of habitat not only reduces the total amount of habitat available, but also isolates blocks of habitat from other blocks of similar habitat. This can reduce movement of habitat-dependent species from one block to another, creating what many ecologists have described as "habitat islands," surrounded by "seas" of other habitat (or, in some cases, the "nonhabitat" of urban or agricultural lands). Additionally, processes that recycle matter and transform energy within the habitat that are effective at larger scales often are disrupted at smaller scales in fragmented habitats.

The worldwide trend of habitat loss and fragmentation combined with the simultaneous emergence of the **theory of island biogeography** in the 1960s and 1970s stimulated numerous early studies in conservation biology on the effects of habitat fragmentation. The theory of island biogeography was developed to explain the processes and patterns of variation in species diversity and abundance on islands. But even the earliest publications on the theory extrapolated it to conceptually similar problems in mainland habitats. The first book on the theory, *The Theory of Island Biogeography* (MacArthur and Wilson 1967) made explicit comparisons between islands and fragmented terrestrial habitats. The authors, Robert H. MacArthur and Edward O. Wilson, said of their theory, "The same principles apply, and will apply to an accelerating extent in the future, to formerly continuous natural habitats now being broken up by the encroach-

ment of civilization" (MacArthur and Wilson 1967:4). Further, the theory's emphasis on the effects of reduced areas, barriers to dispersal, and altered climatic variability associated with islands led to natural and immediate applications to habitat fragmentation. Thus, the theory of island biogeography provided a powerful intellectual paradigm for conservation biology in the early days of its development.

Island biogeography's applications to habitat fragmentation in terrestrial environments initially appeared too good, and too easy, to be true. They were. In particular, the equilibrium theory of island biogeography stated that the species number of an island is constant, but that local turnover changes composition. But the theory's application to terrestrial habitats has been heavily criticized because predicted local extinctions have not been consistently or reliably demonstrated (Simberloff et al. 1995). In fact, the species-area curves often do not account for a high percentage of the variability in data (Boecklen and Gotelli 1984). The theory also has sometimes distracted conservation biologists from more practical and immediate problems. As Simberloff (1986) noted, "unwarranted focus on the supposed lessons of island biogeography theory has detracted from the main task of refuge planners, determining what habitats are important and how to maintain them."

The theory of island biogeography has experienced major revisions even in its original context (island faunas), and many of its hoped-for applications to terrestrial habitats have proved illusory. Despite these problems, the theory of island biogeography has been a seminal intellectual stimulus to conservation biology, particularly on issues of habitat fragmentation. In the chapter devoted specifically to terrestrial habitat conservation (chapter 8, The Conservation of Habitat and Landscape), we will examine the concepts of island biogeography in greater detail and evaluate its continuing applicability to issues of habitat fragmentation and habitat preservation.

Another of the unifying concepts of conservation biology that addresses the problem of habitat conservation, and which is closely related to the problem of habitat fragmentation, is the concept of **habitat heterogeneity.** The term itself is often used to refer to a collection of theories and concepts that model and evaluate the effect of differences in habitats at a variety of spatial scales, and the effect of these differences on the demographics of a population. Heterogeneity has been defined as *any form of variation in environment, including physical and biotic components* (Ostfeld et al. 1997).

A naturally heterogeneous landscape has a rich internal structure of differing habitat patches that promote a corresponding increase in biodiversity in the landscape. Today the fragmentation of habitats, most of which results from human activity in the landscape, creates a pattern that superficially resembles the heterogeneity of a naturally patchy landscape. But the resemblance is just that—superficial. Compared to the rich internal structure of a naturally patchy landscape, a fragmented landscape is only a matrix of simplified and more isolated patches. Other features of fragmented landscapes, such as roads and various human activities, pose specific threats to the viability of individual populations.

The application of these and other concepts in the conservation of habitats leads naturally to the study and efforts of a related specialty, **restoration ecology,** the restoration of degraded habitats. Restoration focuses on identifying and mitigating processes that have led to the habitat's degradation, and then determining realistic goals and measures of success, with the means to achieve them. When such methods are shown to be repeatedly successful, they can be confidently incorporated in various strategies of land-use planning and management.

Problem Four: The management of landscapes through ecosystem processes. The heterogeneity and ecological health of habitats are maintained primarily through ecosystem processes. Populations, like habitats, are often fragmented, and individuals must move from one habitat patch through larger landscape systems. Thus, efforts to restore degraded habitats, preserve habitat fragments, and manage existing landscapes lead naturally to considerations of operative processes in the surrounding ecosystem. A fundamental premise of conservation biology is that the long-term preservation of biodiversity can occur only in and through systems that provide sustainable levels of resources through intrinsic, autogenic processes.

Conservation biology's emphasis on managing ecological processes, such as fire, hydrology, and nutrient cycling, rather than individual species has stimulated development of the concept of **ecosystem management.** A definition of this concept is offered by Noss and Cooperrider (1994) as "any land management system that seeks to protect viable populations of all native species, perpetuates natural disturbance regimes on the regional scale, adopts a planning timeline of centuries, and allows human use at levels that do not result in long-term ecological degradation." Grumbine (1994) defines ecosystem management as management that "integrates scientific knowledge of ecological relationships within a complex sociopolitical and values framework toward the general goal of protecting native ecosystem integrity over the long term."

Historically, ecosystem management developed out of a concern for preserving ecosystems, not merely individual species (Shelford 1933), combined with the realization that parks and preserves were, in their present condition, too small to preserve sustainable populations of many indigenous species, especially large mammals (Wright and Thompson 1935; Craighead 1979). These concerns, which developed within the scientific community, were complemented in the 1970s by the recognition among experts in natural resource policy that ecosystems could and should be used as the basic unit for public land policy (Caldwell 1970). In all its variety of contexts and definitions, the development of ecosystem management has consistently included and stressed three foundational premises. First, *the ecosystem, rather than individual organisms, populations, species, or habitats, is considered the appropriate management unit.* Second, *emphasis is placed on the development and use of adaptive management models, which treat the ecosystem as the subject of study and research, and management activities as experimental and uncertain.* This means that in ecosystem management, management actions should follow careful experimental design, include environmental controls (untreated sites or subjects), and be carefully monitored over time. If the experimental design is sound, the results of the management action should be relatively unambiguous, but must still be interpreted stochastically

(within a range of outcomes with differing probabilities), rather than as deterministic outcomes generated by simple cause-and-effect relationships. A third premise is that *those with vested interests in the health and services of the ecosystem (stakeholders) should participate in management decisions.* The implications of ecosystem management are extensive, and will receive detailed attention in a subsequent chapter (chapter 10, Ecosystem Management).

A variety of authors endorse five goals for ecosystem management. First, ecosystem management aims to maintain viable populations of all native species in the ecosystem. Second, it strives to represent within protected areas all native types of ecosystems across their range of variation. Third, ecosystem management attempts to manage systems through ecological processes and historical patterns of disturbance. Fourth, ecosystem management is oriented to manage over periods of time long enough to maintain the evolutionary potential of species and ecosystems. Finally, ecosystem management is to accommodate human use and occupancy within these constraints (Grumbine 1994).

Conservation biology naturally and appropriately endorses the concept of ecosystem management because both share an emphasis on the value of biodiversity, integrity of community and ecosystem units, and preservation of natural ecosystem processes at historic frequencies. But the deliberate transition from single-species and commodity/utilitarian management to ecosystem management requires more than a change in management approach. To embrace the concept of ecosystem management also requires a significant shift in values, policy, law, and administrative procedures (Grumbine 1994). Such a transition will not occur without intense debate, both within agencies and in public forum. That debate is now underway, but its outcome is still uncertain.

Problem Five: Sustainable development of human economies and human populations. Although conservation biology is fundamentally and historically rooted in more traditional sciences, such as biology and ecology, that have focused on nonhuman subjects, conservation biology must, of necessity, address problems of economic development and the growth of human populations. All economies, from the most primitive to the most sophisticated, are ultimately based on supplies, distribution, and consumption of natural resources. Extraction, use, and processing of natural resources in human economies require energy, create waste, and consume physical space. All of these activities affect other species, the habitats and ecosystems that sustain them, and humans. Thus, conservation biology's cross- and multidisciplinary character is perhaps most evident in its attempts to solve problems associated with economic development.

Economics, for all its uncertainties, remains the most widely used tool for determining valuation of objects, resources, commodities, and services associated with species, habitats, and ecosystems. Further, it is a science that provides well-defined methodologies for risk assessment and cost-benefit analysis in decision making and for comparing present and future values of resources through explicit assumptions. All these evaluations are necessary for scientific facts and recommendations to be transformed into policy decisions.

However, just as economics informs conservation biology and its applications, so conservation biology, especially its emphasis on the intrinsic value of biodiversity, changes the nature and scope of economics. Traditionally, forestry, wildlife management, fisheries management, and range management allowed values to be determined by market behavior and then managed resources according to market valuations. In contrast, conservation biology attributes value to all species and to the biodiversity of natural systems. If such value is real, its presence forces economics to redefine methods of determining valuation. Now, values of species are not determined by the bargaining of individual producers in a private marketplace, but usually through public negotiating and mediating of institutionalized political processes (Funtowicz and Ravetz 1994), especially in democratic societies.

Conservation biology's focus on sustainable development also forces human beings, individually and socially, to reframe traditional arguments about value in nonhuman terms, and to reconsider what humans choose to value. As one observer noted when the city of Los Angeles installed plastic trees in place of real ones along a major boulevard, "the demand for rare environments is . . . learned" and "conscious public choice can manipulate this learning so that environments which people learn to use and want reflect environments that are likely to be available at low cost. . . . Much more can be done with plastic trees and the like to give most people the feeling that they are experiencing nature" (Krieger 1973).

The "problem of plastic trees," as ethicist Lawrence Tribe called it, reveals that human desires and satisfactions are not sufficient criteria for determining what is truly valuable. First, we must determine what we shall value, because what we value determines what we shall be as persons (Tribe 1974). Conservation biology's appropriate focus on sustainable development forces human beings to address two important questions. First, what ought to be sustained? Second, what ought to be developed? The old answers to the first ("the present human standard of living") and the second ("resources needed for economic use and human consumption") are no longer adequate.

POINTS OF ENGAGEMENT—QUESTION 3

If we try to accomplish the goals of conservation strictly through a system of nature preserves, what statement are we making about ourselves in relation to other species? If, in contrast, we attempt to achieve conservation biology's goals through both reserves and restraints on our own behavior where we live, are we making a different statement? In what ways do you think our choices might make a difference in the outcomes of conservation?

Numerous studies have made it increasingly apparent that the largest nature reserves in the world are, by themselves, inadequate to sustain most of the species present in them for long periods of time. If the powerful forces and processes associated with human presence and activity in the landscape are not deliberately modified to permit other species to persist with humans and to permit normal ecosystem processes to occur, the worldwide system of nature reserves alone will not sustain current levels of biodiversity. Unless humans as individuals, cultures, and economies change fundamental patterns of behavior, worldwide biodiversity will continue to decline rapidly. These problems and their analyses will be the subject of subsequent chapters.

STUDYING AND LEARNING FROM THIS TEXT: A PERSPECTIVE

The chapters that follow use these fundamental problems of conservation biology as the controlling ideas and organizing principles to examine and understand this fascinating and growing field. We will not attempt to examine every subject that has been, or is now alleged to be, part of the science of conservation biology, although we will consider examples of problems, paradigms, and concepts, combined with well-chosen case histories, to aid your understanding. Facts have meaning only when they exist in the context of concepts that explain their significance. No professional scientist finds success in work or respect among peers simply because he or she "knows all the facts." Rather, respected scientists are able to command large quantities of information precisely because they understand its *meaning* in relation to essential concepts and theories that define their discipline. Similarly, students do not long remember facts simply because professors require them to be memorized. Rather, students mature to true command of their discipline when the facts they know become vivid examples of theories that can explain repeated examples of related events.

I believe that conservation biology can be presented as a unity of knowledge made meaningful by intimately related concepts and theories that explain what facts mean together with the historical perspectives that led to their discovery. Therefore, I believe that the purpose of a text in conservation biology is not to present every "fact" associated with the discipline, but rather to present a unity of concepts related in such a way as to mark the boundaries of the discipline (what conservation biology is not, as well as what it is), its appropriate subjects of study and problems to be solved, and the examples and data which illustrate the meaning of such concepts and their applications.

The subjects of conservation biology change constantly. Any person pursuing this field, whether personally or professionally, will encounter problems and situations completely unlike any case history we could examine today. Thus, the best preparation is not to memorize the details of the work of others, interesting and helpful as that may be. Rather, it is to understand the principles, with help from previous studies, of investigations that lead to ongoing and informed solutions to the dilemmas of conservation biology, both as it is today, and as it will be in the years to come.

What Makes Conservation Biology a Distinctive Discipline? *—A Directed Discussion*

Reading assignment: (1) Bunnell, F. L., and L. A. Dupuis. 1995. *Conservation Biology*'s literature revisited: wine or vinaigrette? *Wildlife Society Bulletin* 23:56–62. (2) Noss, R. F. 1995. Assessing rigor and objectivity in conservation science. *Wildlife Society Bulletin* 23:539–541.

Questions

1. Bunnell and Dupuis report that, during the period examined, articles in *Conservation Biology* were more theoretical and less likely to have clearly defined, repeatable methods than articles in *The Journal of Wildlife Management* and *Wildlife Society Bulletin*. They assert that this tendency creates obstacles to testing and applying results of *Conservation Biology* articles. Evaluate the validity of their methodology and justification. What are the advantages of publishing studies with repeatable methods? Can you think of any limitations of publishing only work with repeatable methods?

2. Bunnell and Dupuis assert that conservation biology is more value driven than wildlife management and that "a forthrightly value-driven science will find itself much more vulnerable to changes in society than one that strives for value-neutral ground." Their rationale is that society determines values through legislation and that the role of scientists is to address the adequacy of management to achieve legislated values. Is conservation biology more likely to shape social values or to be shaped by them?

3. *The Journal of Wildlife Management* and *Wildlife Society Bulletin* contain more articles about single-species management than *Conservation Biology*. What are the advantages and disadvantages of the single-species approach?

4. In the years since these articles were published, it has become clear that the Society for Conservation Biology and The Wildlife Society will not merge, but will remain separate and distinct. What are the foundational differences in philosophy, values, and scientific/management approaches that maintain the two groups as "separate species"? What, if anything, is the basis for continued and future cooperation between the two societies and their members?

Learning Online

Visit our webpage at www.mhhe.com/conservation for case studies, animations, practice quiz questions, and additional readings to help you understand the material in this chapter. You'll also find active links to the following topics:

General Environmental Sites
General Ecology Sites
History of Environmental Studies
Environmental Philosophy
Land-Use Planning
Species Preservation

Environmental and Ecological
   Organization Sites
Sustainable Agriculture
The Ultimate Ecological Answers
Individual Contributions to
   Environmental Issues

Conservation and Management of
   Habitats and Species
Miscellaneous Environmental
   Resources

Literature Cited

Alvard, M. S. 1993. Testing the "ecologically noble savage" hypothesis: interspecific prey choices by Piro hunters of Amazonian Peru. *Human Ecology* 21:355–87.

Andrewartha, H. G., and L. C. Birch. 1954. *The distribution and abundance of animals.* Chicago, Ill.: University of Chicago Press.

Bennett R., J. S. Dixon, V. H. Calahane, W. W. Chase, and W. L. McAtee. 1937. Statement of policy. *The Journal of Wildlife Management* 1:1–2.

Boecklen, W. J., and N. J. Gotelli. 1984. Island biogeographic theory and conservation practice: species-area or specious-area relationships. *Biological Conservation* 29:63–80.

Boyce, M. S. 1992. Population viability analysis. *Annual Review of Ecology and Systematics* 23:481–506.

Bunnell, F. L., and L. A. Dupuis. 1995. *Conservation Biology*'s literature revisited: wine or vinaigrette? *Wildlife Society Bulletin* 23:56–62.

Caldwell, L. K. 1970. The ecosystem as a criterion for public land policy. *Natural Resources Journal* 10:203–21.

Callicott, J. B. 1990. Whither conservation ethics? *Conservation Biology* 4:15–20.

Carson, R. 1962. *Silent spring.* Boston, Mass.: Houghton Mifflin.

Caughley, G., and A. Gunn. 1996. *Conservation biology in theory and practice.* Cambridge, Mass.: Blackwell Science.

Clements, F. E., and V. Shelford. 1939. *Bio-ecology.* New York: Wiley.

Cox, G. F. 1997. *Conservation biology.* 2d ed. Dubuque, Iowa: William C. Brown.

Craighead, F. 1979. *Track of the grizzly.* San Francisco, Calif.: Sierra Club Books.

Danielson, B. J. 1991. Communities in a landscape: the influence of habitat heterogeneity on the interactions between species. *American Naturalist* 138:1105–20.

den Boer, P. J. 1968. Spreading of risk and stabilization of animal numbers. *Acta Biotheoretica* 18:165–94.

Dudley, J. P. 1995. Rejoinder to Rohlf and O'Connell: biodiversity as a regulatory criterion. In *Readings from* Conservation Biology: *the social dimension—ethics, policy, management, development, economics, education,* ed. D. Ehrenfeld, 102–4. Cambridge, Mass.: Blackwell Science.

Emerson, R. W. 1836. *Nature.* Boston, Mass.: James Monroe.

Errington, P. L., and F. N. Hamerstrom. 1937. The evaluation of nesting losses in juvenile mortality of the ring-necked pheasant. *The Journal of Wildlife Management* 1:3–20.

Frankel, O. H. 1974. Genetic conservation: our evolutionary responsibility. *Genetics* 78:53–65.

Frankel, O. H., and M. E. Soulé. 1981. *Conservation and evolution.* Cambridge, England: Cambridge University Press.

Franklin, I. R. 1980. Evolutionary change in small populations. In *Conservation biology: an evolutionary-ecological perspective,* eds. M. E. Soulé and B. A. Wilcox, 135–50. Sunderland, Mass.: Sinauer Associates.

Frome, M. 1987. *The national parks.* 3d ed. Chicago, Ill.: Rand McNally.

Funtowicz, S. O., and J. R. Ravetz. 1994. The worth of a songbird: ecological economics as a post-normal science. *Ecological Economics* 10:197–207.

Gibbons, A. 1992. Conservation biology in the fast lane. *Science* 255:20–22.

Grumbine, R. E. 1994. What is ecosystem management? *Conservation Biology* 8:27–38.

Hamilton, E., and H. Cairns, eds. 1961. *Plato: the collected dialogues.* Princeton, N.J.: Princeton University Press.

Hanski, I. 1991. Single-species metapopulation dynamics: concepts, models and observations. *Biological Journal of the Linnaean Society* 42:17–38.

Hanski, I. 1997. Habitat destruction and metapopulation dynamics. In *The ecological basis of conservation: heterogeneity, ecosystems, and biodiversity,* eds. S. T. A. Pickett, R. S. Ostfeld, M. Shachak, and G. E. Likens, 217–27. New York: Chapman and Hall.

Hooper, M. D. 1971. The size and surroundings of nature reserves. In *The scientific management of animal and plant communities for conservation,* eds. E. Duffey and A. S. Watt, 555–61. Oxford, England: Blackwell.

Hoopes, M. F., and S. Harrison. 1998. Metapopulation, source-sink, and disturbance dynamics. In *Conservation science and action,* ed. William J. Sutherland, 135–51. Oxford, England: Blackwell Science.

Hunter, M. L., Jr. 1996. *Fundamentals of conservation biology.* Cambridge, Mass.: Blackwell Science.

Jensen, M. N., and P. R. Krausman. 1993. *Conservation Biology's* literature: new wine or just a new bottle? *Wildlife Society Bulletin* 21:199–203.

Kalupahana, D. J. 1987. *The principles of Buddhist psychology.* Albany: State University of New York Press.

Krieger, M. H. 1973. What's wrong with plastic trees? *Science* 179:446–55.

Leopold, A. 1966. *A Sand County almanac with essays on conservation from Round River.* New York: Sierra Club/Ballantine.

Levins, R. 1968. *Evolution in changing environments: some theoretical explorations.* Monograph in population biology, no. 2. Princeton, N.J.: Princeton University Press.

Levins, R. 1969. Some demographic and genetic consequences of environmental heterogeneity for biological control. *Bulletin of the Entomological Society of America* 15:2337–2340.

Levins, R. 1970. Extinction. In *Some mathematical problems in biology,* ed. M. Gesternhaber, 77–107. Providence, R.I.: American Mathematical Society.

MacArthur, R. H., and E. O. Wilson. 1967. *The theory of island biogeography.* Princeton, N.J.: Princeton University Press.

Marsh, G. P. 1965. *Man and nature.* Cambridge, Mass.: Belknap Press.

Meffe, G. K., and C. Ronald Carroll. 1997. *Principles of conservation biology.* Sunderland, Mass.: Sinauer Associates.

Moore, N. W. 1962. The heaths of Dorset and their conservation. *Journal of Ecology* 50:369–91.

Muir, J. 1997. *Nature writings.* New York: Library of America.

Naess, A. 1986. Intrinsic value: will the defenders of nature please rise? In *Conservation biology: the science of scarcity and diversity,* ed. M. E. Soulé, 504–15. Sunderland, Mass.: Sinauer Associates.

Nash, R. 1967. *Wilderness and the American mind.* New Haven, Conn.: Yale University Press.

Nash, R. F. 1989. *The rights of nature.* Madison, Wisc.: University of Wisconsin Press.

Noss, R. F. 1989. Who will speak for biodiversity? *Conservation Biology* 3:202–3.

Noss, R. F. 1995. Assessing rigor and objectivity in conservation science. *Wildlife Society Bulletin* 23:539–541.

Noss, R. F., and A. Cooperrider. 1994. *Saving nature's legacy: protecting and restoring biodiversity.* Washington, D.C.: Defenders of Wildlife and Island Press.

Ostfeld, R. S., S. T. A. Pickett, M. Shachak, and G. E. Likens. 1997. Defining the scientific issues. In *The ecological basis of conservation: heterogeneity, ecosystems, and biodiversity,* eds. S. T. A. Pickett, R. S. Ostfeld, M. Shachak, and G. E. Likens, 3–10. New York: Chapman and Hall.

Petulla, J. M. 1977. *American environmental history.* San Francisco, Calif.: Boyd and Fraser.

Petulla, J. M. 1988. *American environmental history.* 2d ed. New York: Macmillan.

Pimm, S. L., and M. E. Gilpin. 1989. Theoretical issues in conservation biology. In *Perspectives in ecological theory,* eds.

J. Roughgarden, R. M. May, and S. A. Levin, 287–305. Princeton, N.J.: Princeton University Press.

Primack, R. B. 1993. *Essentials of conservation biology.* Sunderland, Mass.: Sinauer Associates.

Pulliam, H. R. 1988. Sources, sinks, and population regulation. *American Naturalist* 132:652–61.

Rohlf, D. J. 1995a. Six biological reasons why the Endangered Species Act doesn't work—and what to do about it. In *Readings from* Conservation Biology: *the social dimension— ethics, policy, management, development, economics, education,* ed. D. Ehrenfeld, 86–95. Cambridge, Mass.: Blackwell Science.

Rohlf, D. J. 1995b. Response to O'Connell. In *Readings from* Conservation Biology: *the social dimension—ethics, policy, management, development, economics, education.* ed. D. Ehrenfeld, 100–101. Cambridge, Mass.: Blackwell Science.

Roosevelt, T. 1926. *State papers as governor and president 1899–1909.* New York: Scribner.

Shaffer, M. L. 1981. Minimum population sizes for species conservation. *BioScience* 31:131–34.

Shelford, V. E. 1933. Ecological Society of America: a nature sanctuary plan unanimously adopted by the Society, December 28, 1932. *Ecology* 14:240–45.

Simberloff, D. 1986. Are we on the verge of a mass extinction in tropical rain forests? In *Dynamics of extinction,* ed. D. K. Elliot, 165–80. New York: Wiley.

Simberloff, D. 1988. The contribution of population and community biology to conservation science. *Annual Review of Ecology and Systematics* 19:473–511.

Simberloff, D. 1997. Biogeographic approaches and the new conservation strategy. In *The ecological basis of conservation: heterogeneity, ecosystems, and biodiversity,* eds. S. T. A. Pickett, R. S. Ostfeld, M. Shachak, and G. E. Likens, 274–84. New York: Chapman and Hall.

Simberloff, D. 1998. Small and declining populations. In *Conservation science and action,* ed. William J. Sutherland, 116–34. Oxford, England: Blackwell Science.

Simberloff, D., J. A. Farr, J. Cox, and D. W. Mehlman. 1995. Movement corridors: conservation bargains or poor investments. In *Readings in* Conservation Biology: *the landscape perspective,* ed. David Ehrenfeld, 74–85. Cambridge, Mass.: Blackwell Science.

Soulé, M. E. 1980. Thresholds for survival: maintaining fitness and evolutionary potential. In *Conservation biology: an evolutionary-ecology approach,* eds. M. E. Soulé and B. A. Wilcox, 111–24. Sunderland, Mass.: Sinauer Associates.

Soulé, M. E. 1985. What is conservation biology? *BioScience* 35:727–34.

Soulé, M. E., ed. 1986. *Conservation biology: the science of scarcity and diversity.* Sunderland, Mass.: Sinauer Associates.

Soulé, M. E., and K. A. Kohm, eds. 1989. *Research priorities for conservation biology.* Washington, D.C.: Island Press.

Soulé, M. E., and B. A. Wilcox, eds. 1980. *Conservation biology: an evolutionary-ecology approach.* Sunderland, Mass.: Sinauer Associates.

Teer, J. G. 1988. Conservation biology: the science of scarcity and diversity. Book review. *The Journal of Wildlife Management* 52:570–72.

Thomas, J. W., and H. Salwasser. 1989. Bringing conservation biology into a position of influence in natural resource management. *Conservation Biology* 3:123–27.

Thoreau, H. D. 1863. *Excursions.* Boston, Mass.: Ticknor and Fields.

Thoreau, H. D. 1975. *The selected works of Thoreau,* ed. Walter Harding. Boston, Mass.: Houghton Mifflin.

Thoreau, H. D. 1995. *Walking.* New York: Penguin.

Tribe, L. H. 1974. Ways not to think about plastic trees: new foundations for environmental law. *Yale Law Journal* 83:1315–48.

Trumbell, W. The Washburn Yellowstone Expedition, parts 1 and 2. May–June 1871. *Overland Monthly* 6(5–6):431–37, 489–96.

Udall, S. L. 1988. *The quiet crisis and the next generation.* Salt Lake City, Utah: Gibbs Smith.

Wiens, J. A. 1997. The emerging role of patchiness in conservation biology. In *The ecological basis of conservation: heterogeneity, ecosystems, and biodiversity,* eds. S. T. A. Pickett, R. S. Ostfeld, M. Shachak, and G. E. Likens, 93–107. New York: Chapman and Hall.

Wilcox, B. A., and D. D. Murphy. 1985. Conservation strategy: the effects of fragmentation on extinction. *American Naturalist* 125:879–87.

Wolfe, L. M. 1946. *Son of the wilderness; the life of John Muir.* New York: Knopf.

Wright, G. M., and B. Thompson. 1935. *Fauna of the national parks of the U.S.* Washington, D.C.: U.S. Department of the Interior.

CHAPTER 2

Science is increasingly criticized not because it is bad, but because it provides inadequate guidance to answer questions posed by legislatures and administrators.

—A. Dan Tarlock, 1994

The Legal Foundations of Conservation Biology

In this chapter, you will learn about:

1 the development and contemporary expressions of conservation law and its relationship to the science of conservation biology.

2 the most important U.S. and international conservation laws and how they define and empower conservation.

3 specific case histories in which national and international conservation laws have influenced the goals and practices of conservation biology.

CONSERVATION AND LAW

Conservation biology is a legally empowered discipline; that is, it represents a scientific community that has received legal, political, and cultural incentives and reinforcements. Indeed, some have gone so far as to call conservation biology a "regulatory science" that "seeks to develop scientific standards that can be applied to regulatory criteria and then to develop management strategies to meet those standards" (Tarlock 1994). Throughout the world, the goals of conservation biology, including preservation of biodiversity, protection of endangered species, and conservation and management of ecosystems, are increasingly established in and enabled by laws. In the United States, most of the major environmental legislation had been established by the mid-1970s, preceding the development of conservation biology as a professional and academic discipline. Although conservation biology might still have developed without national environmental legislation, it would have been substantially less influential.

As demonstrated in chapter 1, the U.S. federal government had been taking an ever-growing role in conservation since the late nineteenth century. During this period, U.S. citizens had become increasingly accustomed to accept federal government authority and "wisdom" in making conservation and environmental laws, and such laws were more likely to be obeyed. However, in the second half of the twentieth century, an increasingly well-educated, affluent, and urban public also began to be more aware of the need for environmental protection, the decline and loss of individual species, and the growing possibility that the activities of both big business and big government might need legislative restraint to protect environmental quality. The body of modern environmental legislation enacted in the last four decades reflected these social changes and has affected conservation biology in the following ways. First, it has given legal incentives and approval for biodiversity preservation. Second, it has affirmed many of the goals of conservation biology and influenced the public to value conservation. Third, such legislation has provided a legal environment that both requires and sustains scientific research, management, and monitoring.

Despite the obvious advantages of this legal empowerment, conservation biology's ties to legislation, policy, and litigation are not always beneficial. The study of law in general, and of conservation law in particular, is paradoxical. On one hand, laws represent current social values. But laws also shape values for future generations, codifying aspirations or preferences into something more lasting and transcendent. Laws empower action, providing political resources and social force to achieve specific goals, but laws also limit action by setting arbitrary and fixed boundaries that may not correspond to the needs of dynamic systems. Because laws are difficult to change or repeal, they provide a sense of permanence to the values they establish. For the same reason, laws can become rigid, unresponsive to changing conditions, and thus no longer effective in solving the problems they were enacted to address.

Traditionally, scientists have avoided involvement in law and policy formulation because they believed that advocacy would undermine their professional objectivity and public credibility; however, many conservation biologists disagree. Reed Noss, a former editor of the discipline's most well-known journal—*Conservation Biology*—said, "I believe that conservation biologists have a responsibility to enter the policy arena and advocate both general principles and specific actions needed to conserve biodiversity" (Noss 1993). The content of *Conservation Biology* has supported this emphasis: from 1987 to 1995, nearly 10% of the journal's articles were devoted to policy issues (Meffe and Viederman 1995). Because conservation biology espouses advocacy in ways and to degrees that other disciplines do not, connections to conservation law and policy are intrinsic to conservation biology's continuing mission, as well as to understanding its historic development. Students of conservation biology who overlook this point will suffer two disadvantages. First, they will be naive in their understanding of their own discipline. Second, they will be ill equipped to use legislation to achieve conservation goals or to participate effectively in the formulation of conservation policies.

Good science and its attendant empirical data are necessary, but insufficient for achieving conservation biology's goals of stemming species extinction and ecosystem degradation (Meffe and Viederman 1995). Conservation biology, as a discipline, asserts that scientists can and should influence environmental policy. But, to do this, they must first comprehend both science and policy. Therefore, this chapter explains the history and practice of conservation law and policy and their effects on conservation biology.

THE DEVELOPMENT OF CONSERVATION LEGISLATION AND POLICY IN THE UNITED STATES

The first attempts at developing environmental legislation in the United States began in the second half of the nineteenth century and included the establishment of Yellowstone National Park (1872) and the Forest Reserve Act (1891). Such legislative actions, and others like them, were designed primarily to preserve natural resources, whether as scenery or commodities, in what were viewed as static ecosystems. The law that established the National Park Service (NPS) in 1916 articulates this view of ecosystems, stating that the mission of the NPS is to "conserve the scenery and the natural and historic objects and the wildlife therein, and to provide for the enjoyment of the same in such manner and by such means as will leave them unimpaired for the enjoyment of future generations."

Whereas the establishment of Yellowstone National Park and the NPS provide examples of laws that contain a bias toward the preservation of "scenery," other laws were passed to preserve commodity resources and to regulate their extraction and use. But, as noted in chapter 1, the environmental problems that accompanied the extraction and use of natural resources, such as air and water pollution, habitat destruction, and species endangerment, were seen as local annoyances, not global threats. Rachel Carson's book *Silent Spring* on environmental problems

demolished this parochial view by demonstrating how pollutants, in this case, pesticides, were transported worldwide through ecological processes and ultimately affected all ecosystem components. Despite the persuasiveness of Carson's science and prose and a groundswell of public and political support, the publication of *Silent Spring* did not result in the banning of Carson's primary target, the pesticide dichlorodiphenyltrichloroethane, more commonly known as DDT. Public appeals and political efforts to ban DDT were frustrated by the combined efforts of the pesticide and agriculture industries. As late as 1972, a decade after the publication of *Silent Spring,* DDT was still in widespread use throughout the United States. Its demise did not come through popular effort or legislation, but through litigation.

A new conservation organization, the Environmental Defense Fund (EDF), was formed in 1968. Lawyers, who made up a disproportionately large share of EDF's small charter membership, brought with them a distinctively different approach for attacking environmental problems. Their strategy was to legally challenge environmentally destructive activities and then support their challenges with scientific evidence. They believed they could succeed by asserting a fundamental, but unprecedented and untested, public right—the right of every citizen to a clean environment. Relying on careful and detailed scientific testimony, the EDF initially achieved a statewide ban of DDT in Michigan. This success led to a national forum for the case and ultimately resulted in the banning of DDT's use in the United States (Dunlap 1981).

Beginning with the signing of the National Environmental Policy Act (NEPA) on January 1, 1970, and the development of NEPA's implementation procedures, the 1970s witnessed the development of a number of innovative structural and legal arrangements in conservation laws. Three such arrangements intentionally developed to promote greater public participation were of particular importance: liberal provisions for public participation, expanded rights for private organizations and individuals to sue public agencies, and provisions for intervenor funding for legal expenses. For example, NEPA provided the legal basis for a series of legal challenges to federal agencies' environmental actions that began in 1971 and led the U.S. Supreme Court to affirm the right of private citizens and nongovernmental organizations to sue these agencies for harmful or potentially harmful environmental actions. Other court decisions established the ability of citizens and nongovernmental organizations to halt proposed actions by federal agencies if the required environmental impact statements were judged to have been improperly prepared.

The EDF's innovative approach to conservation through law and litigation, NEPA's legal requirement for environmental impact statements accompanying proposed actions by federal agencies, and the success of citizen-led litigation demonstrated the combined power of the legal process and public input in achieving conservation goals. Through these forces, conservation was being transformed from a respectable protest movement of the Progressive Era to a body of ethics and values affirmed by legal statutes. And the growing body of environmental legislation became more important in shaping management decisions affecting conservation on both public and private lands.

The United States emerged as a world leader in conservation in the 1960s and 1970s primarily through the passage of a comprehensive battery of environmental and conservation

Table 2.1 A Partial List of Major U.S. Conservation and Environmental Legislation Since 1960

LEGISLATION	YEAR PASSED	DESCRIPTION
Wilderness Act	1964	• Established the National Wilderness Preservation System and defined wilderness as "an area where the earth and its community of life are untrammeled by man, where man himself is a visitor who does not remain"[a] • Established protection of wilderness lands from industrial development by prohibiting timber harvesting, vehicle access, and building of roads and permanent structures on land designated as wilderness[b]
Land and Water Conservation Fund Act (LWCF)	1965	• Established to preserve land for recreational opportunities for all Americans • Designated federal revenues (i.e., property sales, motorboat fuel taxes, and federal recreation user fees) to establish a trust fund for the acquisition of parks and conservation lands[a] • Amended 1968: expanded to include receipts from oil and gas lease revenues from the outer continental shelf[c]
National Environmental Policy Act (NEPA)	1969	• Established environmental quality as a leading national priority of major federal actions • Required that a detailed statement on the ecological consequences of planned actions be written before any action is taken so that the federal agencies can evaluate how their actions may impact the natural and human environment[b] • Required that an environmental impact statement (EIS) be written if the environmental assessment (EA) finds that the planned actions would have a significant impact on the environment
Clean Air Act (CAA)	1970	• Required EPA to set air quality standards with the intent to avoid adverse health effects and decrease particulate matter in the air[d] • Regulated industrial pollution: required industries to apply, and pay for, pollution permits • Required that each state develop a state implementation plan (SIP) describing its efforts to control pollution • Increased spending on air pollution control: $9 million in 1970 to $30 million by 1990
Clean Water Act (CWA)	1972	• Aspired to attain zero discharge of pollutants and focused on improving the quality of navigable waters • Established a permit system and set industrial limits on tolerable pollution discharge • Required municipal waste treatments with federal aid to construct publicly owned sewage treatment works (POTWs) • Amended 1977: created pretreatment requirements for industrial firms that discharge wastes into POTWs and required that states develop management plans for nonpoint source pollution[d] • Amended 1987: established a program to phase out federal grants by 1995 by giving responsibility to the states to raise and distribute funds
Marine Protection, Research, and Sanctuaries Act (MPRSA)	1972	• Designated sites and times for waste disposal • Regulated waste disposal into the ocean beyond the 3-miles-from-shore territorial limit by prohibiting dumping of waste that would adversely affect "human health, welfare, or amenities, or the marine environment, ecological systems, or economic potentialities" such as radiological, chemical and biological warfare agents, high-level radioactive waste, and medical wastes • Established a research program to study the effects waste disposal has on the ocean • Authorized the issuance of permits by the Secretary of the Army for dredged material disposal

Continued

Table 2.1 *Continued*

LEGISLATION	YEAR PASSED	DESCRIPTION

Endangered Species Act (ESA)	1973	• Designed to prevent extinction of species in the United States and worldwide
		• Amended 1978: the act was made more sensitive to economic impacts
		• Authorized the determination and listing of species as endangered and threatened
		  • Prohibited unauthorized taking, possession, sale, and transport of endangered species
		  • Provided authority to acquire land for the conservation of listed species, using land and water conservation funds
		• Authorized establishment of cooperative agreements and grants-in-aid to states that establish and maintain active and adequate programs for endangered and threatened wildlife and plants
		• Authorized the assessment of civil and criminal penalties for violating the act or regulations, and payment of rewards to anyone furnishing information leading to arrest and conviction for any violation of the act
		  • Required shrimp "certification" and turtle excluder devices (TEDs) in fishing nets (Section 609)
		  • Amended 1989: prohibited importing shrimp from countries that do not meet regulations, but applied only to Atlantic/Caribbean region.
		  • Amended 1996: shrimping regulations and prohibitions applied to all countries

National Forest Management Act (NFMA)	1974	• The primary statute governing the administration of national forests • Required the secretary of agriculture to assess forest lands, develop a management program based on multiple-use, sustained-yield principles, and implement a resource management plan for each unit of the national forest system • Authorized $200,000,000 to be appropriated annually to meet the requirements of the act, for reforesting and treating lands in the national forest system • Stressed that forest planning should protect plant, animal, and tree species so that forest biodiversity may be retained • Restrained forest use that would deplete soil and water resources so that permanent impairment of productivity would not occur

Comprehensive Environmental Response, Conservation, and Liability Act (CERCLA)	1980	• Also known as *Superfund*, the name of the trust fund established under this law, its purpose is to clean up sites in order to minimize unfavorable impacts on public health and the environment
		• Required identification of potentially responsible parties (PRPs) to pay for the cleanup of hazardous waste sites
		• Amended 1986 as Superfund Amendments and Reauthorization Act (SARA)
		  • Established purchaser protection: a purchaser would not be liable for remediation costs if the party could demonstrate that it "did not know, nor had reason to know" of the hazardous waste contamination when the party acquired the property
		  • Promoted citizen participation in deciding how a waste site should be cleaned and developed
		• Amended 1995: Brownfield's Action Agenda
		  • Addressed the problem of brownfields (former industrial properties that are now unused because of uncertain conditions)
		  • EPA would not sue new owners for cleanup of contamination that occurred prior to purchase
		  • Stressed that new owners must agree to limit future use to commercial and industrial purposes
		  • Created the Community Reinvestment Act (CRA): made EPA grants available to clean up and redevelop waste sites

aRodgers 1994,   bLoomis 1993,   cEndicott 1993,   dMeiners and Yandle 1993 laws (table 2.1). Although legislators approved a myriad of individual environmental statutes during this period, it was primarily five laws that formed the foundations of modern conservation legislation in the United States. The Land and Water Conservation Fund Act of 1965 (Section 2) established a special fund for the acquisition of parks and conservation lands; the fund's sources included receipts from oil and gas leasing on the outer continental shelf. The Wilderness Act of 1964 led to the establishment of the National Wilderness Preservation System, which now encompasses more than a million acres in the United States. Because the Wilderness Act required attentive monitoring of conditions in wilderness areas, it set a strong research agenda. The established wilderness areas provided researchers the opportunity to study functioning, intact ecosystems and to preserve biodiversity on a landscape scale, and the law effectively demanded that such opportunity be used. The Federal Water Pollution Control Act Amendments of 1972 (commonly known as the Clean Water Act, or CWA) made it unlawful to discharge pollutants into U.S. water supplies. This statute has had more effect than any other on substantially improving the quality of freshwater ecosystems in the United States. The earlier-mentioned NEPA (1969) required the completion of an environmental impact statement to evaluate all actions by federal agencies or private agencies requiring a federal permit that might have a significant effect on the human environment. The Endangered Species Act (ESA) of 1973 prohibited federal agencies from taking action "likely to jeopardize" the existence of a listed species or result in the destruction or adverse modification of its designated critical habitat. These laws have promoted conservation in the United States, have been imitated throughout the world, and have generated continued controversy over their enforcement.

Common Characteristics of Effective Conservation Law

Although U.S. laws addressing environmental and conservation issues are diverse, the most powerful and effective acts share important characteristics. Their common traits include an inspirational and radical message, the potential for growth in influence, an ability to attract and hold the interest of scientists because they raise questions that must be answered by researchers, and a requirement for monitoring (Rodgers 1994).

The inspirational and radical message of the best modern environmental and conservation laws built a strong foundation of moral and social support. Although court interpretation often has been necessary for the message to be clarified and implemented, such a message has been latent within all truly effective conservation legislation. Legal scholar and law professor William H. Rogers, Jr., said of these exceptional environmental laws that "they lack the compromised and ambiguous form normally associated with an act of Congress" (Rodgers 1994). Indeed, the most effective and lasting statutes in U.S. environmental law were almost brazen in their language, values, and goals, and they inspired popular support. The potential for growth in influence allowed such laws to alter social values, and they gained and held scientific and

professional support because they defined tasks for scientists to perform and questions for them to answer.

Several themes of U.S. environmental legislation have become part of conservation biology. Such legislation (1) required that pollution or environmental degradation be evaluated in the context of ecosystem function (CWA, NEPA); (2) endorsed intrinsic and noneconomic values for resources and nonhuman creatures (NEPA, ESA); (3) emphasized the status of individual species and affirmed that extinction is undesirable (ESA); (4) stated that renewable resources were to be managed sustainably, and that managers of nonrenewable resources must take into account the permanent consequences of present management actions (NEPA); (5) made federal funding available for research and habitat acquisition (ESA, LWCF); (6) made polluters liable for damage (CWA); (7) provided private citizens and nongovernmental organizations with avenues for participation in decision making and litigation against federal agencies (ESA, NEPA); and (8) gave additional power to federal agencies to manage and protect resources (ESA, NEPA). However, it is not sufficient for a conservation biologist merely to know the content of particular environmental laws. To be effective, a conservation biologist must understand the process through which laws and policies are formed and how they affect the responses of citizens and government to conservation problems.

THE PROBLEM OF PROCESS: HOW DOES CONSERVATION LAW ARISE?

The formulation of conservation law and its subsequent translation into administrative policy usually follow a predictable pattern: (1) agenda setting (convincing government officials that an issue is of sufficient importance and urgency that it must be addressed); (2) formulation and legitimization (the setting of specific goals with specific plans and proposals to reach such goals); and (3) assessment and reformulation (evaluating the social effects of government policies, judging the desirability of those effects, and communicating with the government and the public). A fourth step, termination, can occur if the policy or the law upon which the policy rests is repealed. Termination is a relatively rare event because government agencies and directives, once initiated, tend to assume a life of their own and are notoriously difficult to kill (Liroff 1976). The most important pieces of environmental legislation have proved to be almost "repeal proof" (Rodgers 1994). Their continuation does not depend on election results. Thus, they are likely to be with us for the foreseeable future.

Throughout this text I distinguish "policy" from "law." The policy is the necessary outcome of all laws that are actually enforced, and can be defined as "a definite course or method of action selected from among alternatives and in light of given conditions to guide and determine present and future decisions" (Webster 1971). Laws originate with issues, but issues do not become law, and laws are not translated into policies without first going through the phases listed above. In all of these phases, public participation can play a critical role.

Public Participation

Numerous statues and regulations in environmental and conservation law and policy stress the importance of public participation in implementation and enforcement of laws and policies, especially those that are regulatory in nature. Such regulations allowed previously excluded groups to enter into the policy-making and decision-making processes, and made it increasingly difficult for government agencies to ignore environmental and conservation concerns or to make decisions without public knowledge or response. For example, in the preparation of environmental impact statements required under the National Environmental Policy Act, responsible agencies are required to solicit public comment through public meetings, to publish notice of such meetings in local newspapers as well as the *Congressional Record,* to record comments received and to explicitly respond to all reasonable public comments and concerns within the draft and final versions of the environmental impact statement.

Public participation in the United States has also increased through litigation. Expanding the right to sue, a key legal victory won in early environmental lawsuits throughout the 1970s, has provided much greater access to the courts. Litigation—or threat of litigation—has become a powerful weapon for conservationists against government actions that are perceived as harmful to the environment. Intervenor funding was made available for the substantial legal fees associated with conservation litigation. This arrangement made it possible for small groups or even a single individual with limited financial resources to challenge well-financed government agencies and business interests, especially if a judge decided that the interests presented by the plaintiff would have been unrepresented by any other party.

Public comment and litigation on proposed government action, although highly effective in specific cases, is necessarily reactive rather than proactive. The public can comment only after a government agency has proposed or initiated a specific action. A lawsuit can be filed only after a government initiative is already underway. In contrast, the most proactive avenue for public input and participation has become the private nongovernmental organization (NGO). Most NGOs are open to anyone. Some require a membership (usually nominal), whereas others, such as The Nature Conservancy, have no required membership charge at all. The largest NGOs, such as the Audubon Society, boast memberships in the millions; others, like The World Wildlife Fund, operate offices in more than 50 countries.

NGOs have become particularly effective at focusing the concern of public citizens at the international level because, apart from international treaties and conventions, there is no comprehensive international legal system for conservation or any single global authority that could enforce it. NGOs are often able to operate effectively as channels of public input at international levels through an array of different activities, including sponsoring independent research, lobbying local and national governments, applying pressure to local and national governments and multinational corporations to enforce or observe conservation law, spending private funds for practical conservation management, monitoring global environmental conditions, building political coalitions to support public conservation policy, and promoting public awareness of environmental problems (McCormick 1999). All such activities affect both the development and enforcement of conservation and environmental law. Many go even further and take steps to influence law and policy directly. In the United States, for example, individual NGOs or NGO coalitions often form political action committees that raise money to support selected candidates in both national and local elections.

Some NGOs, such as The Nature Conservancy, have used private property laws to their own advantage, developing extensive worldwide systems of private preserves, as well as numerous local networks of private landowners who participate in easements that legally restrict or prevent land development. The role of the NGO as a primary avenue of public participation in the formation of environmental law continues to grow. In 1996 the *World Directory of Environmental Organizations* described more than 2,600 NGOs, but this number included only the largest organizations (McCormick 1999).

THE PROBLEM OF APPLICATION: HOW DOES CONSERVATION LAW WORK?

Two Case Histories: NEPA and the ESA

Environmental and conservation laws have provided conservationists with the legal means to stop activities harmful to the environment or to particular species, especially on federal lands or on projects receiving federal funding or requiring federal permits. Of these, the National Environmental Policy Act and the Endangered Species Act, passed and enforced separately but often interacting legally, have radically altered the practice and enforcement of conservation values in the United States and, by imitation, throughout the world. Although previously mentioned in our review of important U.S. environmental laws, we "backtrack" here to examine these two statutes in detail. More than any other legislation, the radical transformation of conservation law achieved by these two acts created the legal environment and social values in which conservation biology operates today.

The National Environmental Policy Act

In 1966 a professor of public administration and a rather unlikely environmental hero, Lynton K. Caldwell, published a paper titled "Administrative Possibilities for Environmental Control" (Caldwell 1966). In his paper, Caldwell suggested that qualitative environmental standards could provide the administrative coherence historically lacking in natural resource policy (Caldwell 1966; Tarlock 1994). Caldwell's paper, published in the book *Future Environments of North America* (Darling and Milton 1966), quickly became one of the most influential publications on environmental policy of the late 1960s.

The U.S. Congress employed Caldwell as the principal drafter of a law that was designed to be the centerpiece of a new era of environmental and conservation legislation, the National Environmental Policy Act of 1969 (Tarlock 1994). In writing NEPA, Caldwell mandated that a "detailed statement" must accompany "proposals for legislation and *other major federal actions significantly affecting the quality of the human environment*" (emphasis added). This requirement led to the development of the now-familiar environmental impact statement (EIS) that describes the possible environmental effects of actions proposed by federal agencies. Ultimately, policies and procedures associated with preparation of the EIS led to pervasive and well-defined procedures for public involvement, as well as for challenging an EIS in court.

NEPA was signed into law by President Richard Nixon on January 1, 1970, a fitting beginning to what would be called "the decade of the environment." NEPA stated a national policy for the environment and formally established environmental quality as a leading national priority. In NEPA's own words,

"It is the continuing responsibility of the federal government to use all practicable means, consistent with other essential considerations of national policy, to improve and coordinate federal plans, functions, programs, and resources to the end that the nation may: (a) fulfill the responsibilities of each generation as trustee of the environment for future generations, (b) assure for all Americans safe, healthful, productive, and esthetically and culturally pleasing surroundings, (c) attain the widest range of beneficial uses of the environment without degradation, . . . (d) preserve important historic, cultural, and natural aspects of our natural heritage, . . . (e) achieve a balance between population and resource use which will permit high standards of living and a wide sharing of the amenities of life, and (f) enhance the quality of renewable resources and approach the maximum attainable recycling of depletable resources."

Robed in such positive, if vague, platitudes, NEPA passed both houses of Congress with relatively little opposition. Indeed, many representatives considered NEPA almost meaningless, a kind of "motherhood and apple pie" act for the environment. In fact, this high-sounding rhetoric did little to change environmental policy in the United States.

Hidden in the more mundane and practical language of the bill, however, were words that would profoundly affect the practices and decisions of every U.S. federal agency. The requirement that all federal agencies develop information, in the form of a "detailed statement," on the ecological consequences of their actions and weigh these impacts in their policy making would become the "teeth" of NEPA's enforcement power. Each such "detailed statement" must describe (1) the environmental impact of the proposed action; (2) any adverse environmental effects that cannot be avoided should the proposed action be implemented; (3) alternatives to the proposed action; (4) the relationship between local, short-term uses of the environment and the maintenance and enhancement of long-term productivity; and (5) irreversible or irretrievable commitments of resources that would be involved in the proposed action should it be implemented. Such a statement, once prepared, was to be circulated among government agencies and other public venues (NGOs, libraries, and private citizen groups) for comment.

NEPA was unique among environmental and conservation legislation in several ways. First, it was proactive rather than reactive, forcing government agencies to consider the environmental effects of proposed actions in advance. Second,

NEPA forced government agencies to explicitly consider the value of noneconomic resources, ensuring that conservation would be considered in evaluating the proposed action. Third, NEPA introduced environmental assessment as a means to guide administrative decision making (Caldwell 1966; Tarlock 1994). Thus, NEPA not only established a mechanism for environmental review, but also stimulated an increased level of citizen involvement in environmental decision making. Policy analyst Richard A. Liroff summarized the true significance of the act when he noted, "Implicit in NEPA was the notion that the public was to be informed of the rationale underlying environmentally impacting administrative actions. NEPA's architects also sought public involvement in decision making, but they did not indicate when it should occur or what form it should take" (Liroff 1976:88). It is also noteworthy that NEPA was strongly linked to the kind of "ideal" role of government in conservation that had first been developed by Theodore Roosevelt (chapter 1). Specifically, NEPA embodied Roosevelt's ideal of environmental protection resting on the foundation of scientifically informed government decisions modified by citizen input and initiative.

These implicit notions of public participation ultimately became explicit directives for public involvement, first addressed by the courts (most notably in the case of *Calvert Cliffs v. Atomic Energy Commission*) and subsequently refined by individual federal agencies. To grasp the scope of NEPA's effect, one must understand what constitutes a "major federal action" and appreciate the extent of federal lands in the United States and their general management directives.

NEPA and U.S. Federal Lands

The U.S. government is the nation's largest landowner, with responsibility for more than 715 million acres, one-third of the total U.S. land area. Many of the country's western states are largely public domain; more than half the land in Alaska, Nevada, Idaho, Oregon, Utah, and Wyoming is federally owned (Rosenbaum 1985). On or underneath this impressive land area lies an equally impressive wealth of natural resources. Perhaps one-third of all remaining oil and gas reserves, 40% of coal reserves, 80% of shale oil reserves, and more than 60% of low-sulfur coal (Rosenbaum 1985), and sites with high potential for geothermal energy generation exist on public lands.

A "federal action" takes place either on federal lands using federal funds, or on private, state, or locally owned land, but with a permit from a federal agency. Any of these situations constitutes the "federal hook" or "federal nexus" that activates the NEPA procedure. The agency involved in the project may fulfill NEPA's requirements of a "detailed statement" by preparing an environmental assessment (EA), that results in a "finding of no significant impact" (FONSI) or requires additional review with an environmental impact statement. Most federal actions, such as routine maintenance, management, and structural repairs, can be considered "categorically excluded" (CatEx) from further review and do not require the development of an EA. Most federal projects are classed as CatEx, or their review is completed with an EA/FONSI.

If an EA is required, the process could be described as a kind of "mini-EIS." An EA may be prepared by an agency as part of a preliminary analysis to determine if a full-scale EIS is required. An EA must contain (1) a clear and concise description of the proposed action; (2) a detailed description of the environment affected by the proposed action; (3) an assessment of the probable effects of the proposed action; (4) an evaluation of the probable cumulative and long-term environmental effects, both positive and negative; (5) an assessment of the risk of credible potential accidents; (6) a description of the relationship of the proposed action to any applicable federal, state, regional, or local land-use plans and policies; and (7) a brief description of reasonable alternatives and their probable environmental effects, one of which is required to be that of not implementing the proposed action, the so-called no action alternative.

An EA differs from an EIS in scope, length, and detail; however, an EA also includes procedures for public input and requires substantial agency investments of time, effort, and money. If the agency determines, based on the EA, that an EIS is not required, it will then publish a finding of no significant impact, which is a brief document that explains why the proposed action has no significant effect on the environment. The FONSI must describe the action, the alternatives considered, and the environmental effects and the reasons why they are not significant. Individuals or groups unsatisfied with the FONSI or with the EA in general, can take the agency to court for not preparing a full-scale EIS.

Preparation of an Environmental Impact Statement

The NEPA process typically begins when an agency publishes a notice of intent (NOI). The NOI identifies the responsible agency (if an action involves two or more agencies, one is designated the "lead agency" and assumes responsibility for the EIS) and describes the proposed action. Invitations, procedures, dates, times, and locations of public meetings, with availability of related documents, also are listed. At minimum, the NOI will be published in the *Federal Register* and mailed to individuals who request it, individuals known to be interested in the proposed action, and national organizations expected to be interested in it. The NOI may also be in local newspapers, publicized through local media, and posted on the site to be affected (Murthy 1988).

As a first step in preparing the EIS, the lead agency will assemble an interdisciplinary team of professionals capable of assessing the scientific, social, and economic issues likely to be addressed in the EIS. A team leader coordinates the group's activities to produce the EIS within specified guidelines and deadlines, and assembles comments from other team members, other agencies, experts, and the public.

The process of EIS preparation also involves regular contact among the lead agency, other cooperating agencies, and the public. Public-issue identification or "scoping" meetings involve the public early in the process. Scoping has been defined as "an early and open process for determining the scope of issues to be addressed and identifying the significant issues related to a proposed action" (Yost and Rubin 1989). After this process is complete, the lead agency will prepare an EIS implementation plan (IP) and use it to produce a draft EIS (DEIS).

The lead agency conducts an internal review of its DEIS and then publishes a "notice of availability" (NOA) in the federal register. Public comment on the DEIS, including comments received at subsequent public meetings where the DEIS

is presented and explained, is then received, considered, and, if appropriate, incorporated into a revision of the EIS. From this effort, a review draft of a final EIS (FEIS) is prepared, reviewed within the agency, and made available to the public. Based on information presented in the FEIS, the responsible official of the lead agency decides whether to implement the proposed action or one of the alternatives (including the possibility of the "no action" alternative) and publishes the decision in the *Federal Register*. This "record of decision" (ROD), like other NEPA-associated documents, is available to the public and other agencies. Anyone who disagrees with the decision has 30 days to file an appeal. If an appeal is granted, the decision may be overturned and the EIS may have to be rewritten.

Policy analyst Richard A. Liroff has provided a key to understanding NEPA's profound effect on national environmental policy by noting that ". . . NEPA laid the groundwork for a series of procedures whereby environmental considerations could be fed into agency decision-making routines" (Liroff 1976:210). These procedures for environmental assessment radically changed the pattern and process of agency decision making with respect to public lands. Most states now have their own versions of NEPA. In addition, procedures for public input established by agencies and by U.S. courts in response to NEPA set the example for public input requirements in most subsequent environmental and conservation legislation. More than any other statute, NEPA made environmental review a permanent part of environmental decision making. This change profoundly affected the development of conservation biology because it made conservation issues relevant and legally mandated considerations in all proposed actions on public lands. In addition, NEPA transformed U.S. environmental and conservation policies into arenas for public participation rather than simply expressions of elected representatives. Informed by such participation, the public in general, and scientists as public citizens, began to see clearer connections between conservation science and conservation law, and to use these connections as conservation advocates.

Shortcomings of the National Environmental Policy Act

It has now been over three decades since NEPA and its grand design for a national environmental policy became law in the United States. Although NEPA has, like all significant environmental legislation, grown in its influence, not all of that influence has been positive. With its successes, NEPA also has had negative consequences that its planners did not anticipate.

One of the most foundational tensions in NEPA was that it assumed an ecosystem management approach (chapter 10) before there were well-developed concepts and procedures of ecosystem management. Specifically, NEPA's intent is to provide for functioning, sustainable ecosystems and long-term environmental quality. However, its highest-level mechanism, the environmental impact statement, is usually prepared by one administrative unit of a single federal agency, such as the staff of a national forest within the U.S. Forest Service, operating within fixed spatial boundaries, limited jurisdiction, and strong vested interests in particular commodities. NEPA procedures demand that the lead agency identify and inform stakeholders, but its procedures do not truly involve stakeholders as full partners in the decision-making process. The public can express concerns at scoping meetings, through letters, or by direct contact with agency personnel, but the actual preparation of the EIS is solely the responsibility of the agency's interdisciplinary team. Although the public can give additional input after reviewing the draft EIS, such input is also strictly one-way communication. The public speaks, the agency listens, but the final EIS remains an internal agency product. As a result, if the public is still dissatisfied with the outcome of the final EIS and the record of decision that accompanies it, citizens have little choice but to litigate. The purpose of such litigation is, regrettably, not to improve the EIS or its decision but to show that the EIS is inadequate on professional and scientific merits as a basis for the management decision, and therefore must be thrown out and done over. This approach necessarily forces the agency into the position of defending its own EIS, if only to save the taxpayers money and their own personnel more work, and an adversarial climate is created between the agency and the public. Thus, NEPA has often multiplied litigation rather than improved decision making. For example, 92 of 94 completed forest plans (mandated by the National Forest Management Act of 1976) were under appeal by 1990. Five were in court and one was declared illegal. A total of 332 active appeals were pending against these plans, brought by conservation groups, commodity interests, off-road vehicle enthusiasts, state and local governments, native American tribes, and private citizens (Behan 1990).

Faced with the daunting prospect of intense adversarial litigation, resource management agencies have responded by diverting more agency resources and personnel solely to the production of environmental impact statements to make their EISs "litigation proof." But the price for administrative prudence is high. Money is diverted from field research and management to salaries for specialists in EIS preparation, fees for consultants who collect data solely for documentation in the EIS, and legal expenses for ongoing litigation of EISs under appeal. Agency administrators and scientists spend less time in the field and more time preparing or defending NEPA documents. And trust between the agency and the public diminishes rather than increases. Some policy analysts have argued for new, more creative approaches in the NEPA process. These have included such novel propositions as the "citizen jury," in which members of the public, rather than the agency, evaluate the EIS and determine the final management decision by consensus, (Brown and Peterson 1993), or the use of informal advisory groups that would have continuing input to the agency's interdisciplinary team (Sample 1993). However, neither these nor other more novel concepts for solving the problems of NEPA have been tested in real cases (Goetz 1997).

Some experts now argue that NEPA will become more effective, and its true intent more manifest, as U.S. resource management agencies mature in their understanding of and commitment to ecosystem management approaches. There is some evidence from individual agencies (particularly the Forest Service) that, in fact, this is the case, with more recent EISs and decisions reflecting the true intent of NEPA in ecosystem protection, not simply the rules of an administrative procedure (Goetz 1997). NEPA and the EIS have unquestionably shaped the landscape of U.S. policy and administration in ways that profoundly affect the perception and practice of conservation biology. But whether NEPA will ripen to bear the fruit of its full intent depends largely on whether agency and public interests mature into working relationships for conservation or remain conflicts of litigation and mistrust between adversaries.

The Endangered Species Act

The Endangered Species Act has been called the "strongest and most comprehensive species conservation strategy" in the world (Rohlf 1995). The ESA affirms the value of biodiversity, and actions authorized under the ESA have contributed to the persistence of many endangered species, and even the complete recovery of a few formerly endangered species, such as the bald eagle (*Haliaeetus leucocephalus*). More than 1,100 species are still listed as endangered or threatened, however, and seven listed species have become extinct under the act's "protection" (Cooper 1999). The Endangered Species Act may be the world's most admired and most powerful piece of conservation legislation, but it is also one of the most controversial. No other statute has so influenced the development and practice of conservation biology or engendered so much hostility and withering criticism.

First passed in 1966 as the Endangered Species Preservation Act, the original act was passed with little controversy or fanfare, but also little power. It was limited to native vertebrates, focused on populations in existing wildlife refuges, created no new programs or legal authority, and was so vague as to be meaningless. Its immediate successor, the Endangered Species Conservation Act of 1969, was not much better, applying only to the importation of a few rare foreign species (Nash 1989). These legally toothless statutes were rewritten in 1972 by E. U. Curtis Bohlen, then undersecretary of the U.S. Department of Interior, in ways that profoundly changed the legal landscape of conservation in the United States. Bohlen's contribution was essentially a new law rather than simply a revision of the former statutes. The new version expanded the jurisdiction of the ESA from vertebrates to all plant and animal species, although some insect pests, viruses, and bacteria were excluded. The 1973 ESA legally defined a "species" as "any subspecies of fish or wildlife or plants, and any distinct population segment of any species of vertebrate fish or wildlife which interbreeds when mature." Although this definition is not scientifically or intellectually satisfying (it assumes an understanding of the very concept it is attempting to define), it is comprehensive in specifying an enormous array of organisms eligible for protection. Although including plants, the ESA does not protect any "distinct population segment" of plants as it does for animals. In practice, plants also tend to be less protected because, under traditional English common law, plants on private lands belong to the landowner, whereas animals belong to the state. Thus a private landowner has no responsibility to protect an endangered plant on his or her own land unless there exists a "federal nexus," such as a government loan subsidizing the landowner's activity on the land.

Bohlen's rewritten ESA also created a new category for legal protection called "threatened species," and even allowed the listing of species that were threatened only in a portion of their range. The 1973 ESA also introduced the concept of "designated critical habitat" into environmental law, creating the legal provisions that require not only the protection of the species, but also the land or water in which it lives. The 1973 ESA gave primary authority for enforcement of the ESA to the Department of Interior's U.S. Fish and Wildlife Service (FWS) for cases involving terrestrial and freshwater species and to the National Marine Fisheries Service of the Department of Commerce for marine species. The U.S. Fish and Wildlife Service has additional authority to identify and purchase such critical habitat, and to stop activities on such habitat that threatened the species, even if the habitat was privately owned. The ESA also offers provision and incentive for the federal government to initiate active cooperation with state programs as well as to cooperate fully with existing state programs to protect species (Section 6). For example, the act states explicitly that the Secretary of the Interior shall "cooperate to the maximum extent practical with the States," enter into management agreements "with any State for the administration and management of any area established for the conservation of endangered species or threatened species," and "enter into a cooperative agreement . . . with any State which establishes and maintains an adequate and active program for the conservation of endangered species and threatened species." In fact, the ESA actually helped stimulate the kind of federal-state cooperation it envisioned by its very existence because, after its passage, many states passed endangered species laws modeled on the ESA.

It was Bohlen's skill and political savvy in rewriting the Endangered Species Act that changed a formerly obscure statute into what Donald Barry, a former vice-president of The World Wildlife Fund, would call "the pit bull of environmental laws. . . . It is short, compact, and has a hell of a set of teeth. Because of its teeth, the act can force people to make the kind of tough political decisions they wouldn't normally make" (quoted in Rosenbaum 1995:334). The 1973 Endangered Species Act passed both houses of Congress with nearly unanimous support.

In practical terms, the ESA gives the FWS primary responsibility for identifying endangered and threatened species and proposing these species for federal protection through the "listing" process. However, actual listing is normally accomplished through interagency consultation, as specified in the ESA's Section 7, because the ESA authorizes *all* federal agencies to "utilize their authorities in furtherance of the purposes of this Act by carrying out programs for the conservation of endangered species and threatened species." The ESA, like NEPA, also provides for review of actions carried out by federal agencies to ensure that agency actions do not "jeopardize the continued existence of any endangered species or threatened species or result in the destruction or adverse modification of the habitat of such species."

In assessing such actions, the act defines an "endangered" species as one that is "in danger of extinction throughout all or a significant portion of its range." A "threatened" species "is likely to become an endangered species within the foreseeable future throughout all or a significant portion of its range." In managing an endangered or threatened species, the FWS also must define "critical habitat" (habitat of special significance to the species' survival; Bean et al. [1991]) and develop a recovery plan that will restore the species to secure population levels. The ESA also provided explicitly for public participation in the listing process. Any citizen or private citizens' group may directly petition the Secretary of the Interior to add a species to the endangered species list, and the Secretary must respond with a determination for or against the petition (in the words of the act, "warranted" or "unwarranted") within 90 days after it has been filed. Negative findings are subject to judicial review, and public participation can continue because the ESA stipulates that judicial review may be obtained by "any person" regarding "any decision" to refuse to list a petitioned species as endangered.

Given its broad powers and its rather uncompromising, nonnegotiable standards, the ESA was in many ways too comprehensive in its protections and prohibitions to go unchallenged indefinitely. The most famous such challenge began in 1978. In *Tennessee Valley Authority v. Hill,* the U.S. Supreme Court ruled that the Tellico Dam on the Little Tennessee River could not be completed because the dam would destroy the habitat of an endangered fish, the snail darter (*Percina tanasi*). Although environmentalists won the battle in the courts, their victory cost them the war through congressional backlash at what many representatives now perceived as an act that was too restrictive and too insensitive to human need and economic welfare. Within a year, Congress had amended the ESA to create a committee that could waive the law's regulations under special economic conditions. Although officially labeled the Endangered Species Committee, this group soon became known as the "God Squad" because of its power to revoke the ESA's protection for selected species. The committee ruled in favor of the fish, but Congress responded by subsequently excluding the snail darter from protection under the ESA. And as for the obscure species that caused such a fuss in the first place, snail darter populations were transplanted and established in other streams, and the Tellico Dam was completed.

The FWS is prohibited from considering economic effects in decisions about the initial listing of a species, but later amendments to the ESA added the requirement that the FWS conduct an economic analysis of the effects of designating critical habitat. Because such designation usually involves the suspension of other activities in the area, including economically profitable ones, the amended ESA includes an "exclusion process" through which all or part of the critical habitat may be excluded from protection if the economic analysis determines that the cost of protection poses too great a hardship in the form of economic or other forms of loss. As in NEPA, a public comment period is required to allow interested parties to provide additional information that can be included in the analysis (Berrens et al. 1998).

Post-1973 amendments made the ESA more flexible in resolving conflicts between endangered species and humans, but also, in the eyes of many conservationists, weakened and betrayed the act's original intent to preserve endangered species regardless of economic cost (Nash 1989). Nevertheless, the ESA remains armed with formidable provisions to protect listed species and is a cornerstone of environmental protection and biological conservation. In some cases significant amounts of critical habitat have been excluded, as in the case of the northern spotted owl (*Strix occidentalis caurina*), but the God Squad exemption process (denying protection to the species itself) has not been invoked (Berrens et al. 1998).

The process of designating critical habitat is the most frequent source of conflict between the federal government's interest in protecting endangered species and the interests of private landowners. Although the ESA provides for "informed consultation" between the federal government and landowners to determine a mutually satisfactory plan to protect the species without undue infringement of personal property rights (Section 7), private landowners have not always been satisfied with the outcome. In fact, many private landowners assert that the ESA prohibits them from deciding how to use their own lands and violates the fundamental rights associated with private property. Critics claim that the ESA's biggest weakness is its punitive approach to dealing with landowners who violate the act's provisions when endangered or threatened species are found on their land. The threat of punishment often promotes landowner behavior that is harmful to the protected species. As Myron Ebell, a property-rights advocate, has said, ". . . if there is an endangered species on your land, the last thing in the world you want to do is provide habitat for it" (Cooper 1999). To avoid, or at least reduce, landowner-government conflicts and their invariably unproductive outcomes, one recent strategy employed by the federal government to protect endangered and threatened species is the habitat conservation plan.

The Endangered Species Act and Landowner Conflicts: Example of the Red-cockaded Woodpecker

The red-cockaded woodpecker (fig. 2.1) inhabits the southeastern United States, where it usually lives in stands of mature longleaf pine woodlands. The woodpecker prefers open forests with minimal understory, a condition that can be maintained only by recurrent fires and active understory management. During the mid-1900s, the red-cockaded woodpecker declined in abundance to fragmented populations of only a few to several hundred individuals, with a total population of fewer than 15,000 birds.

Most of the historical habitat for the woodpecker is on privately owned land. Landowners typically fear the federal regulations that would be imposed on their land and their use of it if red-cockaded woodpeckers were discovered on their property. As a result, landowners often manage their land to make it unattractive to the woodpeckers by harvesting pines before they reach old-growth stages, replacing longleaf pine with shortleaf pine, suppressing fires, and letting the understory grow.

These actions arise from rational economic behavior and from the landowners' fear of the ESA's prohibition against the "taking" of any endangered species. Historically, the term *taking* meant hunting, fishing, collecting, or trapping a creature to kill it or bring it into personal possession. The ESA's definition of taking is much broader. In the ESA, *taking* includes any act that harms or harasses the protected creature in any way, whether intentional or not. Thus, as Bean et al. (1991) note in their analysis of landowners' conflicts with the ESA, "a landowner whose bulldozers crush the larvae of an endangered butterfly on his land commits just as much of a taking as a hunter who deliberately shoots a bald eagle."

Figure 2.1

The red-cockaded woodpecker (*Picoides borealis*), an endangered species that has been the subject of intense management through habitat conservation plans.

If a landowner inadvertently harms a member of an endangered species through normal land-use activities such as farming, logging, or development, criminal prosecution can result. It is this discouraging prospect that leads many landowners deliberately to alter habitat on their land. If an endangered species does inhabit their property, private landowners may resort to the strategy of the three S's—"shoot, shovel, and shut up." Such attitudes and actions are natural outgrowths of the ESA's broad definition of *taking*. The long-term effect of the resulting behavior is a reduction in available habitat for already endangered animals. This example of the red-cockaded woodpecker demonstrates how even legislation designed to protect endangered species can have unintended adverse consequences if it fails to consider economic and personal interests of private landowners.

The Evolution of Habitat Conservation Planning

Struggles arising from conflicts of interest between private individuals and conservation efforts have repeatedly caused what former U.S. Interior Secretary Bruce Babbitt called "environmental train wrecks" (Kaiser 1997). Conflicts of this sort have occurred because early versions of the ESA did not define the concept of critical habitat well and did little to develop the idea of saving species through preserving habitats (Noss et al. 1997). The ESA did prohibit the destruction of the habitat of endangered species, but in practice this has been difficult to enforce (Bean et al. 1991) and sometimes even overruled in court (Noss et al. 1997). To prevent continued loss of habitat for endangered species and reduce conflicts with private landowners, the Clinton administration, under Babbitt's leadership, increasingly resorted to a mechanism known as the habitat conservation plan (HCP).

HCPs arose out of a 1982 amendment to the ESA that allowed the issuance of "incidental take" permits for endangered species. *Incidental take* was defined as take that is "incidental to, and not the purpose of, carrying out an otherwise lawful activity." To be granted such a take permit, the applicant, whether corporate or individual, must first prepare and submit a conservation plan. The plan must explain what the effects of the taking will be on the endangered species, how the effects will be mitigated, and how the species will benefit. Now called habitat conservation planning, this procedure was patterned after the resolution of an environmental/economic conflict over the proposed development of San Bruno Mountain near San Francisco, California.

San Bruno Mountain, immensely attractive as a site for upper-class residential and commercial development, also represented some of the last undisturbed mountain habitat in the San Francisco Bay area and was the home of two endangered species of butterflies (Lehman 1995). Rather than resorting to adversarial litigation, the parties involved in the San Bruno Mountain controversy devised a series of agreements from 1975 to 1983 that allowed for development of one-fifth of the mountain, but protected the remaining 80% (and 90% of the butterflies' habitat).

Congress was so impressed with the San Bruno Mountain example that it codified the process in a 1982 ESA amendment, in hopes that HCPs like the one developed for San Bruno would "encourage creative partnerships between public and private sectors and among government agencies in the interests of species and habitat conservation" (Lehman 1995). The process was intended to foster resolution through negotiation, compromise, and explicit recognition of the interests of all participants. HCPs became the mechanism of choice for handling conflicts over endangered species during the Clinton administration, and the number of approved HCPs has grown exponentially from fewer than 20 per year (1983–1992) (Kaiser 1997) to more than 400 plans completed or in progress through the first half of 1997 (O'Connell 1997).

Supporters of HCPs maintain that this approach involves all vested interests and focuses on protecting the highest-quality and most productive habitats (Lehman 1995). Critics claim that the plans have inadequate scientific guidance, that agreements permit landowners to destroy habitat later if they enhance it initially (Kaiser 1997), that there are few or no opportunities for public participation in formulating the plans, and that HCPs generally have ineffective management provisions and poor oversight of plan implementation. Furthermore, most HCPs are for single areas, single species, and single landowners and critics argue that this approach is overly narrow, restricted, and fragmented (O'Connell 1997).

Despite these criticisms, officials in the Clinton administration worked to make HCPs more attractive to landowners. In 1994, the U.S. Department of Interior and the Department of Commerce jointly issued a new policy titled "No Surprises: Assuring Certainty for Private Landowners in Endangered Species Act Conservation Planning." This revision, which has come to be known as the "no-surprises" policy, requires the responsible federal agency to provide landowners with assurances that they are not responsible for species protection if unforeseen circumstances arise (Walley 1996; Schilling 1997). Under this policy,

after an HCP is approved, federal agencies cannot require any additional mitigation measures from a landowner to conserve an endangered or threatened species until and unless the agencies demonstrate "extraordinary circumstances" that warrant increased protection.

The no-surprises policy was intended to increase landowner cooperation and make the protection of endangered species more effective, but critics were quick to attack it. One hundred sixty-four scientists, including many of the world's leading conservation biologists, wrote letters protesting the policy to members of the U.S. House Committee on Resources (Walley 1996). Their greatest concern was that because uncertainty and change are intrinsic to ecological systems, there will be many surprises, rather than no surprises, in conservation planning. They argued that the no-surprises policy unreasonably and unfairly restricts the ability of agencies to change conservation plans and adapt to changing conditions. The policy also has been criticized because it guarantees no surprises to the landowner as an inherent right, rather than as a privilege earned through proper conservation planning and mitigation efforts in the initial stages. According to the policy, the no-surprises assurance must be given to all landowners whether or not they make conservation commitments (Walley 1996). More generally, other critics decry HCPs as capitulation to economic and development interests and a selling out of the ESA's fundamental intent to preserve and protect endangered species from the sort of "takings" that HCPs permit (Vincent 1999).

Criticisms of the recent increased emphasis on HCPs have led to increased scrutiny of individual plans by conservation biologists. A recent comprehensive review of 44 HCPs, covering a range of land areas, locations, and landowner categories, was completed by a panel representing eight U.S. universities (Mann and Plummer 1997). Directed by the National Center for Ecological Analysis and Synthesis, the study gave mostly favorable reviews to the HCPs examined. Most of the plans were judged to have reliably determined the health of the species' population before being implemented. About half were judged to have made a reasonable prediction about the harm the landowners would cause species, and to have adequately and correctly determined the primary threats to the species. Although the overall review of HCPs was favorable, there were identifiable problems. Most plans did not do a good job of determining how the HCP would affect overall species viability (beyond just the local population), providing for monitoring, or including basic natural history data on species covered in the HCP (Mann and Plummer 1997).

Limitations of HCPs have led to attempts to improve this approach. Increasingly, HCPs are supplemented with "no-take" management plans implemented via memoranda of agreement (MOA) and so-called safe harbor cooperative agreements (Costa 1997). MOA between a federal agency (usually the FWS) and a corporate landowner outline management and conservation actions that the landowner can take to meet or exceed requirements of the ESA for habitat protection. For example, landowners can satisfy their legal ESA obligations by monitoring populations; managing and retaining current and future nesting habitat; producing and maintaining foraging habitat; conducting cooperative research, education, and outreach; and letting the managed population provide donors for other populations (Costa 1997).

One of the first agreements of this kind was signed in 1992 by the Georgia-Pacific Corporation (a lumber company) and the FWS to preserve red-cockaded woodpeckers. By 1997, this MOA was protecting more than 66,000 acres of forest habitat for the woodpecker (Costa 1997).

"Safe harbor" agreements are contracts under which a landowner agrees to actively maintain suitable habitat ("safe harbor") for a predetermined number of individuals of a species equal to the number present on the site when the agreement was formulated. In return, the landowner receives an incidental take permit that authorizes future land-use changes or management on other parts of the site that may be occupied by additional individuals of the endangered species. The major benefit of the safe harbor agreement is that it provides direct habitat improvement and maintenance for all the individuals or population subunits that are enrolled in the original conservation agreement. Once again, the first example of the use of a safe harbor agreement was for protection of the red-cockaded woodpecker. Established in 1995 in North Carolina, an initial agreement succeeded in enrolling 24 landowners and more than 21,000 acres of habitat to be actively managed for the woodpecker. This acreage originally supported 46 woodpecker groups, but is estimated to be able to support up to 107 groups (Costa 1997).

Despite their imperfections, conservation approaches like HCPs, MOAs, and safe harbor agreements acknowledge fundamental truths about the future of conservation. First, habitats must be conserved if species are to be conserved. Second, habitat and species conservation cannot be successful in the long run if they are restricted entirely to public land or to private reserves established by conservation organizations. Habitat and species conservation can be successful only in a landscape context if private landowners are involved and motivated partners. Even the weaknesses of such approaches, specifically their focus on individual species and on small, isolated sites, reveal an important truth about the future of conservation; namely, that efforts of greater landscape scale are needed to preserve populations and their habitats. It is far easier and more cost-effective to protect intact ecosystems and the species they contain than to initiate emergency measures for critically endangered populations on degraded habitat.

The Natural Communities Conservation Planning Program

The California gnatcatcher (*Polioptila californica*), a small songbird endemic to California, lives year-round only in coastal sage scrub habitats. Because of accelerating habitat loss, the northern race of the species (*P. c. californica*) is federally listed as threatened (U.S. Fish and Wildlife Service 1993). Almost all of the gnatcatcher's habitat is on private land, much of it in areas of high real estate value and development potential. State officials did not believe that listing the gnatcatcher as endangered would ensure its survival, and they feared it might even contribute to its extinction if landowners destroyed critical habitat before the classification was finalized. The state preempted federal listing by instituting its own plan, the "natural communities conservation planning program," or NCCP (DeSimone and Silver 1995; Beyers and Wirtz 1997).

Under the NCCP, voluntary participation by counties, cities, landowners, and state and federal wildlife agencies led to the establishment of habitat reserves sufficient to ensure survival of the California gnatcatcher and other species associated with coastal scrub habitat. An interim allocation of 5% of undeveloped scrub habitat was recommended for placement in the reserve program (Beyers and Wirtz 1997), whose overall objective was to study and resolve emerging conflicts in individual land units. Each such land unit was to design its own natural community conservation plan, which would be subject to approval by a state-appointed scientific review panel. Although the state lacked funds to purchase the land designated as reserves, it was authorized to use its permitting authority to enforce approved NCCPs (Tarlock 1994).

The NCCP approach is an example of "adaptive management" because it can be changed as better information on the gnatcatcher is collected and made available. Landowners are less likely to litigate against the process because reserve boundaries can be modified, landowners are involved in reserve planning from the beginning, and all areas still have opportunities for development. Additionally, because the land-use plans are fluid rather than fixed, there is less incentive for landowners to challenge the plan. Entire tracts of land are not completely stripped of their economic value, so it is more difficult for landowners to demonstrate significant or unwarranted financial losses (Tarlock 1994). It is still too soon to assess whether the NCCP approach has achieved its goal of ensuring the California gnatcatcher's survival. Nevertheless, the NCCP merits attention and examination as a constructive alternative to traditional approaches that exacerbate government-landowner conflicts inherent in the ESA.

Private Property and Conservation Law: The Growing Role of Conservation Easements

Although we have focused on habitat conservation planning because of its connection to the ESA, other, less well-known legal procedures, including some operating at state and local levels, have the potential for even more pervasive and long-term effects. One example is the conservation easement.

Conservation easements were developed to make the value of conservation on private land more explicit and more profitable to landowners. In an easement, the landowner agrees, usually with a government entity or a private conservation organization, to restrict some activities or forms of development on his or her land to achieve specific conservation goals, such as habitat or species protection. Such restrictions lower the assessed value of the land, generating a reduction in property taxes for the owner and a reduction in inheritance taxes for the owner's heirs. The owner, however, retains possession, the right to residence and nonprohibited activities, and full legal title to the land under the easement. Thus, conservation easements often succeed where the ESA fails because they provide incentives for conservation by private citizens on their own land, and permit management objectives that focus on biological integrity rather than on individual species.

Conservation easements are becoming increasingly popular, particularly in the western United States. For example, the law firm of Isaacson, Rosenbaum, Woods and Levy P. C. in

Figure 2.2

A landscape in the Loess Hills of northwest Iowa, protected and managed by a conservation easement between the landowner and The Nature Conservancy, a private conservation organization devoted to habitat preservation and management.

Denver, Colorado, has received so much business from conservation easements in recent years that writing easements has become a "professional niche," creating full-time work for 6 of the firm's 40-plus attorneys (Shepherd 2000). According to the Land Trust Association, 1.4 million acres had been placed under conservation easements with local and regional land trusts by 1998 (Shepherd 2000). And conservation easements have become popular with traditional western landowners as well as recent arrivals. The former, often ranchers and farmers who are land rich and cash poor, find conservation easements an effective way to reduce their tax burdens and ease the transition of family land to their children. The latter, typically younger, former easterners who relocated because of their attraction to the West, are often motivated by a desire to do good for the environment while simultaneously earning a tax break on their property (fig. 2.2).

Criticisms of the Endangered Species Act

Beyond criticisms of habitat conservation plans, the Endangered Species Act itself faced mounting criticism in the 1990s. Complaints from private business and development interests are chronic and predictable, but the ESA also has been increasingly subjected to substantive criticisms from conservation biologists.

Many biologists have argued that instead of focusing on individual species, a more appropriate conservation goal is conservation of overall biodiversity and the management and protection of critical habitats and ecosystems (Rohlf 1991). Such critics contend that a narrow, single-species approach is slow, unwieldy, ignorant of the dynamics of real ecosystems, and wasteful of resources and efforts that could benefit multiple populations in the same habitat or ecosystem (Flather et al. 1998). One constructive response to this criticism is habitat- and regional-level analysis of endangered species' distributions, and development of strategies to promote the recovery of multiple species in the same habitat or region (Flather et al. 1998).

Other biologically based criticisms of the ESA include complaints that the law lacks defined thresholds to delineate endangered, threatened, and recovered species; that it does not adequately protect patchily distributed populations ("metapopulations"); that it does not protect habitat reserves sufficiently to sustain recovered populations; and that uncertain or long-term threats to endangered populations are discounted (Rohlf 1991).

Perhaps the most substantive biological criticism of the ESA is that it is reactive rather than proactive, responding only to the needs of species on the brink of extinction (Karr 1995). Maintenance is always easier than repair, especially when "repair" involves crisis management that may be too late. On this point, the ESA is implicated in a broader criticism of conservation biology as a whole. As noted in chapter 1, conservation biologists must reconcile the small-population paradigm (What are the characteristics of small populations, and how can they be managed to ensure their survival?) with the declining-population paradigm (What processes cause population decline, and how can they be altered to stop such decline?) (Caughley and Gunn 1996). The ESA emphasizes the small-population paradigm and may have contributed to a similarly misplaced emphasis in conservation biology. The reactive nature of the ESA also corrupts the listing process of endangered species. Although listing is primarily the responsibility of the U.S. Fish and Wildlife Service and the National Marine Fisheries Service, many recent listings have occurred through the pressure of lawsuits from environmental groups against the FWS over the failure to list particular species. Lawsuits are expensive to combat, and drain money in the endangered species program budget that was intended to acquire habitat and monitor endangered populations. In 2000, the FWS spent its entire budget for the listing and recovery of endangered species on legal fees. In 2001, to combat this problem, the Bush administration proposed new regulations that would severely limit, for one year, the power of environmental groups to bring lawsuits against the FWS over endangered species. But this proposal has only ignited new controversy and criticism that the administration is attempting to squash efforts to protect endangered species.

On economic criteria, environmental economists have criticized the ESA because both the value of species and the value of critical habitat are usually determined by some form of cost-benefit analysis (CBA). CBA is notoriously inappropriate for determining the value of ecosystem complexity and the value of nonmarket and nonutilitarian benefits associated with endangered species and their habitats (Berrens et al. 1998). One proposed solution is to abandon CBA in favor of the safe minimum standard (SMS) approach. As it was originally understood (Ciriacy-Wantrup 1952), SMS attempts to determine, and then preserve, some minimum level or safe standard of a renewable population unless the social costs of doing so are in some way "intolerable," "unacceptable," or "excessive." Proponents argue that the SMS approach is more compatible with the structure and intent of the ESA than is CBA, and they note that the recently developed concept of a minimal viable population (MVP) (chapter 7) provides one quantifiable measure of what constitutes an SMS for a species (Berrens et al. 1998).

The ESA is at the center of two additional controversies. First, it has been the primary legislative force behind reintroduction of the gray wolf (*Canis lupus*) into Yellowstone National Park. Opponents of wolf reintroduction, especially ranchers and farmers, argue that the ESA should not be used to place a dangerous predator in a system so near grazing lands. Second, and equally controversial, are plans to breach four dams along the lower Snake River in Idaho to restore endangered populations of nine species of salmon and trout. The dams in question provide 5% of the region's electricity, and their loss is expected to increase regional energy prices. In addition, the destruction of the dams would affect availability of water for agriculture and would destroy canals that allow barges to carry goods from Lewiston, Idaho, to the Columbia River, and then to the Pacific Ocean (Cooper 1999).

Despite the ESA's shortcomings, it is difficult even for critics to imagine what sort of legislation could replace or improve upon its fundamental legislative virtues. More than any other statute, the ESA affirms that species have intrinsic value, and U.S. courts have interpreted the ESA to give protection to any species listed as "endangered" by the ESA regardless of the economic cost of protection (Rohlf 1995). The ESA has also clearly and explicitly extended legal rights to nonhuman species (Karr 1995). The U.S. environmental historian Joseph Petulla described the ESA as one of the most remarkable, radical, and original laws ever passed because, through its protection, "a listed nonhuman resident of the United States is guaranteed, in a special sense, life and liberty" (Petulla 1977). Overall, the ESA has performed well at the functional level, and there is general agreement that fewer extinctions have occurred under the ESA than would have without it (Committee on Scientific Issues in the Endangered Species Act 1995).

Besieged by controversy, fraught with limitations, and plagued by failures in restoring endangered species or preventing extinctions, the ESA nevertheless has been instrumental in preserving many species, albeit often at small-population sizes. The ESA has operated in the courts more efficiently than many other legal attempts to preserve biodiversity because it contains easily defined biological concepts and goals. In particular, the "species" concept, which is the cornerstone of the ESA's validity, has proved more definable and defensible in legal circles than have concepts such as "biodiversity," "habitat," or "ecosystem" (Karr 1995). Perhaps most important, the ESA remains an important legislative model for efforts to save species worldwide.

THE PROBLEM OF GLOBAL COOPERATION: HOW CAN WE MAKE AND ENFORCE INTERNATIONAL CONSERVATION LAWS AND TREATIES?

General Principles

While environmental legislation was creating a favorable climate for conservation biology in the United States, a number of international treaties had important effects on global biodiversity conservation (tables 2.2 and 2.3). There are, at this writing, over 1,000 international legal instruments, most of them binding, that focus on environmental and conservation issues. The worldwide trend in environmental and conservation treaties and conventions has been one of increasing compliance, but factors affecting compliance are complex and national responses to international conservation efforts are not uniform.

Although every international convention, treaty, or protocol is a unique product of unique issues and circumstances, the development of international instruments in conservation has usually followed a four-step process: (1) issue definition, (2) fact finding, (3) creation of an international body or legislative regime to address the problem, and (4) consolidation and strengthening of the legislative regime.

Weiss and Jacobson (1999) developed a conceptual model, based on real cases, of the success of a variety of international environmental agreements, to show how various factors affect implementation, compliance, and effectiveness of international conservation treaties (fig. 2.3). Compliance with international treaties was affected by the characteristics of the activity (for example, numbers of participants, characteristics of markets, location of the activity), characteristics of the agreement, and the state of the international environment. A general trend was that the smaller the number of participants involved in the activity, the easier the activity is to regulate internationally. Likewise, participants in an activity that dealt with large, global markets also were easier to regulate than participants in smaller firms and more local markets because global corporations and businesses were far more concerned about international image. The most important characteristic of the treaty or convention itself was equitability. Accords perceived by all parties to provide for fair treatment had much higher compliance than those that were perceived to favor some participants over others. The international environment also plays an important role in compliance. The more persistently and publicly the international community focused on a conservation problem, the more compliance with international conservation agreements related to that problem increased. In addition, the clear support of a "leader" country or group of countries, such as the United States or the European Union, for a particular accord also was a critical factor in the level of compliance. Where such leadership was present, international compliance was high. Compliance was affected in individual countries by both intent and capacity. Intent, judged from the behavior of national leaders and political bodies, was a necessary but not sufficient condition for compliance. The country must also possess the capacity to comply, and such capacity requires an efficient and honest environmental bureaucracy, economic resources, technical expertise and public support. Weiss and Jacobson suggest three strategies for strengthening international compliance. The *sunshine approach* focuses on mechanisms to bring the behavior of key parties into the open for public scrutiny, including such actions as regular reporting, peer scrutiny, on-site monitoring, and media access and coverage. *Positive incentive* strategies are most effective where a country or corporation has the intention to comply but lacks the capacity. Here, inputs of money, technical expertise, capital, training programs, or special considerations from other countries may increase compliance. *Coercive measures* are the most effective strategy for parties that have the capacity to comply but lack the

Table 2.2 Some Major International Treaties That Affect Global Biodiversity Conservation

LEGISLATION	YEAR PASSED	DESCRIPTION
International Convention for the Regulation of Whaling[a,b]	1937	• Established International Whaling Commission (IWC) to protect limited whaling stocks • Banned commercial whaling and limited whaling to those species that are best able to sustain exploitation, providing opportunity for recuperation of other whale species of low numbers • Allowed aboriginal subsistence hunting if modern technology is not used to hunt the whales
International Convention for the Protection of Birds[c]	1950	• Addressed the threat of extermination that endangers certain species of birds and the concerns regarding the numerical decrease in other species, particularly migratory species • Protected all birds during breeding season and migrants during migratory period • Provided year-round protection of endangered birds or those of scientific interest • Prohibited buying/selling of protected birds or bird products
Benelux Convention Concerning Hunting and the Protection of Birds[d]	1958	• Classified game species as (a) large game, (b) small game, or (c) wild fowl • Specified appropriate hunting seasons in the three Benelux countries (Belgium, Luxembourg, and The Netherlands) • Prescribed minimum area of habitat allowed for hunting • Regulated importing and exporting of game
The Convention on Wetlands of International Importance Especially as Waterfowl Habitat (Ramsar, Iran)	1971	• Defined wetlands as "areas of marsh, fen, peatland or water, whether natural or artificial, permanent or temporary, with water that is static or flowing, fresh, brackish or salt, including areas of marine water the depth of which at low tide does not exceed 6 meters"; they "may incorporate riparian and coastal zones adjacent to the wetlands, and islands or bodies of marine water deeper than 6 meters at low tide lying within the wetlands"[e] • Recognized wetlands' contribution to biodiversity conservation and the well-being of human communities • Established and regulated the conservation and wise use of wetlands through national action and international cooperation in order to achieve sustainable development throughout the world

Convention on Fishing and Conservation of Living Resources in the Baltic Sea and the Belts	1973	• Established the commitment to preserve and increase the living resources to optimum yield within the Atlantic region of the Baltic Sea and the Belts
		• Established an international Baltic Sea Fishery Commission whose responsibilities include:
		  • Compile, analyze, and distribute records of fish and living resources
		  • Establish proposals for scientific research using analytic data from records
		  • Prepare recommendations for regulations within the Baltic Sea and the Belts

Continued

Table 2.2 *Continued*

LEGISLATION	YEAR PASSED	DESCRIPTION
Convention on International Trade in Endangered Species (CITES)	1973	• Regulated/prohibited commercial trade of globally endangered species or their products by dividing endangered species into three groups: • Appendix I: those in imminent danger—all trade is barred • Appendix II: less-endangered species—controlled trade subject to permits • Appendix III: no imminent danger—trade restricted to a regional level • Amended 1997: Botswana, Namibia, and Zimbabwe allowed to sell hunting trophies, live animals, and animal-derived goods, including ivory, to Japan
Commission on Conservation of Antarctic Marine Living Resources (CCAMLR)[f]	1982	• Established marine life conservation, which allows for "rational use" of resources • Regulated harvesting: harvesting may not cause populations to diminish below a number that will endanger stable recruitment • Excludes seals and whales because they are covered by other laws
United Nations Conference on Environment and Development (UNCED)	1992	• Also known as the Convention on Biological Diversity[g], Earth Summit, or Rio Summit • Protected threatened species and ecosystems • Recognized the need to monitor and conserve genetic diversity • Agenda 21, the Rio Declaration on Environment and Development, and the Statement of Principles for the Sustainable Management of Forests were adopted by more than 178 governments[h] • Agenda 21: • "A program of action for sustainable development worldwide"[i] • Works to establish sustainable alternatives to dwindling resources[j] • Established partnerships between involved countries so that wealthier countries aid poorer countries • Declared a "polluter pays" rule[k]

[a]Johnson, T. January/February 1998. *The whaling trade*. Ottawa: Canadian Geographic.

[b]Spong, P., N.D. Phillips. 1999. International Whaling Commission Report. *Earth Island Journal* 14:13

[c]http://www.tufts.edu/departments/fletcher/multi/texts/BH255.txt

[d]http://www.tufts.edu/departments/fletcher/multi/texts/tre-0140.txt

[e]Blasco, D. *The Ramsar Convention manual: a guide to the convention on wetlands* (Ramsar, Iran, 1971), 2nd ed. Gland, Switzerland: Ramsar Convention Bureau, 1997.

[f]http://www.ccamlr.org/English/e_basic_docs/e_basic_docs_online/e_basic_docs99_toc.htm#Top of Page.

[g]http://www.biodiv.org/chm/conv/default.htm

[h]http://www.un.org/esa/sustdev/agenda21.htm

[i]Brown, N.J., and Peirre Quiblier, ed. 1994. *Ethics and Agenda 21*. New York: United Nations Environment Programme.

[j]Costanza, R., O. Segura, and J. Martinez-Alier. 1996. *Getting down to earth: practical applications of ecological economics*. Washington, D.C.: Island Press.

[k]Primack, R.B. 1993. *Essentials of conservation biology*. Sunderland, Mass.: Sinauer Associates.

Table 2.3   Websites of Major International Conservation Treaties

Sites summarizing major international treaties	
Laws and Treaties for Nature Conservation	http://www.biodic.go.jp/english/biolaw/bio_law_e.html
International Environmental Law and Policy: Treaties	http://www.wcl.american.edu/pub/iel/treaty.cfm
Environmental Treaties and Resource Indicators (INTRI)	http://sedac.ciesin.org/entri/texts/acrc/aff64.txt.html
Major international treaty conventions and organizations	
CITES (Convention on International Trade in Endangered Species)	www.cites.org
Ramsar Wetlands Convention	www.ramsar.org
World Heritage	www.unesco.org
FAO (Food and Agricultural Organization of the U.S. "treaties")	http://www.fao.org/Legal/TREATIES/Treaty-e.htm
CTSF (Conservation Treaty Support Fund)	http://www.conservationonline.com/org/ctsf/genctsf.htm
Convention on the Conservation of Migratory Species of Wild Animals	http://envionment.harvard.edu/guides/intenvpol/indexes/treaties/CMS.html
Major international treaties	
"Rio Cluster" of VN Procedings	http://www.igc.apc.org/habitat/un-proc/index.html
Nearctica "International Agreements"	http://www.nearctica.com/conserve/internat/treaty.html
European Union Treaties including Conservation Treaties	http://www.struiken.ic.uva.nl:88/index1.htm

intention. Sanctions, penalties, revocation of membership in international organizations and communities or loss of privileges in international dealings all may be effective in motivating unwilling parties to comply with established agreements (Weiss and Jacobson 1999).

It is beyond the scope of this chapter, or even a single book, to examine every international conservation treaty, convention, and protocol. We can, however, learn much about their general structure and their effectiveness by looking at specific case histories that provide representative examples for classes of related problems. These classes are instruments that protect migratory species, endangered species, habitats and ecosystems, and commercially valuable species.

International Protection of Migratory Species

One of the oldest problems of species conservation requiring international cooperation is that of managing and protecting species that routinely cross national boundaries during annual migrations or in the course of their normal life-history patterns. Several treaties address conservation of migratory species, including species that breed on the shore or in rivers leading to the ocean, but that spend all or most of their adult life in the sea; migratory marine species that travel over ocean areas across national and international jurisdictional limits; terrestrial species that migrate from breeding to nonbreeding areas across international boundaries; and relatively sedentary terrestrial species that live near international boundaries and routinely cross them in the course of normal movements

(International Union for Conservation of Nature and Natural Resources 1986).

Multilateral and bilateral treaties addressing migratory species date back to the late nineteenth century (International Union for Conservation of Nature and Natural Resources 1986). For example, the Treaty Concerning the Regulation of Salmon Fishery in the Rhine River Basin was signed in 1885 by the Federal Republic of Germany, Switzerland, and The Netherlands. It prohibited the use of fishing methods that blocked more than half a watercourse, prescribed specifications for fishing nets, provided for closed seasons and regulation of fishing hours, and promoted captive breeding.

One of the first, but most effective, treaties protecting migratory birds was the Convention between the United States of America and Great Britain for the Protection of Migratory Birds (1916), which primarily protected migratory birds flying between the United States and Canada. Originally motivated by a desire to conserve dwindling waterfowl populations, the treaty established hunting regulations and closed seasons for hunted species, prohibited hunting of migratory "insectivorous" birds (most songbirds), and established refuges for selected species (International Union for Conservation of Nature and Natural Resources 1986). The Migratory Bird Treaty Act (MBTA) was ratified in the United States in 1918 and prohibited the taking or killing of migratory birds without a permit and imposed strict penalties and liabilities for violations (Means 1998). Indirectly, the MBTA was the forerunner of many future U.S. laws that empowered state and federal agencies to regulate hunting and punish violators of game laws.

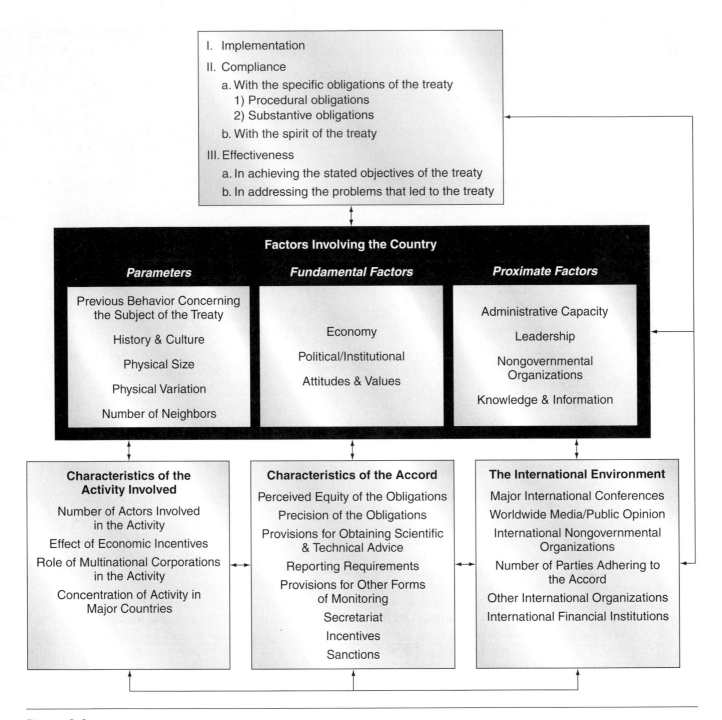

Figure 2.3
A model of factors that affect implementation, compliance, and effectiveness of international treaties and conventions in conservation. After Weiss and Jacobson 1999. Used with permission.

International Protection of Endangered Species: The Convention on International Trade in Endangered Species of Wild Fauna and Flora of 1973 (CITES)

Among modern international agreements that directly affect conservation biology, perhaps the most important has been the Convention on International Trade in Endangered Species of Wild Fauna and Flora of 1973 (CITES), which regulates or prohibits commercial trade in globally endangered species or their products.

Signed in Washington, D.C., in 1973 and going into force in 1975, there has always been disagreement, fueled in part by the treaty's own ambiguous language, about whether CITES is primarily an instrument for species protection or a means to regulate species trade. CITES does not protect all types of species, but only "tradable" species and species products that are bought or sold in transactions involving two or more countries.

The heart of the CITES treaty is found not in the main body of the document, but in three appendices that list categories of species regulated under the terms of the treaty. Appendix I lists species that are endangered and vulnerable to existing or potential trade. Commercial trade in Appendix I species is prohibited, and permits from both the importing and exporting countries must be obtained even for noncommercial transport. Appendix II species are those that either could be threatened by large volumes of trade or that cannot be distinguished from a threatened species. Trade involving species in these categories requires a permit from the exporting country. Appendix III species are not globally endangered, but may be listed at the initiative of an individual state seeking international cooperation for that species' protection. In the case of Appendix III species, nations are asked not to permit importation of the species without an export permit from the listing country. Parties to the treaty meet every 2 years to make amendments to the appendices and develop new species and animal products' lists and identification manuals to improve enforcement (Slocombe 1989).

CITES has proved to be an evolving document, and amendments to original provisions are not uncommon, reflecting changes in perceptions among delegates about the best way to achieve conservation of wildlife. Originally a treaty that equated conservation with strict protectionism in international trade, more recent meetings of CITES participants have shown a growing tendency to permit some trade in formerly protected species if it can be shown that such trade actually enhances their conservation. Thus, attempts to apply CITES to specific conservation dilemmas often have proved problematic. For example, in November 1994 CITES delegates agreed to allow trade in live southern white rhinos (*Ceratotherium simum simum*) from the Republic of South Africa, an action based primarily on the success of rhino conservation programs in that country that had restored a population of 20 individuals (all that remained in the country by 1920) to about 6,300, the largest national population in Africa (Kelso 1995). Sales of white rhinos are actually expected to improve the status of the species in South Africa because proceeds will be spent on further rhino conservation efforts. The rhinos that are sold to other governments are expected to aid in restoring rhino populations currently in decline in other countries.

The same meeting also repealed the 1987 mandate to destroy existing stockpiles of rhino horns, previously sold on the world market as raw material for medicines, aphrodisiacs and, in some Middle Eastern countries, for handles for ceremonial daggers. Although the original mandate was justified as a means to eliminate incentives for national governments to trade in rhino horn products and thus discourage poaching, more recent delegate opinion was that destruction of stockpiles would cause the price of rhino horn to increase, escalating poaching pressure (Kelso 1995). Governments now have been asked to "identify, mark, and secure" their rhino horns in national stockpiles that have, ironically, grown from increasingly effective enforcement of conservation laws, leading to seizures of rhino horns taken by poachers. Although this meeting did not actually approve the sale or trade of horns in such stockpiles, it paved the way to do so at a later time, under strict controls, if current inventory can be carefully marked.

International Protection of Habitats and Ecosystems

Three of the many international agreements designed to protect habitats rather than individual species are: (1) the United Nations Educational, Scientific, and Cultural Organization (UNESCO) Man and the Biosphere Program, (2) the Convention Concerning the Protection of the World Cultural and Natural Heritage (the World Heritage Convention), and (3) the Convention on Wetlands of International Importance Especially as Waterfowl Habitat (1971, commonly called the Ramsar Convention on Wetlands). This last example, the Ramsar Convention, provides an excellent model for understanding the nature of this type of international convention.

The Ramsar Convention, named after its city of adoption in Iran, was the first truly global conservation convention, and it is unique among international conservation treaties in its focus on a particular type of ecosystem—wetlands. Ratified in 1971 and entering into force in 1975, the Ramsar Convention was concerned not only with protecting wetlands for their own sake, but also for the diversity of living creatures that depend upon them. The convention obligates its signers to conduct land-use planning for wetlands and wetland preservation, to identify and designate at least one wetland in their country as a "wetland of international importance," and to establish wetland nature reserves (Koester 1989). One hundred nineteen countries have signed the convention and more than 500 wetlands are on Ramsar's international list. Canada, for example, which holds a disproportionately large share of the world's wetlands (148 million hectares, almost one-sixth of the nation's land area) has designated 13 million hectares (9%) of these wetlands, covering every province, as Ramsar sites. The Ramsar Convention was and remains remarkably farsighted in its recognition of ecosystems as the basis of sustainable populations and the preservation of biodiversity. The success of international conservation efforts demands a greater appreciation of the Ramsar model and a greater application of its design to other ecosystems.

Protection of Commercially Valuable Species: International Conservation Laws of the Seas

The 1972 United Nations Conference on the Human Environment in Stockholm, Sweden, was one of the first international conferences on global oceanic environments. Conservation of the marine environment was an explicit and important priority of the conference. Its proceedings noted, "The marine environment and all the living organisms which it supports are of vital importance to humanity. . . . The capacity of the sea to assimilate wastes and render them harmless and its ability to regenerate natural resources are not unlimited. Proper management is required and measures to prevent and control marine pollution must be regarded as an essential element in this management of the oceans and seas and their natural resources" (Sand 1988).

From 1972 to 1986 regional international treaties were developed for most of the world's oceans, beginning with the Barcelona Convention for the Protection of the Mediterranean Sea Against Pollution. More than 100 nations and 50 international organizations now cooperate in the regional seas programs (Sand 1988). In addition to the regional conventions and protocols, the 1982 Montego Bay Convention, developed in association with the Third United Nations Conference on the Law of the Sea, addressed major issues of ocean conservation on a worldwide basis. In 1977 the United Nations Environmental Program (UNEP) also established a Working Group of Experts on Environmental Law, whose recommendations were endorsed by the UNEP governing council and, in 1982, by the UN General Assembly. Although individual nations were not legally bound to use these guidelines, such so-called soft laws become, over time, increasingly recognized international standards. The most important trend in international marine conservation law has been its increasing shift from a use-oriented to a resource-oriented approach. The use-oriented approach emphasized navigation and fishing, whereas the resource-oriented approach gives far greater attention to sustainable use and development of ocean resources, increasingly defining and enforcing standards of protection, conservation, management, and development (Sand 1988).

Other International Conservation Treaties and Conventions: Patterns and Principles

The 1972 World Heritage Convention, signed one year before the adoption of CITES, established protocols that allow all participating countries to nominate natural and cultural sites to the World Heritage List if they agree to take measures to preserve such sites, thus contributing to a growing network of conservation reserves worldwide (Weiss and Jacobson 1999). In this same year the previously mentioned United Nations Conference on the Human Environment was held in Stockholm, Sweden. The Stockholm Conference was significant in that, for the first time, the United Nations became intentionally involved in world conservation in comprehensive and systematic ways. UNEP was a direct result of the Stockholm conference and made environmental concerns and programs a permanent fixture of the United Nations agenda. Thus, for the first time, a global institution created an international series of programs designed to address environmental and conservation concerns.

United Nations involvement in systematic environmental program development and legislation led directly to a series of subsequent world environmental conferences and to an increasing number of international treaties that addressed species protection and management. Examples of treaties that focused on species protection include: (1) the Commission on Conservation of Antarctic Marine Living Resources (1982); (2) the International Convention for the Regulation of Whaling (1937, which established the International Whaling Commission); (3) the International Convention for the Protection of Birds (1950) and the Benelux Convention on the Hunting and Protection of Birds (1958); and (4) the Convention on Fishing and Conservation of Living Resources in the Baltic Sea and the Belts (1973). More recent treaties protect individual species (e.g., polar bears, vicunas, fur seals), establish refuges and reserves (e.g., Protocol Concerning Protected Areas and Wild Fauna and Flora in the East African Region [Nairobi 1985]), or protect endangered or threatened species in particular regions (e.g., Convention Between the Government of the United States and the Government of Japan for the Protection of Migratory Birds in Danger of Extinction and their Environment [Tokyo 1972]).

The Rio Summit

In June 1992, two global environmental conferences were held in Rio de Janeiro, Brazil, and produced a number of environmental documents signed by most or, in some cases, all of the participating nations. The United Nations Conference on Environment and Development (UNCED), popularly referred to as the Rio Summit or Earth Summit, was a formal conference of official government delegations. Simultaneously, a large gathering of nongovernmental organizations gathered for the Global Forum, a mixture of NGO networking, street shows, trade fairs, and environmental demonstrations (Parson, Haas, and Levy 1992).

The Rio Summit produced five major documents. The best known of these, the Rio Declaration, was originally conceived as a kind of "earth charter" that summarized international consensus on environmental policy and development. The Rio Declaration, signed by all participating nations, affirms environmental protection as an integral part of development.

The Convention on Climate Change primarily addresses emissions limits and standards of "greenhouse gases" associated with fossil fuels. Although the convention does not set specific targets, its ambitious objective is the "stabilization of greenhouse gas concentrations in the atmosphere that would prevent dangerous anthropogenic interference with the climate system . . . within a time frame sufficient to allow ecosystems to adapt naturally" (United Nations 1992). Representatives of 153 countries signed this convention.

The Convention on Biodiversity addresses conservation and sustainable use of biodiversity combined with the fair sharing of genetic resources. The 153 signers pledged to develop plans to protect habitats and species, provide funds and technology to assist developing countries to provide protection, ensure commercial access to biological resources for development, share revenues fairly among sources and developers, establish safety regulations, and accept liability for risks associated with biotechnology development (Parson, Haas, and Levy 1992).

The Statement on Forest Principles was a nonbinding declaration that pledges its signers to keep 17 principles "under assessment for their adequacy with regard to further international cooperation on forest issues" (Parson, Haas, and Levy 1992). Progress toward a formal treaty on forests at the Rio Summit failed primarily because of differences between industrialized countries that wanted a treaty focusing on tropical forests, and developing countries that wanted a treaty including boreal and temperate forests.

The most comprehensive document signed at the Rio Summit was Agenda 21, an 800-page "work plan" addressing social and economic dimensions of environment and development, conservation and management of resources, and means of implementation. Agenda 21's structure is based on key environmental and conservation issues, including the problems of desertification, protecting the atmosphere, and managing toxic wastes. It also addresses social issues with environmental dimensions such as poverty and technology transfer (Greene 1994). In its social and economic dimensions, Agenda 21 affirms the need to eradicate poverty and hunger, to manage resources sustainably, to link human health to environmental and socioeconomic improvements, and to integrate environmental factors into policy making, law, economics, and national accounting. In addressing conservation and management of resources for development, Agenda 21 supports allocation of land that provides the greatest sustainable benefits, and affirms the need for worldwide conservation of biodiversity, the need for proper management of mountain resources, more information on mountain ecosystems, and integrated development of mountain watersheds. Agenda 21 also affirms the importance of freshwater resources, the provision of safe drinking water, and the need for safe management of various kinds of toxic chemicals and hazardous wastes. In its final section on means of implementation, Agenda 21 supports promoting public awareness, establishing a new UN body, the Sustainable Development Commission (SDC) to coordinate pursuit of sustainable development among international organizations and to monitor progress by governments and international organizations toward reaching the goals of the agenda. Agenda 21 concludes with a discussion of the importance of collecting and using information for sustainable development and for implementing Agenda 21. Agenda 21 spurred controversy, and failed to reach agreement, on issues of fish stocks, targets and deadlines for an increase in total official development assistance, and the governance of the Global Environmental Facility (GEF), among others (Parson, Haas, and Levy 1992).

Although not all are legally binding, the Rio documents have influenced international environmental law. Agenda 21 has already found its way into many UN resolutions; the conventions on climate change and biodiversity have increasingly set the standard of international policy, practice, and expectation on the issues they address. The Rio Declaration, although controversial, will contribute to common goals and standards of national behavior informed by environmental principles.

International conservation law has increased and matured rapidly in the last three decades to address issues of international importance. However, in a world of increasing global connection and dependence, initiatives for global conservation by a single nation, even the United States, must increasingly assess their effects upon other nations to be successful. To better understand and appreciate the fascinating complexity of and connections between national and international conservation law, and between governments and nongovernmental organizations, we consider the following examples of legislation designed to protect dolphins from tuna fishers and sea turtles from shrimp trawlers.

THE PROBLEM OF INTERDEPENDENCE: HOW DOES ONE NATION PROMOTE GLOBAL CONSERVATION WITHOUT NEGATIVE EFFECTS ON OTHER NATIONS?

Case History I: Tuna and Dolphins

In 1972, just 2 years after passage of NEPA and only a year before passage of the amended Endangered Species Act, the U.S. Congress enacted the Marine Mammal Protection Act

Figure 2.4

An Atlantic spotted dolphin (*Stenella frontalis*) caught in a tuna net. The "incidental kill" of dolphins in association with tuna fishing has resulted in the deaths of millions of dolphins worldwide.

(MMPA). The MMPA was a relatively minor and noncontroversial piece of legislation that enjoyed broad bipartisan support. The act's clear and simple goal was to protect "certain species and population stocks of marine mammals that are, or may be, in danger of extinction or depletion as a result of man's activities." One of the MMPA's mechanisms to achieve this goal was to reduce "incidental kill or serious injury of marine mammals . . . to insignificant levels approaching a zero mortality and serious injury rate."

The deaths of marine mammals associated with "incidental kill" had increasingly become a cause for scandal and condemnation by the public and the press, particularly in regard to the killing of dolphins by tuna fishers. The problem had been developing since the 1950s, when tuna fishers began to employ purse seine nets in capturing tuna. Such nets captured tuna in large schools when they fed near the surface. After tuna were surrounded by the purse seine net, the bottom of the net was pulled together, trapping the tuna and all other organisms inside (Joyner and Tyler 2000).

Dolphins often travel directly above schools of tuna, so tuna fishers began to track dolphins as an indicator of tuna presence. Thus, it was not surprising, or even "incidental," that dolphins were killed with tuna, either by drowning in the net or being crushed by the harvesting machinery. Since the 1960s, an estimated 6 million dolphins perished in this manner (fig. 2.4).

The MMPA established stringent guidelines for U.S. tuna fishers and all tuna fishing in U.S. waters to assure protection for dolphins and other species. It soon became apparent, however, that other countries, including those harvesting the majority of tuna, were not following standards set by the MMPA. To encourage adoption of such standards on an international level and to protect dolphin populations worldwide, the U.S. Congress twice amended the MMPA. In 1984, the MMPA was altered to require an embargo on tuna imports from any country whose commercial fleets killed more dolphins than U.S. fleets. In 1988, Congress added additional requirements for all tuna-exporting nations attempting to market tuna in the United

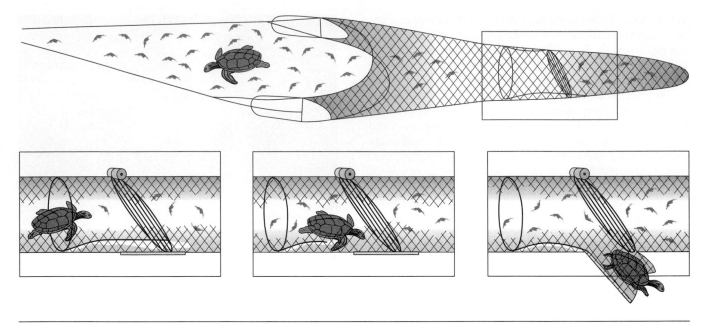

Figure 2.5

A turtle excluder device (TED) that can be installed in a shrimp net to release sea turtles from the net. TEDs, properly installed, can reduce sea turtle mortality associated with shrimp fishing by up to 97%.

States. Tuna-exporting countries were required to reduce incidental kill of nontuna species to the level of the U.S. fishing fleets, and were prohibited from using large-scale drift nets, encircling marine mammals without direct evidence of the presence of tuna, or using purse seine nets after sundown. Failure to comply would lead to a U.S. embargo of the fishery products of that nation. An additional embargo would be placed on tuna products from any nation that imported fishery products from nations under the first embargo (Miller and Croston 1998; Joyner and Tyler 2000).

Case History II: Shrimp and Sea Turtles

In 1989, the U.S. Congress added a provision (Section 609) to Public Law 101-162 that became known as the Sea Turtle Act. The Sea Turtle Act was motivated by concern over continuing worldwide declines in the populations of all seven species of sea turtles and by scientific studies that implicated shrimp nets in sea mortality.

One of the world's largest consumers of shrimp, the United States, was also one of the first nations to begin widespread use of the turtle excluder device (TED), which is a grid trapdoor installed inside a trawling net that keeps shrimp in the net but directs other, larger objects or animals out of the net (fig. 2.5). By the 1980s, TED technology had reached the point that, when devices were properly installed, 97% of sea turtles caught in shrimp nets could be released alive and unharmed without loss of shrimp (Joyner and Tyler 2000).

Earlier legislation had already required the use of TEDs in all shrimp trawlers operating in the Gulf of Mexico and in the Atlantic Ocean off the southeast coast of the United States. The

Sea Turtle Act went even further. It prohibited fish imports from any nation that failed to adopt sea turtle conservation measures comparable to those in the United States. Initially such sanctions were applied only to western Atlantic and Caribbean nations, which eventually complied. However, the largest shrimp importers to the United States were Asian nations that did not use TEDs. As a result, the prohibitions of the Sea Turtle Act were largely symbolic and did little to protect turtles from shrimpers on a global scale.

As these events were taking place, the United States was engaged in negotiations to ratify the General Agreement on Tariffs and Free Trade (GATT). The Clinton administration was reluctant to create controversy with Asian nations over sea turtles that could delay or halt ratification of GATT, and U.S. officials delayed enforcement of the act against its most important shrimp suppliers. Such reticence eventually led to a federal lawsuit by the Earth Island Institute, a U.S. NGO. Earth Island Institute demanded that the provisions of the Sea Turtle Act be enforced uniformly to all nations exporting shrimp to the United States. After a series of appeals, the Earth Island Institute won the case in the U.S. Court of International Trade, forcing the United States to ban imports from nations that had not complied with the Sea Turtle Act, including the largest Asian shrimp exporters.

Legal Challenges to U.S. Conservation Efforts

The tuna and shrimp embargoes, now in full force, led to legal challenges by the sanctioned nations before the World Trade Organization (WTO). In separate but similar cases, the tuna- and shrimp-exporting nations argued that the MMPA and

Sea Turtle Act were violations of the free trade provisions guaranteed by GATT. In the case of tuna and dolphins, the European Union Community also joined in challenging the MMPA because the embargoes prevented them from selling tuna to the United States that they had purchased from Asian nations that did not comply with the MMPA (Miller and Croston 1998). The plaintiffs argued that, under the terms of GATT, an individual nation could not impose restrictions on imports from other nations, even for conservation reasons, that those nations had not been party to developing. Ultimately the WTO agreed and ruled against the United States in the case of both dolphins and sea turtles, agreeing with the plaintiffs that the U.S. laws constituted unfair barriers to free trade. The United States appealed the decisions, but its appeals were not successful (Joyner and Tyler 2000).

Although it pledged to follow the rules of international law, the United States continued its advocacy for the conservation of both marine mammals and sea turtles. In the former case, the United States played a leading role in developing two new international agreements, the La Jolla (California) agreement of 1992, a 10-nation agreement that established a voluntary program to limit dolphin mortality, and the Panama Declaration, which was signed by 12 nations in 1995. The Panama Declaration went beyond the La Jolla agreement in establishing a "permanent" mortality limit for dolphins and stricter enforcement systems. The purpose of both agreements was to foster better methods of harvesting tuna through a voluntary program of setting standards and procedures for dolphin protection, and their collective outcome was the establishment of the International Dolphin Conservation Program. To implement the La Jolla agreement, the U.S. Congress enacted the International Dolphin Conservation Act of 1992. To implement the terms of the Panama Declaration into U.S. law and nationalize the intent of the International Dolphin Conservation Program, Congress passed the International Dolphin Conservation Act of 1997 (Miller and Croston 1998).

The United States maintained its commitment to sea turtle conservation by continuing to sponsor an already existing TED certification program for other nations. In addition, the United States pledged to assist any government seeking help in developing a TED sea turtle protection program of its own. Such commitments appear likely to lead to a regional agreement on sea turtle protection among nations conducting shrimp fishing in the Indian Ocean (Joyner and Tyler 2000).

These difficult cases involving tuna, dolphins, shrimp, and sea turtles offer insight into a world of complex interactions between national and international conservation law, public interest and private industry, and government bureaucracies and NGOs. They illustrate the fine line between conservation leadership and (in the eyes of some) conservation imperialism. Conservation laws of individual countries can no longer be enacted or enforced without first considering the interests of other nations or the likely international response. Today's worldwide commitment to global free trade has created international bodies, such as the WTO, whose decisions have the force of law. Such decisions may override the laws passed by a single nation in matters of international commerce, regardless of that nation's noble intentions for conservation. In the tuna-dolphin and shrimp-turtle decisions, the WTO displayed its own preference for multilateral and international agreements to reach conservation objectives as opposed to unilateral, national initiatives (Joyner and Tyler 2000); however, such decisions by the WTO appear to sacrifice conservation to commerce. The Dispute Settlement Body of the WTO rarely selects panel members and experts for their environmental expertise. Although the Dispute Settlement Body is authorized to seek expert advice on environmental issues, it rarely does so (Miller and Croston 1998). The perception that the WTO favors trade at the expense of conservation was part of the motive behind the anger and violence displayed toward the WTO and its delegates by conservation and environmental organizations, among others, in the large, public, and sometimes violent, demonstrations associated with the 1999 WTO meetings in Seattle, Washington, and the 2000 WTO meetings in Washington, D.C.

U.S. initiatives like the MMPA and the Sea Turtle Act helped provoke the international community to higher standards and greater action on these conservation issues than would have been achieved otherwise. At the same time, it is clear that, in an increasingly global community, the United States will have to increase and improve its efforts to involve other nations in multilateral and international conservation efforts, particularly conservation efforts that affect international trade, if it expects such efforts to be effective and permanent in their effects.

Synthesis

Often environmental regulations and the demands of conservation law press scientists to address and answer questions they may consider "unscientific." Likewise, law and policy require an integrated, interdisciplinary approach that conservation biologists may publicly endorse but are privately unprepared to adopt. And environmental problems, especially on a worldwide scale, may require a much greater level of organization and coordination than has historically been characteristic of the independent nature of science and scientists. For example, the U.S. federal government spends about 5 billion dollars per year to support research on global climate change and toxic waste disposal, yet such research has had only a very limited effect on national environmental decisions (Tarlock 1994). And, as conservation biologists Gary Meffe and Stephen Viederman have noted, "We now understand that much of what we do in conservation biology is worthless if it is not translated into effective policy" (Meffe and Viederman 1995).

In the past, much of the management activity associated with conservation was focused on outcomes that were predictable effects of their management actions. Goals such as sustained yield were based on an expectation of certain return. Today, conservationists are less concerned about certainty of return than about managing risk and determining the boundaries of biological uncertainty. Historically, environmental law has favored policies consistent with our past understanding of the rule of law (i.e., the consistent application of fixed rules that will yield a final, single decision that represents an absolute, moral ideal) (Tarlock 1994). As a result, individual environmental laws have tended to be based on individual scientific premises, and have then continued the application of those premises regardless of what new studies uncovered. Today such legal certainties are inconsistent with the state of our knowledge of ecosystems. Conservation biologists' best estimates of genetic diversity, minimum viable populations, and community ordination are also uncertain estimates. Modern conservation law and policy must mature to the point that they can deal effectively with such uncertainty, rather than simply ignore or reject it.

The development of conservation law and policy demonstrates consistently repeated themes. First, in democratic societies, the scrutiny of a free press and the involvement of a well-educated populace enables private organizations and private citizens to make a great deal of difference in how things turn out. Second, at the international level, even failed attempts at international legislation, such as the Rio Summit, may produce positive results, and must continually be pursued toward the eventual goal of a comprehensive and coordinated system of international conservation legislation. Third, paths to programs of lasting effectiveness in conservation are strongly affected by economic incentives and disincentives, as evidenced by the efforts to save dolphins from tuna fishing and sea turtles from shrimp boats.

For conservation biology and conservation biologists, the new millennium offers two concurrent challenges. On one hand, conservation biologists must become increasingly informed and astute in their understanding of conservation law and policy to make their research and management effective in achieving conservation goals. At the same time, conservation biologists must become increasingly sophisticated in learning how to change laws and policies, and formulate new ones, that will make conservation law more consistent with the scientific findings of how populations, communities, and ecosystems really work. Failure on the first front will make conservation biology an interesting but irrelevant discipline. Failure on the second front will lead to irreconcilable conflicts between the scientific and political communities, and the eventual disconnection of conservation science from conservation law and policy.

Law, Policy, and Science — A Directed Discussion

Reading assignment: Tarlock, A. D. 1994. The non-equilibrium paradigm in ecology and the partial unraveling of environmental law. *Loyola of Los Angeles Law Review* 27:1121–1144.

Questions

1. According to Tarlock, the wave of environmental and conservation legislation that characterized the 1960s and 1970s was based on the equilibrium paradigm in ecology. Environmental legislation was designed to maintain ecosystems at equilibrium and protect them from disturbance. Do major environmental laws such as the National Environmental Policy Act and the Endangered Species Act assume an equilibrium view of nature? If so, what are the ways that this assumption expresses itself in the policies that have arisen from these laws?

2. Modern ecological theory no longer supports the equilibrium view of ecosystems. Today managers often assume and explicitly manage for non-equilibrium states. Do you think environmental legislation of the 1960s and 1970s will "unravel" as Tarlock suggests, or is such legislation general and comprehensive enough to

adapt to a non-equilibrium view of ecosystems and to non-equilibrium management? If it adapts, how might the adaptation be expressed?

3. Consider this quote from Tarlock (p. 1129): "At best, ecosystems can be managed, but not restored or preserved." Is this statement true or false? What kinds of management actions and what management agencies seem to assume that the statement is true? What kinds of management actions and agencies seem to assume that the statement is false?

4. On p. 1131 of the assigned reading, Tarlock states, "Successful conservation biology requires the in-creased production of regulatory science," which he defines as "scientific research directed to provide useful information for regulators." Do you agree or disagree with this statement? If you agree, explain why regulatory science requires a more interdisciplinary approach and greater accountability to societal values than traditional science? If you disagree, explain what conservation biology *does* require, if not "regulatory science." Does regulatory science require *a priori* commitment to normative values to guide decision makers in appropriate use of scientific information?

Learning Online

Visit our webpage at www.mhhe.com/conservation for case studies, animations, practice quiz questions, and additional readings to help you understand the material in this chapter. You'll also find active links to the following topics:

Environmental Policy, Law and
    Planning
Environmental Ethics
Environmental Organizations
Societies and Government Programs
Biodiversity

Global Ecology
Land Use: Forests and Rangelands
Tropical Rain Forests and Land-Use
    Issues
Nontropical Forests and Land-Use
    Issues

Rangelands and Land-Use Issues
Species Preservation
Legislation Regarding Endangered
    Species
Endangered Species Act

Literature Cited

Bean, M. J., S. G. Fitzgerald, and M. A. O'Connell. 1991. *Reconciling conflicts under the Endangered Species Act.* Washington, D.C.: World Wildlife Fund.

Behan, R. W. 1990. The RPA/NEFMA: solution to a nonexistent problem. *Journal of Forestry* 88:20–25.

Berrens, R. P., D. S. Brookshire, M. McKee, and C. Schmidt. 1998. Implementing the safe minimum standard approach: two case studies from the U.S. Endangered Species Act. *Land Economics* 74:147–61.

Beyers, J. L., and W. O. Wirtz. 1997. Vegetative characteristics of coastal sage scrub sites used by California gnatcatchers: implications for management in a fire-prone ecosystem. In *Proceedings of the First Conference on Fire Effects on Rare and Endangered Species and Habitats,* ed. J. M. Greenlee, 81–89. Fairfield, Wash.: International Association of Wildland Fire.

Blasco, D. 1997. The Ramsar Convention Manual: a Guide to the Conservation of Wetlands (Ramsar, Iran, 1971). 2d edition. Gland, Switzerland, Ramsar Convention Bureau.

Brown, T. C., and G. L. Peterson. 1993. A political-economic perspective on sustained ecosystem management. In *Sustainable ecological systems: implementing an ecological approach to land management,* ed. W. W. Covington and L. F. DeBano, 228–35. General Technical Report RM-247. Washington, D.C.: U.S. Department of Agriculture Forest Service.

Brown, N. J., and P. Quiblier, eds. 1994. Ethics and Agenda 21. New York: United Nations Environment Programme.

Caldwell, L. K. 1966. Administrative possibilities for environmental control. In *Future environments of North America,* eds. F. F. Darling and J. P. Milton, 648–71. Garden City, N.Y.: Natural History Press.

Caughley, G., and A. Gunn. 1996. *Conservation biology in theory and practice.* Cambridge, Mass.: Blackwell Science.

Ciriacy-Wantrup, S. V. 1952. *Resource conservation: economics and policy.* Berkeley: University of California Press.

Committee on Scientific Issues in the Endangered Species Act. 1995. *Science and the Endangered Species Act.* Washington, D.C.: National Academy Press.

Cooper, M. H. 1999. Endangered Species Act. *CQ Researcher* 9:851–63.

Costa, R. 1997. The U.S. Fish and Wildlife's red-cockaded woodpecker private lands conservation strategy: an evaluation. *Endangered Species Update* 14 (7–8):40–44.

Costanza, R., O. Segura, and J. Martinez-Alier. 1996. *Getting down to Earth: practical application of ecological economics.* Washington, D.C.: Island Press.

Darling, F. F., and J. P. Milton, ed. 1966. *Future environments of North America.* Garden City, N.Y.: Natural History Press.

DeSimone, P., and D. Silver. 1995. The Natural Community Conservation Plan: can it protect coastal sage scrub? *Fremontia* 23(4):32–36.

Dunlap, T. R. 1981. *DDT: scientists, citizens, and public policy.* Princeton, N.J.: Princeton University Press.

Endicott E. 1993. Land conservation through public/private partnerships. Washington, D.C.: Island Press.

Flather, C. H., M. S. Knowles, and I. A. Kendall. 1998. Threatened and endangered species geography: characteristics of hot spots in the conterminous United States. *BioScience* 48:365–76.

Goetz, P. C. 1997. An evaluation of ecosystem management and its application to the National Environmental Policy Act: the case of the U. S. Forest Service. Ph.D. diss., Virginia Tech, Blacksburg, Va.

Greene, G. 1994. Caring for the earth: The World Conservation Union, the United Nations Environment Programme, and the World Wide Fund for Nature. *Environment* 36(7):25–28.

International Union for Conservation of Nature and Natural Resources. 1986. *Migratory species in international instruments: an overview.* Gland, Switzerland: IUCN.

Johnson, T. January/February. 1998. *The whaling trade.* Ottawa: Canadian Geographic.

Joyner, C. C., and Z. Tyler. 2000. Marine conservation versus international free trade: reconciling dolphins with tuna and sea turtles with shrimp. *Ocean Development and International Law* 31:127–50.

Kaiser, J. 1997. When a habitat is not a home. *Science* 276:1636–38.

Karr, J. R. 1995. Biological integrity and the goal of environmental legislation: lessons for conservation biology. In *Readings from* Conservation Biology: *the social dimension— ethics, policy, management, development, economics, education,* ed. D. Ehrenfeld, 108–14. Cambridge, Mass.: Blackwell Science.

Kelso, B. J. 1995. The ivory controversy. *Africa Report* 40(2):50–55.

Koester, V. 1989. *The Ramsar Convention on the conservation of wetlands: a legal analysis of the adoption and implementation of the convention in Denmark.* Gland, Switzerland: International Union for Conservation of Nature and Natural Resources.

Lehman, W. E. 1995. Reconciling conflicts through habitat conservation planning. *Endangered Species Bulletin* 20(1):16–19.

Liroff, R. A. 1976. *A national policy for the environment: NEPA and its aftermath.* Bloomington: Indiana University Press.

Loomis, J. B. 1993. Integrated public lands management. New York: Columbia University Press.

Mann, C., and M. Plummer. 1997. Qualified thumbs up for habitat plan science. *Science* 278:2052–53.

McCormick, J. 1999. The role of environmental NGOs in international regimes. In *The global environment: institutions, law, and policy,* ed. N. J. Vig and R. S. Axlerod, 52–71. Washington, D.C.: Congressional Quarterly.

Means, B. 1998. Prohibiting conduct not consequences: the limited reach of the Migratory Bird Treaty Act. *Michigan Law Review* 97:823–42.

Meffe, G. K., and S. Viederman. 1995. Combining science and policy in conservation biology. *Wildlife Society Bulletin* 23:327–32.

Meiner's, R. E., and B. Yandle. 1993. Taking the environment seriously. Baltimore, Md.: Rowman and Littlefield.

Miller, C. J., and J. L. Croston. 1998. WTO scrutiny v. environmental objectives: assessment of the international dolphin conservation program. *American Business Law Journal* 37:73–125.

Murthy, K. S. 1988. *National Environmental Policy Act (NEPA) process.* Boca Raton, Fla.: CRC Press.

Nash, R. F. 1989. *The rights of nature.* Madison: University of Wisconsin Press.

Noss, R. F. 1993. Whither conservation biology? *Conservation Biology* 7:215–17.

Noss, R. F., M. A. O'Connell, and D. D. Murphy. 1997. *The science of conservation planning.* Washington, D.C.: Island Press.

O'Connell, M. A. 1997. Improving habitat conservation planning through a regional ecosystem-based approach. *Endangered Species Update* 14(7–8):18–21.

Parson, E. A., P. M. Haas, and M. A. Levy. 1992. A summary of the major documents signed at the Earth summit and the global forum. *Environment* 34:12–15, 34–36.

Petulla, J. M. 1977. *American environmental history.* San Francisco, Calif.: Boyd and Fraser.

Primack, R. B. 1993. *Essentials of conservation biology.* Sunderland, Mass.: Sinauer Associates.

Rodgers, W. H., Jr. 1994. The seven statutory wonders of U.S. environmental law: origins and morphology. *Loyola of Los Angeles Law Review* 27:1009–21.

Rohlf, D. J. 1991. Six biological reasons why the Endangered Species Act doesn't work—and what to do about it. *Conservation Biology* 5:273–82.

Rohlf, D. J. 1995. Response to O'Connell. In *Readings from* Conservation Biology: *the social dimension—ethics, policy, management, development, economics, education,* ed. D. Ehrenfeld,100–101. Cambridge, Mass.: Blackwell Science.

Rosenbaum, W. A. 1985. *Environmental politics and policy.* Washington, D.C.: CQ Press.

Rosenbaum, W. A. 1995. *Environmental politics and policy.* 3d ed. Washington, D.C.: CQ Press.

Sample, V. A. 1993. A framework for public participation in natural resource decision making. *Journal of Forestry* 91(7):22–27.

Sand, P. H. 1988. *Marine environment law in the United Nations Environment Programme: an emergent eco-regime.* London: Tycooly Publishing.

Schilling, F. 1997. Do habitat conservation plans protect endangered species? *Science* 276:1662–63.

Shepherd, R. A. 2000. Conservation is a hot new practice. *National Law Journal* 22:A23.

Slocombe, D. S. 1989. CITES, the wildlife trade, and sustainable development. *Alternatives* 16:20–29.

Spong, P. and N. D. Phillips. 1999. International Whaling Commission report. Earth Island Journal 14:13.

Tarlock, A. D. 1994. The nonequilibrium paradigm in ecology and the partial unraveling of environmental law. *Loyola of Los Angeles Law Review* 27:1121–44.

United Nations. 1992. United Nations Framework Convention on Climate Change. Article 2.

U.S. Fish and Wildlife Service. 1993. *Federal Register* 58(59):16742.

Vincent, B. 1999. Retooling the Endangered Species Act. *Environment* 41:4–5.

Walley, K. K. 1996. Surprises inherent in no surprises policy. *Endangered Species Update* 13(10–11):8–9, 14.

Webster, A. M. 1971. *Webster's seventh new collegiate dictionary.* Springfield, Mass.: G. and C. Merriam Company.

Weiss, E. B., and H. K. Jacobson. 1999. Getting countries to comply with international agreements. *Environment* 41(6):16–31.

Yost, N. C., and J. W. Rubin. 1989. *NEPA deskbook.* Washington, D.C.: Environmental Law Institute.

CHAPTER 3

If you are completely ignorant of values, then you are incapable of making a rational decision, either for or against preserving some species. The fact that you do not know the value of a species, by itself, cannot count as a reason for wanting one thing rather than another to happen to it.

—Elliot Sober, 1986

Values and Ethics in Conservation

In this chapter, you will learn about:

1 characteristics that define and distinguish major categories of value in conservation.

2 methods for determining instrumental and noninstrumental economic values in conservation.

3 philosophical, cultural, and religious traditions that affirm instrumental and intrinsic values of species, biodiversity, and natural objects.

WHAT DOES SCIENCE HAVE TO DO WITH VALUES?

As seen in the previous chapter, conservation biology derives much of its effectiveness and empowerment from conservation law and policy, which are rooted in values and ethics. Laws and policies, however, are not the same as values and ethics. Laws apply to specific situations and events and incorporate only narrow applications of values and ethics to particular circumstances. For example, although the Endangered Species Act protects threatened species and biodiversity, it falls far short of the protection that an encompassing social *value* would offer. As attorney and international environmental negotiator Terry Leitzell notes, "In practice, the ESA does not protect species generally, since thousands of species are expected to become extinct every decade if present rates continue. The ESA protects only a very few species, chosen for protection by a political process using many different values" (Leitzell 1986:250). And the prohibitions written into the Endangered Species Act can protect rare species from extinction only if "they operate within a general climate in which the majority of those affecting the well being of threatened species recognize them as having value" (Carlton 1986:256). Clearly, something more than law is needed to achieve conservation goals.

Many laws would be unnecessary if ethical values were socially and culturally strong enough to ensure that moral ideals like truth, justice, and compassion would always prevail. Historically, laws have been passed only when social norms of value and ethics have collapsed; coercion is society's last resort. Although essential for conservation, laws and policies cannot replace or substitute for values and ethics.

Value refers to a general *basis for an estimation of worth*. Values represent judgments of relative worth, merit, usefulness, importance, or degree of excellence. Values can justify and explain concrete objectives, such as conserving biodiversity, but they are not the same as the objective. *Ethics* are *systematic organizations of values* that establish explicit principles for conduct and behavior by individuals or groups. Laws, policies, and conservation decisions do not merely reflect and implement society's ethics and values; they also shape them. Choices we make today about what is valued will determine the range of experiences in the natural world available to future generations. These choices will also affect future attitudes, preferences, and worldviews. Therefore, conservation biologists need a thorough and sophisticated understanding of values and ethics—two indispensable elements in conservation.

A discussion of values and ethics in the context of a scientific discipline like conservation biology may seem out of place to some because, since the Enlightenment, science has been viewed, by scientists and nonscientists alike, as "value free." Scientists, then, offered humanity knowledge about what and how things are, not a vision of the way things ought to be. In this view of science, facts and truth are not affected by the convictions, commitments, or biases of the investigators. In the words of ethicist C. S. Lewis, such "value-neutral" science separated "the world of facts without one trace of value, and the world of feelings without one trace of falsehood, justice, or injustice . . . and no *rapprochement* is possible" (Lewis 1947:30–31). Just as facts were seen as value free, values were nonfactual, unverifiable, and subjective.

Value-neutral science also implied that scientists should not advocate particular applications of science to specific problems because the role of the scientist was to supply value-neutral information and expertise. Advocacy represented a commitment to particular (and nonfactual) values, a loss of objectivity, and a loss of reliability of scientific information. The application of science to particular problems and the advocacy of particular solutions were left to ethicists, political officials, and religious leaders.

This view of science was fundamentally flawed from its inception and has become increasingly unworkable in producing effective interaction between science and the modern world. The most serious problem of this view was in making objectivity (the accurate perception of reality) synonymous with neutrality (not engaged to, committed, or discerning of value, truth, or worth). As explained in chapter 1, conservation biology emerged as a distinct discipline when a body of scientists deliberately renounced the division between facts and values and rejected the conflation of scientific objectivity with neutrality. Although committed to objective truth, conservation biologists insisted that truth informs decisions of value, and leads to specific commitments of value in the application of science to conservation. In such a context, conservation biology asserted its identity as a "value-laden, mission-driven science" and committed itself to certain core values, which Michael Soulé called "normative postulates" of conservation biology. These postulates, noted in chapter 1, bear repeating here: (1) *diversity of organisms is good,* (2) *ecological complexity is good,* (3) *evolution is good,* and (4) *biotic diversity has intrinsic value, regardless of its utilitarian value.* To be useful, such normative postulates must be informed by specific and practical knowledge of individual conservation dilemmas. Any normative postulate, isolated from all other considerations and pursued exclusively without regard for such considerations, would bear results that would be emphatically "not good." For example, acts of introducing nonindigenous species into a system to achieve a short-term increase in the system's diversity may eliminate native species and severely alter ecosystem structure and function. Normative postulates are valuable guides for identifying the fundamental values of conservation biology. They are not substitutes for intelligent management informed by specific knowledge about particular systems.

Normative postulates are nonempirical statements. That is, they are not statements of physical objectivity, verifiable by experimental analysis. This fact leads some to claim that, therefore, normative postulates cannot be "proven" to be "true" and are really only statements about the personal feelings of those who make them. Such a view of truth, which defines knowledge as consisting solely of descriptions of physical entities, is severely constricted. In fact, it is one in which conservation biology would not be able to practice conservation. In practice, conservation biologists relentlessly seek application of the knowledge they gain from empirical studies. Applications of

knowledge in conservation demand choices and every choice requires a decision about what is "good" or "best" to achieve a given conservation goal. In conservation, management derives its purpose by understanding its outcomes as expressions of value. Adaptive management (chapter 1) developed in a context of conservation biology precisely because of the need to identify values as management ends and then to use scientific knowledge as a means of achieving ends consistent with these values. Thus, to satisfy the inherent drive for application intrinsic to conservation biology and to express management applications in terms of norms, conservationist biologists must possess a coherent system of values that they can articulate universally and persuasively. Further, they must be able to express such values as something more than their own personal preferences, likes, and dislikes about their favorite species or their preferred state of nature. If they fail to recognize the necessity of coherent expressions of value, and to affirm that statements of value (normative postulates) are statements about truth—about how things really *ought* to be—they will be left with no arguments to offer except those that express conservation in terms of ratios of human benefits and costs. Some conservationists do believe that all values should be reduced to this level of analysis, but many would argue that some conservation values exist independent of human welfare and that such values should be pursued regardless of their cost. Indeed, many conservation laws that conservation biologists lobbied for and support, such as the Endangered Species Act, explicitly state that economic considerations may *not* be used in deciding whether to adopt a particular action or policy, such as listing a species as endangered or threatened. Coherent value systems that can guide the applications of conservation biology are diverse. But the persistent effort to discover and understand such systems reveals the genuine need to explain applications and goals of conservation in terms of something more than personal preferences, appetites, and desires.

The values of contemporary conservation biology are deeply rooted in conservation history, particularly in the United States. Conservation biology is unique in that an entire scientific community explicitly embraced a constellation of values that defined its professional identity; however, statements of value, no matter how attractive rhetorically, must have real meaning. People are not persuaded to embrace values by assertion or coercion, but rather by argument. Conservation values cannot simply be plucked out of the air. They must be grounded in thoughtful analysis and social context. They must be informed by facts, not divorced from them. Conservation values, and the ethical systems that contain them, must be treated with intellectual respect, and their associated ethical systems must be viewed with analytical integrity. Conservation biologists must perceive systems of conservation values and ethics as genuine sources of truth and insight, and resist the all-too-common tendency of scientists to treat values as manipulative tools that can be used to dupe the public into adopting "correct" behavior.

Values in conservation biology must be carefully analyzed, measured, and understood if they are to be meaningfully used and persuasively expressed in the work of conservation. Individual statements of value must be integrated into the context of sound ethical systems if conservationists hope to display consistent principles and behavior in their own work or understand the behavior and responses of others to it. Conservation biologists must possess a thorough understanding of the ethical basis of conservation if they aspire to be effective in relating scientific findings to conservation policy. To that end, we now take up the problem of the origins, analysis, and measurement of value in conservation biology.

POINTS OF ENGAGEMENT—QUESTION 1

In recent communication with another scientist on the issue of value, my correspondent wrote, "People choose to study snakes because they love snakes." The correspondent went on to argue that such affection generated its own "mission," and such mission generated its own advocacy for snake conservation. And that was sufficient. Is personal affection a sufficient basis for conservation mission and advocacy? Is it an adequate frame of reference for discourse with other scientists and the public?

THE PROBLEM OF CATEGORIES: HOW DO WE CLASSIFY DIFFERENT KINDS OF CONSERVATION VALUES?

There is no single, universally accepted method of categorizing all values, or even conservation values, but we will begin with a simple dichotomy between *instrumental* and *intrinsic values.* Instrumental values measure the usefulness of a creature or object in meeting a need or providing a service to another, usually a human, and thus facilitating human welfare or happiness. Intrinsic values reside within an object itself. In other words, something has intrinsic value if, in the words of ethicist J. Baird Callicott, it is "valuable *in* and *for* itself—if its value is not derived from its utility, but is independent of any use or function it may have in relation to something or someone else (Callicott 1986:140, emphasis his). Within these larger categories, specific subcategories exist for natural objects in general, and for species in particular (fig. 3.1).

Stephen Kellert, an authority on human attitudes toward wildlife, identifies seven values of wildlife perceived by citizens of developed countries (Kellert 1986). These include:

1. Naturalistic and outdoor recreation values: values that relate to enjoyment from direct contact with wildlife
2. Ecological values: values associated with the importance of a species to other flora and fauna and to the maintenance of ecosystem processes
3. Moral or existence values: values associated with inherent rights or spiritual importance of species
4. Scientific values: actual or potential values associated with a species' contribution to enhancing human knowledge and understanding of the natural world
5. Aesthetic values: values associated with a species' possession of beauty or other qualities admired by humans
6. Utilitarian values: values associated with species as sources of material benefit or use

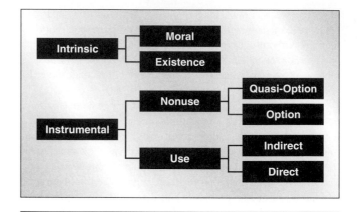

Figure 3.1

Categories of value and their relations. *Intrinsic values* are those that reside in the object itself, either through simply knowing that it exists (existence value) or because it embodies or is associated with something "good" (moral value). *Instrumental values* relate to usefulness or economic value. *Nonuse values* are given to items that are not used by anyone, but have value in the option to use. *Quasi-option values* are given to items that have no use at present, but with further knowledge may prove useful. Items with *option value* (e.g., a recreational area) are important, not because the item is frequently used, but rather because the option to use is valued. *Use value* is associated with consumers. *Direct value* is a measure of consumer demand. An item has *indirect value* if it supports the product with direct value.

7. Cultural, symbolic, or historical values: values associated with strong personal or cultural attachments of human cultural groups to a species, especially associated with symbolic meanings and identification between the species and humans

In Kellert's categories, moral or existence value provides the clearest example of intrinsic value. All the other categories represent some form of instrumental value. As a discipline, conservation biology acknowledges and affirms both instrumental and utilitarian values of species. Therefore, each of these two categories deserves further analysis and examination.

Instrumental Values

All human cultures are sustained, directly or indirectly, through natural capital, including goods and services derived from living organisms; thus, instrumental values can and should play a prominent role in any discussion of the value of living creatures.

The most common method of estimating the instrumental value of goods is through economics. We will examine the intricacies of environmental and conservation economics in chapter 12 (Conservation, Economics and Sustainable Development), but the basic concepts of economic evaluation of instrumental values are introduced here.

Economics is the study of how people allocate scarce resources—including biotic resources—among competing ends to satisfy unlimited human wants. Biotic resources supply all food for humans and, directly or indirectly, most of our fuel and medicines. Plants are the primary source of human clothing and structural materials, and are used to beautify personal property, enhance land values, and reduce soil erosion. In unmechanized cultures and societies, animals provide essential services through their labor in agriculture, transportation, and forestry. Even in mechanized cultures, animals continue to provide such services, although to a lesser extent and in less essential capacities. In human entertainment and recreation, plants and animals play significant roles in increasingly varied and economically important ways. Further, biotic systems perform essential services, such as water purification, soil formation and retention, flood control, oxygen production, and carbon dioxide absorption.

This description is by no means exhaustive, but it illustrates an essential point: biotic resources are instruments of human satisfaction and survival. But biotic resources are also scarce, and their use can be increased only at the cost of forgoing something else that is valued (Randall 1986). All biotic resources are potentially renewable resources, but they are also all potentially degradable and exhaustible. An understanding of these fundamental instrumental values demonstrates an obvious but often overlooked fact. Even though the interests of economics and conservation often appear at odds with one another in specific cases, conservation biology and economics have fundamental interests in common, especially on the issue of species preservation. As economist Alan Randall said, "Since species survival is a precondition for the use of the species as a resource, the preservation problem in principle precedes all other biotic resource issues" (Randall 1986:79).

Many environmental laws specifically require an assessment of value, including noneconomic value, in making environmental decisions. For example, environmental impact statements, required by the National Environmental Policy Act, normally include an economic analysis of the proposed action and all reasonable alternatives. Such an analysis is usually in the form of a cost-benefit analysis that examines the ratio of monetary costs to monetary benefits expected from a particular alternative. Although not part of the Endangered Species Act (ESA) in its original form, economic analyses, including cost-benefit analyses, are now required in the development of recovery plans for endangered species under amendments to the ESA. Thus, economic valuation is deeply ingrained in the fabric of environmental law. Some environmental laws, however, especially regulatory laws like the Clean Air and Clean Water Acts that set environmental standards, specifically prohibit the weighing of benefits against costs in the setting of environmental standards. The purpose of the latter approach is to set some social values, such as clean water or clear air, "beyond the reach" of economic valuation in order to affirm that some environmental values are, in the most literal sense, "priceless," and are not to be compromised by market forces.

Economists attempt to estimate, and, in many cases, predict the market value of resources, including biotic resources, through evaluation of resource scarcity (supply) and the amount of satisfaction the resource provides for humans (demand). Historically, economics has been used in making decisions in conservation, even when such decisions involve judgments about what is "right" and "wrong." Originally, economics was a branch of moral philosophy and only later emerged as an independent discipline (Kelman 1986). Although no longer considered a realm of moral philosophy, contemporary economic evalua-

tions often are influenced by what society believes is "right" in decisions on conservation. Conversely, social and moral choices are informed and influenced by economic valuations.

The economic valuation of biotic resources often can be done through normal market processes. In economics, *market goods* can be exchanged through some form of standard currency in an economic market. Some biotic resources, such as game animals, timber, or rangelands, can be treated and valued as market goods, and their market value can be assigned with relative precision. This valuation is easiest when the value of the resource is measured in utilitarian terms, usually as some type of commodity. In that case, biotic resources often can be treated as *private goods* (i.e., goods that can be bought, sold, and enjoyed solely by the buyer and seller). Natural resources also can serve utilitarian values as *public goods,* goods that are accessible to all and provide benefit to all, usually with the cost shared among the beneficiaries, as in the case of clean air. Every choice for a particular use for a particular resource comes with inherent benefits and costs. Such benefits and costs may be direct (accrue directly to those using the resource) or indirect (accrue to others affected by the way the resource is used).

A resource economist would argue that, theoretically, all instrumental values of resources could be evaluated economically. Evaluation is easiest when biotic resources are valued as private goods, but becomes more complex when they are valued as public goods, and still more complex when future as well as present values must be considered. But these considerations also are part of economic valuation. In economic terms, five categories of value help refine the assessment. **Use value,** the value derived from the actual use of a resource, usually is the easiest to measure and the most amenable to evaluation by market forces. For every use, there is a unique set of *opportunity costs* which represent costs, or losses, associated with the inability to use the resource to produce goods A, B, and C if the resource is being used to produce good D.

There are four major categories of nonuse values. **Option value** refers to the value of a resource's expected future use (i.e., what a person would be willing to pay to guarantee that the resource would be available for future use). **Quasi-option value** is the value of preserving options, given an expectation of growth in knowledge that might lead to a future, but as-yet undiscovered or unrealized use of the resource. Quasi-option value can be conceived as a kind of "speculation value" for what an investor might pay to preserve a resource, such as a rare tropical plant, given the expectation of increasing growth in knowledge of medical applications of all plant species generally. **Bequest value** is the value of knowing that something is preserved for future generations. Finally, **existence value** is the value of knowing that something exists (Randall 1986). In other words, the resource or object is not something that you are going to use now or in the future, but you derive pleasure and satisfaction simply from knowing that it is there. Existence value is unrelated to any actual or potential use of the biotic resource.

To make assessments of what is valued and how strongly it is valued, conservation biologists must have tools for identifying and measuring values, especially the values of the people who will determine conservation policy or be affected by the conservation policies that are determined. A number of tools are used to identify values and determine their relative strength and merit.

Determining Attitudes with Sociological Surveys

Stephen Kellert, speaking of the demise of the Hawaiian avifauna, wrote,

> *"While Hawaii's endemic bird life was biologically vulnerable to these impacts (of exotic introductions and habitat destruction), sociopolitical and economic forces created the contextual basis for this decline, and only the mitigation of these social factors can counter the continuing drift toward extinction [of] the remaining endangered species. Although this assertion may seem obvious, most endangered species efforts pay scant attention to human social factors and perceptions . . . we need to engage in a fundamental reassessment of the relationship between human society and the natural world"* (Kellert 1986).

One way that Kellert and others attempt to assess the underlying attitudes that both reflect and determine human values of wildlife and other natural resources is through surveys. Surveys may consist of "closed-ended" questions (in which the respondent selects a particular answer or a numerical value that best represents her or his answer) (table 3.1) or in-depth interviews involving open-ended questions. Survey design is a major professional discipline within sociology and psychology, and the details and nuances of this complex problem are beyond the scope of the chapter; however, some foundational principles follow: (1) the questions and potential responses of the survey must be carefully worded according to strict protocols to avoid confusion and to clarify categories of response; (2) the sample of individuals surveyed must be representative of the population and sufficiently large to avoid problems associated with

random error; (3) if subgroups of the population differ in important characteristics, representation of the subgroups also must be comprehensive and sufficiently large to ensure accurate sampling; and (4) if some individuals do not respond to the initial survey (as is often the case), there must be follow-up procedures in place to address the problem of "nonresponse bias" because nonrespondents frequently have significantly different characteristics and attitudes than the initial respondents.

If such problems can be successfully addressed, survey data can provide valuable insights into what people value in conservation and why. For example, Kellert investigated basic wildlife values in Japan by creating a typology of wildlife values and attitudes (table 3.2) and then designing multiple questions and scales to measure the strengths of different attitudes in the respondents (Kellert 1991). From these data, Kellert determined that the most common attitudes of the Japanese toward wildlife were humanistic (strong affection for individual animals or particular species) or negativistic (avoidance of animals because of dislike or fear). Compared with attitudes of U.S. citizens determined in earlier studies (Kellert 1979, 1980, 1981, 1985, 1989), Japanese attitudes were more dominionistic (motivated to control animals, especially in sporting activities) and less moralistic (concerned for right and wrong treatment of animals) and ecologistic (concerned for relationships of species within a system) (Kellert 1991).

Coupled with in-depth interviews, survey data may reveal not only attitudes of people toward wildlife and other biotic resources, but also the cultural basis and background of such views. Survey and interview data have revealed that humans around the world are most concerned for creatures that are large, aesthetically attractive, phylogenetically similar to humans, and regarded as possessing capacities for feeling, thought, and pain. Such species represent a category of creatures that have been referred to as "phenomenologically significant animals" (Shepard 1978). Given such human attitudes, it

Table 3.1 **"Closed-ended" vs. "Open-ended" Questions**   By providing a respondent with a list of possible answers (e.g., true/false, or multiple-choice questions), closed-ended questions are useful for directly testing items such as knowledge, awareness, and interaction with a subject. Open-ended questions result in in-depth answers because respondents are not presented with a list of answers from which they may choose. Rather, open-ended questions require an individual to answer thoughtfully and integrate his or her knowledge

QUESTION TYPE	EXAMPLES	RESPONSE
Closed-ended	If a conservation biologist aspires to be effective in relating scientific findings to conservation policy, it is necessary that the basis of conservation ethics be understood. The number of animals is more important than the genetic diversity of the group. The value of an organism should be decided by its supply and demand.	T F T F T F
Open-ended	To what extent should ethics play a role in the process of developing conservation policy? How important is the genetic diversity of a population? What criteria should be used to assign value to an organism?	Variable

Table 3.2   A Typology of Wildlife Values and Attitudes

TYPE OF VALUE OR ATTITUDE	DEFINITION
Naturalistic	Values that relate to enjoyment from direct contact with wildlife
Ecologistic	Values associated with the importance of a species to other flora and fauna and to the maintenance of ecosystem processes
Moral	Values associated with inherent rights or spiritual importance of species
Scientific	Actual or potential value associated with a species' contribution to enhancing human knowledge and understanding of the natural world
Aesthetic	Values associated with the species' possession of beauty or other perceived qualities admired by humans
Utilitarian	Values associated with species as sources of material benefit or use
Dominionistic	Values associated with the mastery and control of animals, typically through sport
Negativistic	Attitudes associated with the avoidance of animals because of dislike or fear
Neutralistic	Attitudes associated with the passive avoidance of wildlife because of a lack of interest
Theistic	Values associated with the belief that a supernatural deity or force creates, sustains, and values wild species

Kellert, S. R. 1991. Japanese perceptions of wildlife. *Conservation Biology* 5:297–308.

is no wonder that animals in this group are often chosen as emblems for major conservation organizations such as The World Wildlife Fund (giant panda (*Ailuropoda melanoleuca*) fig. 3.2), as representatives for government agencies such as the U.S. Forest Service (Smokey Bear, Woodsy Owl), or as a country's national symbol (for the United States, the bald eagle, *Haliaeetus leucocephalus*).

Survey data can also be used to assess human relationships with nature in terms of personal experience. One of the most commonly used methods in this category is a measure called "user satisfaction," which determines the level of personal satisfaction an outdoor recreationist experiences in a particular recreational activity. The higher the level of user satisfaction, the more the experience met or exceeded the expectations and desires of the recreationist. This measure has been used to evaluate the quality of natural environments as well as the preferences and values of those who use them.

Such knowledge can assist conservation biologists in determining what public attitudes are toward wildlife, conservation, and natural areas. In addition, they may be invaluable elements in designing public relations campaigns to win support for conservation goals. However, survey, interview, and user satisfaction data, by themselves, do not help evaluate the fundamental philosophical validity of the attitudes that they identify

and measure, nor do they help conservation biologists argue persuasively for different attitudes or ethical perspectives.

Tools of Economic Valuation: Cost-Benefit Analysis, Safe Minimum Standard Criteria, and Contingency Valuation Analysis

The most common, and sometimes legally prescribed, tool for determining the value of biotic and ecological resources in their natural state versus their value after detrimental acts of development is cost-benefit analysis (CBA). Conceptually simple to imagine, but often practically difficult to achieve, cost-benefit analysis, as its name implies, attempts to assess the benefits of a particular action and compare them to the environmental costs of that action. The preferred alternative is the one in which benefits most outweigh costs. For example, cost-benefit analyses are often associated with environmental impact statements required under the National Environmental Policy Act. "Costs" of the action include both short- and long-term environmental consequences as well as "irreversible and irretrievable commitments of resources" associated with the proposed action. "Benefits" to species, biodiversity, or ecosystem preservation must include both instrumental values—such as game, water,

Figure 3.2

The giant panda (*Ailuropoda melanoleuca*), symbol of The World Wildlife Fund and a classic example of a "phenomenolgically significant animal" (Shepard 1978) that evokes strong feelings of identification, affection, and concern in humans.

range, and timber—and noninstrumental values—such as visual and scenic appeal, religious significance, or historic importance to native peoples or long-term occupants. Cost-benefit analyses also may be associated with the proposed listing of a species as threatened or endangered, with broad changes in environmental or conservation policy, or with the effects associated with new conservation legislation. Government bureaucrats, private consultants or corporations, nongovernmental conservation organizations, or other individuals or groups that have a vested interest in the valuation of biotic resources may perform the actual analysis.

Cost-benefit analyses can be attractive to decision makers in conservation because they attempt to translate all the values associated with a decision into a common currency, usually market value, so that diverse entities (e.g., timber values and wildlife values) can be compared directly. Cost-benefit analysis also has the attraction of forcing decision makers to identify and evaluate the value of all entities affected by their decision. Cost-benefit analysis places the burden of proof on those who value the preservation of biotic resources rather than on proponents of development. In a cost-benefit analysis, defenders of species or biodiversity must show that species or ecosystem preservation is at least as economically valuable as a proposed development that is usually designed from the outset for market valuation and consumption.

Criticisms of cost-benefit analysis as a decision-making tool are numerous. For example, some acts that are morally wrong may have high economic benefits and low costs, whereas some acts that are morally right may have low benefits and high costs. In addition, cost-benefit analysis typically uses currencies most appropriate to private transactions of economic goods as guides for public and social policy, thus equating private preferences with cultural and social preferences. In the case of an endangered species, cost-benefit analysis assumes that preserving a species must be viewed in terms of *human benefits* rather than as an act of *moral obligation*. The distinction between these two

outlooks is critical. An alternative type of economic evaluation, the safe minimum standard (SMS), makes this difference in perspective more explicit.

The SMS approach (Ciriacy-Wantrup 1952) assumes that the desirable or "morally correct" outcome is that the species be preserved. This approach attempts to determine, and then maintain, some minimum level or safe standard of a renewable population unless the social costs of doing so are in some way "intolerable," "unacceptable," or "excessive," and then to use this standard as a guide to regulating economic and development activities that threaten the population or its environment. A human-based value is still inherent in setting the SMS, but it is now based on an existence value (e.g., the minimum population size of a species below which it cannot persist) rather than on an economic value (e.g., the most economically efficient way to harvest the species). The SMS shifts the burden of proof to the proposed developers, who must show, first, that the development will not reduce a biotic resource below a "safe minimum standard" needed for its persistence and, second, that efforts to ensure such persistence will cause "unreasonable," "excessive," or "intolerable" harm to human welfare. Thus, proponents of the SMS argue that it is more compatible than CBA with the structure and intent of the Endangered Species Act and with measures of population and resource sustainability already in use. For example, the concept of the minimum viable population (MVP, chapter 7) provides one quantifiable measure of what would constitute a safe minimum standard for a species (Bishop 1978; Berrens et al. 1998).

As noted earlier, it is easiest to assign values to biotic resources when such resources can be translated into private, market goods as some form of commodity, but only a small percentage of biotic resources can be valued in this manner. It is theoretically possible, however, to assign an economic value to nonmarket goods. To do this, economists often rely on a methodology called contingent valuation analysis (CVA).

Contingent Valuation Analysis

Willingness to Pay (WTP)

Determining the economic value of the satisfaction a person derives from simply knowing that a particular resource or species exists is one of the most challenging problems of economics. The WTP approach attempts to assign monetary worth to existence value by asking, usually through surveys, what a person would be willing to pay in exchange for the preservation of a given entity, such as a rare species, under specific circumstances. To an economist, "benefits" associated with a resource are those things that give it value, and something has value if someone is willing to pay for it, no matter what the reason. To assess benefits of this kind, a typical survey item might propose, "Suppose an undeveloped tract of tallgrass prairie at the edge of your town was found to contain a population of a rare species of butterfly. How much would you be willing to pay to keep this area from being developed and preserve the butterfly population?" The WTP approach is often made more sophisticated by giving the respondent a range of alternatives (e.g., How much would you pay to preserve an endangered bird, snake, plant, butterfly, or beetle?) that allows the person to assign different valuations to different kinds or categories of species. Estimates derived from responses are then used to determine preferences. The WTP approach also can be used to ask more directly how much the person would pay to keep the habitat itself in an undeveloped state, how much to have it turned into a golf course, a housing development, a water treatment plant, or a window factory. To give the question a greater sense of legitimacy and plausibility, the question may be phrased in the form of a potential bond issue, such as a state tax to preserve the habitat of an endangered species. In this form, the researcher can determine: (1) how many people would vote for the bond issue at a given level of taxation, (2) how many people would be unwilling to pay anything at all, and (3) the average cost valuation of those against the bond issue but willing to pay some lesser amount (Hunter 1996). Results are used to infer the cost valuation of a noneconomic good (an endangered species) or state (an undisturbed habitat). The proposed mechanism of payment (higher taxes, government bonds, higher fees for outdoor recreation, or direct cash contribution) may have a significant effect on the answer.

Some conservationists have voiced support for WTP approaches because they may permit normally noneconomic goods, such as endangered species, to stand on equal economic footing with hydroelectric dams, power plants, or upscale subdivisions. Even if the average North American is willing to pay only 10 cents to know that the California gnatcatcher exists, the cost valuation generated from this sentiment in the entire population would run into millions of dollars. Thus, the WTP approach appears to offer the conservation advantage of allowing economists to determine the value of noneconomic goods, such as endangered species, that can lead to assessments of high monetary valuation. Such valuation would permit conservationists to argue that the existence value of a species is worth more than the gains to be realized from its destruction.

Willingness to Accept Compensation

An alternative, related approach to WTP is the willingness to accept (WTA) compensation method. Unlike WTP, which attempts to determine what a respondent would pay for an environmental amenity, WTA attempts to determine what the respondent would accept as compensation for losses suffered as a result of gaining or maintaining such an amenity. Like WTP, WTA typically uses survey methods to determine the average payment affected individuals would accept for losses they incur as a result of conservation practices. WTA can be an effective and necessary method in cases where the achievement of a conservation goal or satisfaction of an environmental amenity comes with a definite and tangible cost to local residents. In fact, determining an acceptable and just level of compensation is often the only way to break otherwise irresolvable value conflicts that may arise in conservation efforts at regional or landscape scales. The reintroduction of the timber wolf (*Canis lupus*) in Yellowstone National Park in the 1990s could not have proceeded without establishing, in advance, a program to compensate ranchers outside the park for livestock losses caused by wolves. Similarly, fair compensation, determined by WTA, is

increasingly viewed as essential in establishing national parks in developing nations in which large numbers of people obtain a living from natural resources in traditional ways, such as hunting, gathering, and pastoral agriculture. For example, the establishment of Mantadia National Park in Madagascar, an area with one of the world's highest densities of endemic species, could not have been successful without a program to compensate local residents for losses associated with changes in land use in and around the park (Shyamsundar and Kramer 1996).

Criticisms of WTP and WTA

Its strengths and successes notwithstanding, criticisms of the WTP and WTA approaches abound from both economists and conservationists. Some economists argue that the WTP and WTA do not measure anything because the respondent's answer is strictly hypothetical and imaginary. In WTP, for example, the respondent gives up no real money or goods to express his or her preference and thus is likely to grossly overestimate a willingness to pay for the existence of the endangered species. This concern is supported by the fact that, in some studies, the public's combined estimated values for individual environmental entities exceeded their aggregate disposable income (Carson, Flores, and Hanneman 1998)!

The WTP approach has been particularly criticized by environmental ethicists and conservationists. Environmental ethicist Mark Sagoff argues that the WTP approach does not measure existence value at all, but instead confuses values with benefits. Valuations generated from WTP analysis often are used as measures of *economic benefits* of preserving endangered species. These measures are then included in cost-benefit evaluations that are used to evaluate policy decisions (whether or not to lease Forest Service land for an oil well, whether to allow housing development in a habitat used by a rare butterfly, whether to allow a wetland to be drained for farming). Most research, however, shows that persons responding to WTP questionnaires are not, in their assessments of valuation, making an estimate of the economic benefits that preserving an endangered species has to them. Rather, they are expressing the relative strength of a moral conviction that the species ought to be preserved. Thus, WTP analysis fails to make a distinction between *what people value because it benefits them* and *what they believe is valuable for ethical reasons or on intrinsic grounds* (Sagoff 2000). In other words, WTP analysis treats all values as expressions of personal preference. This view creates a distortion in value that shifts the focus from the value of the object to an index of the respondent's self-interest. The WTP approach assumes that human welfare is increased when human preferences are satisfied. The more people are willing to pay to have preferences or moral convictions satisfied, the greater the benefit. Ergo, both welfare and benefits are maximized by allocating resources to those who are most willing to pay for them. Or, as Sagoff puts it more sarcastically, "Resources should go to those willing to pay the most for them because they are willing to pay the most for them" (Sagoff 2000). Many believe that this conflation of value with economic benefit is not an accurate measurement of value, but a gross distortion of it.

Such distortion of value can lead to a more systemic and more serious problem. For example, a conservation biologist may feel that she must take up the cause of protecting an endangered species, such as a rare bird, because she perceives value in the species itself and feels an inherent obligation to protect it. But in public discourse, she will feel a strong temptation to frame her reasons in light of human self-interest. For example, she may argue that future generations will be deprived of the value of knowing the species, that the species may have actual or potential market value, or that its loss will negatively affect the value of recreational opportunities, like bird watching, in its habitat. In doing this, the biologist may feel that her end justifies her means. But she may not realize that she is, in the words of ethicist Lawrence Tribe, "helping to legitimate a system of discourse which so structures human thought and feeling as to erode, over the long run, the very sense of obligation which provided the initial impetus for her own protective efforts" (Tribe 1974).

An example of this kind of ethical erosion can be seen in the "economic conservation" displayed in an enterprise called "sea turtle ranching." In this effort to use market forces for the benefit of conservation, eggs of the green turtle (*Chelonia mydas*) are taken from the wild and hatched under controlled conditions. The hatchlings are then raised to market size and used to meet the worldwide demand for sea turtle products such as leather, oil, soup, and meat. Proponents argue that, handled in this way, "surplus" eggs find a useful purpose, some of the domestic hatch can be released into the wild to augment remaining green turtle populations, and the incentive for poaching is reduced.

In his classic essay, "The Business of Conservation," conservation biologist David Ehrenfeld argues that this approach is a pathetic combination of greed and short-sightedness (Ehrenfeld 1992). Ehrenfeld points out that, apart from the myriad of problems associated with keeping and raising sea turtles in captivity, sea turtle ranching actually increases the worldwide demand for sea turtle products. Such demand inevitably not only makes greater demands on world turtle populations through legitimate egg collection, but also makes it much more attractive to poach the eggs of wild turtles. And turtle poaching will always be more profitable than turtle farming because the affluent consumers who have acquired a taste for sea turtle products cannot tell the difference between legal and illegal goods. Ehrenfeld summarizes the problem succinctly: "The power of global demand erodes all safeguards. . . . Thus the commercial ranching of green turtles inevitably brings us around again on the downward spiral—a little closer to the extinction of the remaining populations. By no stretch of the imagination is this conservation" (Ehrenfeld 1992).

Thus, although economic analysis can be useful in an assessment of the value of biotic resources, it is not equally precise in assessing all categories of value. Further, economic analysis is inadequate as a comprehensive ethical system to provide meaningful value for biotic resources. Other measures and methods of value assessment and analysis are necessary to address other dimensions of species' worth, including the most fundamental question, the nature and characteristics of intrinsic value.

THE PROBLEM OF MORAL DIMENSION: HOW DO WE APPROACH INTRINSIC VALUES IN CONSERVATION?

The Search for Intrinsic Value

A fundamental question in any understanding of value is: Are values human intellectual constructs that people design and manipulate, or external realities that the human intellect recognizes and to which it responds? This question can be expressed more simply: "Does intrinsic value exist?" As noted earlier, intrinsic value resides in an object when the object is valuable "on its own," not on the basis of its utility to humans or other species. Some ethicists, not to mention many environmental economists, assert that intrinsic value does not really exist. In their view, all values are anthropocentric, residing in human consciousness and perception. Therefore, all values are human creations and ultimately can be expressed as human preferences and subjected to economic evaluation.

Most ethicists and conservationists would admit that the *locus* of all value is human consciousness, but many would argue that the *source* of all value is not necessarily human consciousness (Callicott 1986). For example, philosopher Donald Regan illustrates one concept of intrinsic value with a natural object—the Grand Canyon—and a hypothetical person—Jones. Regan argues that the intrinsic value of the Grand Canyon is actually a complex of the Grand Canyon itself, Jones's knowledge of the Grand Canyon, and Jones's pleasure in her knowledge of the Grand Canyon. In this example, any natural object can be substituted for the Grand Canyon, and any other person substituted for Jones. But regardless of what object or person is used, Regan's argument contends that, first, there is value in the object itself apart from human perception of it, and, second, that human perception adds value to it. Thus, we are given additional incentive, reason, and moral obligation to preserve objects that we know about in order that we may study them, learn about them, and enjoy our knowledge of them (Regan 1986).

Regan's analytical example is presented in a different way by environmental ethicist Arne Naess. Naess, upon reading an early handbook on the care and treatment of domestic animals, was amazed to discover that the author spoke passionately about caressing pigs. At one point, the author stated, "Those who have experienced the satisfaction of pigs stroked this way, cannot but do it" (Naess 1986:506). At this point, Naess asks the obvious question: how can the author, a human, experience the satisfaction of a pig? The answer, according to Naess, is a simple one. The process that allows us to experience both the satisfaction and worth of a nonhuman creature is that of *identification*, our tendency to see ourselves in everything alive (Naess 1986). Far from being a sentimental mistake that is corrected by formal education, Naess asserts that the sense of identification grows stronger in us as it is increasingly informed by knowledge of a creature or an object. Humans, says Naess, "have the capacity of experiencing the intimate relations between organisms and the nonorganic world . . . the attainment of well rounded human maturity leads to identification with all life forms" (Naess 1986:506).

Thus, Naess asserts that identification with nonhuman life is a kind of intellectual and professional virtue developed through strenuous exercise of intellect and habit, and through an explicit recognition of the intrinsic value of the other life.

Regan again puts the argument more analytically: "What we see is that humans are necessary to the full realization of the 'goods' of other species. The cheetah's speed is good, but it is not good in itself. It needs to be known by a subject who can know it and take pleasure in it in a sophisticated way. The cheetah does not value his speed in the required way. We can and should. That is the proper spelling out of the notion that every creature has 'a good of its own'" (Regan 1986:216).

If such arguments are sound, they lead to the conclusion that the ability to perceive the intrinsic value of other species is not only appropriate, but an important distinction of being human. Indeed, the ability to perceive value in other species does not lower us as human beings, but helps us appreciate a unique role that only humans can fill. The argument that humans hold this special role in appreciating the intrinsic value of other creatures finds support in an unexpected source, the writings of Aldo Leopold:

> *For one species to mourn the death of another is a new thing under the sun. The Cro-Magnon who slew the last mammoth thought only of steaks. The sportsman who shot the last pigeon thought only of his prowess. The sailor who clubbed the last auk thought of nothing at all. But we, who have lost our pigeons, mourn the loss. Had the funeral been ours, the pigeons would have hardly mourned us. In this fact . . . lies objective evidence of our superiority over beasts* (Leopold 1966:117).

Some would assert that Regan's argument does not demonstrate that the objects he describes actually have intrinsic value. For example, if we substitute the smallpox virus in place of the Grand Canyon or the cheetah, most of us, perhaps even most conservation biologists, would have difficulty in feeling any pleasure in our knowledge of the object. Yet most conservation biologists *act* as if they appreciate an inherent value of nonhuman species, and demonstrate it admirably with a career of commitment and a life of personal devotion to the welfare of other creatures. However, if the value inherent in these perceptions is to be shared with others and to be transformed from personal perception or preference into actual "knowledge" about values, it must be expressed in a way that can be understood by everyone. It is not sufficient for conservation biologists to simply "behave" as if nonhuman species have intrinsic value and hope that everyone else will simply imitate their behavior without asking why. If species truly possess intrinsic value, if they have value "on their own," and if such value should change our behavior toward them, what is the basis of that value, and what is its source?

Ecocentrism as a Basis for the Intrinsic Value of Species

Aldo Leopold's land ethic, an important contribution to the ethical framework of conservation biology, assumes and asserts the intrinsic value of nonhuman creatures, even if it does not always articulate a formal argument for its assertions and assumptions. However, one of Leopold's most famous statements reveals much of the basis for his thought: "A thing is right when it tends to preserve the integrity, stability, and beauty of the biotic community. It is wrong when it tends otherwise" (Leopold 1966:262). Inherent in this quotation, and in the context from which it is drawn, are a host of unstated assumptions. Leopold saw the development of a land ethic as a natural extension of an evolving concept of rights, initially applied selectively to free men, then to all people, and finally to the land itself.

Leopold's concept of value for the land and its components was derived from two main sources, ecology and evolution. Leopold saw value in species and ecosystems because each part contributed to healthy overall ecological function. Contrasting intrinsic value and economic value in his home state of Wisconsin, Leopold wrote, "One basic weakness in a conservation system based wholly on economic motives is that most members of the land community have no economic value. Wildflowers and songbirds are examples. Of the 22,000 higher plants and animals native to Wisconsin, it is doubtful whether more than 5 per cent can be sold, fed, eaten, or otherwise put to economic use. Yet these creatures are members of the biotic community, and if (as I believe) its stability depends on its integrity, they are entitled to continuance" (Leopold 1966:246–47).

In these few words, Leopold asserts two points that environmental ethicists would formalize in stricter arguments decades later. First, intrinsic value is based on contribution to ecological stability and integrity. Second, intrinsic value implies a "right to life" of living creatures (note the phrase "entitled to continuance"), a concept that, although radical in Leopold's time, would become a legal reality in the Endangered Species Act.

The second, and closely related, foundation of Leopold's view of concept of the intrinsic value of land and life was evolution; specifically, he viewed the current members of biotic communities as the products of long processes of speciation, adaptation, and change that fitted them for their place in the natural world. This grand investment of natural history and selection imbues these creatures with value, a value that should not be destroyed by momentary and thoughtless acts of a single, recently evolved species, humans. Addressing this concept, Leopold wrote, ". . . man-made changes are of a different order than evolutionary changes and have effects more comprehensive than is intended or foreseen. . . . Can the land adjust itself to the new order? Can the desired alterations be accomplished with less violence?" (Leopold 1966:255–56).

These premises have become the foundations of an ethical paradigm known formally today as *ecocentrism*. Ecocentrism asserts that the fundamental entity to which both values and rights apply is the biotic community, not individual organisms or species. Leopold argued that the value of a species does not reside in itself, but in its value to the integrity, health, function, and persistence of the community of which it is a part. Callicott rephrased Leopold's quote in more modern, ecocentric terms as, "A thing is right when it tends to protect the health and integrity of the ecosystem. It is wrong when it tends otherwise" (Callicott 1994a).

The effects of Leopold's views on conservation biology and related disciplines cannot be overestimated. The Wildlife Society celebrated 1999 as the half-century anniversary of Leopold's *A Sand County Almanac* with a special edition of *The Wildlife Society Bulletin*. In that edition, special features editor and wildlife biologist Richard Knight wrote, "We in The Wildlife Society are fortunate that the founder of our field articulated so clearly and at such length what our discipline is about. Not often does a movement (wildlife conservation) and a scientific discipline (wildlife management) begin with such rich words and clear direction" (Knight 1998).

Conservation biology, no less than wildlife management, bears and acknowledges a great debt to Leopold. In fact, many conservation biologists see themselves as the true intellectual heirs of Leopold's legacy in their efforts to restore ethics, value, and mission to the scientific study of conservation. Fiedler, White, and Leidy (1997:84) assert, "Today the emergence of conservation biology, perceived as a distinct discipline, is a direct result of the failure of resource management fields . . . to fully embrace the values espoused by Leopold." Callicott gives a more comprehensive, if less impassioned, summary of the reasons that Leopold's influence is congenial and attractive to many conservation biologists: ". . . the Leopold land ethic is not based on religious beliefs, nor is it an extension of the ethical paradigm of classic western moral philosophy. It is grounded, rather, in evolutionary and ecological biology. Hence, all nonanthropocentric conservation biologists, irrespective of religious or cultural background, will find the Leopold land ethic intellectually congenial" (Callicott 1994a:44).

Some would argue that Leopold's argument fails to demonstrate an intrinsic value for the land and its components. If species, for example, derive value from their usefulness to community structure and function, then that is an expression of the ecological usefulness of the species to the community, not of the species' intrinsic value. Indeed, many argue that intrinsic value cannot exist apart from a religious tradition in which value is imputed to a resource or species by some higher moral authority. Leopold himself acknowledged that a conservation ethic apart from well-grounded philosophical and religious understanding would be incomplete. Although Leopold's views were not explicitly based on religious traditions, he was nevertheless influenced by such traditions. As a young man, he read Liberty Hyde Bailey's short book, *The Holy Earth,* and in its ideas found stimulus for his early thinking about a "land ethic," a perception of the land's value that existed independent of its use. Leopold recognized that such a view of intrinsic value required validation from philosophical and religious tradition and said so. He stated, "No important change in ethics was ever accomplished without an internal change in our intellectual emphasis, loyalties, affections, and convictions. The proof that conservation has not yet touched these foundations of conduct lies in the fact that philosophy and religion have not yet heard of it. In our attempt to make conservation easy, we have made it trivial" (Leopold 1966:246).

The majority of the world's 6 billion people are "touched by these foundations of conduct," drawing their understanding and applications of values, including conservation values, from religious traditions. We would trivialize conservation ethics by ignoring Leopold's advice, and the majority of human opinion and experience, if we avoided facing the ultimate issues of value that such systems address. Religious traditions, particularly theistic traditions, fully engage the problems and questions inherent in the concept of intrinsic value in comprehensive and diverse ways. They attempt to answer some of the most basic problems of conservation valuation: namely, does intrinsic value exist, what is its source, and how ought we to respond to it?

Intrinsic Value in the Judeo-Christian Tradition

Conservation in the United States developed as a moral movement, undergirded, in part, by a biblical view of the value of the natural world (in biblical terms, "creation"). The early conservationist John Muir was the most explicit in connecting biblical teaching with the value of species. Muir argued for the intrinsic value of species because they were "God's creatures." "All creatures," asserted Muir, "are part of God's family, unfallen, undepraved, and cared for with the same species of tenderness and love as is bestowed on angels in heaven or saints on earth" (Muir 1916:98).

Despite Muir's legacy, the modern era of conservation did not begin auspiciously in its relations between environmental concerns and the Judeo-Christian tradition. In one of the most influential and widely read essays of modern times, "The Historical Roots of Our Ecologic Crisis," published in the journal *Science* in 1967, Berkeley historian Lynn White, Jr., identified the Judeo-Christian tradition, and especially Western Christianity, as the cause of the emerging environmental crisis (White 1967). Specifically, White asserted that, in the Judeo-Christian tradition, nature had no reason to exist except to serve humans and that Christianity established a dualism of humanity and nature and taught that "it is God's will that man exploit nature for his proper ends" (White 1967). Further, White asserted, "By destroying pagan animism, Christianity made it possible to exploit nature in a mood of indifference to the feelings of natural objects" (White 1967).

White's views were widely repeated and reprinted, both with and without acknowledgement, in academic and popular circles throughout the 1970s, and were standard fare in discussions of ecological ethics in the most widely used textbooks on ecology in that era (e.g., Krebs 1972; Colinvaux 1973; Hinckley 1976). The Judeo-Christian tradition was villified in all things environmental, from discussions of landscape architecture (McHarg 1969) to pollution and species extinctions (Ehrlich 1971).

Stung by the criticisms of White and others, and offended by the antireligious climate such claims were creating in environmental circles, Christian scholars responded with rebuttals. Not all were kind. Author and environmentalist Wendell Berry wrote, "The anti-Christian conservationists characteristically deal with the Bible by waving it off. And this dismissal conceals, as such dismissals are apt to do, an ignorance that invalidates it.

The Bible is an inspired book written by human hands; as such, it is certainly subject to criticism. But the anti-Christian environmentalists have not mastered the first rule of the criticism of books: you have to read them before you criticize them" (Berry 1992).

Many ethical scholars of diverse religious and philosophical persuasions took Berry's words to heart. The Bible was read, carefully and in an environmental context. Out of that study and scholarship has emerged what is today referred to as the Judeo-Christian stewardship ethic (Callicott 1994a). Environmental ethicist Bryan Norton lists the Judeo-Christian stewardship ethic as one of the seven dominant worldviews in environmental ethics, and the only one that is derived directly from a religious tradition (Norton 1991). Because of its historical and continuing influence in the development of conservation ethics and values, its activism in conservation, as well as its past and present controversies in interacting with ecological issues, the Judeo-Christian stewardship ethic merits careful examination.

On the central question of the intrinsic value of species, this ethic speaks clearly and without equivocation. In the opening chapter of the Bible's first book, Genesis 1, God repeatedly acknowledges both specific kinds of creatures and natural objects as "good." The goodness acknowledged and perceived by God of his own creation is clearly intrinsic rather than utilitarian because God has no self interests of any kind that would be used to determine the value of the creatures in relation to his own interests.

It is equally clear that God has no need of anything and therefore cannot benefit from or be harmed by his creatures in any way (Callicott 1986). But the goodness of the creatures is also intrinsic and nonutilitarian from the human perspective. The repeated pronouncements of creation's goodness are made by God, not humans, and not in any way in relation to humans. Both nonhuman creatures and humans receive God's blessing to be fruitful and multiply and fill the earth (Genesis 1:22; Genesis 1:28), but the blessing is given first to nonhuman species. In this context, Callicott made the following judgment: "The Judeo-Christian Stewardship Environmental Ethic is especially elegant and powerful. It also exquisitely matches the requirements of conservation biology. The Judeo-Christian Stewardship Environmental Ethic confers objective intrinsic value on nature in the clearest and most unambiguous of ways: by divine decree" (Callicott 1994a:36). Although the Judeo-Christian tradition has usually been referred to as anthropocentric (White 1967), it would be more accurate to describe it as theocentric because the values imparted to created things originate from God and have their locus in God. Although the value of created things can and should be discovered by humans (a concept that some ethicists have termed "inherent" value [Norton 1991]), that value exists independent of humans and human experience. Norton describes such a concept as "strong" intrinsic value, or "intrinsicalism" to distinguish it from inherent value, or "weak" intrinsic value (Norton 1991:234, 235).

Callicott goes on to note that the intrinsic value ("goodness") conferred upon created things is bestowed upon species ("kinds"), not to individual plants or animals. Thus, humans are free to use individual specimens of living things for their own needs, but are not to destroy the goodness of creation's diversity by eradicating entire species. Were this intrinsic goodness the only value conferred upon nonhuman creation by the Judeo-Christian tradition, it would be a substantial contribution. There are, however, five dimensions of biblical teaching that add value to nonhuman creatures and natural objects.

Nonhuman creatures and the nonhuman world are appropriate subjects of active human care and service.

The concept of active care for creation is explicated in the second chapter of Genesis: "The Lord took the man and put him in the Garden of Eden to till it and to keep it." In Hebrew, the word for "till" is derived from the same root as the word normally translated as "serve" (in Hebrew, a form of *'abad*). The idea being conveyed is that the man is to literally "serve" the ground by making it more productive. The word *keep* is derived from the Hebrew *samar,* usually translated elsewhere to signify the idea of persistent loving care, as is the case in a familiar biblical passage that says, "The Lord bless you and keep (samar) you, the Lord make his face to shine upon you and be gracious to you, may the Lord lift his countenance upon you and give you peace" (Numbers 6:24–26). Although White and others imputed the injunction to "subdue the earth and rule over the creatures" in Genesis 1 as a divine license for oppressive behavior by humans toward creation, many modern scholars believe it is more natural to understand this injunction in light of the specific acts given to humans in Genesis 2. Such a view is more consistent with the view of ruling (and of rulers) found in both the Old (Deuteronomy 17:14–20) and New Testaments (Matthew 20:25–28; John 13:3–15). The biblical definition and ideal of ruling is not oppressive or despotic behavior, but rather of wise and loving care expressed through acts of service, even costly personal sacrifice. Thus, the view that the acts of stewardship demonstrated in Genesis 2 represent expressions of the biblical definition of ruling, rather than a separate account of creation (Callicott 1986), is more consistent with the internal coherence of the scripture itself.

Species are the objects of God's covenant protection.

The value of species demonstrated through covenant protection is illustrated in the account of the flood in Genesis 6–9. In this account, God judges human wickedness through a flood that destroys humanity and all living terrestrial creatures, but preserves one righteous man (Noah) and his family, along with representatives of every kind of creature in a great ark. Environmental ethicists and conservation biologists have been alert to see that in this act, Noah is preserving species, not individual specimens, and again affirming the intrinsic value of biodiversity. However, the significance of the story, which is all too often missed, is that God makes a covenant of preservation with all creatures, including humans, following the judgment of the flood. Although Noah acts as creation's representative, the parties of the covenant are not God and Noah, but God and all species. God states, "I will remember my covenant between me and you and all living creatures of every kind. Never again will the waters become a flood to destroy all life" (Genesis 9:15).

Nonhuman creatures and natural objects are manifestations and witnesses of God's creative power and glory. A pervasive theme in the Old Testament, and especially in the Psalms (e.g., Psalm 19) is that created things display and manifest the glory of God and exist in covenant relation to him (Psalm 104). In fact, in the Judeo-Christian tradition, creation's vitality, order, and integrity are viewed primarily as a *response* to God's goodness and wisdom as well as a continuing expression of it (Psalm 95). From this perspective, when creation is destroyed, "God is not so much injured as defied" (Callicott 1986:148) because such wanton human destruction of creation is seen as an act of rebellion against God that diminishes the expression of his glory.

Legal protection is applied to the land as an entity in itself, not as a commodity or as a means of producing a livelihood. In both Exodus and Leviticus, Old Testament books that explicate God's law to his people, God institutes a pattern of rest for the land, a "land Sabbath." "When you enter the land I am going to give you, the land must observe a Sabbath to the Lord" (Leviticus 25:2). God goes on to explain that every seventh year the land shall not be cultivated, but shall receive "a Sabbath rest." The people are admonished not to worry because God will provide sufficient food from the previous year's harvest to ensure a supply until the harvest of the eighth year. Moreover, the land is not to be treated as a commodity at all, but as a personal possession of God. "The land must not be sold, permanently, because the land is mine" (Leviticus 25:23). This concept was to be instituted through the practice of "Jubilee." Every 50 years (the year after every seventh land Sabbath), property acquired by individuals had to be returned to the former owners. Thus land could not be accumulated indefinitely as a possession of the wealthy.

The Bible records that the Jews did not obey these laws either in observing the land Sabbath or in keeping the Jubilee. Breaking such laws is stated explicitly as one of the reasons foreign adversaries deported the Jewish community from the land. Speaking of Nebuchadnezzar, the Babylonian king who deported the people of Judah, the Bible states, "He carried into exile to Babylon the remnant, who escaped from the sword. . . . Then the land enjoyed its Sabbath rests, all the time of its desolation it rested, until the seventy years were completed in fulfillment of the word of the Lord spoken by Jeremiah" (II Chronicles 36:20–21).

Nonhuman creation is included in God's stated plans for the reconciliation and redemption of the world to himself. The Judeo-Christian tradition sees the problem of ecological abuse not as isolated acts of selfishness or ignorance, but fundamentally as a rebellion against God (Hosea 4:1–3). In such a context, the solution is not simply better conservation practices, but a reconciliation between the Creator and his creation. The hope and promise of the reconciliation of God, humanity, and creation are described in numerous Old Testament passages such as Isaiah 11 and Hosea 2, as well as in the New Testament epistles like Romans 8 and Colossians 1. Thus, the idea of environmental stewardship in the Judeo-Christian tradition is not an isolated concept, but an expression of a greater theme of reconciliation between God, humanity, and the created world. In the Christian tradition, this reconciliation is most vividly represented in the incarnation, in which God becomes part of his creation in the person of Jesus Christ, and through his death reconciles all created things to himself (Colossians 1:15–20).

Informed by this understanding, individuals within Jewish and Christian traditions have increasingly modeled expressions of these concepts in the modern world, addressing complex problems of conservation from their faith context. This expression has taken tangible form on many fronts. These have included the production of writings with deliberate implications for conservation policy, the formation of explicitly Jewish and Christian conservation organizations, and efforts to develop conservation policy and influence conservation decisions from a Judeo-Christian perspective.

In September of 1989 the Jewish Theological Seminary of America devoted its High Holiday message, published as a full-page ad in the *New York Times,* to the environmental crisis (Schorsch 1992). In 1991, Jews and Catholic Christians collaborated with the Institute for Theological Encounter with Science and Technology (ITEST) to sponsor the symposium and workshop "Some Christian and Jewish Perspectives on the Creation," which included significant discussion and subsequent publication of Jewish and Christian expositions of biblical teaching on creation stewardship (Brungs and Postiglione 1991). The growth in activity of the religious community on issues of stewardship now makes a list and complete discussion of Jewish and Christian organizations with conservation missions too large to include here. But some organizations have made contributions that are especially noteworthy and comprehensive. These provide examples of the range of conservation activity now taking place.

The National Religious Partnership for the Environment (NRPE), established in 1993, includes leadership from the National Council of Churches, the U.S. Catholic Conference, the Consultation on the Environment and Jewish Life, and the Evangelical Environmental Network. The NRPE's goal is to increase involvement in activities of environmental stewardship by local congregations as well as to demonstrate and address connections between environmental concerns and social justice. Among individual organizations, TargetEarth, formerly The Christian Environmental Association, has undertaken specific policy initiatives opposing tropical deforestation, the development of mining operations near Yellowstone National Park, and the destruction of grizzly bear habitat in Wyoming's Shoshone National Forest, and has initiated efforts to promote the renewal and reaffirmation of the Endangered Species Act. The Evangelical Environmental Network (EEN) was formed in 1993 through a cooperative effort of World Vision and Evangelicals for Social Action (ESA). The EEN also helped initiate and now serves as a secretariat for the more recently formed Christian Environmental Council (CEC), started in 1994, an advisory body of Christian leaders and activists engaged in issues related to the environment. In addition to these and numerous other organizations, many individual denominations have engaged in a variety of activities to promote environmental policies on the basis of a biblical understanding of the stewardship of creation (e.g., Guenthner 1995).

GLOBAL CONTRIBUTIONS TO UNDERSTANDING INTRINSIC VALUE: CONSERVATION ETHICS AND OTHER RELIGIOUS TRADITIONS

The Judeo-Christian tradition, although prominent in the development of values and ethics in conservation, is not unique among world religions in making contributions of ethical and practical significance to conservation efforts. In fact, all major world religions have explored and continue to address significant conservation issues. As conservation becomes increasingly a worldwide effort, such contributions, and their religious motivations, must be more fully understood and appreciated.

Other Western Religious Traditions—Islam

Islam arose as a systematic belief structure in the seventh century A.D. in the Middle East, although, like Judaism, it traces its historical ethnic origins through Abraham. Its adherents, Muslims, believe that Allah (God) communicated to humanity through his representative, the prophet Mohammed. The teachings of Mohammed form the principle distinctions of Islam. The Islamic tradition, like the Judeo-Christian tradition, offers a religious perspective that is monotheistic and a basis of value that is theocentric. Like Judaism and Christianity, Islam perceives the natural world as a creation of God that reveals his glory and attributes: "The seven heavens and the earth and all therein declare His glory: there is not a thing but celebrates His praise" (Qur'an' 17:44). Like Judaism and Christianity, Islam maintains that God is transcendent. Although God values his creation, he is not the same as his creation and is not to be equated with it. God is "totally other," but he fully encompasses his creation and lovingly cares for it. On this point Islam shares with Judaism and Christianity the concept of *immanence*—that God operates within the physical world, always and everywhere intimately present with that world, yet distinct from it. Because God created the world and continues to work within it, the world is not profane, but holy, a place to worship and adore God in any circumstance. To the Muslim, "the whole earth is a mosque" (Manzoor 1984). Thus, all ground is holy ground. The value of the physical world originates from God, not humanity.

The three intellectual pillars of Islam are the concepts of *tawhid* (unity), *khilafa* (trusteeship), and *akhirah* (accountability, or, more literally, the hereafter). These three pillars of Islamic faith are also the pillars of its conservation ethic (Hope and Young 1994).

From its concept of *tawhid,* Islam perceives religion and science, value and fact, as a unity (Hope and Young 1994; Wersal 1995). Within this perspective, Islam is more emphatic and explicit than any other religious tradition in its opposition to any attempt to separate the world into divisions of secular and sacred. This unified perspective makes it natural for Muslims to see the world around them as a creation of God. In such a view they hold a sacramental view of the physical universe. Because Islam sees life and culture as *tawhid,* there is no distinction between religious and secular law, nor any concept akin to "the separation of church and state" that characterizes many Western cultures. Thus, laws on conservation must be grounded in Islamic law and teaching. Further, Islam asserts that an important purpose of religious teaching is to provide a means for independent judgment, correction, and regulation of scientific activity. Many Islamic scholars and scientists are critical of Western science because they see its separation from religious tradition as the cause of its abuses. As Islamic scientist and scholar Seyyed Nasr expresses it, "Western science has become illegitimate because scientists and the rest of society fail to see the need for a higher knowledge into which it could be integrated. The spiritual value of nature is destroyed. We can't save the natural world except by rediscovering the sacred in nature" (Hope and Young 1994).

Islam sees nature as teleological, orderly, harmonious, and dependent. Because such traits represent its original state and God's continued intention for it, Islamic belief leads naturally to an attitude of moral obligation, the concept of the human role as *khilafa* (trustee), whose duty is to maintain these characteristics in natural systems. Thus, the concept that Judeo-Christian thought would describe as stewardship is present in Islam, but is referred to as "viceregency" (Zaidi 1991:41). The words of the Qur'an state the concept plainly: "Behold, the Lord said to the angels: 'I will create a viceregent on earth'" (Qur'an' 2:30).

The third pillar, *akhirah,* arises from the Islamic view that the physical world is a testing ground of human character. Faithful stewardship is one criterion by which God determines the faithfulness of humankind to him, and from this, decides a person's eternal destiny. But, because there is no division of sacred and secular, eternal destiny is enforced and exemplified in present outcomes. For example, if one ceases to manage one's land responsibly, one loses ownership of the land. In Islamic law, land, water, air, fire, forests, sunlight, and other resources are considered common property of all, and not merely humans but of all living creatures (Masri 1992). Expressions of these three principles, and others, in the context of conservation can be studied in detail in a recent comprehensive summary and systematic arrangement of Islamic teachings on stewardship, *The Islamic Principles for the Conservation of the Natural Environment,* a work produced through the collaborative efforts of a number of modern Saudi scholars (Callicott 1994a).

Eastern Religious Traditions and Conservation—Hinduism and Buddhism

Hinduism

Hindus believe that at the core of all being is one reality, but not a "God" in an Islamic, Jewish, or Christian sense of a transcendent, supreme being existing independently of other beings that are his own creations. Rather, in Hinduism all things that appear as individual entities are reflections and manifestations of the one essential being or *Brahman.* Within the reality of Brahman, the Hindu perception of nature is best understood as *prakrti,* the matrix of the material creation. *Prakrti* is significant because it is seen as the expression of the supreme intelligence and physical form of Brahman. Hindu scholars and teachers often refer to Brahman metaphorically as a tree, with its roots "above" (in the spiritual dimension) and its branches "below," in the physical world. The branches of Brahman are conceived as five fundamental elements of *prakrti:* sky (space), air, fire, water, and earth. Before every Hindu worship service, the five elements are purified within the worshippers and in the external environment in which the service takes place. Flowers are offered as purification for the sky, incense for the air, light for fire, water for water, and fragrance for the earth. Thus, it is natural for Hindus to view nature as something internal rather than external, neither alien nor hostile, but inseparable from human identity and existence (Rao 2000).

To the Hindu, every act, willfully performed, leaves a consequence in its wake because human life and action are inseparable from their environment. This is the Hindu concept of *karma,* which is sometimes mistakenly confused with the concept of destiny. A better understanding of *karma* is that every action a person performs creates its own chain of reactions and events that will always be with her, and creates inescapable consequences that must be faced (Dwivedi 2000). Attaining an ideal life depends on choosing right actions that produce good consequences (good *karma*), supported by a purity and balance of the five basic elements within and around a person. The goal is to achieve a life harmonious with nature by creating an ideal environment that is free of pollution, for polluted elements make the human body subject to disease and distortion, no longer an appropriate expression of Brahman. Further, the life in harmony with nature is expressed in a vegetarian diet which consumes plants as appropriate food for a healthy life but does not cause the death of animals. It is also related to one's *karma,* for those who cause violent death will also suffer it (Dwivedi 2000). Likewise, the protection of cattle by Hindus is an expression of the belief that humans are responsible for the care of nonhuman creatures and of belief in the cycle of reincarnation (Rao 2000).

Hinduism has had a long and significant influence on ecology and conservation. Two of the earliest and most influential writers in U.S. conservation, Ralph Waldo Emerson and Henry David Thoreau (chapter 1), were both deeply influenced by Hindu philosophy in their perception and understanding of nature. Hinduism's teaching that all beings are an expression of the one essential being leads naturally to a sense of identity between humans and other living things. Nonhuman species, as well as nonliving objects, are seen as various manifestations of one's own life, and are therefore to be protected and preserved. Hinduism is an important component of ecophilosophies such as "deep ecology" because it strongly identifies humans with other species and natural objects (Naess 1989). More than any other religious tradition, Hinduism is the belief system that most explicitly supports identification with nonhuman life as a genuine perception of reality and as a basis for the care of nonhuman creatures.

Hinduism has been criticized in the context of ecology and conservation because it teaches that the physical world and its diverse entities are derivatives of the undifferentiated and unmanifest Brahman, the supreme reality. The physical world is therefore "less real" and less significant. Callicott, for example, claims that Hinduism teaches that "the empirical world is both unimportant, because it is not ultimately real, and contemptible, because it seduces the soul into crediting appearances, pursuing false ends, and thus earning bad *Karma*" (Callicott 1994b). However, the manifestation of Hinduism in ecological context does not support this criticism. The most famous expression of Hindu belief toward the environment is the *chipko* movement of northern India. The name *chipko* is derived from a Hindi word meaning "to hug" or "to embrace." The movement is usually dated from a protest near the Indian town of Gopeshwar in the Chamoli district of the province of Uttar Pradesh in 1973. There local villagers, protesting government logging policies, went into the forests and physically embraced the trees that were to be cut by the loggers. The loggers, unwilling to inflict harm on the villagers, did not cut the trees and the forest was spared. The chipko movement was borne.

What followed this initial protest was a long and complex struggle over government forest and development practices which continues to this day. The motivations behind the chipko movement were not inspired exclusively by religious conviction. They were also motivated by political issues of self-governance (the right of the villagers rather than the government to determine the fate of local forests), by issues of social justice (the government had denied the villagers a permit to cut trees to make farm implements, but had granted a request in the same forest for a foreign company to cut trees to make sporting goods), and by concerns for local environmental quality (the increase of soil erosion to the detriment of local agriculture and water quality) (James 2000). But they were also inspired by the widespread beliefs of ordinary Hindus that trees were sacred objects. The forest, in local Hindu folk teaching, is seen as the highest expression of the earth's fertility and productivity and is personified by the goddess Vāna Durgā, the tree goddess and earth mother (Shiva 1989). Most of the protesters were, initially, women, and they gave the chipko movement its most dramatic confrontation and its most memorable slogan. In a protest in 1977, a forest officer of the Indian government went into the forests to personally convince the women that the proposed logging was scientifically sound and economically indispensable. He ended his speech by saying, "You foolish women! Do you know what the forests bear? Resin, timber, and foreign exchange!" But far from being intimidated, the women hurled the question back at the forest officer with a very different answer. "What do the forests bear?" they cried. "Soil, water, and pure air! Soil, water, and pure air sustain the earth and all she bears!" (James 2000).

The complexities of the chipko movement must not be underestimated or oversimplified. But chipko does offer a remarkable demonstration of the power of religious values to affect the social and political events that shape the outcomes of conservation. In one sense, Callicott's criticism of Hinduism proved correct. The chipko movement did include an element of negating the world. However, it was not the natural world that was negated in its protests, but rather the world of scientific and economic reductionism that made natural objects worth no more than their value as market goods.

Buddhism

Buddhism is a religious paradox, for it is explicitly agnostic, recognizing no deity. Buddhism focuses on personal mastery of self and the integration of the self with one's surroundings through direct knowledge, discriminating awareness, and deep compassion (Kaza 1990). But Buddhism also has been described as the world's most ecocentric religion (Sponsel and Natadecha-Sponsel 1993). Kaza states that, for the Buddhist, "an environmental ethic becomes a practice in recognizing and supporting relationships with all beings" (Kaza 1990:24).

The fundamental law in Buddhism, noted briefly in chapter 1, is the Law of Dependent Co-Arising, a formalization of the concept that all events and beings are interdependent and interrelated (Kalupahana 1987). On the basis of this law, Buddhism, like Hinduism, stresses a fundamental unity of self and environment. Indeed, the Buddhist ideal of *nirvana*, the awakening into a state of bliss, is reached when the boundary separating the self from its surroundings and all mortal cravings is extinguished (Smith 1958).

Buddhism asserts that the foundation of knowledge is personal experience. Experiential knowledge is especially emphasized in spiritual matters. Thus, personal meditation is accorded high value, and so are natural environments conducive to such meditation. Buddhism teaches respect and compassion for all life, and values undisturbed natural environments as "sacred space" for meditation (Brockelman 1987; Buri 1989).

As a product of such meditation, Buddhism's goal is increasing self-knowledge, and, as a product of self-knowledge, increasing self-mastery and self-restraint. Putting it ecologically, Buddhism does not emphasize that resources are limited, but emphasizes instead that a person should limit his use of resources. Its "middle way" of correct moral behavior emphasizes detachment and moderation from material things and present concerns. Buddhism takes a high view of personal responsibility because it shares, with Hinduism, the doctrine of karma. Buddhists perceive that any future happiness is a direct result of maintaining a satisfactory standard of present conduct. Similarly, wrong actions of the past will produce bad effects in the present. Thus, Buddhism logically values and encourages environmental education and appropriate environmental behavior, not simply for present consequences, but for future ones.

An expression of these values is modeled in current conservation efforts led by Buddhist leaders. The Dalai Lama of Tibet, perhaps the world's foremost Buddhist leader, is also one of the world religious community's foremost conservationists. With his support, the Buddhist Perception of Nature Project, sponsored financially by The World Wildlife Fund, and begun in 1985, identified and integrated environmentally relevant passages from Buddhist scriptures and secondary literature. In Thailand, the Buddhist leader Chatsumarn Kabilsingh contributed to this project by framing Buddhist doctrines into "teaching stories" that have been distributed and used in environmental and conservation education curricula in Buddhist cultures throughout Southeast Asia (Kabilsingh 1990). Many of these stories emphasize Buddhist instruction to not cause harm to others in one's environment, and to protect natural objects such as trees, rivers, and animals (Kaza 1993). Also, in Thailand, the Buddhist monk Pongsak Tejadhammo established the Dhammanaat Foundation in 1985 to preserve forests and create greater environmental harmony and security for local villagers. The foundation is not merely a preserve, but is also a site of ecological restoration (Sponsel and Natadecha-Sponsel 1993). Throughout Thailand, Buddhist monks have adopted a nonviolent strategy of forest protection by ritually ordaining individual trees and wrapping them in sacred orange robes normally worn only by monks. A devout Buddhist would never kill a monk, so the symbolism is obvious in its cultural context. The Buddhist Peace Fellowship, founded in 1978 to address peace and environmental issues, is yet another example of systematic attempts by Buddhists to contribute constructively to worldwide conservation efforts.

CONVERGENCE THEORY: ARE CONSERVATION VALUES MOVING TOWARD A UNIFIED SYSTEM OF POLICIES, VALUES, AND ETHICS?

Our brief examination of the secular and religious traditions that attempt to address conservation issues and articulate conservation values reveals a diverse and, perhaps, fragmented ethical landscape. It also raises a disturbing question: If the worldwide community of conservation biologists represents such divergent ethical backgrounds and cultures, what effect will this have on their ability to work cooperatively toward common goals in conservation? Environmental philosopher and ethicist Bryan G. Norton has addressed this question through what has become known as the "convergence hypothesis" in his book *Toward Unity Among Environmentalists* (Norton 1991), and his work on this problem merits our attention.

Norton argues that, although conservationists differ in the ethical bases of their conservation values, such values, when applied to specific conservation problems, are moving steadily toward a convergence of common, integrated environmental policies. Norton makes this contention on the basis of three premises.

First, speaking in the context of North American environmental issues, Norton asserts that "all environmentalists, regardless of their allegiance to diverging traditions, must seek to manage the entire mosaic that is the American landscape" (Norton 1991:188). Policies of land use are increasingly made at regional or national landscape scales. At such scales, all systems of value must account for and manage the effects of the entire range of human uses of the landscape. For example, agriculture must coexist with efforts to protect biodiversity. Wild and scenic rivers must exist in a context of land that provides a water supply for humans. Human outdoor recreation will exist in and around wilderness. The only successful approach to managing and maintaining such a landscape is to accept the concept of a patchy landscape intelligently arranged (Norton 1991). The realities imposed by regional and national environmental management units make extreme positions, such as pure preservationism or unlimited economic development, unworkable. Thus, according to Norton, environmental policies are not determined by adherence to particular moral principles or ethical systems, but by the effects that human actions in one part of the landscape have upon others. The unity of policy achieved does not come from a common moral basis, but from a common structural one. Land must be used in such a way and according to common policies that sustain and protect natural processes of ecosystems (Norton 1991).

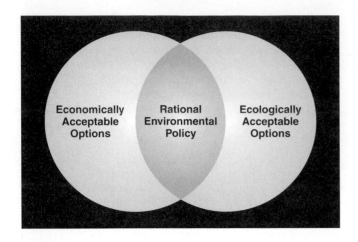

Figure 3.3

A conceptual illustration of how economics and ecology intersect to produce environmental policy. After Norton (1991).

Norton refers to this consensus of policy as *contextualism*, in which rational environmental policy is determined by the intersection of economically feasible policy options with ecologically acceptable policy options (fig. 3.3). In such an intersection, appropriate environmental and conservation behavior is achieved less by legal prohibitions than by positive incentive, reward, and reinforcement. In other words, it is not necessary to make every landowner or land user do the same thing. Rather, at a landscape scale it is enough to provide incentives for *some* landowners and land users to do the right thing to ensure that basic conservation values, such as biodiversity, are preserved. It has now been more than a decade since Norton articulated these views, and some of his ideas have proved prophetic. As noted in chapter 2, habitat conservation plans have been increasingly favored over negative prohibitions to protect endangered species. Conservation easements have become a more common method of protecting habitat than outright purchase of land and exclusion of all human use. These are but two examples that fit the general principle of movement from negative prohibition to positive incentive predicted by Norton.

Norton believes that the growth and effectiveness of contextualization rest upon five axioms of consensus among environmentalists: (1) dynamism (nature is best characterized by dynamic processes, not fixed states or collections of objects); (2) relatedness (all processes are related to other processes); (3) systematicity (processes are not related equally, but are integrated in systems within systems that determine the temporal and spatial expression of the process); (4) creativity (the processes are creative and are the basis of all biologically based productivity); and (5) differential fragility (ecological systems vary in the extent to which they can sustain and recover from human-caused disruptions in their autonomous processes). According to Norton, it is these five axioms that shape the growing consensus of environmental and conservation policy and give common expression to diverse moral and ethical bases for conservation values (Norton 1991).

To say that values and ethics often, or even increasingly, find common expression in management and policy

does not necessarily mean there is an actual convergence of *values*. However, the reality of contextualization and the increasing size and scale of management units inherent in it does mean that moral, aesthetic, and cultural dimensions of the environment must be considered in conservation (Norton 1991). In admitting this, Norton identifies seven major worldviews that, in his judgment, shape modern conservation values (table 3.3). These worldviews affirm vastly different values for the intrinsic value of species, biotic resources, and natural objects, as well as very diverse foundations for those values. Norton is not so optimistic as to predict that these and other worldviews will eventually merge in the cause of conservation, nor does he attempt to answer the vital epistemological question of whether there is a single, correct worldview of conservation values. Norton notes, however, that, operationally, conservationists tend to be ethical pluralists rather than ethical purists. He does not, by this admission, accuse environmentalists of being dishonest. Rather, Norton sees convergence of values and principles occurring through what he refers to as an *integrated worldview* in which moral principles have different domains of application. Ultimately, Norton believes that such integration will not simply show different domains of application, but will establish rules of application in which different values and systems are considered most appropriate to particular types and scales of problems. Such a convergence will lead to a hierarchically organized and integrated system of values, which is necessary because environmentalists and conservationists are not philosophers who operate within a single, closed system of thought. Instead, they are individuals who constantly interact with multiple systems of thought with multiple constituencies, and who gravitate toward similar environmental policies even though they possess diverse value commitments (Norton 1991).

Norton's synthesis is a helpful contribution in seeing how different values and ethical systems work themselves out in determining conservation policies in pluralistic, democratic societies; however, his hypothesis of convergence cannot be accepted uncritically. Norton may be overly optimistic that individuals from all seven major conservation worldviews (and innumerable minor ones) can always work together to achieve common goals. Different worldviews are based on real differences in their perception of truth and the goals that arise from that perception. Adherents to a worldview of constrained economics might easily come to irresolvable disagreements with any worldview, such as deep ecology or Judeo-Christian stewardship, which sees intrinsic value or natural rights in species or natural objects. In practical and contemporary terms, environmental and conservation groups that promote tactics of destroying property or threatening human life have already experienced difficulty in working with conservation organizations whose moral foundations for conservation hold a high view of all life, including humans. Norton also may be overly pessimistic in the belief that a single worldview of conservation values will never emerge. The modern global society has ever-expanding access to information. Worldviews, and the ethical and religious systems on which they are based, will be subjected to increasingly rigorous intellectual scrutiny and comparative evaluation. In addition, conservation biology's explicit

Table 3.3   Seven Major Worldviews that Shape Environmental and Conservation Ethics

WORLDVIEW	TYPE OF VALUE	MOTIVATION FOR CONSERVATION
1. Judeo-Christian stewardship	Theocentric	Preserve the ecological systems that God has commanded humans to care for, as exemplified by the placing of man in the garden to "work it and take care of it"(Genesis 2:15). Humans should respect and not destroy God's handiwork.
2. Deep ecology and related value systems	Ecocentric	The rights or intrinsic values attributed to nonhuman nature place limitations on human prerogatives to use or alter nature and must be respected.
3. Transformationalist/ transcendentalism	Anthropocentric	Respect the spritual value of nature, which provides solace to consider life's deepest questions and can cure human alienation.
4. Constrained economics	Anthropocentric	Resource use is primarily a problem of human economics. Because avoiding irreversible damage to the environment is beneficial, the environment should be preserved when the economic cost is not too great. Low risk taking, common sense, and avoiding irreversible damage to the environment are justification.
5. Scientific naturalism	Science-centered/ ecocentric	Scientific theories of evolution and ecology reveal necessary limits on population growth and violence to the land. Dynamism and contextualism are emphasized.
6. Ecofeminism	Anthropocentric feminism	Because man's domination over nature is symbolic of his domination over women, preserving the environment fights to cure both environmental and social problems.
7. Pluralism/pragmatism	Anthropocentric	Philosophy, although it can serve as a tool to solve moral problems, is not emphasized. Rather, practical problem solving and ethical principles are used to address environmental issues.

*Norton, B. G. 1991. Toward unity among environmentalists, 197–99. New York: Oxford University Press.

commitment to values as a component of its identity and mission means that academic scientists can no longer insulate themselves from discussions of ethics or dismiss them as irrelevant. Conservation biology's commitment to explore the values and ethics of science as an integral part of scientific practice has made the scientific world a more complex place. It means that objective value is a real entity. It is really "out there." If it is found, it must be treated with the respect that truth deserves. The new millennium in conservation biology may see a peaceful convergence toward common values and ethics. But, if there is genuine commitment to treating value as a form of objective truth, it could just as easily witness an intense intellectual and moral battle for the hearts and minds of conservation biologists and the public among a variety of competing ethical systems that attempt to define the value of the biological world.

POINTS OF ENGAGEMENT—QUESTION 2

If value systems in conservation do converge, what forces will lead to their convergence and what forces will hold disparate systems of value together? If they do not converge, what forces are likely to keep different value systems apart, or even lead them to oppose one another?

THE PROBLEM OF PRACTICE: DO CONSERVATION VALUES REQUIRE CONSERVATION VIRTUES?

In 1972, the city of Los Angeles, California, made an innovative proposal for urban beautification. It planned to line the median strip of a major boulevard with plastic trees. The installation of a new box culvert along the strip had not left sufficient soil for real trees to grow, so in their place, city planners proposed the addition of plastic trees constructed of factory-made "leaves" and "branches" wired to plumbing pipes. The trees were "planted" in aggregate rock coated with epoxy, and after their installation, an unknown person or persons added plastic birds (Tribe 1974). Despite the unnaturalness of plastic trees, proponents of the plan could marshal a compelling argument: only plastic trees, they reasoned, could survive the soil-deficient, smog-ridden environment of downtown Los Angeles (fig. 3.4). And a plastic tree, however artificial, would be more appealing to aesthetic values of people than a dead or dying real tree. Urban research planner Martin Krieger, evaluating the problem of plastic trees from the standpoint of cost-benefit analysis, concluded that "the demand for rare environments is a learned one" and "conscious public choice can manipulate this learning so the environments which people learn to use and

Figure 3.4
Downtown Los Angeles, the world's first major city to consider the use of plastic trees as an aesthetic substitute for natural trees in an urban environment.

want reflect environments that are likely to be available at low cost. . . . Much more can be done with plastic trees and the like to give most people the feeling that they are experiencing nature" (Krieger 1973).

In these remarks, Krieger, perhaps unintentionally, succinctly defined a fundamental question to be answered in all debates about the value of natural objects and environments. Namely, is the goal to determine the value of a creature (such as a tree) or a natural object (such as a cavern), or is the goal to teach people to use and want environments that can be made available to them easily and cheaply? In other words, do we address and determine the true value of diversity and rarity, or do we teach people not to miss them when they are gone?

Every attempt to determine the value of resources and species begins with basic assumptions related to our stance on these fundamental questions. For example, in recreation science, the concept of "user satisfaction," noted earlier, is a common measure of the value of landscape and biotic resources associated with outdoor recreational experiences, based on human levels of satisfaction, expectations, and preferences. In a study of recreational user satisfaction in the Apostle Island National Lakeshore in northern Wisconsin, Dustin and McAvoy (1982) reported that "people appear to be growing less sensitive to the environmental degradation that inevitably accompanies crowding . . . as use levels increased, recent visitors became more tolerant of environmental degradation."

In light of Dustin and McAvoy's findings, which are not unusual in modern studies of user satisfaction in natural settings, the words of Aldo Leopold about wilderness become dis-

turbingly prophetic: "Perhaps our grandsons, having never seen a wild river, will never miss the chance to set a canoe in singing waters" (Leopold 1966:116). Biologist Rene Dubos similarly illuminates the meaning of the Los Angeles experience of plastic trees: "Life in the modern city has become a symbol of the fact that man can become adapted to starless skies, treeless avenues, shapeless buildings, tasteless bread, joyless celebrations, spiritless pleasures—to a life without reverence for the past, love for the present, or hope for the future" (Dubos 1965:279).

POINTS OF ENGAGEMENT—QUESTION 3

Conservation biologists have rejected Krieger's vision of trying to provide "environments that people use and want at low cost" as a fundamental objective of their work. Why was this earlier vision of "environmental management" rejected by conservation biology? What vision has replaced it? What does the alternative vision (or visions) reveal about what conservationists value?

From Environmental Values to Ecological Virtues: Categories of Conservation Ethics

We began this chapter by considering the relationship of value to law and policy, and by concluding that, although law and policy influence and are influenced by values, they are no substitute for them. So legal ethicist Lawrence Tribe noted, regarding current efforts to protect the environment, "Policy analysts typically operate within a social, political, and intellectual tradition that regards the satisfaction of individual human wants as the only defensible measure of the good, a tradition that perceives the only legitimate task of reason to be that of consistently identifying and then serving individual appetite, preference, or desire" (Tribe 1974). These words should profoundly disturb students of conservation biology, especially when we consider how much of our discipline's efforts in value determination are based solely on human appetite, preference, and desire.

The medieval theologian Thomas Aquinas asserted that humans surpass all other animals, not in power, but in reason, through which humans are "said to be made in the image of God" (Pegis 1945:27). Although humans can perceive intrinsic and aesthetic values in other species and in natural objects, human appetites, preferences, and desires, cannot be the foundation of value. In such a system the sovereignty of wants becomes the tyranny of whims. If human wants are the measure of all values then reason has been lost, and reason, according to Aquinas, is the means through which humans show the true excellence and distinction of their humanity. Aquinas wrote from a Christian perspective, but his principle has been applied in other contexts. It is only when we link values to purposes greater than individual self-interest that such values become more than expressions of personal preference. And only when values are informed by reason do they attain wisdom, the ability to choose the highest and best ends, together with the surest means of attaining them.

Lawrence Tribe elaborates on this concept: "By treating human need and desire as the ultimate frame of reference, and by assuming that human goals and ends must be taken as externally given, . . . rather than generated by reason, environmental policy makes a value judgment of enormous significance. And once that judgment has been made, any claim for the continued existence of threatened wilderness areas or endangered species must rest on the identification of human wants and needs which would be jeopardized by a disputed development" (Tribe 1974).

Thus, those who are or would be involved in conservation decisions must evaluate whether their conservation values represent only their own individual choices or expressions of personal preference, like items on a dinner menu, or whether they represent something more. In choosing values, we must decide whether our primary obligation as humans is to satisfy our own desires, or to discover and preserve what is valuable, independent of our own desires.

In conservation biology as in all other things, values must mature into ethics that govern behavior if those values would be meaningful. Ethics generally, and conservation ethics specifically, can be grouped into three major categorical systems.

Ethics of *teleology* focus on attention to outcomes and consequences of behavior. Conservation employs ethics of teleology when its arguments focus on the economic values of individual species or total biodiversity, the market value of ecosystem services, or the monetary values of biotic resources like timber or game.

Ethics of *deontology* focus on rules and obligations. Conservation employs arguments of deontology when it addresses our debt to future generations, our need to obey conservation laws, or our duty to respect religious traditions that stress obligations to God, to other people, or to nonhuman life and inanimate objects as manifestations of God's creation that have been placed in our care and to which we are accountable.

The third, and perhaps most neglected, category of ethics, is the category of *areteology*, which focuses on personal virtue. The central claim of areteology is that certain traits of character (virtues) are essential, indeed primary, for correct ethical behavior. These virtues can be thought of as a settled disposition or determination to act excellently. To paraphrase Aristotle, virtue is a state of praiseworthy character, developed over time, made perfect by habit (Bouma-Prediger 1998). Areteology has been a neglected area of conservation ethics, but it addresses significant issues and, perhaps more important, patterns of behavior that affect conservation practice. The development of virtue-based environmental ethics is today an emerging contribution with a radically different approach to ethical dilemmas in conservation.

Conservation Virtues

In virtue-based ethics, correct behavior is not a fulfillment of an obligation or submission to a rule, nor is it an action taken to achieve a desired material outcome. Rather, ethics of areteology view ideal behavior as the habitual pursuit of excellence toward a moral idea. Areteology asserts that we are, and become, what we do. Thus, we become just by doing just acts. We become brave by choosing to act bravely. We become true conservationists and stewards by caring for and preserving species and biodiversity. Christian environmental theologian and philosopher Steven Bouma-Prediger makes the radical assertion that certain traits of character are essential to the "care of creation." These traits include (1) *receptivity and respect* to be able to acknowledge the integrity and interdependence of the nonhuman world and to be able to perceive and appreciate the value of biodiversity; (2) *self-restraint and frugality* to appreciate the limits of creation and not damage its vitality or productivity; (3) *wisdom and hope* to appreciate the fruitfulness of the biological world, enjoy its legitimate pleasures, and confidently expect its future good in God's purposes; (4) *patience, gentleness, and serenity* to be able to allow both ourselves and other creatures to experience Sabbath (rest), either in a time of ceasing from labor or ceasing from taking things from creation and making it labor; (5) *beneficence, benevolence, and love* in order to have the capacity and motive to work for the good of other creatures and empathize with their needs and hardships in a world dominated by the human species. Bouma-Prediger asserts that in cultivating these virtues we can understand and experience what Aldo Leopold meant when he wrote, "One of the penalties of an ecological education is that one lives alone in a world of wounds" (Leopold 1966:197). Finally, (6) *justice and courage* enable one to pursue, in Bouma-Prediger's words, "righteousness" for other creatures, seeking to advance laws, policies, and social norms that treat them with fairness and respect.

Like Aquinas, Bouma-Prediger expresses his ideas in Christian terms, but that does not restrict their applications. Ethics of areteology form the core of Buddhist approaches to environmental and conservation problems. The philosopher Immanuel Kant, writing from a humanist perspective, also stressed the primacy of virtue, the aim to become a better person, as central to our treatment of the natural world. In *The Metaphysics of Morals,* Kant wrote, "A propensity to wanton destruction of what is beautiful in inanimate nature . . . is opposed to man's duty to himself; for it weakens or uproots that feeling in man which, though not of itself moral, is still a predisposition of sensibility that greatly promotes morality or at least prepares the way for it: the disposition, namely, to love something . . . even apart from any intention to use it" (Kant 1991:443).

Areteology contributes a neglected but essential principle for the effective application of conservation values and ethics in all cultural and religious contexts. Namely, it is not sufficient for conservation biologists to simply assert that a particular slate of values is compatible with conservation goals. Conservation biologists also must practice and personify habits of behavior and qualities of character. Such habits and qualities are not only necessary to make the values of conservation biology persuasive to others, but also essential to every conservation biologist who desires to appreciate and enjoy, in other living things, the values that conservation biology promotes.

The idea that conservation requires habits of virtue and not merely choices of values is deeply rooted in the writings of early U.S. conservationists such as Emerson, Thoreau, Muir, and Leopold. Speaking of the conservation of his own day, Leopold wrote, "It defines no right or wrong, assigns no obligation, calls for no sacrifice, implies no change in the current philosophy of values. In respect of land-use, it urges only enlightened self-interest" (Leopold 1966:244). Leopold consummated his point with an allusion to the biblical story of Satan tempting Jesus to turn stones into bread to satisfy his hunger. "In our attempt to make conservation easy," Leopold asserted "we have made it trivial. When the logic of history hungers for bread and we hand out a stone, we are at pains to explain how much the stone resembles bread" (Leopold 1966:246).

Thus, we come full circle. The primacy of virtue, espoused by philosophers like Aquinas and Kant, practiced in conservation by Thoreau and Muir, and articulated in the land ethic of Leopold, is required to supplement what conservation science and law alone cannot provide. Unless people cultivate the practice of conservation virtue based on well-defined ethics and values, as well as the observance of conservation law, conservation is likely to remain a frustrated and limited enterprise.

Synthesis

Values are the engine of action. Although economic valuations can, theoretically, attribute some measure of worth to all types of values, economic assessments often mistakenly assume that value should be based on human preference, need, or desire, rather than on moral obligation, ideal outcome, or the intrinsic worth of the object. Addressing this problem, Aldo Leopold wrote, "The 'key-log' which must be moved to release the evolutionary process for an ethic is simply this: quit thinking about decent land use as solely an economic problem. Examine each question in terms of what is ethically and esthetically right, as well as what is economically expedient" (Leopold 1966:262).

By affirming normative postulates of what constitutes "decent" and "right" in conservation and not merely what is "economically expedient," conservation biology inherently aligns itself with the philosophical position that biotic resources have intrinsic worth. Therefore, one of the fundamental problems of conservation biology is to determine the basis of intrinsic worth, make meaningful assessments of intrinsic value, and then promote policies and behaviors that reinforce such value.

Many different systems of secular ethics and religious belief show convergence upon common values at the core of conservation biology. But values in conservation are not fact-independent sentiments that are proved by assertion, nor are they to be used as tools to manipulate people of different convictions into doing what conservationists want. Values are subject to inspection and analysis, and must be rooted in sound philosophical and intellectual frameworks to be persuasive. In an increasingly global community, it is essential that conservation biologists appreciate the intellectual content and ethical context of expressions of values and their motivations in different individuals and cultures. Conservation biologists must treat these systems with intellectual integrity and respect, and continue to learn from such systems in ways that allow them to better understand, articulate, and practice their own values as conservation biologists. In the same way, conservation biologists must be prepared to change their own behavior as they become informed by traditions of values and ethics from different sources, and even be prepared to make intellectual commitments to particular systems of value. Commitment to an ethical system of values does not lead merely to greater self-understanding and ability to persuade others to take right action, but to becoming a person who has a firm disposition and character to always act for the best and highest good in conservation, an individual possessing a high degree of conservation virtue.

Values and Benefits—A Directed Discussion

Reading assignment: Sagoff, M. 2000. "Environmental economics and the conflation of value and benefit. *Environmental Science and Technology* 34:1426–32.

Questions

1. Sagoff begins with a question: Are environmental and conservation issues problems of design that can be corrected by reconciling "conscience with consumption," or are they problems of economics solved by balancing benefits and costs? How would different answers about the same problem affect the conservation strategy employed to solve the problem?

2. A willingness to pay methodology of assessing the existence value of a natural resource or state of nature assumes that the benefits people get from something are equal to what they are willing to pay for it, and that those who are willing to pay more should determine resource allocation and policy. Is this assumption valid? Why or why not?

3. Sagoff argues that respondents who give a monetary value for protecting an endangered butterfly are not really responding with an assessment of the butterfly's value *to them* but are actually using a monetary value to express the relative depth of their moral conviction that the butterfly *ought to be preserved*. An economist might respond that people's monetary estimates are fair valuation of how much their moral satisfaction is worth to them. Would you side with Sagoff or with the economist? Why?

4. Sagoff advocates an open, democratic discussion mechanism to determine policies based on moral and ethical values instead of using the willingness to pay approach. He argues that environmentalists should resist the temptation to justify their conservation efforts in economic terms. "By joining the CV bandwagon," writes Sagoff, "environmentalists make a pact with economic theory that may gain them the world at the expense of their souls." In this statement, who or what does Sagoff equate with economic theory? What does he seem to regard as the "soul" of the environmentalist? Do environmentalists need to retain these "soul" values in order to participate in democratic, community-based discussions of conservation values?

Learning Online

Visit our webpage at www.mhhe.com/conservation for case studies, animations, practice quiz questions, and additional readings to help you understand the material in this chapter. You'll also find active links to the following topics:

Environmental Philosophy

Conservation and Management of Habitats and Species

Endangered Species Act

Environmental Ethics

Literature Cited

Berrens, R. P., D. S. Brookshire, M. McKee, and C. Schmidt. 1998. Implementing the safe minimum standard approach: two case studies from the U.S. Endangered Species Act. *Land Economics* 74:147–61.

Berry, W. 1992. Christianity and the survival of creation. In *Sacred trusts: essays on stewardship and responsibility*, eds. M. Katakis and R. Chatham, 38–54. San Francisco, Calif.: Mercury House.

Bishop, R. C. 1978. Endangered species and uncertainty: the economics of a safe minimum standard. *American Journal of Agricultural Economics* 60:10–18.

Bouma-Prediger, S. 1998. Creation care and character: the nature and necessity of ecological virtues. *Perspectives on Science and Christian Faith* 50:6–21.

Brockelman, W. 1987. Nature conservation. In *Thailand natural resources profile*, eds. A. Arbhabhirama, D. Phantumvanit, J. Elkington, and P. Ingkasuman, 90–119. Bangkok: Thailand Development Research Institute.

Brungs, R. A., and M. Postiglione, eds. 1991. Some Christian and Jewish perspectives on the creation. *Proceedings of the ITEST workshop March 15–17, 1991*. St. Louis, Mo.: Institute for Theological Encounter with Science and Technology.

Buri, R. 1989. Wildlife in Thai culture. *Culture and Environment in Thailand* 22:51–59.

Callicott, J. B. 1986. On the intrinsic value of nonhuman species. In *The preservation of species; the value of biological diversity*, ed. B. G. Norton, 138–72. Princeton, N.J.: Princeton University Press.

Callicott, J. B. 1994a. Conservation values and ethics. In *Principles of conservation biology*. G. K. Meffe and C. Ronald Carroll and contributors, 24–49. Sunderland, Mass.: Sinauer Associates.

Callicott, J. B. 1994b. *Earth's insights: a survey of ecological ethics from the Mediterranean Basin to the Australian Outback*. Berkeley: University of California Press.

Carlton, R. L. 1986. Property rights and incentives in the preservation of species. In *The preservation of species; the value of biological diversity*, ed. B. G. Norton, 255–67. Princeton, N.J.: Princeton University Press.

Carson, R., N. E. Flores, and W. M. Hannemann. 1998. Sequencing and valuing public goods. *Journal of Environmental Economics and Management* 36:314–23.

Ciriacy-Wantrup, S. V. 1952. *Resource conservation: economics and policy*. Berkeley: University of California Press.

Colinvaux, P. A. 1973. *Introduction to ecology*. New York: Wiley.

Dubos, R. 1965. *Man adapting*. New Haven, Conn.: Yale University Press.

Dustin, D. L., and L. H. McAvoy. 1982. The decline and fall of quality recreational environments. *Environmental Ethics* 4:48–55.

Dwivedi, O. P. 2000. Dharmic ecology. In *Hinduism and ecology: the intersection of earth, sky, and water*, eds. C. K. Chapple and M. E. Tucker, 3–22. Cambridge, Mass.: Harvard University Press.

Ehrenfeld, D. 1992. The business of conservation. *Conservation Biology* 6:1–3.

Ehrlich, P. R. 1971. *How to be a survivor*. Westminster, M.: Ballantine Books.

Fiedler, P. L, P. S. White, and R. L. Leidy. 1997. The paradigm shift in ecology and its implications for conservation. In *The ecological basis of conservation: heterogeneity, ecosystems, and biodiversity*, eds. S. T. A. Pickett. M. Shachak, and G. E. Likens, 78–95. New York: Chapman and Hall.

Guenthner, D. 1995. *To till it and to keep it: new models for congregational involvement with the land.* Chicago, Ill.: Evangelical Lutheran Church in America and The Land Stewardship Project.

Hinckley, A. D. 1976. *Applied ecology: a nontechnical approach.* New York: MacMillan.

Hope, M., and J. Young. 1994. Islam and ecology. *Cross Currents* 44(2):180–92.

Hunter, M. L., Jr. 1996. *Fundamentals of conservation biology.* Cambridge, Mass.: Blackwell Science.

James, G. A. 2000. Ethical and religious dimensions of Chipko resistance. In *Hinduism and ecology: the intersection of earth, sky, and water,* eds. C. K. Chapple and M. E. Tucker, 499–530. Cambridge, Mass.: Harvard University Press.

Kabilsingh, C. 1990. Early Buddhist views on nature. In *Dharma gaia,* ed. A. Hunt-Badiner, 34–53. Berkeley, Calif.: Parallax Press.

Kalupahana, D. J. 1987. *The principles of Buddhist psychology.* Albany: State University of New York Press.

Kant, I. 1991. *The metaphysics of morals.* Trans. M. Gregor. Cambridge, England: Cambridge University Press.

Kaza, S. 1990. Towards a Buddhist environmental ethic. *Buddhism at the Crossroads* 1:22–25.

Kaza, S. 1993. Acting with compassion: Buddhism, feminism, and the environmental crisis. In *Ecofeminism and the sacred,* ed. C. J. Adams, 50–69. New York: Continuum.

Kellert, S. R. 1979. *Public attitudes toward critical wildlife and natural habitat issues.* Washington, D. C.: U.S. Government Printing Office and U.S. Fish and Wildlife Service.

Kellert, S. R. 1980. *Activities of the American public relating to animals.* Washington, D.C.: U.S. Government Printing Office and U.S. Fish and Wildlife Service.

Kellert, S. R. 1981. *Knowledge, affection and basic attitudes toward animals in American society.* Washington, D.C.: U.S. Government Printing Office and U.S. Fish and Wildlife Service.

Kellert, S. R. 1985. American attitudes toward and knowledge of animals: an update. In *Advances in animal welfare science,* eds. M. Fox and L. Mickley, 177–213. Washington, D.C.: Humane Society of the United States.

Kellert, S. R. 1986. Social and perceptual factors in the preservation of animal species. In *The preservation of species; the value of biological diversity,* ed. B. G. Norton, 50–73. Princeton, N.J.: Princeton University Press.

Kellert, S. R. 1989. Perception of animals in American culture. In *Perceptions of animals in American culture,* ed. R. J. Hoage, 5–24. Washington, D.C.: Smithsonian Press.

Kellert, S. R. 1991. Japanese perceptions of wildlife. *Conservation Biology* 5:297–308.

Kelman, S. 1986. Cost-benefit analysis: an ethical critique. In *People, penguins, and plastic trees: basic issues in environmental ethics,* eds. D. VanDeVeer and C. Pierce, 242–49. Belmont, Calif.: Wadsworth.

Knight, R. L. 1998. A celebration of *A Sand County Almanac.* *Wildlife Society Bulletin* 26:695–96.

Krebs, C. J. 1972. *Ecology: the experimental analysis of distribution and abundance.* New York: Harper and Row.

Krieger, M. H. 1973. What's wrong with plastic trees? *Science* 179:446–55.

Leitzell, T. L. 1986. Species protection and management decisions in an uncertain world. In *The preservation of species: the value of biological diversity,* ed. B. G. Norton, 243–54. Princeton, N.J.: Princeton University Press.

Leopold, A. 1966. *A Sand County almanac with essays on conservation from Round River.* New York: Sierra Club/ Ballantine Books.

Lewis, C. S. 1947. *The abolition of man.* Macmillan, New York.

Manzoor, S. P. 1984. Environment and values: the Islamic perspective. In *The touch of Midas: science, values, and environment in Islam and the West,* ed. Z. Sardar, 39–54. Manchester, England: Manchester University Press.

Masri, A. B. A. 1992. Islam and ecology. In *Islam and ecology,* eds. F. Khalid and J. O'Brien, 1–23. London: Cassell Publishers.

McHarg, I. 1969. *Design with nature.* Garden City, N.Y.: Natural History Press.

Muir, J. 1916. *A thousand mile walk to the gulf.* Boston, Mass.: Houghton Mifflin.

Naess, A. 1986. Intrinsic value: will the defenders of nature please rise? In *Conservation biology, the science of scarcity and diversity,* ed. M. E. Soulé, 504–15. Sunderland, Mass.: Sinauer Associates.

Naess, A. 1989. *Ecology, community, and lifestyle.* Cambridge, England: Cambridge University Press.

Norton, B. G. 1991. *Toward unity among environmentalists.* New York: Oxford University Press.

Pegis, A. C., ed. 1945. *Basic writings of Saint Thomas Aquinas.* Vol. 1. New York: Random House.

Randall, A. 1986. Human preference, economics, and the preservation of species. In *The preservation of species; the value of biological diversity,* ed. B. G. Norton, 79–109. Princeton, N.J.: Princeton University Press.

Rao, K. L. S. 2000. The five great elements (*Pañcamahābhūta*): an ecological perspective. In *Hinduism and ecology: the intersection of earth, sky, and water,* eds. C. K. Chapple and M. E. Tucker, 23–38. Cambridge, Mass.: Harvard University Press.

Regan, D. H. 1986. Duties of preservation. In *The preservation of species; the value of biological diversity,* ed. B. G. Norton, 195–220. Princeton, N.J.: Princeton University Press.

Sagoff, M. 2000. Environmental economics and the conflation of value and benefit. *Environmental Science and Technology* 34:1426–32.

Schorsch, I. 1992. Learning to live with less: a Jewish perspective. In *Spirit and nature: why the environment is a religious issue,* eds. S. C. Rockefeller and J. C. Elder, 25–38. Boston, Mass.: Beacon Press.

Shepard, P., Jr. 1978. *Thinking animals.* New York: Viking Press.

Shiva, V. 1989. *Staying alive: women, ecology, and development.* London: Zed Books.

Shyamsundar, P., and R. A. Kramer. 1996. Tropical forest protection: an empirical analysis of the costs borne by local people. *Journal of Environmental Economics and Management* 31:129–44.

Smith, H. 1958. *The religions of man.* New York: Harper and Row.

Sober, E. 1986. Philosophical problems for environmentalism. In *The preservation of species: the value of biological diversity,* ed. B. G. Norton, 173–94. Princeton, N.J.: Princeton University Press.

Sponsel, L. E., and P. Natadecha-Sponsel. 1993. The potential contribution of Buddhism in developing an environmental ethic for the conservation of biodiversity. In *Ethics, religion, and biodiversity: relations between conservation and cultural values*. eds. L. S. Hamilton and H. F. Takeuchi, 75–97. Cambridge, England: The White Horse Press.

Tribe, L. H. 1974. Ways not to think about plastic trees. *Yale Law Journal* 83:1315–48.

Wersal, L. 1995. Islam and environmental ethics: tradition responds to contemporary challenges. *Zygon* 30:451–59.

White, L. 1967. The historical roots of our ecologic crisis. *Science* 155:1203–7.

Zaidi, I. H. 1991. On the ethics of man's interactions with the environment: an Islamic approach. *Environmental Ethics* 3:35–47.

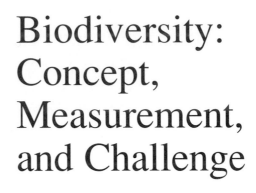

CHAPTER 4

> *If there is danger in the human trajectory, it is not so much in the survival of our own species as in the fulfillment in the ultimate irony of organic evolution; that in the instant of achieving self-understanding through the mind of man, life has doomed its most beautiful creations.*
>
> —E. O. Wilson, 1992

Biodiversity: Concept, Measurement, and Challenge

In this chapter, you will learn about:

1 the concept of biodiversity.

2 methods of measuring and valuing biodiversity.

3 factors that affect levels of biodiversity and global patterns of species abundance.

4 preserving and managing biodiversity.

BIODIVERSITY AND CONSERVATION BIOLOGY

From its inception, the idea of biodiversity in concept, measurement, and conservation has been so central to the study and practice of conservation biology that it cannot be overemphasized. Michael Soulé noted, reflecting on conservation biology's beginnings, that an increasing perception of an accelerating and global loss of species—the "extinction crisis"—was a major factor in conservation biology's emergence as a distinct discipline (Soulé 1986:4).

The shift in focus among conservation biologists from the problem of "endangered species" to the problem of "loss of biodiversity" might, at first, sound like an exercise in biological semantics, but the change is significant. As noted in chapter 1, conservation biology's historical origins were in applied sciences such as forestry, wildlife management, fisheries, and range management. Such disciplines, emerging as distinctive professional communities from the 1930s through the 1950s, were traditionally species specific in their approach to management and in their understanding of species' values. In such a professional environment, studies of species' natural history and habitat requirements received priority. Resource management disciplines began to influence environmental law in the 1960s and 1970s, and the environ-

mental legislation drafted in this period reflected a similar emphasis. The Endangered Species Act (chapter 2), with its emphasis on individual species as the primary targets of conservation efforts, is the best, but not the only example of this type of legislation. Other examples of conservation legislation, such as the Sea Turtle Act and the Marine Mammals Act, displayed a similar focus.

The emergence of conservation biology from this background reflected a shift in emphasis and a break with historic perceptions of the applied sciences about the nature of the "endangered species problem." The traditional view of recent extinctions as a collection of tragic, individual case histories has been replaced in conservation biology with a conviction that the global extinction crisis is caused by fundamental disruptions of ecosystem processes. And the extinctions are now perceived not simply as a parade of passing species, but as a fundamental loss of resources in genetics, community attributes, and ecosystem properties. With this change in perception of *what* was being lost in extinction came a change in the perception of *why* species were being lost. The emphasis on natural histories and habitat requirements of individual species was complemented with studies that sought to identify patterns of species loss and the global, ecological processes that contributed to the emerging patterns of extinction. The extinction crisis created an urgency within conservation biology to develop and define an alternative concept to that of endangered species. That concept was **biodiversity.**

THE PROBLEM OF CONCEPT AND QUANTITY: HOW DO WE KNOW WHAT BIODIVERSITY IS AND HOW DO WE MEASURE IT?

A Conceptual Definition of Biodiversity

Biodiversity, a relatively recent contraction of the term "biological diversity" (Wilson and Peter 1989), has been defined by a multitude of authors and agencies, and not always consistently (table 4.1). One of the best definitions of biodiversity is *the structural and functional variety of life forms at genetic, population, community, and ecosystem levels* (Sandlund, Hindar, and Brown 1992). This definition is especially helpful in that it focuses on the two ideas that make biodiversity a workable concept—that biodiversity is the entire array of biological variety, not simply a collection of individual species, and that the variety that defines biodiversity exists at multiple biological levels.

Biodiversity and the Definition of Species

Historical Origins of the Species Concept

If one were to ask, "What is a fundamental goal of conservation biology?" a good answer would be "to save species." But, to be meaningful, such an answer requires a comprehensive understanding of what species are.

The roots of the species concept extend beyond the history of science into the roots of philosophy itself. The idea of species was inherent and explicit in Plato's concept of the universe as an array of ideal forms. The actual thing observed, including all material objects, even living ones, was considered a shadow of its true form (*eidos*). Because each creature was a representation of a true or ideal form, variation among individuals within a population and among populations was deemphasized and emphasis was placed on the ideal that the creature imperfectly manifested. This formed the basis of the *typological* concept of species that defined species as distinct morphological types.

The **typological concept of species,** subsequently combined with Aristotle's principles of logical divisions of organisms based on "common essence" (a unique attribute that makes the species what it is) was central to biology for many centuries, although modified with John Ray's concept that species "bred true" and with Linnaeus's concept that species were fixed, discrete, and natural entities created by God (Stuessy 1990). With the advent and acceptance of the theory of evolution by natural selection, the view of species as discrete and immutable was replaced with Darwin's view of species as mutable and constantly changing. Darwin stressed the evolutionary integrity of species, in which all individuals were descended from a common ancestor, thus providing the foundations of the modern **evolutionary species concept.**

However, it was in Darwin's day, and remains in this one, difficult to always determine the common ancestor of species with certainty or objective criteria. Biologist Ernst Mayr, considered the father of the biological species concept, proposed a more practical criterion. Mayr's definition of a species was "a group of interbreeding populations that are reproductively isolated from other such groups" (Mayr 1969). To this day, this is still the most widely held and understood use of the term *species* among biologists. In Mayr's definition, the conceptual basis of the species definition is the criterion of reproductive isolation (Stuessy 1990).

Problems inherent in the biological definition of species are many, and the precise demarcation of species remains challenging at best and impossible at worst. Some biologists go so far as to argue that species are only mental constructs that do not really exist in nature (Bessey 1908), but most believe that species possess not only mental reality, but biological and evolutionary reality as well (Stuessy 1990).

Contemporary Issues of the Species Concept

The species concept has undergone and continues to experience a revolution in its definition. Among the seven major types of species concepts extant today (table 4.2), it is the **genetic concept of species** that, more than any other, has revolutionized the definition of species and our understanding of its reality. Modern techniques of molecular genetics now permit direct examination and comparison of the DNA, chromosomes, and gene loci of individual organisms and populations. In science, it is often technique that determines definition, and genetic techniques have both enabled and demanded a new definition of species. The modern genetic concept of species asserts that the way to define a species is through measuring genetic similarities, differences, and distances among populations or groups of populations. There are many difficulties with a genetic concept of species. We rarely really know the actual genetic distances or differences within or among populations, and when we do, we are still not sure when statistically significant differences constitute biologically significant differences. We will examine these problems in detail in chapters 6 and 7.

Our concept of species affects the way we will approach species conservation and the type of problems we will attempt to solve. Conservation geneticist Martha Rojas has framed the problem clearly by asking conservationists whether they are attempting to conserve species as types or species as evolutionary units (Rojas 1995). Our choice on this critical issue strongly affects how we view the value of conserving genetic diversity. If we view species as types, argues Rojas, we will not consider the issue of maintaining their genetic variability. When we have protected an area containing the species of interest, we will consider our work complete.

In contrast, if we view species as evolutionary units, then it is not merely the present group of individuals that concerns us, but the evolutionary potential of the species itself. This perspective invokes evolutionary or cladistic concepts of species

Table 4.1 A Diversity of Definitions for Biodiversity

SOURCE	DEFINITION
Cox (1997)	The richness of the biosphere in genetically distinct organisms and the systems they compose.
Fielder and Jain (1992)	The full range of variety and variability within and among living organisms, their associations, and habitat-oriented ecological complexes. Biodiversity encompasses ecosystem, species, and landscape as well as intraspecific (genetic) levels of diversity.
Hunter (1996)	The diversity of life in all its forms, and at all levels of organization.
Hurlbert (1971)	A function of the number of species present and the evenness with which the individuals are distributed among these species.
International Council for Bird Preservation (1992)	The total variety of life on earth. It includes all genes, species, and ecosystems and the ecological process of which they are part.
Johnson (1993)	The total diversity and variability of living things and of the systems of which they are a part. Biodiversity covers the total range of variation in and variability among systems and organisms at the bioregional, landscape, ecosystem, and habitat levels; at the various organismal levels down to species, populations, and individuals; and at the level of the population and genes.
Magurran (1988)	The variety and relative abundance of species.
McAllister (1991)	The genetic, taxonomic, and ecosystem variety in living organisms of a given area, environment, ecosystem, or the whole planet.
Peet (1974)	The species richness or the number of species in the community, and the equitability or evenness with which importance is distributed among the species.
Reid and Miller (1989)	The variety of the world's organisms, including their genetic diversity and the assemblages they form. It is the blanket term for the natural biological wealth that undergirds human life and well-being. The breadth of the concept reflects the interrelatedness of genes, species, and ecosystems.
Sandlund, Hindar, and Brown (1992)	The structural and functional variety of life forms at genetic, population, species, community, and ecosystem levels.
U.S. Congress Office of Technology Assessment (1987)	The variety and variability among living organisms and the ecological complexes in which they occur. Diversity can be defined as the number of different items and their relative frequency. For biological diversity, these items are organized at many levels, ranging from complete ecosystems to the chemical structures that are the molecular basis of heredity.
Wilson (1992)	The variety of organisms considered at all levels, from genetic variants belonging to the same species through arrays of species to arrays of genera, families, and still higher taxonomic levels; includes the variety of ecosystems, which comprise both the communities of organisms within particular habitats and the physical conditions under which they live.

(table 4.2). For example, the cladistic concept would view a species as a lineage of ancestral-descendent populations that are distinguished by their relative proportion of shared primitive and derived characters (fig. 4.1). If we view species as evolutionary units, we must preserve, not only the organism, but also the organism's ability to respond to environmental change. We must preserve not only the individuals that make up the species, but also the present species' potential to give

Table 4.2 **Six Major Species Concepts**

SPECIES CONCEPT	DEFINITION	REFERENCE
Morphological	Morphospecies are distinct from one another based on morphological characteristics, which can be determined by gross observation. Plant taxonomists most commonly use this concept.	Cronquist (1978) Shull (1923)
Biological	Biospecies consist of a group that is distinct from other populations based on reproductive isolation and typically ecological context. This concept is most commonly used by systematists.	Mayr (1969)
Genetic	Genetic species are separated according to genetic differences such as restriction fragment length polymorphisms, amino acid sequence similarity, or other genetic criteria. The criteria are continually being reconsidered as genetic evaluation technology improves.	Crawford (1983) Gottlieb (1977, 1981) Nei (1972) Palmer and Zamir (1982)
Paleontological	Paleo/chronospecies are distinct from each other by character state gaps that may occur between forms at different time zones, or if character change is gradual, then time is the dividing factor between species. This concept is most commonly used by paleontologists.	Cook (1899) George (1956) Simpson (1961) Sylvester-Bradley (1956)
Evolutionary	An evolutionary species is an ancestral descendant sequence of populations evolving separately from other such sequences and possessing its own unique evolutionary tendencies.	Simpson (1951)
Cladistic	Species are separated according to cladograms. Each new branching unit consists of a new species. Thus, two species could have a common ancestral lineage.	Wiley (1978)

Source: After Stuessy 1990.

rise to future species. We must protect not only the current biological entity and its contribution to biodiversity, but also the potential of the entity to enhance the sum of biodiversity in years to come.

As a discipline, conservation biology has come to embrace the view of species as evolutionary units, and to consider the long-term maintenance of the evolutionary potential of organisms, their genetic diversity, as essential to their survival. Summarizing the importance of genetic diversity, Leoschcke, Tomiuk, and Jain (1994) write, "Genes are the basic material of evolutionary changes and, at the same time, hold in memory the records of events that have occurred in the past. Thus conservation of species is also conservation of the result of an evolutionary process manifested in the genetic and genomic structures of a population."

Ironically, modern insights and techniques regarding the genetics of organisms have led us full circle to something like the old philosophical concept of species. In the view of Plato and Aristotle, there was some unique "essence" of each species as an entity that made it different from other such entities. Modern conservation geneticists would agree. And like Plato and Aristotle, they would agree that the essence is not visible to us,

but an intrinsic component of the creature that is the definition of what it is. In a real sense, the genetic constituency of an organism is the biological equivalent of that essence. It is what makes the creature what it is. And our definition and understanding of what a species is, affects how we will manage and conserve it.

Attractive as the genetic concept of species is, most conceptual understanding, mathematical measurement, legal protection, and management of biodiversity actually make more use of both the biological and morphological concepts of species. Both concepts have many problems, critics, and alternatives. However, the biological concept in particular has proven enduringly useful because it possesses three important characteristics. (1) *It is a testable and operational definition.* The "test" in this case is reproductive isolation; if individuals in the population breed with one another but not with individuals in other populations, then the criteria for being a "species" has been met. (2) *The definition is compatible with legal concepts inherent in conservation laws.* Because species can be identified as independent entities, it is easier to assign protection, rights, values, and duties to species than to assign them to other levels or measures of biodiversity. This is exactly what the -

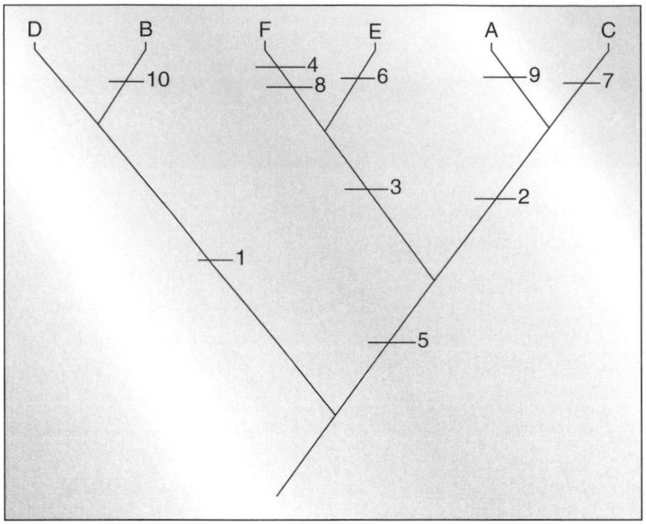

Figure 4.1

A *cladogram* is a tool that can be used to identify and conceptualize species as evolutionary units. This cladogram depicts relationships among species A–F based on characteristics that occur in two discrete states (binary characters), one considered "primitive" (ancestral) and the other "derived" (advanced). The point of change from the primitive to the derived condition for each character is shown by a horizontal bar and number.

After Steussy (1990).

Endangered Species Act does (Petulla 1988). (3) *The definition focuses on a level of biodiversity that fits traditional expressions of conservation.* Management policies and practices designed to increase biodiversity typically aim to increase the number of species in a community, landscape, or ecosystem. Practices such as habitat manipulation, introducing organisms to an area, or controlling mortality through adjustment of hunting, fishing, or trapping regulations are species specific in their applications.

Measuring Biodiversity

To be meaningful to science, biodiversity must be measureable as a mathematical variable. For conservation biologists to measure and express biodiversity in ways that are objective and meaningful to others, they must clarify what level of diversity is under investigation and what dimension of that level is being evaluated. Three levels of diversity—alpha, beta, and gamma—are recognized.

Alpha Diversity

Alpha diversity is the diversity of species within an ecological community, more practically, "the species richness of standard sample sites" (Vane-Wright, Humphries, and Williams 1991), where "richness" is the number of species in the community. A broad definition of a **community** is "all populations occupying a given area at a particular time." In practice, a single site is usually

considered to have multiple communities distinguished from one another by common taxonomic levels or ecological traits. For example, a contiguous block of tallgrass prairie might be described in terms of its plant community, its invertebrate community, its small mammal community, and its avian (bird) community, to name a few. Further, more specialized subdivisions may be used to better identify functional relationships (e.g., the predator community or the detritivore community) or more specialized traits. Alpha diversity in such a community is normally described as a measure of two attributes—species richness and species evenness. Table 4.3 provides an example of data from a community of birds in a tallgrass prairie that can be used to assess different dimensions of alpha (community) diversity.

Species richness refers to the number of species present in a particular community. In our example, site A has 8 species and site B has 11. To compare different kinds of communities, species richness may be expressed in numbers per unit area. Alternatively, the species richness of different kinds of communities may be compared through association measures, such as a coefficient of community (Goodall 1973), in which two or more communities are evaluated on the basis of the percentage of species shared by both. If species richness is the sole measure of diversity, then the community with the larger number of species is considered more diverse.

As a measure of diversity, species richness has a number of positive attributes. The data needed are relatively easy to collect through samples or surveys. Individuals of different species need not be counted; the only data the observer needs to record is whether the species is present or absent. The final re-

Table 4.3   Abundance (individuals/10 ha) and diversity (Shannon index, $H' = -\Sigma(p_i \ln p_i)$) of avian species from two tallgrass prairie sites at DeSoto National Wildlife Refuge, Iowa. Note that site A, with fewer species (8) and two highly abundant species (common yellowthroat and field sparrow), has a lower value of diversity than site B, which has more species (11) that are more equally abundant.

SPECIES	SITE A	SITE B
Common yellowthroat	8.24	1.21
Field sparrow	2.94	2.84
Dickcissel	1.18	2.23
Red-winged blackbird	0.29	0.81
Brown-headed cowbird	2.06	1.82
American goldfinch	1.47	1.02
Ringneck pheasant	0.59	1.63
Mourning dove	1.18	0.61
Eastern kingbird	—	1.60
Grasshopper sparrow	—	4.48
Northern bobwhite	—	2.64
Shannon diversity (H')	1.64	2.25

sult is easy to present, interpret, and compare with other, similar communities, and the number of species present offers a useful first approximation of the biodiversity of the area or habitat.

But using species richness as an index of diversity also has its drawbacks. Because species richness tells us nothing about the relative or absolute abundance of individual species in the community, we do not know whether the species present are equitable in numbers or distribution or whether the community is composed of a few abundant species and many rare ones. Note, for instance, in the data in table 4.3, that the two sites share many species, but have very different densities of these species. What is a reasonable and objective way to incorporate differences in abundance into our estimates of diversity?

In alpha diversity, a second dimension of community biodiversity is evenness. Our two sites, A and B, may have similar numbers and kinds of species; however, if site B has more species and its species are more equally abundant, whereas site A has fewer species and is dominated by just one species (common yellowthroat, *Geothlypis trichas*), then site B will be considered more diverse (table 4.3). This determination has important implications for conservation. When a community is dominated by only one or a few species, it may be that the rarer species are at risk of disappearing from the site. The more common species may even be part of the problem if their behavior is in some way detrimental to rarer species. Additionally, a distribution pattern in which one or a few species are far more abundant than all others may indicate that the habitat lacks a sufficient diversity of structure, patchiness, or resources to allow many species to exist with sustainable populations. An examination of species evenness can be a first step toward generating intelligent hypotheses about possible species interactions, and can lead to greater understanding of more complex processes that influence diversity in the community.

There are dozens of different measures of species evenness (table 4.4). We will not review each individually, but it is helpful to understand how some of the most widely used measures of evenness make assessments of community diversity. One of these is the Shannon index, which calculates diversity (H') as

$$H' = -\Sigma_i p_i \ln (p_i), (i = 1, 2, 3, \ldots S),$$

where p_i is the proportion of the total community abundance represented by the ith species and $\ln (p_i)$ is the natural log of p_i. Numerous mathematical and statistical programs are available to calculate the Shannon index and many other measures of evenness (e.g., Baev and Penev 1995). Any of these indices can be computed on a handheld calculator if the observer knows either the actual number of individuals of each species present or has some measure of each species' abundance or importance. What is important to note is that, in every case, diversity indices involving evenness incorporate quantitative measures of species abundance in relation to the total abundance of all species. Thus, the value of an evenness index increases as the number of species increases and as the species become more equal in abundance.

The diversity of species in a community is not the only level at which biodiversity should be examined. Two other kinds of diversity, beta diversity and gamma diversity, also can be measured and evaluated.

Beta Diversity

Whereas alpha diversity measures the diversity of species *within* a community, *beta diversity* measures the diversity of species *among* communities. Thus, beta diversity provides a first approximation of area diversity or regional diversity. Beta diversity, sometimes called "beta richness," measures the rate of change in species composition in communities across a landscape. Ecologist R. H. Whittaker is credited with the origin of the term, and used it specifically to indicate the change in species composition of communities along a gradient (Whittaker 1975). The "gradient" is normally an environmental variable such as slope, moisture, or soil pH, which can be measured in the same way and at the same scale in all communities. The communities (sometimes designated with the more neutral term *biotic assemblages*) are measured along the gradient. Whittaker's mathematical measurement of beta diversity, today formalized as *Whittaker's Measure*, is

$$S/\alpha - 1,$$

where S is the number of species in the entire set of sites and α represents the average number of species per site, with sites standardized to a common size. In the simplest case, if every site has the same number of species, then $S/\alpha = 1$ and therefore $S/\alpha - 1 = 0$ (i.e., the value of beta diversity is 0 when sites do not change in species composition, indicating a highly homogeneous landscape with respect to a particular environmental gradient). Such a result also would indicate that the species examined had wide tolerances for that particular environmental variable, and thus had broad ecological tolerances and wide niche overlap. At the other extreme of beta diversity, suppose the entire collection of species was equal to 100, but the average number of species found at each site was only 10. Then $S/\alpha - 1 = (100/10) - 1 = 10 - 1 = 9$. Theoretically, there is no upper limit to Whittaker's measure of beta diversity, although empirical values above 10 are rare in studies.

Beta diversity provides insights into three important issues. First, *beta diversity gives a quantitative measure of the diversity of communities that experience changing environmental gradients.* As such, beta diversity provides a way of comparing different communities in a landscape in which important environmental variables change over distance. Second, *beta diversity provides insight into whether species in different communities are relatively sensitive or insensitive to changing environments, and whether associations of species are interdependent (individual species require the presence of other species in the assemblage to persist) or independent (species are added or lost in a more or less random fashion).* Analyses of both trends are important in understanding biodiversity at the landscape level. Finally, *beta diversity can be used to measure nonenvironmental gradients, and thus measure how species are gained or lost relative to other factors.* For example, measures of beta diversity can be used on the same site at different times, thus determining, as the site ages, whether it increases or decreases in number of species. Such measurement is important in evaluating the effects of disturbance and time on a site's biodiversity, and can provide

Table 4.4   Ten Common Estimates of Alpha (Community) Diversity and Their Distinctions*

INDEX	CALCULATION	DESCRIPTIONS, DISTINCTIONS, AND APPLICATIONS	SOURCE
Brillouin's diversity index	$HB = \dfrac{\ln(N!) - \Sigma_i \ln(n_i!)}{N}$	The Brillouin index is recommended for fully censused communities and is therefore free from statistical error. However, its value changes when species numbers increase but proportions remain constant.	Pielou (1969, 1975)
Brillouin's evenness index	$HBe = HB/HB_{max}$	Because this index is based on Brillouin's HB, it is not an estimate but an accurate statistic. Although more difficult to compute than most diversity indices, this index reduces the sensitivity of the estimate to changes in species density.	Pielou (1969, 1975)
Brillouin's maximum diversity index	$HB_{max} = \dfrac{1}{N!}\ln\dfrac{N!}{\left[\left(\frac{N}{S}\right)\right]^{s-1}\left\{\left[\left(\frac{N}{S}\right)+1\right]\right\}^{r}}$	HB_{max} represents the maximum diversity possible with a given sample size and species richness. HB_{max} is used to calculate Brillouin's evenness index	Pielou (1969, 1975)
Hill's diversity index	$N_1 = \exp\left[-\Sigma(p_i \ln p_i)\right]$	Hill's diversity index is an exponential form of the Shannon index. It is widely used, but is sensitive to single-species dominance.	Hill (1973)
Hill's reciprocal of C	$N_2 = 1/C = (\Sigma_i p_i^2)^{-1}$	This reciprocal of the Simpson index (C) is commonly used along with Hill's N_1, but is not as dependent on the number of species as is N_1.	Hill (1973)

*S = number of species at site; N = total number of individuals; p_i = percentage of ith species at site; n_i = number of ith species at site.   (continued)

important insights into succession and its effects on community species composition.

Gamma Diversity

Gamma diversity normally refers to the diversity of species across larger landscape levels. In more measurable terms, gamma diversity is the product of the alpha diversity of a landscape's communities and the degree of beta differentiation among them (Vane-Wright, Humphries, and Williams 1991). Thus, the term *gamma diversity* is used to denote the diversity of different kinds of communities within a landscape. Cody defines gamma diversity as "the rate at which additional species are encountered as geographical replacements within a habitat type in different localities" (Cody 1986). Thus, gamma diversity "is a species turnover rate with distance between sites of similar habitat, or with expanding geographic areas" (Cody 1986:126). Unlike beta di-

versity, gamma diversity is independent of habitat and is calculated as

$$dS/dD[(g + l)/2],$$

which is the rate of change of species composition (S) with respect to distance. *D* is the distance over which species turnover occurs, and *g* and *l* are respective rates of species gain and loss.

The three types of diversity can change independently of one another (fig. 4.2), but in real ecosystems, they are often correlated. High levels of diversity, whether alpha, beta, or gamma, almost always lead to some form of natural rarity (Cody 1986). As species are added to a community, numbers of individuals in individual species typically decline, a phenomenon that has been called *alpha rarity*. *Beta rarity* occurs in habitat specialists; they are abundant in one environment, but rare or absent from environments which manifest even slight changes in one or more

Table 4.4   *Continued*

INDEX	CALCULATION	DESCRIPTIONS, DISTINCTIVES, AND APPLICATIONS	SOURCE
Margalef's diversity index	$DMg = \dfrac{(S-1)}{\ln(N)}$	This widely used index is simple to calculate but best employed on large sample sizes.	Margalef (1968).
Pielou's index of evenness	$E = \dfrac{H'}{\ln S} = \dfrac{-\Sigma_i p_i \ln p_i}{\ln S}$	Pielou's evenness index (the ratio of observed diversity [H'] to the maximum possible diversity of a community with the same species richness [H'_{max}]) is applicable to sample data, but mathematically relates evenness and richness, which are not necessarily related biologically.	Pielou (1969)
Probability of interspecific encounter	$PIE' = 1 - \Sigma_i p_i^2$	PIE' is the complement of the Simpson index ($1 - C$), and estimates diversity instead of dominance.	Baev and Penev (1995)
Shannon index	$H' = -\Sigma (p_i \ln p_i)$	Probably the most widely used diversity index, the Shannon index is employed with both large and small sample sizes.	Shannon and Weaver (1949)
Simpson index	$C = \Sigma_i p_i^2$	The Simpson index is actually a measurement of dominance and assesses the probability that two randomly selected individuals from a community will belong to the same species.	Simpson (1949)

critical variables. *Gamma rarity* describes species that may have large populations in local communities and demonstrate broad environmental tolerances, but are restricted to particular geographic areas and so are lost with increasing distance from their population centers. All these dimensions of biodiversity should be measured and understood before a biologist can interpret the biodiversity of a system and the processes that produce it.

Application and Integration of Diversity Measures to Address Issues in Conservation: A Case Study from Eastern Amazonia

The ability to measure biodiversity with precision allows conservation biologists to monitor changes in biodiversity that are concurrent with human activities in the landscape. With this application in mind, Lopes and Ferrari (2000) investigated the relationship between biodiversity and human activity in the tropical rain forests of eastern Amazonia in Brazil. They measured three dimensions of alpha species diversity that have been previously described mathematically in table 4.4—species richness (Margalef's index), species diversity (Shannon's index), and species

evenness (Pielou's index)—and related each to two elements of human disturbance—forest fragmentation and hunting pressure (table 4.5). Every measure of species diversity declined as disturbance became more intense. More specifically, the diversity of game species declined with increasing hunting pressure, and the diversity of nongame species declined with increasing forest fragmentation. Other, more particular, trends emerged regarding each measure of diversity in relation to each type of disturbance (table 4.6). Thus, this study demonstrates that a precise use of varying measures of biodiversity can be helpful in determining which kinds of disturbance affect which dimensions of diversity in a given community.

Problems of Diversity Indices and Alternative Measures

Although all mathematical measures of diversity provide precise and quantifiable indices of species richness, evenness, or dominance, they possess inherent problems that can obscure rather than enlighten conservation efforts. Controversies about the concept of

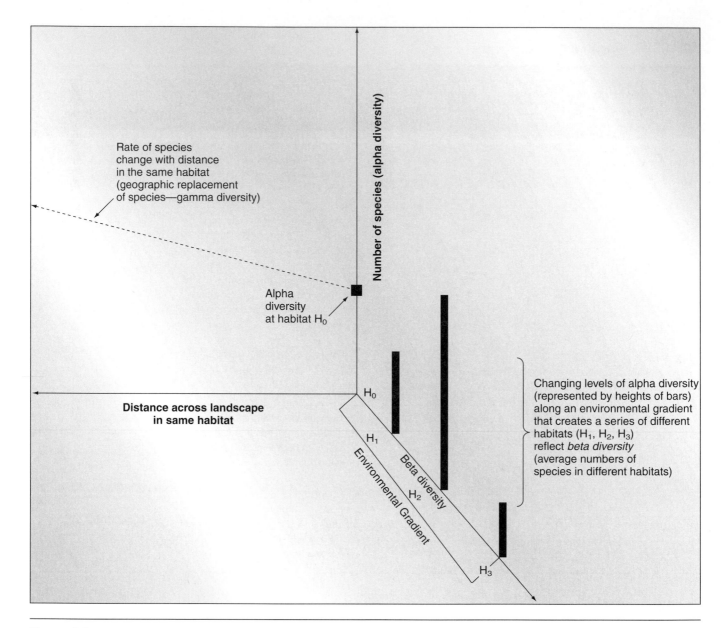

Figure 4.2
The number of species on a given site in one kind of habitat is a measure of alpha diversity (species richness). The average number of species per site along an environmental gradient (number of species per habitat) is a measure of beta diversity. The *rate* of species change over landscape scale distances in the *same* habitat is a measure of gamma diversity (geographic replacement of species).

diversity became intense in the early 1970s, with some mathematical ecologists going so far as to call species diversity a "nonconcept" because of its semantic, conceptual, and mathematical problems, and to claim that "diversity per se does not exist" (Hurlbert 1971). Semantic problems arise when species richness and species evenness are conflated. Richness and evenness, although often correlated, can exhibit inverse relationships along some gradients (i.e., increases in species evenness can be accompanied by decreases in species richness). Mathematical measures can create problems because they do not always, or even usually, correspond to ecological importance in quantifying diversity, so the value of each species in a diversity index is not the same as the value of the species in ecosystem function or conservation priority. Additionally, different diversity measures may yield different val-

ues of diversity for the same community or even change the rank order of diversity in different communities (Hurlbert 1971). Species richness and evenness also tend to be influenced by the number of samples taken and the size of the area sampled. Finally, some values of alpha diversity, such as the Shannon index, can give nearly identical values for very different patterns of species abundance under some conditions (Huston 1994).

Although the mathematical and conceptual problems of measuring and interpreting diversity indices have been resolved or, in some cases, selectively ignored, some problems associated with diversity measurement are persistent. For example, species richness and evenness often increase as a result of human activities that lower the conservation value of the overall landscape. Habitat fragmentation tends to increase the amount

Table 4.5 Species diversity indices and biomass recorded in standard (200 km) samples at five study sites in eastern Amazonia.

	SITE				
VARIABLE	FAZENDA MONTE VERDE, PEIXE-BOI, PARÁ	FAZENDA BADAJÓS, IPIXUNA, PARÁ	FAZENDA SÃO MARCOS, IRITUIA, PARÁ	COMPANHIA REAL AGRO-INDUSTRIAL, TAILÂNDIA, PARÁ	GURUPI BIOLOGICAL RESERVE, MARANHÃO
Area (ha)	200	8,000	10,000	17,485	341,000
Forest disturbance rank	5	4	3	1	2
Hunting pressure rank	4	3	2	5	1
Mammal species	9	9	13	8	15
Total number of individuals	610	281	346	186	343
Shannon's index of diversity (H')	0.984	1.615	1.797	1.697	2.160
Pielou's index of evenness (E)	0.448	0.735	0.701	0.816	0.798
Margalef's index of richness (R_1)	1.247	1.419	2.053	1.340	2.389

Source: After Lopes and Ferrari 2000.

Table 4.6 Overall effects of forest disturbance and hunting pressure on species composition in a tropical rain forest in eastern Amazonia.

	DISTURBANCE EFFECT	
VARIABLE	FOREST DISTURBANCE	HUNTING PRESSURE
Sightings of nongame species	Increase	No effect
Sightings of game species	No effect	Decrease
Species diversity (H')	Decrease	Decrease
Species evenness (E)	Decrease	No effect

Source: After Lopes and Ferrari 2000.

of edge (lengths of habitat borders), and edges are often associated with greater species richness and alpha diversity. However, the sorts of species attracted to edges typically are habitat generalists with large dispersal distances and wide geographic ranges. Habitat specialists, often "interior species," tend to disappear as large, contiguous blocks of habitat are fragmented and edge increases. Thus, even as diversity is increasing, species of greatest rarity and highest conservation value may be disappearing. This problem illustrates why conservation biologists must be well informed about the species composition of a community, and not use diversity indices alone to evaluate the community's ecological health or conservation value.

Conservation biologists are increasingly concerned about the fact that mathematical diversity indices treat all species as equivalent even though this is not appropriate for setting conservation priorities. One species of panda is not the conservation equivalent of one species of rat, even if both are equally abundant in a given community. Vane-Wright, Humphries, and Williams (1991) propose a "taxic diversity measure" that measures diversity not as species richness or species evenness but as the amount of "taxonomic distinctness" present in a community, based on the number and abundance of different taxonomic levels present. We will explore taxic diversity measures further in an example later in this chapter when we examine how to estimate the price of noninstrumental values of biodiversity. Measures of taxonomic diversity also are used in combination with other data to conduct what is called "critical faunal analysis" (Ackery and Vane-Wright 1984), in which fauna of different sites are ranked according to the number of endemic species (species found only in one site or area) present. The site with the highest number of endemics is ranked first and the lowest is ranked last. Then the minimum number of sites is selected that will preserve all endemic taxa (i.e., a complete list of "critical areas") (Vane-Wright, Humphries, and Williams 1991).

THE PROBLEM OF VALUE: WHY SHOULD WE CARE ABOUT BIODIVERSITY?

Ecocentric Values of Biodiversity

Aldo Leopold provided one of the earliest and clearest assertions that biodiversity has value. Writing metaphorically of the value of noneconomic species as parts of the "ecological watch" with which scientists "tinker," Leopold stated, "To keep every cog and wheel is the first precaution of intelligent tinkering" (Leopold 1966:190). Although admirable in its juxtaposition of brevity and elegance, Leopold's statement, without further support, will not stand as an argument for preserving biodiversity. Rarity, by itself, demonstrates neither ecological value nor ecological necessity. Rare species often have little effect on ecosystem processes or structures when present, and are of little consequence to ecosystem persistence when absent (Ehrenfeld 1988). Nevertheless, there are important values of biodiversity that are grounded in the persistence and health of the ecosystem itself, and hence can be called "ecocentric values" of biodiversity. The ecocentric value of biodiversity is rooted in two distinct, but not mutually exclusive, spheres: the effects of species on one another and the effects of species on general ecosystem structures and functions.

Instrumental Values of Biodiversity

As vital as the concept of biodiversity is to the practice of conservation biology, it is difficult to value. Recall one of Michael Soulé's normative postulates of conservation biology: *biotic diversity has intrinsic value, regardless of its utilitarian value.* Although this assertion affirms the value of biodiversity, it does not change the fact that such value, as with all values associated with conservation biology, must be rigorously examined and articulated in a variety of different forms before it gains any persuasive status. The first important question to ask is, intrinsic value aside, does biodiversity have instrumental value and, if so, what kind of instrumental value does it have?

In a comprehensive review of the instrumental values of biodiversity, conservationist Norman Myers identifies three major categories of instrumental values of biodiversity. These are agricultural/genetic resources, medicinal resources, and industrial resources (Myers 1983). Most of the world's major crop plants, including corn, rice, barley, potatoes, tomatoes, and coffee, have their origins in wild, tropical plants. Commercial strains of all these crops are extremely low in genetic diversity, but their wild relatives possess enormous genetic variability that can be incorporated into domestic varieties to increase their vigor and resistance to disease.

In medicine, tropical plants are used extensively and directly as therapeutic agents and as raw materials for thousands of medicinal derivatives. Some of the best known and most widely used examples are *Digitalis* spp. (foxglove), used to treat heart disease and hypertension; *Rauwolfia serpentina*, the source of reserpine, which is used to treat hypertension, anxiety, and schizophrenia; and *Ephedra* spp., which is the source of ephedrine, a drug used to treat amoebic dysentery. Tropical animals are also widely used in pharmaceutical research (Myers 1983).

Tropical plants also contribute raw materials for textiles, fats, oils, resins, rubber, fuels, dyes, and other resources for industrial processes. Many such compounds appear to be promising replacements for petrochemicals. These "phytochemicals" are, unlike petrochemicals, renewable, and considerably less polluting. The total value of such resources is incalculable, but estimates of different components are staggering. The medicinal value of tropical pharmaceutical plants has been estimated in the past at $40 billion (U.S.) annually. Values of genetic benefits to agriculture and raw materials to industry were similar or higher (Myers 1983).

Noninstrumental Values of Biodiversity

Amenity Values of Biodiversity

The economic value of biodiversity, especially the biodiversity of tropical systems, is admittedly impressive, and can form a persuasive argument for preserving biodiversity. The fact that the rosy periwinkle (*Catharanthus roseus*), a species of plant from Madagascar, can provide a cure for a child suffering from leukemia forms a powerful appeal for its value. The truth is, however, that the value of bioprospecting, a form of option value (chapter 3) rarely, if ever, outweighs the value of present, immediate consumptive use. Further, many conservationists are not kindly disposed to this kind of consumptivist approach to the valuing of biodiversity. Resource economist D. H. Meadows, writing with derision of what he calls the "Madagascar periwinkle argument," argues that "even many ecologists hate this argument because it is both arrogant and trivial. It assumes that the earth's millions of species are here to serve the economic purposes of just one species" (Meadows 1990). Meadows's remarks are a fair summary of why some conservationists and ethicists object to evaluating biodiversity in purely economic terms. As J. Baird Callicott has said, "Some things have a price, others have a dignity" (Callicott 1994:33). As a result, some entities, those that we think possess the most dignity, are deliberately removed from economic valuation. Many conservationists argue that biodiversity is such an entity and that the arguments for its conservation ought to be made on the basis on noninstrumental values.

Ironically, many economists agree with this argument and have offered original and creative methods for "pricing" noninstrumental, public values of biodiversity. Environmental and resource economists contend that conservationists and land managers can benefit from "pricing" biodiversity, crass as this may sound to some, because prices convey and summarize information in market economies (Montgomery et al. 1999).

One approach to pricing biodiversity for amenity values has been advanced by economist Martin Weitzman, who describes the problem of conserving biodiversity as the "Noah's ark problem." Is there, he wonders, any rational economic method that a conservationist, like Noah, can use to determine which species to bring "into the ark"? In other words, which species will most increase the value of biodiversity if they are the targets of conservation (Weitzman 1998)? Weitzman suggests that a ranking index, R_I, can be determined through the equation

$$R_I = (D_i + U_i)\,(\Delta P_i / C_i).$$

D_i is a measure of the "distinctiveness" of the ith species, which might be quantified as the distance or difference from its most closely related species. Thus, in Weitzman's system, the most taxonomically distinct species (species with the fewest close relatives) would receive the highest values of D. U_i represents the direct utility of the ith species, or, in Weitzman's words, "how much we like or value the existence of species i," a measure that could be calculated through techniques like contingent valuation (chapter 3). ΔP_i represents the change (increase) in the probability of the survival of the ith species as a result of some conservation action or management strategy. C_i represents the cost of such a strategy for the ith species.

The values of these four terms are difficult to calculate, and the mathematical underpinnings (Weitzman 1998) for doing so are beyond the scope of our discussion here. But such calculations are possible, and understanding the relations in Weitzman's basic equation helps us set a research agenda for how to assess the noninstrumental values of biodiversity. We can use the equation to identify the four fundamental questions that must be answered in setting conservation priorities for biodiversity: (1) How unique is each species compared with other species? (2) What is the utility of this species to humans or to the ecosystem? (3) How much can we improve this species' chances for survival with management efforts? (4) How much will it cost to improve such survivorship? With this information, it is possible to estimate a critical value of R_I, designated R^*, that generates a decision rule for which species to invest management resources in and which species to ignore (i.e., On which species do we invest limited management resources to achieve the highest value of biodiversity?).

Economic evaluation of biological diversity is possible but difficult, often demanding large amounts of precise information to make meaningful predictions and estimates. Economic evaluation is helpful in that it forces conservation biologists to make their assumptions explicit, and to make the most accurate measurements possible of the contributions of different species to biodiversity. If such assumptions are in error, the economic analyses on which they are based will be of little use. For example, Weitzman (1998) gives higher values to species that are distinct taxonomically, but does not assess the ecological contribution of different species to biodiversity. A better evaluation might be to identify possible keystone species in the species assemblages and give them higher values than other species. But to do so would require even more information, as well as additional assumptions.

At this point, we do well to return to Soulé's fourth normative postulate: *biotic diversity has intrinsic value, regardless of its utilitarian value.* Ultimately, it is clarity of thinking and persuasiveness of argument about the intrinsic value of biodiversity, rather than economic analysis, that may determine the success of conservation biologists in preserving it. Thus careful examination and analysis of the arguments for the intrinsic value of biodiversity become even more important.

POINTS OF ENGAGEMENT—QUESTION 1

Can you envision any ways that careful and precise measurement of the *quantity* of biodiversity in an area might help a manager make more coherent public statements and policies about the *value* of diversity in the same area? If so, how, and what form would this take?

Moral Values of Biodiversity

Naturalistic- and Evolution-Based Moral Values

In his well-known essay, "Thinking Like a Mountain," Aldo Leopold describes the experience of killing a wolf, and of his later reflections on the meaning of the wolf's death. "I now suspect that just as a deer herd lives in mortal fear of its wolves, so does a mountain live in mortal fear of its deer. And perhaps with better cause, for whereas a buck pulled down by wolves can be replaced in two or three years, a range pulled down by too many deer may fail of replacement in as many decades" (Leopold 1966:140). In these words, Leopold is arguing that we must change the scale of what we value. "Thinking Like a Mountain" is really a metaphorical argument for thinking in terms of time and process. Specifically, Leopold is arguing for the intrinsic value of biological process.

Understanding the value of ecosystem processes is not simply a theoretical exercise in ethics or an extension of intrinsic value to the point of ecological abstraction. Understanding the role and value of ecological process has enormous practical implications for the successful management of biodiversity. For example, increasing deer densities in eastern U.S. national parks can result in a long-run reduction in the species diversity of plants in such parks. Porter and Underwood (1999) argue that management that does not grasp the value of ecological process under these conditions is certain to fail, not only in achieving a beneficial outcome, but also in communicating a coherent policy to the public. They note that managers must both understand and value ecological process because they must choose the point at which they intervene in the process (what density of deer to allow) and then communicate their reasons to diverse public interests. Porter and Underwood correctly perceive that this choice is a value judgment that has profound practical implications. "Whether we define ecological integrity in terms of species or processes, we must inevitably make a decision as to where in the . . . sequence we choose to intervene. That choice represents a value judgment.

Although the connotation often associated with value judgments is negative, such decisions are the essence of management and cannot be avoided" (Porter and Underwood 1999).

E. O. Wilson writes, "organisms support the world with efficiency because they are so diverse, allowing them to divide labor and swarm over every square meter of the earth's surface. They run the world as precisely as we would wish it to be run" (Wilson 1992:347). Here, Wilson asserts another dimension of the value of processes associated with biodiversity, namely that the intrinsic value of biodiversity is not an intrinsic value of individual creatures, but of *kinds* of creatures, and of the processes of speciation that create them and the ecological functions that they perform. In the scale of time, this view of intrinsic value entails a shift from focusing on the present moment to a focus on generations to come. In space, it requires shifting value from local sites to landscapes and geographic regions.

One basis for making such a shift is the concept that individual species have a right to exist as members of communities to which they belong. As noted earlier, the concept of the right to exist is very close to the actual words of Aldo Leopold in his basic assertion of the intrinsic value of biodiversity: "these creatures are members of the biotic community, and if (as I believe) its stability depends on its integrity, they are entitled to continuance" (Leopold 1966:246–47). The concept of a right to exist generates a corollary that species should be respected because they are elaborate and elegant products of speciation and evolutionary processes that have required long periods of time and selective effort to produce. Those who advocate this dimension of intrinsic value argue that species are to be respected for what they have been able to accomplish in the course of their evolutionary history. E. O. Wilson states, "Every kind of organism has reached this moment in time by threading one needle after another, throwing up brilliant artifices to survive and reproduce against nearly impossible odds" (Wilson 1992:345). In the view of Leopold, Wilson, and others, our respect for other kinds of life is a recognition of what species are as products of extensive, intricate, long-term biological processes. By the same logic, our proper expression of this respect is to allow species to exist as wild creatures without undue human interference (Callicott 1989) and, thus, to help perpetuate processes that allow for speciation. The recognition of this kind of intrinsic value, then, is what justifies our shift in focus to value intergenerational processes, not simply present events.

The shift to value landscape and ecosystem scales over local scale is justified by viewing ecosystem processes as having intrinsic value, not merely for the services such processes provide to us, but because of the value of the processes themselves. Here again, those who argue that the processes have intrinsic value support their assertion by pointing to what the processes produce, namely systems of order, integrity, and stability. Ethicist Bryan Norton writes, "Some of us call the creative drive of ecological systems 'autopoiesis' (from the Greek: 'self-making')—it is the mysterious driving force that creates, through dissipation of energy in open systems, a kind of growth or development, as order is created out of chaos" (Norton 1995).

Obligation- and Virtue-Based Ethics of Biodiversity Preservation

Norton (1995) argues that our recognition of the worth of biodiversity itself, the processes that create biodiversity, and the processes that biodiversity sustains lead naturally to obligations for humans to preserve biodiversity and its associated processes. These obligations are rooted in (1) the obligation humans have to restore the biodiversity of natural landscapes that they have reduced or destroyed; (2) the obligation to preserve the current framework of biodiversity for future generations of humans and nonhuman creatures; and (3) the obligation to preserve biodiversity as a recognition of its inherent beauty, complexity, integrity, and evolutionary achievement (Norton 1995).

Not all moral arguments are based on concepts of obligation or outcomes. Ereteological (virtue-based) ethics for preserving biodiversity are founded on the principle that the act of preserving biodiversity makes us better people (chapter 3). Or, in terms more appropriate to the concept of biodiversity itself, we become a better, less arrogant species by the act of preserving other species, and thus we are ourselves changed and made more receptive and better able to appreciate the value of other creatures. Restoration ecologists, for example, argue that the act of restoring communities and their biodiversity "liberates us from our position as naturalists or observers of the community into a role of real citizenship" in the ecological community (Jordan 1994), an idea that shall receive more detailed attention in chapter 11. Again it is Leopold who articulates the concept well. Speaking of the outcomes of adopting a "land ethic" that enlarges ethical boundaries to include nonhuman creatures and landscape processes, Leopold writes, "a land ethic changes the role of *Homo sapiens* from conqueror of the land community to plain member and citizen of it. It implies respect for his fellow-members, and also respect for the community as such" (Leopold 1966:240).

Religious Traditions and the Value of Biodiversity

As noted in chapter 3, all major world religions make unique and profound contributions to the values and ethics of conservation biology. All contain precepts from which the value of biodiversity could be inferred. Of these traditions, none engages the question, Why should humans care about biodiversity? more directly and more thoroughly than Judaism and Christianity. The Judeo-Christian engagement with the value of biodiversity occurs at four levels: first, the goodness of all created things that is pronounced upon them by God; second, the blessing of all created things to be fruitful and multiply upon the earth; third, the protection of all species as partners in an everlasting covenant with God; and, fourth, that human care for and protection of biodiversity is an expression of being made in the image of God.

The most repeated phrase of the first chapter of Genesis, "God saw that it was good," is stated six times in the first 26 verses, all prior to the creation of human beings. The goodness of created things, as noted in chapter 3, is noninstrumental and nonanthropocentric. In terms of biodiversity, the goodness is pronounced upon the entire created order of life. The goodness of creation is expressed most fully when *all* created forms of life are present ("And God saw *all* that he had made, and, behold, it was *very good*" Genesis 1:31). In the Judeo-Christian tradition, the presence of the full spectrum of biodiversity encourages the development of moral discernment in humans to perceive the goodness of God and of what God has made. When biodiversity is diminished, so is the manifestation of God's goodness and of his good work.

The phrase "be fruitful and multiply" is often associated with God's blessing to humankind. However, this blessing is given twice in Genesis 1, and the first time to nonhuman creatures: "Be fruitful and multiply, and fill the waters in the seas, and let birds increase upon the earth" (Genesis 1:22). The Judeo-Christian tradition affirms that God's intended purpose for the earth is to fill it with life, both human and nonhuman. All living creatures of every kind are equally and unequivocally included in this blessing. The Judeo-Christian tradition perceives that this blessing is most clearly manifested when biodiversity reaches its highest levels.

The third dimension of Judeo-Christian engagement with the question, "Who cares about biodiversity?" is answered with the assertion, "God does," demonstrated through the covenant with creation established in Genesis 9. As noted in chapter 3, environmental ethicists and conservation biologists, regardless of intellectual or faith commitments, have noted that in the preservation of creation in the ark, the ark builder, Noah, is preserving *species,* not individual specimens, and thus affirms the value of biodiversity. More significantly, recall that God makes a covenant of preservation with all creatures, including humans, following the judgment of the flood. (Genesis 9:15). Thus in the Judeo-Christian tradition, the deliberate destruction of species by human actions is contrary to the intent and purposes expressed by God in protecting all species (i.e., biodiversity) under the provisions of this covenant.

Within this perspective, conservation biologist David Ehrenfeld concludes, "Diversity is God's property, and we who bear the relationship to it of strangers and sojourners, have no right to destroy it" (Ehrenfeld 1988). But having no right to destroy biodiversity does not automatically mean that we possess the positive right to preserve biodiversity. This is the fourth dimension of the Judeo-Christian answer to the question, "Why should we care about biodiversity?" Humankind is created in the image of God, and, in that capacity, is given *the right and privilege to care* because humans are uniquely empowered and instructed to perform acts of service and beneficence to other creatures for their good (Genesis 2:15). Care for biodiversity is thus seen as a fundamental expression of the Judeo-Christian assertion that humans are created in the image *of* God, and in that image must express the same care *as* God for what has been created.

THE PROBLEM OF PROCESS AND PATTERN: WHAT FACTORS AND THEORIES EXPLAIN VARIATION IN LOCAL BIODIVERSITY?

Ecological Processes

Ecological Components of Biodiversity

If biodiversity would be cared for and preserved, the processes that sustain it and the patterns that define it must be understood. The two primary components of biological diversity in any community are the number of *functional types* and the number of *functional analogs* (Huston 1994), (fig. 4.3). Among animals, functional types are sometimes referred to as "guilds" and among plants as "life-forms." Recall that, in human societies, guilds were associations of workers of similar skills and interests. In an ecological community, guilds, or functional types, are species that exploit similar resources to survive. In a community of birds, for example, we might have guilds of nectar feeders, fruit eaters, seed eaters, insectivores, omnivores, birds that specialize in killing other birds, birds that specialize in killing mammals, and whatever else the resources of the

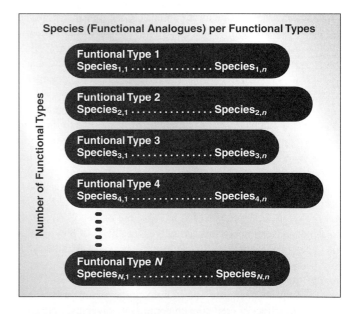

Species (Functional Analogues) per Functional Types

Number of Functional Types

Funtional Type 1
$Species_{1,1}$ . . . . $Species_{1,n}$

Funtional Type 2
$Species_{2,1}$ . . . . $Species_{2,n}$

Funtional Type 3
$Species_{3,1}$ . . . . $Species_{3,n}$

Funtional Type 4
$Species_{4,1}$ . . . . $Species_{4,n}$

Funtional Type *N*
$Species_{N,1}$ . . . . $Species_{N,n}$

Figure 4.3

Total species diversity can be measured as the product of the number of functional types and the number of species per functional type. Two populations may have the same species diversity and still differ. For example, one may have many functional types and few functional analogues, and the other may have many analogues but few functional types. The relative number of functionally analogous species within each functional type is indicated by the width of the oval.

community may provide. Functional types correspond to the concept of "niche," which can be conceptually understood as the ecological role of a species in a community. Determining what causes the number of functional groups or niches to increase is a complex ecological process in which multiple factors are at work. However, a first step in investigating and understanding levels of biodiversity in any community is the recognition that any process that causes an increase in the number of functional groups will result in increased species diversity.

A second key to understanding the processes that shape biodiversity in a community is the recognition that any process that leads to an increase in functionally analogous species (functional analogs) also results in an increase in diversity. In the case of functional analogs, species diversity increases, not because there are more species doing different things in a community, but because there are more species doing the same thing in the community. In this case, all functional analogs are members of the same functional type (fig. 4.3).

Thus, diversity in a community may increase if the number of functional groups increases, or if the number of species within a group increases, or both. But the two components of species diversity, functional groups and functional analogs, are affected by very different processes. When we observe increases in species diversity, such as increasing species diversity from temperate regions to tropical regions, we ought to ask, "Is diversity increasing because the number of functional groups are increasing, or is diversity increasing because the number of functional analogues is increasing?" Interestingly, as we go from temperate to tropical communities, we find that both the number of functional groups and the number of functional analogues within groups increases but, according to Huston (1994), most of the increase in diversity occurs because there are many more species within groups, compared with rather small increases in the number of functional groups.

In many systems, predators strongly influence community biodiversity (alpha diversity) through their responses to differing levels of species abundance. Generalist predators may engage in "switching" responses, concentrating their search efforts on the most abundant prey, thus increasing overall community diversity and reducing the number of individuals of the most abundant prey species. Both generalist and specialist predators tend to reduce average densities of prey species, increase the overall evenness of density in different prey species, and reduce effects of prey species on vegetation. Predators also can increase beta and gamma diversity because their presence causes changes in prey species' ecological niches and habitat choices (Sih 1986; Rosenzweig 1991). Some of the most elegant and illuminating experiments demonstrating the effect of predators on community diversity were conducted by the ecologist Robert Paine on intertidal zone communities (Paine 1966, 1969). Paine systematically removed the predatory sea star, *Pisaster ochraceus*, from some sites but not others. Where *Pisaster* was removed, the sites were eventually dominated by a single species of mussel, *Mytilus californicus*, which crowded out most other species. When *Pisaster* was present, it preyed on this mussel, reduced its numbers, and prevented the establishment of competitive equilibrium, and the competitive exclusion of other, less dominant species. On these sites, the average number of species present was 15, compared with an average of 8 species where *Pisaster* was not present.

The Species-Area Relationship

One of the earliest and most fundamental correlations established by the study of biodiversity has been the direct relation between species richness and area, a relationship first observed by early biogeographers comparing the number of species on islands of different sizes (fig. 4.4) (Darlington 1957). The **species-area relationship** provided a foundational concept for the development of the theory of island biogeography. Island biogeography's most basic and familiar equation, $S = cA^z$, states that the number of species on an island (S) is a constant power of the island's area (A), mediated by two constants, c and z. c is a constant specific to a particular taxonomic group and z is an "extinction coefficient" that integrates the rate of extinction in the group to the number of species associated with the area (chapter 5). Although there are many exceptions to this basic relationship in both insular and mainland communities, it is a well-documented pattern that larger areas contain more species than smaller ones, even if the areas are of similar habitat and landscape features. For example, Nupp and Swihart (2000) demonstrate that the species richness of forest-dwelling mammals in Indiana increases with forest area and is highest in unfragmented sites (fig. 4.5). An assortment of similar, granivorous small mammals can coexist in relatively small areas of forest, but only if such areas are part of a large, contiguous forest.

If this relationship holds, it permits prediction of the loss of species based on the loss of area. Specifically, according to Pimm (1998), if the amount of habitat we have now is less than the amount in some "original" state, then

$$S_{now}/S_{original} = (A_{now}/A_{original})^z.$$

Because species often are lost from habitats when the area of habitat decreases, the worldwide problems of habitat loss and habitat fragmentation (chapter 8) pose a particular threat to the conservation of biodiversity. In this context, Rosenzweig (1999) predicts that biodiversity will be lost from shrinking natural areas in a three-step extinction process (fig. 4.6). In phase one, there will be a loss of endemic species. These species will be the first to go because their entire range of habitat can disappear into human hands. The phase 1 effect of habitat loss on endemic species is well illustrated in Pimm's so-called "cookie cutter model" of habitat loss (Pimm 1998). Consider a model of habitat loss in which an area containing an array of habitats is analogous to a sheet of cookie dough, and the agent of habitat loss (e.g., deforestation), acts as a cookie cutter that randomly removes chunks of similar-size habitat. Then imagine an array of species with different sized ranges using the area (i.e., overlaying the cookie dough). To aid your imagination, figure 4.7 provides a visual image. If the cookie cutter strikes at subarea A, seven species loose habitat but none is exterminated. In contrast, suppose the cookie cutter strikes subarea B, an area containing species with more restricted ranges. Again seven species lose habitat, but this time four species are exterminated. Thus, random habitat loss will cause a disproportionately high rate of extinction in endemic species.

After the loss of endemic species in phase 1, Rosenzweig predicts that a slower but larger phase 2 loss of species will occur through the loss of "sink species," those species "whose remaining populations are all sink populations" (Rosenzweig

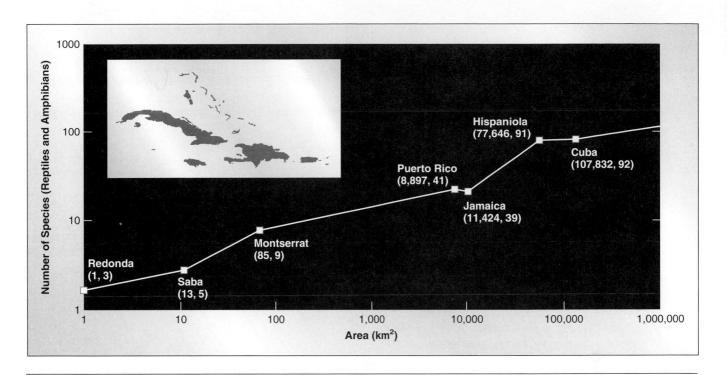

Figure 4.4

A general species-area relationship among some Caribbean islands. Note that species richness on islands increases with increasing area.

Based on data from Darlington (1957:483).

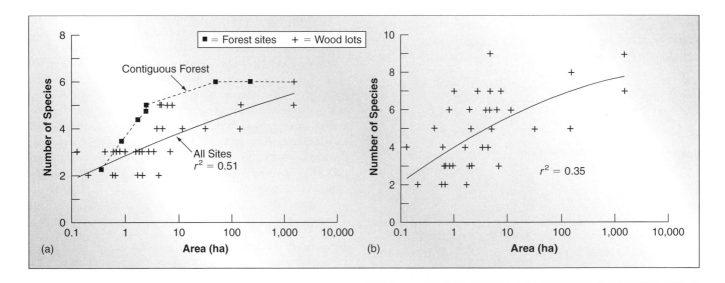

Figure 4.5

An illustration of the relationship between area and species richness of (a) granivores and (b) all small mammal species in woodlots (crosses) and contiguous forest sites (squares). Species richness increases with woodlot area. In (a), note that granivore species richness increases with area more rapidly in contiguous forest than in woodlots. This pattern suggests that species richness not only declines with habitat loss, but also with habitat fragmentation.

After Nupp and Swihart (2000).

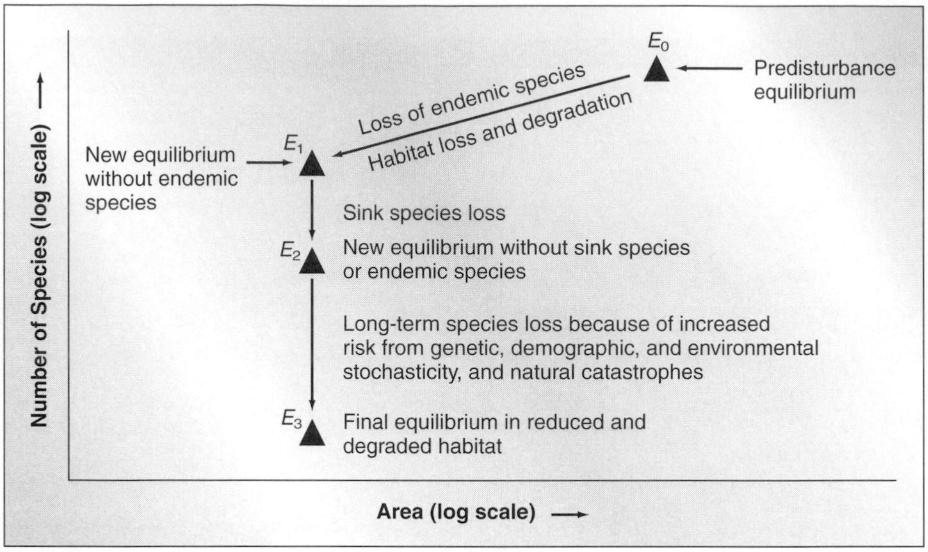

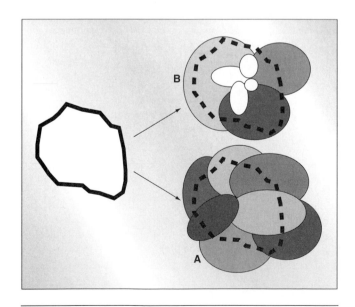

Number of Species (log scale) →

E_0 — Predisturbance equilibrium

Loss of endemic species

New equilibrium without endemic species — E_1

Habitat loss and degradation

Sink species loss

E_2 — New equilibrium without sink species or endemic species

Long-term species loss because of increased risk from genetic, demographic, and environmental stochasticity, and natural catastrophes

E_3 — Final equilibrium in reduced and degraded habitat

Area (log scale) →

Figure 4.6

When the size of a natural area is decreased, the first species lost are endemics. Next, sink species (those that are not reproducing fast enough to replace themselves) go extinct locally. Finally, failure to replace accidental losses fast enough brings the province to a still lower steady state of biodiversity.

After Rosenzweig (1999).

Figure 4.7

The "cookie cutter" model of the effects of habitat loss on endemic species. If the cookie cutter strikes at subarea A, seven species lose habitat but none is exterminated. In contrast, if the cookie cutter strikes subarea B, an area containing species with more restricted ranges, seven species lose habitat, and four species are exterminated. Thus, random habitat loss produces a disproportionately high rate of extinction in endemic species.

After Pimm (1998).

1999). Sink populations (chapter 5) are those that end up in habitats of sufficient quality and size to attract and hold individuals, but in which mortality is greater than recruitment, thereby leading to an overall decline in the population. In phase 3, the diversity of the reduced area will relax to a lower steady state with far fewer species. Species will be lost slowly through "rare accidents" of extinction that accumulate over long time spans.

Most conservation efforts focus on saving habitat rather than saving area. Given the importance of area for maintaining biodiversity, this may be a mistaken emphasis. It is estimated that humans already use or degrade up to 95% of the earth's terrestrial surface (Rosenweig 1999), so simply trying to make nature preserves larger may not be enough. To save biodiversity, conservation biologists must advocate strategies of human habitation and development that use or degrade far less land than current practices.

Habitat Heterogeneity, Productivity, and Disturbance

Biodiversity is often correlated with three different, but ecologically related factors: habitat heterogeneity, primary productivity, and disturbance. Many studies have demonstrated that biodiversity is often directly correlated with habitat and landscape heterogeneity. The more kinds of topography, soil conditions, microclimate, and habitat that occur within a given area, the greater its species diversity. One of the most common and well-established expressions of this correlation is the direct relation between foliage height diversity and bird species diversity. That is, the more layers of vegetation that are present in a community, the greater the number of bird species in that community, usually. Another is the familiar wildlife management concept that species diversity increases as the amount of edge, or habitat border, increases.

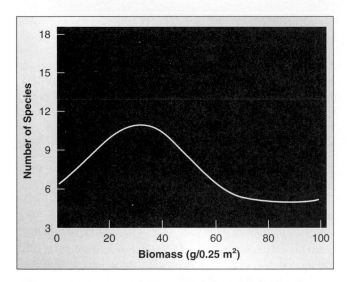

Figure 4.8

The "hump-shaped" relation of plant biomass to plant species richness. Based on data from five habitats in a southwest U.S. desert.

Adapted from Guo and Berry (1998).

When examining the relationship between biodiversity and productivity, many studies have demonstrated a "hump-shaped" relationship between the two variables (fig. 4.8) (Guo and Berry 1998). The relationship is often observed in relationships between biodiversity and biomass, with biomass serving as an index of productivity. Across a range of different habitats, biodiversity is low when biomass is low. Biodiversity rises as biomass rises to a certain level, plateaus, and then falls when biomass increases to still higher levels. For example, Guo and Berry (1998) demonstrated that the number of plant species in 0.25-m² quadrats in five different microhabitats in an Arizona desert environment followed this hump-shaped pattern.

Habitat heterogeneity and productivity are interactive in their effects on biodiversity. Specifically, the range of biomass level from "poor" to "rich" microhabitats strongly influences biodiversity patterns. The greater the range of environmental conditions (i.e., the greater the degree of habitat heterogeneity), the more complete was the development of the hump-shaped relationship (Guo and Berry 1998).

Frequency of disturbance influences both habitat heterogeneity and productivity. Thus, disturbance and diversity are often correlated in such a way as to also produce a hump-shaped curve reflecting the highest levels of biodiversity at intermediate levels of disturbance. This particular correlate has received so much attention in recent literature that it has even acquired its own name, the **"intermediate disturbance hypothesis"** (Huston 1979, 1994). This hypothesis states that ecological disturbance (such as fire, disease, or flood) that occurs too often or in too severe a form, lowers diversity by eliminating too many established species, creating a "species debt." At the other extreme, communities that are rarely, if ever disturbed become dominated by a small number of highly efficient "supercompetitors" that, over time, remove other species from the community, lowering its diversity. A problem with this hypothesis

has been the difficulty of objectively defining what the "intermediate" level of disturbance is. In the worst examples, some researchers simply create a tautology, where the intermediate level of disturbance is the one associated with more species. Some do better, and define "intermediate" according to objective criteria of the frequency of the disturbance, the severity of the disturbance, or a combination of both (Huston 1994).

In a similar way, competition is also assumed to lie behind the hump-shaped relation seen in the biodiversity-productivity relationship. Here, the theory is that the intensity of competition increases as the rate of biomass production increases. After biomass reaches a critical level, competition removes less-competitive species from the community, resulting in lower species diversity. Because disturbance frequency can be a controlling factor in determining community productivity, these factors may often be interactive. The current need in conservation biology research is to test such hypotheses experimentally. One approach might be to selectively remove "supercompetitive" species associated with high levels of biomass or low levels of disturbance and see if there is a postremoval increase in diversity caused by an influx of less-competitive species.

How Much Biodiversity Is There?

Even if biodiversity on a site or within a region can be precisely measured, three obvious questions remain when we examine worldwide patterns: How much biodiversity is there? What factors affect biodiversity? Where is the world's biodiversity distributed? Until the recent crisis of extinction and biodiversity loss, the question, "How many species are there?" was seldom considered important enough to ask, much less answer. Today fewer than 2 million kinds of organisms have been recognized as "species," but the total worldwide estimate is much higher, and much less precise. Scientists estimate that there are anywhere from 5 million to more than 50 million species (May 1988). Why are we so ignorant about such a vital aspect of conservation biology? The answer is that, although the knowledge of how many species exist is essential, it is also difficult to determine and challenging to interpret.

The first difficulty is that we have explored the world very unevenly in its vast array of biological diversity. About two-thirds of all species classified have traditionally come from temperate areas, especially from North America and western Europe, where high human population densities, accessibility of most areas, and detailed scientific exploration all contributed to a relatively high level of knowledge of local species. Insects contribute the greatest number to the worldwide total of identified species, and most of these are from temperate regions. However, in larger terrestrial vertebrates which have been especially well studied all over the world, there are roughly twice as many tropical species as temperate ones. If the same ratio holds for insects, this would mean that there are about two unclassified species of tropical insects for every classified temperate insect species. If this is true, the total estimate of species worldwide would be revised upward to 3 to 5 million (May 1988).

To address the problem of "species ignorance," the International Biodiversity Observation Year (IBOY) 2001–2002, involves a worldwide effort by the scientific community to catalog and monitor global biodiversity in a more coordinated way than ever before. Program Coordinator Jose Saruhan, an ecologist at the Universidad Nacional Autónoma de México, states that the program has two major goals. "One is to increase social and governmental awareness about the transcendental importance of conserving and sustainably managing biodiversity. The other is to bring together global data sets on the status and trends of biodiversity, focusing on those areas that are most amenable to a short-term, intensive program of international scientific cooperation" (Norris 2000). Linked to the IBOY efforts is a preexisting initiative called Species 2000, chaired by Frank Bisby of the University of Reading (England). Species 2000 seeks to link databases of information on the approximately 1.75 million species of animals, plants, fungi, and microorganisms that presently exist, a wealth of information, but one that Bisby describes as still "chaotic" (Norris 2000). The International Biodiversity Observation Year will also attempt to coordinate new and existing monitoring programs that assess current biodiversity, name and describe new species, and monitor threatened and endangered sepcies, and assess the factors contributing to the loss of recently extinct species. Difficulties associated with assessing and monitoring global biodiversity will not be solved by a single scientific effort, but if the IBOY initiative is effective, the quality of data on global biodiversity, and the ability of scientists to use and access this information, will be vastly improved, and the uncertainty associated with the question, "How much biodiversity is there?" will be substantially reduced.

Biodiversity and Rarity

Diversity and rarity are positively correlated in communities and environments. Alpha, beta, and gamma diversity will increase from community to landscape to region if the number of species and their proportional evenness of abundance increase, which normally means that the densities of most species may

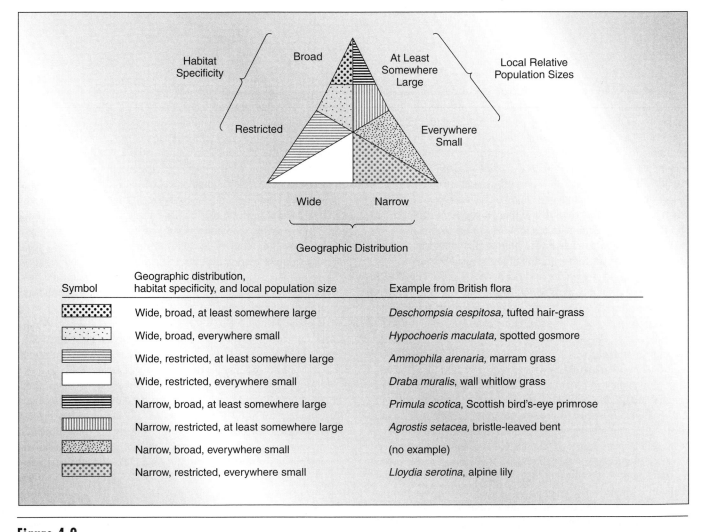

Figure 4.9

Eight categories of species abundance in British plants based on geographic range, habitat use, and relative population size. Note that only one category (broad habitat specificity, wide geographic distribution, and large local population) can truly be considered "common." Species in the other seven categories are rare in one or more dimensions.

Adapted from Rabinowitz, Cairns, and Dillon (1986).

become lower as diversity becomes higher at any and all of these scales. If conservationists desire to maximize biological diversity, then they must learn to manage rarity.

The archetypal rare species is one characterized by small populations, specialized habitat requirements, and a restricted geographic range. Some species, especially the most critically imperiled, do meet all three of these criteria. For example, the endangered Kirtland's warbler has an estimated world population of fewer than 2,000 individuals, breeds only in homogeneous, young, even-aged stands of jack pine (*Pinus banksiana*), and concentrates approximately 60% of its breeding effort in three counties in the state of Michigan in the United States. But rarity is not always so well defined.

A helpful conceptual beginning to understanding factors underlying rarity is offered by Rabinowitz, Cairns, and Dillon (1986). Their typology is based on eight categories created by three dichotomies (Fig. 4.9): (1) Is the population dense or sparse? (2) Does the species use many different habitats or only one or a few habitats? (3) Does the species have a wide geographic distribution or a narrow geographic distribution? Rabinowitz, Cairns, and Dillon (1986) surveyed botanical experts on 177 species of wildflowers in the United Kingdom for which acceptable data on abundance and distribution were available and attempted to place each species in one of the eight categories. Of the species surveyed, 17 were controversial regarding their abundance, but the other 160 produced consensus. Of these, only about a third (36%) were truly "common" in all categories, having large populations, wide distributions, and a broad range of habitats. Only 2% were floral equivalents of the Kirtland's warbler, with small populations, specialized habitat preferences, and restricted geographic ranges. The majority of species (62%), were "rare" in one or two dimensions of their pattern of abundance. Each dimension of rarity needs some concentrated attention to better understand its effect on overall biodiversity.

Habitat Generalists Versus Habitat Specialists

Habitat generalists are species that can exploit a variety of habitats in a given geographic range. Within that range, they are relatively invulnerable to extinction through habitat loss or general changes in land use because, if one habitat is changed or destroyed, they simply move to another. In contrast, **habitat specialists** are typically highly successful and competitive in one or a few types of habitat, but unable to use others. For habitat specialists, loss of the preferred habitat at local or regional levels is catastrophic and typically leads directly to endangered status.

Large Populations Versus Small Populations

Some species almost always occur in large numbers and high densities. Their abundance may be the result of natural-history traits such as high reproductive rates, high rates of juvenile and adult survival, or strong competitive abilities that allow them to dominate other species. Some species, even if widespread geographically, never are abundant anywhere. Natural history traits that contribute to low abundance include low reproductive and survival rates, specialized diets (especially among carnivores), and the need for large areas to find food or complete their life cycles (a particular problem for large-bodied, territorial ani-

Figure 4.10

The mountain lion (*Puma concolor*), an example of a "rare" species with extensive geographic range and wide habitat tolerance, but uniformly low population density. The mountain lion historically had the widest distribution of any American mammal other than *Homo sapiens*, but local densities were almost always less than one individual per 20 km².

mals). The mountain lion, *Puma concolor* (fig. 4.10), historically possessed the largest geographic distribution of any American mammal (Jones et al. 1983), ranging from the southern tip of the Arctic Circle in North America to southern Chile and Argentina in South America, and from the Atlantic to the Pacific in both hemispheres. As a large, territorial carnivore, one resident lion requires a minimum of 16 to 20 km², and some lions have been known to use areas of up to 600 km² when prey are scarce (Hempker 1982). As a result, no single area ever had large numbers of mountain lions. When naturally small populations become isolated, prey availability is reduced, and habitat is restricted. As in the case of the subspecies known as the Florida panther (*Puma concolor coryi*), the population becomes imperiled and the probability of extinction is high.

Widespread Distribution Versus Restricted Distribution

If a species possesses a widespread distribution, like the mountain lion, the probability of persistence at regional or global levels can remain high although individual populations are threatened, or even exterminated, in particular areas. Indeed, if such species have dispersal abilities sufficient to allow individuals to move among regions and populations, they effectively comprise a metapopulation that can periodically recolonize sites that suffer local extinction (chapters 5 and 7). However, if a species has a restricted distribution, it is far more susceptible to local or regional changes associated with land use, human population growth, or climate change. Perhaps the best example of a restricted distribution is the Haleakala silversword (*Argyroxiphium macrocephalum*) (fig. 4.11), a striking plant covered with fine, silvery hair and producing a tall flower stalk at maturity, after which the plant dies. Fifty thousand individuals live on the Hawaiian island of Maui, and only in the crater of the volcano Haleakala.

Figure 4.11
The Haleakala Silversword (*Argyroxiphium macroephalum*), an example of a "rare" species with a dense population of individuals confined to a single site, the crater of Haleakala, a Hawaiian volcano.

The Problem of Endemism

The silversword, in its extremely restricted geographic range, illustrates a pattern typical of other restricted species. This is the phenomenon of **endemism,** in which a species is restricted to a particular area or region. Many tropical species are highly endemic, although there are examples of endemic temperate and polar species as well. The restricted range of an endemic species makes it especially vulnerable to extinction because local changes in land-use patterns or climate affect all individuals. If a species is concentrated in a single area, there is no "reserve" of individuals in another area that might be used to replenish the species if it suffers extinction. Because endemic species are at such high risk, some conservation biologists have argued that the problem of extinction should be studied and understood primarily as the problem of endemism, and endemic species should receive priority in protection (Pimm 1998). Endemism also has attracted attention among conservation biologists because it is logical to form the hypothesis that endemism and biodiversity should be correlates. It has been natural to assume that if we protect regions with high levels of endemism we are simultaneously protecting areas with high levels of biodiversity. Thus, endemism has long held the attraction of being a possible index of biodiversity that could be estimated much more easily that other kinds of biodiversity measurements.

Endemism and Rarity

Kerr (1997) empirically tested the hypothesized relationship between endemism and biodiversity in North America at four different taxonomic levels: a class (Mammalia), a family of butterflies (swallowtails, Papilionidae), a subfamily of moths (Plusiinae), and a genus of bees (*Lasioglossum*). He first divided the area into quadrats of $2.5° \times 2.5°$. He then estimated endemism for each group by counting the number of quadrats in which the species occurred, taking its inverse, and then summing the total for each quadrat through the formula

$$\text{Endemism} = \sum_{i=1}^{s} Q^{-1},$$

where S is the total number of species in the taxon being measured and Q is the number of quadrats within the species' range. By structuring the formula in this way, species with narrow ranges will receive high scores for endemism. Kerr then determined the correlation between a quadrat's rank in endemism for each taxon and its rank in species richness in the same taxon.

Correlations with endemism were high in all cases in these taxa (table 4.7), indicating that areas that protect the endemic species of a taxon are likely to also protect high levels of the biodiversity in that taxon. Mönkkönnen and Viro (1997) note a similar pattern in birds, finding, in the northern hemisphere, that both species richness and endemism are highest in east Asia. However, Kerr found that correlations among different taxons, although still statistically significant, were much weaker (table 4.8). Thus, protecting the endemic species of one taxon will not ensure that endemic species in other taxa are protected in the same area, or that protecting areas with high species richness in one taxon will protect the species richness of other taxa in the area. It is especially noteworthy that Kerr examined this correlation in mammals, whose protection is sometimes justified as an "umbrella" that protects species in other taxa. Kerr's data suggest that the "umbrella" provided by mammal diversity is rather leaky when it comes to the conservation of invertebrate diversity.

In some areas, endemism is characteristic of nearly all species. For example, 90% of Hawaiian plants and 100% of Hawaiian land birds are endemic. In the Fynbos region of southern Africa, known for its unique plant communities, 70% of all plant species are endemic. On a continental scale, 74% of Australian mammals are found only in Australia (Pimm et al. 1995). Not surprisingly, many conservation organizations have made endemism the controlling criterion in prioritizing areas for protection. This focus is the basis of the so-called hot spot concept for ranking areas for conservation. For example, conservationist Norman Myers identified four such hotspot areas for tropical forest plants that were particularly vulnerable to species loss because of high rates of endemism coupled with high rates of deforestation (table 4.9) (Myers 1988). In each of these areas, the number of endemic species is high, and even higher is the ratio of endemics to the area of the forest in which they occur. Unfortunately, geographic endemism is not uniform in different taxonomic categories (van Jaarsveld et al. 1998). This fact makes it hard to apply hot spot criteria universally. Areas with high rates of endemism for birds may not have many endemic amphibians (Pimm 1998). Areas rich in species diversity may be poor in genus or family diversity, and vice versa. Thus the prospect of finding "indicator taxa" that could be used as indices of diversity of other taxonomic categories is not promising (van Jaarsveld et al. 1998). The diversity of different taxa cannot often be protected in the same conservation areas. Conservation strategies to protect endemic species, and so conserve biodiversity, must be

Table 4.7 **Pearson correlations between species richness and endemism within 10 taxa in the conterminous United States and Canada** Pearson correlations can range from 1 to -1. A value of 1 indicates that there is a perfect positive linear relationship between the two variables (they change together by the same amount and in the same direction). A value of -1 indicates a perfect negative linear relationship (the variables are perfectly but inversely correlated; when one increases by x, the other decreases by x). A value of 0 indicates that two variables have no linear correlation. All correlations shown are significant at $P < 0.001$ and the value of endemism is log $(x + 1)$ transformed.

TAXA	CORRELATION OF SPECIES RICHNESS AND ENDEMISM
All mammalia	0.807
Artiodactyla	0.807
Carnivora	0.384
Chiroptera	0.814
Insectivora	0.523
Lagomorpha	0.665
Rodentia	0.773
Lasioglossum	0.851
Papilionidae	0.703
Plusiinae	0.772

Source: Modified from Kerr 1997.

Table 4.8 **Association (Pearson correlations) of species richness and endemism in four taxa** All correlations shown are significant at $P < 0.001$. Note that, although correlations are still significant, there is much less correlation (values are smaller) between levels of endemism or species richness in two unrelated taxa (e.g., endemism in mammals and endemism in Plusiinae) than between the values of endemism *and* species richness *within the same taxonomic group*. Thus, it appears that endemism is strongly associated with species richness, and vice versa, when dealing with the same kinds or organisms, but endemism or species richness in one group is not necessarily associated with endemism or species richness in other groups. Values of endemism are log $(x + 1)$ transformed.[*]

	MAMMALIA	*LASIOGLOSSUM*	PAPILIONIDAE
	Species richness		
Lasioglossum	0.833*		
Papilionidae	0.831	0.676	
Plusiinae	0.514	0.610	0.376
	Endemism		
Lasioglossum	0.805		
Papilionidae	0.594	0.341	
Plusiinae	0.459	0.516	0.238

[*]For further explanation of Pearson correlations, see Table 4.7.
Source: Modified from Kerr 1997.

comprehensive in their approaches, both biologically and geographically, to be successful, and must contain specific information about the distribution of individual species, not just indicator taxa or community diversity (van Jaarsveld et al. 1998).

Endemism and Island Species

Many island species represent extreme cases of endemism that merit conservation priority. By their very nature, island species usually meet at least two of the three requirements for rarity. First, island species have highly restricted ranges. They are

Table 4.9   Areas of Disproportionately High Plant Biodiversity (Hot Spots) in Tropical Forests

HOT SPOT	AREA OF FOREST (KM²) ORIGINAL	AREA OF FOREST (KM²) PRESENT (PRIMARY)*	PLANT SPECIES IN ORIGINAL FORESTS	NUMBER OF ENDEMICS IN ORIGINAL FORESTS (PERCENTAGE)		ENDEMICS AS PERCENT OF EARTH'S TOTAL PLANTS	PRESENT FOREST AREA AS PERCENT OF EARTH'S LAND SURFACE
Madagascar	62,000	10,000	6,000	4,900	(82)	1.96	0.00675
Atlantic Coast Brazil	1,000,000	20,000	10,000	5,000	(50)	2.00	0.0135
Western Ecuador	27,000	2,500	10,000	2,500	(25)	1.00	0.0017
Colombian Chocó	100,000	72,000	10,000	2,500	(25)	1.00	0.0486
Uplands of western Amazonia	100,000	35,000	20,000	5,000	(25)	2.00	0.0236
Eastern Himalayas	340,000	53,000	9,000	3,500	(39)	1.40	0.0358
Peninsular Malaysia	120,000	26,000	8,500	2,400	(28)	0.96	0.0175
Northern Borneo	190,000	64,000	9,000	3,500	(39)	1.40	0.04
Philippines	250,000	8,000	8,500	3,700	(44)	1.48	0.0054
New Caledonia	15,000	1,500	1,580	1,400	(89)	0.56	0.001
Totals	2,204,000	292,000	†	34,400		13.8	0.2
For comparison:							
Hawaii	14,000	6,000	825	745	(88)	0.30	0.004
Queensland	13,000	6,300	1,165	435	(37)	0.17	0.004

*Some, though not many, primary forest species can survive in degraded forests.
†It is unrealistic to sum these figures for plant species, on the grounds that there is some overlap between adjacent regions (e.g., some plants occur in peninsular Malaysia, northern Borneo, and the Philippines).
Source: After Myers 1988.

typically endemic to their islands, unless they can fly, swim, or float for long distances. Not surprisingly, the high rates of species endemism on islands are associated with high rates of biodiversity per unit area (table 4.10) (Whittaker 1998).

The high endemism and per area biodiversity of island flora and fauna are often coupled with extreme vulnerability to extinction. Many island species specialize in the use of habitats unique to their own island or island system and may occupy niches on their island or islands that are occupied by other species in mainland environments. Thus, they are especially vulnerable to invasion from mainland species. For example, the famous woodpecker finch (*Camarhynchus pallidus*) of the Galápagos Islands would be unlikely to retain its feeding niche (removing invertebrates from the bark of trees with a cactus spine) if it had to compete directly with true woodpeckers (family Picidae). Island species have no evolutionary exposure to mainland predators, competitors, and diseases, and so are especially vulnerable to invasions of such species, a subject that will receive more detailed attention in chapter 7. Given these ecological predispositions to extinction, it is not surprising that the rates of island extinctions have been high; that island species have been particularly vulnerable to habitat destruction, predation, and adverse competitive effects from intro-

duced species; and that island species currently form a disproportionately high percentage of the global array of endangered species. Even popular expressions like "dead as a dodo" (a reference to a large flightless pigeon [*Raphus cucullatus*] of the island of Mauritius that had been exterminated by humans and their associated introductions of pigs, rats, and cats by the eighteenth century) underscore how pervasively island species epitomize the extinction process, and how many of these species have already been lost. If island species are to survive, conservation strategies must preserve critical island habitat, eliminate introduced competitors and predators, and protect island species from human hunting, collection, and disturbance.

THE PROBLEM OF DISPERSION: WHERE IS BIODIVERSITY LOCATED?

Global Patterns of Biodiversity

Diversity is a complex phenomenon. Just as diversity cannot be explained by a single hypothesis or theory or measured with a single statistic, it cannot be easily described on a global scale.

Table 4.10 Species Richness and Endemism Among Higher Plants on Selected Islands

ISLAND OR ARCHIPELAGO	TOTAL SPECIES	ENDEMICS	% ENDEMIC
Borneo	20,000–25,000	6,000–7,500	30*
New Guinea	15,000–20,000	10,500–16,000	70–80
Madagascar	8,000–10,000	5,000–8,000	68.4
Cuba	6,514	3,229	49.6
Japan	5,372	2,000	37.2
Jamaica	3,308	906	27.4
New Caledonia	3,094	2,480	80.2
New Zealand	2,371	1,942	81.9
Seychelles	1,640	250	15.2
Fiji	1,628	812	49.9
Mauritius, including Réunion	878	329	37.5
Cook Islands	284	3	1.1
St. Helena	74	59	79.7

*Ten of the 13 selected islands have flora that are composed of 30% or more endemic species.

Source: Compiled by Whittaker 1998 using data from Groombridge 1992 and Davis, Heywood, and Hamilton 1995.

Each major taxonomic group (not to mention most minor ones) shows important and unique patterns in the distribution of its diversity. We will identify some patterns shared by two or more taxa, as well as some important exceptions.

All classes of vertebrates, as well as plants, show marked increases in the number of species as one moves from temperate to tropical latitudes (fig. 4.12) (Reid and Miller 1989; Huston 1994). However, there are exceptions in which diversity increases in temperate, or even polar areas instead of decreasing. These include sea birds, lichens, marine benthic organisms, parasitic wasps, and soil nematodes (Huston 1994).

As with latitude, diversity tends to show an inverse relationship to altitude, particularly in plants. This is not surprising given that increases in altitude produce environmental and climate effects similar to those of latitude. In plants, diversity tends to be highest at low to middle latitudes and lowest at high latitudes and altitudes.

Within these broad patterns, important regional and habitat trends in diversity exist. In marine environments, diversity is highest in coral reef habitats, in coastal zones and estuaries, and overall is higher in tropical marine ecosystems than in temperate ones. In terrestrial habitats, the diversity of tropical rain forests is higher than all other habitats; it is, in fact, legendary. Investigators have found up to 300 different tree species per hectare in study plots in Peru (Wilson and Peter 1989). Biologist E. O. Wilson recorded as many ant species (43) and genera (26) from a single tropical rainforest *tree* in Peru as were present in all of the British Isles (Wilson 1989). In temperate areas, freshwater wetlands contain disproportionately high levels of species diversity at the landscape level. Wetlands may be particularly important in systems of low diversity. For example, boreal swamp forests contribute high biological diversity to otherwise low-diversity boreal forest ecosystems (Hornberg et al. 1998). As noted earlier, island floras and faunas tend to have high rates of endemism, and make disproportionately high contributions to world biodiversity.

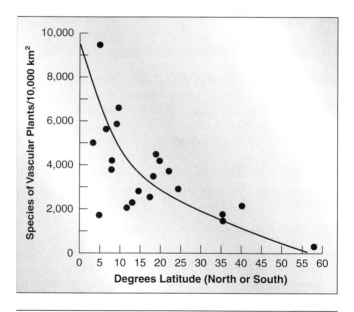

Figure 4.12

Latitudinal patterns in species richness from tropical to temperate regions. In most taxa the number of species increases from temperate to tropical regions.

After Reid and Miller (1989). Reprinted from Huston (1994).

Recent Measurements of Biodiversity: How Can We Identify Hot Spots with Incomplete Information?

There is simply not enough money, labor, and expertise to identify, count, and map the distribution of every species in every taxon at a global scale in time frames that can assist current conservation decisions. Thus, conservation biologists have been engaged for some time in attempting to find noncensus indicator methods that can rapidly and reliably identify areas with disproportionately high levels of biodiversity. The earliest efforts to accomplish this goal relied on the **umbrella species** concept for biodiversity preservation. The umbrella species approach assumes that if an area containing a species or group of species of particular conservation or public interest (for example, large mammals) is preserved, this protection will benefit species in other taxa, thus protecting a high level of biodiversity. As noted earlier in Kerr's (1997) study, empirical data have provided no support for this optimistic outcome. The best data available indicate that taxonomic distributions of organisms are remarkably independent. No one major group of organisms consistently predicts the distributions of other groups (Prendergast et al. 1993). But if biodiversity must be evaluated one taxon at a time, how can we index global biodiversity without censusing species?

One recent approach to this problem has been to measure species richness at higher taxonomic levels, such as families, and then combine the family indices of different taxa to produce an aggregate biodiversity index for arbitrarily defined regional grid cells that can then be mapped worldwide. Williams, Gaston, and Humphries (1997) examined the problem through combining "family richness" in four major taxa: plants, amphibians, reptiles, and mammals. Their choice of these four groups was based on (1) the need to limit the number of groups to keep the index from becoming unmanageable, (2) the popular appeal and already established conservation efforts worldwide to conserve species in these groups, and (3) the availability of reliable regional information for these groups. Williams, Gaston, and Humphries divided the world into grid cells of 611,000 km^2 at intervals of 10° longitude, and then, using a variety of existing data sets, calculated three measures of "family richness" for each cell. *Absolute family richness* was obtained by summing local family richness counts (number of families in the grid cell) for each of the four groups according to the formula

$$\text{Absolute family richness} = f_{p,1} + f_{a,1} + f_{r,1} + f_{m,1},$$

where f is the number of families in each group, designated by subscript (i.e., p for plants, a for amphibians, r for reptiles, and m for mammals). *Proportional family richness* is determined by summing the local *proportion* of family richness in the different major groups through the formula

$$\text{Proportional family richness} = (f_{p,1}/F_p) + (f_{a,1}/F_a) + (f_{r,1}/F_r) + (f_{m,1}/F_m).$$

The new term in this formula, F, represents the total worldwide number of families in each group. Thus it effectively equalizes the contributions of groups with different numbers of families and therefore weights the index to favor areas that have a greater proportion of the families in each group, not simply the areas that have the greatest number of families. A third measure, *proportional family richness weighted for species richness,* is calculated as

$$S_p(f_{p,1}/F_p) + S_a(f_{a,1}/F_a) + S_r(f_{r,1}/F_r) + S_m(f_{m,1}/F_m).$$

Here the new term, S, represents the total number of species in each group (Williams, Gaston, and Humphries 1997).

Using this method, Williams, Gaston, and Humphries found a clear pattern of increasing diversity with decreasing latitude (i.e., biodiversity increased from polar to tropical regions) and close correlation of the three indices with one another (0.949 to 0.991 in Spearman's rank correlation coefficients). Given the correlation of results, it is not surprising that the methods made similar identifications of regions with high biodiversity (fig. 4.13). Central and southern Columbia, Nicaragua, Oaxaca (southern Mexico), and southern peninsular Malaysia all were identified as biodiversity hot spots in at least two of the three methods.

This method has obvious faults, the most blatant of which is its exclusion of groups that are far more species rich than those chosen. It excludes, among other things, all insects, fungi, and bacteria. It is also not known if patterns of biodiversity at the family level are similar at higher levels, such as orders or classes, or at lower levels, such as genera or species. However, the indices derived do provide estimates of diversity that cover a broad range of groups of conservation concern, and the method can be applied to other groups as the reliability of data improves and as conservation efforts expand to be more intentional in including other taxa.

THE PROBLEM OF APPLICATION: HOW DO WE PRESERVE AND MANAGE BIODIVERSITY?

How Can Managers Index and Preserve On-Site Biodiversity?

When it comes to preserving biodiversity in the field, managers must often make quick decisions with incomplete information. What tools can they use to make the best possible decisions in less than ideal circumstances with limited time, resources, and information? Often managers must resort to the use of *indicators* of various kinds to make an assessment of the present status and future trends in biodiversity. Such indicators take various forms and can be assessed by various methods.

Biodiversity Indicator Species

An **indicator species** is one whose status is assumed to reflect the status of other species with which it shares the community. It has been defined precisely by Landres, Verner, and Thomas (1988) as "an organism whose characteristics, such as presence or absence, population density, dispersion, or reproductive success are used as an index of attributes too difficult, inconvenient, or expensive to measure." Indeed, exactly what an indicator species "indicates" varies in different contexts. No less than seven different kinds of "indicator species," each "indicating" something different, have been described by various authors (Lindenmayer, Margules, and Botkin 2000). The problem that managers must solve with respect to indicator species is twofold. First, can they identify an indicator species that indicates anything meaningful about biodiversity? Second, what aspect of biodiversity should they attempt to measure with an indicator species?

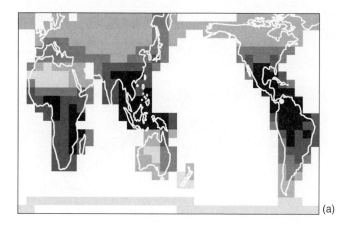

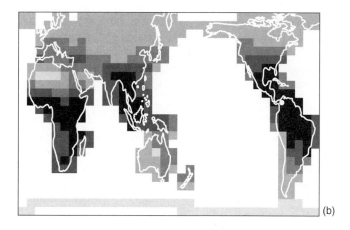

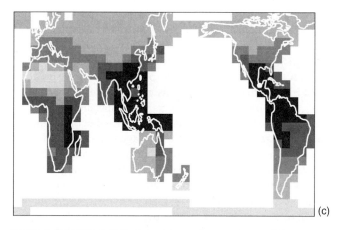

Figure 4.13

Maps of combined family richness of terrestrial and freshwater seed plants, amphibians, reptiles, and mammals worldwide on an equal-area grid map (grid-cell area ca. 611,000 km², for intervals of 10° longitude). Maps produced by summing (a) absolute family richness; (b) proportional family richness; and (c) proportional family richness weighted for species richness. Maximum scores shown in black, other scores divided into five grayscale classes of approximately equal size based on numbers of grid cells. Although units and numerical values differ, frequency classes are comparable among maps.

After Williams, Gaston, and Humphries (1997). Based originally on data from Gaston et al. (1995).

Taxon-Based Biodiversity Indicators

Taxon-based indicator species are those whose presence or absence is correlated with significant changes in biodiversity in a particular taxonomic group. Although an intuitively appealing and attractive concept, the search for such species has met with limited success. For example, in California's coastal sage scrub vegetation, a relatively rare and ecologically sensitive habitat that has diminished as a result of human development, Chase and associates (2000) attempted to determine if the presence of *any* individual bird or small mammal species found in this habitat at a particular sampling point was correlated with increased species richness at those same points. The results were disappointing. For one thing, any species found at all sites, all but one site, or only one site could not serve as an indicator species. These criteria eliminated 21 of 37 potential species, including two species of conservation concern, the loggerhead shrike (*Lanius ludovicianus*) and Stephens' kangaroo rat (*Dipodomys stephensi*). Both were detected at only one site. The species of greatest conservation concern in coastal sage scrub, the California gnatcatcher (chapter 2), also failed as an indicator of biodiversity. Thus, the authors concluded, "species of conservation concern cannot be assumed to be indicators of 'hotspots' of bird and mammal species richness in coastal sage scrub" (Chase et al. 2000). Among more common species, the presence of sage sparrows (*Amphispiza belli*) was correlated with decreased species richness in birds and increased species richness in mammals, but the actual difference in both categories was only one species. The presence of common yellowthroats was associated with higher bird species richness and the presence of the western scrub jay (*Aphelocoma californica*) with lower mammal species richness. However, there was no clear biological rationale as to why such correlations existed. The authors summarized the implications of their findings by saying, "efforts to conserve bird and small mammal biodiversity in coastal sage scrub should not focus exclusively on rare species . . . but instead should focus on a diverse suite of species that are representative of the range of variation in communities found in coastal sage scrub habitats" (Chase et al. 2000). Such a summary may accurately reflect the study's results, but advising managers to assess biodiversity by sampling absolutely everything forgets the original reason for trying to determine an indicator species in the first place. The problem remains. What alternatives exist?

Structure-Based Biodiversity Indicators

Structure-based indicators attempt to index changes in biodiversity through assessing changes in ecological structure (Lindenmayer, Margules, and Botkin 2000). Such indicators are particularly appealing in habitats where one or more elements, such as vegetation, are structurally complex. For example, in forest ecosystems, structural elements such as stand complexity and foliage height diversity have been used as indices of biodiversity. Other indicators, more applicable to a variety of habitats, include measures of connectivity and heterogeneity. Structure-based biodiversity indicators may be appealing because, in many cases, they can be shown to reflect changes in ecosystem processes or patterns. Although some are intuitively sensible (e.g., bird diversity has been repeatedly shown to be a correlate of foliage height diversity), few have been thoroughly tested.

Function-Based Biodiversity Indicators

Function-based indicators of biodiversity assume that, in any community, some species are "drivers" and some are "passengers" in the ecologic process. The drivers are active determinants of the characteristics of the ecosystem in which they live because of ecological functions that they perform in the system. The passengers "ride along" on the effects created by the drivers. The driver species, then, can potentially serve as what could be called *function-based indicators.*

The driver species are analogous to the more familiar concept of "keystone species." **Keystone species** are species that have disproportionate effects on community or ecosystem processes and, as a result, disproportionately affect biodiversity. The American bison (*Bison bison*) provides an excellent case history of the specific mechanisms through which keystone species mediate biodiversity levels by altering ecosystem function (fig. 4.14).

Bison as an Example of a Function-Based Keystone Species

As bison move in herds through a prairie, their urine and feces create elevated levels of nitrogen in distinct patches, altering the gradient of a vital plant nutrient, raising the level of habitat heterogeneity, and increasing overall biodiversity at both alpha and beta levels (Steinauer and Collins 1995). By consuming nitrogen stored as amino acids in plants and returning more labile forms of nitrogen to the soil in the form of urine and feces, bison increase the rate of nitrogen recycling in the system. This increase affects plant growth rates (Risser and Parton 1982), production (Steinauer and Collins 1995), efficiency of water use by plants, and plant community composition (Risser and Parton 1982). Overall, the urine and feces deposition of bison on prairies forms distinct habitat patches recognizable at both species and community levels (Steinauer and Collins 1995).

Bison also exert strong effects on biodiversity through grazing because they preferentially graze dominant C_4 grasses (grasses that use 4-carbon rather than 3-carbon pathways in photosynthesis) such as big bluestem (*Andropogon gerardii*), Indian grass (*Sorghastrum nutans*), switchgrass (*Panicum virgatum*), and little bluestem (*Schizachyrium scoparium*). Reduction in these dominant grasses increases densities and diversities of forb species that are a less important component of the bison's diet. Such effects are particularly important after a fire because C_4 grasses tend to dominate prairie sites after burning if grazing does not occur. In an experimental analysis of these effects, Hartnett, Hickman, and Fischer-Walter (1996) found an average of 40 species on sites grazed by bison, but only 29 species on ungrazed prairie sites.

In addition to effects mediated by grazing, bison affect alpha and beta diversity levels by trampling and wallowing. Wallows are small depressions in grasslands, often associated with moist sites, that are created when bison trample the ground and roll in the exposed soil (Polley and Collins 1984). Wallowing creates openings in grasslands with different soil textures, levels of soil moisture, pH, and available phosphorus,

Figure 4.14

The North American bison (*Bison bison*) is an example of a "keystone species" which, through a variety of effects including grazing, trampling, wallowing, and deposition of feces and urine, changes the biodiversity of a community, as well as its vertical and horizontal structure and heterogeneity.

among other variables, creating conditions that lead to increased beta diversity in the landscape. These environmental gradients occurring over a relatively small spatial scale lead to assemblages of plant species in the wallow that are very different from the surrounding prairie. However, the effects of bison activities favor increased diversity of prairie vegetation rather than invasion by nonprairie vegetation. The physical action of trampling and rolling in the wallow by bison tends to break and crush woody vegetation that could invade such sites, thereby reducing encroachment by woody vegetation into prairie communities (Polley and Collins 1984).

A third effect of bison occurs through significant interactions with other species. Bison and prairie dogs (*Cynomys* spp.) form mutually beneficial grazing associations in mixed-grass prairies. Feeding preferentially at the edge of prairie dog colonies or "towns," bison reduce forage biomass and density, making it more difficult for terrestrial predators to approach the town without being seen. Prairie dogs, through their grazing activities at the edge of the town, create microenvironments of new, more vigorous plant growth preferred by bison. The combined effects of bison–prairie dog grazing lead to higher shoot–nitrogen concentrations in plants than in the surrounding prairie (Kruegar 1986). Bison also use the town's interior, substantially devegetated by prairie dogs, for resting and wallowing (Kruegar 1986). These activities help keep the interior of the town free from vegetation, more amenable to burrowing, and more difficult for predators of prairie dogs to use without being detected.

Identifying keystone species allows conservation biologists to concentrate management efforts on those species that have the greatest effect on overall community and landscape biodiversity and to imitate and influence the processes and actions of keystone species that lead to increased biodiversity. Of course, simply saying that we should concentrate on conserving keystone species again fails to understand a manager's dilemma. Namely, how do we identify a keystone species and its effects so that we will know *what* ought to be conserved?

Ecological Redundancy and Function-Based Biodiversity Indicators

Conservation biologist D. H. Walker offers a four-step approach integrated in the concept of *ecological redundancy*. Walker asserts that the key question to ask is "how much, or rather, how *little*, redundancy is there in the biological composition of systems?" (Walker 1992). First, determine the functionally different kinds of organisms in the ecosystem. That is, what are the rate-limiting processes in the system, and which species are involved in which processes? Second, determine the number of species in each functional group (guild). If a functional group has only one or a few species (low redundancy) it should receive priority over a group that has many species (high redundancy). Third, examine the interactions among species in each guild. There is complete functional redundancy if the loss of one species is completely compensated for by an increase in density of another species. Finally, consider the relative importance of each functional group in ecosystem maintenance. Groups that perform functions considered more essential would be given conservation priority over groups that perform functions deemed less essential.

With appropriate background information, a function-based approach of identifying indicator species can make correct decisions about conservation priority on the basis of objective criteria. The weakness of such an approach is that it requires extensive and detailed information to correctly identify an ecosystem's functional groups and the species in them.

Even when all alternatives are considered, managers have no "magic indicators" that provide assessments of complex variables (biodiversity) with minimal effort (single-species or single-variable measurements). But managers must still manage, even if their information is incomplete. What are possible constructive approaches to the management of biodiversity even when available information is incomplete and imprecise?

POINTS OF ENGAGEMENT—QUESTION 2

It is not possible to assess, manage, or preserve all biodiversity in a system or management unit. It may not even be desirable. Based on available techniques and approaches, what do you now think would be the "best" way for a manager to evaluate biodiversity, and what criteria would you use for the manager to determine when "enough" diversity is being effectively managed and preserved?

Management Approaches to Biodiversity at Landscape Levels

Gathering Appropriate Background Data

A first step in conserving biodiversity is to appreciate a spatial scale perspective for alpha and beta diversity. Specifically, if the goal is to maximize local diversity at a single site (alpha diversity), choose a site with high species richness. But what if the management objective is to maximize biodiversity at a regional level (beta diversity)? If we change scales, the former strategy of simply reserving all the sites with the highest levels of species richness will have little value if the different areas contain the same species. In general, most locations, even sites with high species richness, are dominated by generalist species. Ecological specialists may occur at sites with relatively low species richness. But at the regional level, the key is not simply to select the sites with highest richness, but to select sites in which species compositions are *most dissimilar to one another* (Knopf and Samson 1994). By protecting areas of lower biological similarity (i.e., managing for increased beta diversity), a manager increases the degree of protection given to regional endemic species and to ecological specialist species. It is these species that contribute the most to biodiversity at regional and global levels.

Regional Biodiversity Management—Defining Functional Conservation Areas

Because the distribution of global biodiversity is complex, mapping and protecting key areas requires taxon-specific approaches, careful conceptual methods, and sophisticated technologies for spatial problem solving. A significant contribution in conceptual methods has been offered by Poiani and associates (2000) through the development of "functional conservation areas." By first examining biodiversity at different spatial scales (fig. 4.15), Poiani and associates (2000) define ecosystems and species at four different levels: local, intermediate, coarse, and regional. Within these levels, functional conservation areas (FCA) are identified. An FCA is "a geographic domain that maintains focal ecosystems, species, and supporting ecological processes within their natural range of variability" (Poiani et al. 2000). FCAs are delineated as sites, landscapes, and networks. Functional sites conserve one or more endangered species or rare ecosystems, typically at a local scale. Functional landscapes encompass full terrestrial and aquatic habitat gradients and a diversity of ecological processes needed to maintain those gradients and the species that live within them. Functional networks provide "spatial context, configuration, and connectivity to conserve regional scale species with or without explicit consideration of biodiversity at finer scales" (Poiani et al. 2000).

POINTS OF ENGAGEMENT—QUESTION 3

Recall from chapter 2 that many U.S. environmental and conservation laws are either species based (e.g., Endangered Species Act, Marine Mammal Protection Act) or impact based (National Environmental Policy Act, Clean Water Act). None are designed primarily to protect biodiversity *per se*. What would be the value of a conservation statute written explicitly to protect biodiversity? What form might it take, and why might it require "functional conservation areas" as management units to be effective?

Poiani's conceptual framework, admirably well conceived, is of little value without reliable information on species, habitat, and ecosystem distribution at multiple scales. Thus, conservation of biodiversity increasingly turns to geographic information systems (GIS) and the related Gap Analysis Program (GAP) (chapter 8) to supply information necessary for

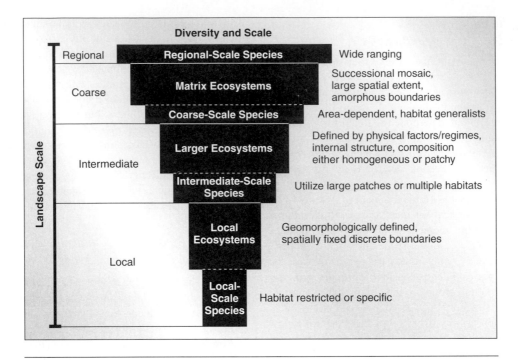

Figure 4.15

Biodiversity and scale. A method of categorizing biodiversity at regional, coarse, intermediate, and local geographic scales.

Modified from Poiani et al. (2000). © 2001 American Institute of Biological Sciences.

intelligent management and conservation decisions. GIS systems can organize and overlay thematic data, such as soils, vegetation, hydrology, and dominant vegetation, within a defined area, and then relate such data, along with land-use patterns and locations of existing nature reserves, to the distribution of endangered or endemic species (Scott et al. 1987). Using such a technique, conservation biologists and land-use planners can determine (1) what proportion of an area's biodiversity is protected in an existing distribution of nature preserves and under existing land-use practices, (2) whether such protection can be expected to permit the persistence of endangered or endemic species, and (3) the best location and arrangement of new nature preserves or the best areas to attempt to change current land-use patterns. Only when conservation biologists effectively integrate global patterns of biodiversity, taxon- and site-specific variations, well-organized conceptual frameworks for biodiversity protection, and technologically advanced data analysis and land-use planning toward the goal of protecting biodiversity at multiple scales is there reasonable hope that biological diversity will persist in any area or ecosystem.

Synthesis

Scientific concepts, measurements, and values are not set by decree. They survive and become accepted only if they are operational, testable, and open to analytical refinement. Biodiversity is one of the core concepts and motivating values for the science of conservation biology, but its persistence and value as a scientific idea are not yet assured. The conceptual definition of biodiversity requires thorough understanding, its mathematical definitions careful measurement, and its valuation rigorous analysis if the concept is to be translated into meaningful ideas that can shape conservation strategies. If biodiversity can be thoughtfully understood and articulated, quantitatively measured and tested, and carefully valued in both instrumental and noninstrumental ways, it will become an increasingly important component of conservation science, conservation law, and conservation policy.

The fundamental processes that control the level of biodiversity in ecological systems are still not well understood. The most effective future research on biodiversity will not be those studies that simply continue to measure it, but those that explore and test hypotheses about the ecological processes that shape it. Biodiversity is a concept that requires further refinement, and it cannot stand apart from other conservation priorities. Current measurements of biodiversity, with their emphasis on species richness and evenness, do not always reveal correlations between diversity and conservation value. New indices that address taxonomic uniqueness and ecological importance must be developed and used in conjunction with traditional measures of diversity if biodiversity is to provide meaningful information about the relative value of different community and landscape assemblages.

Biodiversity and Wilderness—A Directed Discussion

Reading assignment: Sarkar, S. 1999. Wilderness preservation and biodiversity conservation—keeping divergent goals distinct. *BioScience* 49:405-411.

Questions

1. Sarkar argues that biodiversity and wilderness conservation, although often conflated, are distinct conservation goals. What conflicts and potential setbacks might arise when conservationists use a wilderness preservation strategy to protect biodiversity? What are some reasons why wilderness areas might actually be lower in overall biodiversity than nonwilderness areas?

2. Sarkar objects to using a wilderness strategy to protect biodiversity. Would his objection also apply to strategies that depend on nature reserves to protect biodiversity? If not, what is the fundamental difference? If so, what strategies would complement the preservation of biodiversity through nature reserves?

3. Sarkar cites a number of cases in which human presence and influence have been beneficial for maintaining or enhancing biodiversity. In these examples, Sarkar finds " . . . no essential contradiction between social interests and biodiversity conservation." Are there "essential contradictions" between the goals of society and conservation in the United States? If so, identify the contradictions, and describe a conservation management strategy that could reduce them without resorting to reserves and wilderness areas.

4. If the exclusion of humans from biologically diverse areas is an insufficient and inappropriate method of protecting biodiversity on a global scale, conservation biologists must be well versed in disciplines other than biology to design alternate strategies. Which additional disciplines should they study, and with what kinds of professional specialists should they collaborate to envision and achieve conditions in which biodiversity is preserved in environments where humans live and work?

Learning Online

Visit our webpage at www.mhhe.com/conservation for case studies, animations, practice quiz questions, and additional readings to help you understand the material in this chapter. You'll also find active links to the following topics:

Biodiversity
Tropical Rainforests
Refuges and Sanctuaries
Extinction Issues
Tropical Forests and Extinction

North American Habitats and
    Extinction Issues
Conservation and Management of
    Habitats and Species
Endangered Species Act
Endangered Species

Legislation Regarding Endangered
    Species
Refuges and Sanctuaries
Species Abundance, Diversity, and
    Complexity

Literature Cited

Ackery, P. R., and R. I. Vane-Wright. 1984. *Milkweed butterflies.* London, England: British Museum.

Baev, P. V., and L. D. Penev. 1995. BIODIV Version 5.1.: Program for calculating biological diversity parameters, similarity, niche overlap, and cluster analysis, 2d ed. Sofia, Bulgaria: Pensoft.

Bessey, C. E. 1908. The taxonomic aspect of the species question. *American Naturalist* 42:218–24.

Callicott, J. B. 1989. *In defense of the land ethic.* Albany: State University of New York Press.

Callicott, J. B. 1994. Conservation values and ethics. In *Principles of conservation biology,* G. K. Meffe and C. R. Carroll and contributors, 24–49. Sunderland, Mass.: Sinauer Associates.

Chase, M. K., W. B. Kristan III, A. J. Lynam, M. V. Price, and J. T. Rotenberry. 2000. Single species as indicators of species richness and composition in California coastal sage scrub birds and small mammals. *Conservation Biology* 14:474–87.

Cody, M. L. 1986. Diversity, rarity, and conservation in Mediterranean-climate regions. In *Conservation biology: the science of scarcity and diversity,* ed. M. E. Soulé, 123–52. Sunderland, Mass.: Sinauer Associates.

Cook, O. F. 1899. Four categories of species. *American Naturalist* 33:287–97.

Cox, G. W. 1997. *Conservation biology: concepts and application.* Dubuque, Iowa: Brown.

Crawford, D. J. 1983. Phylogenetic and systematic inferences from electrophoretic studies. In *Isozymes in plant genetics and breeding,* Part A, eds. S. D. Tanksley and T. J. Orton, 257–87. Amsterdam, The Netherlands: Elsevier.

Cronquist, A. 1978. Once again, what is a species? In *Biosystematics in agriculture,* ed. L.V. Knutson, 3–20. Montclair, N.J.: Allenheld Osmun.

Darlington, J. P. 1957. *Zoogeography: the geographical distribution of animals.* New York: Wiley.

Davis. S. D., V. H. Heywood, and A. C. Hamilton, ed. 1995. *Centres of plant diversity: a guide and strategy for their conservation.* Vol. 2, *Asia, Australia, and the Pacific.* Cambridge, England: World Wildlife Fund and World Conservation Union.

Ehrenfeld, D. W. 1988. Why put a value on biodiversity? In *Biodiversity,* eds. E. O. Wilson and F. M. Peter, 212–16. Washington, D.C.: National Academy Press.

Fiedler, P. L., and S. K. Jain, ed. 1992. *Conservation biology: the theory and practice of nature conservation.* London, England: Chapman and Hall.

Gaston, K. J. , P. H. Williams, P. Eggleton, and C. J. Humphries. 1995. Large scale patterns of biodiversity: spatial variation in family richness. *Proceedings of the Royal Society of London* B 260:149–54.

George, T. N. 1956. Biospecies, chronospecies and morphospecies. In *The species concept in palaeontology,* ed. P. C. Sylvester-Bradley, 123–37. London, England: Systematics Association.

Goodall, D. 1973. Sample similarity and species correlation. In *Handbook of vegetation science.* V, *Ordination and classification of communities,* ed. R. H. Whittaker, 107–56. The Hague, The Netherlands: Junk.

Gottlieb, L. D. 1977. Electrophoretic evidence and plant systematics. *Annals of Missouri Botany and Gardens* 64:161–80.

Gottlieb, L. D. 1981. Electrophoretic evidence and plant populations. *Progress in Phytochemistry* 7:1–46.

Groombridge, B, ed. 1992. *Global biodiversity: status of the earth's living resources.* London, England: Chapman and Hall.

Guo, Q., and W. L. Berry. 1998. Species richness and biomass: dissection of the hump-shaped relationships. *Ecology* 79:2555–59.

Hartnett, D. C., K. R. Hickman, and L. E. Fischer-Walter. 1996. Effects of bison grazing, fire, and topography on floristic diversity in tallgrass prairie. *Journal of Range Management* 49:413–20.

Hempker, T. P. 1982. Population characteristics and movement patterns of cougars in southern Utah. Master's thesis, Utah State University, Logan.

Hill, M. O. 1973. Diversity and evenness: a unifying notation and its consequences. *Ecology* 54:427–32.

Hornberg, G., O. Zackrisson, U. Segerstrom, B. W. Svensson, M. Ohlson, and R. H. W. Bradshaw. 1998. Boreal swamp forests. *BioScience* 48:795–802.

Hunter, M. L., Jr. 1996. *Fundamentals of conservation biology.* Oxford, England: Blackwell Science.

Hurlbert, S. H. 1971. The nonconcept of species diversity: a critique and alternative parameters. *Ecology* 52:577–86.

Huston, M. A. 1979. A general hypothesis of species diversity. *American Naturalist* 113:81–101.

Huston, M. A. 1994. Biological diversity: the coexistence of species on changing landscapes. Cambridge, England: Cambridge University Press.

International Council for Bird Preservation. 1992. *Putting biodiversity on the map: priority areas for global conservation.* Cambridge, England: International Council for Bird Preservation.

Johnson, S. P. 1993. *The Earth Summit: United Nations Conference on Environment and Development (UNCED).* London, England: Graham and Trotman.

Jones, J. K., Jr., D. M. Armstrong, R. S. Hoffman, and C. Jones. 1983. *Mammals of the northern Great Plains.* Lincoln: University of Nebraska Press.

Jordan, W. R., III. 1994. "Sunflower forest": ecological restoration as the basis for a new environmental paradigm. In *Beyond preservation: restoring and inventing landscapes,* eds. A. D. Baldwin, J. De Luce, and C. Pletsch, 17–34. Minneapolis: University of Minnesota Press.

Kerr, J. T. 1997. Species richness, endemism, and the choice of areas for conservation. *Conservation Biology* 11:1094–1100.

Knopf, F. L., and F. B. Samson. 1994. Biological diversity—science and action. *Conservation Biology* 8:909–11.

Kruegar, K. 1986. Feeding relationships among bison, pronghorn, and prairie dogs: an experimental analysis. *Ecology* 67:760–70.

Landres, P. B., J. Verner, and J. W. Thomas. 1988. Ecological uses of vertebrate indictor species: a critique. *Conservation Biology* 2:316–28.

Leopold, A. 1966. *A Sand County almanac with essays on conservation from Round River.* New York: Sierra Club/Ballantine Books.

Leoschcke, V., J. Tomiuk, and S. K. Jain, ed. 1994. *Conservation genetics.* Basel, Switzerland: Birkhäuser Verlag.

Lindenmayer, D. B., C. R. Margules, and D. B. Botkin. 2000. Indicators of biodiversity for ecologically sustainable forest management. *Conservation Biology* 14:941–50.

Lopes, M. A., and S. F. Ferrari. 2000. Effects of human colonization on the abundance and diversity of mammals in eastern Brazilian Amazonia. *Conservation Biology* 14(6):1658–65.

Magurran, A. E. 1988. *Ecological diversity and its measurement.* Princeton, N.J.: Princeton University Press.

Margalef, R. 1968. *Perspectives in ecological theory.* Chicago: University of Chicago Press.

May, R. M. 1988. How many species are there on earth? *Science* 241:1441–49.

Mayr, E. 1969. The biological meaning of species. *Biological Journal of the Linnaen Society* 1:311–20

McAllister, D. E. 1991. What is biodiversity? *Canadian Biodiversity* 1:4–6.

Meadows, D. H. 1990. Biodiversity—the key to saving life on earth. *Land Stewardship Newsletter* (summer):4–5.

Mönkkönnen, M., and P. Viro. 1997. Taxonomic diversity of the terrestrial bird and mammal fauna in temperate and boreal biomes of the northern hemisphere. *Journal of Biogeography* 24:603–12.

Montgomery, C. A., R. A. Pollak, K. Freemark, and D. White. 1999. Pricing biodiversity. *Journal of Environmental Economics and Management* 38:1–19.

Myers, N. 1983. *A wealth of wild species.* Boulder, Colo.: Westview Press.

Myers, N. 1988. Threatened biotas: "hot spots" in tropical forests. *The Environmentalist* 8:(3)187–208.

Nei, M. 1972. Genetic distance between populations. *American Naturalist* 106:283–92.

Norris, S. 2000. A year for biodiversity. *BioScience* 50:103–7.

Norton, B. G. 1995. A broader look at animal stewardship. In *Ethics on the ark: zoos, animal welfare, and wildlife conservation,* eds. B. G. Norton, M. Hutchins, E. F. Stevens, and T. L. Maple, 102–21. Washington, D.C.: Smithsonian Institution Press.

Nupp, T. E., and R. K. Swihart. 2000. Landscape-level correlates of small-mammal assemblages in forest fragments of farmland. *Journal of Mammalogy* 81:512–26.

Paine, R. T. 1966. Food web complexity and species diversity. *American Naturalist* 100:65–75.

Paine, R. T. 1969. The *Pisaster-Tegula* interaction: prey patches, predator preference, and intertidal community structure. *Ecology* 50:950–61.

Palmer, J. D., and D. Zamir. 1982. Chloroplast DNA evolution and phylogenetic relationships in Lycoperiscon. *Proceedings of the National Academy of Sciences* 79:5006–10.

Peet, R. K. 1974. The measurement of species diversity. *Annual Review of Ecological Systematics* 5:285–307.

Petulla, J. M. 1988. *American environmental history,* 2d ed. Columbus, Ohio: Merrill.

Pielou, E. C. 1969. *An introduction to mathematical ecology.* New York: Wiley.

Pielou, E. C. 1975. *Ecological diversity.* New York: Wiley.

Pimm, S. L. 1998. Extinction. In *Conservation science and action,* ed. W. J. Sutherland, 20–38. Oxford, England: Blackwell Science.

Pimm, S. L., G. J. Russell, J. L. Gittleman, and T. M. Brooks. 1995. The future of biodiversity. *Science* 269:347–50.

Poiani, K. A., B. D. Richter, M. G. Anderson, and H. E. Richter. 2000. Biodiversity conservation at multiple scales: functional sites, landscapes, and networks. *BioScience* 50:133–46.

Polley, W. H., and S. L. Collins. 1984. Relationships of vegetation and environment in buffalo wallows. *American Midland Naturalist* 112:178–86.

Porter, W. F., and H. B. Underwood. 1999. Of elephants and blind men: deer management in the U.S. national parks. *Ecological Applications* 9:3–9.

Prendergast, J. R., R. M. Quinn, J. H. Lawton, B. C. Eversham, and D. W. Gibbons. 1993. Rare species, the incidence of diversity hotspots and conservation strategies. *Nature* 365:335–37.

Rabinowitz, D., S. Cairns, and T. Dillon. 1986. Seven forms of rarity and their frequency in the flora of the British Isles. In *Conservation biology: the science of scarcity and diversity,* ed. M. E. Soulé, 182–204. Sunderland, Mass.: Sinauer Associates.

Reid, W. V. and K. R. Miller. 1989. *Keeping options alive: the scientific basis for conserving biodiversity.* Washington, D.C.: World Resources Institute.

Risser, P. G., and W. J. Parton. 1982. Ecosystem analysis of the tallgrass prairie nitrogen cycle. *Ecology* 63:1342–51.

Rojas, M. 1995. The species problem and conservation: what are we protecting? In *To preserve biodiversity—an overview,* ed. D. Ehrenfeld, 35–43. *Readings from* Conservation Biology.

Cambridge, Mass.: Society for Conservation Biology and Blackwell Science.

Rosenzweig, M. L. 1991. Habitat selection and population interactions: the search for mechanism. *The American Naturalist* 137:S5–S28.

Rosenzweig, M. L. 1999. Heeding the warning in biodiversity's basic law. *Science* 284:276–77.

Sandlund, O. T., K. Hindar, and A. H. T. Brown, ed. 1992. *Conservation of biodiversity for sustainable development.* Oslo, Norway: Scandinavian University Press.

Scott, J. M., B. Csuti, J. D. Jacobi, and J. E. Estes. 1987. Species richness: a geographic information systems approach to the protection of biodiversity. *BioScience* 39:782–88.

Shannon, C. E., and W. Weaver. 1949. *The mathematical theory of communication.* Urbana: University of Illinois Press.

Shull, G. H. 1923. The species concept from the point of view of a geneticist. *American Journal of Botany* 10:221–28.

Sih, A. 1986. Predators and prey lifestyles: an evolutionary and ecological overview. In *Direct and indirect impacts on aquatic communities,* eds. W. C. Kerfoot and A. Sih, 203–24. Hanover, N.H.: University Press of New England.

Simpson, E. H. 1949. Measurement of diversity. *Nature* 163:688.

Simpson, G. G. 1951. The species concept. *Evolution* 5:285–98.

Simpson, G. G. 1961. *Principles of animal taxonomy.* New York: Columbia University Press.

Soulé, M. E. 1986. *Conservation biology: the science of scarcity and diversity.* Sunderland, Mass.: Sinauer Associates.

Steinauer, E. M., and S. L. Collins. 1995. Effects of urine deposition on small-scale patch structure in prairie vegetation. *Ecology* 74:1195–1205.

Stuessy, T. F. 1990. *Plant taxonomy.* New York: Columbia University Press.

Sylvester-Bradley, P. C. 1956. The new palaeontology. In *The species concept in palaeontology,* ed. P. C. Sylvester-Bradley, 1–8. London, England: Systematics Association.

U.S. Congress Office of Technology Assessment. 1987. *Technologies to maintain biological diversity.* Washington, D.C.: U.S. Government Printing Office.

Vane-Wright, R. I., C. J. Humphries, and P. H. Williams. 1991. What to protect? Systematics and the agony of choice. *Biological Conservation* 55:235–54.

van Jaarsveld, A. S., S. Freitag, S. L. Chown, C. Muller, S. Koch, H. Hull, C. Bellamy, M. Krüger, S. Endrödy-Younga, M. W. Mansell, and C. H. Scholtz. 1998. Biodiversity assessment and conservation strategies. *Science* 279:2106–8.

Walker, B. H. 1992. Biodiversity and ecological redundancy. *Conservation Biology* 6:18–23.

Weitzman, M. L. 1998. The Noah's ark problem. *Econometrica* 66:1279–98.

Whittaker, R. H. 1975. *Communities and ecosystems.* New York: Macmillan.

Whittaker, R. J. 1998. *Island biogeography: ecology, evolution, and conservation.* Oxford, England: Oxford University Press.

Wiley, E. O. 1978. The evolutionary species concept reconsidered. *Systematic Zoology* 27:17–26.

Williams, P. H., K. J. Gaston, and C. J. Humphries. 1997. Mapping biodiversity value worldwide: combining higher taxon richness from different groups. *Proceedings of the Royal Society of London* B 264:141–48.

Wilson, E. O. 1989. Threats to biodiversity. *Scientific American* 261:108–16.

Wilson, E. O. 1992. *The diversity of life.* Cambridge, Mass.: The Belknap Press of Harvard University Press.

Wilson, E. O., and F. M. Peter, eds. 1989. *Biodiversity.* Washington, D.C.: National Academy Press.

CHAPTER 5

The Historic and Foundational Paradigms of Conservation Biology

Theories . . . do not evolve piecemeal to fit facts that were there all the time. Rather, they emerge together with the facts they fit from a revolutionary reformulation of the preceding scientific tradition, a tradition in which the knowledge-mediated relationship between the scientist and nature was not quite the same.
—Thomas S. Kuhn, 1970

In this chapter, you will learn about:

1 the scientific paradigms that have shaped conservation biology.

2 the importance of these paradigms to solving major conservation problems, including:

a problems of conserving genetic diversity.

b estimation of minimum viable populations.

c applications of the theory of island biogeography to habitat fragmentation and reserve design.

d the management of metapopulations.

e managing communities as nonequilibrium systems.

WHY PARADIGMS MATTER

In his classic work *The Structure of Scientific Revolutions,* Kuhn describes how **paradigms** within a scientific community become the primary determinants not only of what is studied, but also of how it may be studied. Kuhn uses the term *paradigms* to describe scientific achievements that share two characteristics. Paradigms represent work "sufficiently unprecedented to attract an enduring group of adherents away from competing modes of scientific activity," and paradigms are "sufficiently open-ended to leave all sorts of problems for the redefined group of practitioners to solve" (Kuhn 1970:10). By these criteria, a small number of ideas and achievements in conservation biology qualify as paradigms. These guide the discipline's current efforts and direct the course of its future development. In Kuhn's words, paradigms define the "puzzles" a discipline must solve. The great asset of scientific paradigms is their ability to give coherence and focus to a variety of data that may at first appear unrelated. Their great liability is a tendency to exclude data and questions that the paradigm is not designed or equipped to answer.

In many ways, conservation biology represents a radical departure from antecedent disciplines in its outlook, methods, and questions; Kuhn would describe its emergence as a "paradigm shift." Yet conservation biology remains deeply, if sometimes paradoxically, indebted to its intellectual predecessors. As ecologist Daniel Simberloff put it, "conservation science did not spring fully armed from genetics and ecology as Athena did from the head of Zeus" (Simberloff 1988). When we examine the development of conservation biology as a scientific discipline, we see a pattern of conceptual transitions that are still embedded in a context of older scientific achievements.

The major paradigms in conservation biology are associated with conservation genetics, island biogeography, metapopulations (spatially subdivided populations of a single species), habitat heterogeneity, and the nonequilibrium theory of ecosystems. The purpose of this chapter is to show how the development of specific scientific paradigms in these areas contributed to the formation of conservation biology, how such paradigms determined and defined the important questions of conservation, and why questions of conservation changed when paradigms changed.

PROBLEMS OF GENETIC DIVERSITY

Conservation Genetics and the Rise of Conservation Biology: Historic Connections

The recent and historic crisis of species extinction, which was first perceived with alarm in the 1960s, soon became more focused on concerns about genetic diversity and genetic viability of populations. Specifically, biologists concerned with species conservation saw the fundamental question as, "Do populations suffer a significant genetic deterioration as a consequence of a sudden or gradual decrease in numbers?" (Frankel and Soulé 1981). In other words, is there a relationship between population size and long-term fitness or evolutionary potential?

Even as genetics was emerging as a major concern in conservation science, it was also thrust into a more prominent role through the effects of key conservation legislation. Until the 1970s, the science of taxonomy generated little notice outside academia. Species, subspecies, and other taxonomic units were determined using the morphological characteristics of living or preserved specimens, and often from small samples. Systematic and taxonomic uncertainties were great, but of little particular interest or relevance outside professional circles. All this changed with the passage of the Endangered Species Act (ESA) of 1973 (chapter 2). The act extended legal protection to species and subspecies, making taxonomic determination a matter of life and death, as well as the basis of legal protection and the criterion for legal challenge. Geneticist Stephen J. O'Brien, reflecting on the "innocence" of taxonomic science in the years before the ESA, wrote, "When taxonomic distinctions became the basis for legal protection afforded by the Endangered Species Act of 1973, this innocence was lost forever. Disagreements over taxonomic status fueled legal assaults on the Act, and misclassification led to inappropriate conservation measures resulting in losses of some species" (O'Brien and collaborators 1996). The rise of taxonomy as a legal basis for conservation was accompanied by a number of simultaneous advances in molecular genetics that provided more exact and quantifiable means of species and subspecies identification. These coincident developments made genetics a cornerstone of designating and protecting endangered species.

Molecular Genetics and the Assessment of Variation

Geneticist R. C. Lewontin, writing in 1991 about the changes he had observed in the field of genetics during his own career, said, "When I entered Th. Dobzhansky's laboratory as a graduate student in 1951, the problematic of population genetics was the description and explanation of genetic variation within and between populations. That remains its problematic 40 years later in 1991. What has changed is our ability to characterize variation at the genic and nucleotide level and, linked to the ability to give detailed descriptions of variation, the development of a theory of population genetics that takes into account the full implication of historical ancestries in real populations" (Lewontin 1991).

The technique that began the revolutionary transition from Mendelian genetics to molecular genetics was gel electrophoresis of proteins, the details of which will be covered in greater depth in chapter 6 (The Conservation of Genetic Diversity). It had long been known that a point mutation in the coding sequence of a molecule of DNA would result in an amino acid substitution in the protein synthesized from that sequence. Theoretically, that substitution should be detectable by analyzing the protein. The protein could then be compared with nonmutated proteins from other individuals and used as an index of genetic variation among them. But the laboratory methods of protein analysis were so time consuming and laborious as to be impractical for genetic surveys.

Protein electrophoresis made such analysis feasible. It was already known that a single amino acid substitution in a protein could change the physical configuration and charge of a protein enough to cause it to move at detectably different rates in a charged field. This movement could be observed visually in a gel by treating proteins with different types of stains (fig. 5.1). Differences in movement rates could be used to infer differences in amino acid sequences in the protein, and such differences could be treated as a legitimate index of genetic variation in different individuals. When such methods were adapted to large-scale surveys of genetic variation in natural populations, science had a powerful tool in hand for practical assessment of genetic variation at molecular levels (Lewontin 1991). More recently, such methods have been extended to direct analysis of DNA variation among individuals (fig. 5.1)

By itself, such a development need not have produced any important implications for conservation. Prior to the development of allozyme electrophoresis, geneticists debated the role and average levels of variation in natural populations. The

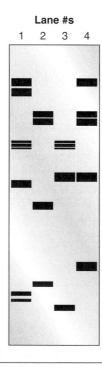

Lane #s

1   2   3   4

Figure 5.1

An example of 4 "DNA fingerprints" from different individuals generated by gel electrophoresis. Each lane represents one individual.

so-called balance school asserted that individuals in sexually reproducing populations were heterozygous at most loci, whereas the classical school believed that most loci were homozygous (Lewontin 1991). Experiments in Mendelian genetics had failed to resolve the argument, but electrophoresis confirmed high levels of variation and heterozygosity in most populations. This outcome had two important implications: (1) precise detection of genetic variation among individuals and populations, and (2) empirical evidence that such variation was common, or "normal" in most populations. Thus, electrophoresis supported the idea that "healthy" populations maintained high levels of genetic variability, apparently as a means of coping with recurrent environmental variation. Electrophoresis and other genetic techniques also played important roles in more precisely determining the genetic and taxonomic identities of endangered species and their relationships to more abundant, closely related species. Such genetic techniques, findings, and implications set the stage for genetics to rise to a preeminent role in the developing science of conservation biology.

Theoretical Population Genetics as a Basis for Conservation Ethics

In a paper remarkable for its prescience in foreseeing the explicit recognition of value as a basis for conservation and conservation biology, geneticist O. H. Frankel argued in "Genetic Conservation: Our Evolutionary Responsibility," that conservation of plant species was a necessary expression of social responsibility by the scientific community (Frankel 1974). According to Frankel, such responsibility was to express itself by accomplishing two goals, both genetic in nature: (1) achieve the "highest possible evolutionary potential" (i.e., genetic variation) in world plant gene pools, and (2) establish an "evolutionary ethic" that would "make it acceptable and . . . inevitable for civilized man to regard the continuing existence of other species as an integral part of his own existence" (Frankel 1974).

Although explicitly anthropocentric and, by today's standards of environmental ethics, rather shallow in its analysis and development of values, Frankel's remarks are notable because they advocated a clear connection between science and values in the pursuit of conservation and perceived this connection primarily in the field of genetics. Geneticists equated the preservation of genetic diversity with preservation of evolution, a process, Frankel argued, that was essential to the continuance of the human race. Thus, genetics was now equipped with the scientific technology, legal imperative, and value rationalization to assume a leading role in the development of conservation biology.

Effective Population Size and the Small Population Paradigm

As noted in chapter 1, conservation biology developed with two different, and sometimes antagonistic, paradigms regarding populations. The **"small-population paradigm"** focused on the characteristics associated with small populations and the factors that keep them small. The **"declining-population paradigm"** focused on factors responsible for population decline, irrespective of population size (Caughley and Gunn 1996). In the 1960s

and early 1970s, even before conservation biology emerged as a distinct discipline, genetic concerns about small populations were growing. Moore (1962) and Hooper (1971) considered problems associated with inbreeding depression that could arise in populations confined to refuges. However, Hooper's conclusions minimized the risk of inbreeding by asserting that small amounts of immigration could stem inbreeding depression and that inbreeding itself facilitated the adaptation of local population subunits (demes) to particular environments. Small and declining populations also raised increasing concern over the potentially deleterious effects of **genetic drift,** the random fluctuations in gene frequencies that occur as a result of nonrepresentative combinations of gametes created during breeding. Geneticist R. J. Berry raised the concern that loss of genetic variation through drift could limit future adaptation to environmental change, but ultimately concluded that natural selection would have sufficient strength to overcome the adverse effects of genetic drift (Berry 1971).

Despite such sanguine views about the potential problems of loss of genetic variation in small populations, later investigators examined the connections between genetics and conservation and came to radically different conclusions. Some of these were summarized in 1973 in Michael Soulé's classic review, "The Epistasis Cycle: A Theory of Marginal Populations," published in the *Annual Review of Ecology and Systematics*. Soulé noted six factors that account for loss of genetic variation in marginal populations: inbreeding, reduced gene flow with other populations, genetic drift, problems associated with **effective population size** (the size of an "ideal" population that would undergo the same amount of genetic drift as the actual population), reduced variation in niche width, and directional selection (Soulé 1973). Soulé demonstrated that, in populations of fruit flies (*Drosophila*), marginal populations rarely possessed novel or unique gene arrangements and many had reduced allelic diversity. Although Soulé made no explicit connections to conservation, he laid the foundation for what would become the principal genetic paradigm for concerns about small populations. Soulé wrote, "Marginal populations are . . . prone to severe reduction in numbers and can experience intermittent drift. Just a trickle of gene flow can, however, restore lost alleles; but, if the organism has poor dispersal powers, then marginal, isolated demes are expected to be allelically depauperate" (Soulé 1973).

Frankel and Soulé (1981) later collaborated to make a compelling case for the preeminence of genetic concerns in conservation in their landmark book *Conservation and Evolution*. There the authors provide detailed theoretical arguments to demonstrate that small populations experience reductions in genetic variation and allelic diversity and increased rates of inbreeding, leading to a loss of heterozygosity and fitness. Such problems are exacerbated in a demographic event known as a **"bottleneck"** (Frankel and Soulé 1981) (fig. 5.2). After a bottleneck, the remaining individuals represent only a sample of the original source population. The smaller the sample, the more likely that it may not be representative of the source from which it was taken, and the more certain that some alleles, especially the rarer ones, may have been lost. This loss of genetic variation can mean a loss of heterozygosity in the population, which may be correlated with a loss of overall

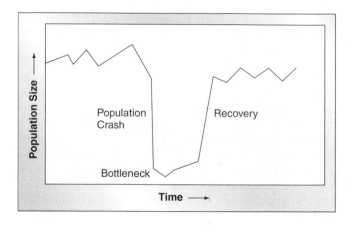

Figure 5.2

A graphical representation of population size before, during, and after a population bottleneck.

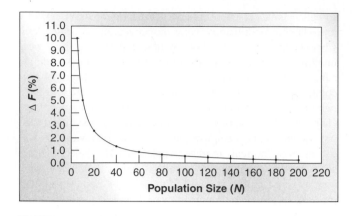

Figure 5.3

Percent change in the inbreeding coefficient (ΔF) at different population sizes. Note that the value of the inbreeding coefficient increases as population size declines.

After Frankel and Soulé (1981).

fitness (Frankel and Soulé 1981) because it exposes a greater proportion of recessive genes in a homozygous condition, and traits previously masked are now expressed. Many of these recessive genes have deleterious or even lethal effects on organisms. But other genetic problems also are associated with small populations. Little genetic variation is lost if the reduction in population size is temporary, but significant genetic variability will be lost if the population remains small for many generations (Frankel and Soulé 1981) (table 5.1). Additionally the less common alleles are in the original population, the more likely they are to be lost in a population bottleneck.

A second problem associated with the genetics of small populations is genetic drift. Recall that genetic drift represents the random fluctuations in gene frequencies that occur as a result of nonrepresentative combinations of gametes created during breeding. The proportion of such nonrepresentative matings tends to increase in small populations. Geneticist Ian Robert Franklin provides one of the clearest definitions of **genetic drift:** "In a finite population, the array of genotypes is formed by sampling gametes from the previous generation; virtually all

of the genetic effects which arise in small populations are an unrepresentative consequence of sampling, a process known as genetic drift" (Franklin 1980). Thus genetic drift is to genetics what sampling error is to statistics. The smaller the population, the greater the probability that the sample (random matings of individuals) may represent neither the average nor the range of characteristics found in the population.

A third concern regarding the genetics of small populations is inbreeding, the mating of individuals of above-average genetic relatedness. On the basis of theoretical arguments regarding inbreeding, Frankel and Soulé derived "the basic rule of conservation genetics," expressed as a percent change in the **inbreeding coefficient** (the probability that two alleles at the same locus in an individual are identical by descent). The "rule" asserted by Frankel and Soulé (1981:73) is that natural selection for performance and fertility can balance inbreeding depression if the change in the inbreeding coefficient (ΔF) is no more than 1% per generation (fig. 5.3). The significance of this

Table 5.1   Percent Change in Genetic Variation and Proportion of Rare Alleles Lost from a Population at Bottlenecks (Minimum Sizes) of Different Magnitudes. Values of p represent proportions of each of 4 alleles

NUMBER OF INDIVIDUALS IN SAMPLE (N)	PERCENT CHANGE IN GENETIC VARIATION	PROPORTION OF RARE ALLELES LOST	
		$p_1 = 0.70, p_2 = p_3 = p_4 = 0.10$	$p_1 = 0.94, p_2 = p_3 = p_4 = 0.02$
1	50.0	0.6300	0.7200
2	25.0	0.4950	0.6925
6	8.3	0.2125	0.5900
10	5.0	0.0925	0.5000
50	1.0	0.0025	0.1000
∞	0.0	0.0000	0.0000

Source: After Frankel and Soulé 1981.

rule is well stated by the authors themselves: "We refer to the 1% rule as the *basic rule of conservation genetics* because it serves as the basis for calculating the irreducible minimum population size consistent with the short-term preservation of fitness" (Frankel and Soulé 1981:73). Such short-term fitness preservation was considered safely achieved in most populations with an effective population size of 50 (Franklin 1980; Frankel and Soulé 1981). In contrast, long-term fitness was based on adaptation, measured most objectively in the ability of a population to speciate. Comparing long-term adaptive potential to short-term considerations regarding inbreeding, geneticist Ian Robert Franklin (1980) wrote, "In the long-term, genetic variability will be maintained only if population sizes are an order of magnitude higher" (i.e., 500). Franklin based his "500 rule" on data compiled by geneticist Russel Lande from mutation rates associated with *Drosophila,* maize, and mice. Lande (1988) later noted that the figure of 500 may be roughly correct to "maintain typical amounts of heritable variation in selectively neutral quantitative characters," but should never be used as a "blanket application to species conservation" and should not be incorporated into species' survival plans when other factors have not been considered that might require larger populations for effective persistence. Even as early as 1981, Frankel and Soulé, while citing Franklin's rule, noted that to accommodate continuing evolution, "the actual number of individuals that satisfy this criterion may be several times greater than 500" for a variety of reasons. They went on to argue against rule-of-thumb estimates and for continued genetic monitoring of endangered populations to determine if genetic variation was remaining at or above critical minimum levels.

Frankel and Soulé (1981) took genetic considerations into another critical dimension of conservation biology by addressing the relationship of genetics and nature preserves. Frankel and Soulé concluded that "it appears doubtful that existing nature reserves are large enough to support the process of speciation" based on estimates of speciation in different taxa on islands of varying sizes. Small mammals, for example, showed evidence of speciation in areas as small as 110,000 km^2, whereas larger vertebrates appeared to need a minimum of 600,000 km^2. Thus, early genetic paradigms in conservation biology addressed the question not only of how large a population must be to avoid extinction, but also of how large a reserve must be to permit speciation.

Despite its limitations, the "50/500 rule" was important to the development of conservation biology because it provided one of the first specific estimates of what constituted a "minimum viable population." Although the term **minimum viable population,** or MVP, would not appear in published literature until a year after Franklin's estimates (Shaffer 1981), it is clear, in retrospect, that Franklin's efforts represented a first attempt to answer the question of minimum numbers needed for population persistence. Today the 50/500 rule is of historical interest in understanding the conceptual development of conservation genetics, but is of little practical value. Subsequent studies revealed increasingly complex issues of genetics and demography, preventing any single rule of thumb from being used with certainty. In the area of genetics, long-term inbreeding depression has been demonstrated in populations with effective sizes of 50 to 500, and may occur in larger populations as well (Latter

Figure 5.4
Genetic samples of cheetahs (*Acinonyx jubatus*) have shown little genetic variation at sampled loci. Various studies have shown the cheetah to suffer higher-than-average rates of infant mortality, infertility, sperm abnormalities, and susceptibility to disease, all characteristics associated with high rates of inbreeding and low genetic variability.

and Mulley 1995; Frankham 1995). As geneticist R. Frankham (1995) stated, "No finite population appears to be immune from inbreeding depression in the long term." Thus genetics provided the first paradigms, albeit initially little more than rough guidelines, for estimating minimum populations needed for species persistence.

Genetics was not established as a preeminent paradigm in conservation biology only on the basis of theoretical argument. Soulé's earlier (1973) concern about some demes becoming "allelically depauperate" was dramatically supported with the publication of O'Brien and associates' 1983 *Science* paper titled, "The Cheetah Is Depauperate in Genetic Variation" (fig. 5.4). Most articles in scientific periodicals begin with objective, descriptive (and rather dull) titles, but the choice of such a vivid declarative sentence as a title in this investigation illustrated the authors' convictions about the veracity and significance of their findings. Specifically, an examination of 47 allozyme loci in 55 cheetahs (*Acinonyx jubatus*) from two populations revealed no polymorphic loci and an average heterozygosity of 0.0 (O'Brien et al. 1983). The authors attributed this genetic uniformity in cheetahs to past population bottlenecks followed by severe inbreeding, and supported their explanation with data showing that sperm counts were 10 times lower in cheetahs than in related felid species, and that 70% of the sperm were morphologically aberrant. Because the cheetah was a species with a wild population estimated at only 1,500 to 25,000 individuals, scientific interest was high and concern grew about loss of genetic diversity in wild populations.

In captive-bred cheetahs, infant mortality was nearly 30%, greater than for most other animal species. In skin tissue grafts among 14 cheetahs, 12 of which were between unrelated animals, all grafts were accepted beyond the rapid rejection stage, suggesting that the major histocompatibility complexes (MHC) of individual cheetahs were identical (i.e., the grafts were the equivalent of receiving tissue from a genetically identical individual) (O'Brien et al. 1985). The authors asserted that such lack of genetic variation contributed to high infant mortality and increased susceptibility to disease in cheetahs (O'Brien et al. 1985). Similarly, isolated and inbred population of lions (*Panthera leo*) in the Ngorongoro Crater in Tanzania underwent a population crash because of poor reproductive performance and high susceptibility to epizootics, providing further support for the possibility that inbreeding and loss of genetic variation were causes of population decline (Packer et al. 1991).

The findings of low or zero genetic variation in cheetahs raised understandable alarm. Some conservation biologists, such as Graeme Caughley and Anne Gunn, warned that linking low heterozygosity to low reproduction and high susceptibility to disease in cheetahs was "a risky conclusion" (Caughley and Gunn 1996) in light of subsequent analysis of the originally published data (Caughley 1994; Caro and Laurenson 1994). However, warnings about flawed experimental designs, inconsistent theoretical applications of ideas, and inappropriate measurement techniques and comparisons for heterozygosity (Caughley 1994) did not deter the growing conviction in the scientific community that low heterozygosity in declining populations was a significant factor in such decline. Genetic concerns thus emerged as a driving force in the development of conservation biology in the 1980s because of such empirical studies and because of more theoretical concerns associated with the risk of extinction in small populations that were aggravated by loss of genetic diversity and inbreeding depression (Caughley 1994). Today some population biologists argue that problems associated with inbreeding depression and other genetic factors have been overrated, and that some small populations show excellent long-term viability with no sign of negative genetic effects (Simberloff 1988; Walter 1990). Populations of concern to conservation biology, such as populations on islands, populations in fragmented habitats, and populations in zoos were examples that made such a paradigm compelling because it readily defined critical applications as well as important theoretical puzzles to solve.

POINTS OF ENGAGEMENT—QUESTION 1

What factors gave genetic concerns such an important role in the development of conservation biology? Did other related disciplines, such as wildlife biology, address these concerns in the same way? Why or why not?

MINIMUM VIABLE POPULATIONS AND POPULATION VIABILITY ANALYSIS

Because the problem of species extinction was a compelling concern in the genesis of conservation biology, a critical question emerged: in a world of limited resources for conservation and many sociopolitical boundaries, what was the minimum number of individuals needed in a population to ensure its survival? As noted earlier, the first attempts to answer this question were rules of thumb based on genetic considerations. Franklin (1980) asserted that inbreeding is kept to a tolerable level in populations with an effective size of 50 or more individuals and that effective populations of 500 or more tend to retain acceptable levels of heterozygosity. Franklin stated, "in randomly mating populations, such as are found in most mammals and birds, inbreeding considerations alone require that population numbers should be not less than 50 individuals." Other investigators found that greater threats to small populations lay in problems associated with random variation in birth rates, death rates, and other demographic variables (demographic stochasticity) and the effects of random environmental variation on the population's rate of increase (environmental stochasticity). Both demographic and environmental stochasticity often had increasingly deleterious effects as population size declined.

The lack of reliability, precision, and sensitivity to the demographics and environments of individual populations inherent in genetic rules of thumb motivated the search for more precise and comprehensive estimates of minimum population thresholds needed for population persistence. Out of this need arose the concept of "minimum viable populations" (Shaffer 1981) as estimates of the minimum number of individuals needed for the population to survive for a given period of time with a specified probability of persistence. A common convention that emerged for minimum viable populations was a 95% probability of persistence for 100 or 1,000 years. Concurrent with development of the concept of the minimum viable population, there was a systematic examination of the process of extinction as a series of interacting factors. These factors were perceived to have increasingly stronger effects in small populations, a perception that was eventually formalized in an interactive conceptual model that came to be called "the extinction vortex" (Gilpin and Soulé 1986).

The extinction vortex models of Gilpin and Soulé (1986) illustrated four routes of interactive factors that might lead to extinction. Two of these, the so-called F vortex and A vortex, are genetically based paths in which low population sizes lead to increased inbreeding, inbreeding depression, genetic drift, accumulation of deleterious mutations, and decreasing adaptive correspondence to the population's environment, all of which combine to lead to still lower population size (N_e) (fig. 5.5).

The F vortex is driven by the consequences of inbreeding depression in small populations, especially in the rise in so-called **"lethal equivalents"** in a population associated with deleterious homozygous recessive gene combinations causing direct mortality in some individuals (Ralls, Brugger, and Ballou 1979; Ralls, Ballou, and Templeton 1988). The increased expression of lethal equivalents and other consequences of inbreeding, including reduced reproduction, increased susceptibility to disease, and reduced growth rate lower the population's rate of increase (r), causing further and accelerating declines, and increasing the rate of inbreeding still further because of a smaller population size (N).

The A vortex is also a consequence of genetic drift and loss of genetic variation, but is manifested in a reduced ability of the population to adapt itself to a changing environment. As the effective population size declines, the efficacy of stabilizing and directional selection also drops. The result is a decreasing

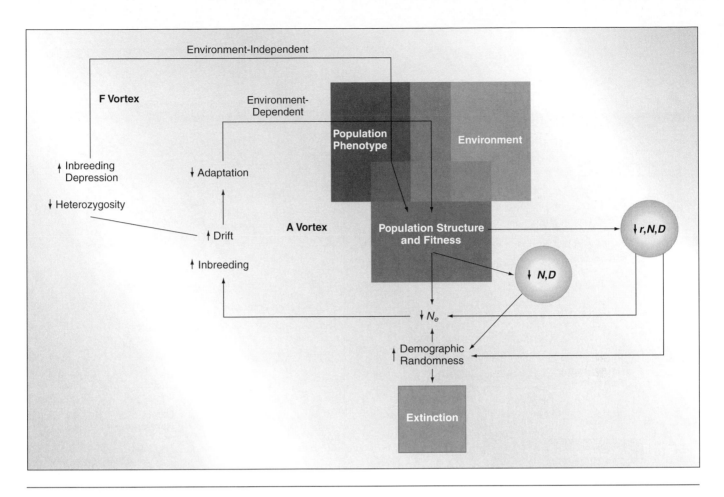

Figure 5.5

The F vortex and A vortex, two accelerating and degenerative cycles of population decline driven by an increasing level of inbreeding depression (F vortex) or a decreasing ability of the population to adapt to a changing environment (A vortex). Both are exacerbated in small populations. N is the population size, D is the population distribution, r is the population's instantaneous rate of increase, and N_e is the effective population size.

After Gilpin and Soulé (1986).

correspondence between the population's **phenotype** and its environment. As a result of this loss in fitness, both r and N decline further, increasing the effects of genetic drift and accelerating the loss of genetic variation to further exacerbate the lack of fit between the population and its environment. Thus begins a cycle of decline from which the population cannot recover.

The R vortex (fig. 5.6) reflects the importance of variations in the population's instantaneous rate of increase (r). Events associated with the R vortex begin when a chance lowering of population size (N) is coupled with increases in the variance of the population's instantaneous rate of increase. Now reduced to smaller numbers, the population is increasingly vulnerable to **environmental stochasticity,** which causes even greater variation in r. As Gilpin and Soulé (1986) put it, "the severity of the impact of a disturbance may be exacerbated by the current states of r and N, with a series of similar disturbances having progressively more serious consequences on the population."

The D, or "discontinuity," vortex (fig. 5.7) occurs when conditions described in the R vortex (lower N and increasing

variance of r) cause increased spatial fragmentation of a population, increasing rates of extinction in the now-divided population subunits. This fragmentation produces effective population sizes that may be a full order of magnitude lower than the census population size (Gilpin and Soulé 1986).

Although conceptual rather than mathematical, these models were useful in systematizing and categorizing the potential effects of many factors in the process of population decline. The extinction vortex models also provided a means for evaluating case histories of population decline and extinction and for understanding common, foundational principles that can be applied to specific cases. The associated concepts and estimation of minimum viable populations, equally important to the development of conservation biology, are explained more fully in chapter 7. Here we simply note that the MVP concept became more useful with the development of the technique called **"population viability analysis,"** or PVA, in which analytical or simulation models made precise estimates of the probability of species persistence within a defined

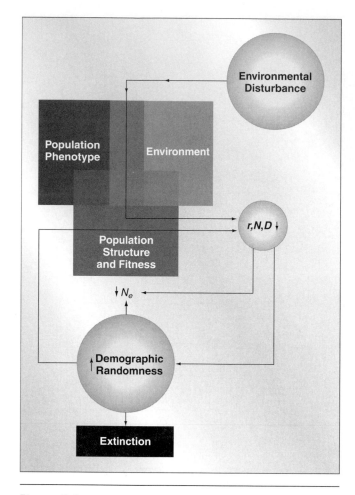

Figure 5.6
The R Vortex, an accelerating and degenerative cycle of population decline driven by increasing vulnerability to environmental disturbance at low population sizes. N is population size, D is population distribution, r is the population's instantaneous rate of increase, and N_e is the effective population size.

After Gilpin and Soulé (1986)

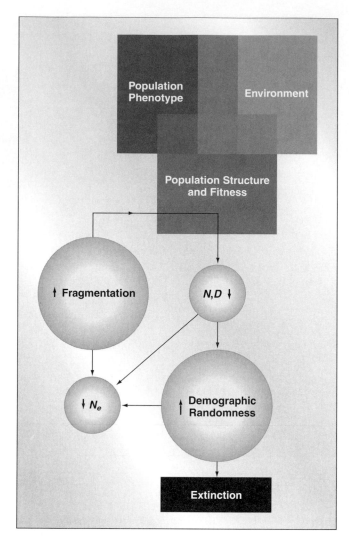

Figure 5.7
The D or discontinuity vortex, an accelerating and degenerative cycle of population decline driven by the fragmentation of the population into smaller and smaller subunits. N is population size, D is population distribution, and N_e is the effective population size. A lowering of N and an increase in demographic randomness can alter the spatial distribution of a population, introducing or increasing fragmentation. More fragmented distributions increase the likelihood of local extinctions.

After Gilpin and Soulé (1986).

time period at a given level of probability (i.e., uncertainty). PVA models, especially computer simulation models for individual species, quantified and evaluated the various factors of extinction described in the conceptual extinction vortex models, and ultimately replaced extinction vortex models because they were able to provide quantitative estimates of the probabilities of essential stochastic events in populations (Groom and Pascual 1998).

Genetics remains today at the forefront of conservation research and management, but it has been increasingly complemented by greater emphasis on the effects of environmental variation and population demography in determining the causes of scarcity, decline, and extinction of populations. However, conservation biology began with an emphasis on communities, not populations. To understand the shift of emphasis from communities to populations, we must review the development of the paradigm most responsible for the emergence of conservation biology as a distinct discipline. That paradigm was the theory of island biogeography.

THE THEORY OF ISLAND BIOGEOGRAPHY

Foundational Concepts of Island Biogeography Theory

In 1967, ecologist Robert H. MacArthur and zoogeographer Edward O. Wilson collaborated to produce a contribution to the Monographs in Population Biology series titled *The Theory of Island Biogeography* (MacArthur and Wilson 1967). Their stated purpose was "to examine the possibility of a theory of

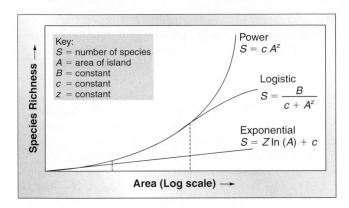

Figure 5.8

Species-area curves and their relationship. More than one equation can be used to develop species-area curves. Presented are three equations that can be used to generate a graphical representation of a species-area relationship, forming a species-abundance curve.

After Fangliang and Legendre (1996).

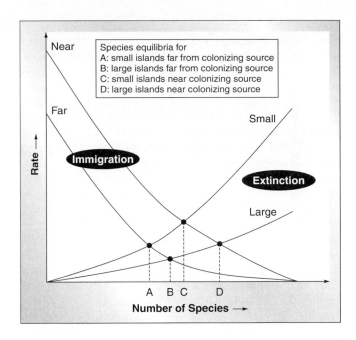

Figure 5.9

The equilibrium model of island biogeography predicts that numbers of species on an island represent an equilibrium between rates of immigration and extinction. Immigration rates increase with decreasing distance from an island's colonizing source. Extinction rates increase with decreasing area of the island. The four equilibria shown (A, B, C and D) depict different combinations of island size and distance from its colonizing source. The equilibrium theory of island biogeography predicts that large islands near a colonizing source will have more species than small islands far from a colonizing source.

Adapted from MacArthur and Wilson (1967).

biogeography at a species level" (MacArthur and Wilson 1967:5), and their work quickly became one of the most widely read and oft-cited publications of its time.

MacArthur and Wilson began with an assertion that was neither radical nor original, namely that the area of suitable habitat and the species diversity in a habitat vary directly. A graphical representation of a species-area relationship is called a **species-abundance curve** (fig. 5.8). Data from island flora and fauna demonstrated that the relation in such a curve could be summarized as the equation $S = cA^z$ (chapters 1 and 4). As described in chapter 4, this equation states that the number of species on the island is equal to the area of the island (A) multiplied by a taxon-specific constant (c) and raised to the power of a second constant, z. The constant z, often called the "extinction coefficient," is a factor that adjusts the slope of the species-area curve to integrate the rate of extinction of species on the island.

Building on this relationship MacArthur and Wilson demonstrated, with some actual data and considerable theoretical mathematics, not only that the species-area relationship could be predicted, but also that the number of species present could be estimated as a relation between immigration and extinction rates and the distance from the island to a colonizing source (mainland) (fig. 5.9). Increasing distance lowered the immigration rate, whereas decreasing distance raised it. Note that the curve falls down and to the right because, as more species become established, fewer immigrants belong to new species. Simultaneously, the extinction curve rises because the more species present, the more likely it becomes that any given species will become extinct, either because of low population size or ecological or genetic accident. MacArthur and Wilson envisioned that, in such a situation, the rate of extinction would increase exponentially with increasing numbers of species because "the combination of diminishing population size and increasing probability of interference among species will have an accelerating detrimental effect" (MacArthur and Wilson 1967:22). In such a view of island communities, immigration and extinction are counteracting forces that lead to a dynamic equilibrium of

species abundance determined by the size of the island and its distance from the mainland. Hence, MacArthur and Wilson's theory is properly called the **equilibrium theory of island biogeography.** Not surprisingly, the theory predicts that the number of species will be greatest on large islands near mainland colonizing sources.

Predictions of Island Biogeography Theory

The equilibrium theory of island biogeography made specific predictions about (1) expected times to extinction for established populations and for colonizing populations of different sizes (MacArthur and Wilson 1967:74–78), and (2) demographic traits that make particular species good colonizers (MacArthur and Wilson 1967:81). MacArthur and Wilson's theory also addressed the concept of invasibility (the ability of a colonizing species to "penetrate" an established community) and the role of the "variable niche" that can determine species abundance in island faunas. The latter concept was expressed in what the authors termed the **"compression hypothesis,"** which asserts that, on islands, as more species invade and are packed into the landscape, the niche occupied by each species becomes smaller. Island biogeography theory also addressed the role of "filtering" and "stepping stones" in explaining the diversity of island faunas. **Filtering** refers to the reduction in the number of species, genera, or higher taxonomic categories during dispersal. Filtering occurs when a certain group is unable to cross the barrier between the island and the mainland or when such a group arrives on the island but can find no suitable habitat. Filtering reduces the absolute diversity of the island fauna compared with the associated mainland because not all categories have an equal probability of success in reaching the mainland. **Stepping stone islands** are smaller islands that lie between the main recipient island and its mainland colonizing source. Such stepping stone islands can increase the rate of exchange between the recipient island and the mainland "by many orders of magnitude" (MacArthur and Wilson 1967:144).

The equilibrium theory of island biogeography held out the promise of being able to answer fundamental questions of essential conservation concern: (1) How many species can a given land area hold, assuming constant and known rates of immigration and extinction? (2) What is the estimated time to extinction of a given, established population within the designated area? (3) What is the likely persistence time of a colonizing population of a given size? (4) What will be the effects of barriers, distance, and established species on the diversity of the area's fauna? and (5) What will happen if smaller areas of suitable habitat exist between colonizing sources and recipient areas? By all Kuhnian criteria, island biogeography qualified as an important scientific paradigm. Its theoretical and management applications became a driving force in the early development of conservation biology.

Applications of Island Biogeography

Habitat Fragmentation

Although the equilibrium theory of island biogeography could not answer with equal precision all the questions that it raised, the fact that it could define such questions well and suggest how they could be answered made it attractive to biologists attempting to predict changes in species composition in fragmented habitats. The relationship of island biogeography to the problem of fragmented habitats was explicit from its inception. Even before the coining of the term **habitat fragmentation,** MacArthur and Wilson clearly had such a concept and its applications in mind. The authors wrote, "The same principles apply, and will apply to an accelerating extent in the future, to formerly continuous natural habitats now being broken up by the encroachment of civilization" (MacArthur and Wilson 1967:4). This statement is followed by a display of maps of woodland areas in southern Wisconsin from 1831 to 1950 (Curtis 1956) that implicitly but clearly depict the relict woodland fragments as "islands." One section titled "Habitat Islands on the Mainland," makes the analogy even more explicit (MacArthur and Wilson 1967:114–15).

Applications of the theory to terrestrial habitats followed quickly, especially in the design of nature reserves. Simberloff (1997) summarizes these developments well: "The original proposal of the equilibrium theory . . . envisioned 'island' as a metaphor for any insular habitat—a lake, a forest surrounded by fields, and the like. Thus it is hardly surprising that habitat

islands were prominent in the equilibrium theory literature. And because reserves are usually habitat islands, it is small wonder that the equilibrium theory was applied to conservation."

The equilibrium theory has been applied widely, from woodlots in agricultural landscapes (Temple and Cary 1988) to montane mammals living in isolated mountain ranges (McDonald and Brown 1995). Conservation biologists embraced island biogeography theory because it offered a coherent and comprehensive way of integrating the problems of habitat fragmentation and species extinction into predictive mathematical models. Island biogeography defined, in the truest Kuhnian sense, the puzzles conservation biologists ought to solve and how they ought to solve them. But nowhere was the equilibrium theory of island biogeography embraced more confidently than in the design of nature reserves.

Reserve Design

During the 1970s, the equilibrium theory of island biogeography experienced a continual growth in influence. Part of the theory's appeal was that it clearly addressed the growing concern over habitat fragmentation, a problem that has been called "*the central issue in conservation biology*" (Wiens 1996). As ecologist John Wiens wrote, "Island biogeography theory, by postulating a simple relationship between island (or fragment) isolation, area, and species number, offers a way to anticipate the magnitude of species loss as habitat patches become smaller and increasingly isolated by fragmentation. These relationships led to the formulation of several principles of reserve design . . . some of which have been implemented in practice" (Wiens 1996). As Wiens implies, a growing number of conservationists used the theory to suggest "rules" for the design of nature reserves. Such applications further strengthened the theory's status as the controlling paradigm of conservation biology. Ultimately, a comprehensive collection of such rules was adopted in the World Conservation Strategy (IUCN 1980). With such adoption, the theory of island biogeography became the foundation of reserve design and conservation planning.

The rules are easiest to understand and explain if displayed visually (fig. 5.10), but can be summarized as follows: (a) larger reserves are better than smaller reserves, (b) a single large reserve is better than a larger number of smaller reserves of equal total area, (c) closely spaced reserves are better than widely spaced reserves, (d) reserves arranged at equal distances from one another are better than reserves arranged in a linear sequence of increasing distance from the first reserve to the last reserve, (e) connected reserves are better than unconnected reserves, and (f) a circular reserve is better than a noncircular reserve (IUCN 1980).

These rules, though intuitively appealing, are not entirely consistent with or grounded in island biogeography. Rule a is actually derived from the historic species-area relationship— not from island biogeography theory per se. Likewise, the theory does not dictate rules b, c, d, or f. Rule f is based on the belief that minimizing the perimeter-to-interior ratio of a habitat is good, whereas rule c presupposes the development of the theory of metapopulations (to be discussed further in this chapter and in chapter 7). Rule e, the linking of reserves with corridors, is predicted by the theory because it assumes that

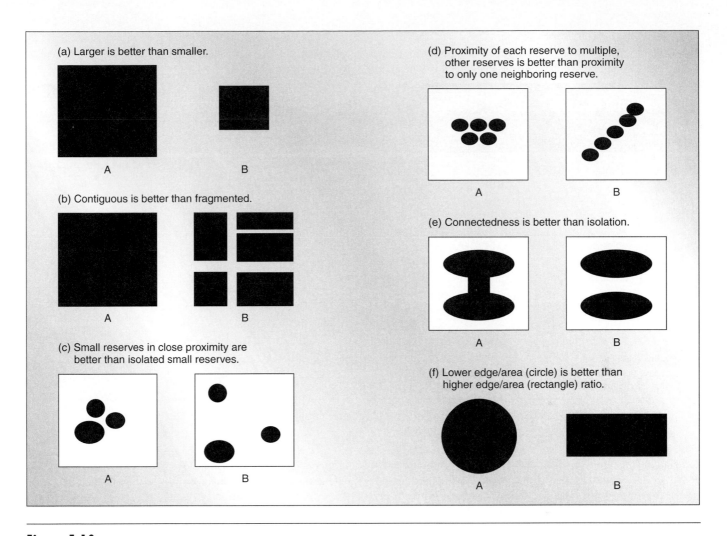

(a) Larger is better than smaller.

A   B

(b) Contiguous is better than fragmented.

A   B

(c) Small reserves in close proximity are better than isolated small reserves.

A   B

(d) Proximity of each reserve to multiple, other reserves is better than proximity to only one neighboring reserve.

A   B

(e) Connectedness is better than isolation.

A   B

(f) Lower edge/area (circle) is better than higher edge/area (rectangle) ratio.

A   B

Figure 5.10

A graphical representation of the "rules" of island biogeography applied to nature reserves. In each case, design A is considered superior to design B.

immigration rates will be raised if corridors are present, and was anticipated by MacArthur and Wilson's (1967) analysis of stepping stone islands (Simberloff 1997). Corridors will receive more detailed analysis in chapters 7 and 8.

Disaffection with Island Biogeography Theory in Conservation

By the 1980s, a growing number of studies had shown that local population extinction was not as frequent as island biogeography theory predicted. The contribution of new individuals and increased genetic diversity from conspecific immigrants tended to reduce extinction rates below what island biogeography theory had forecast, a phenomenon that came to be known as the "rescue effect" (Brown and Kodric-Brown 1977). The rules for reserve design proved to be unsupported by data in some cases and simply untrue in others. For example, single large sites did not always have more species than groups of small sites. Sometimes there was no difference at all. Sometimes the opposite was true. Simberloff and Gotelli (1984), for example, examined the species richness of plant communities in remnant prairies

and forests in the midwestern United States. Specifically, through iterative computer simulations, they made repeated comparisons of the plant species composition of a single large prairie remnant to a matching, randomly grouped set of smaller remnants of approximately equal area. Simberloff and Gotelli made the application to island biogeography theory explicit by calling the small remnant sets "archipelagoes" (an archipelago is a group of clustered islands) and by referring to the isolation of the remnants as "insularization." Although species richness was significantly related to area, groups of smaller sites consistently contained more species than single large remnants of the same size (table 5.2). No species was excluded from groups of small sites. And, when actual results were compared to results of a "random colonization model" that predicted the number of species that should be found in remnants of different sizes—smaller sites had more species than predicted while larger remnants had fewer (Simberloff and Gotelli 1984).

Inconsistencies between expected and observed results were not the only problem that emerged in applications of island biogeography theory. More troubling were things that the theory did not predict at all. For example, the theory of island biogeography made predictions about numbers of species

Table 5.2   Comparisons of species numbers (*S*) in single large remnants and clusters of smaller remnants (archipelagoes) of forests and prairies of approximately equal total area (ha).

SINGLE REMNANT		MULTIPLE REMNANTS		
AREA	*S*	Σ AREA	NUMBER OF REMNANTS	*S*
0.066	7	0.059	3	11*
0.096	8	0.092	3	14
0.329	16	0.322	2	16
1.750	17	1.749	8	22
3.473	24	3.203	2	28
96.000	23	93.062	8	31
3.810	22	3.650	2	47
8.300	32	8.210	2	49
10.890	39	10.460	3	49
12.270	20	11.970	2	49
30.660	33	30.670	4	54
75.200	27	73.600	9	76
0.790	10	0.780	2	11
1.600	10	1.600	3	18
3.800	16	3.780	3	21

*Note that in 14 of 15 comparisons, the cluster of small remnants contained more species than the corresponding single large remnant.
Source: Data from Simberloff and Gotelli 1984.

associated with areas of different sizes, but did not predict which species would be found in new areas, a key question in any conservation effort. The theory provided no information or predictive ability in regard to individual species and habitats and ignored important information regarding the natural history and autecology of particular species (Lack 1976). Likewise, it became increasingly clear that terrestrial habitats, even when severely fragmented, were not islands, and their populations were not entirely isolated. Enthusiastic applications of island biogeography theory to montane mammals (McDonald and Brown 1995), for example, were later severely criticized in subsequent reviews and analyses that revealed that the populations were not isolated and that some of the assumptions of previous analyses were based on incorrect information about population distribution and dispersal (Skaggs and Boecklen 1996; Simberloff 1997). There was a growing perception that a new paradigm was needed to account for population persistence in small areas, and to understand the role of small sites and small populations in conservation.

Continuing analysis of island faunas cast growing doubt on island biogeography theory's prediction of widespread local extinctions (Hanski and Simberloff 1997). Some biologists began to argue that the absence of particular species from individual islands did not represent an extinction event, but rather evidence of the occasional presence of fugitive populations characterized by high rates of temporal and spatial turnover. Detailed comparisons of the theory's predictions with empirical results of field studies were increasingly at variance. The growing number of mismatches between the expected and the observed—what Kuhn would describe as "anomalies"—eventually produced a Kuhnian "crisis" in the science of

conservation biology. Island biogeography was replaced by an alternative view of populations, dispersal, and habitats—the theory of **metapopulations.**

POINTS OF ENGAGEMENT—QUESTION 2

How does the theory of island biogeography view small populations? How does it view vacant islands or vacant fragments of habitats? How would these perceptions affect the theory's influence on the problem of reserve design? How would these perceptions affect the influence of the theory on determining conservation priorities?

THEORIES OF METAPOPULATION BIOLOGY

Origins and Development of Metapopulation Theory

As noted in chapter 1, population ecologists H. G. Andrewartha and L. C. Birch stated that "a natural population occupying any considerable area will be made up of a number of . . . local populations" (Andrewartha and Birch 1954). The accompanying schematic illustration depicted a series of spatially subdivided populations of a species with different densities in each subunit (fig. 1.8). This seemingly innocuous statement and illustration initially generated little attention, and less controversy, but it is probably one of the first expressions of the concept that a population may exist as spatially disjunct subunits at different densities in habitat patches of varying carrying capacity. Andrewartha and Birch also noted that individual subunits suffered periodic extinction, followed by recolonization by individuals dispersing from neighboring subunits. By making these population traits explicit, Andrewartha and Birch had functionally defined a *metapopulation,* even though the word would not appear in scientific literature for more than a decade. This concept of spatially subdivided populations, and the study of their properties of demography and dispersal, would become a foundational paradigm of conservation biology—the paradigm of metapopulations.

By the 1950s, the realities of habitat fragmentation and species extinction began to create a professional climate more receptive to a concept of fragmented populations, but the metapopulation paradigm did not immediately appear to address these changing conditions and problems. The view of populations as subdivided by interacting units was made more explicit by the biologist C. B. Huffaker and his coworkers in an elegant series of experiments involving mites and oranges in the late 1950s and early 1960s (Huffaker 1958; Huffaker, Shea, and Herman 1963). Huffaker's work, designed to evaluate dynamics of predator-prey relationships, used the six-spotted mite (*Eotetranychus sexmaculatus*) as the prey species, and another species of mite (*Typhlodromus occidentalis*) as the predator. Both species can sustain large populations on the skin of an orange, so Huffaker created "habitats" of oranges and "nonhabitats" of rubber balls placed in various combinations on a tray. Mites could not leave the tray, but could move among the balls and oranges (fig. 5.11). When the prey species was forced to feed in habitats (oranges) concentrated in large areas and grouped at adjacent, joined positions, predators exterminated prey within two weeks, whereupon all the predatory mites consequently starved. When the oranges were widely dispersed, prey survived longer, and predator and prey populations followed regular and predictable cycles of increase and decrease (fig. 5.12) (Huffaker, Shea, and Herman 1963).

To put this study in the proper context, we should note that Huffaker was not attempting to investigate the dynamics of metapopulations. There was no such paradigm, nor was Huffaker attempting to create one. Even the word *metapopulation* had not yet been invented. Huffaker and his colleagues designed their experiments to investigate predator-prey oscillations, specifically as elaborations of Gause's (1934) studies of predator-prey interactions predicted by the classic Lotka-Volterra equations (Berryman 1992). The Lotka-Volterra equations predict that predator and prey persist in the same environment through stable numerical oscillations, one rising as the other is falling (fig. 5.13). Gause had falsified such predictions with an experiment in a simplified laboratory environment that contained only one species of predator and prey. In Gause's test tube universe, the predator always exterminated the prey unless the prey had recourse to refuge habitat that the predators could not penetrate, in which case the predators starved (Gause 1934). Huffaker and his colleagues, seeking to determine if a stable oscillation between predator and prey populations was possible, created a more heterogeneous environment in which prey and predators were widely dispersed. In such an environment, persistence of both predator and prey, and the stability of their oscillating cycles of abundance, increased.

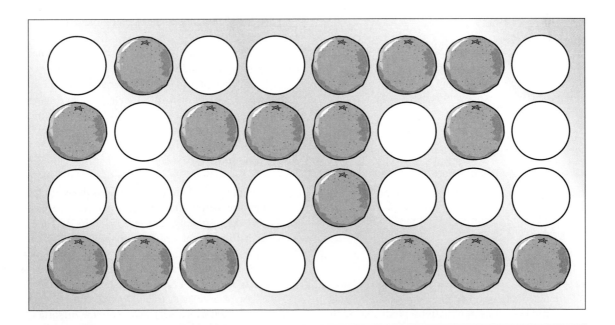

Figure 5.11

A diagrammatic representation of Huffaker's experiment on the persistence of a predator-prey system of two species of mite. Dark circles represent oranges that mites could colonize and white circles represent rubber balls of "nonhabitat" that they could not colonize.

After Huffaker (1958) and Huffaker, Shea, and Herman (1963).

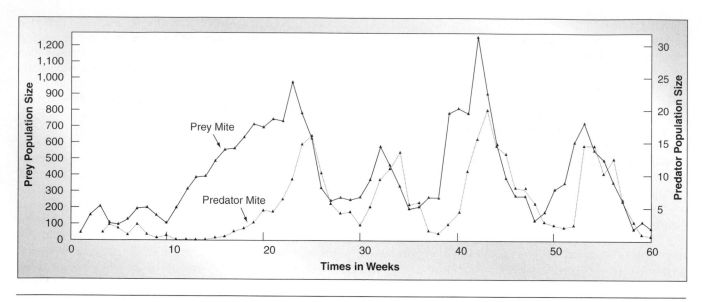

Figure 5.12

Oscillations in the densities of the predatory mite, *Typhlodromus occidentalis*, and its prey, the six-spotted mite, *Eotetranychus sexmaculatus*, in an experimental system of oranges (habitat for *Eotetranychus sexmaculatus* which feeds on oranges) and rubber balls (nonhabitat) over a period of 60 weeks. Note that predator and prey populations follow a series of regular, synchronized fluctuations.

After Huffaker, C. B., K. P. Shea, and S. G. Herman (1963).

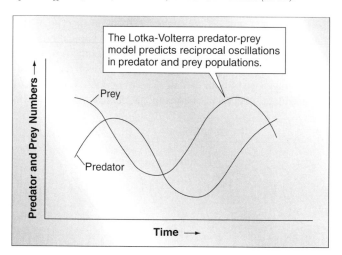

Figure 5.13

Stable oscillations of prey and predator populations predicted by the Lotka-Volterra equations. Note that prey increase when predators decline. Predators respond numerically to increasing prey densities with population growth, initiating declines in prey, which then lead to subsequent declines in predators. The cycle then iterates continuously.

Such experiments, and others that followed, demonstrated the importance of environmental heterogeneity in maintaining the stable predator-prey interactions predicted by the Lotka-Volterra model. They also demonstrated (somewhat inadvertently) that populations can, under certain circumstances, persist as "subpopulations" that occupy small or fragmented habitats on a temporary basis, and may move regularly from one habitat subunit to another. In such an arrangement, the population persists even though individual subpopulations

suffer extinction and only a portion of all available habitats is occupied at any one time. Unlike island biogeography theory, which defined area and distance from colonizing sources as the key variables, the universe created by Huffaker made distances and dispersion patterns among habitat fragments, along with species' dispersal abilities, the most important variables determining population persistence. Such a change in emphasis anticipated the shift from island biogeography to metapopulation theory and is crucial to understanding the reasons why that shift occurred.

While Huffaker was creating fragmented habitats in his laboratory by the random placement of oranges and rubber balls, less benign human activities in the outside world were creating fragmented habitats on a global scale. Habitat fragmentation was rapidly becoming the leading cause of species extinction and, as noted earlier, the central concern of the growing extinction crisis, which it remains to this day.

The process of fragmenting habitats inevitably divided formerly contiguous populations into spatially discrete population subunits. Recall rule c of reserve design: closely spaced reserves are better than widely spaced reserves. This rule acknowledges spatially discrete subunits and the effect that spatial distribution of associated habitat subunits may have on population dynamics. Rule d (reserves arranged at equal distances from one another are better than reserves arranged in a linear sequence of increasing distance from the first reserve to the last reserve) and rule e (connected reserves are better than disconnected reserves) also imply the importance of spatial arrangement in populations. Conservation biologists increasingly perceived that the generalities implied in these rules needed to be made more explicit and better defined. The explication of these rules into more precise conceptual and mathematical constructs led to the development of the metapopulation paradigm.

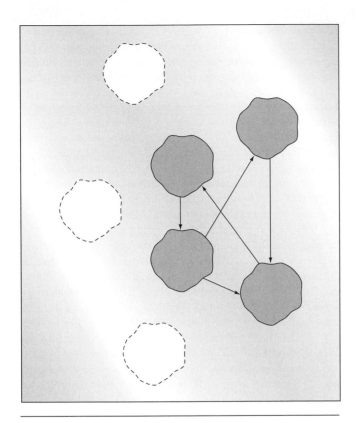

Figure 5.14

Levins' metapopulation model portrays extinction and migration patterns of populations existing as discrete subunits. Without recolonization of the habitat, each local population is in danger of becoming extinct. Arrows represent population recolonization. Solid circles are occupied habitats. Dashed circles represent populations that have become extinct because of the lack of immigrants.

After Harrison (1991).

The concept of metapopulations was an alternative to the traditional view of populations as panmictic, demographically homogeneous units. This alternative emerged as an explicit population biology model in the late 1960s and early 1970s. Levins (1970) offered the first definition of a **metapopulation**: "any real population [that] is a population of local populations which are established by colonists, survive for a while, send out migrants, and eventually disappear" (fig. 5.14).

Levins' theory of metapopulations arose from an examination of habitat heterogeneity and problems associated with the control of insects that damaged crops. In a paper presented at the symposium "Genetics in Biological Control" at the 1968 meeting of the Entomological Society of America, Levins stated that his purpose was "to show that the pattern of environmental variation in space and time can be utilized in the control of pests and to indicate the information which is needed for the selection of the most promising predator" (Levins 1969). Indeed, Levins practical objective was to determine the optimum properties of the predator population that could control the pest and then to produce such a population through genotypic selection. His purpose was to move entomologists away from the concept of thinking about "average" conditions and concentrate

instead on using specialized predators that would not be uniformly effective in all environments (Levins 1969).

In retrospect, it is ironic that what would eventually become one of the most important paradigms of conservation biology arose from concerns about agricultural pest control. But understanding its original context also helps us understand why the concept of the metapopulation, as Levins defined it, received little attention from conservation biologists until the early 1980s; it simply was not designed to address a problem of importance to conservation biology.

Conceptually, Levins envisioned a population separated into spatially discrete subunits (habitat islands). Individual population subunits suffered periodic and predictable extinction, but were recolonized by dispersers from neighboring subunits. Thus, population size was determined by the relationship between extinction and migration rates. If N represents the total number of local populations at a given time, T is the total number of sites that could support populations, and m is the migration rate (the probability that migrants from any given population can reach another site), then populations would be established according to the migration rate multiplied by the probability that the site reached is vacant:

$$mN\left(1 - \frac{N}{T}\right).$$

For example, if the probability that migrants can reach another site is 50% per year ($m = 0.5$) and there are 100 populations ($N = 100$) living in an environment with 200 suitable sites ($T = 200$), then the annual number of migrations (50) times the probability of encountering a vacant site ($1 - 100/200$, or 0.5) is 25 (i.e., 25 new populations will be established annually). However, some populations are eliminated by local extinctions with a probability of E and a rate of EN. Thus, the change in the number of individuals in the population over time will be

$$\frac{dN}{dt} = mN\left(1 - N/T\right) - EN.$$

N reaches equilibrium, N_{eq}, when the right side of the equation is 0. Therefore the population will reach equilibrium at

$$N_{eq} = T\left(1 - \frac{E}{m}\right).$$

Persistence of the population requires that $m > E$, and equilibrium is reached at a population size at which E and m are equal. When the population is large, changes in the rate of extinction of subpopulations have relatively little effect. But if the population experiences a more general, overall decline, changes in E begin to have significant effects on population size. The value of E is almost certain to be variable because it will be affected by random environmental fluctuation. If E varies over time, then the value of N will never reach an equilibrium, but rather will fluctuate within some range of values according to a given probability distribution. And if the extinction rate ever exceeds the migration (colonization) rate, the population will disappear (Levins 1969).

Like the theory of island biogeography that preceded it, Levins' metapopulation model meets Kuhn's essential criteria for being a paradigm. Specifically, the Levins model took existing data and gave them a new framework of interpretation. By treating space as a discrete entity, it defined an important fundamental question to be answered: How does one distinguish habitat from nonhabitat (the matrix) for the species of interest? (Hanski and Simberloff 1997). In practical terms, the Levins metapopulation model asserted that a key preliminary task of an empirical metapopulation study is to determine the important differences between habitat and nonhabitat and delimit the number and area of suitable habitat patches within a study area.

Levins' model also assumed that local (subunit) population dynamics were density dependent, that population dynamics in different patches were independent of one another, and that there was limited dispersal linking population subunits. Additionally, the original model assumed that all patches were of similar size and quality. There was no spatial correlation (clumping) of the patches, all patches were equally available to dispersers, the number of patches was very large, local populations were not affected by dispersal, and patches were modeled as either "occupied" (at carrying capacity) or "unoccupied" (no individuals in the patch) (Wiens 1996). The last assumption is the reason that Levins' model eventually came to be called the **"occupancy model"** (Gilpin 1996) to distinguish it from other types of metapopulation models that developed later.

The original Levins model represents what is today referred to as a **"spatially implicit" model** of metapopulations. That is, habitat patches and local populations were discrete, but all were assumed to be equally connected to one another (Hanski and Simberloff 1997). Spatially implicit models, because of their elegance and simplicity, facilitated mathematical and conceptual analysis of how metapopulations might work. Unfortunately, spatially implicit models were unrealistic, and their dependence on other assumptions about populations limited the questions that could be asked.

More recent efforts in the modeling of metapopulations have increasingly relied on **spatially explicit models,** which assume differing degrees of connectedness between population subunits and feature "localized interactions." Localized interactions are those in which population subunits interact primarily or exclusively with neighboring subunits, not with all subunits. A further refinement in metapopulation modeling has been the development of **"spatially realistic models"** (Hanski and Simberloff 1997) that include considerations of the specific geometry of particular patches (especially on issues of size, shape, and arrangement of patches). Other types of metapopulation models and their assumptions will be considered in detail in chapter 7.

The assumptions of the metapopulation paradigm are not obviously different from those of island biogeography. Both paradigms view extinction and colonization of population subunits (whether islands or habitat "patches") as fundamental population processes determining population persistence. In addition, both paradigms see the world as a collection of patches of suitable habitat surrounded by a matrix of hostile nonhabitat and assume that a given individual uses only one patch, and that each patch has a distinctive local population (Andrén 1994).

Despite its similarities with island biogeography, and despite the fallacies of some of the assumptions of early metapopulation models, the metapopulation concept proved a more compelling alternative to conservation biologists for several reasons. Fundamentally, metapopulation theory shifted emphasis from communities to populations as the fundamental units of conservation, a view that many conservationists found more appropriate and more manageable than the community-based emphasis of island biogeography. Metapopulation theory's conceptualization of populations as spatially discrete subunits in fragmented, yet still connected, habitats offered a picture of what biologists perceived to be the case in nature. Biologists realized that any plan for maintaining extant populations would have to incorporate the preservation of many habitat fragments, rather than rely exclusively on the large, contiguous habitat blocks. Further, conservationists perceived that population persistence also would depend on the ability of individuals to disperse successfully among habitats (McCullough 1996).

Perhaps the most important difference between the two paradigms, and the greatest reason that metapopulation theory eventually prevailed over island biogeography theory, was the difference in the way the two paradigms viewed the significance of local extinctions and small populations. Biologists observed that as real habitats were fragmented, some species disappeared from the fragments. Although the equilibrium theory of island biogeography acknowledged that extinctions on islands could be replaced by recolonization from a mainland source, such recolonizations might be relatively rare events widely separated in time. In contrast, metapopulation theory suggested that vacant habitats might be recolonized on a regular basis as part of normal patterns of dispersion and population growth. On this point, metapopulation theory has significant differences from island biogeography. First, metapopulation theory suggests that unoccupied habitat is as important as occupied habitat in long-term population persistence. Metapopulation theory provides incentives for maintaining suitable habitat, even if fragmented, and even if some fragments are not occupied by particular species. In addition, metapopulation theory holds out the hope that a fragmented group of population subunits could actually enhance population structure and persistence. Finally, metapopulation theory acknowledges an important role for small and fragmented habitats and their associated populations in overall population persistence. Because of its emphasis on the species-area relationship, the theory of island biogeography devalues small sites. In contrast, metapopulation theory rescues them (Simberloff 1997).

The theory of metapopulations articulated a process that could stem the growing wave of extinctions, and initial modeling results were encouraging on this point (Hanski and Gilpin 1991). Spatial structure, underemphasized in island biogeography theory, became a key concept of metapopulation theory and modeling (Hanski and Simberloff 1997). Modelers of metapopulations also found that they could relax the initial assumptions of the Levins (1970) model and make subsequent metapopulation models more realistic (Gilpin 1996). By the early 1990s, the metapopulation paradigm had effectively replaced the equilibrium theory of island biogeography as the governing perspective in conservation biology (Hanski and Simberloff 1997).

Refinements of Metapopulation Theory

As metapopulation theory and modeling continue to develop, they increasingly rely on two key premises: (1) populations are

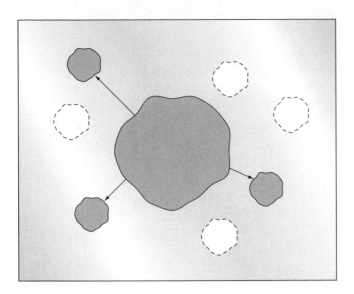

Figure 5.15

The mainland-island metapopulation model as proposed by Boorman and Levitt (1973). The Boorman-Levitt metapopulation consists of individuals inhabiting a large source area and several small sinks. The large "mainland population" is capable of supporting its own population in addition to supplying immigrants to smaller islands. Thus, the mainland serves as the primary source of inhabitants and is relatively extinction resistant. Arrows represent recolonization of sinks by individuals from the mainland. Dashed circles represent sinks in which extinction occurs without recolonization.

After Harrison (1991).

spatially structured into assemblages of locally breeding populations, and (2) migration among local populations has some effect on local population dynamics, including the possibility of population reestablishment following extinction (Hanski and Simberloff 1997). Subsequent to the development of the occupancy model of metapopulations (Levins 1970), Boorman and Levitt (1973) produced an alternative metapopulation model sometimes referred to as the "mainland-island metapopulation" model (fig. 5.15). In this model, one population subunit is significantly larger and more permanent than all others, and serves as the primary "source" population for smaller subunits. The "mainland" population never goes extinct; therefore, the metapopulation never suffers extinction (Hanski and Simberloff 1997). Frequent dispersal from an extinction-resistant mainland to extinction-prone "island" populations prevents all small populations on the "islands" from suffering extinction at the same time (Harrison 1991).

Harrison (1991) elaborated a classification scheme for metapopulation models in four categories (fig. 5.16). Besides Levins' model (renamed the "classical model" by Harrison, fig. 5.16a) and the mainland-island model previously discussed, Harrison proposed two other types of metapopulations. One is the "patchy" model, in which migration among subunits is so frequent that the patches function as a single demographic unit. The other is the "nonequilibrium" model, in which movement among

the subunits is so limited that each subunit functions as a separate population unit. Extinction is not offset by recolonization, and the population suffers a long-term decline.

A further contribution to the metapopulation paradigm came in the form of the concept of source-sink habitats, originally developed by Pulliam (1988) (fig. 5.17). Recall that in Levins' original model, all habitat patches were of equal quality. In source-sink models, patches differ in their ability to support populations; therefore, subunit populations will exhibit habitat-specific demographic rates. More specifically, source-sink theory proposes that populations exist in heterogeneous habitats that include areas in which population surpluses are produced (**sources**) and areas in which the population cannot replace itself without immigration (**sinks**). The sink populations occupy habitat patches that do not support them for long, especially in the face of environmental variation. Source populations not only exist longer, but fare well enough for individuals to recolonize the other patches. If all this is true, it means that the relative quality of each habitat and the distribution of individuals among habitats of differing quality are the major determinants of overall population dynamics (Pulliam 1988; Kadmon 1993).

Thus, source-sink metapopulation models draw conservationists' attention to variable habitat quality and to the fact that population survival may be dependent on a few key habitats, not always those in which the species is most abundant (Caughley and Gunn 1996; Hoopes and Harrison 1998). Examinations of source-sink systems in theoretical modeling have been enlightening (Danielson 1991), but the concept presents unique challenges for experimental design and testing in the field. One noteworthy attempt to examine the process with real populations in variable habitats was the effort of Ronen Kadmon on the desert annual grass *Stipa capensis* in Israel. *Stipa capensis* grows on slopes, depressions, and wadis (desert drainage channels that are usually dry). The vast majority of its habitat is on slopes, but most of its seed production occurs in wadis. Further, the grass's ability to produce seed and the proportion of the seed that germinates are highly rainfall dependent, and the environmental response is habitat dependent. In average (dry) years, seed production in the wadis accounts for more than 90% of the abundance of *Stipa capensis* on slopes. Increased rainfall has the greatest effect on slope populations and the least effect on wadi populations, leading to far higher contributions of seed production and germination from slope habitat in wet years than in dry ones. Thus, rainfall has a determining effect on whether a particular habitat is a "source" or a "sink" in a given year. Reflecting on the complexities of these populations in a heterogeneous environment, Kadmon noted, "a very small portion of the total area may maintain a large and disproportionate portion of the whole population. This finding points to the importance of identifying the types of habitats available, their relative abundances, and the between-habitat distribution of the individuals when studying the dynamics of natural populations" (Kadmon 1993). Thus, although the source-sink concept may be a fundamentally sound way of looking at habitat distribution in natural populations, the environmental interactions associated with source-sink population dynamics may be extremely complex.

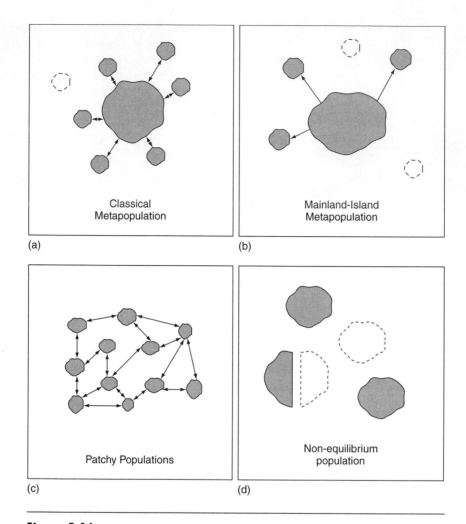

Figure 5.16

Types of metapopulation models. In a *classical metapopulation,* (a) some colonies may not exhibit high rates of movement for long periods of time. Also, colonization may unite several patches within a larger patch as a single entity that contributes to other sinks. Colonies farthest from the source are most prone to extinction. The *mainland-island metapopulation* (b) depicts local extinctions occurring mainly among a subset of populations. The mainland/source, resistant to extinction, functions as the major provider of colonists. The island and sink metapopulations have little effect upon regional persistence. In *patchy populations* (c), because of the high levels of emigration and immigration, the patches function as a single unit. It is rare that discrete local populations become extinct. The absence or insufficiency of recolonization to balance extinction distinguishes *nonequilibrium populations* (d). Extinction of metapopulations occurs as part of an overall regional decline (i.e., a product of the reduction, fragmentation, or deterioration of a habitat).

After Harrison (1991).

Models of source-sink habitats illustrate certain key concepts not explicit in earlier expressions of metapopulation theory. One of these, clearly evident in Kadmon's study, is the idea that certain habitats may be disproportionately important to populations, and that this may not be immediately evident from species density or abundance within the habitat (Hoopes and Harrison 1998). Source-sink models also have demonstrated that the qualities of poor habitat can affect the viability of populations in good habitat. Finally, source-sink models have shown that relatively small changes in the ratio of good to bad habitat can cause relatively large changes in species viability and population persistence. This theory, and the models and experiments derived from it, reveal that the relative abundance and dispersion of source and sink habitats may be strong determinants of population density, and even of the outcome of species interactions, such as competition (Danielson 1991).

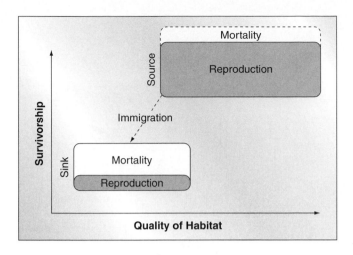

Figure 5.17
A visual representation of the source-sink model of habitat distribution. In source habitats, reproduction produces a population surplus (i.e., mortality does not decrease the number of individuals because of overcompensation through reproduction). Surplus individuals move to sink habitats where mortality exceeds survivorship. Sink habitats cannot be maintained by reproduction, but depend on immigration to maintain a population.

Figure 5.18
The Florida scrub jay (*Aphelocoma coerulescens*) demonstrates a pattern of spatial distribution and population demography corresponding to the concepts and predictions of metapopulation theory.

A Metapopulation Case History: The Florida Scrub Jay

One species that has been used to classify, develop, and test metapopulation models is the Florida scrub jay (*Aphelocoma coerulescens*), Florida's only endemic bird (fig. 5.18). This species provides an excellent opportunity for metapopulation modeling because of its habitat requirements and natural history. Scrub jays are habitat specialists that prefer low-growing, scrub oak vegetation (*Quercus* spp.) interspersed with bare openings on sandy, nutrient-poor soils. Such habitat can be maintained only by frequent fires, and tends to be patchily distributed among large areas of other types of vegetation, unsuitable and uninhabited by jays (Woolfenden and Fitzpatrick 1984).

Scrub jays are monogamous, cooperative breeders that establish well-defined territories in their preferred habitat. It is common for younger males to remain with the parents as "helpers" for one or more years rather than dispersing immediately. Such helpers typically increase the family's overall breeding success, help expand territory borders, and increase their own chances of "inheriting" part or all of the parental territory in subsequent years (Woolfenden and Fitzpatrick 1984). Given this combination of reproductive behaviors and habitat preferences, Florida scrub jays are a textbook example of a metapopulation, a group of spatially discrete population subunits with high costs of dispersal among habitat patches.

Data on population distributions of the Florida scrub jay have revealed more complexity than traditional classification systems could incorporate. Smith and associates (1996) delineated 42 Florida scrub jay populations (fig. 5.19). Twenty-one of these were classified as nonequilibrium (small and extinction prone subunits), three conformed to the classical (Levins) model, three fit the patchy model of Harrison, and five matched the mainland-island model. To deal with new permutations seen in the remaining 10 populations, Smith and associates (1996) added the category of "midland" subpopulations, groups that were not necessarily prone to extinction but were not invulnerable to extinction. With the addition of the midland category, Smith and associates (1996) categorized 9 of the remaining 10 populations as midland-island populations and one as a mainland-midland population. Further studies of *in situ* populations may benefit from insights gained by metapopulation modeling, but are also likely to provide additional examples of complex population structures that do not fit neatly into model classifications.

Metapopulation Genetics

Prior to the articulation of the concept of spatially divided populations by Andrewartha and Birch, a genetic basis for the same concept had been proposed by Sewall Wright in the 1930s (Wright 1931). Although Wright did not use the term *metapopulation,* he did propose a "shifting balance theory" of natural selection in which small, subdivided populations (demes) achieved high levels of local adaptation to changing environments through (local) natural selection, genetic drift, migration (among subunits), and interdemic selection (i.e., local extinctions of less-fit demes and colonization of new or vacated areas by more-fit demes). The problem that Wright was attempting to solve was how novelty arose in a constantly changing environment. The

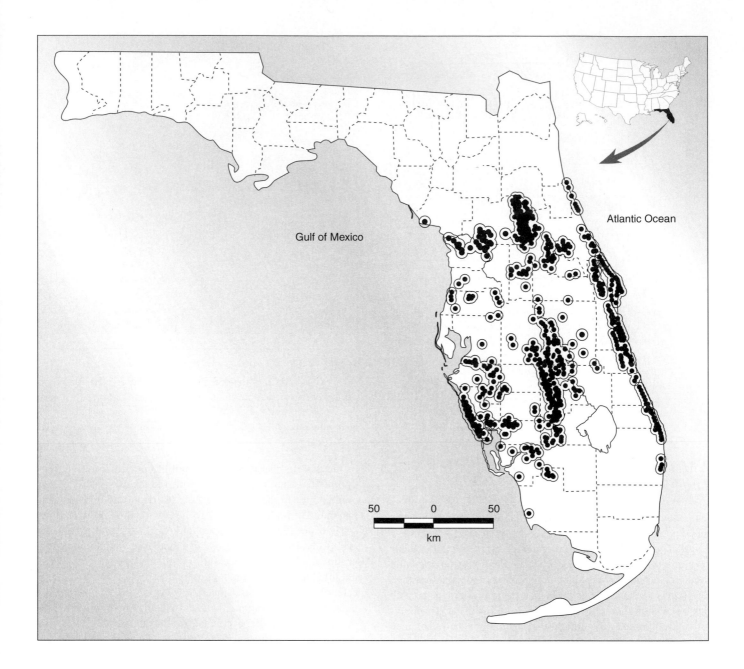

Gulf of Mexico

Atlantic Ocean

50   0   50
km

Figure 5.19

Florida scrub jay metapopulation distribution map. Dark areas represent scrub jay subpopulations. The outer lines designate separate metapopulations.

Smith et al. (1996:203).

small, subdivided populations served as "nature's many small experiments" (Wade and Goodnight1998), and speciation resulted as a by-product of local adaptation when it produced reproductive isolating mechanisms. It was this subdivision that prevented the averaging of environmental variation into a single genetic optimum (one size fits all) for the entire population. In other words, genetic optima shifted spatially and temporally in different population subunits.

Historically, the scientific alternative to Wright's shifting balance theory was R. A. Fisher's theory of large population size (Fisher 1958), which was an attempt to explain how existing adaptations were refined in a slowly changing environment (Wade and Goodnight 1998) (table 5.3). Fisher saw the world as a collection of large, interbreeding (panmictic) populations in which adaptation occurred primarily through mutation and natural selection. Genetic drift and migration were at best inconsequential, and at worst counterproductive to adaptation. Whereas Wright saw genetically subdivided populations of multiple and varied fitness peaks, Fisher envisioned fitness as a global average for the entire population.

Table 5.3 **Essential Differences Between Wright's Shifting Balance Theory (SBT) and Fisher's Theory of Large Population Size (LPS)**

CONCEPT	WRIGHT (SBT)	FISHER (LPS)
Central Problem of Evolutionary Theory	The origin of adaptive novelty in a continually changing environment.	The refinement of existing adaptation in a stable or slowly changing environment.
Major Process of Evolutionary Change	A combination of local natural selection, random genetic drift, migration, and interdemic selection contribute to evolutionary change.	Both mutation and natural selection contribute to evolutionary change.
Ecological Context of Evolutionary Change	Change occurs in small, subdivided populations.	Change occurs in large, randomly breeding populations.
Genetic Basis of Evolutionary Change	Epistasis and pleiotropy; allelic effects are context dependent.	Additive genetic effects; allelic effects are context independent.
Process of Speciation	An inevitable by-product of local adaptation in epistatic systems.	Selection is disruptive or locally divergent.

Source: After Wade and Goodnight 1998.

Although both theories found support, the weight of opinion traditionally favored Fisher's theory as the simpler explanation for empirical data from natural populations, which were assumed to be panmictic. If populations had apparent separation and spatial diversity, they still faced no real barriers to migration, exchange, or gene flow. However, by the 1960s, the realities of habitat destruction and fragmentation, combined with increasing concern over growing rates of species' extinctions, led researchers to question the generality of contiguously distributed, panmictic populations.

Wright's shifting balance theory has profound implications for the maintenance of biodiversity as well as for the dynamics of metapopulations. Wright proposed that evolution might proceed rapidly in spatially structured populations, especially if local extinctions and recolonizations occurred (Hanski and Simberloff 1997). Wade and Goodnight (1998) note that "Wright imagined that the membership of most species was distributed into small, semi-isolated breeding groups." In such a system, random genetic drift and selection within demes become strong evolutionary forces. Far from being detrimental to the population, random genetic drift and epistasis become potential sources of additional genetic variance by fueling additional adaptation to local conditions within metapopulations (Wade and Goodnight 1998) and can create more genetic variance per generation than mutation (Wade 1996).

The patterns of recurrent local extinctions and recolonizations characteristic of metapopulations significantly affect the genetic structure within and among local population subunits (demes), and additional genetic models of metapopulations have been formulated to explain and predict the intricacies of metapopulation genetics. The nature of the genetic effect depends upon specific recolonization patterns. The propagule pool model, a genetic analogy to the previously discussed mainland-island model of metapopulations, assumes that all colonists are drawn from a single extant deme in the metapopulation. The migrant pool model assumes that colonists to a new deme are drawn randomly from the entire metapopulation (Slatkin 1977). In both models, local extinctions cause decreases in the genetic diversity of the metapopulation, both within and among demes, but genetic diversity is maintained at higher levels in the migrant pool model (Pannell and Charlesworth 1999). Such higher levels of diversity occur because sites colonized by individuals from different demes approximate the genetic diversity of the entire metapopulation, rather than the genetic diversity of only one deme.

The predictions of these models have been supported by studies of real metapopulations, reinforcing the assumption that genetic diversity within and among demes is strongly influenced by characteristics and methods of dispersal. Hydrothermal vents offer a classic metapopulation example because the vents are highly specialized, discrete, patchily distributed, and surrounded by nonhabitat of more typical ocean floor. Deep-sea invertebrates often disperse along rifts in the ocean floor where vent activity is prominent. Vrijenhoek (1997) found that indices of gene flow and genetic differentiation declined with increasing distance between populations in some species that had poor dispersive abilities, such as the giant tube worm (*Riftia pachyptila*), a result consonant with island biogeography and metapopulation theory. Distance had less effect on species with greater dispersive abilities, such as the mussel *Bathymodiolus thermophilus*, whose mobile larvae disperse widely through ocean currents.

POINTS OF ENGAGEMENT—QUESTION 3

Are genetic considerations more important in metapopulation theory or island biogeography theory? How might genetic considerations be addressed in an actual conservation effort involving a metapopulation?

Habitat Implications of Metapopulation Theory: Theories of Habitat Heterogeneity

Just as island biogeography theory tended to devalue small sites and small populations, it also devalued the importance of habitat in population persistence and ignored habitat considerations in its rules for reserve design. This omission was a contributing factor in its demise. As conservation biologist Daniel Simberloff complained years earlier, "unwarranted focus on the supposed lessons of island biogeography theory have detracted from the main task of refuge planners, determining what habitats are important and how to maintain them" (Simberloff 1986). Metapopulation theory envisions a biological world of spatially discrete populations and habitat patches, surrounded by inhospitable and unoccupied areas of "nonhabitat." It is small wonder, then, that the development of metapopulation theory has placed a premium on the measurement and prediction of *habitat heterogeneity* (differences in habitat). With the increasing interest in metapopulation dynamics and the development of source-sink models, theories of habitat heterogeneity have become one of conservation biology's most active research fronts.

Heterogeneity can be defined as *any form of variation in the environment, including physical and biotic components.* Variation can be spatial or temporal, fixed or dynamic, and steady state or transient. For example, the classic "shifting mosaic" model of forest succession (fig. 5.20) described by Bormann and Likens (1979) predicts that a forest reaches a steady state (defined as no net change in biomass over time) not

Figure 5.20

A "shifting mosaic" forest of uneven age distribution created by the differing rates of death among the dominant age class of trees.

Figure 5.21

The California roach (*Hesperoleucus symmetricus*, not pictured) and steelhead (*Oncorhynchus mykiss*, pictured), two species whose ecological effects as predators change markedly according to the relative heterogeneity of aquatic habitat, even over very fine spatial scales.

because the forest is homogeneous, but because it is heterogeneous. On individual sites, the death of old, dominant trees takes place constantly but unpredictably. The successional process on each such site may be at a different stage, but, when seen in the landscape perspective of the entire forest, the overall effect is a dynamic equilibrium in which there is no net change (accumulation) in biomass. Bormann and Likens called such a state a "shifting mosaic" because the forest is composed of different "textures" (i.e., age classes) of vegetation on individual sites that change in distribution over time.

As ecologists and conservation biologists focused greater attention on the issue of heterogeneity, they discovered that it played an important role in ecological function. For example, in stream ecosystems, changes from boulder to gravel substrate altered the relative success of predatory fish, such as California roach (*Hesperoleucus symmetricus*) and steelhead (*Oncorhynchus mykiss*) (fig. 5.21). On gravel substrates, predation by these species had relatively little effect on ecological function. Prey species achieved greater success in escaping from predation by using the interstitial spaces present in gravel that are absent in boulders, reducing the foraging efficiency of the predators. In addition, prey species in gravel tended to be more cryptic than those in boulders. Over boulder substrates, predatory fish suppressed densities of damselfly nymphs and other small predators, which in turn released algae-feeding chironomids (*Pseudochironomus richardsoni*) from predation. The chironomids, increasing in density, then greatly reduced standing crops of algae. These effects could be documented over a distance of centimeters in the stream substrate (Power 1992).

The reality of habitat heterogeneity may change long-held views of some of the most basic processes of population regulation, such as density-dependent population regulation. As the density of a population increases, the incidence of intraspecific agonistic encounter and interference is assumed to rise, resulting in a hostile social environment that leads to reduced breeding success or lower survivorship of offspring,

Figure 5.22

The imperial eagle (*Aquila adalberti*), a species whose population demography is affected by the degree of habitat heterogeneity in its environment.

particularly among territorial species (Lack 1966; Fretwell and Lucas 1970). In this situation, it is assumed that all breeders are equally affected and, although average fecundity is depressed, the amount of variation of fecundity among individuals does not change. But what if density dependence is driven by habitat heterogeneity? Then breeders in low-density populations would all have access to high-quality breeding habitat, but as the population increased, more and more breeders (especially younger individuals or new arrivals) would be forced to breed in lower-quality habitats. Breeders on high-quality sites would continue to have high fecundity, but those on poor sites would have much lower fecundity, and the variation in fecundity among individuals would increase.

Ferrer and Donazar (1996) evaluated these distinctive predictions in a growing population of imperial eagles (*Aquila adalberti*) (fig. 5.22) in Spain's Doñana National Park. The population of eagles grew for 16 years, but remained stable during the next 16 years. The number of fledglings per pair declined over the entire period, and there was an inverse relationship between population size and productivity. Productivity on the first six (original) territories in the population remained high and stable throughout the study, but productivity on more recently established territories was lower and much more variable. Overall, variation in productivity increased as average productivity declined (Ferrer and Donazar 1996). In years of poor overall breeding success, pairs in "good" territories contributed 90% to 100% of all new birds to the population. Thus, the data evaluated by Ferrer and Donazar support the hypothesis that habitat heterogeneity, not an increased frequency of agonistic encounters, creates a density-dependent population response. The large contributions of new birds from good sites, particularly in poor years, also suggest that a removal of a portion of high-quality habitat would affect the eagle population size in both the good habitat and the poorer habitat because the latter may depend on the production of individuals in the former (Ferrer and Donazar 1996).

Habitat heterogeneity also may be useful in estimating the variability and extinction probability of metapopulation subunits. In a study of bush cricket (*Metrioptera bicolor*) populations in southern Sweden, Oskar Kindvall determined that the variability of cricket population subunits in grassland patches separated by pine forests was negatively correlated with the degree of grassland habitat heterogeneity (fig. 5.23). In other words, the more heterogeneous the habitat, the less variable the number of bush crickets from year to year. In addition, over the course of six years, local (site-specific) extinctions of population subunits were concentrated on sites with low habitat heterogeneity, small area, or both (Kindvall 1996). Similar results have been observed for the bay checkerspot butterfly (*Euphydryas editha bayensis*) found in the San Francisco Bay area of California. In this population, extinctions have been observed on large habitat patches that lack topographic heterogeneity. Ehrlich and Murphy (1987) suggested that butterfly larvae survive best on wetter, north slopes during most (normal) years, but do better on drier, south-facing slopes in wet years. Similarly, Kindvall (1996) suggested that bush cricket populations do better in dense, tall grass in dry years, but survive better in sparse vegetation during wet years. If the interpretations of these studies are correct, the best choice for nature preserves may not necessarily be the largest sites, but

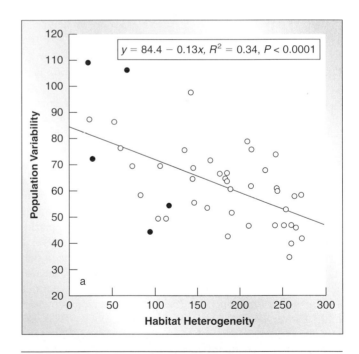

$$y = 84.4 - 0.13x, R^2 = 0.34, P < 0.0001$$

Figure 5.23

Populations of bush cricket (*Metrioptera bicolor*) subunits exemplify that population size is less variable as heterogeneity increases. Dark circles indicate patches where local extinctions occurred. White circles indicate patches with extant populations. Population variability was measured by the coefficient of variance (cv) of local population size, and habitat heterogeneity was measured using digitized infrared aerial photographs. Each patch was assigned values according to how much the patch deviated from the standard level of gray in the photographs (SD-hue).

After Kindvall (1996).

the sites that provide the greatest heterogeneity within occupied habitats and the greatest diversity of total habitats.

To understand heterogeneity, scientists must pay attention to the problem of scale, which was once a nonissue in biology. Ecologists in general, and conservation biologists in particular, now realize that most of the processes that determine species abundance and population dynamics, particularly in metapopulations, are scale dependent. To a wood thrush (*Hylocichla mustelina*) flying through an even-aged temperate deciduous forest, forested sites may appear quite uniform, and the habitat relatively homogeneous. But to a wood-boring beetle living in a rotting log beneath the thrush's flight path, habitat is uniform and homogeneous only within the log. Where the log ends, suitable and unsuitable habitat are demarcated by an abrupt boundary (the border of the log and the forest understory). Failure to perceive and respect such boundaries would lead, from the beetle's scalar perspective, to a fatal error in judgment, and to a new residence in the belly of the thrush. Thus, as conservation biologists strive for increased precision in understanding habitat needs of threatened and endangered species, the study of ecological scale has become and is likely to remain an issue of vital concern in conservation biology for the foreseeable future.

THE NONEQUILIBRIUM PARADIGM IN ECOLOGY AND CONSERVATION

Patchiness and heterogeneity are compelling scientific concepts, especially when integrated into theories of metapopulation dynamics, because they draw attention to the spatial matrix of ecological processes, the fluxes of organisms and materials across boundaries, the dynamics of an entire mosaic as well as its parts, and the importance of scale in understanding the effects of ecological processes (Ostfeld et al. 1997). Patchiness and heterogeneity are controlling ideas in conservation biology for practical as well as scientific reasons; conservation is increasingly practiced in a fragmented world where human land use, ownership, and jurisdictional boundaries impose patchiness on a variety of systems (Ostfeld et al. 1997).

The increased importance of patchiness and heterogeneity in ecological studies was one manifestation of a broader shift in fundamental views of how ecosystems functioned. In a 1992 paper in *BioScience,* Eugene P. Odum, director emeritus of the University of Georgia's Institute of Ecology and considered by many the father of modern ecology, listed 20 "great ideas" in ecology (Odum 1992). His first "great idea" is the concept, "An ecosystem is a thermodynamically open, far from equilibrium system." This is an excellent, concise statement of what has come to be called the nonequilibrium paradigm.

Throughout most of the twentieth century, the equilibrium paradigm governed the science of ecology. According to Frederic Clements, ecological communities were the equivalent of "superorganisms" that always followed predictable patterns of population replacement to arrive at similar, stable associations (chapter 1) (Clements 1916, 1920, 1936). Another early ecologist, A. G. Tansley, described ecosystems as existing in "relatively stable, dynamic, equilibrium." Tansley wrote, "There is in fact a kind of natural selection of incipient systems,

and those which can attain the most stable equilibrium survive the longest." Tansley continued, "the normal autogenic succession is a progress towards greater integration and stability. The **'climax'** represents the highest stage of integration and the nearest approach to perfect dynamic equilibrium. . . . The equilibrium attained is however never quite perfect; its degree of perfection is measured by its stability" (Tansley 1935).

In the view of Clements and Tansley, equilibrium was synonymous with perfection, evolutionary progress, stability, and integration. Aldo Leopold popularized this idea as a core concept of his famous "land ethic." In *A Sand County Almanac,* Leopold wrote, "A thing is right when it tends to preserve the integrity, stability, and beauty of the biotic community. It is wrong when it tends otherwise" (Leopold 1966:262). The message of such a paradigm was unmistakably clear: left to themselves, ecosystems attain equilibrium. Equilibrium equals perfection. Let nature be.

The equilibrium paradigm found a receptive audience among science and the public because of the match between the "balance of nature" and popular and scientific visions of a perfect universe. It was similarly attractive to legislators who perceived it as the foundational premise for good environmental legislation (Tarlock 1994). Wilderness was valued because it was "untrammeled by man." Endangered species were to be "preserved" because their loss represented a potential loss of equilibrium in the system. Environmental impact statements were required of federal actions that might have a significant effect on the environment and create a departure from equilibrium.

As early as the 1920s, rigorous experimental studies and mathematical modeling of ecosystems began to discredit the equilibrium paradigm, and the erosion of its credibility accelerated through the 1960s and 1970s. As early as 1926 ecologist Henry Gleason had presented theoretical and empirical evidence contradicting the equilibrium paradigm in plant ecology (Gleason 1926). Pioneering work on the role of disturbance in forest ecosystems by F. Herbert Bormann and Gene E. Likens demonstrated a "shifting mosaic" pattern of plant associations in forests, not a uniform climax state (Bormann and Likens 1979). The development and substantiation of the climax theory of succession revealed that climax communities of vegetation represented patterns of populations corresponding to and changing with patterns of environmental gradients, rarely if ever achieving equilibrium with their environment (Whittaker 1953). The discovery that numbers of predators and prey, even in a wilderness, are neither constant in numbers nor predictable in their oscillations (Mech 1966, 1970) and the redefinition of traditional equilibrium concepts of carrying capacity (numbers of herbivores that could be supported by a given association of plants) (Caughley 1979) also demonstrated that there was not a single equilibrium point between herbivores and vegetation.

The common themes of nonequilibrium emerging from diverse studies eventually led to an overall synthesis that has been described as the nonequilibrium paradigm (Botkin 1990) and to a growing interest in the effects and implications of nonequilibrium systems for conservation problems, especially threatened species (Pimm 1991). Increasingly, scientists found that multiple equilibria could be achieved through management. Current research on ecosystems consistently reveals that such systems are neither fixed nor balanced. Rather, ecosystems represent patches or collections of conditions that exist for finite periods of time. The disturbances created by human activities, as well as other, nonhuman forces, are both predictable and random. That is why ecosystems can be managed, but they cannot be restored to some perfect or balanced state, nor can they be preserved indefinitely in a kind of ecological stasis (Tarlock 1994). Present conditions in an ecosystem are retained only by dynamic inputs of matter and energy, and such inputs can be altered through management actions.

In a strict sense, the shift from an equilibrium view of ecosystem processes to a nonequilibrium view was essentially complete by the time conservation biology emerged as a distinct field; however, the emergence of the nonequilibrium paradigm facilitated two critical developments in conservation biology. One was the development of the adaptive management concept (Hilborn and Walters 1992; Lee 1993). Traditionally, management actions were viewed as deterministic, and their outcomes understood in relationships of cause and effect. In contrast, adaptive management treats management actions as an ongoing means of learning about how the system functions. Thus, management efforts, viewed as experimental in nature, should follow careful experimental design, include environmental controls (untreated sites or subjects), and be carefully monitored over time. If the experimental design is sound, the results of the management action should be relatively unambiguous, but must still be interpreted stochastically (within a range of outcomes with differing probabilities), rather than as deterministic outcomes generated by simple cause-and-effect relationships. Such a shift, with its accompanying rise in uncertainty, has implications for population viability analysis and minimum viable population estimation. Because different equilibria are possible between populations and their environments, different-size minimum populations may be viable at different points in time, depending on the state of the system. Thus, to be useful, population viability analysis for nonequilibrium systems must be increasingly time and condition specific.

The shift to a nonequilibrium view of ecosystems also changed the focus of conservation from the protection of sites to the management of ecosystems, thus facilitating the development of ecosystem management (chapter 10) as a conservation strategy. As long as ecosystems were viewed as relatively static equilibria, the emphasis in conservation was on the preservation of *states* of nature. Preserved sites were often, even deliberately, small because the object of conservation was a particular entity (such as a species or habitat) of natural or ecological heritage that existed in a particular place of fixed boundaries. Management was either entirely passive or focused on the exclusion of humans and other kinds of disturbances from the site. As the nonequilibrium view took hold, emphasis shifted from preserving species or habitats to managing ecosystem processes such as fire, water regimes, and nutrient cycling at the landscape scale. Because the systems were now seen as open rather than closed, boundaries were more difficult to define and cooperation was required among various interests over much larger areas. The emphasis on fixed natural areas was replaced by an emphasis on heterogeneous landscapes that could sustain and create essential ecosystem processes (Barrett and Barrett 1997). This change in emphasis is evident in the increased attention now given to the study of disturbance and patch dynamics in ecosystems.

A "disturbance" may be defined as any relatively discrete event in time that disrupts ecosystem, community, or population structure and changes resources, substrate availability, or the physical environment (White and Pickett 1985). A heterogeneous distribution of habitat patches can occur in a landscape for a variety of reasons, including patterns of external and internal disturbances and the patterns of vegetational change that follow disturbances. Disturbance and vegetational change determine the size, density, frequency, and longevity of patches of habitat in a landscape (Pickett and Thompson 1978). The internal structure of the patches themselves will be determined by species composition, population densities, population dispersions, and the size and geometry of the organisms in them (Pickett and Thompson 1978). To generate patches of habitat, disturbances must directly affect existing vegetation. Fire is perhaps the most common example, but other patch-creating disturbances include wind, flooding, landslides, disease, death of dominant vegetation, and the activity of animals, especially grazers (McNaughton 1984).

Endogenous Disturbances

Endogenous disturbances, also called "autogenic successional processes," are generated by processes intrinsic to the ecosystem (Bormann and Likens 1979). One of the best, most important, and most widespread examples of endogenous disturbance in forest ecosystems is the random, but continual, death of large, dominant trees in a forest and their subsequent fall. Treefall, with associated uprooting at the base of the tree, is a key process in the population and community ecology of many forests (Lertzman 1992). The processes of **gap-phase replacement,** through which gaps in the forest floor and the forest canopy are created and subsequently filled with new species, play a central role in species coexistence in many kinds of forests. Gap-phase replacement processes also are a primary means of maintaining species richness and diversity (Schaetzl et al. 1989). When a gap is created in a forest canopy or on a forest floor, increased availability of light and decreased competition, with changes in moisture, heat, and nutrient availability, release resources previously unavailable to many species. This is why canopy gaps in a forest, and individual microsites on a forest floor such as stumps and logs, can lead to accelerated growth, increased species richness (Dyrness 1973; Thompson 1980; Spies and Franklin 1989), and increased survivorship of individuals compared to closed-canopy sites (Monk 1961; Dunn, Guntenspergen, and Dorney 1983; Lertzman 1992) (fig. 5.24). In forests, the distribution of gaps of different sizes, shapes, and ages determines the species composition of the community, and is an essential element in maintaining community biodiversity.

In grassland ecosystems, one of the most pervasive forms of endogenous disturbances is grazing. Native grazing animals, especially large ungulates, actively modify grassland systems by their individual and collective behavior. Trampling by the grazers breaks up litter, making it more susceptible to microbial decomposition and increasing germination rates of new seedlings (McNaughton, Ruess, and Seaqle 1988). Grazing physically stimulates increased regrowth and a higher rate of nutrient movement in plants, and a grazer's saliva can cause chemical reactions in plants that lead to increased plant productivity (McNaughton 1979). Large groups of grazers create "grazing lawns" where frequent, intensive grazing selects for small-leafed, dwarfed ecotypes of grass species that are easier for grazers to exploit. Grazing can enhance the nutritional quality

Figure 5.24
An example of a treefall gap in a deciduous forest in the eastern United States. Differences in light availability between the gap and the surrounding forest create corresponding differences in species composition of the understory and increase community biodiversity.

of plants, particularly through stimulating plants to produce higher levels of nitrogen and protein and a higher forage mass per unit volume of plant material (McNaughton 1984).

Browsers (animals that feed on woody vegetation) constitute another form of endogenous disturbance that can actively modify shrub and forest communities. In African savannas and grasslands, browsers prevent saplings from growing large enough to resist fire, thereby inhibiting the development of a closed-canopy forest (McNaughton, Ruess, and Seagle 1988). Like grazing, browsing creates flushes of regrowth in plants that lead to community productivity two to three times higher than in similar, unbrowsed systems (Teague 1985). Mixed feeders, such as elephants, regulate the balance between woody vegetation and its grassland understory in African savannas (Laws 1970). All feeding classes of large ungulates in African ecosystems increase nutrient availability, alter plant patterns of nutrient uptake and mobility, and enhance primary productivity of vegetation through the addition of their dung and urine.

Exogenous Disturbances

Exogenous disturbances are caused by forces outside the ecosystem (Schaetzl et al. 1989) and include events such as fire, flooding, thunderstorms, tornadoes, hurricanes, and short- and long-term climate change. Although exogenous disturbances are not generated by the system itself, they are often regular and recurrent, and plant species in such disturbed systems typically show adaptation to, even dependence upon, such disturbances. If these disturbances are suppressed, such species may decline or disappear.

Fire is perhaps the most pervasive and influential of all exogenous disturbances; it has been a major factor in the development of the temperate northern hardwood forests (Bormann and Likens 1979), the boreal forest (Heinselman 1973; Rowe and Scotter 1973), Mediterranean chaparral communities, and all types of grasslands (Barbour, Burk, and Pitts 1987). In conif-

erous forests of North America, some populations of jack pine (*Pinus banksiana*) and its western ecological equivalent, lodgepole pine (*P. contorta*) possess "serotinous" cones that will not shed their seeds unless exposed to temperatures associated with fire. Seeds of such species also germinate at higher rates if exposed to fire. Fire-resistant seeds are common in many species of grassland plants, and the high silica content in the stems of many species of grasses makes them highly resistant to fire (Barbour, Burk, and Pitts 1987). One of the best examples of communitywide fire adaptation is seen in the chaparral community of California, in which most plants are not only able to withstand fire, but actively spread it. Chaparral plants exhibit high flammability, intricate branching patterns, large surface-to-volume ratios, periodic dieback that leaves a residue of highly flammable twigs, and production of volatile, flammable oils (Barbour, Burk, and Pitts 1987) (fig. 5.25).

Fire can, in some settings, be controlled as an agent of human management. In contrast, wind, in all settings, is independent of human control, but is no less pervasive in its effects on succession in some areas, especially forest ecosystems. A single wind event can affect an entire region. For example, one hurricane in the northeastern United States in 1938 damaged trees in 4.5 million hectares out of 5.3 million hectares in the states of Connecticut, Rhode Island, Massachusetts, and Vermont (Bormann and Likens 1979).

The effects of wind, fire, flooding, pestilence, or other exogenous disturbances are mediated by topography and existing vegetation. Like endogenous disturbances, they also affect the landscape unevenly and create habitat heterogeneity, but at a different scale. Exogenous disturbance, because it occurs less frequently, but with greater destructive intensity and over greater areas than endogenous events, tends to increase the temporal, spatial, and intensity scales at which succession proceeds.

The development of disturbance theory, coupled with an emerging non-equilibrium view of ecosystems, created an

Figure 5.25
A chaparral community in California. Chaparral plants are adapted to fire through possession of volatile oils and bodies with large surface-to-volume ratios. Note the intricate branching patterns of many of the plants, a growth form that permits fast and efficient conduction of fire.

intellectual climate within conservation biology that could embrace the concept of ecosystem management through direct manipulation of ecological processes. Disturbances, whether exogenous or endogenous, often act as natural "removal experiments," eliminating some species from the community. In nature, as in contrived experiments, conservationists must remember that the effect of such removals and other disturbance-induced changes will change with time. Thus, it is vital for conservationists to monitor and understand disturbance effects in a context of temporal scale and, informed by this understanding, to set management objectives that are time specific, as well as habitat and species specific, as well as to manage disturbance effectively toward conservation goals.

Synthesis

The emergence of modern conservation biology as a scientific discipline began with concerns about the genetics of small and declining populations. Problems associated with population bottlenecks, genetic drift, and inbreeding rose in prominence concurrent with increasingly sophisticated molecular techniques for genetic assessment. This confluence of growing concern and increasingly precise methods to assess and measure genetic parameters was a strong contributing factor in the preeminence genetic paradigms gained in the early development and definition of conservation biology.

The rise and fall of island biogeography, as well as the ongoing reevaluation of classical genetic concerns associated with small populations, illustrates a danger inherent in scientific theories that arise during crises—the use of broad generalizations to address complex and specific problems is likely to lead to profound disillusionment. Metapopulation theory faces the same danger, and so must continue to grow and mature in its grasp of the complexities of real populations, or suffer the same fate.

Many species now exist as metapopulations because human activity has fragmented their habitats. By definition, every declining population will reach some critical threshold in its loss of habitat when simple, additive habitat loss becomes compounded by habitat fragmentation and patch isolation (Andrén 1994). When this happens, we have perhaps added a new "metapopulation" as a potential research target that may advance our understanding of our new paradigm.

But the use of a sophisticated name for the demographic effects of our relentless shredding of the natural world will be no comfort to the species that receives metapopulation status, and no surety that the new mantle it wears will help it to persist. Unless individuals in the metapopulation have the dispersal abilities to move from patch to patch in an increasingly broken landscape, its subunits will not be examples of enhanced spatial demographics and "rescue effects." They will only too soon become another statistic of contemporary extinction.

We perceive now that rates of dispersal, immigration, and emigration, although hard to measure and long overshadowed by our fascination with population birth and death rates, may hold the keys to conservation. Our knowledge of how well species disperse, and when and where, may become the means through which we understand not only the dynamics of metapopulations, but also whether or not any species can long endure the new paradigm we have created for it. Today, even the existence and apparent survival of metapopulations in fragmented landscapes may be deceptive. Contemporary metapopulations may be far from equilibrium because the landscape has changed so quickly (Hanski 1997). Some current "metapopulations" may already be on the road to extinction. Their conservation will increasingly depend not only on our knowledge of metapopulation dynamics, but also on our understanding of habitat heterogeneity and ecological scale in their population processes.

Minimum Population Sizes—A Directed Discussion

Reading assignment: Shaffer, M. L. 1981. Minimum population sizes for species conservation. *BioScience* 31:131–134.

Questions

1. Why do you think the more well-established disciplines of ecology and wildlife management had not articulated the "minimum viable population" concept before the publication of Shaffer's article? What events made this concept appealing or useful to some scientists by the 1980s?
2. Did the concept of a minimum viable population strengthen the foundation of the small-population paradigm or the declining-population paradigm?

Explain how this concept contributed to the development of conservation biology as a scientific discipline distinct from ecology or wildlife management.

3. The minimum viable population concept is inextricably linked to population viability analysis. What question does a population viability analysis answer? Do you agree with population ecologist Graeme Caughley that this is a trivial question? If so, what question should be asked instead? If not, what makes the question important?

4. Should the calculation of a minimum viable population size be used to describe a population's current state, diagnose what caused the population to reach its current state, or prescribe management actions for

recovery? Defend your choice, and explain why the minimum viable population estimate and associated population viability analysis should not be used in the other two ways.

Learning Online

Visit our webpage at www.mhhe.com/conservation for case studies, animations, practice quiz questions, and additional readings to help you understand the material in this chapter. You'll also find active links to the following topics:

Legislation Regarding Endangered
  Species
Endangered Species Act

Ecosystem Management
Endangered Species
Species Preservation

Refuges and Sanctuaries
Miscellaneous Environmental
  Resources

Literature Cited

Andrén, H. 1994. Effects of habitat fragmentation on birds and mammals in landscapes with different proportions of suitable habitat: a review. *Oikos* 71:355–66.

Andrewartha, H. G., and L. C. Birch. 1954. *The distribution and abundance of animals*. Chicago: University of Chicago Press.

Barbour, M. G., J. H. Burk, and W. D. Pitts. 1987. *Terrestrial plant ecology*. 2d ed. Menlo Park, Calif: Benjamin/Cummings Publishing.

Barrett, N. E., and J. P. Barrett. 1997. Reserve design and the new conservation theory. In *The ecological basis of conservation*, eds. S. T. A. Pickett, R. S. Ostfeld, M. Shachak, and G. E. Likens, 236–51. New York: Chapman and Hall.

Berry, R. J. 1971. Conservation aspects of the genetical constitution of populations. In *The scientific management of animal and plant communities for conservation*, eds. E. Duffey and A. S. Watt, 177–206. Oxford, England: Blackwell.

Berryman, A. A. 1992. The origins and evolution of predator-prey theory. *Ecology* 73:1530–35.

Boorman, S. A., and P. R. Levitt. 1973. Group selection on the boundary of a stable population. *Theoretical Population Biology* 4:85–128.

Bormann, F. H., and G. E. Likens. 1979. Catastrophic disturbance and the steady state in northern hardwood forests. *American Scientist* 67:660–69.

Botkin, D. B. 1990. *Discordant harmonies : a new ecology for the twenty-first century*. New York: Oxford University Press.

Brown, J. H., and A. Kodric-Brown. 1977. Turnover rates in insular biogeography: effect of immigration on extinction. *Ecology* 58:445–49.

Caro, T. M., and M. K. Laurenson. 1994. Ecological and genetic factors in conservation: a cautionary tale. *Science* 263:485–86.

Caughley, G. 1979. What is this thing called carrying capacity? In *North American elk: ecology, behavior and management*, eds. M. S. Boyce and L. D. Hayden-Wing, 2–8. Laramie: University of Wyoming Press.

Caughley, G. 1994. Directions in conservation biology. *Journal of Animal Ecology* 63:215–44.

Caughley, G., and A. Gunn. 1996. *Conservation biology in theory and practice*. Cambridge, Mass: Blackwell Science.

Clements, F. E. 1916. Plant succession: an analysis of the development of vegetation. Publication 242. Washington, D.C.: Carnegie Institution of Washington.

Clements, F. E. 1920. Plant indicators: the relation of plant communities to process and practice. Publication 290. Washington, D.C.: Carnegie Institution of Washington.

Clements, F. E. 1936. Nature and the structure of the climax. *Journal of Ecology* 24:252–84.

Curtis, J. T. 1956. The modification of mid-latitude grasslands and forests by man. In *Man's role in changing the face of the earth*, ed. W. L. Thomas. Chicago: University of Chicago Press.

Danielson, B. J. 1991. Communities in a landscape: the influence of habitat heterogeneity on the interactions between species. *The American Naturalist* 138:1105–20.

Dunn, C. P., G. R. Guntenspergen, and J. R. Dorney. 1983. Catastrophic wind disturbance in an old-growth hemlock-hardwood forest, Wisconsin. *Canadian Journal of Botany* 61:211–17.

Dyrness, C. T. 1973. Early stages of plant succession following logging and slash burning in the western Cascades of Oregon. *Ecology* 54:57–68.

Ehrlich, P. R., and D. D. Murphy. 1987. Conservation lessons from long-term studies of checkerspot butterflies. *Conservation Biology* 1:122–31.

Fangliang, H., and P. Legendre. 1996. On species-area relations. *The American Naturalist* 148:719–21.

Ferrer, M., and J. A. Donazar. 1996. Density-dependent fecundity by habitat heterogeneity in an increasing population of Spanish imperial eagles. *Ecology* 77:69–74.

Fisher, R. A. 1958. *The genetical theory of natural selection*. Toronto: Dover Publications.

Frankel, O. H. 1974. Genetic conservation: our evolutionary responsibility. *Genetics* 78:53–65.

Frankel, O. H., and M. E. Soulé. 1981. *Conservation and evolution*. Cambridge, England: Cambridge University Press.

Frankham, R. 1995. Conservation genetics. *Annual Review of Genetics* 29:305–27.

Franklin, I. R., 1980. Evolutionary change in small populations. In *Conservation biology: an evolutionary-ecological perspective*, eds. M. E. Soulé and B. A. Wilcox, 135–49. Sunderland, Mass.: Sinauer Associates.

Fretwell, S. D., and H. L. Lucas. 1970. On territorial behavior and other factors influencing habitat distribution in birds, theoretical development. *Acta Biotheoretica* 19:16–36.

Gause, G. F. 1934. *The struggle for existence*. Baltimore, Md.: Williams and Wilkins.

Gilpin, M. 1996. Metapopulations and wildlife conservation: approaches to modeling spatial structure. In *Metapopulations and wildlife conservation*, ed. D. R. McCullough, 11–27. Washington, D.C.: Island Press.

Gilpin, M., and M. E. Soulé. 1986. Minimum viable populations: processes of species extinction. In *Conservation biology: the science of scarcity and diversity*, ed. M. E. Soulé, 19–34. Sunderland, Mass.: Sinauer Associates.

Gleason, H. A. 1926. The individualistic concept of the plant association. *Bulletin of the Torrey Botanical Club* 53:1–20.

Groom, M. J., and M. A. Pascual. 1998. Saving species through population viability biology and viability analyses: a morass of math, myth or mistakes? In *Conservation biology: for the coming decade*, 2d ed., eds. P. L. Fiedler and P. M. Kareiva, 4–27. New York: Chapman and Hall.

Hanski, I. 1997. Metapopulation dynamics: from concepts and observations to predictive models. In *Metapopulation biology: ecology, genetics, and evolution*, ed. I. Hanski and M. E. Gilpin, 69–91. San Diego, Calif.: Academic Press.

Hanski, I., and M. Gilpin. 1991. Metapopulation dynamics: brief history and conceptual domain. In *Metapopulation dynamics: empirical and theoretical investigations*, eds. M. Gilpin and I. Hanski, 3–16. London, England: Academic Press.

Hanski, I., and D. Simberloff. 1997. The metapopulation approach, its history, conceptual domain, and application to conservation. In I. Hanski and M. E. Gilpin, eds. *Metapopulation biology: ecology, genetics, and evolution*, 5–26. San Diego, Calif.: Academic Press.

Harrison, S. 1991. Local extinction in a metapopulation context: an empirical evaluation. *Biological Journal of the Linnean Society* 42:73–88.

Heinselman, M. L. 1973. Fire in the virgin forests of the Boundary Waters Canoe Area, Minnesota. *Quaternary Research* 3:329–82.

Hilborn, R., and C. J. Walters. 1992. *Quantitative fisheries stock assessments: choice, dynamics, and uncertainty*. New York: Chapman and Hall.

Hooper, M. D. 1971. The size and surroundings of nature reserves. In *The scientific management of animal and plant communities for conservation*, eds. E. Duffey and A. S. Watt, 555–61. Oxford, England: Blackwell.

Hoopes, M. F., and S. Harrison, 1998. Metapopulation, source-sink, and disturbance dynamics. In *Conservation science and action*, ed. W. J. Sutherland, 135–51. Oxford, England: Blackwell Science.

Huffaker, C. B. 1958. Experimental studies on predation: dispersion factors and predator-prey oscillations. *Hilgardia* 27:343–83.

Huffaker, C. B., K. P. Shea, and S. G. Herman. 1963. Experimental studies on predation: complex dispersion and levels of food in an acarine predator-prey interaction. *Hilgardia* 34:305–30.

International Union of the Conservation of Nature and Natural Resources. 1980. *World conservation strategy*. Gland, Switzerland: IUCN.

Kadmon, R. 1993. Population dynamic consequences of habitat heterogeneity: an experimental study. *Ecology* 74: 816–25.

Kindvall, O. 1996. Habitat heterogeneity and survival in a bush cricket metapopulation. *Ecology* 77:207–14.

Kuhn, T. S. 1970. *The structure of scientific revolutions*, 2d ed., enlarged. Chicago: University of Chicago Press.

Lack, D. 1966. *Population studies of birds*. Oxford, England: Clarendon.

Lack, D. 1976. *Island biology illustrated by the land birds of Jamaica*. Oxford, England: Blackwell.

Lande, R. 1988. Genetics and demography in biological conservation. *Science* 241:1455–60.

Latter, B. D. H., and J. C. Mulley. 1995. Genetic adaptation to captivity and inbreeding depression in small laboratory populations of *Drosophila melanogaster*. *Genetics* 13:287–97.

Laws, R. M. 1970. Elephants as agents of habitat and landscape change in East Africa. *Oikos* 21:1–15.

Lee, K. N. 1993. *Compass and gyroscope: integrating science and politics for the environment*. Washington, D.C.: Island Press.

Leopold, A. 1966. *A Sand County almanac*. New York: Sierra Club/Ballantine.

Lertzman, K. P. 1992. Patterns of gap-phase replacement in a subalpine, old-growth forest. *Ecology* 73:657–69.

Levins, R. 1969. Some demographic and genetic consequences of environmental heterogeneity for biological control. *Bulletin of the Entomological Society of America* 15:237–40.

Levins, R. 1970. "Extinction." In *Some mathematical problems in biology*, ed. M. Gesternhaber, 77–107. Providence, R.I.: American Mathematical Society.

Lewontin, R. C. 1991. Twenty-five years ago in GENETICS: electrophoresis in the development of evolutionary genetics: milestone or millstone? *Genetics* 128:657–62.

MacArthur, R. H., and E. O. Wilson. 1967. *The theory of island biogeography*. Princeton, N.J.: Princeton University Press.

McCullough, D. R. 1996. Introduction. In *Metapopulations and wildlife conservation*, ed. D. R. McCullough, 1–10. Washington, D.C.: Island Press.

McDonald, K. A., and J. H. Brown. 1995. Using montane mammals to model extinctions due to global change. In *To preserve biodiversity—an overview*, ed. D. Ehrenfeld, 215–21. Readings in *Conservation Biology*. Cambridge, Mass.: Blackwell Science and The Society for Conservation Biology.

McNaughton, S. J. 1979. Grazing as an optimization process: grass-ungulate relationships in the Serengeti. *The American Naturalist* 113:691–703.

McNaughton, S. J. 1984. Grazing lawns: animals in herds, plant form, and coevolution. *The American Naturalist* 124:863–86.

McNaughton, S. J., R. W. Ruess, and S. W. Seagle. 1988. Large mammals and process dynamics in African ecosystems. *BioScience* 38:794–800.

Mech, L. D. 1966. *The wolves of Isle Royale.* U.S. National Park Service Fauna Series, no. 7. Washington, D.C.: U.S. Government Printing Office.

Mech, L. D. 1970. *The wolf: the ecology and behavior of an endangered species.* Garden City, N.Y.: Natural History Press.

Monk, C. D. 1961. The vegetation of the William L. Hutcheson Memorial Forest, New Jersey. *Bulletin of the Torrey Botanical Club* 88:156–66.

Moore, N. W. 1962. The heaths of Dorset and their conservation. *Journal of Ecology* 50:369–91.

O'Brien, S. J., and collaborators. 1996. Conservation genetics of the felidae. In *Conservation genetics: case histories from nature,* eds. J. C. Avise and J. L. Hamrick, 50–74. New York: Chapman and Hall.

O'Brien, S. J., M. E. Roelke, L. Marker, A. Newman, C. A. Winkler, D. Meltzer, L. Colly, J. F. Evermann, M. Bush, and D. E. Wildt. 1985. Genetic basis for species vulnerability in the cheetah. *Science* 227:1428–34.

O'Brien, S. J., D. E. Wildt, D. Goldman, C. R. Merril, and M. Bush. 1983. The cheetah is depauperate in genetic variation. *Science* 221:459–62.

Odum, E. P. 1992. Great ideas in ecology for the 1990s. *BioScience* 42:542–45.

Ostfeld, R. S., S. T. A. Pickett, M. Shachak, and G. E. Likens. 1997. In *The ecological basis of conservation: heterogeneity, ecosystems, and biodiversity,* eds. S. T. A. Pickett, R. S. Ostfeld, M. Shachak, and G. E. Likens, 3–10. New York: Chapman and Hall.

Packer, C., A. E. Pusey, H. Rowley, D. A. Gilbert, J. Martenson, and S. J. O'Brien. 1991. Case study of a population bottleneck: lions of the Ngorongoro Crater. *Conservation Biology* 5:219–30.

Pannell, J. R., and B. Charlesworth. 1999. Neutral genetic diversity in a metapopulation with recurrent local extinction and recolonization. *Evolution* 53:664–76.

Pickett, S. T. A., and J. N. Thompson. 1978. Patch dynamics and the design of nature reserves. *Biological Conservation* 13:27–37.

Pimm, S. L. 1991. *The balance of nature? Ecological issues in the conservation of species and communities.* Chicago: University of Chicago Press.

Power, M. E. 1992. Habitat heterogeneity and the functional significance of fish in river food webs. *Ecology* 73:1675–88.

Pulliam, H. R. 1988. Sources, sinks, and population regulation. *American Naturalist* 132:652–61.

Ralls, K., J. D. Ballou, and A. Templeton. 1988. Estimates of lethal equivalents and the cost of inbreeding in mammals. *Conservation Biology* 2:185–93.

Ralls, K., K. Brugger, and J. Ballou. 1979. Inbreeding and juvenile mortality in small populations of ungulates. *Science* 206:1101–3.

Rowe, J. S., and G. W. Scotter. 1973. Fire in the boreal forest. *Quaternary Research* 3:444–64.

Schaetzl, R. J., S. F. Burns, D. L. Johnson, and T. W. Small. 1989. Tree uprooting: review of impacts on forest ecology. *Vegetatio* 79:165–76.

Shaffer, M. L. 1981. Minimum population sizes for species conservation. *BioScience* 31:131–34.

Simberloff, D. 1986. Design of nature reserves. In *Wildlife conservation evaluation,* ed. M. B. Usher, 315–17. London, England: Chapman and Hall.

Simberloff, D. 1988. The contribution of population and community biology to conservation science. *Annual Review of Ecology and Systematics* 19:473–511.

Simberloff, D. 1997. Biogeographic approaches and the new conservation biology. In *The ecological basis of conservation: heterogeneity, ecosystems, and biodiversity,* eds. S. T. A. Pickett, R. S. Ostfeld, M. Shachak, and G. E. Likens, 274–84. New York: Chapman and Hall.

Simberloff, D., and N. Gotelli. 1984. Effects of insularization on plant species richness in the prairie-forest ecotone. *Biological Conservation* 29:27–46.

Skaggs, R. W., and W. J. Boecklen. 1996. Extinctions of montane mammals reconsidered: putting a global warming scenario on ice. *Biodiversity and Conservation* 5:759–78.

Slatkin, M. 1977. Gene flow and genetic drift in a species subject to frequent local extinction. *Theoretical Population Biology* 12:253–62.

Smith, B. W., J. W. Fitzpatrick, G. E. Woolfenden, and B. Pranty. 1996. Classification and conservation of metapopulations: a case study of the Florida scrub jay. In *Metapopulations and wildlife conservation,* ed. D. R. McCullough, 187–215. Washington, D.C.: Island Press.

Soulé, M. 1973. The epistasis cycle: a theory of marginal populations. *Annual Review of Ecology and Systematics* 4:165–87.

Spies, T. A., and J. F. Franklin. 1989. Gap characteristics and vegetation response in coniferous forests of the Pacific Northwest. *Ecology* 70:543–45.

Tansley, A. G. 1935. The use and abuse of vegetation concepts and terms. *Ecology* 16:284–307.

Tarlock, A. D. 1994. The nonequilibrium paradigm in ecology and the partial unraveling of environmental law. *Loyola of Los Angeles Law Review* 27:1121–44.

Teague, W. R. 1985. Leaf growth of Acacia karroo trees in response to frequency and intensity of defoliation. In *Ecology and management of the world's savannas,* eds. J. C. Tothill and J. J. Mott, 220–22. Canberra: Australian Academy of Science.

Temple, S. A., and J. R. Cary. 1988. Modeling dynamics of habitat-interior bird populations in fragmented landscapes. *Conservation Biology* 2:340–47.

Thompson, J. N. 1980. Treefalls and colonization patterns of temperate forest herbs. *American Midland Naturalist* 104:176–84.

Vrijenhoek, R. J. 1997. Gene flow and genetic diversity in naturally fragmented metapopulations of deep-sea hydrothermal vent animals. *Journal of Heredity* 88:285–93.

Wade, M. J. 1996. Adaptation in subdivided populations: kin selection and interdemic selection. In *Adaptation,* eds. M. R. Rose and G. Lauder, 381–405. Sunderland, Mass.: Sinauer Associates.

Wade, M. J., and C. J. Goodnight. 1998. Perspective: the theories of Fisher and Wright in the context of metapopulations: when nature does many small experiments. *Evolution* 52:1537–53.

Walter, H. S. 1990. Small viable population: the red-tailed hawk of Socorro Island. *Conservation Biology* 4: 441–43.

White, P. S., and S. T. A. Pickett. 1985. Natural disturbance and patch dynamics: an introduction. In *The ecology of natural disturbance and patch dynamics,* eds. S. T. A. Pickett and P. S. White, 3–13. Orlando, Fla.: Academic Press.

Whittaker, R. H. 1953. A consideration of climax theory: the climax as a population and pattern. *Ecological Monographs* 23:41–78.

Wiens, J. A. 1996. Wildlife in patchy environments: metapopulations, mosaics, and management. In *Metapopulations and wildlife conservation,* ed. D. R. McCullough, 53–84. Washington, D.C.: Island Press.

Woolfenden, G. E., and J. W. Fitzpatrick. 1984. *The Florida scrub jay.* Princeton, N.J.: Princeton University Press.

Wright, S. 1931. Evolution and Mendelian populations. *Genetics* 16: 97–159.

PART TWO

Concepts

The five chapters in this section examine the conceptual and theoretical basis for four fundamental problems of modern conservation: (1) How do we preserve genetic diversity? (2) How do we preserve and manage populations and species, especially species that face imminent threats of extinction? (3) How do we preserve habitat needed to support such populations in terrestrial and aquatic environments? and (4) How do we manage arrays of populations and habitats in a functioning ecosystem in which humans may play important roles?

These chapters will approach their subjects by first introducing the theories that lie behind the problem, for such theories also lie behind the solution. By making the effort to understand concepts and theories first, we create a framework in which examples of particular management problems presented to us can be categorized and comprehensively understood. I believe that there is no conservation problem that cannot be solved if a conservation scientist will first place the problem in its appropriate theoretical context, and then, with skill, apply both theoretical and practical principles to its solution.

Thus, in examining the conservation of genetic diversity (chapter 6), we begin with the context of genetic theory that explains the pervasive problems of inbreeding depression, loss of genetic variation, accumulation of harmful mutations (mutational meltdown), hybridization and introgression, and outbreeding depression. Similarly, chapter 7 offers a careful, conceptual definition of what populations are, followed by the theoretical context that explains their decline and predicts their persistence. Chapters 8 and 9 examine theories of habitat use, the theoretical interaction of habitat and population processes, and theories of habitat change (succession) and patch distribution (heterogeneity). Chapter 10 examines one of the most challenging of all modern conservation problems: the problem of managing an entire ecosystem. Because even the concept of ecosystem management is problematic, we begin with a thorough exploration of what ecosystem management actually means. Equipped with an operational definition of ecosystem management, we then explore the scientific basis for being able to treat ecosystems as units of management in conservation.

CHAPTER 6

The Conservation of Genetic Diversity

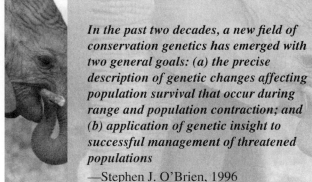

In the past two decades, a new field of conservation genetics has emerged with two general goals: (a) the precise description of genetic changes affecting population survival that occur during range and population contraction; and (b) application of genetic insight to successful management of threatened populations

—Stephen J. O'Brien, 1996

In this chapter you will learn about:

1 genetic concerns associated with small populations.

2 techniques for making precise genetic assessments of individuals and populations.

3 measurement of genetic attributes of individuals and populations.

4 Applications of genetics to captive breeding and conservation of wild populations.

THE GENETICS OF SMALL POPULATIONS

Conservation biology is a science concerned with the fate of populations, which are defined and identified by their genetic constituency. This unique genetic makeup not only distinguishes one population from others, but also determines a population's capacity to adapt to changing conditions and, potentially, to produce new species. Many conservationists would argue that the conservation of genetic diversity is the fundamental and foundational basis of all conservation efforts because genetic diversity is requisite for evolutionary adaptation, and such adaptation is the key to the long-term survival of any species (Schemske et al. 1994). To ensure such survival, conservation biologists have two primary goals in the area of genetics. One is to preserve significant amounts of heritable variation in plant and animal populations, particularly small populations threatened with extinction. The other, simultaneous goal is to prevent the fixation of deleterious alleles, a fixation that can contribute to inbreeding depression and accumulation of harmful mutations (Lynch 1996). The first goal has the positive thrust of maintaining a population's evolutionary potential, its capacity to adapt to environmental change over the long term. The second goal, oriented to more short-term effects, is designed to prevent declines in survivorship and fecundity that often occur in small populations as a result of reduced genetic diversity.

Meeting these goals is designed to address five primary genetic threats to small populations: (1) **inbreeding depression,** (2) loss of genetic variation and reduction in heterozygosity, (3) accumulation of harmful mutations, (4) **introgression** and **hybridization** with larger populations, and (5) **outbreeding depression**. All of these factors, acting alone or in combination, increase the risk of extinction in small populations. Before evaluating the factors directly, we examine some unique and even more foundational genetic concerns that present distinctive threats to small populations and that may interact with the above factors in ways that reduce fitness.

Bottlenecks and Genetic Drift

As noted in chapter 5, a population bottleneck is a "minimum population size as a result of a crash" (Frankel and Soulé 1981). The remaining individuals, and their genes, possess only a sample of the genetic variation present in the original source population. This loss of genetic variation can lead to a loss of heterozygosity in the population, which, in some studies, has been correlated with a loss in overall fitness (Frankel and Soulé 1981). The correlation exists because loss of heterozygosity allows a greater proportion of recessive alleles to occur in a homozygous condition so that traits previously masked are expressed. Many of these recessive genes have deleterious or even lethal effects on an organism.

Small populations that suffer a prolonged bottleneck may experience *genetic drift*, the random fluctuations in gene frequencies that occur because matings of individuals with uncommon genotypes may make a greater proportional contribution to the population's total reproductive output. Conversely, there is also a greater risk that some genotypes will not be represented in matings that do occur and subsequently will be lost from the gene pool.

Genetic drift can lead to a loss of heterozygosity or a fixation of deleterious alleles. These outcomes can cause random changes in the phenotype, and can lead to a decline in genetic variability (Franklin 1980). Such effects are exacerbated in small populations, particularly if they are closed to migration. In such a state, there is a decrease in the number of different alleles at a single locus in the population and in heterozygosity (Caughley 1994). The degree of decline in heterozygosity is a function of the population size, N, over the number of generations, t (Wright 1931). For example, over one generation, the amount of heterozygosity, H, changes this way:

$$H_1 = H_0 [1 - 1/(2N)],$$

where H_0 represents the original level of heterozygosity (usually expressed as a proportion) and H_1 represents the new level of heterozygosity after one generation. Generalizing the equation so that it can be applied for any number of t generations, heterozygosity declines in this manner:

$$H_t = H_0 [1 - 1/(2N)]^t.$$

Note that the smaller the value of N, the greater the decline in heterozygosity. For example, a population of 50 individuals that began with a 0.5 level of heterozygosity would lose 1% of its heterozygosity in each generation (from 0.5 to 0.495). In contrast, a population of 10 individuals with the same initial heterozygosity would lose 5% of its heterozygosity (from 0.5 to 0.475) (table 6.1). Mutations can and do occur, and they increase genetic variability and heterozygosity, but this change in heterozygosity, ΔH, also is affected by population size:

$$\Delta H = -H/(2N) + mH = H\left(m - \frac{1}{2N}\right),$$

where m is the addition of heterozygosity through mutation, typically expressed as a rate. Populations reach an equilibrium level of heterozygosity ($\Delta H = 0$) at

$$H^* = 2Nm.$$

Thus, the smaller the size of the population, the lower its equilibrium heterozygosity.

Genetic Drift and Effective Population Size

The theoretical consequences of genetic drift are normally calculated for an "ideal" population in which each individual contributes gametes equally to a genetic pool from which the next generation is formed. Real populations rarely conform to this ideal genetic vision. However, the effective population size, N_e, represents the size of a randomly mating population that is subject to the same degree of genetic drift as a particular "real" population. The effective population size or, more correctly, the "variance effective size" (Franklin 1980) of a population is

Table 6.1   Heterozygosity and the Effect of Population Size.   The heterozygosity (H) of smaller populations declines at a faster rate than that of larger populations. Shown here are two populations: Population A with a starting size of 50 and population B with a starting size of 10. Within one generation, population B has declined to a level of 0.475. In contrast, it takes population A five generations to decline to that level.

H_t	POPULATION A (50)	POPULATION B (10)
H_0	0.500	0.500
H_1	0.495	0.475
H_2	0.490	0.451
H_3	0.485	0.429
H_4	0.480	0.407
H_5	0.475	0.387

Source: After Caughley 1994.

affected by a number of variables, including variance in progeny number (brood or litter size), differential sex ratios, fluctuations in total numbers, and deviations from random mating systems. The first three problems can be evaluated separately if we assume that mating is random.

To examine the effect on N_e of variation in the number of progeny, let N equal the population's actual size (census size) and σ^2 the variance in progeny number. Then

$$N_e = \frac{4N}{2 + \sigma^2}.$$

Thus, if the size of the population is 100, the brood size ranges from 0 to 8 and the variance is 4, then the effective population size (N_e) is 400 divided by 6, or 67, which is one-third less than the census population size. Experimental studies of *Drosophila* have shown that populations subjected to equalization of family size had greater genetic variation, greater reproductive fitness, and slower rates of inbreeding than populations in which family size was not equalized (Boriase et al. 1993).

For populations with unequal sex ratios, the formula to determine the effective population size is

$$N_e = \frac{4N_m N_f}{N_m + N_f},$$

where N_m is the number of males and N_f is the number of females. Consider a population of 100 elk (*Cervus elaphus*) (fig. 6.1). If there are 50 bulls and 50 cows, if each bull mates with one cow, and if each pair represents a unique association of individuals, then the effective population size is 10,000 divided

Figure 6.1

Elk (*Cervus elaphus*) are an example of a species with a harem mating system that reduces the effective population size.

by 100, or 100. But there is no wild elk population anywhere with such a sex ratio, nor are there any that use such a mating system. Through both natural selection and the effects of sexually differential hunting pressure, all wild elk populations have more females than males. In autumn, during the breeding period or "rut," males gather groups of females (harems) that they defend against other males for exclusive breeding privileges. Suppose, in such a setting, that the population of 100 elk is actually composed of 10 males and 90 cows, and that each male takes a harem of 9 females and successfully defends it from other males. Such a scenario is still a gross oversimplification of what really happens, but it is a little closer to real elk life. In such harem mating systems, which are common in large ungulates, the relatedness of offspring born to females within a harem is much higher than the relatedness of offspring from females of different harems. Therefore, in this revised scenario, the effective population size for the elk is 4 × 10 × 90 divided by 10 + 90, or 3,600 divided by 100, producing a result of 36.

Thus, the effective population size of a group of animals with a biased sex ratio is only about one-third that of a monogamous population with a balanced sex ratio. In other words, the sampling error (genetic drift) associated with random mating in a population of 36 individuals is equivalent to the sampling error associated with mating in a population of 100 individuals with the sex ratio and mating system just described, and would lead to increased rates of inbreeding. Studies by Briton and colleagues (1994) confirmed this prediction. Polygamous mating systems associated with unequal sex ratios did increase rates of inbreeding and loss of genetic variation, leading Frankham (1995a) to assert that harem breeding structures should be avoided or circumvented whenever possible in captive breeding programs.

Finally, effective population size changes when population size fluctuates. If population size varies from generation to generation, then the effective number is the *harmonic mean* (the reciprocal of the arithmetic mean of the reciprocals of a finite set of numbers):

$$\frac{1}{N_e} = \frac{1}{t}\left(\frac{1}{N_1} + \frac{1}{N_2} + \dots \frac{1}{N_t}\right),$$

where N_t is the effective size of the population at generation t.

POINTS OF ENGAGEMENT—QUESTION 1

Work out the mathematics of the previous equation for 10 generations with a population of 100 in every generation, and then repeat the calculation a second time, but let one generation "crash" to only 10 individuals. What happens to the effective population size?

Like the effects of unequal family sizes and unequal sex ratios, unequal population sizes over time should lead to increased levels of genetic drift and loss of heterozygosity. This prediction, like those above, has been verified experimentally (Woodworth et al. 1994).

All of these formulas assume random mating, but that assumption is often violated in real populations. More complex mathematics are required to determine effective population sizes where there is significant deviation from random mating. Even if random mating is approximated, however, most populations will have a lower genetically effective population size than their census size. The problem of genetic drift becomes especially important when the effective population size is very small. When the effective population size is large, the expected variation in a typical genetic character is determined mainly by the strength of selection for or against that character (i.e., genetic variation is determined by mathematical probability). However, when the effective population size becomes less than a few hundred individuals, the expected variation of the character becomes largely independent of the strength of selection and is determined primarily by the balance between mutation and genetic drift (i.e., variation is determined by random events).

Bottlenecks, Small Populations, and Rare Alleles

Although bottlenecks have relatively little effect on genetic variability in a population unless the population remains small for a long time, the effect of size reductions on rare alleles is a different matter. Rare alleles can be lost quickly in small populations that experience a sudden decline or that remain at low levels for extended periods. The expected number of alleles, $E(n)$, remaining after a genetic bottleneck is equal to

$$m - \Sigma(1 - P_j)^{2N_e},$$

where m is the number of alleles prior to the bottleneck, P is the frequency of the jth allele, and N_e is the effective number of individuals at the bottleneck. Suppose that $m = 4$ and that one allele is common, but the other three are rare. Look what happens to the average number of alleles (table 6.2) as the effective number of individuals drops from 50 to 1. The rarer the alleles are, the more likely that they will be lost during a bottleneck (Frankel and Soulé 1981).

As noted in chapter 5, the earliest paradigms of modern conservation biology arose from concerns about long-term loss of genetic variation in small populations. One of the earliest stated goals of conservation was the retention of 90% of a population's genetic variability for 200 years (Soulé et al. 1986).

Table 6.2 **Decreasing Population Size Influences the Average Number of Alleles.**   In this case, four alleles are observed—one with a high frequency and three with lower frequencies. Rare, less common alleles are more likely to be lost during a bottleneck. These rare alleles are typically not essential in the initial environment; however, as the environment changes they might become crucial for survival.

EFFECTIVE NUMBER OF INDIVIDUALS (N_e)	AVERAGE NUMBER OF ALLELES RETAINED, GIVEN THE ORIGINAL FREQUENCY OF ALLELE	
	$p_1 = 0.70, p_2 = p_3 = p_4 = 0.10$	$p_1 = 0.94, p_2 = p_3 = p_4 = 0.02$
	4.00	4.00
50	3.99	3.60
10	3.63	2.00
6	3.15	1.64
2	2.02	1.23
1	1.48	1.12

Source: After Frankel and Soulé 1981.

The loss of genetic variation is reduced (and the probability of meeting this goal improves) as the effective population size grows to about 1,000 individuals. Beyond this level, further increases in effective population size do not usually increase the amount of genetic variability in the population (Lynch 1996). But effective population size is often only one-tenth to one-third the number of breeding adults in the population for reasons noted above, including unequal family sizes, unequal sex ratios, and unequal population sizes over time. Thus the $N_e >$ 1,000 criterion suggests the need for a stable population of 3,000 to 10,000 breeding adults in each generation to prevent long-term loss of genetic variation.

Inbreeding Depression

As if the problem of finding a mate were not enough, individuals in small populations may suffer just as much or more as a result of finding the wrong mates. More specifically, they are likely to mate with close relatives with whom they share many genes. This situation is known as **inbreeding.** Inbreeding depression occurs when historically large, outcrossing populations suddenly decline to only a few individuals. The population then experiences reduced survival and fecundity. As explained in chapters 1 and 5, inbreeding is a major concern in conservation because it may contribute to extinction.

Inbreeding depression increases as relatedness increases (fig. 6.2). When the degree of relatedness of individuals in the population (so-called inbreeding-by-descent or paternity inbreeding) is regressed against one or more traits affecting fecundity or survival, the resulting regression can be used to calculate the degree to which increased mortality or lower

fecundity is associated with increased relatedness. Inbreeding depression is an especially well-documented problem in captive populations of vertebrates (Frankham 1995b). Of the 45 captive, inbred vertebrate populations examined by Ralls and Ballou (1983), 42 had reduced juvenile survival compared with outbreeding populations of the same species. In wild populations, inbreeding depression has been documented in fish, snails, lions, shrews, white-footed mice, and plants (Frankham 1995a and references therein).

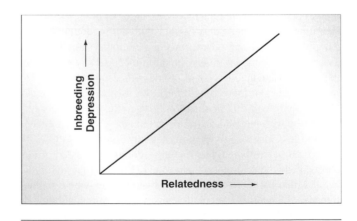

Figure 6.2

Relationship of relatedness in mating to levels of inbreeding depression, which is the decline in fitness (reduced survival and fecundity) associated with increased frequencies of mating among closely related individuals. The probability of an individual mating with a relative increases as the population size decreases.

(b)

(a)

Figure 6.3

(a) The squinting bush brown butterfly (*Bicyclus anynana*) and (b) the Sonoran topminnow (*Poeciliopsis occidentalis*) are two species whose fitness declines with inbreeding.

Some inbreeding occurs in all populations, no matter how large. However, inbreeding has disproportionately detrimental effects on small populations. Caughley (1994) summarized a disastrous sequence of events that inbreeding can initiate:

1. The frequency of mating between close relatives rises.
2. Heterozygosity is reduced in offspring, reducing the ability of the population to respond to environmental change.
3. Semilethal recessive alleles are expressed in a homozygous condition.
4. As a result of this expression, fecundity is reduced and mortality is increased (inbreeding depression).
5. The population becomes even smaller, amplifying the sequence initiated in step 1.

Caughley referred to this sequence as the "extinction vortex" of a positive feedback loop, or as Caughley put it, "the worse it gets, the worse it gets" (Caughley 1994).

Some of the effects of inbreeding can be evaluated by measuring the rate of juvenile survival, which is calculated theoretically as

$$\ln(S) = A + BF,$$

where S is the rate of juvenile survival, A is the instantaneous rate of juvenile mortality in progeny of unrelated parents, B is the same rate when the line is completely inbred ($H = 0$), and F is the coefficient of inbreeding, the probability that an individual receives at a given locus two genes that are identical by descent (Ralls, Brugger, and Ballou 1979). For a self-compatible population, the value of F is equal to

$$\frac{H_0 - H}{H_0}.$$

H_0 is the observed proportion of heterozygous genotypes in the population and H is the actual frequency. Thus, F is functionally a measure of how much heterozygosity is reduced as a result of inbreeding compared to a randomly mating population with the same frequency of alleles.

If one knows the F coefficient and a measure of the individuals' fitness in the population, it is possible to estimate the severity of inbreeding depression. Conventionally, the magnitude of inbreeding depression is expressed as the average reduction in mean value per 10% increase in the F coefficient (Van Oosterhout et al. 2000). Given this information, it becomes possible to estimate the cost of inbreeding, i, as

$$1 - \frac{\text{survival rate at } F = 0.25}{\text{survival rate at } F = 0.0}.$$

For example, suppose that the survival rate at $F_{0.25} = 0.2$ and at $F_{0.0} = 0.8$. Then the cost of inbreeding is $1 - (0.2/0.8)$, or $1 - 0.25 = 0.75$. Conceptually, the cost of inbreeding is the proportional decline in survival that can be attributed to inbreeding of a given magnitude. In 38 species of mammals, Ralls, Ballou, and Templeton (1988) estimated the actual average cost of inbreeding to be 0.33.

Other investigators also have documented measurable fitness costs in inbred populations. In the squinting bush brown butterfly (*Bicyclus anynana*) (fig. 6.3a), lifetime female fecundity was inversely proportional to the inbreeding coefficient of the female parent (Van Oosterhout et al. 2000). Smaller and more inbred populations also had more sterile egg clutches. In addition, zygote mortality, juvenile mortality, adult male and female longevity, and male development all were detrimentally affected by inbreeding (Van Oosterhout et al. 2000).

Genetic variability, which often declines with inbreeding, is correlated with fitness in some populations. For example, the Sonoran topminnow (*Poeciliopsis occidentalis*) (fig. 6.3b), a small fish once widespread in the southwestern United States, exists today as a group of isolated populations in springs in the state of Arizona. One of these, the Monkey Spring population,

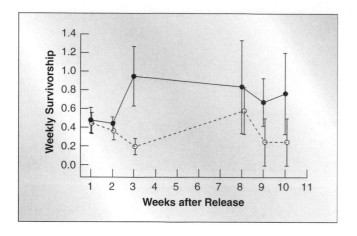

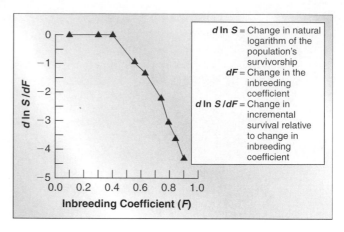

Figure 6.4

Survivorship of inbred (dotted line) and noninbred (solid line) white-footed mice (*Peromyscus leucopus novaboracensis*) in a mixed deciduous forest in Illinois. Bars represent standard errors of estimates. Noninbred animals had higher survivorship than inbred animals in all time intervals. Although none of the differences between groups was significant for any one estimate, when estimates were used as repeated measures of survivorship, the difference was statistically significant.

After Jiménez et. al. (1994).

had a mean heterozygosity of 0.0%, the lowest of all populations studied. The Monkey Spring population also had the highest mortality, slowest growth rate, poorest fecundity, and weakest developmental stability of all populations (Vrijenhoek 1994). In a controlled experiment, Jiménez and associates (1994) estimated the survivorship of inbred and noninbred white-footed mice (*Peromyscus leucopus novaboracensis*) in a mixed deciduous forest in Illinois. Mice in both treatments had been raised in a laboratory but were all descendants of wild mice. Mice were released into the field and recaptured at regular intervals, and trapping was done on surrounding adjacent habitat to estimate the proportions of mice that left the release site. Most mice were captured within 50 m of their release sites, and emigration rates were not different in the two treatments. Based on release-recapture ratios of marked mice in both inbred and nonbred treatments, Jiménez and associates determined that, over a 10-week period, noninbred mice had consistently higher survivorship than inbred mice (fig. 6.4). Some differences were sex specific. For example, male mice in both treatments lost weight in the first few days after release. Noninbred males regained their weight loss, but inbred males continued to lose weight throughout the experiment.

Low survivorship is one explanatory factor in extinction, and there is a clear threshold relationship between rates of inbreeding and rates of extinction in many species, with sharply increasing rates of extinction beginning at intermediate levels of inbreeding (Frankham 1995b). In one model of inbreeding, Frankham (1995b) demonstrated that changes in the inbreeding coefficient (F) produced little change in survivorship, at low levels of inbreeding, but large decreases in survivorship as inbreeding increases. Figure 6.5 depicts this relationship by comparing the value of the inbreeding coefficient with the ratio

Figure 6.5

The threshold effect of inbreeding on fitness. This model suggests that inbreeding has little effect at very low levels, but begins to cause an increasing rate of decline in fitness when it passes a "threshold" level at about $F > 0.45$.

After Frankham (1995b).

of the change in the natural log of survivorship ($d \ln S$) to change in the inbreeding coefficient. Note that when the value of F is relatively low (0.00–0.45), increases in F produce almost no changes in survivorship (the slope of the line is near zero). But when the inbreeding coefficient passes a critical "threshold" level (in this model, above about 0.45) survivorship declines rapidly with even small increases in the value of F. Although Frankham's model demonstrates that high levels of inbreeding lower survivorship, it should be noted that even very low rates of inbreeding have been associated with extinction at fixed population sizes (Latter et al. 1995). Inbreeding is almost always likely to contribute to overall population decline, especially in species with low reproductive rates (Mills and Smouse 1994).

Aside from the obvious and disastrous effects of increased mortality and decreased fecundity associated with inbreeding depression, inbreeding also appears to have significant effects on the genetic structure of populations (Sullivan 1996). Although inbreeding is not the direct cause of a loss of genetic variability, it increases the effect of genetic drift, which is likely to reduce a population's genetic variability (Templeton and Read 1994). Sullivan (1996) found that among populations of Colorado chipmunks (*Tamias quadrivittatus*) in the southwestern United States, inbreeding depression was more serious in relict and isolated populations with limited gene flow, even if the overall population was large and geographically widespread. The most isolated populations had little or no genetic variation at up to 30 loci (Sullivan 1996).

In plant populations, inbreeding may show extremes from no effect to complete reproductive failure. The latter outcome can occur in plants that are "self-incompatible," in other words, plants that are unable to generate viable seeds when they receive their own pollen. When inbreeding is high, different plants in the population may not recognize pollen from other individuals as "different," and reproduction fails. For example, in the rare lakeside daisy (*Hymenoxys acaulis*), native to the shoreline areas of the southern portions of the Great Lakes, the Illinois population was reduced to a small number of individuals that failed to set seed for 15 years. Researchers found that all individuals in the population were members of a single compatibility type, so pollen produced by an individual in this population could not fertilize any other individual in the same population. The introduction of plants from an Ohio population may enable the daisy to persist in Illinois, but only at the cost of the loss of the genetically unique Illinois population (Holsinger, Mason-Gamer, and Whitton 1999). At the other extreme are self-compatible plants whose populations may show no reduction in reproductive output even when inbreeding is at high levels. For example, Groom (1998) compared the numbers of seeds produced by self-pollinated and outcrossed pollinated individuals of the annual herb *Clarkia concinna concinna,* a species found only along the northern portion of the inner coastal range of California. In this self-compatible plant, there was no difference in seed production between self-pollinated and outcrossed individuals within or among patches (Groom 1998).

Geneticists today continue to emphasize the dangers associated with inbreeding (Frankham 1995a), but some conservation biologists have asserted that genetic concerns in small populations are of less importance than historically believed and that "loss of diversity is more likely to be a symptom of endangerment than its cause" (Holsinger, Mason-Gamer, and Whitton 1999). As Holsinger, Mason-Gamer, and Whitton (1999) put it, "those alleles most likely to be lost as a result of genetic drift—rare alleles—are also the least likely to contribute to any immediate response to natural selection."

As the results of more studies accumulate, the effects of inbreeding depression and loss of genetic diversity have become the subject of intense debate. Cheetahs provide a notorious case history of a population with low genetic diversity. As noted in chapter 5, sampled individuals from southern Africa had a complete absence of genetic diversity at 47 allozyme loci, and further studies of more loci showed similar trends (O'Brien et al. 1983; O'Brien et al. 1985; Caro and Laurenson 1994). Reciprocal skin grafts between pairs of unrelated cheetahs were either accepted or showed only slow rejection, indicating that the histocompatibility complex, a highly linked group of loci responsible for rejection of allogenic skin grafts, had little genetic variation (O'Brien et al. 1983; O'Brien et al. 1985; Caro and Laurenson 1994). In the wild, cheetahs suffer high mortality between birth and emergence from their lairs, a condition consistent with the high juvenile mortality of inbred species. However, long-term field studies of cheetahs determined that the most important cause of mortality during this period was lion predation (Caro and Laurenson 1994). Further mathematical analysis also revealed some suspicious results. Using the formula for loss of heterozygosity in small populations, $H_t = H_0 [1 - 1/(2N)]^t$, Amos (1999) estimated that the loss of 99% of the cheetah's genetic variability would require a population bottleneck of either 16 generations with a population size of 2, or 227 generations at a population size of 25. Both scenarios are unlikely. It is more plausible that cheetahs have historically had low levels of genetic diversity for reasons we still do not fully understand, or that patterns of cheetah dispersal and reproductive behavior, characteristic of metapopulations, may account for the observed low variability. These hypotheses are not mutually exclusive. Low genetic diversity in geographically widespread populations is not unique to the cheetah. The Eurasian badger (*Meles meles*) and Eurasian otter (*Lutra lutra*), both of which have large populations ranging over extensive areas, have lower than expected genetic variation (Amos 1999; Cassens et al. 2000), although the otter may have undergone a recent genetic bottleneck.

Mutational Meltdown

Along with genetic drift and inbreeding, the accumulation of deleterious mutations may also disproportionately affect small populations. When this occurs, it is known by the colorful name of **"mutational meltdown."** Suppose that the maximum number of progeny that each individual in a population can produce is equal to some value, R. If the reduction in the fitness of individuals because of accumulated deleterious mutations lowers the probability of survival of the offspring to maturity to less than $1/R$, then the per capita reproductive rate becomes less than one. At this point, the population would begin to decline. As population size declines, random genetic drift becomes proportionately more important than selection as an evolutionary force and the rate of accumulation of deleterious mutations increases, producing further declines in fitness. Such declines in fitness accelerate the rate of population decline still further, causing further increases in the force of drift and accumulation of further deleterious mutations, ultimately leading to the population's extinction.

To understand this phenomenon more analytically, imagine a population divided into "mutation classes." Individuals in the same class share the same number of mutations (although not necessarily mutations at the same loci). Assume that, initially, one class has no mutations. This is called the "zero mutation" class. However, if this class is lost (its members experience at least one mutation), it cannot be restored from other mutation classes (because all these carry at least one mutation). Once the zero mutation class is lost, it is lost forever.

After the loss of the zero mutation class, the class carrying one mutation is considered the "least-loaded" class. This class also will be lost if mutations accumulate. Assuming that mutations decrease fitness, then, with the loss of each least-loaded class, the mean fitness of the population will decline. This process of gradually lowering population fitness through an increasing concentration of mutations is called "Muller's ratchet," after the geneticist who first described the process (Muller 1964).

Assume that the population experiences a mutation rate of μ. The speed of Muller's ratchet depends on the size of the least-loaded class, C_0, in relation to the size of the population, N. If the order of events in a population is reproduction, mutation, selection, then the ratio of C_0/N is

$$\frac{C_0}{N} = e^{\mu(1-s)/s}.$$

Assume that there is selection against mutants, and that s is a selection coefficient equal to the fractional reduction in viability caused by a single mutation. Then the probability p of losing the least-loaded class in the next generation is proportional

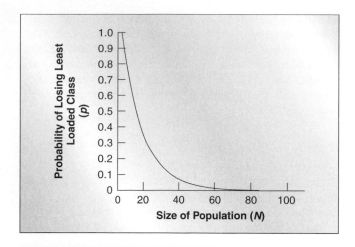

Figure 6.6

"Muller's ratchet." The smaller the population, the larger the probability that the least-loaded class will be lost as mutations increase in concentration in individuals. Given a constant rate of mutation and a constant selection coefficient, the key variable determining the rate of accumulating mutations is once again population size.

After Gabriel and Bürger (1994).

to the mutation rate, the strength of selection against mutation, and the size of population in this relation:

$$p = 1 - \exp\left[-\frac{\mu(1-s)}{s}\right]^N.$$

The higher the probability, the greater the speed of the ratchet, and the faster the mutations accumulate. Given a constant rate of mutation and a constant selection coefficient, the key variable is again population size. Muller's ratchet is depicted graphically in figure 6.6. Note that the smaller the population, the larger the probability that the least-loaded class will be lost (mutations will increase in concentration in individuals). The probability of losing the least-loaded class increases exponentially in populations with fewer than 40 individuals (fig. 6.6). Given that there is selection against mutants such that mutations reduce fitness, this means that increasing amounts of mutation will decrease population size, and that the effect will generate a positive feedback loop of increasingly more rapid population decline. Geneticists W. Gabriel and R. Bürger summarize where the population is now headed: "If the accumulation of deleterious mutations reduces the actual population size, then subsequent mutations accumulate faster because the chance of losing the least-loaded class is enhanced at reduced population size. Each loss of the actual fittest mutation class further reduces population size. This again speeds up the ratchet. Therefore, mutation accumulation and random genetic drift synergistically intensify each other and drive the population to extinction" (Gabriel and Bürger 1994). Lande (1994) predicted that such accumulation of deleterious mutations is as important as environmental stochasticity as a cause of extinction in populations of up to 1,000 individuals, and more important than demographic stochasticity. Not all scientists agree that the accumulation of unfavorable mutations contributes to a

significant risk of extinction. Although small populations are at greater risk for the accumulation and fixation of unfavorable mutations, leading to an overall loss of fitness in the population, the probability of this type of event is small. Thus, Holsinger, Mason-Gamer, and Whitton (1999) argue that "mutation accumulation is less likely to pose a threat to the continued persistence of most populations than typical levels of environmental variation."

Hybridization and Introgression—A Liability in Endangered Animal Populations

Hybrids are the offspring of matings between individuals of different species. In animals, hybrids typically suffer a number of disadvantages compared with nonhybrid individuals. Animal hybrids may be infertile, or even if fertile, they may have reduced mating success because nonhybrid individuals in both their parental species may not recognize them as potential mates. These and related problems make hybridization in animals a waste of reproductive effort by the parent individuals, and wasted reproduction is not something individuals in small populations can afford. The problem is compounded in a practical way because the Endangered Species Act (ESA) offers no protection to hybrids. Thus, hybrid animals may suffer two significant liabilities: lower fitness and no legal protection.

These liabilities become significant when hybrids occur between rare species, or between a rare species and a more common species. That is, when two closely related species coexist, and one of them is very rare, it may be genetically swamped, and its unique genome effectively exterminated, by interbreeding and hybridization with the more common species. A related problem, known as introgression, is the long-term acquisition and incorporation of genetic material from one species into the genome of another species. Introgression makes it difficult to identify, establish, and maintain the genetic integrity of a species in relation to other, closely related species. One of the best, and most complex, examples of the problems that can arise from inbreeding, hybridization, and introgression involves the case of the red wolf (*Canis rufus*) of the south-central United States (fig. 6.7).

Attempts to reestablish populations of the endangered red wolf, which became extinct in the wild about 1975 (Wayne 1996), were controversial because of the possibility of past hybridization and are threatened by the possibility of current hybridization. Some scientists have argued that the red wolf is not a true species because of past hybridization and introgression. Recent genetic studies have found no definitive genetic features in the red wolf and have been unable to establish any definitive pattern of genetic phylogeny for the red wolf compared with coyotes (Wayne and Jenks 1991; Roy et al. 1994; Roy, Girman, and Wayne 1994). This has led some to argue that the ongoing captive breeding program and proposed reintroduction of the red wolf should be abandoned (Ezzell 1991). Current reintroductions could be hindered by the potential for hybridization between red wolves and coyotes (*Canis latrans*) or domestic dogs (*Canis familiaris*). This hybridization could hinder recovery because it may occur before the population of red wolves reaches sufficient size to maintain breeding within species.

Figure 6.7
The red wolf (*Canis rufus*), a species threatened with genetic extinction through hybridization with and introgression from gray wolves (*Canis lupus*), coyotes (*Canis latrans*), and dogs (*Canis familiaris*).

Despite these criticisms, morphological examination of museum specimens supports the view of the red wolf as a species distinct from the coyote, and early twentieth-century specimens of the gray wolf (*Canis lupus*) from central Texas are completely distinct from red wolf–coyote hybrids that were known and collected from that area at the same time (Novak 1999). Additionally, fossil remains of wolves from Florida have cranial measurements identical to red wolves, not gray wolves, leading to the theory that the red wolf represents a surviving line of "small primitive wolves that once occurred throughout the Holarctic and that formed an evolutionary stage between the coyote and the modern gray wolf" (Novak 1999). Whether the genetic similarities of red wolves to other canids represent recent introgression or the acquisition of DNA from other species through hybridization in the more distant past, the case of the red wolf vividly demonstrates the dangers of genetic introgression for a species at low numbers when confronted by sympatric, closely related, and reproductively compatible species that exist around it at high numbers.

The complexities of hybridization and introgression are not confined to the Canidae. The Florida panther (*Puma concolor coryii*) also faces issues of genetic integrity that have both legal and biological ramifications. The Florida panther is an endangered subspecies of mountain lion (also known as cougar, catamount, or puma) that once ranged throughout the southeastern United States but is today confined to the Everglades and Big Cypress Swamp of southern Florida. In the late 1980s researchers discovered two distinct genetic strains in the remaining individuals of the Florida panther population. One strain was associated with the main population in the Big Cypress Swamp, and the other was identified in two family groups that had appeared in the Everglades. Genetic analysis demonstrated that the Everglades individuals were distinct not only from the Big Cypress panthers,

but also from western U.S. mountain lions. Their genetic characteristics were closer to subspecies from South and Central America (O'Brien and collaborators 1996). Inspection of the archives of the Everglades National Park revealed that between 1957 and 1967, seven animals from a captive stock had been released into the Everglades with National Park Service cooperation. This stock was derived from a mixture of Florida panthers and individuals from South American subspecies (O'Brien et al. 1990). Unfortunately, as noted earlier, the Endangered Species Act does not protect hybrids between endangered taxa because such hybrids are not considered to protect or help the recovery of listed species, and could jeopardize its genetic integrity. This legal understanding of the ESA, established through court cases, has become known as the "hybrid policy," and meant that the Florida panther was no longer subject to legal protection under the ESA. However, the hybrid policy was subsequently suspended to avoid penalizing species because of "a bureaucratic precedent that did not anticipate the resolving power of molecular genetics" (O'Brien and collaborators 1996).

Hybridization in Plants—Conservation Threat or Conservation Asset?

Among animals, hybridization is relatively rare and hybrids are often infertile, but plants hybridize freely and hybrid offspring are often capable of mating with one another and with either parental species. Thus, plant ecologists have long considered hybridization a normal phenomenon in plant populations and an important contributor to adaptation and subsequent speciation (Stebbins 1950; Grant 1971; Lewis 1980). In plants, hybridization can alternatively present a threat to endangered species or, in other cases, a stimulus to increased biodiversity.

Plant conservationists working in the field suspect that the hybridization of rare species with closely related, common species is threatening a large number of rare plants with extinction. For example, land stewards of The Nature Conservancy in the United States cited a number of rare or endemic species in the western United States that are believed to be in danger of being lost through hybridization with other species. These include endangered, threatened, or candidate endangered species such as the white firewheel (*Gaillardia aestivalis*), the Bakersfield saltbush (*Atriplex tularensis*), the western bog lily (*Lilium occidentale*), Nelson's sidalcea (*Sidalcea nelsonii*), and the peacock larkspur (*Delphinium pavonaceum*). However, the accounts of hybridization in these species exist largely in the form of unpublished government or agency reports, and none is supported by molecular experiments or evidence (Rhymer and Simberloff 1996). This lack of publication or molecular data does not mean that these concerns are unwarranted. It does, however, suggest that potential threats from hybridization in plants can occur rapidly, and can potentially produce extinction before conservationists are able to enjoy the luxury of careful scientific experimentation and study.

Interestingly, field and experimental data indicate that zones of active plant hybridization are also zones of high levels of biodiversity in nonplant taxa, especially insects and fungi (Whitham et al. 1999). For example, in zones of hybridization among different species of *Eucalyptus* in Australia, 29 of 40

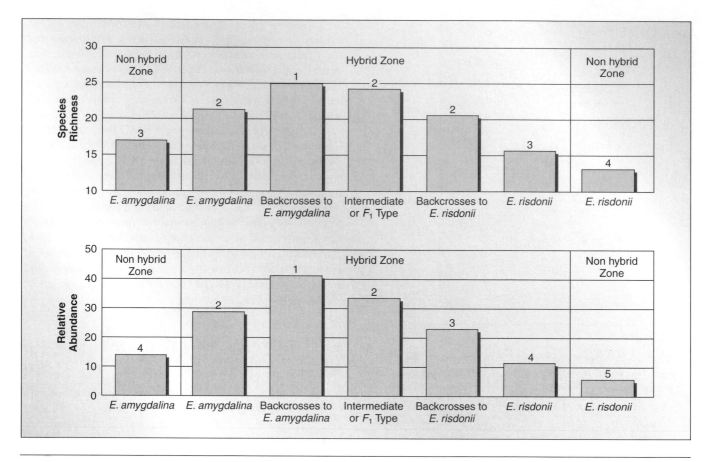

Figure 6.8

Average species richness and relative abundance of 38 insect and 2 fungal taxa were higher on hybrid *Eucalyptus* trees than on parental trees in pure stands or intermixed with hybrids. Groups with different numbers are significantly different in value. Relative abundance was calculated by averaging standardized abundance measures for each taxon.

After Whitham et al. (1999).

insect and fungal taxa examined were more abundant in hybrid zones than in "pure" zones (Whitham et al. 1999). This occurred primarily because hybrid trees accumulated insect and fungal species that were otherwise unique to each parent species, thus producing the genetic equivalent of an "edge effect" (fig. 6.8). More controlled experiments in specially created "hybrid gardens" supported these field observations. Thus, hybridization in plants may be an asset to conservation rather than a liability because it is positively correlated with, and may provide an underlying genetic basis for, higher levels of community biodiversity. After reviewing both field and experimental evidence, Whitham and colleagues (1999) concluded that, in plants, "hybrid zones can be centers of biodiversity."

Outbreeding Depression, Self, and Intrinsic Coadaption

Although usually less significant in its effects than inbreeding depression, the problem of outbreeding depression deserves mention, both to define the concept and to note the characteristics of populations that may suffer from its effects. As the external environment may mold local adaptations by natural selection, the internal genetic environment of a population may lead to the pro-

duction of local subunits of genes that interact in a mutually favorable manner. Thus, population subunits that are highly adapted to local conditions (a condition common in metapopulations) and that have low vagility may evolve coadapted gene complexes, in which genes must be inherited together to produce appropriate adaptive effects. When individuals from such normally inbreeding populations breed with individuals from other populations of the same species (outbreed), they may decline in fitness as their uniquely coadapted genetic combinations are broken up. Outbreeding depression is most common in plants (Frankham 1995a), especially in populations that have evolved high levels of self-pollination (selfing). In fact, selfing has evolved repeatedly in plant taxa that were previously outcrossing (Stebbins 1957; Grant and Grant 1965; Raven 1979). Selfing is particularly adaptive where pollinators are at low density and the accompanying probability of being pollinated is low. Under these conditions, selfing populations may show little if any effects of inbreeding depression as noted earlier (Groom 1998). For example, in the annual and perennial herb genus *Epilobium*, selfing and outcrossing species are sympatric, often growing on the same site. *Epilobium ciliatum* is mostly self-pollinating, while the closely related species *E. angustifolium* is usually outcrossing. In forest clear-cuts of the U.S. Pacific Northwest, both species often

Figure 6.9
The collared lizard (*Crotaphytus collaris*), a species in which local, highly inbred populations are common.

invade the same site after trees are removed. Parker, Nakumura, and Schemske (1995) compared the relative levels of inbreeding depression in these sympatric species and found, not surprisingly, that there was less inbreeding depression in the selfing *E. ciliatum* than in the outcrossing *E. angustifolium*. However, poor reproductive performance (outbreeding depression) was also documented in some lines of outcrossing of *E. angustifolium*, suggesting the presence of local populations with co-adapted gene complexes that may have been disrupted by outcrossing (Parker, Nakumura, and Schemske 1995).

Although far less common than in plants, outbreeding depression is also a potential genetic problem in animals. One of the most fascinating examples is found in populations of the collared lizard (*Crotaphytus collaris*) (fig. 6.9) of the Ozark Mountains of the south-central United States (Templeton 1986). Today the collared lizard lives primarily in dry grasslands and deserts in the southwestern United States, but its range previously extended further east. When, in recent centuries, the eastern portion of its range became wetter and more forested, the collared lizard declined in these regions. Declines were further exacerbated by fire suppression that reduced or eliminated tallgrass prairies in the southeast where the lizard still survived. Today the only eastern remnants of the species are populations in the Ozarks (mainly in the state of Arkansas). There lizards survive in small openings (glades) associated with rocky soils and outcrops, but will not disperse through the surrounding oak-hickory forests to other glades. As a result of an estimated 2,000 generations of isolation among populations averaging fewer than 50 individuals, intraglade populations have become genetically identical. A small number of genes also appear to exist together in a coadapted gene complex, with the genetic structure of the complex slightly different in each glade. (Templeton 1986).

Although outbreeding depression has not been explicitly documented in the collared lizard, this species illustrates two forces that can allow outbreeding depression to occur. One is the phenomenon of *local adaptation* and accompanying genetic differentiation. The second phenomenon is known as *intrinsic coadaptation,* in which genes in a local population primarily

adapt to the genetic environment defined by other genes. This latter phenomenon can occur in a species that becomes subdivided into small, isolated populations. The outbreeding depression associated with intrinsic coadaptation is normally a temporary phenomenon rapidly eliminated by natural selection but, in the collared lizard, intrinsic coadaptation has become a more permanent fixture of these glade-specific populations.

The collared lizard presents unique genetic considerations for conservation efforts that attempt to establish new populations in currently unoccupied glades. In new glades, individuals from different populations must be used in the founding group. Although outbreeding depression was initially a serious consideration in these conservation efforts, it is no longer expected to have any deleterious effects on founding populations. The coadapted gene complex is expected to either be preserved in the founding population (because the number of genes in the complex is small) or selection is expected to quickly and efficiently establish a new parental genotype of high fitness (Templeton 1986).

POINTS OF ENGAGEMENT—QUESTION 2

Design an experiment to determine if outbreeding depression has a significant effect on populations of collared lizards. What would be your research hypothesis, your experimental design, and your test consequence? Even if no effects are documented, would genetic considerations still matter in a management program to establish new populations? If so, how would such considerations affect the management plan?

GENETIC TECHNIQUES: SOLVING THE PROBLEM OF ASSESSING GENETIC STATUS AND CHANGE

It is not merely increased concern over genetic variation that has driven the science of genetics to its current prominence in conservation biology, but also the increasing precision and sophistication of techniques associated with genetic analysis, particularly at the molecular level. It is beyond the space and scope of this chapter to review all the genetic techniques currently used to assess genetic variation in individuals and populations. Interested readers should study the recent excellent review of these developments in the book *Molecular Genetic Approaches in Conservation* (Smith and Wayne 1996) for more detail. In this chapter we will examine the principles, procedures, strengths, and weaknesses of the techniques that are used most widely in conservation and that appear most often in current conservation biology literature. By appreciating what individual techniques can and cannot do, one can not only choose the right technique to solve a particular problem, but also avoid misinterpretations and inappropriate applications of results that such techniques provide.

Allozyme Electrophoresis: Genetic Variation at Molecular Levels

Allozyme electrophoresis was one of the first molecular techniques to offer a sensitive assay of genetic variation, and is still one of the most widely used. Using electrophoresis, protein variants of different alleles can be identified, and the frequencies of different alleles in different populations can be used as indices of potential relatedness and extent of gene flow. This technique can allow many different genes in a population to be analyzed (Cronin 1993). From 1991 to 1994, more genetics papers published in the journal *Conservation Biology* used allozyme electrophoresis than all other methods combined (Leberg 1996), but the importance of analyzing allozymes is likely to decline as other molecular techniques become more available and precise.

In allozyme electrophoresis, enzyme proteins with different charges are separated through movement in a chemical medium (gel) to oppositely charged poles in an electric field. Allozymes, different allelic variants at a single locus, respond to electrophoresis because they are differentially charged particles that move through the gel at different rates. Charge differences are influenced by their amino acid composition. Because amino acid composition is directly controlled by the sequence of nucleotides in DNA, charge differences are assumed to represent differences in DNA sequences (Leberg 1996). The common way to visualize an allozyme in a gel exploits the enzymatic nature of the allozyme itself. The gel is incubated in a colorless mixture of reactants that can be modified by the allozyme. The modification of the reactants produces a colored product that stains the gel, producing discrete, visible bands of color.

Allele frequencies determined by allozyme analysis are based on the assumption that different allozymes are unique because of differences in their nucleotide sequences. Genetic analyses of populations can determine the frequency of these alleles. Through these genetic analyses, the relationships between populations that are inferred from the allozyme data are based on the similarity of the allele *frequencies,* not on the similarity of the alleles themselves.

Allozymes are of interest to conservation biologists because they are the products of genes and their polymorphisms permit comparisons among individuals, populations, or species that can be useful in showing degrees of genetic relatedness. And because allozymes can reveal codominant inheritance patterns in genotypes, allele frequencies can be determined directly. Allozymes themselves can be obtained for study from almost any type of tissue, although tissues rich in protein are generally used. Leberg (1996), in a comprehensive review of allozyme electrophoresis, notes nine potential and actual applications of this method in conservation genetics. These include the estimation of genetic variation, determination of association of genetic diversity with fitness and population viability, determination of effective population size, rates of gene flow, types of mating systems (degree of inbreeding avoidance) used in the population, population structure, degree of hybridization, identification of paternity and species determination, and resolution of phylogenetic relationships. Most of these applications could also be addressed by other genetic techniques, which will be treated in detail later in this chapter.

Information about allozymes gathered by electrophoresis makes possible comparisons of genetic characteristics and variation that cannot be made with other techniques. For example, Hartl and Pucek (1994) demonstrated, using allozyme data, that species that had experienced past population reductions had reduced levels of allozyme diversity. The extensive base of allozyme data makes it possible to survey genetic information on a large number of independent loci in many individuals of a population, an important task in many conservation efforts. The procedure is of relatively low cost and, for a trained individual, easy to apply (Leberg 1996).

The Polymerase Chain Reaction (PCR): A Noninvasive Method for Genotyping Endangered Species

The development of the **polymerase chain reaction (PCR)** as a standard genetic technique revolutionized conservation biology and management beginning in the late 1980s. This noninvasive technique for determining genotypes of individual animals is based on a relatively simple reaction, in which a short region of a DNA molecule, even as small as a single gene or smaller, is copied repeatedly by a DNA polymerase enzyme. In standard PCR, any region of a DNA molecule can be chosen as long as the borders (beginning and ending sequences) are known. To carry out the PCR, two **oligonucleotides** (short pieces of DNA) hybridize to the DNA molecule, one to each strand of the double helix. These act as primers for the subsequent DNA synthesis and delimit the region to be amplified. An enzyme, DNA polymerase 1, is added from the bacterium *Thermus aquaticus,* an organism native to hot springs whose enzymes, including its DNA polymerase (known as *Taq*), resist denaturization when exposed to heat (Brown 1995). The Taq enzyme is incubated in the solution and facilitates the production of new complementary DNA strands. The mixture is then heated so that the new strands detach from the original DNA, then the strands are cooled, allowing more primers to hybridize at their respective positions, including positions on the newly synthesized strands. The Taq enzyme carries out a second round of DNA synthesis and the cycle can be repeated many times, eventually resulting in the synthesis of several hundred million copies of the amplified DNA fragment, which can then be analyzed in various ways (Brown 1995). The most common and useful analysis for conservation biology is direct sequence analysis of the PCR products, resulting in an identification of the genotype of the organism from which the material was obtained. The products also can be seen visually by electrophoresis in an agarose gel after staining with ethidium bromide (Fritsch and Rieseberg 1996) (fig. 6.10).

Random Amplified Polymorphic DNA (RAPD) Analysis

A different genetic analysis technique, but one that makes use of PCR, is **random amplified polymorphic DNA (RAPD) analysis.** In RAPD, only a single random oligonucleotide primer is used. RAPD can generate essentially unlimited numbers of loci

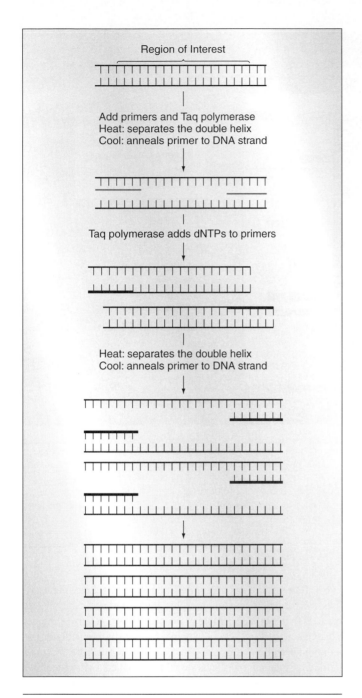

Figure 6.10

A schematic representation of the mechanism for a polymerase chain reaction (PCR), which is used to multiply (i.e., amplify) the segment of DNA of interest.

for analysis using a variety of types of genome samples. Only a small amount of tissue is needed, and this tissue can be stored indefinitely prior to sampling. The same DNA primer can be used on any organism (Fritsch and Rieseberg 1996). But what renders the RAPD technique especially useful to conservation biologists, particularly in field studies, is that it makes noninvasive genetic sampling possible (Morin and Woodruff 1996). Small amounts of discarded hair, feathers, feces, urine, fins, scales,

eggshells, antlers, or other material from a living creature can be used without capturing or handling the organism (Morin and Woodruff 1996). In fact, RAPD analysis can be used even with DNA collected from materials that animals have chewed and then spat out (Inoue and Takenaka 1993; Takenaka et al. 1993).

PCR and RAPD as Tools of Taxonomic Assessment

PCR analysis can be used as a postmortem technique, such as on museum specimens (Morin and Woodruff 1996), and only minute amounts of DNA are required for analysis. The analysis can include genetic fingerprinting to identify particular individuals. Using reverse transcription PCR (RT-PCR), RNA can be amplified and the amount of messenger RNA (mRNA) can be determined in the sampled tissues, providing an accurate index of the activity of the parent gene. From this index, it is possible to infer many aspects of the physiological status of the organism, including reproductive and nutritional conditions (Morin and Woodruff 1996).

PCR and RAPD analyses also have important applications in phylogenetics. For example, the banding pattern observed in RAPD analysis, is a reflection of sequence variations of the template DNA taken from the sample. Differences in sequence variation of two or more individuals can be assessed for relatedness because closely related organisms would be expected to have banding patterns more similar than distantly related organisms. In RAPD, the banding patterns can be broadly classified into two groups: variable (polymorphic) and constant (nonpolymorphic). Products that qualify as constant or variable are relative to the taxonomic group. For example, an RAPD analysis can identify certain fragments that are always the same for all species within a genus, and also other fragments that differ among species. Similarly, some fragments are the same for all populations within a species, but other fragments differ among populations. The variable fragments can be used to make a positive identification of the species or population that contributed the genetic material (Hadrys, Balick, and Schierwater 1992) (fig. 6.11). This technique is an increasingly powerful and common tool in resolving taxnomic and phylogenic ambiguities in related species or in identifying relatedness of different populations. Such analysis enables conservation biologists to become more certain in determining which populations and taxa to manage and protect. Because PCR and RAPD can be used not only on tissues from living organisms, but also even from those long dead, such as museum specimens, it provides a means to compare past populations to current ones. As a result, PCR and RAPD have played an important role in the previously discussed studies evaluating the comparative genetics of gray wolves, red wolves, and coyotes in historic and extant populations, and in evaluating past and current levels of hybridization and introgression (Wayne and Jenks 1991; Roy et al. 1994). Such analyses can be taken still further to determine paternity and kinship relationships as well because, in principle, RAPD markers can be treated like Mendelian alleles, and then used in single-locus "fingerprint" profiles to determine, with a high degree of probability, the genetic relatedness between individuals (Hadrys, Balick, and Schierwater 1992). Such fingerprinting applications have increasingly widespread application in conservation biology.

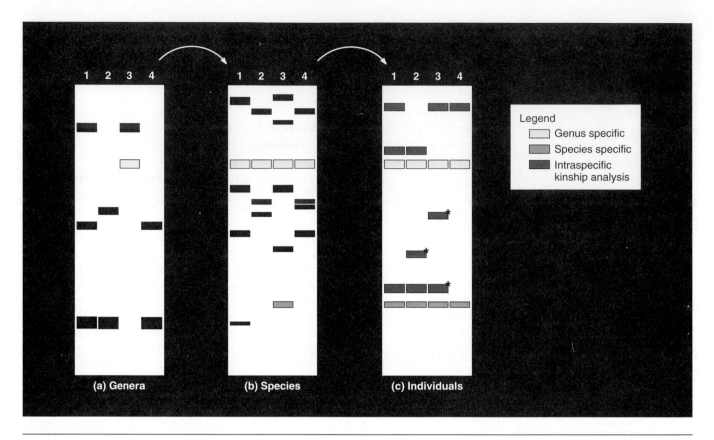

Figure 6.11

An example of the use of polymorphic and nonpolymorphic RAPD fragments to achieve different levels of taxonomic resolution. In (a), the presence of a genus-specific band identifies individual 3 as a member of a particular genus. In (b), all individuals have the same genus marker, but a species-specific fragment present in individual 3 identifies it as a particular species within the genus. In (c), three polymorphic fragments (indicated with *) can be used to distinguish different individuals of the same species.

After Hadrys, Balick, and Schierwater (1992).

DNA Fingerprinting: The Use of Satellite Markers

Many sections of an organism's genome consist of short sequences of DNA that may be repeated up to 1 million times. Such repeating segments of DNA often have different sequences of bases than other forms of DNA, and therefore different molecular densities. Such highly repetitive sequences are known as **"satellite DNA"** (Sudbery 1998).

Some types of satellite DNA are interspersed throughout an organism's genome. There are two classes of such DNA, respectively called **minisatellites** and **microsatellites**. The minisatellites consist of sequences of 10 to 100 base pairs in length, repeated in tandem arrays that vary in size from 0.5 to 40 kb (kilobase, a unit of 1,000 base pairs). Some of the loci on homologous chromosomes within minsatellites are highly variable in length. Hence, minisatellites are also sometimes referred to as variable number tandem repeats (VNTRs), and their variability forms the basis for DNA fingerprinting (fig. 6.12). Differences in length are unique to particular individuals, and thus provide a basis for certain identification.

Compared with minisatellites, microsatellites are much smaller, consisting of short tandem repeats (STRs) of units only two to four nucleotides in length (Sudbery 1998). Like minisatellites, microsatellites are polymorphic and provide valuable genetic markers, but they have a more uniform distribution in the genome. This makes them especially valuable in determining pedigrees of individuals and in forensic applications to determine the origin of individual animals or (illegal) wildlife products. Microsatellites also can be evaluated with less expensive and less labor-intensive techniques. The high variability of mini- and microsatellite markers makes both useful for many applications in conservation biology because allozyme (protein-coding) DNA markers are invariant in many species (Hedrick 1999).

Although extraordinarily valuable in distinguishing genetic differences among individuals and populations, minisatellite and microsatellite DNA analysis must be interpreted with caution. The high variability that gives mini- and microsatellites an advantage over allozyme markers can also lead to serious errors of interpretation. Precisely because both mini- and microsatellites usually possess very high within-population heterozygosity, the magnitude of differentiation measured between populations may be small. For example, the commonly used value of allelic differentiation in alleles among multiple subpopulations, G_{ST}, can be small because it is derived from the difference between the proportion of heterozygous individuals in the population (H_T) and the

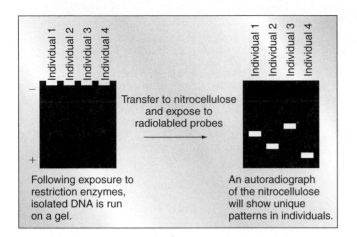

Figure 6.12
A schematic example of a DNA "fingerprint" produced by electrophoresis of DNA associated with microsatellite markers. Restriction enzymes cut the DNA at specific sequences. The amount of DNA between sequences differs between individuals; therefore, unique patterns will appear. Because DNA has a slight negative charge, it will move toward the positive electrode. Smaller pieces move faster and are found near the base of the gel. Following separation, the DNA double strands are denatured to single strands using heat or chemical treatment. A DNA print of the gel is transferred and fixed onto nitrocellulose paper. The single-stranded DNA is subjected to a probe specific for a sequence within the microsatellite. Unique banding patterns of the radioactive probe can be observed using autoradiography.

averaged weighted heterozygosity within subpopulations (H_S) divided by H_T. In other words,

$$G_{ST} = \frac{H_T - H_S}{H_T}$$

If both H_T and H_S are large (and they can be if measured with micro- or minisatellite DNA), then the value of G_{ST} will be small. For example, if $H_T = 0.80$ and $H_S = 0.75$, then $G_{ST} = 0.05/0.8$, or 0.0625. Thus, the method can give a (false) impression of low genetic variability within the population. The same problem also occurs with other commonly used measures of genetic variability in populations, so measures chosen to estimate differences should be variation independent (Hedrick 1999). A second problem with satellite DNA is that bottlenecks can generate large genetic distances in a short time in these loci. Mini- and microsatellite data from two populations can be used to estimate time since divergence because genetic distance is assumed to increase linearly with time since divergence (Nei 1972). However, if one or both of the groups have gone through a substantial reduction in population size, the observed genetic distance may not accurately reflect time since divergence. The problem is further accentuated in comparisons of multiple groups (three or more populations) if those populations have experienced historic differences in effective population size. This problem can be avoided by using statistical tests to determine from genetic data whether or not a bottleneck has occurred (Cornuet and Lukart 1996).

A third problem is perhaps the most basic and important of all: genetic markers are indices of genomic variation, and the genes they contain are not necessarily the ones that are the most important targets of natural selection, or the most important determinants of an organism's fitness (O'Brien 1994). In other words, the patterns of variability present in the most adaptive loci (i.e., loci that exert more direct influence on fecundity and survival) may not be closely correlated with variability in the highly variable loci of minisatellites and microsatellites (Hedrick 1999). Practically, this means that statistically significant differences in genetic data associated with satellite markers may have little biological significance. To avoid this problem, Hedrick (1999) recommends using tests of statistical power to evaluate a known biological effect. For example, one can determine the relationship between the number of loci needed for evaluation to detect bottlenecks of different sizes at different levels of probability. This relationship can be seen in figure 6.13, where each of the curves displayed represents a statistical "isobar" of equal probability of detection for a combination of number of loci and bottleneck size. For example, with only five loci, the probability of detecting a population bottleneck of 30 is less than 0.80. But if the number of loci is increased to 20, the probability of detecting a bottleneck of the same size increases to more than 0.95. Such an analysis demonstrates that it is relatively easy to detect bottlenecks if many loci are evaluated. However, if such bottlenecks last only one generation, they may have little biological significance.

Alternatively, there may be a lack of statistically significant differences in satellite DNA even when there is a biologically meaningful difference between the two groups. Hedrick (1999) notes that Scotch pine (*Pinus sylvestris*) in Finland shows no significant genetic distance differences when satellite markers are assessed between northern and southern populations, yet these populations differ in a number of important adaptive traits and do not transplant well from one region to another. The opposite error arises when there are statistically significant differences between genetic markers in populations, but no measurable adaptive differences. This is a particular problem in cases with many highly polymorphic loci where very small differences may be statistically significant but do not affect adaptive traits.

To solve these kinds of problems, analyses of satellite markers should be combined and complemented with analyses of allozyme markers that are more likely to include genetic material more closely correlated with fitness. However, future advances in genetics may solve this problem in a more meaningful way. Just as the Human Genome Project is now successfully mapping the location of every human gene and determining its function, so concurrent research is now mapping the genomes of other species. This research has so far revealed many similarities in basic organization and a highly conservative genetic structure among related organisms. Thus, geneticists may one day map the genomes of other nonhuman organisms, perhaps including those most imperiled with extinction, using the human genome as a starting point.

Mitochondrial DNA

Mitochondria are organelles found in all eukaryotic cells that function primarily in converting chemical energy into metabolic energy that can be used directly by the cell. Unlike most

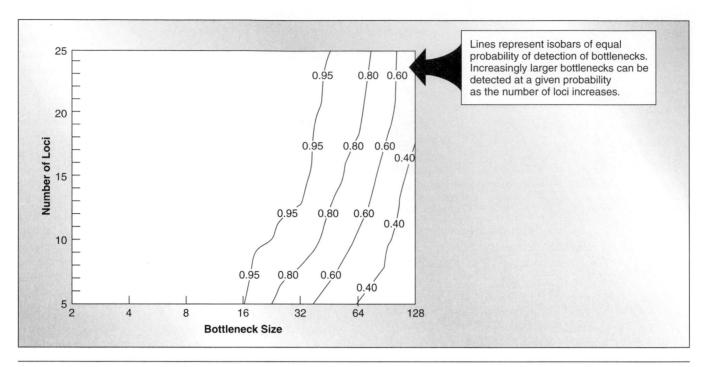

Figure 6.13
Number of loci and bottleneck size affect the probability of detecting past bottlenecks through genetic analysis. As the number of loci examined increases, the probability of detecting a past bottleneck increases. As the size of the bottleneck increases, the probability of detection decreases.

After Hedrick, P. W. (1999).

other organelles, mitochondria have their own, nonnuclear complement of DNA. Each mitochondrion contains between 2 and 10 copies of a circular genome, much smaller than the corresponding nuclear DNA genome. Mitochondrial DNA is specific for particular cellular functions associated with the mitochondria, including the synthesis of subunits that function in cellular respiration, units that code for the synthesis of the transfer RNA (tRNA) of each amino acid, and DNA involved in the synthesis of ribosomal RNA (rRNA) that functions in protein synthesis (fig. 6.14).

Mitochondrial DNA is also unique in that it is maternally inherited as a linked set of genes, passed on to progeny in the cytoplasm of the egg cell (Cronin 1993; Sudbery 1998). Thus, there is no recombination between maternal and paternal genomes. This mode of inheritance results in rapid mtDNA differentiation relative to nuclear genes and makes the construction of phylogenetic trees straightforward because, without recombination, the number of nucleotide differences between the mtDNA genomes of different individuals, populations, or species can be assumed to be a direct measure of phylogenetic relatedness (Sudbery 1998). Theoretically, phylogenies derived from mtDNA are not affected by historic changes in effective population sizes, unequal sex ratios, or unequal family sizes. Many taxonomists consider phylogenetic analysis of mtDNA to be a superior, more objective method of determining phylogeny because it is based entirely on quantitative characters. Because it is maternally inherited, mtDNA can be used to determine a maternal lineage in an individual or group, but not a paternal one.

The small size of the circular mtDNA genome means that its sequences have long been known, and it is easy to identify polymorphisms in mtDNA. This characteristic is complemented by the fact that certain regions of mtDNA evolve at faster rates than single-copy nuclear genes in mammals (Wilson et al. 1985), so they can be especially useful for studying differences at the population level. The mtDNA molecule as a whole evolves at a rate of about 2% per 100,000 years, but some areas on the molecule can change at up to five times this rate (Cann, Stoneking, and Wilson 1987; Greenberg, Newbold, and Sugino 1983). Overall, mtDNA evolves 5 to 10 times faster than nuclear DNA, and thus permits studies of recent evolutionary change that nuclear DNA would not detect. Mini- and microsatellite analysis, RAPD analysis, protein electrophoresis, and restriction enzyme analysis are commonly used on mtDNA.

As powerful, and as useful, as mtDNA analysis can be in conservation biology, it is critical to remember that mtDNA normally evolves independently of and at a much faster rate than nuclear DNA. Additionally, mtDNA may not be as closely correlated to the organism's fitness or to its true phylogenetic divergence from other populations as nuclear DNA. Thus, interpretations and conclusions about phylogeny and taxonomy based on mtDNA results have led to questionable interpretations, erroneous conclusions, and imprudent proposals. For example, the recent discovery that mtDNA in Sumatran and Bornean orangutans (*Pongo pygmaeus*) was different led to proposals that they be classed as separate species (Xu and Arnason 1996; Zhi et al. 1996), as did similar findings in geographic races of North American brown bears (*Ursus arctos*). However, such populations have not long been geographically separated and show little or no consistent morphological distinctions. In these cases, it is unlikely that important conserva-

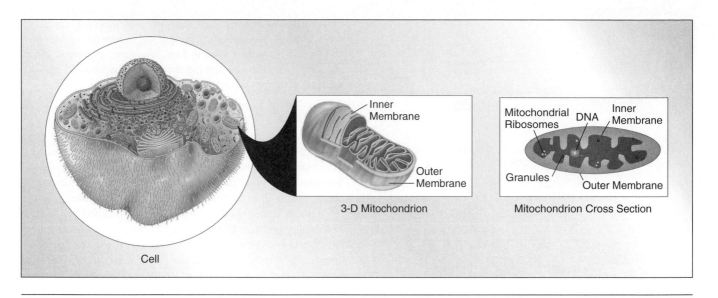

Figure 6.14

Schematic diagram of a mitochondrial cross section. Unlike nuclear DNA, mitochondria DNA is not enclosed within a membrane. Rather, it is found within the matrix of the organelle.

tion goals are advanced by treating every genetically distinct population as a separate entity, and such a strategy could lead to inefficient and ineffective efforts to conserve species.

Although some related species and subspecies have phylogenetically distinct mtDNA, other entities long known and recognized as separate species show little or no divergence in mtDNA. Cronin (1993) notes that the mule deer (*Odocoileus hemionus hemionus*) has mtDNA that is divergent from its conspecific subspecies, the Sitka black-tailed deer (*O. h. sitkensis*) and the Columbian black-tailed deer (*O. h. columbianus*), but not distinct from a different species, the whitetail deer (*O. virginianus*). Likewise, there is more similarity of mtDNA in brown bears and polar bears (*U. maritimus*) than among geographic races of brown bears, and there is more similarity between some populations of mallards (*Anas platyrhynchos*) and black ducks (*Anas rubripes*) than among different populations of mallards. No one would seriously propose, however, that brown bears and polar bears, or mallard ducks and black ducks, be classified as the same species, receive the same management treatments, or be subject to the same conservation strategies. Thus, Cronin concludes that mtDNA alone does not provide a sufficient assessment of overall genetic differentiation, but should be used only when it can be complemented by an analysis of other, nuclear genetic material (Cronin 1993). The variable mtDNA genotypes in ancestral populations may be sorted independently in descendent populations, some of which may differentiate into new species, although others may not. After speciation, nuclear genes in the two groups may diverge greatly, but ancestrally shared mtDNA genotypes may not. As a result, perfectly recognizable and distinct species may have nearly identical mtDNA sequences. As conservation biology continues to use genetic techniques and paradigms as a basis for determining appropriate populations and strategies for conservation, care and discernment are needed to ensure that the power of a genetic technique does not run ahead of sound biological judgment.

Restriction Fragment Length Polymorphism (RFLP): A Technique for Assessment of Genetic Variation Among Individuals

To measure variations of nuclear DNA among individuals, one of the most common and effective techniques uses **restriction fragment length polymorphisms** or RFLPs, which are variations in the length of restriction fragments produced from identical regions of the genome (Lodish et al. 2000). Variations in DNA sequence (DNA polymorphisms) may create or destroy restriction enzyme recognition sites. As a result, the patterns of restriction fragment lengths produced from exposure to restriction enzymes may vary in homologous chromosomes among individuals. To detect the differences, a radioactive marker (probe) is used to bind to a specific sequence associated with the fragments in order to detect the presence of the fragments. If there are no differences in the sequences of two homologous chromosomes in a particular DNA region, a restriction enzyme that cuts that region will produce fragments of identical length and sequence, and the probe will recognize this as a single fragment that will appear as one band (the restriction site occurs at the same place in both chromosomes). However, if a mutation has occurred that has destroyed a recognition site for a restriction enzyme in one individual compared with another, fragments of differing lengths will be produced, and these will be recognized as different fragments by the marker and will appear as two bands (fig. 6.15).

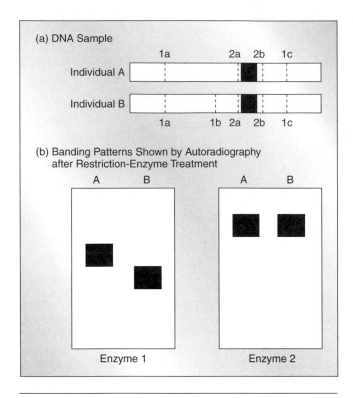

(a) DNA Sample

Individual A

1a   2a 2b   1c

Individual B

1a   1b 2a 2b   1c

(b) Banding Patterns Shown by Autoradiography after Restriction-Enzyme Treatment

A   B   A   B

Enzyme 1   Enzyme 2

Figure 6.15

A schematic diagram of an RFLP result. The black bands shown in the DNA samples represent the area to which a radioactive probe will bind. One of the restriction-enzyme recognition sites for enzyme 1 is mutated in individual A. As a result, different banding patterns will appear when the radioactive probe is detected with autoradiography. Treatment with enzyme 2 gives the same pattern in both individuals because they have identical restriction-enzyme recognition sites for enzyme 2.

After Lodish et al. (2000).

POINTS OF ENGAGEMENT—QUESTION 3

Can you identify a specific question that a particular molecular genetic technique allows a conservation biologist to ask that, prior to the technique's development, could not have even been formulated as a question? What value is added to a conservation effort by being able to ask and to answer the question you have identified?

THE PROBLEM OF MEASUREMENT: HOW DO WE DETERMINE GENETIC CHARACTERISTICS OF INDIVIDUALS AND POPULATIONS?

Captive breeding programs provide opportunities to develop and test precise measures of the genetic characteristics and traits of individuals and populations and to manage captive pop-

ulations in ways informed by such measures. With refinement, some of these measures may be applicable to wild populations and their management as well. Here we examine some of these measures, methods for making the measurements, and how to interpret and apply the information they provide.

How Do We Measure Genetic Diversity?

Some measures of genetic diversity are identical to the measures of community diversity described in chapter 4. For example, the Shannon index (a measure of species diversity in a community) can be used with equal efficacy as a measure of genetic diversity if the proportional abundance of alleles is substituted for the proportional abundance of species. Likewise, the Simpson index (chapter 4), used as a measure of dominance in the assessment of community structure, can be used as a measure of expected heterozygosity with the same substitution of allelic frequencies for species abundance (Vida 1994). These measures provide indices of genetic diversity in the population, and thus are a test consequence that can be used to evaluate the efficacy of different kinds of captive breeding strategies. However, the most meaningful genetic measurements in a population are often not of diversity, but of the strength of different forces affecting the population's genetic makeup. The most important of these forces are inbreeding and genetic drift, and the accurate measurement of their effects is critical to effective conservation and management.

How Do We Measure Inbreeding and Genetic Drift in Small Populations?

Geneticists Templeton and Read distinguish three biological meanings of inbreeding: (1) a measure of shared ancestry in the maternal and paternal lineages of an individual, (2) a measure of genetic drift in a finite population, and (3) a measure of a system of mating in a reproducing population. Each of these three dimensions of inbreeding grows stronger in its effects as population size declines, and such effects can be measured to make informed and appropriate management decisions regarding breeding strategies for small populations (Templeton and Read 1994).

The first concept of inbreeding, the measure of the shared ancestry of an individual in its maternal and paternal lines, has been called "inbreeding by descent" (Templeton and Read 1994). This type of inbreeding is quantified as the inbreeding coefficient, symbolized by F_p (pedigree inbreeding). The value of F_p, which varies from 0 to 1, can be calculated only for an individual of known pedigree. It measures the amount of ancestry an individual shares with its maternal and paternal lines. Pedigree inbreeding intensifies as the size of a population decreases.

In the second case, inbreeding is a measure of genetic drift in a population. If we knew the individual values of F_p for every individual in a population, added these values together, and divided the sum by the number of individuals, the resulting quotient would be the average probability of inbreeding by descent, symbolized as F_d. Here the subscript d is meant to signify that this value of F is a measure of the averaged inbreeding by descent of all members of the local population, or deme (Templeton and Read 1994). This value represents the average probability of inbreeding by descent, a measure of the effect of

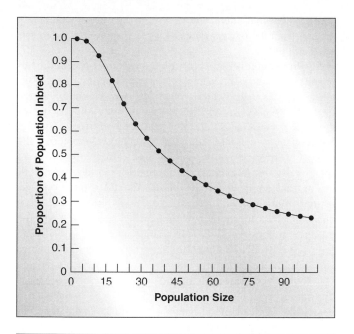

Figure 6.16

The relation between average pedigree inbreeding in a population (F_d) and population size. The smaller the population, the less time it will take to become completely inbred.

genetic drift on a population relative to an "ideal" population experiencing completely random mating.

Remember that the first type of inbreeding, **pedigree inbreeding,** increased in magnitude as population size declined. Similarly, the value of F_d increases as population size decreases and, for a given population size, F_d increases over time. This is expressed by the relationship

$$F_{d(t)} = 1 - [1 - 1/2N]^t,$$

where t is equal to time in generations. Note that the larger t becomes, the closer $F_{d(t)}$ comes to 1 (a completely inbred population). Thus, $F_{d(t)}$ will eventually reach a value of 1, and how fast it does so is a function of population size (fig. 6.16). The smaller the population, the faster it will become inbred. In this scenario, genetic drift causes the average probability of inbreeding by descent to increase and genetic variation to decrease. This means that inbreeding and loss of genetic variation are *correlates,* but inbreeding of this type is not the *cause* of a loss of genetic variation.

Finally, inbreeding can be used as a measure of a system of mating in a population, quantified as a value called the **panmictic index**, f. The panmictic index measures inbreeding as a deviation from a reference population, which has a system of mating in which alleles at a locus are paired in proportion to their frequencies in the overall population (by definition, random mating). The panmictic index thus evaluates deviations from the heterozygosity frequencies expected under random mating, so

$$f = 1 - H_O/H_e,$$

where H_e is the expected heterozygosity under random mating and H_o is the observed heterozygosity. Recall that in a randomly

mating population, the frequency of heterozygosity is defined by the Hardy-Weinburg equation. For two alleles, p and q, that frequency is

$$(p + q)^2 = p^2 + 2pq + q^2.$$

Thus, the expected frequency of the heterozygote is $2pq$. For example, if the frequency of allele p is 0.6 and the frequency of allele q is 0.4, then the H_e is $2 \times 0.6 \times 0.4 = 0.48$. Observed heterozygosity (H_O) can be calculated from genetic measurements of sampled individuals in the population. If observed heterozygosity is greater than expected, $f < 0$, the population has a reproductive system that avoids inbreeding. If observed heterozygosity is less than expected, $f > 0$, inbreeding is not avoided. Thus, the value of the panmictic index can be used to quantify the degree of avoidance of inbreeding in a population. Allelic frequencies generated from allozyme data, previously described, are frequently used to compute this index.

Measuring these three aspects of inbreeding separately and accurately provides a powerful array of information from which to make intelligent management decisions. Using the values of F_p in a captive breeding program, for example, a manager can determine which individuals are most and least inbred. If the goal is to minimize inbreeding, the manager can choose to mate individuals who are least inbred. Using the value of F_d, the manager can determine the effect of genetic drift on a population and the degree of heterozygosity present in that population and, from these data, make an intelligent decision about whether the current population size is sufficient to maintain an acceptable level of heterozygosity. Using the value of f, a manager can determine if current mating systems in the population lead to avoidance or encouragement of inbreeding, and then act accordingly.

Captive Breeding and Population Subdivision

Another strategy of captive breeding, which may be used in conjunction with those already discussed, is to subdivide the captive population into separate breeding units. In the wild, conservation biologists work actively against such subdivision, often the result of habitat fragmentation and, if subdivision occurs, work to provide means of connectivity and ensure dispersal among population subunits. However, in captivity, if the survival of the (small) subunits can be assured in a carefully controlled environment, population subdivision can preserve genetic variability more effectively than a single breeding unit. Geneticist Russell Lande explains, "Subdivision of a population and random genetic drift within the subpopulations converts the original genetic variation within the base population in genetic variation between subpopulations. Population subdivision also allows genetic variation between populations to accumulate by random genetic drift and fixation of new mutations. Once alternative alleles at a locus are fixed in different subpopulations, this component of genetic variability is permanently maintained and cannot be lost as long as the subpopulations persist. Splitting a population into separate subpopulations is therefore a powerful way of maintaining genetic variability, even though the total population size may be small" (Lande 1995).

Theoretically, migration between subdivided populations should reduce inbreeding depression and loss of heterozygosity.

Such theory predicts that as little as one migrant per generation should retard fixation of alleles in local populations (Wright 1969). Experimental tests on house flies (*Musca domestica*) have demonstrated that, in fact, fitness is maintained, even at these low levels of migration, over 24 generations of captive breeding in populations founded by only four individuals (Backus et al. 1999).

THE PROBLEM OF APPLICATION: HOW DO WE USE GENETIC INFORMATION AND TECHNIQUES IN CONSERVATION?

The applications of genetic techniques to conservation are varied, but can be grouped into five broad categories: (1) identifying relatedness among populations and determination of meaningful taxonomic units, (2) determining rates of gene flow and dispersal among populations, (3) determining the time since past genetic bottlenecks in a population or the time since significant differentiation between populations, (4) understanding patterns of reproductive ecology, and (5) locating sources of wildlife products, an important issue in the enforcement of laws protecting wildlife from commercial exploitation.

Genetics Can Clarify Relatedness, Taxonomy, and Phylogeny

The clarification of relatedness and taxonomy has profound implications for one of the most basic question in conservation biology: How many species are there? An excellent example of this application can be seen in a recent study of the Colorado chipmunk (*Tamias quadrivittatus*) of the southwestern United States. Here Sullivan (1996) used genetic information from different chipmunk populations to identify and describe a new subspecies, *T. q. oscuraensis,* that occurs only in the Oscura Mountains of New Mexico. The new subspecies used habitats that were drier, steeper, and less vegetated than other populations, and its adaptability to these new and more extreme conditions permitted an expansion of the overall species range (Sullivan 1996).

Similarly, recent molecular genetic studies among sea turtles compared the Kemp's ridley turtle (*Lepidochelys kempi*) and the similar olive ridley turtle (*L. olivacea*), vindicating the claim that the Kemp's ridley deserved recognition as a separate species (Avise 1998). Studies of the genetics of minke whales (*Balaenoptera acutorostrata*) have led investigators to advocate that the Northern and Southern Hemisphere populations be treated as distinct species (Hoelzel and Dover 1991a). The same is true for sympatric populations of killer whales (*Orcinus orca*) (Hoelzel, Dahiheim, and Stern 1998; Hoelzel and Dover 1991b). The case of killer whales is particularly interesting, because it suggests that observed differences in behavior in sympatric populations, so-called resource polymorphisms, may be genetically based (Hoelzel 1998).

Using PCR amplification techniques, Smith and colleagues (1991) described a new species of African shrike from a single specimen. DNA from one individual was amplified, compared to homologous sequences from other known species, and the new individual was determined to be as genetically distinct from the other species as those species were from one another.

Proper identification of relatedness can prevent the hybridization, and sometimes genetic extinction, of "look-alike" species in zoos and collections. For example, captive breeding of gazelles and dik-diks that were supposedly of the same species has sometimes produced infertile offspring. Subsequent cytogenetic study and analysis revealed that the parents were genetically distinct and, effectively, different species (Benirschke and Kumamoto 1991). Such interbreeding of genetically different individuals and populations has further implications when individuals are transferred from one collection to another. Commenting specifically on the problem in dik-diks, Benirschke and Kumamoto (1991) note that "not only were animals of supposedly identical species chromosomally different, but also hybrids between Kirk's (*Madoqua kirkii*) and Guenther's (*Madoqua rhyncotragus*) dik-diks were found in 300 collections." The authors conclude that "a cytogenetic analysis should be mandatory before captive populations are established" because "without such planned investigations one can confidently predict that unnecessary hybridization and reduced fertility will take place in captive groups. It is similarly important that of those animals that are to be released again into nature, only cytogenetically similar animals be used" (Benirschke and Kumamoto 1991).

Genetic Techniques Can Determine Rates of Gene Flow Among Populations

Determining rates of gene flow among populations has been an important component in evaluating conservation strategies for many species. In Rocky Mountain bighorn sheep (*Ovis canadensis canadensis*), Lukart and Allendorf (1996) used mtDNA to infer rates of gene flow and genetic differentiation in a number of populations in the western United States, and found a wide distribution of mtDNA groups (**haplotypes**) over large geographic areas, suggesting that, in the past, gene flow among populations occurred over a wide geographic area. However, many populations of bighorn sheep are now highly differentiated, and some have become genetically fixed for a single haplotype, suggesting more recent isolation and fragmentation of individual populations.

Measurement of gene flow is also an important consideration in plant conservation, for it is often low in small, rare, and endemic plants. For example, Dolan and associates (1999) used gel electrophoresis techniques previously described to measure isozyme variation in three species of rare perennial shrubs in Florida scrub vegetation. The level of interpopulation gene flow (N_m) was estimated by calculating differences among populations (Wright 1951). Although levels of gene flow were different in each species, one species, *Hypericum cumulicola,* was determined to be experiencing almost no gene flow among populations. This species also had the lowest proportion of species- and population-level polymorphic loci, the fewest number of alleles per polymorphic locus, and the lowest level of heterozygosity (table 6.3) (Dolan et al. 1999). In a population

Table 6.3   Comparison of Genetic (Isozyme) Variation in Three Species of Endemic Shrubs of Lake Wales Ridge, Florida. Numbers in parentheses are standard errors

	Eryngium cuneifolium	*Hypericum cumulicola*	*Liatris ohlingerae*	OTHER ENDEMICS[a]
Number of extant populations[b]	20	90	115	—
Number of populations surveyed	16	34	30	—
Number of loci	21	18	12	
Mean number of plants/locus	30.0	28.1	22.0	—
Species-level % loci polymorphic	43.8	28.0	50.0	43.8
Population-level[c] % loci polymorphic	16.0 (1.6)	6.2 (0.9)	31.4 (1.4)	29.2
Mean number of alleles/polymorphic locus	1.61 (0.05)	1.25 (0.03)	1.93 (0.06)	2.6
Observed heterozygosity	0.041 (0.004)	0.006 (0.001)	0.095 (0.005)	—
Expected heterozygosity	0.054 (0.004)	0.023 (0.003)	0.121 (0.005)	0.074
Estimated rate of gene flow (N_m)	0.31	0.09	1.83	

[a]Listed in Godt and Hamrick 1996.
[b]Based on databases of the Florida Natural Areas Inventory and field surveys.
[c]Averaged across all populations.
Source: After Dolan et al. 1999, with additional data from Godt and Hamrick 1996.

with these characteristics and little or no gene flow among populations, conservationists must protect many individual populations, not just a few in designated reserves, if they intend to preserve genetic diversity in the species. And because genetic diversity in this species is primarily found *among* populations rather than *within* populations, even small populations are worth preserving (Dolan et al. 1999).

Genetic Techniques Can Estimate the Time Since Past Population Bottlenecks

Inferring the length of time since past population bottlenecks is based on two simple yet plausible assumptions. First, the relatively clocklike and monophyletic accumulation of genetic change in mitochondrial DNA is observed in most organisms, although at different, taxon-specific rates. The second assumption is that a bottleneck renders mtDNA loci monomorphic at the time of the bottleneck. Thus by knowing the rate of accumulated genetic change and the present level

of genetic variation at mtDNA loci, one can estimate backward to determine how much time has been needed to achieve the present level of variation starting from a monomorphic condition.

This kind of analysis has been used on populations of Stellar sea lions (*Eumetopias jubatus*) (Bickham, Patton, and Loughlin 1996). In this case, investigators found no evidence of a recent bottleneck in the species, despite the fact that the northern population was at one time considered nearly extinct. The failure to find evidence of a genetic bottleneck suggests that large numbers of individuals may have survived undetected during population lows, that the lows were of such short duration that genetic diversity was not lost (no real bottleneck), or that mtDNA analysis alone is inconclusive and potentially misleading in determining historic occurrences of genetic events in populations. As described earlier in this chapter, precisely because mtDNA evolves at a different rate than nuclear DNA, conservation strategies should not be based solely on mtDNA analysis (Cronin 1993; Lukart and Allendorf 1996).

Genetic Techniques Can Determine Patterns of Reproductive Ecology

Conservationists have long been curious about, as well as frustrated by, the question of whether adult female sea turtles return to or near their natal areas to breed. Although field observations and tagging experiments have not been conclusive, analysis of mtDNA of green turtles (*Chelonia mydas*), loggerheads (*Caretta caretta*), and hawksbills (*Eretmochelys imbricata*) has shown that individuals in particular rookeries show fixed genetic differences in the frequencies of matriline lineages (patterns of female descent). This finding supports the hypothesis that adult females do consistently return to their natal areas for breeding (Avise 1998). Because female sea turtles ultimately govern the reproductive output of a rookery, this discovery has profound implications for sea turtle conservation. It means that other, physically equivalent habitats would not be used by adult females. If the natal breeding area itself were not protected, reproductive success would diminish or be eliminated altogether, and the loss of reproductive output in one rookery would not likely be compensated by reproduction in others.

Variation of mitochondrial DNA (mtDNA) in elephants has been used to determine the sources of ivory from populations in different African countries (Bischof 1992). Although there has been a ban on the sale of commercial ivory, the ban has been lifted so that international law now allows some harvest of elephants in areas with large or destructive populations. Recall that mtDNA is inherited maternally as a linked set of genes (Cronin 1993). Because elephants have a matriarchal social system, females remain in stable family groups and several generations of related females may be present in the same group. Thus, unique mtDNA signatures may characterize populations, and even individual groups within them (Bischof 1992).

Captive Breeding: Using Genetic Techniques to Recover Genetic Diversity and Population Size

Over 20 years ago conservation geneticist William Conway offered this assessment of the role and purpose of captive breeding programs in conservation. "Captive propagation has had a simple-minded directness: an immediacy that offers attractive reinforcement to the complex geo-political tasks of habitat preservation. It has been a kind of 'ark.' Such programs pretend to offer no overall cure for the epidemic of extinction but provide topical treatments of the symptoms expressed by the loss of higher animals. Simply expressed, captive propagation offers another way of fighting the continuing reduction of earth's diversity—an opportunity to preserve options" (Conway 1980).

These "topical treatments" of captive breeding programs have taken and continue to take a variety of forms. But, for all their diversity of expression, they have been remarkably unified in their pursuit of four primary conservation functions. These programs are designed to provide (1) substitutes for wild individuals and populations that can be used in basic biological research, (2) test cases for the development of care and management techniques, (3) demographic and genetic reservoirs that will augment existing wild populations, and (4) a last remnant of those species for which there is no immediate opportunity to survive in the wild (Conway 1980).

The captive breeding of rare plants and animals was once a collection of haphazard, highly individualized programs of particular zoos and preserves around the world. Today such efforts are highly coordinated, international efforts. By 1996, there were established pedigrees for over 250 species being managed for recovery in captivity according to principles of population genetics and demography (Rodríguez-Clark 1999). Many of these are so-called closed populations, populations which must be managed to be self-sustaining in captivity with no further inputs from wild stock for the foreseeable future. The goals of such programs are to minimize changes in the genetic constitution of these populations (i.e., preserve the original level of genetic diversity) and to avoid, or at least reduce, problems associated with inbreeding depression. The first goal ultimately looks forward to a day when offspring of captive individuals may be reintroduced into the wild to augment natural populations. For such efforts to have hope of success, the captive-bred individuals must have as much genetic diversity as possible remaining from their wild-caught ancestors in order to cope with their natural (and unfamiliar) environment. The second goal, the reduction of inbreeding depression, is aimed primarily at maintaining high survivorship and vitality among captive populations for as long as they may remain in captivity.

Because captive breeding programs often begin with a small number of individuals or "founders," inbreeding may be unavoidable. Historically, conservationists have been concerned that such founder events would enhance genetic drift and lead to an eventual reduction of genetic variation in the resulting population, and numerous theoretical and empirical studies (Frankham 1995a) have supported that concern.

Given these very real problems, there is a lack of agreement about what constitutes an optimal breeding strategy for captive populations. Because captive populations are often small, the first overall objective in breeding strategy is to maximize the ratio of the effective population size, N_e, to the total (census) population size, N. If the ratio of N_e/N can be maximized, the influences of genetic drift and inbreeding depression can be decreased. Basic methods to maximize this ratio are to (1) have as many founding individuals as possible, (2) grow the population to its captive carrying capacity as quickly as possible, (3) maximize the number of breeders in each subsequent generation, (4) equalize family sizes, (5) equalize the sex ratios of breeders, and (6) reduce subsequent fluctuations in the captive population size. Such strategies, if successful, not only maximize the N_e/N ratio, but also help retain existing heterozygosity in the population.

Historically, captive breeders have used a variety of approaches. So-called active selection is one of the oldest such strategies. In active breeding, breeders either brought the "best" animals together for mating or used those that simply bred on their own, assuming that breeding animals were the fittest individuals. But "best" in captivity often translated into either the most docile animals or those that most quickly adapted to a captive environment. Neither trait adds to the survivorship of wild populations. As an active selection strategy was repeated in subsequent generations, it led to increased inbreeding, loss of genetic variability, and fixation of traits adaptive for a captive environment, but of limited value in the wild. In his analysis of a failed attempt to establish a population of hybrid domestic X wild turkey (*Meleagris gallopavo*) in Missouri, Aldo Leopold attributed the failure to specific differences in behavior between the hybrid turkeys (raised in captivity) and wild turkeys. He noted that the domestic turkeys were more tranquil and docile and bred earlier in the season, and that the chicks scattered when the hen sounded a note of alarm. Chicks of wild turkeys hide. Commenting on this failure in conservation, Leopold remarked, "Wild turkeys are wary and shy, which are advantageous characteristics in eluding natural and human enemies. They breed at a favorable season of the year. The hens and young automatically react to danger in ways that are self-protective. Reproductive success is high. . . . Birds of the domestic strain, on the other hand, are differently adapted. Many of their physiological reactions and psychological characteristics are favorable to existence in the barnyard but many preclude success in the wild" (Leopold 1944). Such attempts at turkey reintroduction using captive-raised turkeys were repeated widely throughout the United States. With the exception of northern Michigan where a domestic-raised population successfully established itself in the wild, these efforts were all unqualified failures of active selection.

As an alternative to active selection, another approach to captive breeding has been to minimize inbreeding and maximize heterozygosity. Such a strategy requires a substantial founder population to be successful. For example, Frankham (1995a) argues that initiating populations with at least 20 to 30 unrelated founders are necessary to maximize initial heterozygosity. One method for retaining the heterozygosity and genetic diversity of the founder population has been to breed individuals that minimize relatedness (kinship). Mathematical models by Ballou and Lacy (1995) predict that minimizing kinship is the best method of managing the breeding of small founder populations for maximum retention of heterozygosity and allelic diversity. Average mean kinship (MK) of individuals in a population can be estimated as

$$MK = 1/2 \sum p_i^2/r_i,$$

where p_i is the expected proportion of the present gene pool that is descended from the ith founder and r_i is the "allelic retention" of that founder, in other words, the proportion of founder i's alleles that have survived to the present. Using the MK value, one could determine the mean kinship of each reproductive individual in the population to the founders, and then rank all individuals in the population from lowest to highest in their MK score. Those with low scores underrepresent the founding individuals, whereas those with high scores overrepresent them genetically. The male with the lowest-ranked MK would be mated with the lowest-ranked female, the male with the second lowest MK with the second lowest female, and so on so as not to mix over- and underrepresented lines. The MKs of offspring are then determined and, over time, added to the population to recalculate MK as the offspring reach reproductive age. The ranking process is iterated, the lowest values are paired, and matings continue until carrying capacity is reached. At this point, the expectation is that sufficient numbers exist to generate the types of matings and numbers of offspring that maximize the population's MK, which is the same as maximizing its retention of genetic diversity from the founder population. Compared with alternative strategies, such as random breeding or strategies to maximize avoidance of inbreeding, strategies to maximize mean kinship produce higher levels of heterozygosity in captive-bred populations (Ballou and Lacy 1995). And because heterozygosity is directly correlated with heritability, the evolutionary potential of a captive-bred population should be high if the heterozygosity can be maintained over the long term (Ballou and Lacy 1995; Rodríguez-Clark 1999).

Synthesis

Genetics has played a multidimensional role in the development and practice of conservation biology. Increasing precision and quantification of the genetic characteristics of individuals and populations has given force, along with occasional dilemmas and confusion, to the Endangered Species Act and other laws that address the protection of species. In cases where taxonomic distinctness is an important criterion for setting conservation priorities, genetic analyses take pride of place in decisively answering questions of taxonomic status (Schemske et al. 1994).

The accelerating pace of development in genetic analysis and techniques has elevated captive breeding from being the art of a few specialists to a science that can be practiced with an increasingly uniform set of standards and protocols. Further, improved genetic techniques and analysis have changed captive breeding from an effort that once seemed a last desperate gamble into carefully coordinated programs with real potential for successful long-term species preservation and reintroduction of animals to natural areas. Genetic techniques also have made possible noninvasive and less labor-intensive sampling of wild populations, vastly enhancing the potential of genetic management.

The danger inherent in a period of such achievement and optimism in conservation genetics is that, improperly applied and incorrectly understood, genetic analyses can lead to inappropriate conclusions, incorrect management decisions, and disillusionment because of overinflated expectations. An overemphasis on genetic analysis, to the exclusion of other essential considerations of a population's identity or persistence, can lead to conclusions that may be not only inappropriate and wasteful, but also even absurd. In addition, genetic research in populations is at a stage in which highly sophisticated and widely available techniques make it possible to collect enormous amounts of data without any clear understanding of what such data mean in regard to population modeling, long-term survivorship, or reproductive performance. This problem is particularly acute in plant conservation efforts. For example, the U.S. Fish and Wildlife Service (1990) recovery plan for the endangered plant, Peter's mountain mallow (*Iliamna corei*), calls for collection of data on genetic variation and subsequent development of a population genetics model to determine the number of populations and effective population sizes required for long-term survival. Botanist Douglas Schemske and his coworkers, commenting on the plan, noted, "This emphasis on the genetics of conservation rarity is a clear case of overkill, as only four individuals of the mallow are known to exist in the wild" (Schemske et al. 1994). To avoid the inanity of such "conservation genetics overkill," conservation biologists must take pains to not misappropriate genetic techniques, to not seek excessive amounts of genetic information when it has little value for the immediate needs of conservation planning, and to understand the distinctive concepts and context of genetic techniques so that their results are properly understood and fully integrated with other sources of information and insight. Similarly, conservation biologists must choose their fights carefully, and not make the loss of every genetic variant a conservation crisis. As Rhymer and Simberloff (1996) put it regarding the threat of hybridization, "one cannot be exercised over every situation in which new genes are flowing into a distinctive population, or economic and emotional resources will be insufficient to win most of these battles." Finally, conservation biologists must understand that genetic analyses do not, in themselves, make management decisions, create conservation plans, or provide essential management data. Alone, genetic information will always be insufficient for achieving conservation goals. Nongenetic dimensions of a species' life history, environmental constraints, or population demography may be more important to the persistence of a species than the present state of its genetic diversity. Sophisticated techniques are no substitute for sound judgment and discerning biological insight into conservation efforts.

Empirical Evidence that Inbreeding Causes Extinction—A Directed Discussion

Reading assignment: Van Oosterhout, C., W. G. Zilstra, M. K. Van Heuven, and P. M. Brakefield. 2000. Inbreeding depression and genetic load in laboratory metapopulations of the butterfly *Bicyclus anynana*. *Evolution* 54:218–25.

Questions

1. After reviewing the role of genetics in the development of conservation biology, Caughley (1994) concluded that issues of genetics were over-rated as significant factors in the extinction of wild populations. He asserted that "no instance of extinction by genetic malfunction has been reported." Do the results of Van Oosterhout et al. make a convincing case that extinction could occur by genetic malfunction? Based on their experimental results, describe the circumstances under which such malfunction could occur.

2. In the experimental design used by Van Oosterhout et al., subpopulations 1 and 4 experienced no migration (i.e., no gene flow). Recall that Levins (1970) defined a metapopulation as "any real population [that] is a population of local populations which are established by colonists, survive for a while, send out migrants, and eventually disappear." How would you describe the subpopulations that experience no gene flow? Do they form a metapopulation, or are they better classified in another way? Should this affect the interpretation of Van Oosterhout et al.'s results?

3. Identify the advantages and disadvantages of conducting this study with captive, experimentally manipulated populations rather than with natural populations of different sizes and different rates of migration.

4. Can you name one or more real populations (wild or captive) with characteristics similar to these experimentally created populations? Based on the results, suggest what steps managers of such populations might take to reduce the probability of extinction. How would you design a logical next step in experimental research to build on the results of Van Oosterhout et al.?

Learning Online

Visit our webpage at www.mhhe.com/conservation for case studies, animations, practice quiz questions, and additional readings to help you understand the material in this chapter. You'll also find active links to the following topics:

Species Abundance, Diversity, and
    Complexity
Population Biology

Genetics
Refuges and Sanctuaries

Legislation Regarding Endangered
    Species

Literature Cited

Amos, W. 1999. Two problems with the measurement of genetic diversity and genetic distance. In *Genetics and the extinction of species: DNA and the conservation of biodiversity,* eds. L. F. Landweber and A. P. Dobson, 75–100. Princeton, N.J.: Princeton University Press.

Avise, J. C. 1998. Conservation genetics in the marine realm. *Journal of Heredity* 89:377–82.

Backus, V. L., E. H. Bryant, C. R. Hughes, and L. M. Meffert. 1999. Effect of migration or inbreeding followed by selection on low-founder-number populations: implications for captive breeding programs. *Conservation Biology* 9:1216–24.

Ballou, J. D., and R. C. Lacy. 1995. Identifying genetically important individuals for management of genetic variation in pedigreed populations. In *Population management for survival and recovery: analytical methods and strategies in small population conservation,* eds. J. D. Ballou, M. Gilpin, and T. J. Foose, 76–111. New York: Columbia University Press.

Benirschke, K., and A. T. Kumamoto. 1991. Mammalian cytogenetics and conservation of species. *Journal of Heredity* 82:187–91.

Bickham, J. W., J. C. Patton, and T. R. Loughlin. 1996. High variability for control-region sequence in a marine mammal: implications for conservation and biogeography of Steller sea lions (*Eumetopias jubatus*). *Journal of Mammalogy* 77:95–108.

Bischof, L. 1992. Genetics and elephant conservation. *Endangered Species Update* 9 (7–8):1–4, 8.

Boriase, S. C., D. A. Loebel, R. Frankham, R. K. Nurthen, and D. A. Briscoe. 1993. Modeling problems in conservation genetics using captive *Drosophila* populations: consequences of equalization of family sizes. *Conservation Biology* 7:122–31.

Briton, J., R. K. Nurthen, D. A. Briscoe, and R. Frankham. 1994. Modeling problems in conservation genetics using *Drosophila:* consequences of harems. *Biological Conservation* 69:267–75.

Brown, T. A. 1995. *Gene cloning, an introduction.* London, England: Chapman and Hall.

Cann, R. L., M. Stoneking, and A. C. Wilson. 1987. Mitochondrial DNA and human evolution. *Nature* 325:31–36.

Caro, T. M., and M. K. Laurenson. 1994. Ecological and genetic factors in conservation: a cautionary tale. *Science* 263:485–86

Cassens, I., R. Tiedemann, F. Suchentrunk, and G. B. Hartl. 2000. Mitochondrial DNA variation in the European otter (*Lutra lutra*) and the use of spatial autocorrelation analysis in conservation. *Journal of Heredity* 91:31–35.

Caughley, G. 1994. Directions in conservation biology. *Journal of Animal Ecology* 63:215–44.

Conway, W. G. 1980. An overview of captive propagation. In *Conservation biology: an evolutionary-ecological perspective,* eds. M. E. Soulé and B. A. Wilcox, 199–208. Sunderland, Mass.: Sinauer Associates.

Cornuet, J. M., and G. Lukart. 1996. Description and power analysis of two tests for detecting population bottlenecks from allele frequency data. *Genetics* 144:2001–14.

Cronin, M. A. 1993. Mitochondiral DNA in wildlife taxonomy and conservation biology: cautionary notes. *Wildlife Society Bulletin* 21:339–48.

Dolan, R. W., R. Yahr, E. S. Menges, and M. D. Halfhill. 1999. Conservation implications of genetic variation in three rare species endemic to Florida rosemary scrub. *American Journal of Botany* 86:1556–62.

Ezzell, C. 1991. Conserving a coyote in wolf's clothing? *Science News* 139:374–75.

Frankel, O. H., and M. E. Soulé. 1981. *Conservation and evolution.* Cambridge, England: Cambridge University Press.

Frankham, R. 1995a. Conservation genetics. *Annual Review of Genetics* 29:305–27.

Frankham, R. 1995b. Inbreeding and extinction: a threshold effect. *Conservation Biology* 9:792–99.

Franklin, I. R. 1980. Evolutionary change in small populations. In *Conservation biology: an evolutionary-ecological perspective,* eds. M. E. Soulé and B. A. Wilcox, 135–49. Sunderland, Mass.: Sinauer Associates.

Fritsch, P. and L. H. Rieseberg. 1996. The use of random amplified polymorphic DNA (RAPD) in conservation genetics. In *Molecular genetic approaches in conservation,* eds. T. B. Smith and R. K. Wayne, 54–73. New York: Oxford University Press.

Gabriel, W., and R. Bürger. 1994. Extinction risk by mutational meltdown: synergistic effects between population regulation and genetic drift. In *Conservation genetics,* eds. V. Leoschcke, J. Tomiuk, and S. K. Jain, 69–89. Basel, Switzerland: Birkhäuser Verlag.

Godt, M. J. W., and J. L. Hamrick. 1996. Genetic structure of two endangered pitcher plants, *Sarracenia jonesii* and *Sarracenia oreophila* (Sarraceniaceae). *American Journal of Botany* 83:1016–23.

Grant, V. 1971. *Plant speciation.* New York: Columbia University Press.

Grant, V., and K. A. Grant. 1965. *Flower pollination in the phlox family.* New York: Columbia University Press.

Greenberg, D. B., J. E. Newbold, and A. Sugino. 1983. Intraspecific nucleotide sequence variability surrounding the origin of replication in human mitochondrial DNA. *Gene* 21:33–49.

Groom, M. J. 1998. Allee effects limit population viability of an annual plant. *American Naturalist* 151:487–96.

Hadrys, H., M. Balick, and B. Schierwater. 1992. Applications of random amplified polymorphic DNA (RAPD) in molecular ecology. *Molecular Ecology* 1:55–63.

Hartl, G. B., and Z. Pucek. 1994. Genetic depletion in the European bison (*Bison bonasus*) and the significance of electrophoretic heterozygosity for conservation. *Conservation Biology* 8:167–74.

Hedrick, P. W. 1999. Perspective: highly variable loci and their interpretation in evolution and conservation. *Evolution* 53:313–18.

Hoelzel, A. R. 1998. Genetic structure of cetacean populations in sympatry, parapatry, and mixed assemblages: implications for conservation policy. *Journal of Heredity* 89:451–58.

Hoelzel, A. R., M. E. Dahiheim, and S. J. Stern. 1998. Low genetic variation among killer whales in the eastern North Pacific, and genetic differentiation between foraging specialists. *Journal of Heredity* 89:121–28.

Hoelzel, A. R., and G. A. Dover. 1991a. Mitochondrial D-loop DNA variation within and between populations of the minke whale. *Report of the International Whaling Commission* 13:171–82.

Hoelzel, A. R., and G. A. Dover. 1991b. Genetic differentiation between sympatric killer whale populations. *Heredity* 66:191–95.

Holsinger, K. E., R. J. Mason-Gamer, and J. Whitton. 1999. Genes, demes, and plant conservation. In *Genetics and the extinction of species: DNA and the conservation of biodiversity,* eds. L. F. Landweber and A. P. Dobson, 23–46. Princeton, N.J.: Princeton University Press.

Inoue, M., and O. Takenaka. 1993. Japanese macaque microsatellite PCR primers for paternity testing. *Primates* 34:37–45.

Jiménez, J. A., K. A. Hughes, G. Alaks, L. Graham, and R. C. Lacy. 1994. An experimental study of inbreeding depression in a natural habitat. *Science* 266:271–73.

Lande, R. 1994. Risk of population extinction from fixation of new deleterious mutations. *Evolution* 48:1460–69.

Lande, R. 1995. Breeding plants for small populations based on the dynamics of quantitative genetic variance. In *Population management for survival and recovery: analytical methods and strategies in small population conservation,* eds. J. D. Ballou, M. Gilpin, and T. J. Foose, 318–40. New York: Columbia University Press.

Latter, B. D. H., J. C. Mulley, D. Reid, and L. Pascoe. 1995. Reduced genetic load revealed by slow inbreeding in *Drosophila melanogaster. Genetics* 139:287–97.

Leberg, P. L. 1996. Applications of electrophoresis in conservation biology. In *Molecular genetic approaches in conservation,* eds. T. B. Smith and R. K.Wayne, 87–103. New York: Oxford University Press.

Leopold, A. 1944. The nature of heritable wildness in turkeys. *Condor* 46:133–97.

Levins, W. H. 1980. *Polyploidy: biological relevance.* New York: Plenum.

Lewis, R. 1970. Extinction. *Lectures on Mathematics in the Life Sciences.* 2:75–101.

Lodish, H., A. Berk, S. L. Zipursky, P. Matsudaira, D. Baltimore, and J. E. Darnell. 2000. *Molecular cell biology,* 4th ed. New York: Freeman.

Lukart, G., and F. W. Allendorf. 1996. Mitochondrial-DNA variation and genetic population structure in Rocky Mountain bighorn sheep (*Ovis canadensis canadensis*). *Journal of Mammalogy* 77:109–23.

Lynch, M. 1996. A quantitative genetic perspective on conservation issues. In *Conservation genetics: case histories from nature,* eds. J. C. Avise and J. L. Hamrick, 471–501. New York: Chapman and Hall.

Mills, L. S., and P. E. Smouse. 1994. Demographic consequences of inbreeding in remnant populations. *American Naturalist* 144:412–31.

Morin, P. A., and D. S. Woodruff. 1996. Noninvasive genotyping for vertebrate conservation. In *Molecular genetic approaches in conservation,* eds. T. B. Smith and R. K. Wayne, 298–313. New York: Oxford University Press.

Muller, H. J. 1964. The relation of recombination to mutational advance. *Mutation Research* 1:2–9.

Nei, M. 1972. Genetic distance between populations. *American Naturalist* 106:283–92.

Novak, R. M. 1999. Red wolf *Canis rufus.* In *The Smithsonian book of North American mammals,* eds. D. E. Wilson and S. Ruff, 143–46. Washington, D.C.: Smithsonian Institution Press.

O'Brien, S. J. 1994. Genetic and phylogenetic analyses of endangered species. *Annual Review of Genetics* 28:467–89.

O'Brien, S. J., and collaborators. 1996. Conservation genetics of the felidae. In *Conservation genetics: case histories from nature,* eds. J. C. Avise and J. L. Hamrick, 50–74. New York: Chapman and Hall.

O'Brien, S. J., M. E. Roelke, L. Marker, A. Newman, C. A. Winkler, D. Meltzer, L. Colly, J. F. Evermann, M. Bush, and D. E. Wildt. 1985. Genetic basis for species vulnerability in the cheetah. *Science* 227:1428–34.

O'Brien, S. J., M. E. Roelke, N. Yuhki, K. W Richards, W. E. Johnson, W. L. Franklin, A. E. Anderson, O. L. Bass Jr., R. C. Belden, and J. S. Martenson. 1990. Genetic introgression within the Florida panther *Felis concolor coryii. National Geographic Research* 6:485–94.

O'Brien, S. J., D. E. Wildt, D. Goldman, C. R. Merril, and M. Bush. 1983. The cheetah is depauperate in genetic variation. *Science* 221:459–62.

Parker, I. M., R. P. Nakumura, and D. W. Schemske. 1995. Reproductive allocation and the fitness consequences of selfing in two sympatric species of *Epilobium* (Onagraceae) with contrasting mating systems. *American Journal of Botany* 82(8):1007–16.

Ralls, K., and J. Ballou. 1983. Extinction: lessons from zoos. In *Genetics and conservation: a reference for managing wild animal and plant populations,* eds. C. M. Schonewald-Cox, S. M. Chambers, B. MacBryde and W. L. Thomas, 164–84. Menlo Park, Calif.: Benjamin/Cummings.

Ralls, K., J. D. Ballou, and A. Templeton. 1988. Estimates of lethal equivalents and the cost of inbreeding in mammals. *Conservation Biology* 2:185–93.

Ralls, K., K. Brugger, and J. Ballou. 1979. Inbreeding and juvenile mortality in small populations of ungulates. *Science* 206:1101–03.

Raven, P. H. 1979. A survey of reproductive biology in the Onagraceae. *New Zealand Journal of Botany* 17:575–93.

Rhymer, J, A., and D. Simberloff. 1996. Extinction by hybridization and introgression. *Annual Review of Ecology and Systematics* 27:83–109.

Rodríguez-Clark, K. M. 1999. Genetic theory and evidence supporting current practices in captive breeding for conservation. In *Genetics and the extinction of species,* eds. L. F. Landwebber and A. P. Dobson, 47–73. Princeton, N.J.: Princeton University Press.

Roy, M. S., E. Geffen, D. Smith, E. Ostrander, and R. K. Wayne. 1994. Patterns of differentiation and hybridization in North American wolf-like canids revealed by analysis of microsatellite loci. *Molecular Biological Evolution* 11:553–70.

Roy, M. S., D. J. Girman, and R. K. Wayne. 1994. The use of museum specimens to reconstruct the genetic variability and relationships of extinct populations. *Experimentia* 50:551–57.

Schemske, D. W., B. C. Husband, M. H. Ruckelshaus, C. Goodwillie, I. M. Parker, and J. G. Bishop. 1994. Evaluating approaches to the conservation of rare and endangered plants. *Ecology* 75:584–606.

Smith, E. F. G., P. Arctander, J. Fjeldsa, and O. G. Amir. 1991. A new species of shrike (Laniidae: *Laniarius*) from Somalia, verified by DNA sequence data from the only known individual. *Ibis* 133:227–35.

Smith, T. B., and R. K. Wayne, eds. 1996. *Molecular genetic approaches in conservation*. New York: Oxford University Press.

Soulé, M., M. Gilpin, W. Conway, and T. Foose. 1986. The millennium ark: how long a voyage, how many staterooms, how many passengers? *Zoo Biology* 5:101–14.

Stebbins, G. L. 1950. *Variation and evolution in plants*. New York: Columbia University Press.

Stebbins, G. L. 1957. Self-fertilization and population variability in the higher plants. *American Naturalist* 91:337–54.

Sudbery, P. 1998. *Human molecular genetics*. Essex, England: Addison-Wesley Longman.

Sullivan, R. M. 1996. Genetics, ecology, and conservation of montane populations of Colorado chipmunks (*Tamias quadrivittatus*). *Journal of Mammalogy* 77:951–75.

Takenaka, O., H. Takasaki, S. Kawamoto, M. Arakawa, and A. Takenaka. 1993. Polymorphic microsatellite DNA amplification customized for chimpanzee paternity testing. *Primates* 34:27–35.

Templeton, A. R. 1986. Coadaptation and outbreeding depression. In *Conservation biology: the science of scarcity and diversity*, ed. M. E. Soulé, 105–16. Sunderland, Mass.: Sinauer Associates.

Templeton, A. R., and B. Read. 1994. Inbreeding: one word, several meanings, much confusion. In *Conservation genetics*, eds. V. Leoschcke, J. Tomiuk, and S. K. Jain, 91–105. Basel, Switzerland: Birkhäuser Verlag.

U.S. Fish and Wildlife Service 1990. Peter's mountain mallow (*Iliamna corei*) recovery plan. United States Fish and Wildlife Service, Twin Cities, Minnesota.

Van Oosterhout, C. V., W. G. Zijlstra, M. K. Van Heuven, and P. M. Brakefield. 2000. Inbreeding depression and genetic load in laboratory metapopulations of the butterfly *Bicyclus anynana*. *Evolution* 54:218–25.

Vida, G. 1994. Global issues of genetic diversity. In *Conservation genetics*, eds. V. Leoschcke, J. Tomiuk, and S. K. Jain, 9–19. Basel, Switzerland: Birkhäuser Verlag.

Vrijenhoek, R. C. 1994. Genetic diversity and fitness in small populations. In *Conservation genetics*, eds. V. Leoschcke, J. Tomiuk, and S. K. Jain, 37–53. Basel, Switzerland: Birkhäuser Verlag.

Wayne, R. K. 1996. Conservation genetics in the Canidae. In *Conservation genetics: case histories from nature*, eds. J. C. Avise and J. L. Hamrick, 75–118. New York: Chapman and Hall.

Wayne, R. K., and S. M. Jenks. 1991. Mitochondrial DNA analysis supports extensive hybridization of the endangered red wolf (*Canis rufus*). *Nature* 351:565–68.

Whitham, T. G., G. D. Martinsen, K. D. Floate, H. S. Dungley, B. M. Potts, and P. Keim. 1999. Plant hybrid zones affect biodiversity: tools for a genetic-based understanding of community structure. *Ecology* 80:416–28.

Wilson, A. C., R. L. Cann, S. M. Carr, M. George, U. B. Gyllensten, K. M. Helm-Bychowski, R. G. Higuchi. S. R. Palumbi, E. M. Prager, R. D. Sage, and M. Stoneking. 1985. Mitochondrial DNA and two perspectives on evolutionary genetics. *Biological Journal of the Linnaean Society* 26:375–400.

Woodworth, L. M., M. E. Montgomery, R. K. Nurthen, D. A. Briscoe, and R. Frankham. 1994. Modeling problems in conservation genetics using *Drosophila*: consequences of fluctuating population sizes. *Molecular Ecology* 3:393–99.

Wright, S. 1931. Evolution and Mendelian populations. *Genetics* 16:97–159.

Wright. S. 1951. The genetic structure of populations. *Annals of Eurogenics* 15:323–54.

Wright, S. 1969. *Evolution and the genetics of populations*. Vol. 2, *The theory of gene frequencies*. Chicago: University of Chicago Press.

Xu, X., and U. Arnason. 1996. The mitochondrial DNA molecule of Sumatran orangutan and a molecular proposal for two (Bornean and Sumatran) species of orangutan. *Journal of Molecular Evolution* 43:431–37.

Zhi, L., W. B. Karesh, D. N. Janczewski, H. Frazier-Taylor, D. Sajuthi, F. Gombek, M. Andau, J. S. Martenson, and S. J. O'Brien. 1996. Genomic differentiation among natural populations of orangutans (*Pongo pygmaeus*). *Current Biology* 6:1326–36.

7

I have seen something else under the sun: the race is not to the swift, or the battle to the strong, nor does food come to the wise, . . . but time and chance overtake them all

—Ecclesiastes 9:11

The Conservation of Populations

In this chapter you will learn about:

1 what populations are.

2 what causes populations to decline and become extinct.

3 how to develop a plan for the conservation of a small and declining population.

4 how to understand and manage problems associated with non-indigenous species.

5 how metapopulations can be conserved.

6 a conceptual approach to making management decisions for individual populations.

Populations are defined by genetic integrity and distinctiveness (chapter 6), but conserving populations and species requires more than conserving their genetic integrity. In this chapter we move beyond genetics to address intrinsic characteristics of population growth and decline (population demography), effects of extrinsic environmental forces on demographic characteristics, and ways of conserving populations through the management of external forces. We begin with clarification of the word **population.**

DEFINING POPULATIONS

The traditional definition of *population* is "all coexisting individuals of the same species." As noted in chapters 4 and 6, the species concept has been subjected to new interpretations, particularly as genetic techniques provide increasing precision on issues of the genetic similarities and differences of organisms. Wells and Richmond (1995) argue that spatial structure, genetic structure, and demographic structure define a group, and that a population is a group of individuals showing a clear disjunction from other groups in at least one of these characteristics (Wells and Richmond 1995).

Spatial disjunction refers to a distribution pattern in which groups of individuals, although they might be physically or genetically similar, are separated from one another by their

location. Consequently, individuals from one group are not normally able to interact with individuals in other groups. To be considered disjunct, each group must occupy a particular and identifiable area relative to other groups, the space between the groups must not contain individuals of any group, and individuals must not normally travel from group to group (fig. 7.1). Spatial disjunction is one criterion for defining populations, and is the easiest to detect.

In contrast, *genetic disjunctions* occur when all the individuals in one group have a common set of genetic attributes that are not shared by individuals in other groups. If two groups share two genes, 1 and 2, but the first group contains alleles A and B at gene 1 and C and D at gene 2, whereas the second group contains alleles W and X at gene 1 and Y and Z at gene 2, then the groups are genetically disjunct and, hence, distinct populations.

A *demographic disjunction* occurs when a group of individuals shares common demographic properties that differ in value from the same properties in other such groups. The best conceptual explanation of this kind of disjunction is provided by Cole (1957), who defined a population as "a biological unit at the level of ecological integration where it is meaningful to speak of a birth rate, a death rate, a sex ratio, and an age structure in describing the properties of the unit." Cole's definition identifies populations as groups that have common rates of birth and death, characteristic sex ratios and age structures, and other

Figure 7.1

An example of spatial disjunction can be observed in bighorn sheep (*Ovis canadensis*) in the Intermountain West of the United States. Each group can be considered a population because, although similar in morphology and genetics, they occupy distinctly different areas that are not contiguous.

demographic properties. To produce such a discontinuity, different groups must experience too few exchanges of members to significantly affect one another's demography. If immigrants from one group do markedly change the birth rate, death rate, age structure, or sex ratio of another group, then the discontinuity is broken and the two groups are, from a demographic standpoint, really one.

Among these three types of disjunctions or discontinuities, spatial discontinuity is the most important factor. If the populations are not separated spatially, the other criteria will not exist. For example, if we find a variety of alleles in many individuals of the same species, but we cannot delineate in space how the alleles are segregated, then we have no genetic discontinuity, only genetic variety. As any good conservation geneticist knows, the first step in identifying genetic differences is to look for spatial disjunctions (Wells and Richmond 1995). Although spatial discontinuity could occur without

genetic or demographic discontinuity, it is difficult to imagine how genetic or demographic discontinuity could occur without spatial separation.

Whenever we refer to a group of individuals as a "population," we ought to identify immediately the discontinuity—spatial, genetic, or demographic—that exists between this group and other groups. If we find no discontinuity, then the group of individuals we are examining do not constitute a population. Instead they form a part of some larger population that is in some way disjunct from other groups.

An important goal of all conservation biologists is to develop a common understanding of what populations are and how to identify them. Populations are the fundamental units of conservation and the primary targets of management and policy directives that provide meaningful protection to groups of organisms that are declining or small in number, or facing the imminent threat of extinction. Having defined populations, the next step is to determine what factors determine the size and persistence of populations. And, of special relevance to conservation, what causes populations to decline and become small?

WHY DO POPULATIONS DECLINE?

"Extinction," the quality of "ceasing to be," marks the termination of a population's existence. This is an event that conservation biologists seek to avoid. But few populations suddenly disappear when they are large, vigorous, and growing. In almost all cases, extinction is preceded by decline. The genetic dangers faced by small populations have been examined in chapters 5 and 6. To these we now add other factors that can cause population decline and eventual extinction.

Population Demography

Traditionally, population growth is defined by birth, death, immigration, and emigration. In a group of individuals born at the same time (cohort), the sum of the probabilities of survival of each individual to a particular age (survivorship) influences the trajectory of population change over time. Because losses of individuals represent subtractions from the population, more individuals are added by the births of new individuals, a process known as **recruitment**. Recruitment is driven by **fecundity**, usually defined as the number of young or eggs produced per female (animals) or seeds per individual (plants) per unit time. Other increases that accrue to the population through immigration, and additional losses that are incurred through emigration, are functions of *dispersal*, the permanent movement of an organism from its area of birth to a new area. Dispersal must be measured in terms of both *rate* (proportion of individuals that leave the natal area) as well as *distance* (how far an organism travels from the natal area before it resumes a settled existence).

Simple models of population growth integrate the complexities of these multiple factors with relatively few mathematical concepts. The simplest model of all is that of **exponential population growth**, which is defined solely by the population's size, N, and its rate of increase, r, which is defined as the

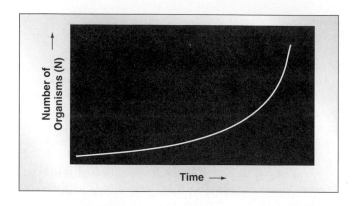

Figure 7.2

The exponential growth curve, a graphical depiction of a population increasing at an ever-increasing rate over time.

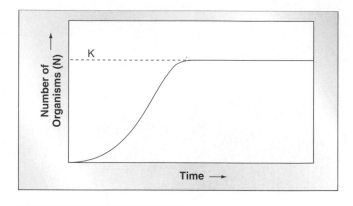

Figure 7.3

The logistic growth curve, a graphical depiction of a population's growth as it approaches an environmental limit, or carrying capacity (K).

difference between the population's rate of birth (*b*) and its rate of death (*d*). Immigration and emigration are either ignored, considered inconsequential, or added to the values of *b* and *d*, respectively. In exponential growth, the change in numbers (*dN*) in the population over change in time (*dt*) is determined by the equation

$$\frac{dN}{dt} = rN.$$

Viewed graphically, exponential growth is a J-shaped curve showing a population growing over time at an ever-increasing rate (fig. 7.2).

Many populations experience exponential growth under certain conditions for limited periods of time. An algal bloom in a freshwater pond, the growth of bacteria on a nutrient-rich substrate, or the introduction of a non-native species into a new and suitable habitat are all common cases in which populations may grow exponentially. However, exponential growth is not a realistic, long-term phenomenon in any population that is influenced or affected by its environment. As biologist Mark Boyce remarked bluntly about the exponential growth model, "it has no ecology" (Boyce 1992). But a real population does have an ecology, and an accurate model must reflect, at least to some degree, the interaction of a population with its environment.

The simplest model of population growth that is affected by the environment is **logistic population growth,** which includes an environmental limit on the population size (the carrying capacity) that slows population growth as *N* approaches the carrying capacity, *K*. Logistic growth is mathematically defined as

$$\frac{dN}{dt} = rN \left(\frac{K - N}{K} \right).$$

When *N* is small relative to *K*, the expression (*K* − *N*/*K*) approaches unity, and the population grows at an exponential rate. But as *N* approaches *K*, the value of the expression approaches zero, and so does the rate of growth, *dN/dt*. Thus, a visual representation of the model depicts a population growing at a nearly exponential rate during early stages of growth, but gradually slowing in growth until it reaches a stable equilibrium (fig. 7.3).

Many variations, some extremely sophisticated, have been developed to increase the correspondence between the

logistic growth model and real populations, most of which do not move smoothly or uniformly toward equilibrium or necessarily stay there if they achieve it. The simple model described above assumes that (1) each new individual has an instantaneous effect on the population's growth rate, which is almost always false; (2) individuals are added to the population at a constant rate, a condition that is not met by many populations; and (3) carrying capacity is constant, when in reality it often changes in response to environmental variation and interaction with the effects of the population on available resources. It is more accurate to define carrying capacity not as a constant value limit that a population cannot exceed, but as a population size reflecting an equilibrium between a population and its resources. Thus the value of *K* varies according to environmental variation, and populations can lower the value of *K* by damaging renewable resources, such as plants, in ways that reduce the plants' long-term productivity (Caughley 1979).

The logistic model also assumes that all individuals in a population are essentially demographic equivalents of one another in reproduction and survivorship. This is rarely true, especially in populations with long-lived individuals, where population growth is often defined and determined by **age structure** and **sex ratio**. With sufficiently detailed information, one can construct a **life table** and compute age-specific rates of birth, mortality, survivorship, fecundity, and other parameters that determine the growth of a given population (table 7.1; Appendix 7A).

Although life tables are valuable tools for identifying the specific traits of populations that determine patterns of growth over time, they are burdened with a number of important limitations. One is the assumption that the demographic parameters in the life table remain constant over time, or at least from the time the data were collected to the period of interest. Unless such data were very recently collected, this assumption is not likely to be true. A second limitation is that life tables focus on a population's past, but conservation biologists are primarily interested in the future. Instead of simply asking about the past or present demographic parameters of a population, a conservation biologist would ask, "What is the likelihood of this population persisting if these parameters remain unchanged, and for how long?" Although the questions are related, they are also pro-

Table 7.1   **A Life Table for Belding's Ground Squirrel (*Spermophilus beldingi*).**   Life tables, properly constructed from appropriate data, provide important summaries of age-specific demographic characteristics of plant and animal populations: *n* is the actual number of individual squirrels alive in each age interval; *d* is the number dying during the interval; *l* is the proportion of the original cohort alive at the beginning of the age interval; *q* is the mortality rate from interval *x* to inteveal *x* + 1; *e* is the life expectancy of individuals in the age interval; and *x* is the age interval to which the value refers. Calculations of *l* do not include individuals first marked as adults.

AGE (YEARS)	FEMALES					MALES				
	n_x	d_x	l_x	q_x	e_x	n_x	d_x	l_x	q_x	e_x
0–1	337	207	1.000	0.61	1.33	349	227	1.000	0.65	1.07
1–2	252*	125	0.386	0.50	1.56	248†	140	0.350	0.56	1.12
2–3	127	60	0.197	0.47	1.60	108	74	0.152	0.69	0.93
3–4	67	32	0.106	0.48	1.59	34	23	0.048	0.68	0.89
4–5	35	16	0.054	0.46	1.59	11	9	0.015	0.82	0.68
5–6	19	10	0.029	0.53	1.50	2	0	0.003	1.00	0.50
6–7	9	4	0.014	0.44	1.61	0	—	—	—	—
7–8	5	1	0.008	0.20	1.50	—	—	—	—	—
8–9	4	3	0.006	0.75	0.75	—	—	—	—	—
9–10	1	1	0.002	1.00	0.50	—	—	—	—	—

Source: Sherman and Morton 1984.
*Includes 122 females first captured as yearlings.
†Includes 126 males first captured as yearlings.

foundly distinct. To understand how conservation biologists evaluate the status of a population, we take up the second question in more detail.

Stochastic Perturbations

To determine the probability that a population will persist in time, and for how long, we must evaluate the factors that affect the population's size. Such factors can be broadly categorized as "deterministic" or "stochastic." **Deterministic factors** are those that affect the population in a constant relation to the population size. For example, if predators consistently removed 10% of all individuals in a population year after year, regardless of variations in the population's size, we would be justified in calling predation a deterministic factor. If deterministic factors are the primary cause of population decline, specific management actions and strategies may alleviate them. Rarely, however, is any significant factor affecting a population such a simple, direct, cause-and-effect relationship that can be counted on to produce a determined and predictable result. In fact, factors that significantly

influence population size are almost always **stochastic.** That is, their effects can vary randomly, although the variation is usually within a limited range.

In a much quoted paper titled "Minimum Population Sizes for Species Conservation," U.S. Fish and Wildlife biologist Mark Shaffer identified four "sources of uncertainty" that can affect the size of a population: **genetic stochasticity, demographic stochasticity, environmental stochasticity,** and **natural catastrophes** (Shaffer 1981). All of these can be considered stochastic rather than deterministic in that their effects are not certain, but rather come from a random distribution of events whose probabilities are unique to particular populations and their environments. If a population is large, the outcomes associated with these sources of uncertainty (stochastic variation) follow the law of averages. But if the population is small, its success or failure may deviate drastically from the average because it often hinges on chance events that affect only a small number of individuals, sometimes with devastating results. For example, the heath hen (*Tympanuchus cupido cupido*) (fig. 7.4), a bird similar in appearance and behavior to the prairie chicken (*Tympanuchus* spp.), was once common throughout the

Figure 7.4

The heath hen (*Tympanuchus cupido cupido*), once abundant throughout the northeastern United States, was reduced to a relict population on the island of Martha's Vineyard off the coast of Massachusetts. Despite complete protection, it went extinct by 1932 as a result of the random, detrimental, and combined effects of environmental and demographic stochasticity, inbreeding, and natural catastrophes.

northeastern United States. By 1876, overhunting and habitat destruction had restricted its range to the island of Martha's Vineyard in Massachusetts (Shaffer 1981). By 1900 there were fewer than 100 survivors, and a refuge was established for the population on the island in 1907. By 1916, the population had increased to around 800 individuals and seemed to be headed for recovery. Then a series of environmental and demographic "bad luck" befell the survivors. In the year that they reached their peak population, a fire devastated the island and destroyed most of the remaining habitat and nests. A high winter concentration of goshawks (*Accipiter gentilis*), an efficient avian predator, deepened the disaster. After a minimal recovery in 1920, disease swept through the population, eliminating all but 100 birds. Decline continued from this point on, with increasing numbers of birds experiencing sterility. Worse, the proportion of males increased until, in the final years, there were no females at all! The population was extinct by 1932.

The factors of environmental stochasticity, demographic stochasticity, genetic stochasticity, and natural catastrophe—the "Four Horsemen of the Extinction Apocalypse"—were all active in the demise of the heath hen. Although they are complex and interactive, we will examine each factor briefly and individually.

Genetic Stochasticity

Problems of genetic stochasticity were explained in detail in chapters 5 and 6, but are briefly reviewed here. Small populations tend to suffer from increased rates of inbreeding, increased effects of genetic drift, and the accumulation of unfavorable mutations. Many small populations in the wild, such as the Florida panther (*Puma concolor coryi*) (O'Brien 1996), and captive-bred populations of endangered species have shown measurable detrimental effects associated with inbreeding depression, mutational meltdown, and genetic drift. These effects accumulate in small populations and can lead to further population declines and eventual extinction.

Environmental Stochasticity

Environmental stochasticity refers to fluctuations in the probability of birth and death in a population because of the temporal variation of habitat parameters; populations of competing, parasitic, or predatory species; and incidence of disease. The importance of environmental stochasticity can be best understood relative to the average rate of increase of the population. Let r_{av} represent the average rate of increase of the population and let V_e represent the variance in population growth attributable to environmental variation. If r_{av} is greater than V_e, then the expected persistence time of a population increases directly with increasing population size at an ever-increasing rate (fig. 7.5). In this scenario, environmental stochasticity is unlikely to cause extinction as long as the population is not very small. On the other hand, if V_e is greater than r_{av}, the shape of the population persistence curve is profoundly different. Persistence time still increases as the size of the population increases, but it reaches an upper asymptote, beyond which further increases in population size do not significantly increase expected time of population persistence. This second case describes a population with large, environmentally induced population fluctuations and a relatively small rate of increase. In such a case, even large populations would be very vulnerable to extinction, and the best protection against extinction would not necessarily be to generate the largest population, but to ensure that the total population did not experience the same environmental variations at once (Simberloff 1998).

Demographic Stochasticity

Demographic stochasticity refers to random fluctuations in birth and death rates, emigration and immigration, or sex ratio and age structure of a population. Such processes are still stochastic even if the observed rates remain constant. Biologist Robert Lacy notes that "with the exception of aging, almost all events in the life of an organism are stochastic" (Lacy 1993). In large populations, variation among individuals rarely matters; in small populations, however, such variation matters very much. The loss of a pregnant female, a new generation with a highly skewed sex ratio, or the accidental death of a few prime breeding adults can have enormous impacts on small populations. Like those of environmental stochasticity, the effects of demographic stochasticity diminish rapidly with increasing

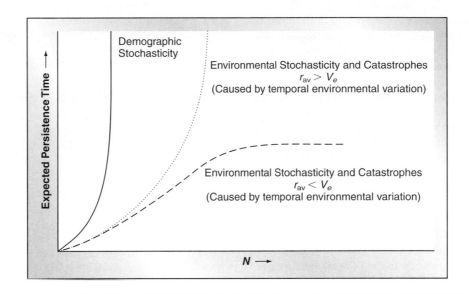

Figure 7.5

The effects of environmental and demographic stochasticity on the persistence time of a population at different population sizes (N). Although the probability of extinction from such forces is very low in large populations, these forces create a high probability of extinction at low populations.

After Caughley (1994).

population size, and population persistence is all but assured (fig. 7.5). But at small population levels the effects of demographic stochasticity alone makes extinction almost certain.

A further consideration of demographic stochasticity that weighs heavily on small populations is the so-called **Allee effect,** named for the British ecologist, W. C. Allee, who first described it (Allee et al. 1949). Compared with large populations, individuals in small populations may "suffer reduced fitness from insufficient cooperative interactions with conspecifics" (Lande 1999). The outcome of such "insufficient cooperative interactions" can trigger a number of different mechanisms, acting alone or in concert, that reduce the population's fitness. In populations below a certain size or density, individuals, as geneticist Russell Lande puts it, "may have difficulty encountering potential mates. These effects can render population growth negative in small populations, creating an unstable equilibrium at small population size below which the population tends to decline to extinction" (Lande 1988). There are other detrimental effects of small population size on fitness. In some populations, social groups stimulate mating activity. For example, some gallinaceous birds (order Galliformes) like the sage grouse (*Centrocercus urophasianus*) (fig. 7.6) gather for mating on communal display and breeding grounds known as leks. If numbers are insufficient to promote lek formation, displays and breeding may not take place. Also, in many species, groups may deter predators through cooperative defense and increased vigilance or increase foraging efficiency by altering the vegetation community itself (McNaughton 1984). When density drops below a level at which such groups can form, the ability to detect predators decreases, vegetation can no longer be altered for optimal foraging efficiency, and survivorship may decline. In other species, groups of individuals may benefit one another through various means of physically or chemically conditioning the environment (such as huddling for warmth or

Figure 7.6

The sage grouse (*Centrocercus urophasianus*), a gallinaceous bird of the western United States, gathers for mating on communal display and breeding grounds known as leks. If numbers are insufficient to promote lek formation, displays and breeding may not take place.

making trails through heavy snow that can be used repeatedly to save energy), or through communal nesting (for example, weaver finches, family Ploceidae). At very low densities, animals may be unable to take advantage of these and other benefits they formerly enjoyed when living in larger groups. Because of the Allee effect, there may be a minimum threshold density in small populations below which the population may be unable to recover.

Natural Catastrophes

Some (e.g., Simberloff 1998) have argued that natural catastrophes are simply extreme cases of environmental stochasticity. There is some sense in which this is true, as natural catastrophes are often extreme forms of normal environmental variation (e.g., extreme and prolonged drought or flash floods resulting from intense, heavy rain). But catastrophes also may be considered a separate category because they occur so infrequently as to lie outside the normal probability distribution of random events associated with environmental variation. Further, catastrophes may be qualitatively as well as quantitatively different in their effects. In the understated words of biologist Robert Lacy, "a forest fire is not just a very hot day" (Lacy 1993). Catastrophes, although rare, pose special threats for small populations because they have the potential to eliminate all individuals in a small group. For small populations, the most viable protection against catastrophes is spatial dispersion.

Population Viability Analysis

Given that small populations are more vulnerable than large populations to decline and extinction from all of the above forces, it is less important to examine every aspect of their demography than it is to determine the relative threat of extinction from each of these four forces. Such an assessment is called a **population viability analysis (PVA).**

Population viability analysis has been defined as *the estimation of extinction probabilities by analyses that incorporate identifiable threats to population survival into models of the extinction process* (Lacy 1993). Ralls and Taylor (1997) describe PVA as a method of analysis that uses data in a model to estimate the risk of extinction for a population. The goal of a PVA is to provide an evaluation of whether a population will fail or prosper in response to specific circumstances and, more particularly, an assessment of the risk of extinction for a population over a specific time horizon under a given set of circumstances.

As noted earlier, small populations are at much higher risks of extinction from random events associated with environmental and demographic stochasticity, genetic drift, and natural catastrophes. Therefore, PVAs normally do not allow a population to reach a level of zero (actual extinction), but rather specify a lower limit or extinction threshold, below which extinction would be nearly certain because of these same forces. This "quasi-extinction threshold" is then built into the model as a fixed and limiting parameter.

The output of a PVA can be in one of four forms. If the quasi-extinction threshold and the time horizon are both allowed to vary, the output takes the form of a three-dimensional quasi-extinction surface in which time to threshold, extinction threshold, and probability of dropping to or below the threshold are plotted simultaneously. If time is fixed, the output produced is a two-dimensional quasi-extinction curve in which different numerical thresholds are plotted against different quasi-extinction probabilities. Or, if the quasi-extinction threshold is fixed (e.g., the manager determines that the population must never drop below 200 individuals), then the output generated is a probability distribution of different extinction times. This kind of output is used to find the mean or median value of the probability distribution. A fourth approach to output is to run multiple simulations with defined thresholds, defined times, and defined probabilities. Such an approach produces a series of curves known as "quasi-extinction contours" that reveal, for a given quasi-extinction

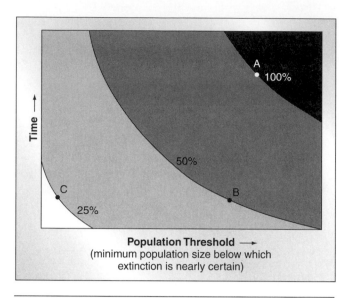

Figure 7.7

An example of quasi-extinction contours generated from a PVA. Each point on a given curved line (contour) represents a combination of time and population size having a given probability of quasi-extinction (reduction of the population to such a low level that extinction is nearly certain) such that all points on the same line have equal quasi-extinction probabilities.

After Groom and Pascual (1998).

probability, a combination of time and threshold associated with it (fig. 7.7) (Groom and Pascual 1998).

The use of PVAs has increased dramatically during the past decade. Groom and Pascual reviewed 58 PVAs published in major conservation journals from 1987 to 1996. In the first five years, there were only 10 PVA publications (17%), whereas the second five years saw 48 such publications (83%) (Groom and Pascual 1998). The National Research Council of the United States has recommended even greater reliance on PVAs for the management and conservation of endangered species (National Research Council 1996).

PVAs also have shifted strongly in their expressed purpose. The majority of the earliest publications (1987–1989) were designed to determine optimum harvest levels, but since 1990 the most common uses have been determining the viability of populations, particularly of threatened and endangered species, and for conservation management (Groom and Pascual 1998). In the United States through 1996, 56 species had been the subject of at least one PVA published in a major scientific journal (Groom and Pascual 1998). Several species, such as the northern spotted owl (*Strix occidentalis caurina*), had been the subject of multiple PVAs, and many other species had been the subject of unpublished PVAs submitted as reports to federal agencies.

PVAs can be categorized in multiple ways. A broad dichotomy exists between analytical-deterministic and stochastic PVAs. Analytical-deterministic models are usually also "structured" PVAs, in that they divide (structure) the population into ages, classes, stages, or sizes and assign values of demographic variables specific to such divisions (Groom and Pascual 1998). Such structured PVAs make a deterministic projection of the population's growth, size, and trajectory into the future.

In theory, deterministic or analytical PVAs can provide the most accurate estimates of population viability, provided that all needed data are available and measured with little or no error; however, these two conditions are rarely met, especially in small and endangered populations. In addition, remember that precision is not the same as accuracy. Most biological processes are stochastic in nature, and outcomes are inherently uncertain. Models that fail to express the degree of uncertainty are prone not only to error but also to serious misinterpretation. As J. Maynard Smith, speaking of biological modeling long before the advent of PVAs, stated with no equivocation, "the use of deterministic rather than stochastic models can only be justified by mathematical convenience" (Maynard Smith 1974). Thus, the second category of PVAs, stochastic simulation models, have increasingly become the approach of choice and widest acceptance by the professional community. Stochastic simulation models also may be structured, in which case a different probability distribution for basic demographic variables (reproduction, growth, mortality) may be assigned to each group or class.

PVAs also may be spatial or nonspatial. That is, the model may assume that the population enjoys a common distribution throughout a given environment (nonspatial), or that it exists as separate subpopulations that are spatially disjunct from one another (spatial). Nonspatial PVAs are more common, primarily because they are easier to model. Nonspatial PVAs do not, for example, require information on migration rates, but spatial PVA models do, and their predictions are strongly affected by even small errors in this variable. Spatial PVAs also require subgroup-specific data on population demography in order to model the persistence of subgroups as independent events. Obviously, subdividing a single population into multiple subunits means that each subunit consists of a smaller number of individuals, making the subunits more susceptible to extinction-causing processes, especially to the effects of demographic stochasticity. Such subdivision tends to decrease effective population size and have a negative effect on the projected viability of the population. Thus, most spatial models tend to predict lower population viability than nonspatial models of the same population (Groom and Pascual 1998). Spatial PVAs, when properly designed and given accurate data, can make more accurate projections about populations that experience complex interactions with their environment and can evaluate the persistence of metapopulations. However, they demand labor-intensive collection of field data, as well as increased computer space and running time, making them a minority among published PVAs to date.

VORTEX: A Model Used in Population Viability Analysis

In the past decade, one of the most widely used generic simulation PVAs has been the computer model VORTEX (Lacy 1993), which originated from earlier models developed for and used by the Captive Breeding Specialist Group of the International Union for the Conservation of Nature (IUCN). Although there are now many different computer programs that can produce a population viability analysis, an examination of the details of one such program helps explain what PVAs do, the data they require, and what assumptions and limitations affect interpretation of their output.

VORTEX is a model that simulates the effects of deterministic and stochastic forces affecting a population, including environmental, demographic, and genetic stochasticity. VORTEX can provide calculations of the effects of inbreeding depression, migration rates between two populations, the mortality attributable to environmental variation in each age class, frequencies and effects of catastrophes, the effects of harvesting, and life table calculations of expected growth.

VORTEX models demographic stochasticity by determining the occurrence of probabilistic events such as reproduction, litter size, sex determination, and death with a "pseudo-random" number generator (Lacy 1993). The number generator is referred to as pseudo-random because the process that generates the numbers is itself deterministic, not random. Working with a carrying capacity also specified by the user, VORTEX will remove individuals from the population if the population size rises above this carrying capacity.

VORTEX models environmental stochasticity primarily through annual, environmentally induced variations in the carrying capacity, and secondarily through variations in demographic variables that are attributable to environmental variation. Catastrophes, the extreme examples of environmental variation, are treated as random events that occur with specified probabilities, and multiple types of catastrophes can be modeled. The effect of a catastrophe on the population is determined using a "severity factor" specific to a particular demographic event, such as survival or reproduction.

VORTEX models genetic effects by assigning unique alleles to each individual. The user can choose a recessive alleles model or a heterosis model to simulate genetic effects (Lacy 1993). In the recessive alleles model, each founder individual has one unique recessive allele. If such an allele occurs in subsequent generations in a homozygous condition (through inbreeding), it is lethal for the individual that carries it. Deaths remove some of these lethal alleles, but also reduce the genetic variation of the population. In contrast to the recessive alleles model, the heterosis model assumes that the cause of inbreeding depression is reduced levels of heterozygosity; thus all homozygotes are assumed to have reduced fitness compared with heterozygotes. In this model, deleterious alleles are not fully recessive and are not removed by natural selection. All alleles are advantageous in the heterozygous condition and deleterious in the homozygous condition. Because deleterious alleles are not removed by selection in this model, the effects of inbreeding do not decline over time as they would in the recessive alleles model.

VORTEX models a number of distinct deterministic processes and their effects. Reproduction can be modeled as a density-dependent effect, either direct (fewer individuals and lower reproduction, i.e., an Allee effect) or indirect (fewer individuals, but greater reproduction). Individuals can

Table 7.2   Assumptions associated with the computer simulation model VORTEX, a widely used population viability analysis model for long-lived, low-fecundity species. Summarized from Lacy 1993.

1. When population size is less than carrying capacity, survival probabilities are density independent. Additional mortality imposed when the population exceeds K affects all age and sex classes equally.
2. The relationship between changes in population size and genetic variability is examined for only one locus. Therefore, potentially complex interaction between genes located on the same chromosome (linkage disequilibrium) are ignored. Such interactions are typically associated with genetic drift in very small populations, but it is unknown if, or how, they would affect population viability.
3. All animals of reproductive age have an equal probability of breeding. This ignores the likelihood that some animals within a population may have a greater probability of breeding successfully, and breeding more often, than other individuals. If breeding is not random among those in the breeding pool, then loss of genetic variation and increases in inbreeding will occur more rapidly than in the model.
4. The life-history attributes of a population (birth, death, migration, harvesting, supplementation) are modeled as a sequence of discrete and therefore seasonal events. However, such events are often continuous through time and the model ignores the possibility that they may be aseasonal or only partly seasonal.
5. Inbreeding is assumed to depress only one component of fitness: first-year survival. Effects on reproduction could be incorporated into this component, but longer-term impacts such as increased disease susceptibility or decreased ability to adapt to environmental change are not modeled.
6. The probabilities of reproduction and mortality are constant from the age of first breeding until an animal reaches the maximum longevity. This assumes that animals continue to breed until they die.
7. A simulated catastrophe will have an effect on a population only in the year that the event occurs.
8. Migration rates among populations are independent of age and sex.
9. Complex, interspecific interactions are not modeled, except in that such community dynamics might contribute to random environmental variation in demographic parameters. For example, cyclical fluctuations caused by predator-prey interaction cannot be modeled by VORTEX.

be added to or removed from populations in specific sex and age classes, and trends in the carrying capacity can be modeled as an annual percentage change in a linear relationship (Lacy 1993). VORTEX also can simulate a metapopulation by modeling up to 20 different populations (subpopulations) with distinct population parameters, among which individuals move. In this case, VORTEX tracks occurrence of both extinctions and recolonizations of the populations.

For a given number of runs (e.g., 100), VORTEX will determine the proportion of runs in which the population survived for specified time periods, such as 10 years, 100 years, 500 years, 1,000 years, or other time increments, as well as the size of the population at the end of the time period, its observed and expected heterozygosity, and the number of extant alleles as a measure of surviving genetic variation. VORTEX can also calculate the same results for metapopulations and, with these, determine the number of individual recolonizations and extinctions that occur in the metapopulation during a specific time period. If more than 50% of the simulated runs end in extinction, VORTEX calculates the median time to extinction and the mean time to extinction of the populations that become extinct.

VORTEX makes several explicit assumptions (table 7.2) about the populations being modeled. These are, at best, simplifications of the way real populations behave and, at worst, inaccurate representations of population behavior. Among other things, the model ignores the long-term effects of catastrophe, relationships of migration rates to sex and age, and more

complex interactions that often occur in populations. Although these explicit assumptions are serious concerns, the implicit assumptions are even more important. VORTEX assumes that the user inputs accurate data on all requested demographic, environmental, and genetic parameters about the population, but in reality, few populations are so well studied as to have reliable data for every value demanded by VORTEX. Many key variables must be entered as "best guesses," with a consequent drop in the reliability of the conclusions that they generate.

Population ecologists Graeme Caughley and Anne Gunn, after working with VORTEX, offered this critique:

VORTEX and like programs do exactly what they are told to do, as constrained by the static single-species models that provide their structure. They can be useful for various purposes so long as the user understands what the programs are doing and how the ecology implied by the model differs from the ecology of real populations. Even if the parameter values fed into them are correct, their structure may be inappropriate and, hence, their output may be of dubious usefulness. A further problem is that the detailed information they seek is almost never available from an endangered species. Hence they often get fed with guesses, or "consensus views of experts familiar with the species," which amounts to the same thing. This is a fault not of the software but of the user (Caughley and Gunn 1996:208–9).

DECISION ANALYSIS IN MANAGING SMALL POPULATIONS

PVA, MVP, and the Analysis of Risk

It may be presumptuous to think that modeling can define a minimum viable population (Boyce 1992), but PVA, when combined with techniques of adaptive management and risk analysis, can contribute to early recognition of problems associated with small populations and provide an accurate assessment of the nature and extent of these problems and even suggest an appropriate solution. PVA also provides standards with which to evaluate solutions to see if they are effective when implemented. Finally, PVA offers the ability to continue evaluation of the problem over time, as well as to evaluate the effectiveness of management efforts to solve it (Lindenmayer et al. 1993).

Boyce (1992) advocates combining PVA with techniques of adaptive management in managing for different MVPs in spatially disjunct populations. Such experiments, combined with risk assessment, can enable managers to evaluate the effectiveness of different conservation strategies (Lindenmayer et al. 1993). This synthesis of approaches and techniques is necessary because the management of any population, with or without PVA, is a problem of decision analysis. For every hypothesis we form about the factor or factors that limit a small population, there is a corresponding management option appropriate to the hypothesis. For every hypothesis-management combination, there is some probability of different states of the population or its environment that might occur as a result.

For example, suppose a manager determines that a small, forest-dwelling population of endangered animals, now consisting of 100 individuals, is most threatened by the potentially catastrophic effects of forest fire. A manager may have the choice of maintaining the endangered population as a single unit at one location, which will maximize the effective population size, or translocating half the animals to a different site to reduce the risk that a fire could destroy the entire population. Suppose that the probability of a fire over the next 100 years is estimated at 0.10. Further, suppose that the probability of extinction over this time under noncatastrophic environmental conditions for a population size of 100 is 0.05, but the probability of extinction for a group of 50 is 0.15. Which strategy minimizes the risk of extinction?

If we display our choices and the associated probabilities of these events as a decision tree (fig. 7.8), we can evaluate all possible outcomes and their probabilities systematically. If the population remains as a single unit and there is no fire, the extinction probability is 0.05. If a fire occurs, their extinction is certain ($pE = 1.0$). Their expected probability of extinction is then the sum of the probabilities of these events, or $(0.05 \times 0.9) + (0.1 \times 1) = 0.045 + 0.1 = 0.145$. If the population is managed as two units, there are four possibilities: (1) a fire occurs in

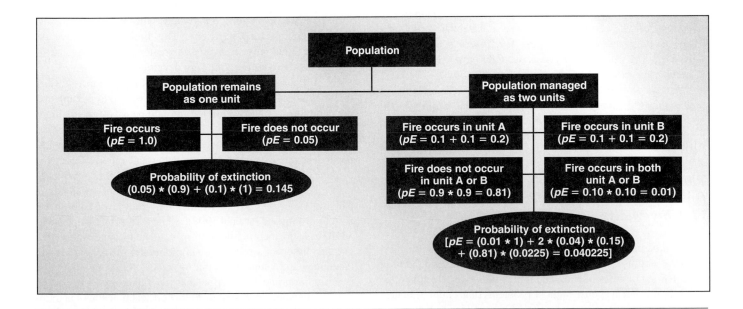

Figure 7.8

A decision-making tree of risk analysis for the probabilities of extinction (*pE*) for a population under two different management strategies. When combined with results gained from a PVA, risk analysis can provide valuable insight into the relative risks of different management strategies for a population of given characteristics.

both units, (2) a fire does not occur in either unit, (3) and (4) a fire occurs in one unit but not in the other. The probability of a fire in both units is $0.10 \times 0.10 = 0.01$; and the probability that a fire does not occur in either unit is $0.9 \times 0.9 = 0.81$. The probability of a fire in at least one unit but not in both is $0.1 + 0.1 = 0.2$. Then the extinction probability of the two subunits is $(.01 \times 1) + 2(0.04 \times 0.15) + (0.81 \times 0.0225) = 0.01 + 0.012 + 0.018225 = 0.040225$. In this particular case, the risk of extinction is more than three times as high for the single large population as for the separated populations.

Risk analysis, like PVA, makes many assumptions. For one thing, it assumes that the probabilities it derives from the PVA are accurate estimates of the probabilities of these events. Further, it assumes that catastrophes like forest fires are independent events in the two areas (they may not be if the two areas are not sufficiently far apart). Using PVA for endangered species is especially uncertain because it means that the very data needed to make the PVA more reliable, namely population parameters obtained by direct measurement or manipulation of the population, probably cannot be obtained because the population's small size may make direct handling or experimental manipulation an unacceptable risk. Yet it is often in cases of endangered species that a reliable PVA is most needed. Thus, far from solving all dilemmas of population management, PVA may create some of its own.

All forms of analysis are limited by both their assumptions and their procedures. These limitations are not reasons to avoid using the analysis, but rather are constraints to be appreciated so that the analysis is not misunderstood or misapplied. Caughley and Gunn are among the most explicit of conservation biologists in their assessment of the limits of PVA and its correct applications. Their remarks are worth repeating:

> Viability analysis has often been used as if it were a diagnostic tool, it has been used to justify management recommendations on how to cure a problem that it cannot in logic diagnose, and its use has sometimes led to status quo management when a recovery operation was urgently needed. The viability analyses are at their most useful in estimating the vulnerability of small populations to stochastic events. Populations in reserves are often small and face the risks of stochastic events, and thus it follows that viability analyses are useful to estimate the size of reserves (Caughley and Gunn 1996:219).

The Problem of PVA Application: How Do We Use and Interpret Population Viability Analyses?

PVA can be used to improve many aspects of conservation policy. PVA permits the problems associated with small populations to be defined in greater detail, and it may facilitate selection of a more appropriate response to those problems than could be made through other means of analysis. PVA also can lead to better appraisal of management efforts (i.e., Did the policies and management strategies employed solve the problem?).

Like all models, PVA forces conservation biologists to make explicit what is known about the population, as well as current assumptions about unknown population parameters. Thus, PVA serves a purpose in assembling known facts in a meaningful way, while also identifying areas of research needed to make such analysis more reliable and information needed for management planning (Lindenmayer et al. 1993). PVA can synthesize interacting factors and identify trends in population behavior, and it can identify processes that threaten the population. It can even be used in defining the "minimum critical area" for a species, a valuable concept in designing reserves.

PVA is limited in numerous ways and dimensions, however. First and foremost, PVA is a data-intensive technique; the more data, the better the analysis. In a strict sense, this is a problem of the state of scientific knowledge, not a limitation of the technique per se, but effective techniques must work with the data that are available. PVA's heavy dependence on, and the connection of its predictions to, accurate data means that, if such data are not available, the predictions produced from a PVA may not be reliable. The more realistic the model, the more data it requires. This problem is of special concern for the spatial models that could accurately represent a metapopulation.

PVA is typically used in a single-species approach to conservation, which is understandable because a PVA's unit of interest is a population, not a community. In a strict sense, this limitation also is less a criticism of the method than of the lack of PVA-equivalent techniques for larger biological units. It would be a poor carpenter who criticized his saw because it cut wood well but was useless in driving nails. Nevertheless, it is important to recognize that PVA was developed for and should be used primarily on a population-by-population basis. It is not an appropriate technique for evaluating the viability of communities or ecosystems, nor is it equally well suited to all *types* of populations. Most specific models for estimating PVA, such as VORTEX, are designed for species with low fecundity and long life spans, such as some birds and mammals. Such models perform poorly in evaluating populations with high fecundity and short life spans, such as insects. This limitation, however, is likely to be gradually removed as a greater and more diverse array of PVA analysis programs are developed.

A more serious limitation of PVAs is that most still use linear assumptions of population growth (i.e., they assume exponential growth and do not incorporate density-dependent effects). Of 58 PVAs reviewed by Groom and Pascual (1998), only 16% (9 studies) incorporated any kind of density-dependent mechanism. Groom and Pascual reflect, somewhat cynically, on this state of affairs, remarking, "Since we teach undergraduates in ecology that simple linear models are patently false, one has to wonder why they are standard for PVA" (Groom and Pascual 1998). Some PVA programs, such as VORTEX, do incorporate a form of density dependence, but one that assumes a relatively simple model of logistic growth (Caughley and Gunn 1996) that is truncated exponentially and is not appropriate to describe many populations. One partial, but still meaningful, response to this criticism of PVAs is to integrate the quasi-extinction threshold with known Allee effects in given populations. Recall that Allee effects are a form of density dependence that could be called "threshold density dependence," because they only begin to affect populations negatively when numbers drop below critical minimum sizes. The role of this kind of density dependence can be incorporated into PVAs by making the extinction threshold of the model the point at which Allee effects would begin to occur.

Overall, PVA simplifies the dynamics of populations as well as interactions among different environmental and demographic parameters. Such simplification reduces the correspondence between the model and the real population. Acknowledging this limitation helps identify specific areas in which models for PVA can be improved and their correspondence to real populations increased.

A major limitation of current PVAs is that there are no standard protocols or objective criteria that define a valid PVA. Judgments in particular cases can be arbitrary. Associated decisions about acceptable probabilities of extinction or time of persistence are not subject to any established standard; thus, it is difficult to compare the quality of one PVA to that of another. This problem is exacerbated by the fact that most published PVAs to date offer only one model as a solution and do not compare the solutions of models with alternative assumptions or criteria. After completing their review of recently published PVAs and noting this deficiency, Groom and Pascual argue "for the further step of building and analyzing multiple models to obtain the best possible portrait of future outcomes of a management strategy or a particular disturbance . . . by stressing the selection of a single model we risk discarding reasonable alternatives, producing a distorted assessment of actual risks associated with management procedures" (Groom and Pascual 1998). The authors argue for a more candid admission of uncertainty, a simultaneous evaluation of several alternative models, and a search for management alternatives that are robust to the choice of the model.

Other serious criticisms of PVA arise from rigorous mathematical analysis of its inputs and outputs. Mathematician and biostatistician Donald Ludwig calculated quasi-extinction probabilities of natural populations from time series of census data or estimated abundance in several different animal populations. Even when errors in estimates were ignored, confidence intervals associated with extinction probabilities were large. In three species of birds, the Laysan finch (*Telespyza cantans*), palila (*Loxoides bailleui*) and snow goose (*Anser caerulescens careulescens*), most confidence intervals included extinction probabilities of 0 and 1, rendering them meaningless. When errors in estimates were included, things got worse. Ludwig demonstrated that neglecting observational errors produced a bias in the extinction probability that significantly lowered its value, thus giving an overly optimistic estimate of the probability of population persistence (Ludwig 1999). Ludwig concluded "A proper population viability analysis should include estimates of likely ranges in parameter estimates based upon available data and also include the corresponding ranges for quantities related to extinction. The results should not be regarded as reliable unless the ranges are small. In view of the difficulties cited above, it is difficult to imagine reliable estimates of small extinction probabilities for populations that appear to be threatened" (Ludwig 1999).

PVA is primarily a description of the existing demographic and environmental stochasticity of a population projected into the future. Like all models, constructing a model for a PVA forces us to make explicit all assumptions about the population and to identify the processes and parameters that influence our conclusions. But this is an advantage of model building, not model output. PVA can, rightly constructed and interpreted, help managers make more informed judgments, but neither its assumptions nor its predictions can be empirically tested. PVA is expected to be defensible against legal challenges. But given its uncertainties, it may not be able to withstand them. Thus, biologists who perform PVA may not be able to meet the expectations of their employers (Ludwig 1999), particularly if the employer is a government agency whose decisions are accountable to the public and potential targets for litigation.

Unfortunately, many conservationists forget the limitations of PVA and incorrectly use it as a diagnostic tool to determine why populations are declining and how to save them from extinction. PVA has no power to diagnose the causes of decline or prescribe a remedy for it, but PVA may provide, when combined with human discernment, clues that lead the investigator to correctly identify which factors pose the greatest risk to the population's persistence. It does not, however, identify what made the population small in the first place or what must be done to make it bigger. Similarly, PVA cannot be presented in public or in court as an estimate of certainty. Its level of uncertainty may be large, even unacceptably so. Again Ludwig addresses the issue perceptively, noting that "an exaggeration of our capabilities carries a high risk of failure and subsequent disillusionment. It would be better to be more modest about our understanding and achievements, and to help decision makers understand the complex, realistic arguments that pertain to most conservation decisions" (Ludwig 1999).

Although PVA will remain an attractive technique in conservation biology because it holds out the promise of assessing and categorizing the causes of population decline in generalized categories, PVAs are notoriously weak in their assessment of the specific ecological causes of decline and extinction in particular populations, and typically ignore fundamental factors such as habitat and predation (Boyce 1992). This is one of the most serious criticisms that can be made of PVA, precisely because the ultimate causes of extinction are almost always ecological. For example, Temple (1986) notes that, among endangered birds, 82% are threatened by habitat destruction, 44% by overexploitation, 35% by introductions of nonnative species, and 12% by chemical pollution or the consequences of natural events. To conserve populations successfully, specific ecological pressures and threats must be understood, evaluated, and managed in detail, not masked in modeling generalities of stochastic factors.

The critical question for conservation that the PVA attempts to answer is summarized in the famous query of former President Ronald Reagan: "Just how many spotted owls do we need?" That is, how many individuals are necessary to achieve a population that is relatively safe from the assaults of environmental and demographic stochasticity, genetic drift, and natural catastrophe? This key question of conservation biology is answered in the determination of a value closely tied to the practice of population viability analysis. This is the concept of minimum viable populations.

Minimum Viable Populations

The origin of the term **minimum viable population (MVP)** is uncertain. It may have ties to the directive of the National Forest Management Act that the U.S. Forest Service shall maintain *viable populations* of native vertebrates on national forest lands (Gilpin and Soulé 1986). Boyce defines a minimum viable population as "an estimate of the minimum number of organisms of a particular species that constitutes a viable population" (Boyce 1992). The operative phrase in the definition—"viable population"—cannot always be consistently or objectively determined. According to Shaffer, one of the originators of the MVP concept, "a minimum viable population for any given species in any given habitat is the smallest isolated population having a 99% chance of remaining extant for 1,000 years despite the foreseeable events of demographic, environmental, and genetic stochasticity, and natural catastrophes" (Shaffer 1981).

Although not all conservation biologists would agree with the probabilities of persistence and time spans specified by Shaffer, his use of specific quantities for both variables illustrates one of the chief attractions of the MVP concept—it permits an estimate of population persistence at a specified level of probability for a specified period of time. Thus, it holds out the promise of answering an important question of conservation biology (How many individuals must a population have to persist?) with a response that is at once specific (for how many years) and straightforward about the degree of uncertainty in the answer (the probability of persistence). However, this is a promise that cannot always be fulfilled.

There are five main avenues for determining an MVP: experiments, biogeographic patterns, theoretical models, genetic considerations, and simulation models. Not all are equally useful, practical, or precise in the task of MVP estimation, especially for small populations. Experiments are often inappropriate, not to mention extremely risky, to conduct on small-populations; the outcome of such an experiment could well be another extinct species. The second approach, an examination of biogeographic patterns of disjunct populations of a species over a large area, does provide some first-order approximation of the minimum sizes of areas that populations can use, and the (apparent) minimum number of individuals that sustainable populations must have. However, as in all descriptively oriented data, the inferences drawn may be unreliable. Various factors, all uncontrolled in a simple examination of the biogeographic pattern, may affect the population and area sizes used by species because population characteristics vary widely in different areas and habitats. Populations with minimum numbers and minimum areas may not be stable populations. They may, in fact, be already declining to extinction. Thus, the use of a minimum for decision rules in management and conservation could doom a population. A biogeographic approach will not provide worthwhile results unless information is available on dispersal, migration, and colonization rates of new habitats. And, perhaps most obviously, a biogeographic approach will not work well with species that do not inhabit sufficiently large areas or that do not inhabit insular or patchy habitats.

Some theoretical models have been employed to determine minimum population sizes. One is diffusion theory, in which the movement, dispersion, and growth of a population are assumed to follow the principles of diffusion observed at a molecular level. However, most organisms do not behave precisely like molecules of gas, even in an entirely novel environment, and the theory requires environments completely unpredictable to the organism for its assumptions to be valid. Many other models, although elegant in their simplicity, also contain unrealistic assumptions or unresolved mathematical difficulties (Shaffer 1981). Because every species is unique, general theoretical models often fail to achieve realism when applied to the particulars of individual species' environments and demography.

As noted in chapter 5, genetic considerations were the original source of estimates of minimum viable populations in conservation biology. Franklin (1980) suggested that effective population sizes of at least 50 individuals were needed to prevent deleterious effects of inbreeding, and a minimum effective size of 500 was required to maintain sufficient genetic variation to be able to respond to continuing environmental variation. We have already noted that such "rules" were too general to be of value to specific populations. In many cases, much larger numbers may be required. Of greater importance is the fact that such an estimate of MVP reflects only genetic considerations. Real populations, especially those that are declining or small, may be threatened by nongenetic factors. If nongenetic factors are ignored, the estimated effective population size may deviate from the population size needed to alleviate threats from other factors, such as environmental stochasticity, demographic stochasticity, and catastrophes.

Managing real populations is rarely as simple as identifying a single cause of decline, eradicating one exotic competitor, or forming a speculative hypothesis based on a correlation between environmental and demographic variables. Causes of population decline are complex problems that may require management of multiple threats over long time spans. To better appreciate the complexity and integration of such threats and how they can be managed to achieve a desirable outcome, even under adverse conditions, we consider the case history of one small and declining population, the Lord Howe Island woodhen, (*Tricholimnas sylvestris*).

THE LORD HOWE ISLAND WOODHEN: A CASE STUDY IN MANAGING MULTIPLE THREATS TO A SMALL AND DECLINING POPULATION

The Lord Howe Island woodhen (fig. 7.9) was not described scientifically until 1869 (Sclater 1869), but had been noted by European explorers visiting Lord Howe Island (between Australia and New Zealand) in the late eighteenth century. The earliest accounts describe it as a rather fearless bird that would often approach human visitors; indeed, biologists conducting a more recent investigation of the Lord Howe woodhen described it as "suicidally inquisitive" (Miller and Mullette 1985). Combined pressures of hunting, habitat destruction, and the spread of introduced feral pigs on the island caused drastic reductions

Figure 7.9
The Lord Howe Island woodhen (*Tricholimnas sylvestris*), a flightless rail found only on Lord Howe Island between Australia and New Zealand. The woodhen was nearly exterminated through a combination of hunting, habitat destruction, and predation by introduced feral pigs (*Sus scrofa*). An aggressive management program of woodhen protection, pig eradication, and release of captive-bred individuals to new locations has increased numbers and long-term prospects for the persistence of the woodhen population.

in woodhen numbers and range. Pigs were the most significant problem, because they destroyed ground cover in preferred habitat (an essential element for nesting success), and preyed on nests, young birds, and adults.

By 1978 the population had been reduced to 10 breeding pairs located on the most remote areas of the island's highest mountain. Within two years, only three healthy breeding pairs remained, and circumstantial evidence suggested that pigs were the primary limiting factor in woodhen distribution even though the ranges of pigs and woodhens did not overlap, but often abutted one another. As recent investigators noted, "In several places, the distributions were separated only by a low rock face" that the pigs could not climb (Miller and Mullette 1985).

Using the evidence of the geographic distribution of pigs and woodhens as support for the hypothesis that pigs were the primary factor limiting range expansion, the Australian government began a pig eradication program in 1979 and a woodhen captive breeding and reintroduction program in 1980. Because it was believed that woodhen distribution was indicative of its preferred habitats, captive-reared woodhens were released on mountainous slopes at lower elevations. Although they survived and established territories, recovery was slow. A more serendipitous event provided the insight that would prove critical to the woodhen's recovery.

In 1979, a male woodhen appeared in a lowland area near the southern tip of Lord Howe Island. He defended a territory that included a large outdoor garden of the King family, long-term residents of the island. Soon known as the "King garden woodhen," the male accepted a captive-bred female released into his territory in 1981. Now "adopted" by the King family and given protection and some supplemental food, the pair laid their first clutch of eggs in 1982, but successfully raised only one chick. However, more clutches followed in

quick succession, and by the end of the year the pair had laid nine clutches and raised 15 chicks in 11 months, plus two chicks that were raised in captivity! Biologist P. J. Fullagar, who observed the process firsthand, noted, "Multiple broods were observed in which young birds from one brood assisted in defense and feeding of a subsequent brood" (Fullagar 1985). Supplemented with additional birds from the captive breeding program (and the additional food provided by the King family), this colony of woodhens prospered to the extent that birds established territories throughout the southern end of Lord Howe Island in lowland habitats. Within two years "it was becoming difficult to obtain accurate figures on their numbers" (Fullagar 1985). Up to 20 pairs were thought to be present, exceeding the numbers of the original wild mountain population. Continued growth of the King colony, successful reintroduction of captive-raised birds into other areas, and the growth of the mountain population (now unconstrained by pigs) brought the woodhen population to over 200 individuals by 1984.

The Lord Howe woodhen illustrates several important components of successful rehabilitation of a small population. First, evaluating the distribution of the woodhen and its predators helped construct a hypothesis to explain its decline (pigs limit woodhens), which could be used to formulate a specific prediction (woodhens should increase in number if pigs are removed). That prediction could be evaluated with a management plan (eradicate pigs). This sequence of clear thinking and appropriate action saved limited resources (there was no attempt at rat, owl, or cat control) and produced rapid results that could be compared directly to the prediction derived from the hypothesis.

Second, small populations can, in some cases, be helped dramatically by careful, well-planned supplementation of individuals from captive breeding programs. The release of captive-bred birds in both the mountain area and the areas surrounding the King colony made significant additions to the growth of both populations. As a general principle, individuals raised in wild populations are always preferable to captive-reared individuals, but captive breeding programs are sometimes required to supplement natural reproduction in populations that have reached critically low levels.

Third, the unexpected and fortuitous development of the King colony demonstrates a fact that might have otherwise been overlooked. Small, remnant populations of what was once a widespread species do not necessarily occupy their optimal habitat as they approach extinction. Indeed, forces causing their decline (in this case, pigs) may be driving the last individuals into marginal habitats where survivorship and fecundity are low. In this case, it appears the high-elevation forests were not optimal habitat for the woodhen, but refuge habitat. It is a common and repeated observation of studies of animal behavior that habitat preferences change in the presence of predators, and that habitat that is optimal for breeding is usually not optimal for refuge and escape (Rosenzweig 1991). Thus, conservation biologists should not necessarily assume the habitat occupied by the last remaining individuals in the population is the best habitat for reintroduction. In addition to consulting historical data on the population's distribution, a better method may be to use an adaptive management approach and attempt reintroductions in multiple habitat types and monitor the results carefully over time.

THE PROBLEM OF NONINDIGENOUS SPECIES: HOW DO WE MANAGE AN INVASIVE POPULATION?

In a world of increasingly mobile human populations, it is not surprising that such human movement has been accompanied by increasing numbers of invasions of other, **nonindigenous species** (NIS), to many parts of the world where they were not previously found. Today many conservationists consider effects of nonindigenous species second only to habitat destruction as a cause of extinction and endangerment of species worldwide.

Regrettably, the human role in spreading such species has often been deliberate in its intent while ignorant of its consequences. In the nineteenth century, the Naturalization Society in New Zealand attempted, and for the most part succeeded, in re-creating the ambience of an English countryside in some parts of New Zealand by releasing common British songbirds, often to the detriment of the native birds of New Zealand (Godfray and Crawley 1998). But even more introductions are accomplished simply by thoroughgoing human ignorance, as people, along with their goods, clothes, ships, planes, and automobiles, serve as the conduits for plant and animal stowaways carried unnoticed throughout the world. Invasions of nonindigenous aquatic organisms may be facilitated by whole communities of creatures attached to the bottom of ships or living in the ballast water of such vessels (Ruiz et al. 1997). Multiple species of forest-dwelling insects may move throughout the world in processed or unprocessed wood products, while others that feed on vegetables and fruits, such as the Mediterranean fruit fly or "medfly" (*Ceratitis capitata*) often move worldwide as adults or larvae in produce shipments (Carey 1996). Plants may be dispersed long distances as spores or seeds, or actively collected and planted under cultivation, only to later escape to the wild.

It is difficult to assess the total number of nonindigenous species worldwide or their rates of invasion, because estimates vary according to phylogenetic group and reference time frame. In the United States, the U.S. Congressional Office of Technology Assessment estimated that there are at least 4,500 established nonindigenous species (OTA 1993). From their own research and reviews of other sources, forest scientist Pekka Niemelä and insect ecologist Willliam J. Mattson have, together, concluded that nearly 2,000 species of insects and 2,000 species of weedy plants have invaded North America in the last 500 years (Niemelä and Mattson 1996). Godfray and

Crawley (1998) estimated that at least 20,000 nonnative plant species have been introduced into Great Britain, nearly 1,200 of which have become naturalized. Among aquatic organisms, Carlton and Geller (1993) list 46 species of NIS that have been introduced around the world from ballast water discharges, just since the 1970s.

The majority of introduced species fail to establish persistent populations in their new environments, and most of the successful ones that do, live benign and inconspicuous lives among the natives. A small number experience enormous population growth and range expansion, often with devastating ecological and economic effects on the native communities. In the United States, it has been estimated that a small number of nonindigenous species (pest species) have caused nearly 100 billion dollars in damages from 1906 to 1991 (OTA 1993), a figure that does not take into account the biological damage to ecosystems or the cost of extinction or endangerment to native species caused by these nonindigenous species.

Whether benign or pestilent, such invasions cause changes in community composition, structure, and function. Indeed, it is impossible to understand existing communities worldwide without an understanding of species' introduction and invasion. Thus Godfray and Crawley (1998) assert that "the composition of many (perhaps even most) communities is determined by the history of introduction."

An understanding of nonindigenous species takes on special urgency in conservation because such exotic introductions, invasions, and establishments often displace or even eradicate native species, especially species with small populations, specialized habitat requirements, or limited range. Such events occur not only through competition, but also through predation. Predation is often most devastating to a prey species when an introduced predator encounters native prey species that have evolved few or no defenses against it. A classic example of extinctions caused solely by the actions of an introduced predator is the case of the brown treesnake (*Boiga irregularis*) (fig. 7.10), which reached the Pacific island of Guam in 1967. The snake's spread coincided with the disappearance and extinction of bird species native to the island; research confirmed that the snake—the only predator unique to Guam—was indeed the cause. Some species are neither competitive nor predatory, but create such pervasive changes in vegetation and habitat that indigenous species cannot survive. For example, pigs (*Sus scrofa*) and goats (*Capra hircus*) from Europe have devastated native flora and fauna of tropical islands throughout the world where they have been introduced.

Such stories of nonindigenous species invasions and their effects make for fascinating reading, and comprise the entire content of many excellent books (e.g., Elton 1958; Groves and Burton 1986; Mooney and Drake 1986; Drake et al. 1989). However, our purpose here is not to tell every story of alien invasion, but to explicate patterns, principles, and theories regarding such invasions in general that we may better understand individual case histories in particular, and thus better respond to and manage NIS populations such that native species are conserved. We begin with a general review of common trends that emerge from repeated patterns of invasion.

Figure 7.10
The brown treesnake (*Boiga irregularis*), an introduced predator on the Pacific island of Guam, has exterminated all species of native birds. Here, a brown treesnake is caught in the act of consuming a bird.

Characteristics of Successful Invading Species

No single comprehensive theory adequately explains all patterns of invasion or all characteristics of invasive species. In general, however, successful invaders often show three consistent characteristics: (1) The invading species can deliver seeds, breeding individuals, or other types of propagules at a high rate at an opportune moment for invasion and at a high density to an opportune site or sites. (2) The invading species is able to persist for extended periods at low densities under unfavorable conditions until favorable conditions permit it to grow to higher densities. (3) The invading species is a good "ecological match" for the environment, and is able to exploit local conditions and abiotic factors that favor completion of its life cycle as well or better than native species.

Many highly successful invaders are especially adept in category one. For example, Rejmánek and Richardson (1996) examined the invasive characteristics of 24 species of pine (*Pinus* spp.), 12 of which they classed as noninvasive (planted on at least three continents but never reported as spreading) and 12 classed as invasive (spreading on at least two continents). After evaluating 10 different life history traits in both groups via discriminant analysis, they found only three that were significant in classification. These traits were the square root of mean seed mass, the square root of the minimum juvenile period, and the mean interval between large seed crops. Invasive species had low mass of individual seeds, short juvenile periods, and short intervals between large seed crops. The second and third traits allowed invaders to achieve early and consistent reproduction once established, whereas the first contributed to higher numbers of more widely dispersed seeds. Interestingly, the invasive species were all concentrated in the same subgenus

(*Diploxydon*), and noninvasive species were all members of a different subgenus (*Strobus*). This dichotomy suggests that, at least in this taxon, membership in a subgenus might be used as the first indication of the possible invasiveness of a species (Rejmánek and Richardson 1996).

Other principles also help explain the success of invaders in multiple contexts. Specifically, in planned introductions, an introduced species is more likely to be successful if (1) more individuals are released rather than few, (2) more release sites are used rather than fewer, and (3) the releases are repeated many times (Veltman, Nee, and Crawley 1996). Thus, nonindigenous species that can invade in large numbers at multiple sites in repeated efforts will have higher probabilities of success.

Although the ability to persist at low densities is undoubtedly important, we know relatively little about the abilities of invasive species in this regard because, at such low densities, they are often undetected. For example, recent evidence suggests that the Mediterranean fruit fly (*Ceratitis capitata*) has been able to persist at extremely low population levels in the Los Angeles Basin of California and slowly spread to other locations in the region despite intense efforts to eradicate it (Carey 1996). Along with low densities, some data also suggest that invasive species are adept at enduring long periods of unfavorable conditions. For example, in Great Britain, alien plant species were more likely to show protracted (> 20 year) seed dormancy than were native species (Godfray and Crawley 1998). Further, successful invaders must be able to increase in numbers when rare, overcoming the Allee effects that normally cause the decline of small populations.

Many invasive species demonstrate the importance of "ecological match" or "preadaptation" to a novel environment. Some aspects of ecological match occur at extremely broad levels. For example, among herbaceous plants, the best indicator of ability to invade a new area appears to be latitudinal range (Forcella, Wood, and Dillon 1986; Rejmánek 1995). The greater the spread of latitude (and, by inference, climatic conditions) that an herbaceous plant can tolerate in its indigenous range, the more likely it is to invade new areas. Paralleling this pattern in plants, Niemelä and Mattson (1996) determined that one reason European phytophagous insects are more likely to be successful invaders of North American forests than vice versa is because they possess broader capacities to accommodate new host species.

Structural habitat components also appear to play a major role in creating an ecological match that facilitates invasion success in some groups of organisms. In an extensive review of nonindigenous fishes in California, Moyle and Light (1996) determined that abiotic conditions in streams, rather than biotic characteristics of native stream communities, were the most important determinants of successful invasion (Moyle and Light 1996). The most successful invaders were those adapted to the local hydrologic regimes, specifically to patterns of seasonal changes in water flow. The investigators concluded, "the most important factor determining the success of an invading fish is the match between the invader and the hydrologic regime" (Moyle and Light 1996).

Another dimension of ecological match is the ability of a species to alter the habitat itself, effectively creating a niche where one did not previously exist. An example of this can be seen in the nitrogen-fixing tree *Myrica faya,* an invasive species on the island of Hawaii. *Myrica faya* is adept at colonizing volcanic ash and open native forests, both of which are nutrient-limited systems. In these systems, *Myrica faya* can increase inputs of nitrogen up to four times (Lodge 1993). One would think that such nitrogen additions might be beneficial to native species in these nutrient-limiting environments. However, for many species, the effects of shading and high rates of litter accumulation under and around *Myrica* outweigh these benefits and lead to their decline (Lodge 1993). Thus, *Myrica faya* alters the habitat by changing rates of nutrient cycling as well as the physical structure and light penetration of open forests, creating a new niche favorable to itself, but one in which native species cannot survive.

The Problem of Prediction: Can We Construct Models of Invasive Patterns to Understand the Invasive Process?

The identification of common trends and traits in invading species is helpful in understanding the processes through which NIS become established in new environments, and may assist a conservation biologist in making a preliminary assessment of which potentially invasive species have the greatest probability of success, and the greatest potential for harm. Better still would be the ability to model the invasion process in a systematic manner, and thus gain greater understanding of its mechanisms and greater ability to make more specific predictions about its outcomes.

Many invasions show patterns that can be modeled and described as wave motions, in which individuals at a particular point or site move outward in concentric circles of ever-increasing radii. This view of biological invasion was recognized intuitively and described conceptually over 40 years ago. In his classic book, *The Ecology of Invasions by Animals and Plants,* the British ecologist Charles Elton (Elton 1958), included maps with concentric circles depicting the spread of such introduced species as the European starling (*Sturnus vulgaris*), muskrat (*Ondatra zibethica*), Chinese mitten crab (*Eriocheir sinensis*), and other invasive organisms.

Even before Elton's work, biometrician J. G. Skellam, in his paper "Random Dispersal in Theoretical Populations" (1951), mathematically described the spread of an invading organism as a type of reaction-diffusion equation. This **reaction-diffusion model** predicts that the advancing front of the invading organism should travel as a wave at a velocity (V) described by the equation

$$V = 2\sqrt{rD},$$

where r is the population's intrinsic rate of increase and D is the "diffusion coefficient," equal to one-half the mean squared distance moved in a time unit by an organism (Godfray and Crawley 1998). For example, D is often expressed in km^2/year (e.g., Grosholz 1996).

For all its simplicity, the predictions of this equation do closely match observed results in many species (Grosholz 1996; Godfray and Crawley 1998). The speed of advancement in a

muskrat invasion, for example, was locally constant, but influenced by topography and habitat preference (i.e., the muskrat's affinity for wetlands) (Skellam 1951). The model, however, can be modified to account for spatial heterogeneity and the patchy distribution of habitats by taking the form

$$V = 2\sqrt{r_a D_h},$$

where r_a is the arithmetic mean of the population's intrinsic growth rate across patches and D_h is the harmonic mean of the diffusion coefficient across patches (Shigesada, Kawasaki, and Teramoto 1986). Both values can be calculated if one knows the rates of growth and spread in a sufficient number of habitat patches to calculate a reliable average. This model has also performed well when subjected to more stringent experimental tests in which the value of r was calculated from a life table and the value of D from controlled experiments of dispersal, then compared to data associated with an actual invasion. Andow and associates (1990) subjected the wave model to such tests with an invasion of muskrats, the cereal leaf beetle (*Oulema melanopa,* a European insect introduced in the United States in 1958) and a butterfly (*Pieris rapae,* also European and introduced to the United States several times in the nineteenth century). The model performed well for the muskrat and butterfly, but drastically underestimated the rate of spread in the beetle. Andow and associates (1990) believed that the model failed in the case of the beetle because it did not incorporate the effects of long-distance dispersal by a few individuals. Such events, although rare, can profoundly affect the rate of spread. More sophisticated recent models have attempted to incorporate such effects, along with effects of density dependence and carrying capacity as inner circles become saturated with individuals. For example, Veit and Lewis (1996) constructed a model with such elements to describe and explain the spread of the house finch (*Carpodacus mexicanus*) from a small population of about 250 birds in New York to a population that now covers most of the United States. Their model predicted that range expansion would be slower than expected by traditional models because of Allee effects, but that speed of distribution would increase at an increasing rate as the population grew. Both of these predictions are confirmed in historical data. The model's incorporation of long-distance dispersal by a few individuals also matched well with the rather jagged and erratic pattern of historic expansion.

Such elements as long-distance dispersal and density-dependent rates of spread are incorporated more explicitly into a category of invasion models known as **stratified diffusion models**. In these models, populations at different distances from the source of invasion are assigned different values of r and D, thereby creating different layers or strata in the dispersing population. It is worth noting that stratified diffusion models ultimately resemble classic age-structured models of population growth, except that the founding of each colony in successive strata takes the place of birth and colony growth takes the place of aging. If new colonies are established near existing colonies and coalesce, the rate of spread changes from accelerating to linear. This pattern of invasion has been referred to as "starburst" to distinguish it from the traditional "traveling wave" form. Godfray and Crawley (1998) note, "Both types of spread (traveling wave and 'star-

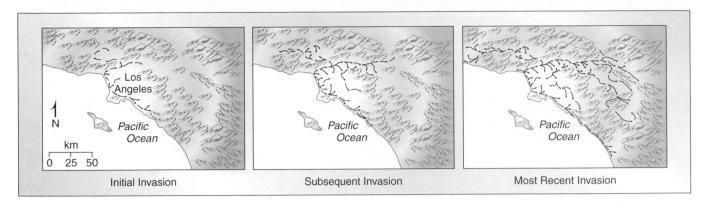

Figure 7.11

Schematic diagram of the "invasion-stream" model of NIS invasion, exemplified by the Mediterranean fruit fly, or "medfly" (*Ceratitis capitata*), in the Los Angeles Basin of California. Note that the spread of the medfly follows configurations of the valleys.

After Carey (1996).

burst') are seen as two ends of a single continuum; the key parameter is the distance between successive foci of establishment. When this distance is small, the assemblage behaves like a traveling wave, but when it is large starburst effects predominate."

The reaction-diffusion model has proven useful in describing, explaining, and, in some cases, predicting patterns of movement associated with invasions of nonindigenous species (NIS). However, it has the drawback of blinding conservationists to other types of patterns, patterns that may suggest factors other than population growth and diffusion coefficients as the controlling agents of spread. In his analysis of the invasion of the Mediterranean fruit fly or medfly in California, James Carey proposes that the spread of the medfly does not follow the pattern of a traveling wave, but the flow of a stream. Like a stream's flow, movement is determined primarily by topography, with dispersing individuals seeking paths of least resistance to new locations (fig. 7.11). Further, Carey argues that biologists were predisposed to look for a concentric wave pattern of spread in their analysis of medfly distribution. When no such pattern was found, reports of medflies were treated as separate invasions from foreign sources, with increased emphasis on inspection at ports of entry. In fact, Carey demonstrates that such reports are evidence of established populations spreading via "linear channeling," directed primarily by the locations of mountains, valleys, and shorelines. Carey notes, "This linear channeling occurs as organisms follow the path of least resistance, and produces a pattern more like a tentacle than a moving wave. The stream-like movement of an invasion explains how two separate outbreaks a great distance apart can still be the same infestation, and the diffusion process in regions where the infestation first started explains why medflies are found repeatedly in the same neighborhoods over a multiple-year period" (Carey 1996).

The Problem of Practical Response: How Do We Prevent or Control Invasions by Nonindigenous Species?

Although a few invasions by nonindigenous species have proven beneficial, and many have been benign or at least incon-spicuous, the enormous economic and ecologic damage done by some nonindigenous species, as well as the direct extermination and endangerment of native species by nonindigenous species throughout the world, suggest that further invasions of nonindigenous species should be prevented. Thus, prudent management should follow a path of preventing nonnative species from entering; of identifying, controlling, and where possible, eradicating nonindigenous populations if they are established, especially when they are still small; and of strategically controlling their effects on native species when their populations are large, widespread, well established, and beyond hope of eradication. Management practices, therefore, change according to the status of the nonindigenous species being considered.

Step 1: Preventing Entry of Nonindigenous Species

To prevent the initial entry of nonindigenous species, managers should carefully answer the following questions: (1) To what nonindigenous species is my preserve, community, or ecosystem most vulnerable? (2) Which of these nonindigenous species has the greatest potential or opportunity to invade, based on its current distribution and pattern of dispersal? (3) What are the most likely points and methods of entry? Can these be closed or monitored in such a way as to prevent the entry of the nonindigenous species?

Answering the first two questions requires a thorough knowledge of both the system being protected and the ecology of invaders that would be most adept at becoming established in it. Thus, conservation biologists must remain well informed about current invasive species and their present distribution. In some cases, it may be possible to alter environmental variables in ways that make a community less prone to invasion. For example, suppression of fire in grassland and prairie ecosystems makes such systems more likely to suffer encroachment from nonprairie species, including nonindigenous plants. In aquatic systems, unmanaged human disturbance and disruption of systems seems to increase the success of invaders. In the previously cited work of Moyle and Light (1996) on fish communities, the authors note, "In aquatic systems with high levels of human disturbance, a much wider range of species can invade than in systems with low levels of human disturbance." For plants, open disturbed communities are more invasible than closed, less disturbed communities (Crawley 1987). Godfray and Crawley (1998) formalize this as a rule of predictable patterns of invasion. Namely, "the rate of establishment of alien species will be proportional to the frequency and intensity of disturbance of the habitat." Further, in such areas of disturbance, species are more likely to be transported by humans into other places. Thus any management actions that can reduce unplanned human disturbance to the habitat reduce its invasibility and its rate of transmission of organisms into nonnative habitats.

Determining the most likely points and methods of entry is the basis for intelligent and strategic efforts to control entry into the protected system by intensive monitoring at key locations, or by changing the practices of persons, vehicles, or vessels moving through such locations. It is not surprising that alien species richness is positively correlated with proximity to centers of human transport (docks, cities, railways, roads, trails) and inversely related to a habitat's degree of isolation (Godfray and Crawley 1998). This correlation suggests obvious procedures that could lower the transmission and invasion of alien species, although all would be challenging to enforce. For example, changes in disposal and treatment of ballast water from foreign sources would, if implemented, achieve significant reductions in invasion rates of aquatic organisms throughout the world. At a regional level, cleaning and inspection of boat surfaces, and stringent removal of biotic material from boats, especially aquatic plants, entering and leaving vulnerable bodies of water also would reduce invasion probability. Although such an effort would be overwhelming from an enforcement perspective, it could succeed through voluntary cooperation achieved via public education. In terrestrial environments seeds and spores of invaders are often carried on tires of autos or on the shoes or clothing of hikers. Road and trail closures in strategic areas can thus reduce the probability of invasion. Despite all these precautions, many invasions have been the result of escapes from cultivation or captivity. Thus the long-term solution is for more stringent controls on the importation and planting of nonnative plants as ornamentals, and on the keeping, sale, and distribution of exotic animals for pets, display, or commercial use.

Step 2: Controlling Initial Infestations of Nonindigenous Species

Small populations of nonindigenous species are the easiest to control, and potentially can be completely eradicated. However, such small populations are also the most difficult to detect. Regular inventories of protected systems, with intentional and focused effort to detect nonindigenous species considered most likely to enter the system, are essential to discover populations of nonindigenous species while such populations are still small, dispersed, and not well established. Once detected, strenuous efforts to reduce or eradicate such small populations are better than delaying action until populations have grown too large.

There are several methods of eradicating small populations of nonindigenous species. The most direct, and potentially the most effective, is physical removal. Because sexually reproducing animal populations are limited by the number of females, killing or removing all females will eradicate such populations. In plants, physical removal may require pulling up or cutting down every individual of an invading species. Direct removal methods are effective if the invaders are few and confined to a small number of specific locations. However, direct removal is labor intensive and, therefore, often expensive.

In plants, some invertebrates, some fish, and some small mammals, a less labor-intensive method of direct removal is chemical control, the application of pesticides. Pesticides, properly applied, can remove invasive species, but they present serious risks. Most pesticides are not species specific. They often kill individuals of native, nontargeted species in addition to (or sometimes, instead of) the targeted nonindigenous species. Thus pesticides should never be used where they pose a potential threat to native, small, or declining populations. Pesticides can spread out of the target zone via air and water, killing individuals in nontarget areas. Some pesticides can persist in the environment for an extended length of time, posing a longer-term threat even after the nonindigenous species is removed. In plants, some herbicides may enter nutrient cycling pathways and be spread throughout the system, killing individuals of some species directly and interfering with nutrient uptake or metabolic pathways in others.

The alternative to chemical control is biological control, the introduction of a new species for the specific purpose of controlling the alien invader, often through predation and parasitism. There are cases in which biological control successfully reduces, or even eliminates, the targeted nonindigenous species. But there are also many failures, and there are always risks. The greatest risk is that of unforeseen effect. Biological control usually requires the introduction of yet another nonindigenous species to control the first nonindigenous species. Like the old lady in the children's song who swallowed a fly, and then swallowed a spider to catch it, attempts at biological control can set in motion a chain of events that can rapidly escape management constraints, usually with very negative consequences. Mongoose (*Herpestes* spp.) introduced in Hawaii and Caribbean islands to control rats have proven themselves much more efficient at devastating populations of native island birds and reptiles, particularly ground-foraging skinks (a type of lizard, family Scincidae) and snakes (Whittaker 1998). Further, using biological control is a concession that the original nonindigenous species probably cannot be completely eradicated. Although predators and parasites may reduce the abundance of a prey or host species, they rarely eliminate it, especially if the prey or host species is well established at high densities.

Step 3: Controlling Negative Effects of Established Nonindigenous Species on Native Populations

Throughout the world, many nonindigenous species are now, even after only a few decades, so well established that their eradication in the new environment is no longer a realistic goal. It is highly unlikely, for example, that the introduced Eurasian species cheat grass (*Bromus tectorum*) will be eliminated from western range communities in the United States by any amount of management effort, or that starlings will cease to be an abundant species among North American birds in the foreseeable future. Rather than commit further resources to increasingly futile efforts at eradication, managers and conservationists should determine the ways in which the distribution and abundance of nonindigenous species can be reduced and how their negative effects on native species can be mitigated. In grasslands, many introduced species, such as cheat grass, tend to withstand heavy grazing better than native species. Managing grazing levels to create conditions that favor natives over exotics does not eliminate introduced species, but can shift community distribution in favor of natives. In aquatic communities, the alteration of water levels and flow rates, where possible, can be used in ways that favor native plants and animals over nonnative invaders. In forest ecosystems, harvesting practices and methods can create environments that favor native species over introduced ones.

When eradication is no longer possible, managers and conservationists must accept the nonindigenous species as part of the community and determine what interactions they may have with other species and with the physical environment, and what resources they provide, that could provide specific benefit and value. If such interactions and resources can be correctly identified, properly managed, and prudently used, positive features and components of the natural community can be conserved even in the presence of infestations of the nonindigenous species.

THE PROBLEM OF METAPOPULATIONS: HOW DO WE MANAGE POPULATIONS OF SPATIALLY DISJUNCT SUBUNITS?

In chapter 5, we examined the concept of metapopulations, one of the controlling paradigms of conservation biology. To review, metapopulations are "spatially disjunct groups of individuals with some demographic or genetic connection" among individuals in different groups (Wells and Richmond 1995). What creates such a connection is normally movement (migration) of individuals among groups along with gene flow that occurs as individuals arriving in a new group breed with those already present.

Classic metapopulations, or so-called Levins model metapopulations (Levins 1969, 1970), assumed that groups had essentially independent, rather than interdependent, demographics, or at least that extinction and recolonization were independent events for all groups. Original models assumed that although extinction and recolonization events were independent, rates of extinction and recolonization were the same for all patches. This last assumption, patently unrealistic in actual populations, can be relaxed, as more recent modeling efforts have shown. However, the attractiveness and popularity of the concept of metapopulations has led to tendencies to call any population with individuals in different places a metapopulation, a misuse of language that confuses rather than clarifies population processes and conservation efforts.

Metapopulation theory has made a number of distinctive contributions to conservation biology, although its long-term value remains a matter of debate. A fundamental theoretical contribution of metapopulation theory is that the persistence of a species depends on its existence as a set of local populations, "largely interdependent yet interconnected by migration" (Harrison and Taylor 1997). The concept is compelling to conservationists for several reasons. First, the idea of spatially disjunct populations appears to match, at least superficially, the distributions that result from habitat fragmentation. Second, small, disjunct subpopulations hold out the promise that not all will experience the same set of environmental variations, reducing (theoretically) the potential for environmental stochasticity to be the primary cause of extinction in small and declining populations. Third, metapopulation theory attributes value to small sites (and, by extension, small reserves) as significant units for population persistence. Finally, metapopulation theory asserts that the disappearance of a species from a local site may not represent a permanent loss or extinction, but only a temporary absence in a broader context of local, site-specific extinction and colonization in a landscape context. Thus, unoccupied sites of apparently suitable habitat have value in being preserved if they exist within a framework of other patches of suitable habitat, at least some of which is currently occupied by the species.

Empirical Studies of Metapopulations

Despite strenuous effort and high motivation to find classical metapopulations in nature, the search has not been entirely fruitful. A metapopulation, in contrast with a traditional population having a dispersed distribution in a patchy environment, must meet two essential criteria: (1) all local populations must be prone to extinction and (2) the persistence of the entire population (the metapopulation) must require recolonization of individual sites (Harrison and Taylor 1997). Two species that meet these criteria are the pool frog (*Rana lessonae*) of coastal Sweden and the butterfly (*Melitaea cinxia*) of southern Finland. Both species are characterized by disjunct population subunits. Extinction on individual sites occurs frequently, migration is limited, and vacant sites are recolonized by individuals from other sites (Harrison and Taylor 1997).

However, many recent studies have invoked metapopulation concepts almost faddishly, with little analysis of whether the population met the definition of a metapopulation. Valverde and Silvertown (1997), for example, describe a common English primose (*Primula vulgaris*) as a metapopulation because it occurs in forest canopy gaps that are separated from one another by intervening closed canopy forest. However, *Primula vulgaris* can disperse over long distances; that is, movement of seeds among groups of individuals in different gaps is frequent, and most recruitment within a gap is not local. Additionally, when forest disturbance is high, some subpopulations in individual gaps never become extinct, and this violates one of the basic requirements of the metapopulation concept. The situation in this species is further complicated because gaps are ephemeral. Extinctions are less a result of independent demographic events in the subpopulations than of habitat loss that occurs as forest succession proceeds in the gap (the gap becomes dark and *Primula vulgaris* cannot survive). Thus, it would be more accurate to describe the changing distribution of *Primula vulgaris* in a forest as a pattern of colonization of new sites, not as a metapopulation process of extinction and recolonization of the same sites.

The extension of the metapopulation concept to include the population of a "habitat tracking" species like *Primula vulgaris* is one among many examples that support the assessment reached by Harrison and Taylor (1997) after an extensive review of empirical studies of metapopulations. They conclude, "classical metapopulations form a minority, even among the modest number of systems that have been well studied in metapopulation terms. We find only a few examples of a single species existing in a balance between the extinction and colonization of populations and almost none of the systems in which multiple species coexist through tightly coupled metapopulation dynamics at comparable spatial scales."

Metapopulations, Landscape Features, and Disease

Because the persistence of metapopulations depends heavily on the ability of individuals in spatially discrete units to move from one unit to another, landscape features and their effects on individual movement are of critical importance to metapopulations. Most spatially explicit metapopulation models of all categories use interpatch distance and migration rates as primary determinants of the probability that a given patch will be colonized. As a result, metapopulation theory must interface with landscape ecology (chapter 10) so that we can assess **connectivity** (Wiens 1996).

Connectivity is at issue in the debate over the use of corridors to increase the persistence of populations in general, and metapopulations in particular. We know, from Levins' classical model of metapopulations, that the rate of colonization must exceed the rate of extinction in a metapopulation for that metapopulation to persist (chapter 5). If corridors increase rates of colonization by increasing the rates of movement of individuals among sites, then, theoretically, such corridors should increase the persistence of the population. Additionally, if corridors increase opportunities for breeding among individuals of different sites, this should increase the effective population size and help avoid initiating the so-called discontinuity vortex. However, if connectivity makes all groups susceptible to the same degrees and kinds of environmental stochasticity (e.g., a fire spreads via corridors, a disease agent moves through corridors, a predator decimates successive patches by using corridors), then one of the primary advantages of having a spatially disjunct population—the assurance that different populations will not be simultaneously subjected to the same environmental variations and catastrophes—is lost.

Disease is a particularly troubling problem facing fragmented populations, and its effects have been modeled extensively. Hess (1996) developed and evaluated an epidemiological model of disease transmission in metapopulations using four spatial arrangements of connectivity among population subunits:

island (patches all equally accessible to one another), necklace (patches accessible only through step-by-step movement through adjacent patches), loop (a circular version of the necklace model), and spider (individuals can travel from a central patch to any other patches). Hess found that highly contagious diseases of moderate severity spread widely in metapopulations, and increased the probability of metapopulation extinction. Of greater importance was Hess's finding that the type of spatial configuration in a metapopulation affected the rate of transmission of a disease as well as the possibility of managing its spread. At low levels of migration more typical of classical metapopulations, the probability of extinction was significantly different in different spatial arrangements. The necklace arrangement had the lowest extinction probabilities in cases of moderately severe diseases (life span reduced 50 to 60%), but for more severe diseases (life span reductions of 70%), the spider arrangement had the lowest extinction probabilities. Interestingly, spatial arrangement had no effect on extinction probabilities for mild (<50% life span reduction) or severe diseases (>70% life span reductions). Extremely virulent diseases killed so many individuals that few dispersers survived. In contrast, mild infections had little effect on the growth rates of populations in different patches. In necklace and island arrangements, increasing the rate of migration in the presence of disease increased the probability of extinction under some conditions.

The effects of spatial arrangement were more pronounced when, in the presence of disease, a single population was quarantined, something that a manager might actually be able to do in a real-world situation. In the necklace arrangement, the quarantine of a single patch dramatically reduced the transmission of the disease and the probability of metapopulation extinction. The same was true in the spider arrangement if the quarantine was applied to the central patch (Hess 1996) because it severely limited movement of infected individuals, but did not impede movement of uninfected individuals in outer patches; thus, high levels of recolonization were maintained. Increasing movement among population subunits in metapopulations, such as might be achieved through corridors, may carry a greater risk than is commonly appreciated (Hess 1996).

PRACTICAL STEPS IN MAKING MANAGEMENT DECISIONS: A CONCEPTUAL FRAMEWORK

As was noted earlier (chapter 5), conservation biology has traditionally suffered in its treatment of small populations because of tension between two relevant paradigms, the small-population paradigm and the declining-population paradigm. The small-population paradigm provides theoretical insights into what makes populations small and keeps them small. Although it addresses particular dangers—genetic, demographic, environmental, and catastrophic—to which small populations are uniquely susceptible, the small-population paradigm does not always illuminate what caused the population to decline and what ought to be done to restore it to viable levels.

The declining-population paradigm originated in the applied sciences, and it is helpful in identifying the causes of species decline and their remedies. When applied in conjunction with theories from the small population paradigm, it can offer constructive insight for strategies that may lead to a population's recovery. What follows is a possible conceptual approach to managing a small population for recovery to viable levels.

Step 1: Determine the Cause of the Decline

Populations may decline for many reasons. In conservation as in medicine, a correct remedy is impossible without correct diagnosis. The cause of decline may appear obvious, but the "obvious" cause may, in fact, have little to do with the observed decline. To determine the real reason for the population decline, one should examine available data that compare the population in its present state with the same population in the past, particularly if records exist that describe the population when it was larger. Comparisons should focus on critical variables of interest:

1. *Geographic range:* What are the differences between the past and present geographic ranges of the population?
2. *Habitat use:* What are the differences between past and present habitat use in this population?
3. *Competitors, predators, parasites, and disease:* Are there differences in the current and historical types, species, or intensities of competitors, predators, parasites, or diseases that interact with this population?
4. *Environmental conditions:* Are environmental conditions for the population today the same as or different than those faced by this population in the past?
5. *Integrative comparisons:* If data are available to answer the above, they can be integrated to answer more complex questions. Some of the more important integrative questions are:
   a. How do distributions of competitors, predators, parasites, and disease organisms compare with present and historical distributions of the population? Are these obvious overlaps or disjunctions in particular interactions?
   b. How has the availability of preferred habitat changed in the population's geographic range over time?
   c. Are changes in range and habitat use, if any, associated with changes in environmental conditions experienced by the population?

In addition to comparisons to historic conditions, an examination of present conditions should determine the status of two forces that, if present, must be stopped immediately if the population is to survive.

6. *Direct exploitation:* Does the population experience any form of direct exploitation by humans today, legally or illegally?
7. *Habitat destruction:* Is the critical habitat of the population, especially breeding habitat, stable in quantity and quality?

Important Factors Regarding Present Conditions

Historical data are not always available and, even if they are, there are dimensions of a population's current status that merit examination independent of historical analysis. These are categories that can be examined in some detail through PVA simulation models, but they are repeated here because they can be examined through other PVA techniques, as well as some kinds of non-PVA analysis.

1. *Environmental stochasticity:* What level of environmental variation is experienced by the population, and does such variation affect population numbers?
2. *Demographic stochasticity:* What is the current demographic status of the population (birth rates, death rates, age-specific survivorship, mortality and reproduction, and immigration and emigration)?
3. *Genetic constraints:* What is the population's current level of inbreeding and heterozygosity? Does the population show any obvious signs or effects of inbreeding, including morphological distinctives, deformities, sterility, or abnormally high juvenile mortality?
4. *Susceptibility to natural catastrophes:* Is the population currently susceptible to any types of natural disasters? If so, what kind, and at what frequency and severity? Is it feasible to protect the population from such disasters?

Step 2: Formulate a Hypothesis About the Cause of Population Decline

Even after a careful examination of all available data (which will be imprecise at best and absent at worst), the facts alone will not unequivocally reveal the cause of the population's decline and its persistent low numbers, nor will they necessarily offer an obvious management solution. To advance further, one must make the most informed guess possible about the cause of the population decline, and frame this guess as an explicit hypothesis. For example, as was the case with the Lord Howe Island woodhen, examination of historical comparisons of numbers and distributions, combined with an examination of the distributions and numbers of predatory species, led to the conclusion that nest and habitat destruction by pigs were the most important agents of population decline. This idea can be framed more rigorously as a research hypothesis: The distribution of Lord Howe Island woodhens is limited by pigs.

This hypothesis leads to a specific prediction: If pigs are eradicated from the island, woodhens should expand their range and habitat use. Where possible and appropriate, the prediction could be framed as an experiment with appropriate control and replication—areas accessible to woodhens that contained pigs and other, similar areas of equal accessibility from which pigs were removed. A significant increase in woodhen distribution and abundance in pig-free areas would support the hypothesis. No difference between pig-free and pig-infested areas would negate the hypothesis and lead to formulation of a new hypothesis with appropriately different predictions and experimental design. The hypothesis and its predictions should be subsequently incorporated into management actions. In the case of the Lord Howe Island woodhen, the appropriate management action was a pig eradication program, a practical, albeit labor-intensive effort that had the potential for effective implementation and success.

The urgency of the problems associated with small and declining populations often make rigorous scientific experiments impossible to perform in a timely manner, and the small numbers and limited distribution of endangered populations may make replications impossible, but urgency is still no excuse for unclear thinking. Without hypotheses and predictions, there is no way to know if a management action is successful or to discern the real cause of a population's decline.

Step 3: Determine Potential Avenues for Increasing Population Size

Whether experiment manipulation is possible or not and whether the results of such experiments are clear or not, managers will have to try something to lead the endangered population toward recovery. Ultimately, management actions to increase the size and persistence of small populations fall into one of three categories:

1. *Intensive ecological and environmental management of the species in its natural habitat:* Using this strategy, managers depend primarily on natural reproductive capabilities and adaptations for survivorship of a wild population, but enhance the environment in such a way as to maximize favorable environmental conditions, minimize detrimental environmental variation, and optimize population demography toward maximum growth through removing competitors, predators, and parasites; controlling disease; creating favorable habitat; increasing the quality of available habitat; improving opportunity for migration and movement, and, in some cases, translocating animals and offering supplemental feeding. In terms of habitat enhancement, this kind of effort has been more formally described as the **optimal niche gestalt** approach to habitat management. The optimal niche gestalt invokes the idea that there are structural features of an environment that allow a species to thrive over and above those that allow it to merely persist (James et al. 2001). Thus managers should first identify these features by identifying correlations between environmental features and high-density populations or subpopulations, experimentally test hypotheses about the underlying causes that lead to these correlations, and then, informed by the results, manage selected sites intensively so as to favor processes and structures that create the features associated with these high-density populations rather than simply managing to create average environmental conditions found over the species' range (James et al. 2001).

2. *Supplementing wild populations with additions of captive-reared individuals:* To employ this strategy, managers must have or create a captive-bred population, the offspring of which they release into the wild at favorable sites to supplement existing populations or start new ones. Without such additions, the Lord Howe Island woodhen population might have perished. However, managers must first ask astute questions to evaluate the risks associated with supplemental additions from captive stock and determine if such a strategy is wise. Captive-bred individuals often have low rates of survivorship and fecundity in natural environments. Captive breeding increases the opportunity for inbreeding, and the release of highly inbred individuals into the wild population may perpetuate unfavorable genetic traits, or inbred individuals may carry an "opportunity cost" to the population by mating with healthy wild individuals, reducing reproductive success and removing opportunities for mating by unrelated individuals. Even if not inbred, captive-reared individuals may be genetically different from wild populations. These differences may spread traits that are not environmentally adaptive, or that genetically contaminate the population as a true phylogenetic species. Finally, if a captive-bred population does not already exist, the decision to start one means that some individuals must be taken from the wild population to do it. Animal capture and handling is risky even under the best of conditions. Some animals will die from trapping, handling, or transport. Some will not survive in a captive environment. Thus the wild population will be rendered even smaller and more prone to extinction.

3. *Removal of all remaining wild individuals to preserve the population in captivity:* This scenario is admittedly the most extreme, but it is not strictly hypothetical. Faced with an extremely small number of individuals in the wild, the U.S. Fish and Wildlife Service captured all remaining California condors (*Gymnogyps californianus*) and vested all efforts for the survival of this species in captive breeding programs in Los Angeles and San Diego zoos, with the hope that, one day, a wild population of condors could be reestablished from captive-reared birds. Similarly, individuals of the last wild population of black-footed ferrets (*Mustela nigripes*) were captured and placed in a captive breeding program just before canine distemper eliminated that population. Black-footed ferret populations have already been reestablished in Wyoming, Montana, South Dakota, and Arizona using captive-bred individuals, and condors are beginning to be reestablished in California.

Managers of small populations must assess which strategy carries the greatest potential for recovery and which the lowest risk of extinction. Unfortunately, those may be two different strategies! Small populations leave little room for error. Managers must examine all available data carefully, formulate clear hypotheses regarding causes of population decline, and make testable predictions that can be evaluated in management actions if they are to be effective in managing small and declining populations.

Synthesis

Conservation biology has rightly sought to distance itself from the *single-species management* approaches that characterized applied resource management sciences in the past, an approach that led to entire communities and ecosystems being managed for the benefit of one or a few species. However, conservation biology must not confuse single-species management with *species-specific conservation,* an approach that should be valued and practiced if conservation biologists have any hopes of seeing the recovery of threatened and endangered populations. Populations are the primary currency and concern of conservation biology, but the need persists to define populations more rigorously in order to avoid misunderstanding and mismanagement. This is particularly critical at a time when more and more populations are threatened with extinction, and when many species have been reduced to only a small number of disjunct populations with relatively few individuals.

Conservation biologists have increasingly focused on the question, "What is the minimum number of individuals that are needed for a population to persist through time?" Not surprisingly, the precise estimator of the answer (the minimum viable population) and the analytical tool that estimates it (population viability analysis) have become essential to the practice of conservation biology. However, conservation biologists must mature in their recognition of PVA as an analytical technique rather than a diagnostic tool, and thus better appreciate its limitations.

The estimate of an MVP should not be made solely through PVA, but should include a detailed assessment of the ecology of the species at risk, and conservation biologists should incorporate risk analysis and adaptive management procedures into assessments of PVA before determining final management strategies to restore small populations to viable numbers.

Confusion about what constitutes a metapopulation leads to misunderstandings and misjudgments about the nature of spatially disjunct population units. Empirical evidence suggests that classical metapopulations may be relatively rare in nature, and conservation biologists must be careful not to call all patchy populations, dispersed populations, or fragmented, nonequilbrium populations "metapopulations." Doing so could lead to management solutions that are inappropriate to actual population problems.

The dreadful urgency of attempting to save many small and declining populations from imminent extinction compels conservation biologists to implement management strategies quickly. However, this combination of concern and rapid response must not tempt conservation biologists to be careless in their systematic analysis of the causes of a population's decline. Each assessment of cause must be framed as a carefully constructed hypothesis that leads to specific predictions, a clear and practical management strategy, and a measurable way to test consequences that determine its veracity.

Predicting the Success of Invasive Species—A Directed Discussion

Reading assignment: Reichard, S. H., and C. W. Hamilton. 1997. Predicting invasions of woody plants introduced into North America. *Conservation Biology* 11:193–203.

Questions

1. Evaluate the four approaches for screening potentially invasive plant species (p. 194) based on the relative merits and disadvantages of each. Which approach is most similar to the method used in the United States? Which approach seems best suited for achieving conservation goals? If this is not the approach used in the United States, suggest other criteria that may have determined the choice.
2. The authors used a discriminant analysis to classify woody plant species as invasive or non-invasive based on their attributes. Given sufficient data, this technique could be used to predict characteristics other than invasiveness. What is another conservation concern for which species could be classified based on their attributes? What are the benefits of applying the technique in this new case?
3. If a discriminant analysis incorrectly classifies an invasive species as non-invasive, what might be the cost of this error? What might be the ramifications of the opposite error—labeling a non-invasive species as invasive? Given the potential for misclassification, predict how interest groups (other than conservationists) would respond to the implementation of this technique in the permit-granting procedure for importing non-native plants.
4. Identify the elements of this approach that would be attractive to managers and policy makers. What obstacles do you envision to the adoption of this approach in management, law, or policy? Suggest ways of overcoming the obstacles by changing the procedure or constraining its use.

Learning Online

Visit our webpage at www.mhhe.com/conservation for case studies, animations, practice quiz questions, and additional readings to help you understand the material in this chapter. You'll also find active links to the following topics:

Fisheries and Conservation Issues
  Concerning Teleosts
Conservation Issues Concerning
  Amphibians

Conservation Issues Concerning
  Reptiles
Conservation Issues Concerning Birds
Refuges and Sanctuaries

Conservation Issues Conservation of
  Mammals
Legislation Regarding Endangered
  Species

Literature Cited

Allee, W. C., A. E. Emerson, O. Park, T. Park, and K. P. Schmidt. 1949. *Principles of animal ecology.* Philadelphia: Saunders.

Andow, D. A., P. M. Kareiva, S. A. Levin, and A. Okubo. 1990. Spread of invading organisms. *Landscape Ecology* 4:177–88.

Boyce, M. S. 1992. Population viability analysis. *Annual Review of Ecology and Systematics* 23:481–506.

Carey, J. R. 1996. The incipient Mediterranean fruit fly population in California: implications for invasion biology. *Ecology* 77:1690–97.

Carlton, J. T., and J. B. Geller. 1993. Ecological roulette: the global transport and invasion of nonindigenous marine organisms. *Science* 261:78–82.

Caughley, G. 1979. What is this thing called carrying capacity? In *North American elk: ecology, behavior, and management,* eds. M. S. Boyce and L. D. Hayden-Wing, 2–8. Laramie: University of Wyoming.

Caughley, G. 1994. Directions in conservation biology. *Journal of Animal Ecology* 63:215–44.

Caughley, G., and A. Gunn. 1996. *Conservation biology in theory and practice.* Oxford, England: Blackwell Science.

Cole, L. C. 1957. Sketches of general and comparative demography. *Cold Spring Harbor Symposium on Quantitative Biology* 22:1–15.

Crawley, M. J. 1987. What makes a community invasible? In *Colonization, succession, and stability,* eds. A. J. Gray, M. J. Crawley, and P. J. Edwards, 429–53. Oxford, England: Blackwell Scientific.

Drake, J. A., H. A. Mooney, F. di Castri, R. H. Groves, F. J. Kruger, M. Rejmanek, and M. Williamson, eds. 1989. *Biological invasions: a global perspective.* New York: Wiley.

Elton, C. S. 1958. *The ecology of invasions by animals and plants.* London, England: Methuen and Company.

Forcella, F., J. T. Wood, and S. P. Dillon. 1986. Characteristics distinguishing invasive weeds within *Echium* (Bugloss). *Weed Research* 26:351–64.

Franklin, I. R. 1980. Evolutionary change in small populations. In *Conservation biology: an evolutionary-ecological perspective,* eds. M. E. Soulé and B. A. Wilcox, 135–49. Sunderland, Mass.: Sinauer Associates.

Fullagar, P. J. 1985. The woodhens of Lord Howe Island. *Aviculture Magazine* 91:15–30.

Gilpin, M. E., and M. E. Soulé. 1986. Minimum viable populations: processes of species extinctions. In *Conservation biology: the science of scarcity and diversity,* ed. M. E. Soulé, 19–34. Sunderland, Mass.: Sinauer Associates.

Godfray, H. C. J., and M. J. Crawley. 1998. Introductions. In *Conservation science and action,* ed. W. J. Sutherland, 39–65. Oxford, England: Blackwell Science.

Groom, M. J., and M. A. Pascual. 1998. Saving species through population viability biology and viability analyses: a morass of math, myth or mistakes? In *Conservation biology: for the coming decade,* 2d ed., eds. P. L. Fiedler and P. M. Kareiva, 4–27. New York: Chapman and Hall.

Grosholz, E. D. 1996. Contrasting rates of spread for introduced species in terrestrial and marine systems. *Ecology* 77:1680–86.

Groves, R. H., and J. T. Burton, eds. 1986. *Ecology of biological invasions: an Australian perspective.* Canberra: Australian Academy of Science.

Harrison, S., and A. D. Taylor. 1997. Empirical evidence for metapopulation dynamics. In *Metapopulation biology: ecology, genetics, and evolution,* eds. I. Hanski and M. E. Gilpin, 27–42. San Diego, Calif.: Academic Press.

Hess, G. 1996. Disease in metapopulation models: implications for conservation. *Ecology* 77:1617–32.

James, F. C., C. A. Hess, B. C. Kicklighter, and R. A. Thum. 2001. Ecosystem management and the niche gestalt of the Red-cockaded woodpecker in longleaf pine forests. *Ecological Applications* 11:854–70.

Lacy, R. C. 1993. VORTEX: a computer simulation model for population viability analysis. *Wildlife Research* 20:45–65.

Lande, R. 1988. Genetics and demography in biological conservation. *Science* 241:1455–60.

Lande, R. 1999. Extinction risks from anthropogenic, ecological and genetic factors. In *Genetics and the extinction of species: DNA and the conservation of biodiversity,* eds. L. F. Landweber and A. P. Dobson, 1–22. Princeton, N.J.: Princeton University Press.

Levins, R. 1969. Some demographic and genetic consequences of environmental heterogeneity for biological control. *Bulletin of the Entomological Society of America* 15:237–40.

Levins, R. 1970. Extinction. *Lectures on Mathematics in the Life Sciences* 2:75–107.

Lindenmayer, D. B., T. W. Clark, R. C. Lacy, and V. C. Thomas. 1993. Population viability analysis as a tool in wildlife conservation policy: with reference to Australia. *Environmental Management* 17:745–58.

Lodge, D. M. 1993. Biological invasions: lessons for ecology. *Trends in Ecology and Evolution* 8:133–37.

Ludwig, D. 1999. Is it meaningful to estimate the probability of extinction? *Ecology* 80:298–310.

Maynard Smith, J. 1974. *Models in ecology.* Cambridge, England: Cambridge University Press.

McNaughton, S. J. 1984. Grazing lawns: animals in herds, plant form, and coevolution. *American Naturalist* 124:863–86.

Miller, B., and K. J. Mullette. 1985. Rehabilitation of an endangered Australian bird: the Lord Howe Island woodhen *Tricholimnas sylvestris* (Sclater). *Biological Conservation* 34:55–95.

Mooney, H. A., and J. A. Drake. 1986. *Ecology of biological invasions of North America and Hawaii.* New York: Springer-Verlag.

Moyle, P. H., and T. Light. 1996. Fish invasions in California: Do abiotic factors determine success? *Ecology* 77:1666–70.

National Research Council. 1996. *Science and the Endangered Species Act.* Washington, D.C.: U.S. Government Printing Office.

Niemelä, P., and W. J. Mattson. 1996. Invasion of North American forests by European phytophagous insects. *BioScience* 46:741–53.

O'Brien, S. J. 1996. Conservation genetics of the felidae. In *Conservation genetics: case histories from nature,* eds. J. C. Avise and J. L. Hamrick, 50–74. New York: Chapman and Hall.

Office of Technology Assessment (U.S. Congress). 1993. *Harmful nonindigenous species in the United States.* OTA Publication OTA-F-565. Washington, D.C.: U.S. Government Printing Office.

Ralls, K., and B. L. Taylor. 1997. How viable is population viability analysis? In *The ecological basis of conservation: heterogeneity, ecosystems, and biodiversity,* eds. S. T. A. Pickett, R. S. Ostfeld, M. Shachak, and G. E. Likens, 228–35. New York: Chapman and Hall.

Rejmánek, M. 1995. What makes a species invasive? In *Plant invasions,* eds. P. Pysek, K. Prach, Rejmánek, M. and P. M. Wade, 3–13. The Hague, The Netherlands: SPB Academic Publishing.

Rejmánek, M., and D. M. Richardson. 1996. What attributes make some plant species more invasive? *Ecology* 77:1655–61.

Rosenzweig, M. L. 1991. Habitat selection and population interactions: the search for mechanism. *American Naturalist* 137:S5–S28.

Ruiz, G. M., J. T. Carlton, E. D. Grosholz, and A. H. Hines. 1997. Global invasions of marine and estuarine habitats by nonindigenous species: mechanisms, extent, and consequences. *American Zoologist* 37:621–32.

Sclater, P. J. 1869. *Ocydromus sylvestris,* sp. nov. *Proceedings of the Zoological Society, London:* 472–73.

Shaffer, M. L. 1981. Minimum population sizes for species conservation. *BioScience* 31:131–34.

Sherman, P. W., and M. L. Morton. 1984. Demography of Belding's ground squirrel. *Ecology* 65:1617–28.

Shigesada, N., K. Kawasaki, and E. Teramoto. 1986. Traveling periodic waves in heterogeneous environments. *Theoretical Population Biology* 30:143–60.

Simberloff, D. 1998. Small and declining populations. In *Conservation science and action,* ed. W. J. Sutherland, 116–34. Oxford, England: Blackwell Science.

Skellam, J. G. 1951. Random dispersal in theoretical populations. *Biometrica* 38:196–218.

Temple. S. A. 1986. The problem of avian extinctions. *Current Ornithology* 3:453–85.

Valverde, T., and J. Silverton. 1997. A metapopulation model for *Primula vulgaris,* a temperate forest understory herb. *Journal of Ecology* 85:193–210.

Veit, R. R., and M. A. Lewis. 1996. Dispersal, population growth, and the Allee effect: dynamics of the house finch invasion of eastern North America. *American Naturalist* 148:255–74.

Veltman, C. J., S. Nee, and M. J. Crawley. 1996. Correlates of introduction success in exotic New Zealand birds. *American Naturalist* 147:542–57.

Wells, J. V., and M. E. Richmond. 1995. Populations, metapopulations, and species populations: What are they and who should care? *Wildlife Society Bulletin* 23:458–62.

Whittaker, R. J. 1998. *Island biogeography: ecology, evolution, and conservation.* Oxford, England: Oxford University Press.

Wiens, J. A. 1996. Wildlife in patchy environments: metapopulations, mosaics, and management. In *Metapopulations and wildlife conservation,* ed. D. R. McCullough, 53–84. Washington, D.C.: Island Press.

Appendix A.   *Calculation of Columns in a Cohort Life Table*

x column

Age is presented as an interval $(x–x')$, typically in years.

n_x and d_x columns

d_x is the number dispersing or dying in each interval. By knowing this number, and the initial number of individuals in the cohort (N), we can calculate the number surviving to the beginning of the next age interval (n_x).

$N - d_x = n_x$ for the second age interval

$n_x - d_x = n_x$ for each subsequent age interval

l_x column

Survivorship (l_x) is equal to the proportion of the original cohort surviving to the *beginning* of each age interval. By definition, survivorship is 1.0 for the original cohort and goes to 0 during the life span of the longest-lived individuals in the cohort.

$$\frac{n_x}{N} = l_x$$

q_x column

The age-specific disappearance rate (q_x) is the proportion of the population that dies or disperses during a particular age interval.

$$\frac{d_x}{N} = q_x$$

e_x column

e_x is the future life expectancy, usually expressed in years. To calculate life expectancy, two additional statistics are necessary: L_x and T_x. L_x is the average time units lived by all individuals in each category within the population. Lx is found by summing the number alive at age interval x and the number at age $x + 1$, and dividing the sum by 2. T_x is the number of time units left for all individuals to live from age x onward. To calculate T_x, simply sum all of the values for L_x from the last age interval up the column until the age interval of interest. e_x is then calculated by dividing T_x for the particular age class x by the survivors for that age, as given in the n_x column.

CHAPTER 8

The Conservation of Habitat and Landscape

Nature reserves, like time capsules, are successful to the degree that their contents retain their integrity. They fail to the degree that their contents are destroyed

—Michael E. Soulé and Daniel Simberloff, 1986

In this chapter, you will learn about:

1 concepts and definitions of habitat and landscape and the role of habitat conservation in conservation biology.

2 definitions and importance of habitat heterogeneity and patch dynamics.

3 specific mechanisms through which habitat loss, fragmentation, and isolation threaten biodiversity.

4 principles of reserve design and their role in habitat conservation.

5 means of reducing the effects of human disturbance to conserve habitat and habitat-dependent populations in nonreserve environments.

THE DEFINITION, CONCEPT, AND IMPORTANCE OF HABITAT

Habitat can be simply but accurately defined as *the physical and biological surroundings of an organism* (Bolen and Robinson 1995). With reference to habitat, **landscape** can be defined as *a large area that comprises more than one type of habitat distributed in numerous patches* (Danielson 1991), or, to include the effect of human influence, *an aggregate of different but interacting landforms, sometimes united by a cultural attribute* (e.g., an agricultural landscape of cultivated fields, pastures, stockponds, and hedgerows) (Bolen and Robinson 1995). These well-framed definitions are important because the fundamental goal of conservation biology—the preservation of biodiversity—rests largely upon the conservation, preservation, and management of habitat and landscape. Habitats and landscapes possess intrinsic and utilitarian values worthy of conservation. Habitat distribution is a primary determinant of species' abundance. In fact, it is often the degree of habitat specificity of a species that determines whether it is common or rare (Rabinowitz, Cairns, and Dillon 1986). Populations become more susceptible to extinction in the face of environmental variation. For example, local variations in regional rainfall

may result in local losses of wetland habitat, threatening the persistence of individual populations of wetland species. Such environmental variation cannot be eliminated even in the largest preserves, but the preservation of habitat (for example, the preservation of wetlands at a regional scale) can reduce the adverse effects of environmental variation and conserve essential, habitat-specific resources needed by habitat-dependent species.

Today many conservation biologists consider habitat alteration the single greatest threat to species and ecosystems worldwide (Soulé 1991; Noss and Cooperrider 1994). In the United States, 88% of species protected by the Endangered Species Act are considered endangered because of habitat destruction and degradation (Wilcove et al. 1996). Habitat alteration includes the physical conversion of natural habitat to an unnatural habitat (habitat loss); the breaking of large, contiguous blocks of habitat into smaller patches (habitat fragmentation); the increasing separation of blocks of habitat from one another (habitat isolation); and the changes in a habitat that affect its composition, structure, or function (habitat degradation) (Noss, O'Connell, and Murphy 1997). The problems associated with habitats today are both extensive and complex. But before examining the problem, we must first consider why habitats matter.

Figure 8.1
Moose (*Alces alces*) are habitat generalists that use a variety of available habitats in varying proportions.

Habitat Preservation and Conservation: Basic Principles

Faced with an array of habitats in a particular landscape, a mobile species, such as a moose (*Alces alces*) (fig. 8.1) can choose which habitats to use. Figure 8.2 gives a summary of these choices, using a sample of radio-collared moose from a larger population (Van Dyke, Probert, and Van Beek 1995). Habitat classification such as this requires the creation of arbitrary categories, but by creating such categories we can analyze the distribution of animals in a matrix of complex vegetation. These data and their interpretation offer some insights into principles of how animals use habitat, and suggest several reasons why understanding species-specific habitat relationships is important for conservation.

First, note that moose, like most species, use more than one kind of habitat. This is because a single habitat seldom provides all necessary resources for long-term survival and reproduction. Second, habitat *use* is often different from habitat *availability.* Some habitats are used at rates greater than expected, whereas other habitats are used less often than expected relative to their distribution in a landscape. These differences provide clues that can lead conservationists to make well-informed hypotheses about why patterns of apparent habitat preference or avoidance appear in particular species. If these hypotheses are tested through careful experimental design and manipulation, the results of the experiments can provide insight about reasons animals choose habitats, and about the resources that habitats provide to the animals. Third, these data also demonstrate the principle that habitat use changes over time (in this case, seasonally). Such changes should be expected because the availability of resources in habitats changes temporally. Although obvious, this fact has often been overlooked in the design of parks and

nature reserves. If the full array of habitat and landscape structure that animals require in different seasons does not exist within the reserve, then they must leave the protection of the reserve to find it or perish in the reserve without it. Because habitats, like organisms, have life spans, provision must be made to create new habitat as well as to preserve existing habitat. Fourth, habitat use differs among individuals in a population. Sex- or age-related differences in body size, social organization, or investment in reproductive effort and parental care of offspring may lead individuals of the same population to select different habitats during the same period of time. Failure to consider these differences can lead to inappropriate choices of which habitats to conserve and result in ineffective conservation efforts.

Looking beyond the insights we can gain from this single data set, there are additional reasons why habitat and landscape are important to conservation biology. (1) Arrangements of habitat patches and distances between patches of the same type of habitat, along with the presence or absence of connecting corridors, may influence populations. (2) Movement of conspecifics among habitat patches can affect population dynamics, social behavior, rates of extinction, and genetic composition of populations. (3) Different species respond differently to a given array of habitats. (4) Movements of other species into, out of, and among habitat patches may be important to a species of conservation interest. (5) Spatial configurations of patches can affect processes of populations and communities (Lidicker 1995).

Determining Species' Preferred Habitats

A simple algorithm for identifying the strength of preference or avoidance for habitats is Ivlev's selection index (Ivlev 1961). Although originally developed to assess selection for food items, it can be effectively applied to measure selection for habitat, as well (Yeo and Peek 1992). Ivlev's index of selection (SI) for a given type of habitat is determined by the formula

$$SI = (U - A)/(U + A),$$

where U is the proportion of the animal's use of the habitat out of all use and A is the proportion of the same habitat available in the landscape. For example, suppose it can be determined that an animal spends 80% of all its time in stands of aspen (*Populus tremuloides*) trees ($U = 80$) within a home range in which aspen stands make up only 10% of the landscape ($A = 10$). In this case

$$SI = (80 - 10)/(80 + 10) = 70/90 = 0.78.$$

Note that, in Ivlev's formula, the selection index will be zero whenever the animal uses the habitat in the same proportion as its availability (no selection). It will approach the value of 1.0 when use is proportionately much higher than availability, as in our example, suggesting strong preference for that habitat. Alternatively, it will approach a value of −1.0 when use is far less than availability, suggesting avoidance of the habitat. If many samples of habitat use are made from multiple individuals in the same area, standard statistical techniques can be used to see if the average value of the selection index is different from zero (Neu, Byers, and Peek 1974; Marcum and Loftsgaarden 1980).

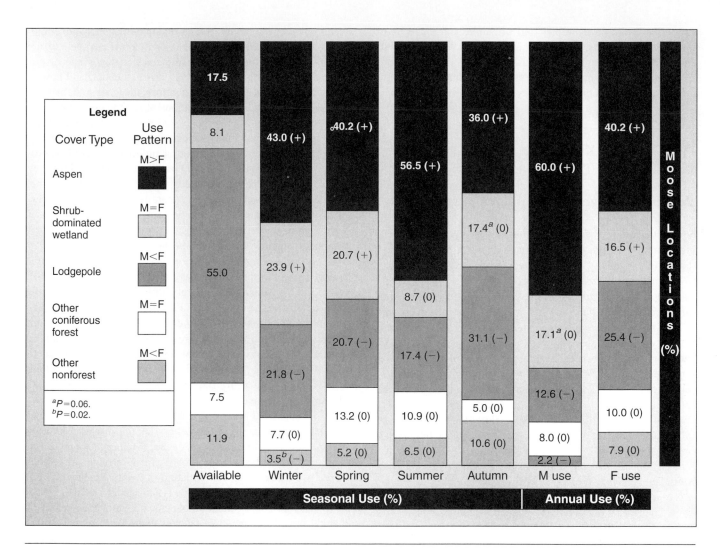

Figure 8.2

Seasonal habitat selection by 3 male (M) and 10 female (F) moose in the Fiddler and Fishtail Creek drainages of south-central Montana, USA1989–1993. Numbers indicate percentages of available habitat or percentages of moose locations in a habitat. Symbols in parentheses indicate selection for (+), selection against (−), a habitat, or no selection (0). $P < 0.01$ for all cases of selection and for differences between sexes, except where noted. Note that, among these individuals, aspen habitat is always selected for and lodgepole habitat is always selected against. Further, note that males and females differ in habitat use in three of five habitat types, and that strength of habitat selection changes seasonally.

Adapted from data in Van Dyke, Probert, and Van Beek (1995).

To acquire data on the availability of habitat in an area, one can use existing vegetation maps, aerial photos, or files of geographic information systems. If none of these exist, availability can be determined from sampling numerous random points within the area until an appropriate number of samples have been made. Use of habitats by an animal can be determined from direct observation, radio-telemetry data, systematic trapping, or determining the locations of songs, tracks, or sign (scat or markings). Regardless of method, there will always be some feasible way of obtaining some estimate of habitat use relative to availability, and then using such information to make a more informed decision about which habitats to preserve for particular species.

Use/availability data have limitations. They can be prejudiced by investigator biases. Investigators make decisions about which habitat is actually "available," which may be influenced by their method of measuring an animal's home range or by their judgment about which habitat an animal is actually capable of using. Methods and sampling intervals for determining both use and availability also can affect estimates of preference or avoidance of habitat. Use/availability ratios do not illuminate the processes and mechanisms that led the animal to choose one habitat over another. Higher than expected use of a habitat does not prove that it is critical or required for an individual or species. That would be demonstrated only by a controlled experiment in which a particular habitat was removed from the landscape and the population showed a significant decline. But an examination of use and availability of habitat, however limited, keeps us from overlooking the obvious when it comes to habitat conservation.

These simple examples and principles introduce us to some of the most basic aspects of animal-habitat relations. But the actual structure of habitat, its use by organisms, and its conservation require a more complex understanding of many issues. We begin with the complexity of habitat distribution.

THE COMPLEXITIES OF HABITAT: HETEROGENEITY AND PATCH DYNAMICS

Habitat Heterogeneity and Patchiness

A persistent quality of habitats is that they are highly variable in space and time, a quality referred to as **habitat heterogeneity**. Habitat heterogeneity can be defined as *any form of variation in the environment, including physical and biotic components. Such variation may appear as spatial or temporal patterns, and may be fixed or dynamic. If dynamic, heterogeneity may appear as a steady or transient state, depending on both the nature of the process and the scale of observation* (Ostfeld et al. 1997). Species richness and species diversity are correlates of habitat heterogeneity and the related attribute of patchiness. Heterogeneity exists because habitats occur in **patches**, which can be defined as *areas, smaller than a landscape, that contain only one type of habitat* (Danielson 1991). **Patchiness**, a quality of habitat arrangement, can be defined as *a form of spatial heterogeneity in which boundaries may be discerned. Patchy heterogeneity appears as contrasting discrete states of physical or biotic phenomena. An array of patches may be seen as patchiness at a particular scale* (Ostfeld et al. 1997).

The realities of heterogeneity and patchiness can cover a wide spectrum of conditions. At one extreme, patches of habitat can be entirely discrete and perfectly discernible from one another. At the opposite extreme, the habitat, although heterogeneous, may have subtle and almost imperceptible gradations from one kind of patch to another. Most undisturbed landscapes are naturally patchy. A patchy landscape often has a rich internal structure of different habitats, but often with only gradual differences between adjacent patches. In any form or at any scale, patchiness creates a discontinuity of resource distribution in the environment that provides for a variety of niches and a variety of species to exploit them. Thus, heterogeneity and patchiness are important controlling ideas not only in our study of habitat and landscape conservation, but in the conservation of biodiversity in general.

Habitats and Landscapes: Understanding Scales of Time, Space, and Intensity

All habitats, whether consisting of discrete patches or of patches with indiscernible boundaries, are ultimately heterogeneous. Perceiving their heterogeneity depends on the spatial and temporal scales used to measure it. Habitat and landscape structure are not constant but change over time and space. The higher the rate of landscape change, the lower the probability of regional population survival (Fahrig and Merriam 1995). Environmental heterogeneity associated with and, to an extent,

determined by the landscape is a function of spatial and temporal heterogeneity. In turn, spatial and temporal heterogeneity are functions of the intensity and frequency of landscape disturbance. Each must be considered in order to understand how to conserve and manage habitat. Therefore, the analysis of habitats and landscapes, and the applications of how to conserve them, must include what Meentemeyer and Box (1987) called "a science of scale," which treats scale as an explicitly stated variable.

Habitats may be classified at multiple scales of heterogeneity, but organisms have only particular scales at which they can respond to such heterogeneity. Begin with a scale of time. To a casual observer, habitats give an impression of being permanent landscape features, but they are not. Just as we speak of "population dynamics," it is just as appropriate to speak of "habitat dynamics." **Temporal scale** can be defined as *habitat life span relative to the generation time of the organism* (Fahrig 1992). If the life span of the habitat is short relative to the life span of the organism (ephemeral habitats, such as an intermittent stream or a seasonal wetland formed during a spring flood), organisms must enter periods of dormancy to endure nonhabitat environments until the habitat reappears, or they must employ nomadic behavior or high rates and distances of dispersal to reach other suitable habitat. If the life span of the habitat is long relative to the organism, sedentary behavior and more limited dispersal are favored.

In contrast to temporal scale, **spatial scale** is a measure of habitat patchiness. It can be defined as *the distance between habitat patches relative to the dispersal distance of the organism* (Fahrig 1992). When we speak of the spatial structure of a landscape, we mean the spatial structure of habitat patches and the matrix in which they are embedded (Fahrig and Merriam 1995).

Lenore Fahrig created a stochastic simulation model of habitat-organism interactions that evaluated the relative importance of temporal and spatial scale on population persistence. Imagine a two-dimensional model universe of habitat "cells" of two categories, usable habitat cells in which organisms can survive and reproduce, and nonusable habitat cells that are inhospitable. The usable cells are clustered as habitat patches. They are not only clustered but transient, existing only for a limited time. New habitat patches appear at random at discrete time intervals (steps). Fahrig's model contains only six constants (fraction of grid in usable habitat, per capita birth rate per time step, death rate per time step, age of organism at maturity, maximum population per habitat cell, and fraction of population dispersing per time step) and three variables (life span of patches L, size of patches S, and mean dispersal distance Z) (Fahrig 1992). Patch size can, in turn, be used to compute average distance between patches because the total area of the grid is constant. Thus, patch size can be used as an index of spatial scale. In each time step, four events occur in sequence: (1) habitat patches are created, (2) some organisms survive and reproduce while others die, (3) some survivors disperse to new patches, and (4) some patches disappear.

By holding the fraction of usable habitat constant but varying the values of patch life span, patch size, and species dispersal distance in different combinations, Fahrig was able to evaluate independently the effects of temporal and spatial scale on population size. The model's outcomes showed that populations increased with increasing temporal scale (i.e., densities of

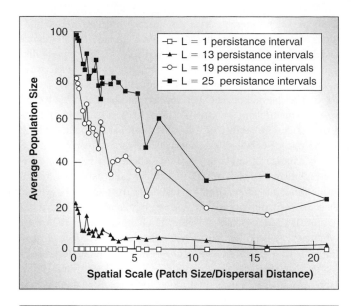

Figure 8.3

The effect of spatial scale on mean population size at four different levels of habitat persistence. Note that as habitat persistence (temporal scale) increases, average population size rises.

Adapted from Fahrig (1992).

organisms increased the longer a given patch of habitat persisted) and decreased with increasing spatial scale (the ratio of distances between patches to the organism's average dispersal distance). In other words, as distances between patches increased relative to an organism's ability to disperse to patches, population densities declined. In relative importance, temporal scale had a far greater effect on population stability than did spatial scale (Fahrig 1992). At any given spatial scale, populations were larger when habitat patches persisted longer (fig. 8.3).

Important insights emerge from this model. If the goal of a conservation effort is to preserve a particular species, conservationists must determine the temporal scale of its preferred habitat. Is its favorite habitat persistent relative to the life of the organism (high temporal scale), or are preferred habitats ephemeral, appearing unpredictably and disappearing frequently? Because temporal scale may be more important than spatial scale, "the size of reserve fragments may be less important than their persistence over time" (Fahrig 1992). Fahrig goes on to recommend that "since habitat continually undergoes modification due to processes such as disturbance . . . more attention should be placed on the duration of reserve fragments as habitat for particular species of interest."

To better understand temporal scale in conserving habitat, conservationists must measure and, if possible, manage disturbance intensity because it is often the governing factor of temporal scale. Disturbance intensity is a measure of the severity, frequency, and extent of environmental disturbance. It incorporates an examination of space and time, but must examine the strength of the disturbance agent. For example, a forest subjected to a slow-moving ground fire is affected differently than one subjected to a fast-moving crown fire. The responses to the two disturbances by plants and animals will differ, as well.

Habitats arise at random through the interaction of environmental disturbance and successional processes, exist for a period of time, and then disappear, being converted, through ongoing population replacement or subsequent disturbance, to new and different kinds of habitat. To determine if a habitat will remain stable, one must determine how long and over what area the present collection of species can persist with a given intensity of disturbance (Connell and Slatyer 1977). For a group of populations of plants and animals to remain viable, an area must be large enough to ensure sufficient area and habitat diversity to provide niche space for all species. If some habitats are short lived and created by disturbance, then the area must be repeatedly disturbed if those habitats are to persist. Such disturbance can occur through natural processes or deliberate management actions, but it must occur with sufficient intensity and frequency to maintain the desired habitat. Disturbance also may be necessary for many long-lived species. Some species cannot survive without catastrophic events. For example, coast redwoods (*Sequoia sempervirens*) and giant sequoias (*Sequoiadendron giganteum*) (fig. 8.4) may require extensive and severe floods or fires to remove older

Figure 8.4

The giant sequoia (*Sequoiadendron giganteum*) is established only by severe, but infrequent fires.

Figure 8.5

The white-footed mouse (*Peromyscus leucopus*) a species that increases its home range size in fragmented habitats.

trees that suppress younger individuals (Stone and Vasey 1968), but with relatively long time intervals between catastrophic events.

Even when habitats are relatively stable, they are likely to be patchy. In individual patches, extinctions may be common. Recolonization after extinction depends upon the distances between occupied and unoccupied patches (landscape spatial structure), the rates and distances of dispersal by the organism (dispersal characteristics), and the length of time that habitat patches persist (temporal changes). The stability of the community will depend on whether a species is likely to replace itself after a disturbance or whether it is likely to be replaced by another species. If changes in landscape structure occur at an unnaturally high rate (as in some anthropogenic changes), neither dispersal nor adaptation can keep up, and the result is a high rate of local and regional extinction.

Consider some field studies that illustrate the effects of landscape and dispersal scale. In unaltered habitats of woods or brush in the eastern United States, the white-footed mouse (*Peromyscus leucopus*) (fig. 8.5) uses home ranges of less than 0.5 ha. When agricultural clearing fragments woodland habitat,

the home ranges increase to tens of ha and mice may move hundreds of meters in one night (Merriam and Lanoue 1990; Wegner and Merriam 1990). Similarly, the red fox (*Vulpes vulpes*) disperses farther in more fragmented urban habitats than in less fragmented boreal forest habitat (Lindstrom 1989; Hansson 1991).

The interaction between the rate of change in landscape spatial structure and the rate of change in dispersal behavior determines the probability of a species' regional survival. As long as the rate of change in dispersal behavior is greater than the rate of change in landscape spatial structure, it is possible for the organism to survive in the changing landscape by moving around in it and integrating the resources over space. However, there will be a maximum possible rate of change in dispersal behavior. If the landscape structure is changing faster than this, organisms will be unable to recolonize areas where local extinctions have occurred at a sufficient rate and the regional population itself will become extinct (Fahrig and Merriam 1995). Unlike animals, plants cannot move about freely in a landscape once they are established. They must cope with environmental change either through high resistance or tolerance to stress or through seeds that can disperse faster than, and in advance of, unfavorable environmental change.

Conservation biologists frequently ask the questions, "How large is the habitat?" and "How far are the habitat remnants from one another?" Too seldom do they ask the question, "Do patches of this habitat remain in the same location for many generations, or do patches appear and disappear frequently and randomly?" Both traditional and contemporary efforts in conservation have focused on issues of spatial scale (i.e., preserving local populations in local places). But such efforts, without considerations of temporal scale, may not conserve habitats or populations. Because rates of change in habitat and landscape structure have larger effects on populations than arrangements and distances between habitats, preserving habits and species alone is not a sufficient condition for long-term population and habitat persistence. Only conservation efforts that incorporate temporal, as well as spatial, management of landscape processes will be effective in the long run.

Habitat patches in landscape context support the world's wealth of biodiversity. Such habitats are repeatedly disturbed and, in responding to disturbance, change over time. This pattern of change in habitat and landscape is called succession. The ability to manage succession and its effects is critical for effective conservation efforts.

Managing Succession

In an ecological context, **succession** refers to a pattern of continuous, directional, and nonseasonal change (replacement) of plant populations on a site over time. Succession of vegetation in a habitat can be managed toward the goals of conservation in four ways. The first and most obvious is for managers to alter the frequency, extent, and intensity of disturbance events. This can be done by prescribed burning, flooding, cutting, applying herbicides, or using mechanical methods of vegetation removal at varying intervals.

A second approach is to manage succession by altering plant and animal interactions. One means to this end is to manipulate the density of herbivores, especially large grazing and

Gardiner River (top left) and *Sand in the Canyon* (top right) are two examples of the dozens of watercolor sketches of the Yellowstone area (Wyoming, U.S.A.) produced by the American artist Thomas Moran during his studies of the area in 1871 with the U. S. Government's Hayden expedition. As the expedition's official artist, Moran used rough field sketches like these as the basis for brilliant watercolor masterpieces like *Hot Springs of Gardiner's River* (bottom, 1872). Moran's genius in capturing Yellowstone's spectacular beauty, along with the photographic images of expedition photographer William Henry Jackson, played a decisive role in capturing public imagination and political support for the radical vision of an American national park. *Gardiner River* and *Sand in the Canyon* courtesy of Yellowstone National Park, U.S. National Park Service. *Hot Springs of Gardiner's River* reprinted by permission, private collection, Washington, D.C., U.S.A.

Looking north from Island Lake, Wyoming, into the backcountry of the Absaroka-Beartooth Wilderness Area of Wyoming and Montana, U.S.A. The Wilderness Act of 1964 (Chapter 2) protects legally designated wilderness areas from human development and settlement as "areas untrammeled by man." Wilderness areas provide conservation biologists with baseline data to determine normal patterns of energy and matter transfer in functioning ecosystems, and may provide refuge for threatened or endangered species. In this scene, yellow globeflower (*Trollius laxus*), a species of alpine buttercup, blooms in the foreground. In the background, solitary Lonesome Mountain (slightly left of center) dominates the landscape. Photo by author.

The American alligator (*Alligator mississippiensis*), the official "state reptile" of Florida (USA) exemplifies a "function-based biodiversity indicator," more commonly described as a "keystone species," (Chapter 4), whose activities create habitat for other species and thus increase overall community biodiversity. Alligators may excavate depressions or "gator holes," in wetlands, thus increasing the depth and water-holding capacity of their immediate surroundings. Many species use or reside in such depressions, which can become particularly important for the persistence of many species during drought. © McGraw-Hill Companies, Inc./Carlyn Iverson, photographer.

Tropical rainforests (pictured), concentrated in equatorial regions of South America, Africa, and Indo-Malaysia, are centers of world biodiversity (Chapter 4). Large areas of rainforests are annually destroyed through slash and burn agriculture, land conversion practices, logging, and fire. © McGraw-Hill Companies, Inc./Barry Barker, photographer.

The pitcher plant (*Sarracenia purpurea*) is a habitat specialist found in wetland environments, particularly on nutrient poor, slightly acidic soils, such as those associated with bogs. Pitcher plants compensate for low nutrient levels, especially low N levels, in its surrounding soil by carnivory, capturing and digesting invertebrates, and occasionally small vertebrates, that fall into its "pitcher" (a modified leaf). The pitcher plant and other types of wetland-dependent habitat specialists exemplify "beta rarity," the tendency of some species to be abundant in one habitat but completely absent in others. Preserving a diversity of habitats is an effective means of protecting high levels of beta diversity (Chapter 4). © Corbis/Volume 46.

Fire can be a naturally occurring disturbance that creates habitat heterogeneity at landscape levels (Chapter 5). Here we fly over a fire scar in Montana's (USA) Gallatin National Forest. Such fires, burning unevenly and creating disjunct burn areas in a matrix of unburned forest, create a "mosaic" of different tree stand age classes that contribute to habitat heterogeneity. Photo by author.

Orangutans (*Pongo pygmaeus*) from Sumatra and Borneo, long considered to comprise the same species, were recently documented to have significant genetic differences, leading some conservationists to suggest that they be classified and treated as separate species (Chapter 6). The increasing refinement of genetic techniques to determine the presence of genetic variability in wild populations will raise new and difficult questions about species classification and associated legal protection and conservation strategies. © Corbis/Volume 5.

The black-footed ferret (*Mustela nigripes*) had been reduced to a single wild population near Meeteesee, Wyoming (USA) before intensive captive breeding efforts began. The wild population was subsequently wiped out by distemper, but the number and genetic diversity (Chapter 6) of captive ferrets was sufficient to eventually provide individuals for release and establishment of new wild populations (Chapter 7) in South Dakota, Montana, Wyoming, Utah, Colorado, New Mexico, and Arizona. © Corbis/Volume 244.

A biologist dresses in a whooping crane (*Grus americana*) suit to feed juvenile whooping cranes being raised in captivity. The use of the suit prevents the young cranes from becoming acclimated to humans and helps prepare them to recognize adult cranes when they are eventually released into the wild. Restricted in recent years to a single wild flock summering in Wood Buffalo National Park, Northwest Territories (Canada) and wintering in Aransas National Wildlife Refuge in Texas (USA), conservation biologists are now attempting to establish a second flock in Florida (USA). This project represents one of many conservation efforts to restore populations of threatened and endangered species through captive breeding and reintroduction (Chapter 7). U.S. Fish and Wildlife Services, NCTC.

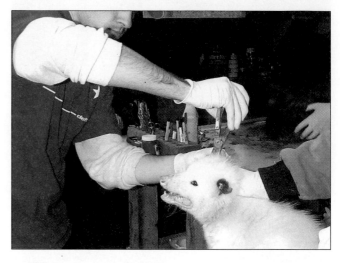

Conservation biologists attach a numbered ear tag to a Virginia opossum (*Didelphus virginiana*). In addition to making it possible to identify the individual later, such marking is vital for population estimation and management (Chapter 7). Biologists can use marked animals to estimate population sizes by comparing ratios of marked to unmarked animals in later sampling efforts in which large numbers of both marked and unmarked animals are captured. Photo by L. K. Page. Used by permission.

Foresters attempt to eradicate invading velvet trees (*Miconia calvescens*) trees with herbicides in a native Hawaiian forest. First trees are cut down, then stumps must be sprayed with herbicides to prevent regrowth. Velvet tree, native to South America, is an example of a non-indigenous species (NIS, Chapter 7) that can supplant native species by forming a dense canopy under which native species cannot grow, but velvet tree seedlings can. Several avenues of biological control, including insects and plant diseases, are being considered. © Frans Lanting/Minden Pictures.

The western prairie fringed orchid (*Platanthera-praeclara*), once widespread across eight midwestern U.S. states, is today a federally threatened species surviving in remnant wet prairies, sedge meadows, and roadside ditches. Growing up to four feet in height, the prairie fringed orchid is pollinated only by certain species of hawkmoths (Family Sphingidae) that possess tongues of suitable length to reach the orchid's recessed nectarines. Although these hawkmoths are not themselves endangered, the fragmentation of prairies by crops and urban development creates barriers to the moths that reduce rates of pollination in orchid populations. Photo courtesy of T. Gouveia, Henry Doorly Zoo, Omaha, Nebraska, U.S.A.

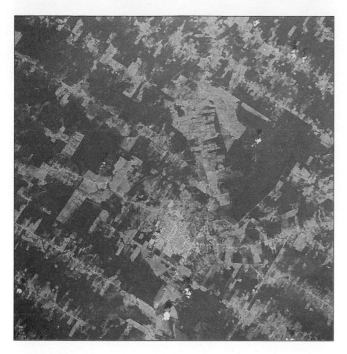

Coral reefs built up by many generations of deposits from coral polyps form important structures for many creatures. When coral reefs and their associated waters are protected as marine reserves, populations of many species, including large predatory fish like this grouper (Family Serranidae), shown swimming over a coral formation, increase in density and biomass (Chapter 9). © Corbis/Volume 64.

A satellite image reveals the reality and extent of tropical deforestation and habitat fragmentation in Brazil's Amazon Basin. Dark areas represent remaining forests. Brown areas have been deforested. Note how formerly contiguous blocks of forest have been isolated into disconnected fragments. As this type of deforestation occurs, species are lost from combined effects of habitat loss, habitat isolation, and habitat fragmentation (Chapter 8). Fragmentation, in particular, increases the ratio of edge (perimeter) to area in habitat, making interior species more vulnerable to "edge effects," which include changes in microclimate, increased competition with "edge" species, and increased vulnerability to predators that can now "penetrate" from the edge to the interior habitat. NASA.

Coral polyps such as these are the building blocks of coral reefs, centers of marine biodiversity in tropical and subtropical marine waters (Chapter 9). Mineral deposits from former generations of coral provide a permanent structure for the present generation of living corals to grow, as well as habitat for a great diversity of other forms of marine life. Photo by N. C. Rorem. Used by permission.

Kelp (Division Phaeophyta) "forests" create heterogeneity in marine environments because their large size and extent alters current flow. The great biomass offers an abundant source of food, and their bodies create physical structure that many marine creatures use for support or concealment. Here a Garibaldi fish (*Hypsypops rubicundus*) swims among a stand of kelp. © Corbis/Volume 64.

The water hyacinth (*Eichornia crassipes*) is a non-indigenous species in the United States that can multiply rapidly and produce choking infestations in slow flowing, warm water canals and streams (Chapter 9). In such cases, control may be attempted through habitat manipulation, chemicals, physical removal, or biological control (for example, manatees eat water hyacinth), but complete eradication is difficult once the infestation is extensive. © Steven P. Lynch.

Lentic systems (Chapter 9) are aquatic habitats characterized by non-flowing or slow-flowing water, like this pond. Under such conditions, rooted aquatic emergents such as water lily (*Nymphaea* spp.) may be established. Such emergents provide physical structure for a variety of invertebrates and smaller vertebrate species (especially amphibians), as well as a food source for larger herbivores, but may also increase rates of sedimentation and biomass accumulation, hastening the pond's succession to a terrestrial habitat. Photo by L. K. Page. Used by permission.

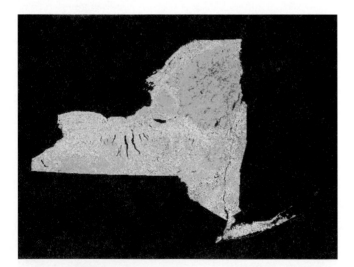

The state of New York (USA) as it appears in a satellite image that can provide source data for geographic information system (GIS) analysis (Chapter 10). This image, originally provided by the Environmental Protection Agency, was enhanced by the Quantitative Studies Laboratory of the College of Environmental Studies and Forestry-State University of New York (Syracuse) so that adjusted colors correspond visually to land use classes. Green indicates forested land, yellow is agricultural land, gray represents urban land, and black represents water (lakes). Such classifications offer a first-order approximation for identifying major land use classes and habitat types (Chapter 8).

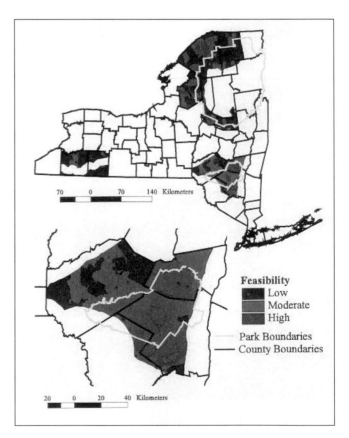

The same image (at left) can be subjected to further analysis in GIS to answer specific questions in spatial analysis. Here the GIS is applied to the problem of determining potentially acceptable areas for re-introduction of elk (*Cervus elaphus*), historically common in the state but exterminated in the nineteenth century. Using a habitat suitability model, the GIS applies rules that define suitable habitat for elk to identify areas of appropriate vegetation characteristics and sufficient size for a re-introduction effort. From Didier and Porter 1999. Used by permission of The Wildlife Society.

A "wave" of fire advances across a tallgrass prairie at the DeSoto National Wildlife Refuge in Iowa (USA). Historically, recurrent fire was common in prairies. Fires enhanced native biodiversity and retarded encroachment by woody, non-prairie vegetation. Today fire suppression in prairies contributes to loss of prairie habitat and prairie species of plants and animals. Here a prescribed fire is set and directed by U.S. Fish and Wildlife Service personnel to rejuvenate native prairie. Prescribed fire is a natural process that can be used to meet specific management objectives in many ecosytems (Chapter 10). Photo by S. Van Riper. Used by permission.

Not only plant species, but many animal species increase densities in areas subjected to burning. For example, the dickcissel (*Spiza americana*) a common bird of tall and mixed grass prairies in North America, has shown recent evidence of regional population decline with the continuing loss of prairie habitat. These maps show the location and configuration of dickcissel territories in a management unit of tallgrass prairie on the DeSoto National Wildlife Refuge, Iowa, shortly after burning (1995) and one year later (1996). Note that immediately after burning, 8 dickcissels established territories on the site. Territories were relatively small, suggesting high densities of resources for dickcissels, and all parts of the prairie were used. One year later, only four territories are established, all are much larger in size, and approximately half of the habitat is no longer used. Maps prepared by author based on unpublished data.

A compass plant (*Silphium perfoliatum*) blooms in a restored prairie in the Morton Arboretum in Lisle, Illinois (USA) near Chicago. The compass plant, so called because its basal leaves often align in a north-south direction, is a species that is quickly lost in degraded or disturbed prairies and may be slow to re-establish in restoration efforts. Comprehensive ecological restoration (Chapter 11) of degraded sites seeks to restore both the components (species) of the ecosystem as well as its ecological functions. Photo by L. K. Page. Used by permission.

Even sites severely degraded by human activity or polluted by toxic wastes can, with skill and patience, be restored to former ecological function and components (Chapter 11). Likewise, soils contaminated by toxic wastes can be detoxified through bioremediation using appropriate management techniques and appropriately chosen plant and soil communities. This plant community is growing on soil taken from a reclaimed mine in Oklahoma. © McGraw-Hill Companies, Inc./Gary Hannan, Eastern Michigan University, photographer.

Ecotourism generates considerable revenue in Zimbabwe's CAMPFIRE program for sustainable development (Chapter 12). However, ecotourism can have harmful effects on wildlife and their habitats. Where demand is high and not properly regulated, or where sites are not properly "hardened" to withstand high densities of tourists, degradation of habitats and reductions in population numbers can occur. Here we see a photo safari in Kenya that reveals the high demand for wildlife viewing, and its potentially degrading effects on wildlife habitat. © McGraw-Hill Companies, Inc./Barry Barker, photographer.

A male Galapagos tortoise (*Geochelone elephantopus vandenburghi*, top) copulates with a female near Volcano Alcedo on Isabela Island in the Galapagos Islands. The Galapagos tortoise exemplifies conservation biology's historic emphasis on conserving rare species (Chapter 1). The tortoise is "rare" and "threatened" in multiple dimensions. It is an endemic species with a narrow geographic range (the Galapagos Islands), is characterized by small population sizes and relatively slow rates of population growth (tortoises are long-lived and do not normally begin breeding until 15-20 years of age), and is threatened by habitat destruction by introduced species (goats) and humans, as well as predation by introduced species such as pigs, rats, dogs, and cats. The government of Ecuador has set the Galapagos Islands aside as a national park, and biologists have undertaken active measures for the tortoise's conservation. Photo by L. K. Page. Used by permission.

The American goldfinch (*Carduelis tristis*) is the state bird of both New Jersey and Iowa (USA). Here a male goldfinch perches on the head of a thistle (*Cirsium arvense*) in an Iowa tallgrass prairie. The goldfinch's striking colors contribute to its choice as an official state symbol. Species that display qualities of visual beauty or other attributes admired by humans exemplify the concept of *aesthetic value* (Chapter 3). Photo by S. Van Riper. Used by permission.

browsing mammals, on the site following disturbance. Manipulations can range from no control, which often leads to heavy use of vegetation by herbivores and significant effects on the amount and composition of vegetation, to total exclusion of herbivores, a strategy that often produces communities of high plant biomass but low net productivity.

A third approach is to change the availability of species that can invade a disturbed site. For example, range managers restoring a degraded prairie may first burn the site, and then seed it with native species that will function best on that site, or that are of particular value for conservation. In a forest, sites disturbed by logging may be arranged close to abundant seed sources of desired species, or such species may be planted directly on the site when logging is completed. Similarly, wetlands created by deliberate flooding may be seeded with plant species of special value to conservation or that have high value as food and cover for wetland wildlife. The same approach can be employed, in a negative way, through species- or type-specific herbicide application after the disturbance event, thus selectively altering the pattern of plant succession on the site.

A fourth approach to managing succession is to manipulate the availability of resources at a site, and so alter the interactions of plants with the environment and with other species. Adding specific nutrients and fertilizers to logged sites in forests or to cleared or burned areas in prairies to favor the establishment of particular species is a common management practice. Removing undesirable species from sites after disturbance may be done to favor species considered more beneficial to system function, more valuable for animals, or of greater value in conservation. Leaving specific resources, such as snags, in place following a logging operation can increase the availability of nest sites for cavity-nesting birds.

One example of a conservation initiative using multiple approaches in succession management is Johnson and Leopold's (1998) effort to manage habitat for the endangered eastern massasauga rattlesnake (*Sistrurus catenatus catenatus*). The eastern massasauga is associated with nonforested wetlands, especially minerotrophic peatlands, throughout its range in the eastern United States and Canada. The massasauga prefers early successional communities and suffers habitat loss as succession proceeds, especially from the encroachment of woody plants into wet meadows and old fields.

Johnson and Leopold applied the first approach, directly altering the type, frequency, and intensity of disturbance events, by cutting and burning all woody vegetation in selected experimental plots in the Cicero Swamp Wildlife Management Area in central New York, a peatland complex inhabited by massasaugas. They then made use of the second approach, altering plant and animal interactions, by enclosing selected plots in wire fences, thus excluding herbivores (primarily white-tailed deer, *Odocoileus virginianus*). Finally, by applying herbicides specific for woody vegetation, combined with cutting, Johnson and Leopold employed the third approach and altered the pattern of plant invasion subsequent to the disturbance.

Over three years, Johnson and Leopold were successful, to varying degrees, in altering patterns of succession through these approaches, particularly in achieving reductions in the density, basal area, and height of shrubs in treated plots compared with untreated areas, and, with herbicides, increasing the rate of mortality and decreasing the rate of resprouting among woody species. Although data on the response of the massasauga to the treatments were inconclusive and limited by a small sample size (nine radio-marked massasaugas), massasaugas did use treated areas at more than four times their availability (10.1% of locations in treated areas that made up only 2.5% of the total area).

Understanding and managing succession is vital to successful conservation. In fact, ecological models based on habitat succession have been shown to be more accurate at predicting the distribution of many organisms than null models (models based on random environmental fluctuation) or so-called isolation models (models based on island biogeography theory and metapopulation theory) (Kareiva, Skelly, and Ruckelshaus 1997).

Increasingly, habitat fragmentation and human influence will disrupt both the endogenous and exogenous processes that historically created disturbed sites. It is the creation of such sites, through disturbance, that continually reinitiates the successional process and provides diversity and heterogeneity of habitat in the landscape. Few conservation reserves can hope to preserve biodiversity simply by trying to preserve an existing state of species and habitats without disturbance. Increasingly, the conservation of biodiversity will depend not on preserving the status quo but on managing processes, and the most important process to be managed will be succession.

POINTS OF ENGAGEMENT—QUESTION 1

How would the perspectives and activities of a reserve manager change if he or she changed the goal from achieving a *state* (e.g., a proportional abundance of a desired habitat or a targeted population level of a particular species) to enhancing a *process* (an exchange of matter or energy in the system that created resources or habitats in the landscape)? What specific changes can you imagine in allocations of time, questions of research interest, and desired outcomes if such a change occurred?

THE PROBLEMS OF HABITAT LOSS, ISOLATION, AND FRAGMENTATION

Overview: Understanding the Problem

The detrimental influence of humans on biodiversity often does not occur through the deliberate killing of plants and animals. Today the most significant, if not always obvious, threats to the conservation of biodiversity are the related problems of habitat loss, habitat isolation, and habitat fragmentation. Anthropogenic (human-induced) effects on a landscape, including the processes of urbanization, agriculture, and industrialization, invariably have the effect of dividing blocks of once-contiguous habitat into smaller blocks of isolated and fragmented habitat.

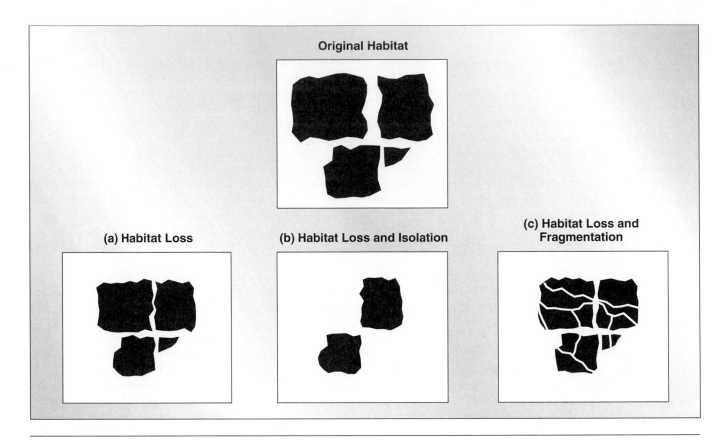

Original Habitat

(a) Habitat Loss

(b) Habitat Loss and Isolation

(c) Habitat Loss and Fragmentation

Figure 8.6

A conceptual illustration of habitat loss, isolation, and fragmentation. In (a), all patches are consistently smaller. In (b), habitat fragmentation is actually decreased because there are fewer patches, but habitat isolation increases. In (c), in addition to increasing patch separation, fragmentation decreases patch size.

Adapted from Fahrig (1997).

Fragmentation occurs when a large expanse of habitat is transformed into a number of smaller patches of smaller total area, isolated from one another by a matrix of habitats unlike the original (Wilcove, McLelland, and Dobson 1986). The resulting fragments of native habitat are then defined as any patch of vegetation around which most or all of the original vegetation has been removed (Saunders et al. 1987). The edges that are created when habitats are fragmented are a junction or border between two different landscape elements, such as plant communities, successional stages, or land uses (Yahner 1988).

The process and effects of habitat fragmentation follow a predictable sequence. (1) A block of contiguous habitat is broken into numerous smaller patches. (2) The remaining habitat patches decrease in size and number because of increased vulnerability to disturbance and invasion by other species and increased ease of alteration or removal by humans. (3) Distance between patches increases (increased habitat isolation). (4) Quality of the patches decreases. (5) Importance (and detrimental influences) of edge effects increases (Oksanen and Schneider 1995).

Habitat loss and fragmentation have significant effects on populations. Because species abundance and diversity are often area dependent, both decline as the amount of habitat is

reduced. The overall loss of habitat and the isolation of the patches can reduce population sizes to such low levels that indigenous species may become extinct in individual patches, or even in an entire group of patches (Gaines et al. 1997; Rosenberg, Noon, and Meslow 1997). Population decline occurs through several mechanisms. Some species require home areas or breeding territories of minimum critical sizes. Although such critical sizes may vary according to the densities of needed resources, there will be some lower limit, below which the species cannot exist. When habitat fragmentation reaches the point that insufficient area remains for these species to maintain a breeding population, these species will be lost from the fragments. Another problem is that large habitat fragments may contain both common and uncommon species, but small fragments often have only common species. Additionally, small fragments are more likely to lose species over time because their small populations are more susceptible to extinction.

It is often difficult to separate the effects of habitat loss, habitat fragmentation, and habitat isolation because such events often occur together and their outcomes are confounded. Figure 8.6 displays these distinctions graphically. Note that in (a) the amount of habitat is reduced, but the number of habitat patches remains constant. This scenario represents simple habitat loss, not additional fragmentation. In (b), two of the four patches are removed, but the two that persist remain constant in size. This scenario is a case of habitat loss *plus* increased habitat isolation (increasing distance between habitat patches), although fragmentation of the remaining habitat is actually decreased. In (c), the four original habitat patches are broken into 14 smaller patches. Here, and only here, do we have a condition of habitat loss *and* habitat fragmentation. How do we determine the relative contribution of each effect to declines in populations and in the diversity of communities? The question is not merely academic. It is of great practical significance to conservation. If declines are due primarily to habitat loss, then the primary goal of conservation is to prevent habitat loss and, where possible, add more habitat. If declines are due primarily to the effects of fragmentation and/or isolation, then the spatial arrangement and connectivity of remaining patches become the primary concern. If we misdiagnose the problem, then we risk applying an inappropriate solution that will waste valuable resources and effort.

Consider a spatially explicit, individual-based stochastic model in which the amount of breeding habitat, represented by the variable COVER, and the amount of fragmentation of the breeding habitat (the variable FRAG) can be varied independently (Fahrig 1997). Other model variables specific to the life history and movement characteristics of the hypothetical organism are also included. The landscape in which the organism moves is a two-dimensional grid of 900 cells arranged in a 30 × 30 matrix. The landscape contains only two categories of habitat, breeding and nonbreeding. Reproduction, movement, and mortality of organisms occur in discrete time steps. Reproduction occurs only in the breeding habitat, and its probability is 0.5 per individual per time step. Every reproduction event produces one offspring per reproducing individual. The probability of movement of

an organism is 1.0 (certainty) per time step if it is in nonbreeding habitat and 0.5 in breeding habitat (i.e., the model assumes that organisms in breeding habitat do not "want" to leave). Movement direction and distance are random. Cells have carrying capacities. Mortality is density dependent below carrying capacity, but individuals in a cell are "killed" at random if their numbers exceed carrying capacity.

The model begins with an arbitrary value of COVER that is equal to the number of cells of breeding habitat. Specifically, a cell is selected at random and assigned a random value between 0.0 and 1.0. If the selected cell has a bordering cell that is designated as breeding habitat, or if the random number is less than the assigned value of FRAG, then the cell is assigned as breeding habitat. This procedure is iterated until all cells are assigned. Note the effect of the value of FRAG on this assignment. If FRAG is assigned a low value such as 0.1 (representing a low level of habitat fragmentation), then most cells receive a larger value for their random number and will not be designated as breeding habitat *unless* they border a cell of breeding habitat. Thus, with low values of FRAG, almost all breeding habitat is clumped and fragmentation is minimal.

Fahrig (1997) ran 2,000 simulations of this model at various levels of COVER and FRAG, each with an initial population of 500 individuals, for 500 time steps or until the population became extinct, whichever came first. The outcomes of these simulations demonstrated that the amount of breeding habitat (COVER) had a much greater effect on the probability of the population's extinction than did the degree of fragmentation. In fact, as long as the value of COVER was greater than 0.2 (more than 20% of the total area in breeding habitat), all populations survived regardless of the degree of fragmentation (fig. 8.7). Only when the amount of breeding habitat dropped below 0.2 did the effects of fragmentation become important.

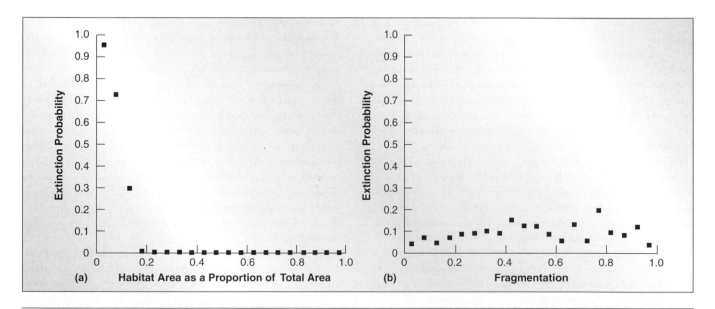

Figure 8.7

The independent effects of (a) habitat loss and (b) habitat fragmentation in a simulation model. Note that the probability of extinction begins to rise when remaining habitat is less than 20% of the total landscape, but fragmentation alone has little effect on extinction probability.

After Fahrig (1997).

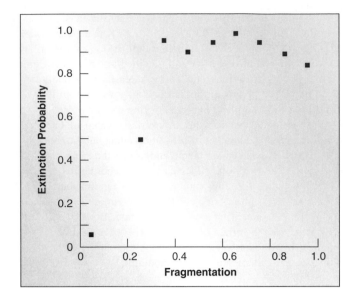

Figure 8.8

The importance of habitat fragmentation on extinction probability in a simulation model when the proportion of breeding habitat in the landscape drops to 10%. At this level of habitat loss, extinction becomes nearly certain when fragmentation indices exceed 0.3. Results based on 200 simulations.

After Fahrig (1997).

Even when the amount of breeding habitat was reduced to 10% of the landscape, most populations persisted until the value of FRAG became greater than 0.3 (fig. 8.8). Fahrig concluded, "when breeding habitat covers more than 20% of the landscape, survival is virtually assured no matter how fragmented the habitat is. Details of how habitats are arranged is unlikely to mitigate risks of habitat loss" (Fahrig 1997).

The value of Fahrig's model, like all well-constructed models, is not that it perfectly mimics the real world. Rather, its value is that it makes explicit our assumptions about such a world. And it is precisely such explication that permits us to separate effects of habitat loss from those of habitat fragmentation. Most field studies and observations cannot do this. However, some studies do support the 20% rule of habitat cover in a landscape that the model suggests (Lande 1987; Lamberson, McKelvey, and Noon 1992; Lawton et al. 1994; Hanski, Moilanen, and Gyllenberg 1996). The model's results also suggest that, where habitat loss is substantial, effects of fragmentation become critically important to population persistence. Thus, where habitat loss is great, fragmentation and isolation do matter, and conservation efforts to achieve optimal spatial arrangement of remaining habitats become increasingly important.

When we assess the importance of fragmentation and loss of habitat, we should note that the relative effects of each vary according to species requirements. Specifically, any given amount of habitat loss has a far greater effect on so-called interior species (species dependent on habitat conditions unique to one habitat) than on so-called edge species (species that are adapted to habitat borders). In fact, habitat fragmentation can, under certain conditions, even increase the amount of available

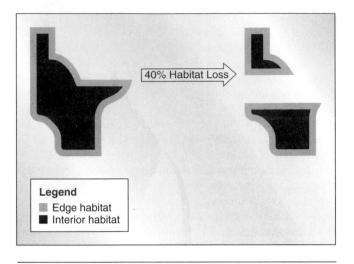

Figure 8.9

The effect of habitat fragmentation. When 40% of habitat is removed, the amount of interior habitat decreases by approximately 60%. In this scenario, edge species do not experience a significant increase or decrease of habitat.

habitat for edge species. Consider, for example, a forest in which 40% of original habitat in a single block is destroyed and the remainder exists as two blocks (fig. 8.9). In this scenario, edge species suffer no loss of habitat, but interior species lose 60% of their habitat.

Given the pronounced effect of habitat fragmentation on a habitat patch's area-to-edge ratio, it is not surprising that some of the most well-known and well-studied aspects of habitat fragmentation are *edge effects*. Consider what happens when a habitat is fragmented. Suppose we begin with a 1 km^2 (100 ha) block of contiguous habitat (fig. 8.10). Assume that changes associated with the edges penetrate 100 m at each border and are the same on all sides. This reduces the amount of "core" habitat (habitat unaffected by edge effects and processes) to an area 800 m long and 800 m wide, or 64 ha. Now suppose that this block is bisected by one road (north-south) and one power line (east-west). Assuming that edge effects continue to penetrate 100 m at each boundary (which now include the new road and the power line) the amount of habitat that is unaffected by edge is reduced from a single block of 64 ha to 4 separate blocks, each only 300 m in length on each side, or 9 ha in area. Together, the four blocks now contribute only 36 ha of habitat that is not affected by edge effects and processes, little more than half (56%) of the unaffected habitat present in the original, unfragmented block.

This example illustrates two important principles. First, as long as fragmentation continues, edge habitat in any given area increases while core habitat decreases. Second, when fragmentation occurs, core habitat is moved closer to edge habitat and to the processes and effects associated with the edge environment. From this general understanding, we can begin to examine the details of edge effects with more precision and gain more insight into why increasing the amount of habitat exposed to edge can be the same as direct habitat loss.

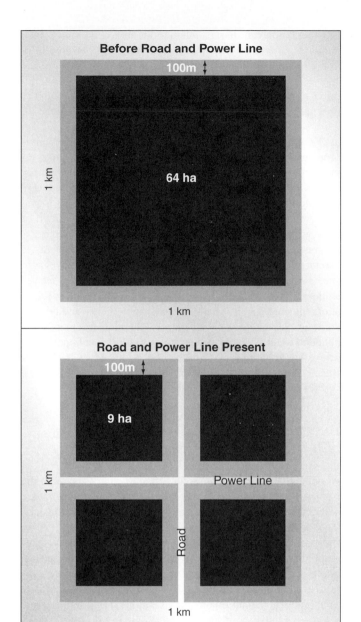

Figure 8.10

The effect of habitat loss on interior and edge species. Note that, in this scenario, when a road and power line intersect in the habitat, edge species experience a net habitat increase of 36 ha to 64 ha, whereas interior species lose 44% (28 ha of the original 64 ha) of previously available habitat.

Characteristics of Edge Environments

Broadly, edges can be classified as inherent or induced. Inherent edges represent long-term landscape features that form because of differences in soil type, topography, microclimate, geomorphology, or disturbance history of the site (Thomas, Maser, and Rodiek 1979). In contrast, induced edges are usually short-lived features at the junctions of different vegetation types or land uses (Yahner 1988). Induced edges may be the result of natural processes, such as fire, or the result of deliberate management actions.

Traditional wildlife management has viewed edges as desirable landscape elements and the addition of induced edges as beneficial. This view arose because species diversity tends to increase as the amount of edge increases, and many game species, including most upland game birds, white-tailed deer, and cottontail rabbits (*Sylvilagus* spp.) make disproportional use of edges (Leopold 1933). Species diversity often increases at edges because edges attract plants and animals that are adapted to the edge environment. These edge species may form a unique community in their own right. Although edges may increase diversity by increasing the abundance of edge species, this increase often comes at the expense of other species that require habitats unaffected by edge features and processes.

Environmental Characteristics

The creation of edge changes important features of a habitat's physical environment. Edges usually receive more direct insolation, and thus are typically warmer and have lower relative humidity than interior areas, especially in forest habitats. Reradiation at night is often greater, and frosts are more frequent and extensive. Radiation and moisture fluxes also tend to be greater at edges. Increased radiation and insolation associated with edges tend to increase soil temperatures and affect rates of invertebrate and microbial activity (Klein 1989; Parker 1989) and associated processes of nutrient decomposition. Increased soil temperatures may also reduce the retention of water in the soil and alter the growth rates and phenology of vegetation.

Wind behaves differently at edges than in the interior of a habitat. As wind moves over a landscape and flows from one type of vegetation to another, the upper part of the wind profile retains the characteristics formed over the previous vegetation type, while the lower portion takes on profile characteristics of the new vegetation. The two profiles do not fully equilibrate with one another for some distance, increasing turbulence and wind shear at their boundaries. The distance required for equilibration is normally 100 to 200 times the height of the vegetation (Saunders, Hobbs, and Margules 1995). Although not well studied, the effects of edge on wind behavior clearly have important and observed consequences. Trees near an edge are more susceptible to damage, wind pruning, uprooting, and other forms of physical damage. Increased wind speeds also can lead to increased evapotranspiration and desiccation. Secondary effects of wind at edges can include increased transport of material into the interior from surrounding, but different, vegetation types, including soil, seeds, insects, and dead organic matter.

Habitat fragmentation and its associated creation of edge also alters local water regimes. Rates of interception and evapotranspiration are changed by removal of native vegetation and by modification of native vegetation at the edges of remnants (Saunders, Hobbs, and Margules 1995). Replacement of deep-rooted perennials, which are more typical of native vegetation, with herbaceous crops, pasture, or nonvegetative surfaces at edges leads to greater runoff and increased surface and groundwater flows, with accompanying increases in rates of erosion and transport of particulate matter.

Biological Characteristics

Biological interactions also change at edges, usually to the detriment of species adapted to interior habitats. Recent studies have shown that edge habitats have a variety of detrimental effects on interior species of both plants and animals, including increased predation, increased parasitism, and increased herbivory (Laudenslayer 1986; Alverson, Waller, and Solheim 1988; Harris 1988; Temple and Cary 1988; Yahner 1988). For example, the presence of edges can alter and intensify interspecific interactions. In bird communities, some species that use edges, such as the brown-headed cowbird (*Molothrus ater*), parasitize the nests of interior species, particularly forest species. Other edge species, such as blue jays (*Cyanocitta cristata*) and house wrens (*Troglodytes aedon*), may engage in nest predation (blue jays) or the destruction of eggs of other species (house wrens).

Overall, increased amounts of edge result in less secure habitat for interior species (Temple 1986) because nest predators and parasites not only reduce nesting success of interior species at edges, but can penetrate more deeply into the interior of other habitats from edges. For example, among forest birds, increased rates of nest predation may extend up to 600 m into the forest from the edge (Wilcove 1985). The outcome is that there is less habitat in which resident species are relatively unaffected by edge processes and effects. Many predators associated with edges of habitats have characteristic "penetration depths" into blocks of homogeneous habitat (Wilcove, McLelland, and Dobson 1986). Thus, when habitat is fragmented and the ratio of edge-to-interior increases, many species dependent on a particular habitat become more vulnerable to predation because they have little or no habitat left that edge predators cannot penetrate. In North America, species of insectivorous songbirds with high vulnerability to predation during the breeding season have declined significantly in recent years, while those species with low vulnerability to predation during the breeding season have increased (Robinson et al. 1995). This finding has led some to conclude that habitat fragmentation and edge creation in temperate forests, with its associated increases in penetration depths of predators, may be more important in songbird declines than the more publicized tropical deforestation.

Effects of Habitat Fragmentation on Population and Community Processes

Habitat fragmentation can lead to extinction of local populations through a number of individual or combined effects (Wilcove, McLelland, and Dobson 1986). First, fragmentation can, if sufficiently severe, reduce fragments to sizes smaller than the minimum critical area needed for the home ranges of some species. Second, fragmentation can lead to extinction through a loss of habitat heterogeneity. When heterogeneity of habitat is lost, so are those species adapted to patch conditions. Further, species that require two or more habitat types suffer if fragmentation makes it impossible for them to move among habitats. Third, changes in environmental variables or increases in populations of edge species associated with fragmentation, previously discussed, may be detrimental to interior populations. A fourth concern is the problem of secondary extinctions.

Fragmentation can disrupt interactions in a community, such as plant-pollinator relations, mutualisms, parasite-host relations, or predator-prey relations. Then the extinction of one species may lead to additional extinctions (secondary extinctions). Fifth, fragmentation reduces area of available habitat, thus reducing population densities. Smaller populations are more susceptible to extinction because they are more subject to demographic instability, inbreeding depression, and environmental stochasticity (chapter 7, Wilcox 1980). Thus, reduction in area, acting through extinction, reduces the number of species that can and will occur in fragments of formerly contiguous habitat.

All of the above factors are affected by the isolation of the fragments from one another, and by the time since the isolation occurred. When fragmentation occurs and the remnants of habitat are isolated, each remnant is likely to initially have more species than it can maintain in the long term. Over time, species are lost from fragments. This process, known as "species relaxation," is predicted by island biogeography theory and considered an inevitable consequence of habitat fragmentation, although exactly which species are lost will depend upon characteristics of both species and habitat. Species that are habitat specialists in native vegetation, species with large territories, and species at low densities are most likely to disappear.

Field and Experimental Studies of Habitat Fragmentation

Although habitat fragmentation is often hypothesized to be the cause of population declines, careful experimental tests are needed to determine if such a hypothesis is valid. Fortunately, there are field studies that shed light on these effects.

In a formerly forested area of Australia now fragmented by agriculture, the size of remaining forest patches proved to be the single best predictor of species richness of terrestrial mammals (Bennett 1987). In this area, the effects of habitat fragmentation on small mammal communities have been especially well documented (Bennett 1990). First, among eight investigated species, six native and two introduced, the introduced black rat (*Rattus rattus*) and house mouse (*Mus musculus*) proved much more tolerant of fragmentation than the native species. The introduced species were more common in small forest fragments (2 to 10 ha) than in larger fragments (20 to 80 ha). In contrast, native species were far less tolerant of fragmentation, occurring less frequently in smaller fragments than in larger ones.

Larger fragments consistently had more species, and the frequency of occurrence of native species consistently increased in samples of larger-size fragments. The frequency of native mammals increased with increasing-size class of forest patches, and the pattern of increase followed a relatively predictable sequence. For example, four species occurred in over 50% of fragments 41 to 100 ha in size, but no species occurred in over 50% of fragments less than 2 ha in size. Two species of bandicoots (*Isoodon obesulus* and *Perameles nasuta*) were rarely found in fragments of less than 40 ha, and the long-nosed potoroo (*Potorous tridactylus*) (fig. 8.11) was rarely found in fragments of less than 8 ha. The general prin-

Figure 8.11
The long-nosed potoroo (*Potorous tridactylus*) of Australia, a marsupial negatively affected by habitat fragmentation.

ciple that fragmentation leads to smaller resident populations is borne out in this study. The bandicoots, which used only the larger fragments, were of particular concern because, even in the larger fragments, their populations were extremely small (Bennett 1990).

Different species respond differently to fragmentation. In the Australian study, variation in body size among species explained 83% of their variation in tolerance of fragmentation. The larger the average body weight of a species, the lower its tolerance. This pattern also has appeared in studies of small mammals in other areas. Diffendorfer and colleagues (1995) examined the effect of habitat fragmentation on three species of grassland rodents, the hispid cotton rat (*Sigmodon hispidus*), the prairie vole (*Microtus ochrogaster*), and the deer mouse (*Peromyscus maniculatus*), by experimentally fragmenting grassland habitat to varying degrees. The largest of the three rodents, the cotton rat, had higher densities in contiguous habitat, but the two smaller species had higher densities in fragmented habitat. The smallest species, the deer mouse, had the most positive response to fragmentation, was most abundant in the smallest areas, and actually used the interstitial areas that were created to separate habitats. Diffendorfer and colleagues (1995) believed that the differences arose from a combination of differences in habitat quality and changes in competitive interactions. The cotton rat suffered from habitat fragmentation because smaller sites had insufficient resources for long-term survival. Its decline in fragmented habitats created competitor-free space for the two smaller species. The deer mouse had the most positive response to fragmentation because the greater the degree of fragmentation, the greater the area of interstitial border areas, which it could use as well as the original habitat. For this species, fragmentation increased the amount of habitat available.

Bennett demonstrated differences in tolerances to fragmentation on the basis of species occurrence. Diffendorfer and colleagues went further and suggested hypotheses, based on population characteristics, about the potential mechanisms through which habitat fragmentation created differences in species abundance. It is even more illuminating to make direct comparisons of population densities or population processes, such as rates of birth, death, predation, or parasitism, in fragmented and unfragmented habitats. Robinson and associates (1995) accomplished this by examining a number of such processes directly on forest birds nesting in fragmented forests of the midwestern United States. Examining landscapes that ranged from 6% to 95% forest cover, Robinson and associates (1995) found that cowbird parasitism and nest predation were negatively correlated with the proportion of forest cover. In areas with less than 55% forest cover, wood thrush (*Hylocichla mustelina*) nests contained more cowbird eggs than wood thrush eggs. But in heavily forested landscapes, cowbird parasitism was so infrequent that it was no longer a significant cause of reproductive failure. Similarly, rates of nest predation on ovenbirds (*Seiurus aurocapillus*) and Kentucky warblers (*Oporornis formosus*) were so high in severely fragmented forests that the investigators concluded that such areas were population sinks for these species (Robinson et al. 1995). Overall, the amount of forest cover (an index of habitat fragmentation) had significant effects on rates of both brood parasitism and nest predation.

APPLICATION PROBLEMS IN HABITAT CONSERVATION: HOW DO WE INTEGRATE HABITAT STRUCTURE WITH RESERVE DESIGN?

The Problem of Function: What Is a Reserve For?

The function of a reserve defines its objective, but unless the relationship between function and objective is clearly understood, establishing reserves will not necessarily advance any of the important goals of conservation. Recall our earlier question about whether managers should focus on preserving states or managing processes. This question anticipates a growing debate in conservation about what the proper objective of a reserve is. One view is that reserves should preserve *states* (species and assemblages of species) (Simberloff 1988). This goal is commonly expressed as a goal of preserving species, sometimes even a single species. This is one of the oldest and most common goals for establishing refuges in the United States. Examples of species whose endangered status led to refuge establishment include the California condor (*Gymnogyps californianus*), the saguaro (*Carnegia gigantia*), the Kirtland's warbler (*Dendroica kirtlandii*), and the redwood (*Sequoia* spp.) (Soulé and Simberloff 1986). More recently, the goal of preserving states is increasingly to preserve entire habitats and functioning communities or biodiversity generally (Soulé and Simberloff 1986).

A more recently developed view of reserves is that they should preserve *processes,* especially ecosystem function (Simberloff 1988). If the objective is to preserve states, then that goal is best achieved by maximizing the density of a targeted species, or the overall richness of species or habitats within the reserve. But if the goal is to preserve processes, then that goal is best achieved by maintaining in perpetuity a highly complex set of ecological, genetic, behavioral, evolutionary, and physical systems and the coevolved, compatible populations that participate in the exchange of matter and energy in these systems. The nature of the objective determines how the reserve should be designed as well as the type of management that should be employed on the reserve.

If the aim is to preserve a specific association of habitats and species, then regular and frequent intervention (management) is necessary in the processes (burning, grazing, predation, competition) that could alter these states. In other words, preserving states requires managing processes. On the other hand, if the aim is to preserve the processes themselves, little or no intervention is required (Simberloff 1988). But preserving processes, rather than states, in a reserve means that the habitats, landscapes, and associated species composition of the reserve will change over time.

Historically, most reserves have been established to maintain particular assemblages of species and habitats. Therefore, management and intervention are essential to achieve this objective. Ironically, parks and preserves often are among the least managed areas where natural habitat exists, and may be the places where popular opinion and political pressure are most strongly entrenched against such management. It is a popular but tragic misconception that parks and nature preserves will sustain populations and habitats without effective management of habitats. For example, an analytical mathematical model of 19 reserves in east Africa indicated that, without active management and supplementation of populations, the smaller parks will lose 10 to 20% of all large mammal species within 50 years following isolation, and a typical reserve will lose almost half its large mammal species within 500 years of isolation (Soulé, Wilcox, and Holtby 1979). As environmental policy analyst A. Dan Tarlock noted perceptively, "At best, ecosystems can be managed, but not restored or preserved" (Tarlock 1994). This is because habitats and ecosystems represent patches or collections of conditions that exist only for finite periods of time (Botkin 1990). An ever-accelerating rate of human interaction with natural habitat makes it impossible for us to return to any "ideal" state of a habitat's past. Instead, we must choose an objective and manage to achieve it.

Some Methodologies for Reserve Design

Two fundamental tasks of a refuge planner are determining what habitats are important to preserve (managing states) and deciding how to maintain them (managing processes) (Simberloff 1986). An early attempt to establish criteria for habitat preservation was the so-called hot spot approach advocated by Norman Myers of the World Wildlife Fund (Myers 1988) (chapter 4). In this method, areas are prioritized for preservation according to their rates of species rarity and endemism. The greater the number of rare and endemic species, the higher the area's priority for protection. This straightforward procedure permits the world's areas with the greatest densities of rarity ("hot spots" of endemic and endangered species, to use Myers' terminology) to be identified quickly and with little resort to technical expertise (fig. 8.12)

because a knowledge of species' ranges and population status is sufficient to employ the method. Using this method, consistently identifiable hot spots include such areas as Brazil's Atlantic coast, western Ecuador, the uplands of western Amazonia, the eastern Himalayas, peninsular Malaysia, northwestern Borneo, New Caledonia, and the Philippines (Myers 1988).

Intuitive, conceptually simple, and inherently appealing, the hot spot methodology is not without problems. For one thing, endemism and rarity are not necessarily consistent across taxa. For example, an area that ranks high in the density of rare birds may rank low in its density of rare mammals, and vice versa. An application of the method on British plants and animals found that species-rich areas frequently did not coincide for different taxa (Prendergrast et al. 1993). A second problem is that the degree of rarity also may be inconsistent at different levels of taxonomy. For example, areas with a high density of rare species may be low in densities of rare genera or rare families. A related problem is that areas with a high density of rare or endemic species may actually have low levels of overall species diversity. For example, Stolz and colleagues (1996) assert that areas of high bird diversity, such as the Amazon Basin, are areas of low endemism, with most of the bird species being widespread, abundant, and in no immediate danger. In contrast, many areas of lower overall diversity in the same region, such as the Brazilian cerado, the midmontane forests of the Andes and the Caribbean, and the dry forests of western South America, face much greater and more immediate threats of extinction to particular avian species (Stolz et al. 1996).

In some cases, an area's reputation for high biodiversity may even be exaggerated or illusory. For example, Mares (1992) found that, among neotropical mammals, South American habitats had relatively low diversity compared with other regions and habitats. He argued that if only one macrohabitat could be chosen to preserve the greatest biodiversity, it should be deserts, scrublands, and grasslands (Mares 1992). These problems and criticisms reveal that the hot spot approach, although not without merit, can be misapplied. Even when used with the most complete and accurate data, the hot spot method still requires arbitrary judgments about which groups of species to protect, and at which taxonomic level.

In the United States, one of the most recent and comprehensive efforts in reserve design and conservation planning is the ongoing Gap Analysis Program, now often referred to simply as GAP or GAP analysis. Originally developed in the early 1990s by J. M. Scott and others at the University of Idaho, GAP uses satellite imagery and geographic information system (GIS) technology to make computer-generated maps of the distribution of dominant vegetation and then relates this distribution to both the distribution of vertebrates and to existing conservation reserves (Scott et al. 1993). Specifically, GAP's sequential tasks are to: (1) map existing vegetation to the level of dominant or codominant species (from satellite imagery); (2) map predicted distributions of native vertebrate species (primarily using museum and agency collection records in conjunction with existing general range maps of each species); (3) map public land ownership and private conservation lands; (4) show the current network of conservation lands (the combined distribution of public lands and private conservation lands); (5) compare the

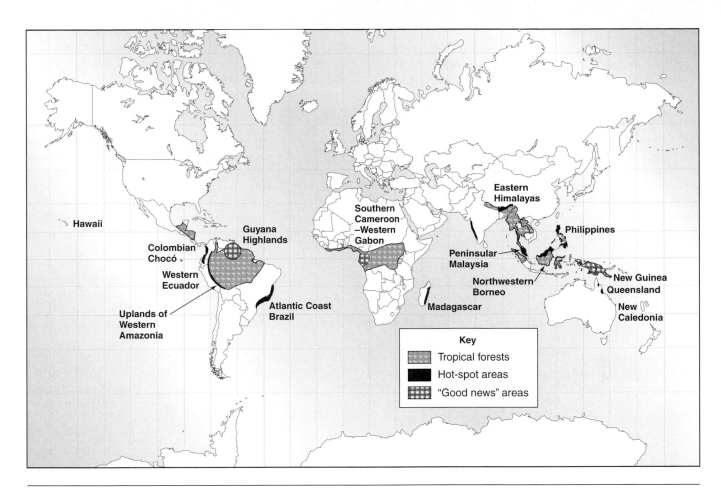

Figure 8.12

Examples of "hot spots" of biodiversity in the tropics. Concentrations of high biodiversity and endemism suggest priority areas for habitat conservation. "Good news" areas refer to regions where species loss due to deforestation is less than anticipated.

After Myers (1988).

distributions of native vertebrates and vegetation communities with the network of conservation lands; and (6) from this comparison, provide an objective basis of information for local, state, and national options in managing biological resources.

Overall, GAP is designed to provide an information base that will enable managers and planners to make the best and most efficient use of land in establishing reserves by showing where conservation efforts should be focused to achieve maximum biodiversity or maximum protection for endangered species. Currently the GAP analysis, sponsored and coordinated by the Biological Resources Division of the U.S. Geological Survey, involves 445 contributing organizations and 44 states. With the aid of its GIS applications and technology, GAP can display relationships of interest at varying cartographic scales, display distributions of individual species or entire suites of species, and overlay maps of species distributions with different jurisdictions of land ownership and land management objectives.

For all its extensive support and technological power, GAP is not without its critics. The assessments that can be made from GAP analysis are only as good as the resolution of the maps used to produce them. Some have argued that the inventory and monitoring of biotic resources on a national scale requires a sampling universe composed of broad, landscape-scale assessments. GAP, in contrast, relies mainly on small-scale, low-resolution assessments of biodiversity, and so only infrequently identifies areas that should become candidates for biological protection (Short and Hestbeck 1995).

Determining Appropriate Reserve Size

The question of how large a reserve should be to achieve its desired objectives remains a critical issue in conservation biology, closely related to habitat and landscape preservation. If the purpose of the reserve is to ensure the persistence of a particular species or group of species, then the needs of such species become the operative criteria for reserve size. If time, expertise, and money permit, the species within the proposed refuge, or at least those considered of highest priority for conservation, should be subjected to a population viability analysis (PVA) (chapter 7). Such an analysis will reveal the average minimum viable populations (MVPs) for important refuge species. These MVPs, divided by estimated population densities, should yield the quotient of minimal area for population persistence (Simberloff 1988).

With minimal area estimates in mind, Soulé and Simberloff (1986) offer a three-step approach to the practical problem of estimating the optimal size of a conservation reserve or collection of reserves. The first step is to identify species whose disappearance would significantly decrease the value of the reserve or its diversity, including threatened and endangered species, species of high public visibility or aesthetic appeal, species whose abundance provides an index of habitat quality (indicator species), and species that create habitat or perform functions that enhance populations of other species (keystone species). Second, determine the minimum number of individuals in a population needed to guarantee a high probability of survival for these species. Third, using known densities, estimate the area needed to sustain this minimum population.

The dilemmas of these choices emerge in practical ways when planners can, or must, choose between making the reserve a single large area or several smaller reserves of approximately equal area. Determining the best choice depends upon several factors. First, knowing that smaller reserves will support smaller populations, one must determine the difference between extinction probabilities associated with large and small populations of the most important species. Large differences favor a single large reserve, while small differences argue for several small reserves (chapter 7). A second key question is: How many populations will a series of small refuges preserve, because, presumably, each small reserve will not contain every species that might be present in the single large reserve? Third, what is the correlation in the year-to-year fluctuation of the environments of the populations in the proposed small reserves? If environmental variation is independent (uncorrelated), multiple reserves, even if small, provide a measure of protection against chance environmental disturbances or catastrophes that could reduce or exterminate a single population. On the other hand, if the separate reserves have a high degree of environmental correlation, they confer no such advantage. If a series of small reserves is considered, what is the probability of recolonization of one of the reserves following a local extinction? If the probability for recolonization is high (individuals disperse well and frequently from one reserve to another), there is an advantage to having multiple reserves. But if the probability of recolonization is low, then local extinctions may be more permanent events, and a single large reserve holds the advantage of a larger population less prone to extinction (Simberloff 1988).

Determining reserve size also must involve a consideration of factors associated with the reserve's habitat heterogeneity and patch dynamics. To effectively preserve both species and habitats, reserves must be larger than the size of the largest disturbance-created patch ("minimum dynamic area"), even including the rarest kinds of disturbance-created patches, and they should contain separate minimum dynamic areas of each habitat type. Additionally, the reserve should include internal sources for repopulating local extinctions, and should include disturbance-created patches of different ages (Pickett and Thompson 1978).

For all the abstract intricacies that theories and problems of reserve design may generate, conservation biologists must never stray far from the realities of managing reserves in the real world. In that world, the most common size for existing reserves is 10 to 30 km^2 (Bolton 1997). Whether or not such reserves are theoretically adequate for conserving biodiversity, they represent valuable resources and opportunities for the species they contain. With the benefit of intensive, careful, and intelligent management, such small reserves can, and indeed must, make an important contribution to worldwide conservation efforts.

Size alone is never a sufficient criterion for reserve design. Species may use only one or a few types of habitats, and these habitats invariably follow a patchy distribution. Even in its preferred habitats, a species will not occur in every patch. In an area with relatively high levels of biodiversity, the problem is intensified. The richer the fauna, the lower will be the average population density of each species and the narrower the range of habitats occupied by each species. Because of such patchy distributions, it may not suffice to set aside a single large piece of habitat to preserve diversity. This distributional patchiness argues against attempting to preserve all local biodiversity in a single large reserve (Diamond 1980).

Even very large reserves may not preserve diversity of habitat, especially if the area selected is insensitive to the realities of patchy habitat distribution. In Idaho, Wright, MacCracken, and Hall (1995) examined four areas that had been proposed by various interest groups as future national parks (fig. 8.13). Although large (averaging 220,000 ha each) the four proposed areas added little to the number of vegetation types already under protection, and none met even the modest goal of protecting 10% of the vegetation types in the ecoregion (Wright, MacCracken, and Hall 1995). Even if the proposals were expanded, they did little to increase the preservation of habitat and vegetation types. In fact, the proposed areas could have been made considerably smaller and still reached the same goals in this regard (Wright, MacCracken, and Hall 1995). Smaller reserves would have had negative effects on the population levels of some species, but they would not have reduced the variety of habitats protected. Alternatively, the current proposals could have increased the number of habitats preserved with relatively few hectares added to their land area. This study demonstrates yet another way in which refuge design and establishment, if uninformed by ecological data, may have little real value in preserving biodiversity and associated habitat diversity.

Connecting Isolated Reserves and Fragmented Habitats with Corridors

Theoretical Basis of Corridors

A quality of populations that increases their ability to persist is connectivity. **Connectivity** is a parameter of landscape function that measures the processes by which subpopulations of organisms are interconnected into a functional demographic unit (Merriam 1984) (fig. 8.14). A related, but distinct, concept is **connectedness.** Connectedness refers to a physical linkage between landscape elements. Although connectedness is necessary for populations to achieve connectivity, connectedness does not guarantee connectivity. Connectivity is achieved only if individuals actually move between connected units. If they do not, connectedness does nothing to increase the persistence of the population.

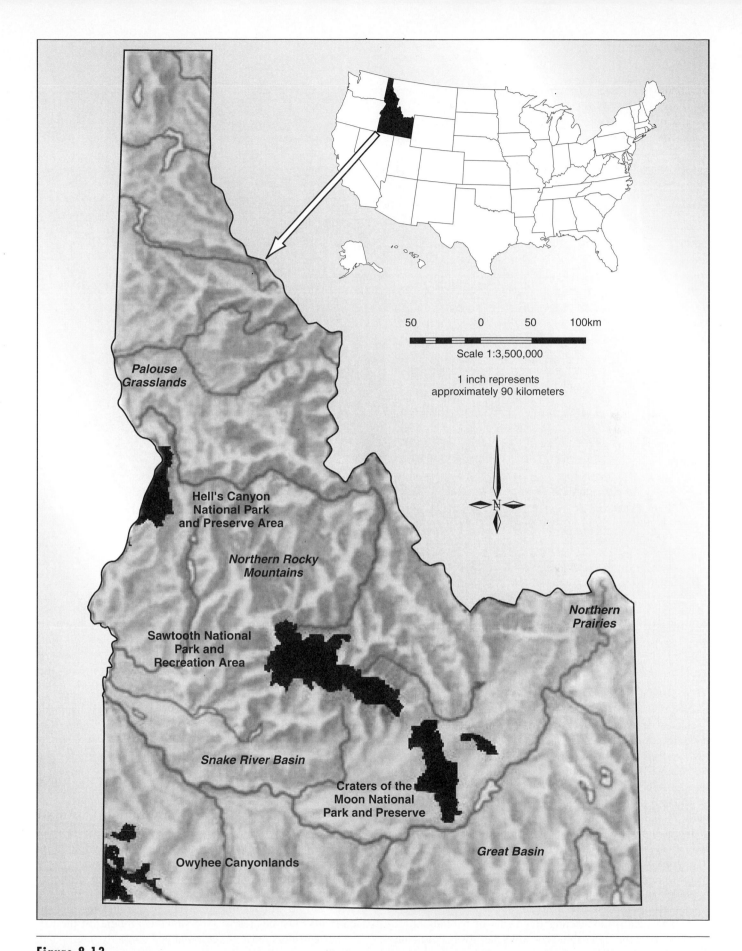

Figure 8.13

Four areas in Idaho proposed for protection as future national parks. Although large, the areas add little to the number of vegetation types already under protection and protect less than 10% of the vegetation types in the ecoregion.

After Wright, MacCracken, and Hall (1995).

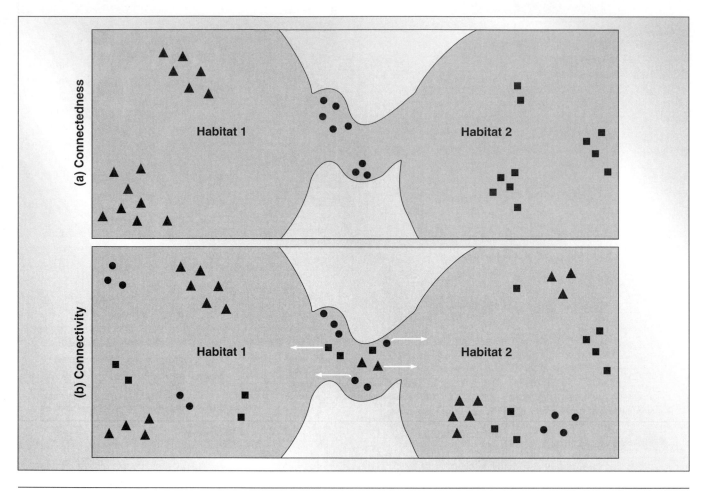

Figure 8.14

A conceptual illustration of the difference between connectedness and connectivity. Corridors create connectedness, but only movement of individuals from patch to patch creates connectivity. In (a), each species (represented by different symbols) occupies a distinct area and there is no movement through the corridor; however, one species (●) uses the corridor as habitat. In (b), individuals of all three species move through the corridor, creating connectivity between habitats.

Habitat corridors are often proposed as a means of linking isolated reserves or fragmented patches of habitat. A **corridor** can be defined as *a linear landscape element that provides for movement between habitat patches, but not necessarily reproduction* (Rosenberg, Noon, and Meslow 1997). The concept of connecting fragments of habitat with corridors originally arose out of predictions of reserve design associated with the theory of island biogeography (MacArthur and Wilson 1967) (chapter 5). The rationale was that corridors would provide avenues of movement and immigration between patches of habitat, leading to increased stability of populations in remaining fragments. The subsequently developed theories of metapopulations attributed much of the dynamics of population persistence to characteristics of the surrounding landscape, and to the ability of individuals in different subpopulations to move from patch to patch.

The idea of connecting fragments of habitats or isolated reserves through habitat corridors has always been intuitively appealing and plausible. Unfortunately, the appeal and plausibility of corridors have outpaced definitive studies and rigorous interpretation to assess the real roles and values of corridors. Most advocates of corridors assert that: (1) animals will make use of corridors to move from one fragment of habitat to another; (2) fragments of habitat connected by corridors will have higher levels of species diversity than isolated fragments; (3) populations in fragments connected by corridors will have lower rates of species replacement (species turnover, a measure of community stability) than isolated populations; and (4) populations connected by corridors will have higher population densities, higher rates of growth, and longer persistence times than populations in similar, but isolated habitats. These assertions are not necessarily well supported by data. To understand and apply the conservation potential of corridors more effectively, three steps are needed: (1) The concept, form, and function of corridors must be clearly understood. (2) Studies of the values of corridors must be reviewed and correctly interpreted. (3) The principles these studies provide must be applied to problems of habitat fragmentation and reserve design.

Conceptual and Definitional Problems of Corridors

The first problem that confronts a systematic study and understanding of corridors is that multiple definitions of corridors exist. Simberloff and associates (1992) noted six different definitions or conceptions of the term published in recent literature. (1) Some habitats constitute corridors and thus deserve protection as a distinct habitat. An example would be vegetation adjoining waterways (riparian habitat). These habitats are unique and valuable in their own right, whether or not they provide opportunity for organisms to move to other habitats. (2) Greenbelts and buffers are called corridors, but their function is amelioration of the human environment, and their value is primarily one of aesthetic amenities. (3) Biogeographic land bridges are called corridors, but their relevance to maintaining viable populations in refuges is obscure. (4) A series of discrete refuges, such as for migratory waterfowl, is called a corridor, although this would be best understood as a complement to a corridor. (5) Underpasses and tunnels to allow animals to cross highways are called corridors. (6) Strips of land that facilitate movement between larger (and usually similar) habitats are called corridors. The last definition is the usual and most appropriate.

Beyond simple definitions, confusion increases when one confronts the vagueness regarding the concept of corridors. For example, landscape ecologists Forman and Godron (1986) define corridors as *narrow strips of land which differ from the matrix . . . on either side.* In contrast, Merriam (1984) defines corridors as *narrow patches of vegetation that facilitate movement among habitat patches, thereby preventing isolation of populations.* The first definition stresses the form of the corridor and the second stresses its function. Such confusion, which is common in the literature of conservation biology, not only makes the definition of a corridor ambiguous, but often reduces the clarity of experimental design and interpretation of results.

Additional confusion stems from the problem of equating a corridor with the more general concept of a **linear habitat**. Linear habitats are long, narrow strips of habitats that provide resources for plant and animal populations. Many species of animals not only move through linear habitats, but may reside in them permanently. Linear habitats have two primary functions for vertebrates. They provide many species with resources and space to live (a habitat function), and they may provide pathways for movement from place to place (a corridor function). But linear habitats and corridors are not necessarily the same (i.e., form does not always equal function). Sometimes corridors provide a habitat function and sometimes they do not. Sometimes so-called corridors provide only a habitat function, and do not facilitate movement between larger patches. Clearing up the confusion is essential if one wants to study the role of corridors or use them effectively in conservation.

Experimental Studies of Corridors

Confusion about the definition of corridors often leads directly to errors of interpretation associated with experimental studies of corridors. In fact, relatively few studies have demonstrated that corridors actually provide connectivity; that is, increase the rate of successful movement of animals between patches. Instead, most studies of corridors have documented the presence of animals in the corridor (a linear patch of habitat lying between larger patches) and have inferred that the linear patches were acting as corridors (Rosenberg, Noon, and Meslow 1997). For example, Merriam and Lanoue (1990), in a study of the use of fencerows by the white-footed mouse, demonstrate that most of the movements and 90% of all activity by radio-tagged mice occurred in fencerows. The authors consistently refer to the fencerows as "corridors" rather than "habitat," implying that the mice were moving through the fencerows to other habitats, but they present no data to substantiate this assumption.

Such errors of interpretation are not unusual. Many published studies of corridors have failed to demonstrate that corridors are, in fact, used for movement between habitat fragments. Of 36 research papers in the book *The Role of Corridors* (Saunders and Hobbs 1991), only 5 presented new data on animal movement in and through corridors. Of these 5, only 1 gathered data on movement between habitat patches without corridors. Only 3 of the 5 studies concluded that corridors had a role in conserving a particular taxon (Simberloff et al. 1992).

Some field studies also have provided evidence that corridors can encourage higher population densities, growth rates, and persistence times in fragmented habitats. For example, populations of white-footed mice in fragmented woodlots had higher rates of growth in woodlots connected by corridors than those in isolated woodlots of the same size (Fahrig and Merriam 1985). Among bird communities in boreal forests, fragmented sites that were connected to one another had lower rates of species turnover (i.e., more stable composition of bird communities) than isolated fragments, but isolated and connected fragments did not differ in species richness, species diversity, or species abundance (Schmiegelow, Machtans, and Hannon 1997). However, neither of these studies demonstrated animal movement through habitat corridors.

A few studies have provided indirect evidence of movement through corridors to adjoining habitats. For example, Haas (1995) demonstrated that American robins (*Turdus migratorius*) were more likely to disperse between shelterwood habitat fragments connected by wooded corridors than to adjacent, equally distant, shelterwood stands that were not connected by wooded corridors. Dmowski and Kozakiewicz (1990) demonstrated that nonlittoral species of birds living in a pine forest near a lake were more likely to visit and use the littoral zones of the lake connected to the forest by a shrub corridor than to visit equally distant littoral zones separated from the forest by an open meadow (fig. 8.15). These authors also were able to demonstrate that movements of these nonlittoral (primarily forest) species of birds to the littoral zone occurred mainly in and through the shrub corridor.

One of the few studies that was successful in combining an examination of both the habitat and movement functions of corridors was the previously cited work of Bennett (1990) on populations of small mammals in Australia. Investigating eight species of small mammals, six native and two introduced, that used both forest fragments and corridors

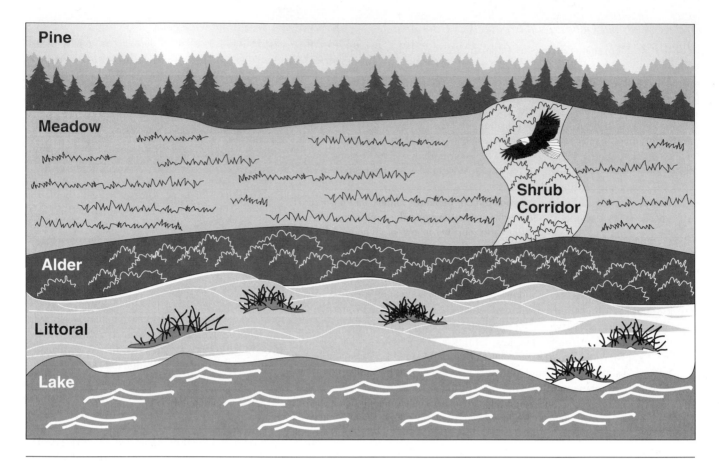

Pine

Meadow

Shrub Corridor

Alder

Littoral

Lake

Figure 8.15

Corridors between a forest and lakeshore in landscape context (model aerial view). Littoral zones were more likely to be used by nonlittoral species of birds (nonshorebirds) if the littoral zone was connected to the forest by a shrub corridor.

After Dmowski and Kozakiewicz (1990).

between the fragments, Bennett documented a number of key corridor functions and values. First, all species occurred in corridors, but use of corridors differed by species. Some species, such as the bush rat, were primarily breeding residents in corridors, using the corridors as permanent habitat. Other species, such as the introduced house mouse and the native brown antechinus (*Antechinus stuarti*), were primarily transients who used the corridors for movements between fragments of forest habitat. Several species had approximately equal proportions of transient and resident individuals in corridors, and several individuals of different species had home areas that encompassed portions of both the corridor and an adjacent forest habitat. Even species and individuals using the corridors as transients followed a variety of movement patterns, some traversing the corridor from one forest fragment to another in a single movement, whereas others moved between adjacent patches via an extended series of gradual movements through the corridor (Bennett 1990).

Bennett's (1990) study illuminated a number of key points by demonstrating that corridors are used in dispersal and as permanent habitat, and that individual species differ in their use of and movement patterns within corridors.

Although all documented dispersal movements of the examined species occurred through corridors, no examination of dispersal without corridors was sought (Simberloff et al. 1992). This omission undermines the credibility of Bennett's study in unequivocally demonstrating that corridors increase rates of dispersal or enhance patterns of dispersal in these species.

Despite the imperfections and limitations of these and other studies of corridor function, there is growing evidence that corridors can add value to linear landscape elements. But such studies also demonstrate that three important qualifiers are needed when discussing the value of corridors. First, the role of corridors in both animal movement and population persistence is complex. Simple generalizations, however intuitive and appealing, are unhelpful, probably erroneous, and likely to lead to bad decisions in landscape management. Second, studies of corridors must include appropriate controls in the experimental design to determine if dispersal, survival, and movement of animals through and within corridors are really any different than without corridors. Third, the value of corridors differs according to species and landscape scale. The above studies and others have demonstrated some specific values of corridors in connecting closely spaced fragments of habitats (typically less than 5 km between adjacent fragments). As yet, no studies have demonstrated the value of corridors at large scales (e.g., 5 to 50 km between habitat fragments), yet this scale is more appropriate to conserving the processes and diversity of functioning ecosystems and is often the scale at which corridors are proposed to link large reserve complexes (Hunter, Jacobson, and Webb 1988). Clearly, much important research still lies ahead.

A review of available research on corridors permits certain generalizations. First, given a choice between habitat and nonhabitat, individual animals are more likely to select movement pathways that include components of their habitat, particularly if these individuals are moving within the boundaries of their home range. However, at the level of local populations, most individuals are not adverse to moving through nonhabitat. They may compensate by moving through nonhabitat more quickly than they would through habitat, but they are not necessarily "cut off" from other fragments of habitat simply because there is nonhabitat in between (Rosenberg, Noon, and Meslow 1997). Second, the probability of successful movement through the corridor depends on three things: the probability of finding the corridor, the probability that the animal selects the corridor as the route for movement, and the probability of the animal successfully traversing the corridor. Thus, the corridor values may be dynamic, reflecting variation in the degree of contrast between patch and matrix environments. Given a choice of moving through the matrix or the corridor, the animal's response is conditional, not obligate (Rosenberg, Noon, and Meslow 1997).

Potential Disadvantages of Corridors

Just as there can be benefits of linking habitat fragments through corridors, so there can be negative effects (Simberloff et al. 1992). Isolation of a population or its habitat is not always bad, particularly if the population or habitat is especially susceptible to environmental variation and disturbance. Because corridors connect habitat fragments, they have the potential to "import" negative effects from one fragment to another. Parasites or agents of disease may move more easily through corridors from one habitat fragment to another than they could between isolated fragments. Nonnative species also may use corridors as routes of dispersal from fragment to fragment. Mortality, especially predation, may be heavier on individuals residing or dispersing within corridors because predators indigenous to other, surrounding habitats may be able to penetrate corridors more easily than remaining habitat fragments. This effect could make corridors population sinks that could have negative effects on the overall sizes of populations in the region.

Corridors could have indirect negative effects on population processes. The persistence of a metapopulation is enhanced by the fact that population subunits are sometimes nonsynchronous in their population demographics. That is, cyclic patterns of population increase and decrease are independent events in different subunits. Dispersal of individuals from different subunits through corridors can reduce or elim-

inate such independence, making the population subunits more vulnerable to regional environmental variation and its effects on population processes. Thus, an environmental event that affects only low-density populations would eliminate some subunits that had asynchronous demographics, but could reduce or eliminate all populations if subunits were synchronous. Another threat that corridors pose to basic population processes is the problem of habitat quality. If patches of habitat are connected by corridors that are similar in quality to the habitats they connect, then persistence of subpopulations in the connected habitats increases (Fahrig and Merriam 1985). But simulation modeling has demonstrated that if the corridors are of lower-quality habitat, dispersers suffer higher mortality in corridors and overall population levels decline (Henein and Merriam 1990). Such modeling is supported in one field study of mammals in a fragmented tropical forest where 42% of variations in species richness in different forest fragments was explained by the area of the fragment, and an additional 40% was explained by the degree of isolation of the fragments and the quality of habitat in connecting corridors (Laurance 1995). Many factors can contribute to corridors being of lower quality than the habitat patches they connect. Even if vegetation in corridors is identical to connected patches, corridor shape (long and narrow) tends to increase edge and the associated edge effects of increased predation and parasitism on interior species within the corridor.

Finally, corridors represent both economic and opportunity costs. Corridors do not magically appear between habitat fragments. Land for corridors must be purchased and then managed to provide value in conservation. Because time, money, and expertise are never unlimited, acquiring and managing corridors is, in fact, a decision *not* to purchase and manage additional isolated fragments or to purchase habitat to enlarge existing fragments. Thus, corridors represent an opportunity cost in habitat management as well as a potential conservation value. The effectiveness of corridors will vary with species, and with the characteristics of the corridor, especially its complexity of vegetation structure, its length from fragment to fragment relative to the dispersal abilities of the species targeted for conservation, and its width. Unless these factors are considered, the purchase and management of corridors, and the effects they have on linking previously isolated fragments, may be counterproductive to conservation goals.

There is often no more evidence to support the potential dangers and disadvantages of corridors than there is to support their value, and sometimes there is a good deal less. But any consideration of corridor design must make careful consideration of all issues. Most important, unless corridors lead to increased connectivity between population subunits, they offer no real advantage, and they are less beneficial to the population than a direct restoration of lost habitat. Regardless of the value corridors have been shown to have, or may yet prove to have for isolated and fragmented habitats and their associated populations, they are not a final solution to the problems of habitat loss and fragmentation. The landscape conditions and processes that put populations in jeopardy must be corrected (Merriam 1995).

MANAGEMENT OF HABITAT ON NONRESERVE LANDS: BIODIVERSITY AND MULTIPLE USE

Mitigating Habitat Disturbance on Nonreserve Lands

Ecologist Gary E. Belovsky, after making a critical review of estimates of minimum viable population (MVP) sizes for individual species and taxa, noted, "The concept of MVP . . . does not inspire optimism. By comparing the areas required for mammal persistence with the sizes of parks in the world . . . we can estimate the future success of man's conservation efforts if we assume management by 'benign neglect.' The largest mammalian carnivores (10 to 100 kg) can be expected to persist 100 years in 0 to 22% of the current parks but no park is large enough to guarantee persistence for 1,000 years. For larger mammals (>50 kg) to persist in evolutionary time (10^5–10^6 years), regions of 10^6 to 10^9 km^2 are required, assuming that major climatic variations do not occur. . . . Obviously, man must re-evaluate his conservation efforts and management schemes in light of these results if the larger mammals are to be preserved in nature without human intervention" (Belovsky 1987:50).

Conservation of species and their habitats cannot be achieved entirely through conservation reserves, no matter how well designed. Conservation also must incorporate careful management of human activity on nonreserve lands. It must manage and restructure human activity and land use that permits coexistence with other life, especially larger animals, over large areas where people also live and work. In this section, we examine some general considerations of habitat management on nonreserve lands and how to reduce the effects of humans and their activities on resident plants and animals. These considerations anticipate some aspects of ecological restoration that we will examine in further detail in chapter 11 (restoration ecology), but must be considered here, as well, to understand how to conserve habitat, and associated resident populations, subjected to human activity. We illustrate these considerations with a case history.

In a remote area of south central Montana near the Absaroka-Beartooth Wilderness Area, a local population of elk, known as the Line Creek herd, winter in sagebrush (*Artemisia tridentata*) covered foothills on land that is a mosaic of private, state, and federal ownership. The herd's name comes from a stream flowing through its range that closely follows the boundary line between Montana and Wyoming for several miles. On this range the U.S. Forest Service granted a lease for oil ex-

ploration and development to the Phillips Petroleum Company. Phillips planned to conduct directional drilling to extract oil from the high-elevation Line Creek Plateau directly to the west. Originally Phillips had planned to drill on the plateau, but had withdrawn that application because of the plateau's highly sensitive alpine vegetation. However, on the lower-elevation winter range, other state and federal agencies, private conservation organizations, and local residents voiced concern that the drilling activity would displace the elk population during its most stressful season. Because of the proximity of the herd to Wyoming, there was particular worry that the animals might move permanently south across the state line, depriving Montana of considerable revenue from license and hunting fees.

Phillips agreed to a comprehensive series of rules to attempt to minimize the adverse effects of the drilling on the elk. The company used the drilling site only in summer and early fall, after elk had moved to other ranges at higher elevations. Access to the site was limited to a single road used only by vehicles of the company and the Forest Service and closed after drilling was completed. After each drilling season ended, the drilling rig was lowered from a vertical to a horizontal position to make it impossible to see from adjacent drainages, thus minimizing its visual impact. When all drilling activity had ended, the site was reseeded to native grasses.

The Line Creek elk had been monitored through radio telemetry prior to drilling, and such monitoring was continued during the drilling period and after the drilling had been completed. Of interest to all, and surprising to many, the population did not leave its original range. The size, shape, and position of the herd's home range remained essentially unchanged during and after the drilling compared to the pre-drilling period. There was evidence that the elk avoided the site of the drilling rig itself and that they increased their use of forested (cover) habitats when they were near the well. However, they did not change their overall patterns of use in the range, even when such patterns were measured at very fine spatial scales (Van Dyke and Klein 1996) (fig. 8.16).

The story of the Line Creek elk is one example of an attempt at disturbance **mitigation**. The verb *mitigate* is defined as "to cause to become less harsh or hostile, to make less severe or painful, to alleviate." Although legal protection and environmental policies sometimes provide sufficient authority to end all disturbance that might cause harm to a critically endangered population, nonendangered populations may be no less valuable in their own right, and may be even more important to functioning ecosystems. In the context of managing habitats at regional levels, it is unrealistic to think that all disturbances can be eliminated, but it is not unrealistic to believe that many disturbances can be mitigated.

The Line Creek example provides several illustrations of certain principles of mitigation. The first is the principle of **timing limitations**. If animals use habitats on a seasonal basis, sites potentially can be used for some human activities when the animals are absent, without harm to the animals. A second is the principle of **limited access**. If roads must be built for human activity, detrimental effects can be reduced if their use and access is limited to essential activities, and if such roads are closed after the activities are completed. A third is the principle of **visual minimization**, exemplified in the practice of lowering the oil drilling rig to a horizontal position. Reducing the distance at which an object associated with disturbance can be seen by

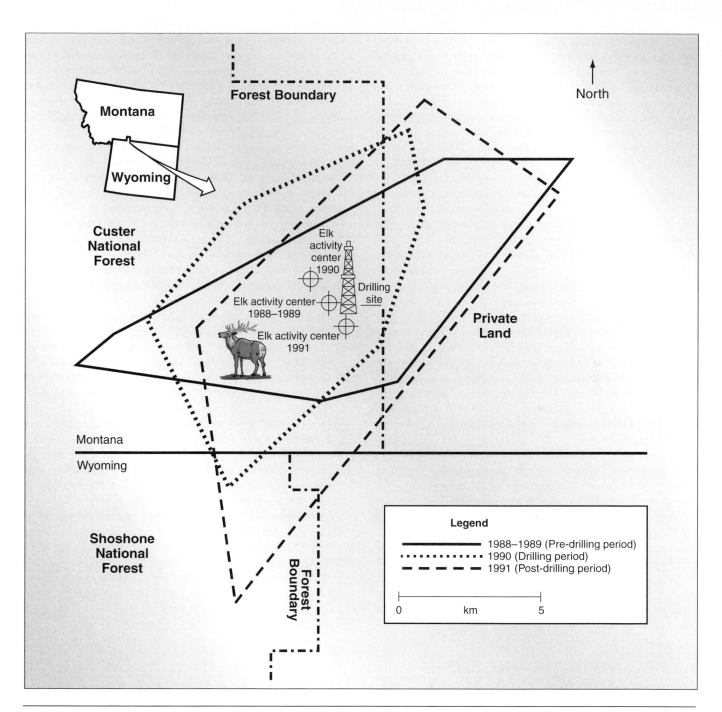

Figure 8.16

Seasonal ranges and home range activity centers of an elk population before, during, and after oil-drilling activity on the population's winter range. After drilling began, elk avoided the drilling site and altered patterns of habitat use. However, with a variety of mitigation measures in place (see text) elk did not significantly alter the size or boundaries of their surrounding range.

Adapted from Van Dyke and Klein (1996).

animals reduces the animals' response to the object, and makes more of the area available for the animals' use. A fourth principle, to be developed in more detail in chapter 11, is that of **reclamation**. By immediately reseeding the disturbed site, the oil company began successional activity on the actual drilling site quickly, and acted in a way that could more rapidly return the site to its previously undisturbed state. A fifth principle is that of **no surface occupancy**. In this case, Phillips demon-

strated this principle by choosing not to occupy a more fragile alpine site and drilling instead on a lower-elevation site where vegetation was less sensitive and more resilient to disturbance. Through improving technology and techniques like directional drilling, mining and drilling on remote sites of low-quality habitat can extract resources from beneath sites covered by high-quality habitat. This practice can reduce or eliminate detrimental effects of these activities on sensitive populations.

Historically, conservationists have viewed only complete preservation of habitats and populations as the "real" goal of conservation efforts, and any concession to resource extraction or human use of an area has been seen as an undesirable compromise. Although complete elimination of human presence and disturbance may be necessary for some species at all times, and for many species in some times and places, the idea that it is the only acceptable conservation strategy is unrealistic, unnecessary, and hypocritical. The strategy of conservation through elimination of the human presence is unrealistic because human activity will increasingly disturb natural environments in the modern world. Human activities and their disturbances can never be completely eliminated, but they can be managed. Successful management of such activities can create productive habitat for plants and animals even when human activity is not completely excluded. The complete exclusion of human activity is often unnecessary because animals do not use all elements of a landscape equally. Even the home range of an animal, regardless of how it is measured, contains large portions of space where the animal has never been. This is why appropriate conservation of habitat rests on an understanding of the habitat preferences of species and their natural history. The elimination of human activity and presence can be hypocritical because it can blind us to the fact that human needs are met through environmental resources. Even efforts to study and conserve species requires the extraction and use of resources from our environment.

Strategies of mitigation force us to admit these needs and then, having made that admission, determine ways to extract needed resources with minimal harm and disturbance to other living creatures and their physical environments. The exclusion of human disturbance from habitats is a strategy that often is possible only in the most pristine environments. Mitigation is a strategy that can be practiced with positive effects even on highly impacted and degraded environments. To be effective in a world in which the human presence will continue to grow, conservation biologists must employ mitigation strategies to maintain viable habitats and healthy landscapes in the midst of human presence and disturbance.

Preserving Habitats Through Landscape Planning: The Multiple-Use Module

Site-specific habitat conservation and disturbance mitigation have long histories and admirable records of accomplishment, but also possess serious weaknesses. Conservation of habitat at individual sites, whether on multiple-use lands or reserves, places the management of habitats at a site-specific level. However, the creation of habitat, as well as the loss of habitat (through disturbance, succession, climate change, and human activity) occurs through forces that operate at a landscape level. Unless habitats can be managed and conserved through the same processes that create and destroy them, habitat management and conservation cannot be successful.

One approach that attempts to solve this problem and to integrate the conservation of habitat on both managed and reserve lands is the **multiple-use module** (MUM) (Harris 1984; Noss and Harris 1986; Noss 1987). First proposed by Harris

(1984) for protecting old-growth stands of Douglas fir (*Pseudotsuga menziesii*) in managed forests of western Oregon, the MUM system envisions a multiple-use landscape unit (module) consisting of a fully protected core area surrounded by concentric zones of natural areas used in progressively more intense fashion for recreation and commodity production (fig. 8.17). Individual modules are then connected through landscape corridors to one another to allow movement of animals among modules while allowing for additional uses of resources within the landscape. The buffer zones that surround the core areas of each MUM are intended to (1) insulate the most sensitive elements in the preserve in a core area free from intensive land use and human disturbance; (2) provide supplementary habitat for animals inhabiting the core area, hence increasing the effective size of the reserve; and (3) provide for a variety of human use and activity in the landscape while minimizing conflicts with other species (Noss 1987). Ultimately, the goal of the MUM approach is to combine applications of corridors and multiple-use zoning strategies to create an integrated network of clustered reserves (Noss 1987).

Successful creation and management of an integrated system of MUMs would require additional applications of ecological restoration, including road closures, reintroduction of extirpated species, removal of human structures and settlements, and restoration of natural disturbances and hydrologic processes (Noss 1987, chapter 11). Although greater human presence and activity, with more extractive use of resources, would be allowed in buffer zones surrounding the core area, management of buffer zones is intended to be consistent with the overall goals of conserving biodiversity and protecting particular species of interest (Noss and Harris 1986). Noss and Harris envision the buffer zones as areas where habitat might be deliberately manipulated to benefit wildlife, but core areas would be considered "inviolate preserves." Proposals for using the MUM concept to integrate conservation landscape planning have been proposed for the north Florida–south Georgia region, for the entire state of Florida (Noss 1987), and for the Ohio Valley in southeastern Ohio (Noss 1987). To date, none of these proposals has been fully implemented.

Potential Problems of the MUM Concept: An Assessment

In many ways, the MUM concept is essentially a large-scale application of the corridor concepts previously discussed and, as such, is subject to many of the same criticisms and concerns. For example, the MUM concept depends heavily on the assumption that extensive corridors connecting nodes of biodiversity (core areas) over long distances would effectively facilitate movement of individual animals among such nodes. The MUM concept also assumes that such movement will increase biodiversity in existing reserves and enhance population persistence. Both assumptions are suspect. Most species, with the exception of some of the larger vertebrates, have small dispersal distances relative to proposed distances between adjacent reserves (Simberloff et al. 1992). In addition, long-distance dispersal through corridors is one of the least-investigated aspects of corridor use. The potential risks

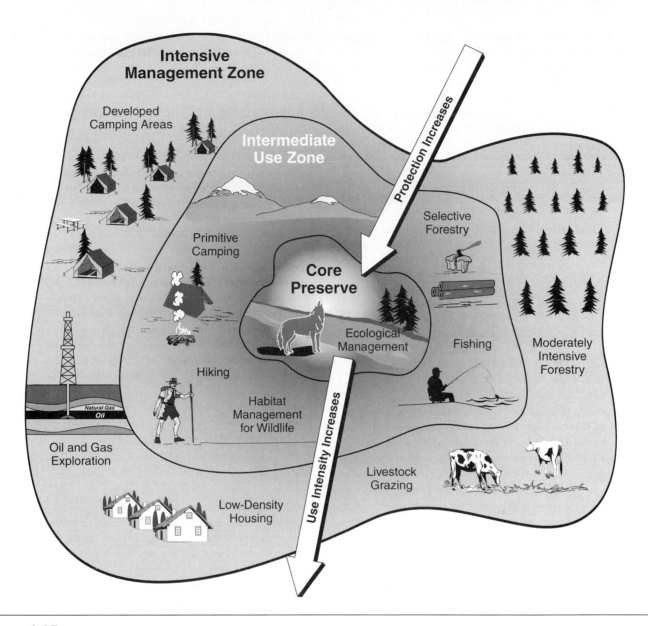

Figure 8.17

The MUM for habitat and landscape conservation uses a combination of corridors and varying intensities of land use to increase available habitat for species.

Adapted from Noss and Harris (1986).

and costs of such long-distance movement for dispersing organisms raises doubts about the efficacy of the MUM approach.

Another potential weakness of the MUM concept is its conceptual dichotomy between "preservation" (in the core area) and "management" (in buffer zones). The MUM concept, in the words of Noss and Harris (1986) "resolves the traditional conflict between 'hands on' conservation/management and 'hands-off' preservation" by physically separating management and preservation into different areas of the module. Although the MUM concept does accept that intensive

management will be necessary to preserve biodiversity, including habitat diversity, even in core areas (Noss 1987), it still tends to assume that management is defined by its activity and its sphere of influence, rather than by its objective. If, instead, conservation management were defined by its objectives rather than by its activity or its location in the MUM, a wider array of management strategies might be used to protect biodiversity and address human use of the same landscape in both core areas and buffer zones.

The MUM approach also suffers because its emphasis on large landscape units invariably affects human settlement, recreation, and resource extraction in the managed area, but its goals are expressed only in terms of conservation values. This weakness contributed to the lack of support for the Ohio Valley proposal by both the principal government conservation agency (the Ohio Department of Natural Resources) and by local chapters of private environmental groups in Ohio (The Sierra Club, The Audubon Society, and The Ohio Nature Conservancy) (Noss 1987). Until the MUM approach becomes more effective at incorporating objectives of human use, settlement, and recreation into its management strategies, and until it can better define management actions by their objectives instead of by their physical location, it is unlikely that the MUM strategy will be consistently applied or consistently supported on a regional basis.

Synthesis

Habitat conservation is the foundation of population conservation. Yet conservation biologists still stuggle, often in an apparent state of conceptual confusion, to understand the separate and interactive effects of habitat loss, habitat fragmentation, and habitat isolation on plant and animal populations, to communicate these effects coherently to the public, and to translate their understanding into meaningful policies that effectively manage habitat and the processes that shape it. The importance of habitat is recognized in concepts like that of "critical habitat," written into the Endangered Species Act (chapter 2), yet often ignored in many other conservation laws. Despite the vital role of habitat in all aspects of conservation, few major federal conservation laws in the United States and few international conventions are written specifically to protect or restore habitat. Likewise, most conservation organizations present their missions in terms of species, not landscapes. A notable exception, The Nature Conservancy, offers a commendable example of how to articulate an alternative conservation vision. The Nature Conservancy states that its mission is "to preserve the plants, animals, and natural communities that represent the diversity of life on Earth by protecting the lands and waters they need to survive," a complement to its memorable organizational slogan, "Saving the last great places."

More conservation organizations, more conservation laws, and more conservation efforts must come to The Nature Conservancy's level of awareness of the importance of habitat conservation if species are to survive. Perhaps humans have been reticent to embrace habitat conservation as enthusiastically as species conservation because habitat conservation is fundamentally an issue of *land use*. Natural habitats cannot be conserved in zoos. They can persist only if people choose to use and occupy less land, and to use the land they occupy in less destructive and degrading ways. To commit to habitat conservation is to commit to radical and pervasive changes in practices of human residence and land use.

Some habitat conservation can and is being achieved through the establishment of refuges and reserves. But not all habitats can be saved by setting them apart from the human presence. Humans must become more intentional in determining what critical habitat components are, and then find ways to occupy and use landscapes in ways such that their presence and activity do not destroy these components, but preserve them, at least in part, in place and function. If refuge design is not complemented by the presence of sufficient functional habitats in the surrounding and vastly greater array of nonreserve lands, no population, especially of a large, mobile species, has any real hope of long-term persistence.

Advantages and Disadvantages of Conservation Corridors—A Directed Discussion

Reading assignment: Simberloff, D., J. A. Farr, J. Cox, and D. Mehlman. 1992. Movement corridors: conservation bargains or poor investments? *Conservation Biology* 6:493–504.

Questions

1. List some of the landscape features that have been called "corridors." What constitutes a corridor? Why is a clear definition essential for determining the value and function of corridors?

2. Simberloff et al. (1992) note that many published studies purportedly supporting the conservation value of corridors lacked data to make such a conclusion or were flawed in their design. Once published, however, these studies themselves can be cited as documenting the value of corridors. Suggest ways to address the problem of continuing, but unsupported, claims about the use of corridors for conservation.

3. Evaluate the relative economic and opportunity costs of a corridor (in this case, the opportunity cost is the cost of *not* acquiring alternative habitat because available funds were used to purchase corridor habitat). What studies, analyses, or assessments ought to be done before a decision is made to purchase corridor

habitat rather than land that could serve as a new, independent reserve?

4. In your opinion, does an emphasis on corridor acquisition focus attention on populations in reserves or on nonreserve lands? Is such an emphasis conducive or detrimental to conservation? If conducive, how can corridor management become more widely employed in conservation? If detrimental, suggests ways of creating less emphasis on corridors and more on alternative management approaches.

Learning Online

Visit our webpage at www.mhhe.com/conservation for case studies, animations, practice quiz questions, and additional readings to help you understand the material in this chapter. You'll also find active links to the following topics:

Conservation and Management of
   Habitats and Species
Biomes and Environmental Habitats
Freshwater Habitats
Marine Habitats

Estuaries and Wetlands
Tropical Rainforests
Temperate Forests
Tundra
Taiga

Savannah
Grasslands
Deserts

Literature Cited

Alverson, W. S., D. M. Waller, and S. L. Solheim. 1988. Forests too deer: edge effects in northern Wisconsin. *Conservation Biology* 2:348–58.

Belovsky, G. E. 1987. Extinction models and mammalian persistence. In *Viable populations for conservation,* ed. M. E. Soulé, 35–57. Cambridge, England: Cambridge University Press.

Bennett, A. F. 1987. Conservation of mammals within a fragmented forest environment: the contributions of insular biogeography and autecology. In *Nature conservation: the role of remnants of native vegetation,* eds. D. A. Saunders, G. W. Arnold, A. A. Burbidge, and A. J. M. Hopkins, 41–52. Sydney, Australia: Surrey Beatty.

Bennett, A. F. 1990. Habitat corridors and the conservation of small mammals in a fragmented forest environment. *Landscape Ecology* 4:109–22.

Bolen, E. G., and W. L. Robinson. 1995. *Wildlife ecology and management.* Englewood Cliffs, N.J.: Prentice Hall.

Bolton, M., ed. 1997. *Conservation and the use of wildlife resources.* New York: Chapman and Hall.

Botkin, D. B. 1990. *Discordant harmonies: a new ecology for the twenty-first century.* New York: Oxford University Press.

Connell, J. H., and R. O. Slatyer. 1977. Mechanisms of succession in natural communities and their role in community stability and organization. *American Naturalist* 111:1119–44.

Danielson, B. J. 1991. Communities in a landscape: the influence of habitat heterogeneity on the interactions between species. *American Naturalist* 138:1105–20.

Diamond, J. M. 1980. Patchy distributions in tropical birds. In *Conservation biology: an evolutionary-ecological perspective,* eds. M. E. Soulé and B. A. Wilcox, 135–50. Sunderland, Mass.: Sinauer Associates.

Diffendorfer, J. E., N. A. Slade, M. S. Gaines, and R. D. Holt. 1995. Population dynamics of small mammals in fragmented and continuous old-field habitat. In *Landscape approaches in mammalian ecology and conservation,* ed. W. Z. Lidicker, Jr., 175–99. Minneapolis: University of Minnesota Press.

Dmowski, K., and M. Kozakiewicz. 1990. Influence of a shrub corridor on movements of passerine birds to a lake littoral zone. *Landscape Ecology* 4:99–108.

Fahrig, L. 1992. Relative importance of spatial and temporal scales in a patchy environment. *Theoretical Population Biology* 41:300–14.

Fahrig, L. 1997. Relative effects of habitat loss and fragmentation on population extinction. *Journal of Wildlife Management* 61:603–10.

Fahrig, L., and G. Merriam. 1985. Habitat patch connectivity and population survival. *Ecology* 66:1762–68.

Fahrig, L, and G. Merriam. 1995. Conservation of fragmented populations. In *Readings from* Conservation Biology: *the landscape perspective,* ed. D. Ehrenfeld, 16–25. Cambridge, Mass.: Blackwell Science.

Forman, R. T. T., and M. Godron. 1986. *Landscape ecology.* New York: Wiley.

Gaines, M. S., J. E. Diffendorfer, R. H. Tamarin, and T. S. Whittam. 1997. The effect of habitat fragmentation on the genetic structure of small mammal populations. *Journal of Heredity* 88:294–304.

Haas, C. A. 1995. Dispersal and use of corridors by birds in wooded patches on an agricultural landscape. *Conservation Biology* 9:845–54.

Hanski, I., A. Moilanen, and M. Gyllenberg. 1996. Minimum viable metapopulation size. *American Naturalist* 147:527–41.

Hansson, L. 1991. Dispersal and connectivity in metapopulations. *Biological Journal of the Linnaean Society* 42:89–103.

Harris, L. D. 1984. *The fragmented forest: island biogeography theory and the preservation of biotic diversity.* Chicago: University of Chicago Press.

Harris, L. D. 1988. Edge effects and conservation of biotic diversity. *Conservation Biology* 2:330–32.

Henein, K., and G. Merriam. 1990. The elements of connectivity where corridor quality is variable. *Landscape Ecology* 4:157–70.

Hunter, M. L., G. L. Jacobson, and T. Webb III. 1988. Paleoecology and the coarse-filter approach to maintaining biological diversity. *Conservation Biology* 2:375–85.

Ivlev, V. S. 1961. *Experimental ecology of the feeding of fishes.* New Haven, Conn.: Yale University Press.

Johnson, G., and D. J. Leopold. 1998. Habitat management for the eastern massasauga in a central New York peatland. *The Journal of Wildlife Management* 62:84–97.

Kareiva, P., D. Skelly, and M. Ruckelshaus. 1997. Reevaluating the use of models to predict the consequences of habitat loss and fragmentation. In *The ecological basis of conservation: heterogeneity, ecosystems, and biodiversity,* eds. S. T. A. Pickett, R. S. Ostfeld, M. Shachak, and G. E. Likens, 163–81. New York: Chapman and Hall.

Klein, B. C. 1989. Effects of forest fragmentation on dung and carrion beetle communities in Central Amazonia. *Ecology* 70:1715–25.

Lamberson, R. H., K. McKelvey, and B. R. Noon. 1992. Reserve design for territorial species: the effects of patch size and spacing on the viability of the northern spotted owl. *Conservation Biology* 6:505–12.

Lande, R. 1987. Extinction thresholds in demographic models of territorial populations. *American Naturalist* 130:624–35.

Laudenslayer, W. F., Jr. 1986. Summary: predicting effects of habitat patchiness and fragmentation. In *Wildlife 2000: modeling habitat relationships of terrestrial vertebrates,* eds. J. Verner, M. L. Morrison, and C. J. Ralph, 331–33. Madison: University of Wisconsin Press.

Laurance, W. F. 1995. Extinction and survival of rainforest mammals in a fragmented tropical landscape. In *Landscape approaches in mammalian ecology and conservation,* ed. W. Z. Lidicker, Jr., 46–63. Minneapolis: University of Minnesota Press.

Lawton, J. H., S. Nee, A. J. Letcher, and P. H. Harvey. 1994. Animal distributions: patterns and processes. In *Large-scale ecology and conservation biology,* eds. P. J. Edwards, R. M. May, and N. R. Webb, 41–58. Boston, Mass.: Blackwell.

Leopold, A. 1933. *Game management.* New York: Scribner.

Lidicker, W. Z., Jr. 1995. The landscape concept: something old, something new. In *Landscape approaches in mammalian ecology and conservation,* ed. W. Z. Lidicker, Jr., 3–19. Minneapolis: University of Minnesota Press.

Lindstrom, E. 1989. Food limitation and social regulation in a red fox population. *Holarctic Ecology* 12:70–79.

MacArthur, R. H., and E. O. Wilson. 1967. *The theory of island biogeography.* Princeton, N.J.: Princeton University Press.

Marcum, C. L., and D. O. Loftsgaarden. 1980. A nonmapping technique for studying habitat preferences. *The Journal of Wildlife Management* 44:963–68.

Mares, M. A. 1992. Neotropical mammals and the myth of Amazonian biodiversity. *Science* 255:976–79.

Meentemeyer, V., and E. O. Box. 1987. Scale effects in landscape studies. In *Landscape heterogeneity and disturbance,* ed. M. G. Turner, 16–34. New York: Springer-Verlag.

Merriam, G. 1984. Connectivity: a fundamental ecological characteristic of landscape pattern. In *Proceedings on the first international seminar on methodology in landscape ecological resources and planning,* eds. J. Brandt and P. Agger, 5–15. Roskilde, Denmark: International Association for Landscape Ecology.

Merriam, G. 1995. Movement in spatially divided populations: responses to landscape structure. In *Landscape approaches in mammalian ecology and conservation,* ed. W. Z. Lidicker, Jr., 64–77. Minneapolis: University of Minnesota Press.

Merriam, G., and A. Lanoue. 1990. Corridor use by small mammals: field measurement for three experimental types of *Peromyscus leucopus. Landscape Ecology* 4:123–31.

Myers, N. 1988. Threatened biotas: "hotspots" in tropical forests. *Environmentalist* 8:187–208.

Neu, C. W., C. R. Byers, J. M. Peek. 1974. A technique for analysis of utilization-availability data. *The Journal of Wildlife Management* 38:541–45.

Noss, R. F. 1987. Protecting natural areas in fragmented landscapes. *Natural Areas Journal* 7:2–13.

Noss, R. F., and A. Cooperrider. 1994. *Saving nature's legacy: protecting and restoring biodiversity.* Washington, D.C.: Defenders of Wildlife and Island Press.

Noss, R. F., and L. D. Harris. 1986. Nodes, networks, and MUMs: preserving diversity at all scales. *Environmental Management* 19:299–309.

Noss, R. F., M. A. O'Connell, and D. D. Murphy. 1997. *The science of conservation planning.* Washington, D.C.: Island Press.

Oksanen, T., and M. Schneider. 1995. The influence of habitat heterogeneity on predator-prey dynamics. In *Landscape approaches in mammalian ecology and conservation,* ed. W. Z. Lidicker, Jr., 122–50. Minneapolis: University of Minnesota Press.

Ostfeld, R. S., S. T. A. Pickett, M. Shachak, and G. E. Likens. 1997. Defining the scientific issues. In *The ecological basis of conservation: heterogeneity, ecosystems, and biodiversity,* eds. S. T. A. Pickett, R. S. Ostfeld, M. Shackak, and G. Likens, 3–10. New York: Chapman and Hall.

Parker, C. A. 1989. Soil biota and plants in the rehabilitation of degraded agricultural soils. In *Animals in primary succession, the role of fauna in reclaimed lands,* ed. J. D. Meier, 423–38. Cambridge, England: Cambridge University Press.

Pickett, S. T. A., and J. N. Thompson. 1978. Patch dynamics and the design of nature reserves. *Biological Conservation* 13:27–37.

Prendergrast, J. R., R. M. Quinn, J. H. Lawton, B. C. Eversham, and N. W. Gibbons. 1993. Rare species, the coincidence of diversity hotspots and conservation strategies. *Nature* 365:335–37.

Rabinowitz, D., S. Cairns, and T. Dillon. 1986. Seven forms of rarity and their frequency in the flora of the British Isles. In *Conservation biology: the science of scarcity and diversity,* ed. M. E. Soulé, 182–204. Sunderland, Mass.: Sinauer Associates.

Robinson, S. K., F. R. Thompson, T. M. Donovan, D. R. Whitehead, and J. Faaborg. 1995. Regional forest fragmentation and the nesting success of migratory birds. *Science* 267:1987–90.

Rosenberg, D. K., B. R. Noon, and E. C. Meslow. 1997. Biological corridors: form, function, and efficacy. *BioScience* 47:677–87.

Saunders, D. A., G. W. Arnold, A. A. Burbidge, and A. J. M. Hopkins. 1987. The role remnants of native vegetation in nature conservation: future directions. In *Nature conservation: the role of remnants of native vegetation,* eds. D. A. Saunders, G. W. Arnold, A. A. Burbidge, and A. J. M. Hopkins, 387–92. Chipping Norton, Australia: Surrey Beatty.

Saunders, D. A., and R. J. Hobbs, eds. 1991. *The role of corridors.* Chipping Norton, Australia: Surrey Beatty.

Saunders, D. A., R. J. Hobbs, and C. R. Margules. 1995. Biological consequences of ecosystem fragmentation: a review. In *Readings from* Conservation Biology: *the landscape perspective,* ed. D. Ehrenfeld, 1–15. Cambridge, Mass.: Blackwell Science.

Schmiegelow, F. K. A., C. S. Machtans, and S. J. Hannon. 1997. Are boreal birds resilient to forest fragmentation? An experimental study of short-term community responses. *Ecology* 78:1914–32.

Scott, J. M., F. Davis, B. Csuti, R. Noss, B. Butterfield, C. Groves, H. Anderson, S. Caicco, F. D'Erchia, T. C. Edwards, Jr., J. Ulliman, and R. G. Wright. 1993. Gap analysis: a geographic approach to protection of biological diversity. *Wildlife Monographs* 123.

Short, H. L., and J. B. Hestbeck. 1995. National biotic resource inventories and GAP analysis. *BioScience* 45:535–39.

Simberloff, D. 1986. Design of nature reserves. In *Wildlife conservation evaluation,* ed. M. B. Usher, 315–17. London, England: Chapman and Hall.

Simberloff, D. 1988. The contribution of population and community biology to conservation science. *Annual Review of Ecological Systems* 19:473–511.

Simberloff, D., J. A. Farr, J. Cox, and D. W. Mehlman. 1992. Movement corridors: conservation bargains or poor investments? *Conservation Biology* 6:493–504.

Soulé, M. E. 1991. Land use planning and wildlife maintenance: guidelines for conserving wildlife in an urban landscape. *Journal of the American Planning Association* 57:313–23.

Soulé, M. E., and D. Simberloff. 1986. What do genetics and ecology tell us about the design of nature reserves? *Biological Conservation* 35:19–40.

Soulé, M. E., B. A. Wilcox, and C. Holtby. 1979. Benign neglect: a model of faunal collapse in the game reserves of east Africa. *Biological Conservation* 15:259–72.

Stolz, D. F., J. W. Fitzpatrick, T. A. Parker III, and D. K. Moskovits. 1996. *Neotropical birds: ecology and conservation.* Chicago: University of Chicago Press.

Stone, E. C., and R. B. Vasey. 1968. Preservation of coastal redwood on alluvial flats. *Science* 159:157–61.

Tarlock, A. D. 1994. The nonequilibrium paradigm in ecology and the partial unraveling of environmental law. *Loyola of Los Angeles Law Review* 27:1121–44.

Temple, S. A. 1986. Predicting impacts of habitat fragmentation on forest birds: a comparison of two models. In *Wildlife 2000: modeling habitat relationships of terrestrial vertebrates,* eds. J. Verner, M. L. Morrison, and C. J. Ralph, 301–4. Madison: University of Wisconsin Press.

Temple, S. A., and J. R. Cary. 1988. Modeling dynamics of habitat-interior bird populations in fragmented landscapes. *Conservation Biology* 2:340–47.

Thomas, J. W., C. Maser, and J. E. Rodiek. 1979. In *Wildlife habitats in managed forest: the Blue Mountains of Oregon and Washington,* ed. J. W. Thomas, 48–59. Agriculture Handbook Number 553. Washington, D.C.: U.S. Forest Service.

Van Dyke, F., and W. C. Klein. 1996. Response of elk to installation of oil wells. *Journal of Mammalogy* 77:1028–41.

Van Dyke, F., B. L. Probert, and G. M. Van Beek. 1995. Seasonal habitat characteristics of moose in south-central Montana. *Alces* 31:15–26.

Wegner, J., and G. Merriam. 1990. Use of spatial elements in a farmland mosaic by a woodland rodent. *Biological Conservation* 54:263–76.

Wilcove, D. S. 1985. Nest predation in forest tracts and the decline of migratory songbirds. *Ecology* 66:1211–14.

Wilcove, D. S., M. J. Bean, R. Bonnie, and M. McMillan. 1996. *Rebuilding the ark: toward a more effective Endangered Species Act for private land.* Washington, D.C.: Environmental Defense Fund.

Wilcove, D. S., C. H. McLelland, and A. P. Dobson. 1986. Habitat fragmentation in the temperate zone. In *Conservation biology: the science of scarcity and diversity,* ed. M. E. Soulé, 237–56. Sunderland, Mass.: Sinauer Associates.

Wilcox, B. A. 1980. Insular ecology and conservation. In *Conservation biology: an evolutionary-ecological perspective,* eds. M. E. Soulé and B. A. Wilcox, 95–117. Sunderland, Mass.: Sinauer Associates.

Wright, R. G., J. G. MacCracken, and J. Hall. 1995. An ecological evaluation of proposed new conservation areas in Idaho: evaluating proposed Idaho national parks. In *Readings from* Conservation Biology: *the landscape perspective,* ed. D. Ehrenfeld, 224–33. Cambridge, Mass.: Blackwell Science.

Yahner, R. H. 1988. Changes in wildlife communities near edges. *Conservation Biology* 2:333–39.

Yeo, J. J., and J. M. Peek. 1992. Habitat selection by female Sitka black-tailed deer in logged forests of southeastern Alaska. *The Journal of Wildlife Management* 53:210–13.

CHAPTER 9

... a river is more than an amenity. It is a treasure. It offers a necessity of life that must be rationed among those who have power over it.

—Oliver Wendell Holmes, 1931

The Conservation of Aquatic Ecosystems

In this chapter, you will learn about:

1 the ecological properties of aquatic habitats.

2 types of freshwater and marine ecosystems.

3 conservation problems, goals, and management strategies associated with freshwater and marine ecosystems.

The majority of literature in conservation biology, as in the rest of biology, focuses on terrestrial environments and the creatures that inhabit them. Yet 71% of the globe is covered by oceans, not land. Freshwater and marine environments may hold the majority of all earth's species, but because they are foreign and threatening to humans, and more difficult to investigate, they are not as well studied as terrestrial sites. The resources of aquatic habitats are vast and essential, but even those we use most frequently are mysterious to us. We often receive them, or exploit them, without truly understanding their value or the processes that sustain them.

Aquatic creatures are important in the diets of most people throughout the world, yet we have no real idea of the sizes of the populations that support most fisheries, especially in the oceans. Our lakes, rivers, and seas are repositories for all types and quantities of human and industrial refuse, yet we do not know the capacity of these systems to hold such waste, or its effects on ecosystem functions. The majority of our commercial fisheries are fully exploited, overexploited, or in decline, yet we go on taking. The oceans of the world have long been one of the principal regulators of its climate, yet, as human activity alters the climate, we are only beginning to appreciate how such changes will affect ocean systems. Subsurface ocean topography and structure determine the abundance of many creatures on which humans depend for food, yet humans alter ocean topography and structure in harvesting food and other resources. Such alterations leave us with less food to harvest and fewer resources to use. Because aquatic habitats are so different from terrestrial ones, the problems associated with

their conservation also are vastly different. Their uniqueness deserves special attention.

HETEROGENEITY IN AQUATIC ENVIRONMENTS

Terrestrial habitats are defined and described primarily by their dominant vegetation and landform or topographic characteristics. In aquatic habitats, there may be little or no vegetation structure, and physical structures may play only a minor role in habitat characteristics. Unlike air in a terrestrial environment, which is rarely considered in a study of "habitat," water in an aquatic environment *is* the habitat. That is, the physical and chemical properties of water often must be considered as the dominant habitat features. Differences in light, temperature, oxygen, and nutrients make aquatic habitats highly heterogeneous, resulting in heterogeneity in the abundance and distribution of aquatic populations.

Heterogeneity in water, as in all environments, may be spatial or temporal. In an aquatic environment, spatial heterogeneity is created less by underwater topography and the structure of physical objects than by depth. Changes in levels of light, temperature, oxygen, pressure, and nutrient availability, among other variables, create different conditions and associated niches at different depths, with corresponding changes in the communities of organisms. The greater the range of depths,

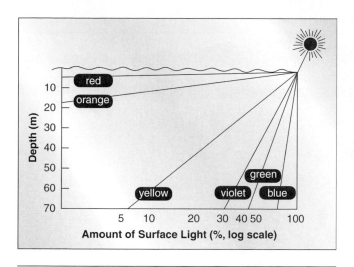

Figure 9.1

The absorption of different wavelengths of light by pure water as a function of depth. Red light is absorbed in the first few meters, whereas blue light can penetrate to depths greater than 70 m. An increase in the turbidity of the water, typical of real aquatic systems, causes light at each wavelength to be absorbed at lesser depths.

Adapted from Brönmark and Hansson (1998).

Figure 9.2

The deep sea angler fish (*Melanocetus johnsoni*), a photoluminescent marine creature of extreme depths, uses a luminescent "lure" at the end of the long stalk on its head to attract prey.

the greater the diversity of conditions, and the greater the diversity of organisms.

Light, for example, travels through air with comparatively little change, but is radically altered when transmitted through water. Depth of light penetration depends on wavelength, and different wavelengths ("colors") are selectively removed with increasing depth (fig. 9.1). Most of the red light entering water is absorbed in the first few meters, depending on temperature, density, turbulence, turbidity, and the presence of organisms. Blue light, in contrast, can penetrate to more than 70 m (Brönmark and Hansson 1998). Below 100 m, there is little or no light. In marine environments, many creatures that live constantly at these depths are bioluminescent or have "lights" on various body parts to serve them in navigation, reproduction, and predation (fig. 9.2).

Marine Habitats

Heterogeneity may be created in oceans by global forces of atmospheric heating, cooling, and air movement. Zonal winds, driven by the uneven heating of the earth's surface, produce circulation patterns that divide oceans into distinct hydrographic regions differing markedly in horizontal and vertical motion, nutrient fluxes, and seasonality (Barry and Dayton 1991). These patterns are determined by the earth's rotational forces and lead to the formation of prevailing currents at global and regional scales (fig. 9.3).

Energy from mesoscale features, tidal and regional wind forcing, and interactions of wave movement with bottom topography or continental margins generate upwellings, fronts, eddy currents, continental shelf waves, filaments, internal waves, and other phenomena that affect the productivity and distribution of aquatic organisms. Whereas surface waves move material over the sea surface, internal waves are generated by tidal and wind stress forces. If the water column is stratified, internal waves will travel horizontally along the layer of ocean water that exhibits the most rapid change in density. The orbital motion of water associated with these internal waves causes the formation of zones of convergence and divergence at the ocean's surface, altering the distribution of populations of zooplankton and phytoplankton (Barry and Dayton 1991).

Undersea topography is also important in creating heterogeneity. Water moving along the ocean's surface is deflected as it encounters continental shelves and continents, forming cells of water called **gyres**, which define the provinces of animals in a marine community. Below the surface, the interaction of currents with undersea pinnacles and seamounts generates intensified currents and attracts filter-feeding organisms at higher densities than at similar depths over flatter subsurface terrain. Plankton communities also increase in density and diversity near these same features.

In marine habitats, the actions of living organisms are among the most important determinants of habitat characteristics and availability. For example, in benthic communities, tube-building species make major modifications in ocean floor habitats that may enhance or restrict the abundance and diversity of other species. Tube builders bind sediment in place and thus stabilize the seafloor. They also change concentrations of key nutrients and so create unique communities associated with their activities (fig. 9.4) (Barry and Dayton 1991). Kelp also make major modifications in habitat, growing to huge sizes (up to 700 feet for a single individual) and occurring at high densities in kelp beds and even kelp "forests." A large kelp bed creates a drag on coastal currents, causing the currents to sweep around the bed. Internal waves,

Global Ocean Circulation Pattern

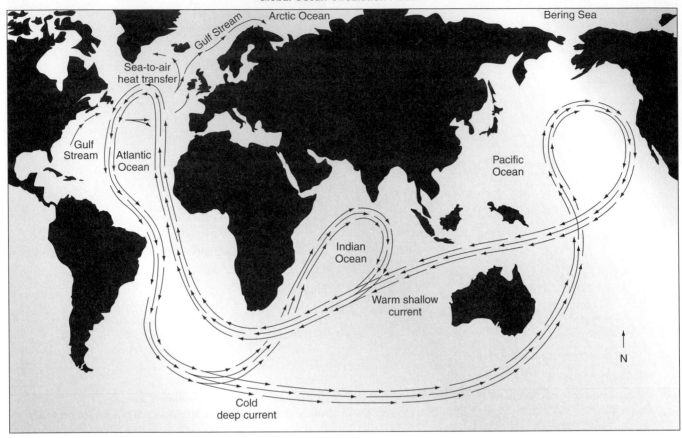

Figure 9.3

Global circulation patterns of marine currents established by prevailing winds and the earth's rotation.

After Broecker (1991).

Figure 9.4

A community of tubeworms (phylum Polychaeta) living in sediment on the floor of the Atlantic Ocean. Tubeworms and other sediment-dwelling species may, through differential use of sediment and sediment nutrients, create microhabitats with unique nutrients and communities of organisms.

transporting pelagic larvae and nekton (free-swimming animals) shoreward, are slowed and disrupted by kelp beds, causing them to lose energy and deposit organisms at disproportionately high numbers at the leading edge of a bed. As a result, fish congregate around the edges of kelp beds where prey organisms accumulate.

Lotic Systems

In streams (technically, *lotic* [flowing water] environments), characteristics of the watershed's climate, topography, vegetative cover, soil, land-use patterns, and bedrock are the primary determinants of habitat characteristics. Streams and rivers have been described as "nothing else than functional parts of higher units: of landscapes . . . of geosynergies . . . or biogeocoenoses . . . , on whose existence they depend" (Sioli 1975:276). Climate, specifically temperature and precipitation, determines the amount of moisture input to the watershed and its rate of evapotranspiration. Topography determines the rate and erosive force of surface runoff entering the stream channel. Vegetative cover determines the rate of erosion of soil in the watershed, and thus the rate at which sediment enters the stream. Soil type also affects the rate of

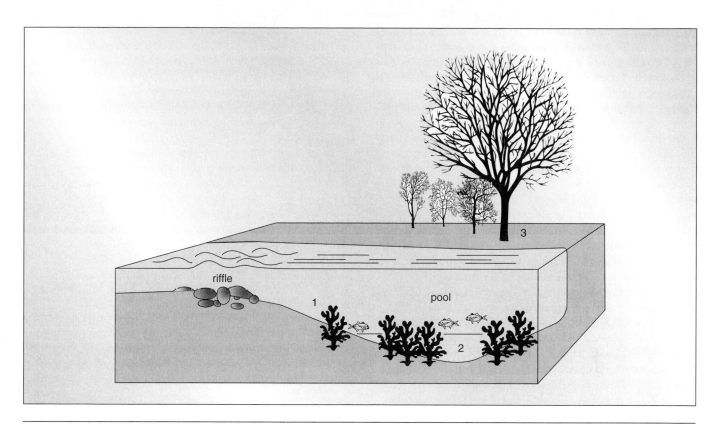

Figure 9.5

Principal categories of material input into a stream. Within the stream flow, the majority of the organic matter is typically dissolved. Organic matter enters the stream ecosystem from autochthonous (stream-derived) and allochthonous (derived from the terrestrial system) sources and may exit via respiration of aquatic organisms. Autochthonous material (1 and 2) may be from sources such as algae and aquatic plants, or from organisms living in stream sediments. Allochthonous substances (3) originate from sources such as throughfall and litter fall.

erosion and the chemical characteristics, turbidity, and bottom features of the stream. Land-use patterns influence rates of erosion and types of material entering the stream through surface runoff and other sources. Characteristics of a stream's watershed also determine input of coarse particulate matter, fine particulate matter, and dissolved organic matter (fig. 9.5).

Within a stream, two dominant habitat features are **riffles** and **pools** (fig. 9.5). These are particularly prominent where water flows at rates of 50 cm/sec or more (fast-water streams). Riffles are sites of primary production in streams where *periphyton,* a community of organisms composed of diatoms, blue-green and green algae, and various aquatic mosses, dominate. Periphyton is ephemeral in nature, constantly being transported through the stream channel by flowing water. Thus, the quantity and composition of the periphyton community within the riffles changes temporally and seasonally, especially in streams in temperate latitudes.

Above and below riffles are pools, catchbasins in which stream topopgraphy permits the velocity of water to decrease. For a given volume of water, pools will form if the stream gradient decreases, the volume of the stream channel increases, or both. Velocity is a measure of energy, so, as the velocity of water decreases, so does its ability to do work. In pools, the ability of the current to carry its load declines, and heavier material is deposited, so pools are sites of biomass accumulation and decomposition.

Larger consumers, such as fish, may congregate in pools at the edge of the upstream riffle because this is often the location where food items appear in highest quantity.

Substrate type is another important determinant of heterogeneity in lotic habitats. Sand and silt substrates generally are the poorest habitat for stream organisms because they offer few attachment sites for periphyton or larger organisms. Bedrock substrate is more solid, but exposes an organism directly to the full force of the stream's flow; few organisms can afford the constant expenditure of energy against a swift current. Gravel, rubble, and boulder bottoms generally characterize the most productive stream habitats because they (1) provide large surface areas and many points of attachment for periphyton, (2) provide cover and refuge for organisms of varying sizes, and (3) divert the force of the stream's current from organisms positioned behind pieces of gravel, rubble, and boulders, thus allowing them to conserve energy.

Lentic Systems

Heterogeneity in lake habitats (**lentic** environments) is little affected by internal currents (flow rates in lakes may be extremely slow) and more by the effect of prevailing surface winds. The two most common effects of wind activity on circulation in lakes are seen in **Langmuir rotations** or **cells** and **seiches.** Langmuir rotations are established by prevailing winds and extend vertically throughout the water column, and are most often observed as "foamlines" at the downwelling boundary of the cell. The circulation of water in such cells is an important mechanism for transporting matter and energy vertically in the lake's water column, affecting the quality of aquatic habitat from the surface to the bottom. Perhaps even more important are external and internal *seiches*. Defined as oscillations of a body of water around points or nodes, seiches are recurring, rocking water movement patterns within a lake basin generated by prevailing winds (external seiches) or by differences in density of water in different layers (internal seiches). In external seiches, prevailing winds "push" water to the far side of the lake, causing a buildup of water volume on the lake's leeward side. When the wind recedes, the displaced water flows back to its original position, initiating a rocking motion of the water in the lake's basin.

Temporal and spatial heterogeneity in aquatic environments are most strongly determined by seasonal variation in temperature, but other factors also change seasonally. Water density changes with temperature, and in temperate lakes, density differences create the most dramatic examples of temporal heterogeneity in aquatic environments. Mobile aquatic creatures respond to changes in temperature and density in the same ways that terrestrial organisms respond to changes in vegetative cover in a landscape. Organisms optimize foraging, reproduction, and growth rates by moving to different strata on seasonal and daily schedules. For example, juvenile sculpins (a bottom dwelling fish, family Cottidae) feed in benthic environments during the day, but move to the warmer waters of the epilimnion during the night where they digest their food more quickly and experience increased growth rates (Neverman and Wurtsbaugh 1994).

Wetlands

Wetlands have been defined as *lands transitional between terrestrial and aquatic systems where the water table is at or near the surface or the land is covered by shallow water* (Cowardin et al. 1979). Wetlands make disproportionately large contributions to global biodiversity and primary productivity. Wetlands often harbor high numbers of endangered species, game species, and other economically important species. But because many wetland areas are transitory or ephemeral in nature, both their definition and their dynamics make it difficult to estimate the exact extent of wetlands in the world today. Not surprisingly then, estimates of global wetland area vary from 5.3 million km^2 (Matthews and Fung 1987) to 8.6 million km^2 (Maltby and Turner 1983).

As habitats and ecosystems, wetlands provide services and products far in excess of the approximately 6% of the earth's surface that they cover (Matthews and Fung 1987; Gosselink and Maltby 1990). Like an economic entity whose value is made up of assets, services, and attributes, wetlands have corresponding values in their structural components, environmental functions, and system organization (Barbier 1995).

Included among the structural components of wetlands are species that form the basis of many sport and commercial fishing industries, hunting, and agriculture (e.g., various forms of domestic and wild rice), as well as wildlife products (especially fur and meat), wood, and water (Barbier 1995). In fact, most game and fur-bearing animals in temperate regions, and many species of game fish spend at least part of their life cycle or at least one season of the year in wetlands, even if they are not "wetland species." A disproportionate number of threatened and endangered species also are wetland dependent.

Wetland functions are varied and essential. For example, because wetlands have the capacity to absorb large inputs of water from surface runoff or upstream sources and yet release relatively little of these inputs downstream in the short term, intact wetland systems protect downstream landscapes, natural systems, and human communities from storm and flood damage. Also because wetlands contain dense, highly productive plant communities, they can absorb large quantities of waste and nutrient runoff. Wetlands also provide opportunities for many types of recreation and water transport. Other wetland services are provided by "constructed wetlands," which are the products of human engineering for specific purposes. Constructed wetlands are created on sites where wetlands did not previously exist or where the original wetlands were destroyed or degraded (Mitsch, Mitsch, and Turner 1994). The most common type of constructed wetland is designed for wastewater treatment (Brix 1994; Kadlec 1994), but wetlands also are constructed for wildlife habitat, for research, and as compensation for loss of natural wetlands under "no net loss of wetlands" statutes in various states and countries.

Wetland organizational characteristics support high levels of primary productivity and biomass. Because water is shallow throughout the wetland environment, all parts of the system can be photosynthetically active, unlike deepwater environments where light cannot penetrate below certain depths. Because water levels vary spatially and temporally (seasonally) within a wetland, the wetland experiences strong moisture and other environmental gradients that support a variety of plant species, including plants of diverse life and growth forms. Such plant diversity creates physical heterogeneity and complexity greater than most terrestrial environments, and often supports a more diverse biotic community.

CONSERVATION CHALLENGES OF FRESHWATER HABITATS

Freshwater habitat quality is degraded worldwide by a small constellation of common factors and processes. The most important threats to freshwater streams and lakes are physical habitat alteration, chemical alteration or pollution of the water, introduction of exotic species (Abell et al. 2000), and, in streams and rivers, alteration of flow regimes, especially as a result of dams (Benke 1990). In the United States, from 1972 to 1982, four times more lake acreage was degraded than was improved in quality (Karr 1991). The only large U.S. stream (more than 1,000 km in length) that has not been severely altered in its flow regimes for hydropower and navigation is the Yellowstone River of Wyoming and Montana (Benke 1990). Along with alteration of flow rates and habitat, pollutants and exotic species have rendered many

$$CaCO_3 + H_2CO_3 \rightleftharpoons CaCO_3 + CO_2 + H_2O \rightleftharpoons Ca(HCO_3)_2 \rightleftharpoons Ca^{2+} + 2HCO_3^-$$

Figure 9.6

Normal buffering reactions that take place in fresh water. Note that a source of carbonate ions, such as calcium carbonate ($CaCO_3$) must be present for hydrogen ions (H^+, from acid sources such as H_2CO_3) to be removed from solution and stabilize the system's pH. Basic carbonate rocks found within the stream or from surrounding terrestrial rocks and soils provide a source of calcium carbonate for the reaction. If carbon dioxide or bicarbonate (HCO_3^-) is removed, calcium carbonate will precipitate and hard water will form.

rivers unfit for most human uses. A recent survey of the majority of U.S. rivers, covering some 643,000 miles of waterways, found that only 56% could support multiple uses such as drinking water, fish and wildlife habitat, recreation, and agriculture. In the 44% of rivers that could not support multiple use, the most important problems were chemical alternation or pollution of the water, specifically sedimentation, nutrient overloading (also known as eutrophication [Farrell 1998]), and acidification. Fifty-six percent of U.S. streams suffer reduced fishery potential because of chemical contamination (Karr 1991).

Eutrophication

The process of **eutrophication** occurs when nutrients, particularly phosphorus, are released into rivers from upstream or surrounding agricultural areas (in the form of fertilizer runoff) or from towns and cities (in the form of human waste) (Brönmark and Hansson 1998). Higher levels of phosphorus trigger a chain of events that begins with a massive increase in the growth of primary producers (usually limited by a scarcity of phosphorus in fresh waters). Periphytic (attached) algae and submersed macrophytes increase in biomass at the beginning of the process, but then decline as phytoplankton and cyanobacteria (blue-green algae) increase in abundance and reduce the amount of light that filters through the water. Dead organisms accumulate as sediment and the bacteria that remove minerals from decaying organic matter extract large amounts of oxygen from the water. Fish kills of some species may follow as oxygen is depleted, but cyprinid fishes (family Cyprinidae, carps and minnows) typically increase in abundance because they can survive in poorly oxygenated waters and are efficient predators of zooplankton, whose numbers increase in the initial stages of eutrophication. As a result of the cyprinid predation, grazing zooplankton decrease. Levels of phytoplankton, the prey of zooplankton, then increase, further increasing the turbidity of the water (Brönmark and Hansson 1998). As eutrophication progresses, the biological community is radically altered, and the lake declines in value as a source of drinking water, recreation, and food.

Acidification

Acidification is a process through which the pH of surface fresh waters, especially lakes, declines (becomes more acidic) because of inputs of acidic precipitation in the form of rain, snow, or fog. Emissions of hydrogen sulfide (H_2S), most commonly associated with the burning of coal to generate electricity, and nitrous oxide (NO), an exhaust waste from cars, can combine with atmospheric water vapor to form weak concentrations of sulfuric acid and nitric acid that fall as precipitation into water bodies or their surrounding drainage areas.

Acidic inputs generally do not affect pH in areas where soil and rock substrates contain significant amounts of calcium carbonate ($CaCO_3$) or other carbonate compounds. These compounds react with water to form carbonate and bicarbonate ions that can buffer a system against acidic inputs (fig. 9.6). In areas without such buffering capacities, however, such as those with granitic substrates or granitic-derived soils, the same inputs of acid precipitation can have disastrous effects on aquatic communities.

The sequence of events begins with a lowering of pH in the aquatic system due to acidic inputs, especially during periods of heavy rain or during spring snowmelt. The most common and immediate effect of lower pH is a lowering or cessation of reproductive effort in many species of fish, amphibians, and aquatic invertebrates, and some species may suffer direct mortality. An indirect, but often more devastating effect of the lower pH is a change in the chemical reactions occurring in the aquatic system, especially those involving metallic ions such as aluminum, lead, or cadmium. Such metals usually remain in solution at higher pH (7 or above), but begin to precipitate out of solution at lower pH levels. Aluminum is deadly to fish because it binds to their gills and impedes respiration (Brönmark and Hansson 1998). When fish populations are reduced in acidified lakes, many invertebrates are released from predation pressure and invertebrate populations may then grow (especially predatory invertebrates) (Brönmark and Hansson 1998). In addition, once aluminum begins to precipitate out of solution, it binds with phosphorus, producing aluminum phosphate. Such a reaction takes phosphorous out of the system and makes it unavailable as a nutrient for organisms.

Habitat Alteration by Nonindigenous Species

Although the problem of nonindigenous species is now a worldwide concern affecting all types of environments (chapter 7), aquatic habitats are especially sensitive to alterations by foreign invaders. Aquatic environments are particularly vulnerable to invasion if a disturbance occurred recently, if predators are absent, or if effective competitors of the invader are absent (Ashton and Mitchell 1989).

Aquatic plants follow a predictable pattern of invasion characterized by four distinct stages: introduction, dispersal, adaptation, and colonization of new habitat (fig. 9.7). Available evidence suggests that, although many aquatic plants can

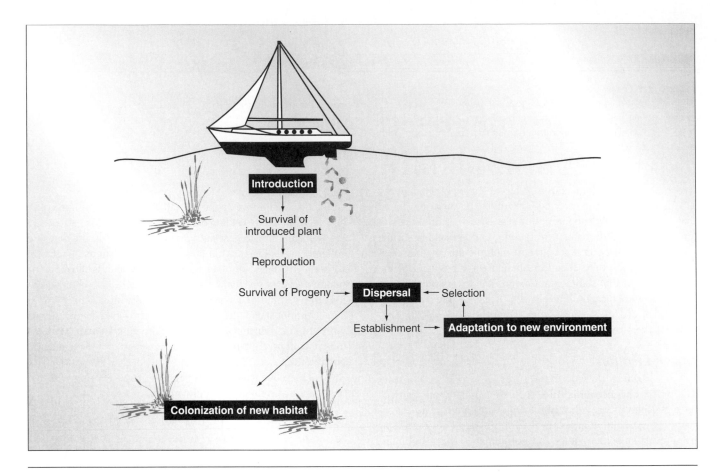

Figure 9.7
Processes and stages associated with the invasion and establishment of an aquatic plant species.

survive unfavorable conditions for extended periods of time, and are readily transported by biological agents such as birds, fish, and insects, almost all invasions of plants that have caused significant habitat alteration and other problems have been human mediated. Assessing the complete history of invasions by aquatic plants, Ashton and Mitchell (1989:117) commented, "we have seen that few aquatic plants are dispersed between unconnected water bodies by natural mechanisms. Indeed, more initial introductions of aquatic plants to new continents have been deliberate in that the introduced species was perceived to have some special attraction and/or intended use for humans. . . . In every case, man has been implicated in their deliberate or accidental introduction to continents outside their native range."

The list of such invaders, including Eurasian water milfoil (*Myriophyllum spicatum*), purple loosestrife (*Lythrum salicaria*), and water hyacinth (*Eichornia crassipes*), and their natural and introduced histories, are beyond the scope of this chapter. But invasive aquatic plants tend to have certain traits in common. First, vegetative reproduction is their common, if not exclusive method of propagation. Second, human activity and transport are their main means of dispersal. Third, all are species capable of extremely rapid reproductive rates. The majority of successful invaders also have free-floating life forms (Ashton and Mitchell 1989). Aquatic invaders, such as purple

loosestrife, may rapidly invade shallow water habitats, especially wetlands, forming dense stands that choke out native species. Water hyacinth, in contrast, is an emergent species that can form dense mats in deeper water, but with the same result. Eurasian water milfoil is a perennial aquatic herb with a slender, elongate floating stem. Often reproducing vegetatively, Eurasian water milfoil can disperse long distances by floating, and may cling to boats or other manufactured structures, facilitating its distribution.

We typically think of animals as being dependent upon plants and their physical environment. But in fact, some aquatic animals may radically alter the physical environment itself or even the properties of the surrounding ecosystem. For example, the zebra mussel (*Dreissena polymorpha*) is a classic invasive species with high reproductive rates, wide environmental tolerances, and large dispersal distances. A native of the Black and Caspian Seas in Eurasia, the mussel spread throughout Europe in the nineteenth century. It had reached Lake St. Clair (shared by the U.S. state of Michigan and the Canadian province of Ontario) by 1986, probably arriving via discharges of ballast water from European ships using the Great Lakes via the St. Lawrence Seaway. Downstream dispersal was rapid. By 1991, the zebra mussel was present in New York's Hudson River and in the St. Lawrence River in Quebec. Upstream dispersal, facilitated by commercial shipping, also occurred. The species reached the Mississippi River by 1992 via the Chicago Sanitary and Ship Canal. From there, the zebra mussel has spread through the Mississippi to Louisiana and has begun to move upstream into the Mississippi's major tributaries (Johnson and Carlton 1996).

The zebra mussel exemplifies the third quality of successful invaders, that of ecological match. Unlike any native species of bivalve in North America, the zebra mussel possesses a tuft of filaments (*byssal threads*) that allows it to attach to any stable surface, including other living creatures. This trait gives the mussel not only access to niches that native clams cannot exploit, but a rapid means of dispersal as well. The zebra mussel produces large numbers of plankton-feeding larvae (*velligers*) that are easily, rapidly, and widely dispersed by prevailing currents.

The zebra mussel is a relatively long-lived species that can actively pump the water it filters while feeding, thus making it better suited than short-lived, passive filter feeders, like insect larvae, to exploit calmer waters associated with lakes and slow-moving rivers. Efficient and voracious filter feeders on phytoplankton, zebra mussels at high densities can exceed the combined filtering activities of all zooplankton (Johnson and Carlton 1996). At densities now found in western Lake Erie, zebra mussels may remove up to 25% of the system's primary production in phytoplankton *daily*! Taken together, these traits make the zebra mussel a unique harvester of planktonic primary productivity. Although such feeding may increase water clarity, it also removes nutrients, energy, and biomass from the pelagic portion of the lake community and shunts it to the benthic zone in the form of increased mussel biomass and feces (Brönmark and Hansson 1998). This shift of matter and energy can radically change community composition and species diversity. The zebra mussel's physiological traits suggest that it is capable of colonizing the fresh waters of most of the United States and southern Canada. The effects of the zebra mussel on native species are of particular concern because 70% of the 297 species of freshwater mussels native to North America are listed as extinct, endangered, threatened, or of special concern (Johnson and Butler 1999). At extremely high densities, the zebra mussel not only displaces similar species from a habitat or substrate, but will even use the bodies of other living creatures *as substrates*. Mollusks and slow-moving crustaceans, such as crayfish (*Orconectes* spp.), are particularly vulnerable, and some individuals die after becoming completely encrusted with zebra mussels (Abell et al. 2000). Economic losses from structural damage and clogging of underwater structures, such as pipes, are estimated in the millions of dollars. Initial infestations may reduce water turbidity because of the enormous amount of water collectively filtered by the population, but the zebra mussel's combination of high reproductive rate and short life span can eventually lead to the accumulation of large numbers of dead mussels that foul the water (Hayes 1998).

Larger nonindigenous species also can radically alter aquatic habitat. The carp (*Cyprinus carpio*), a bottom-feeding fish native to Europe, was brought to the United States in the 1830s and was the subject of massive, intentional introductions to freshwater rivers and streams by the 1890s. Such

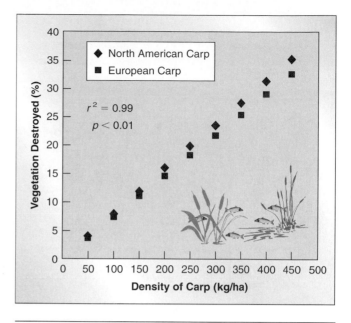

Figure 9.8

The effect of carp (*Cyprinus carpio*) on aquatic vegetation in experimental enclosures. The durations of the North American and European experiments were 92 and 71 days, respectively. Vegetation loss increased linearly with increasing carp biomass.

Original data from studies by Robel (1961) in Utah (U.S.A.) and Crivelli (1983) in France.

introductions were celebrated with high hopes for the carp as an outstanding game fish. Bands played. Politicians made speeches. The outcome, however, was less pleasant than the day's happy events. Tolerant of turbid, poorly oxygenated, even chemically polluted waters, carp proliferated as prophesied, but not to many anglers' delight. Among their other undesirable habits, carp routinely destroy emergent wetland vegetation through their rooting action in the sediment. In controlled experiments in which carp were confined in enclosures, they destroyed up to one-third of all submergent aquatic vegetation. The variation in the proportion of vegetation destroyed can be almost completely explained by variation in the biomass of carp in the exclosure (fig. 9.8). More remarkably, the pattern of plant destruction was almost identical in experiments performed on two different continents, North America and Europe (Robel 1961; Crivelli 1983).

Some nonindigenous species do not change the habitat itself, but may cause profound changes in the use of habitat by other species. The Nile perch (*Lates niloticus*), a large and voracious predatory fish, was introduced into Lake Victoria in east Africa in 1954 as a food source for human populations to supplement dwindling supplies of native fish. Its populations remained low for nearly two decades, but exploded in the 1980s, to the detriment of many endemic species. Lake Victoria's rich biodiversity of haplochromine (*Haplochromis* spp.) cichlids, species found nowhere else in the world, experienced

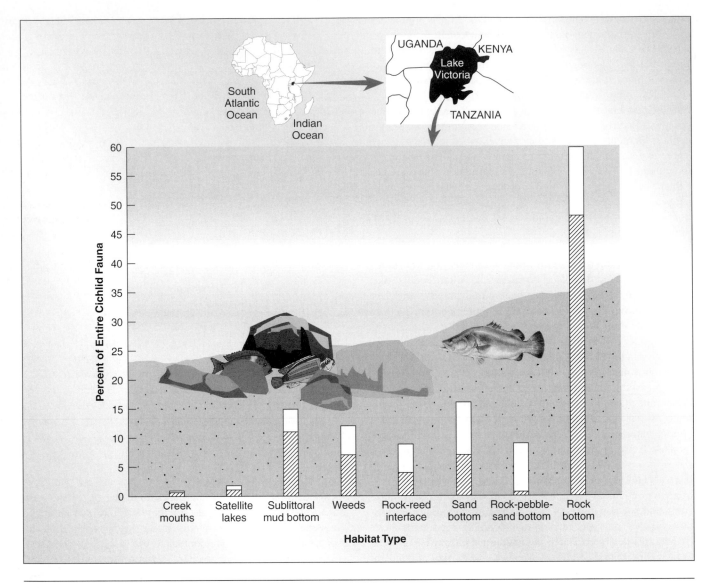

Figure 9.9
Proportions of extant haplochromine (cichlid) fauna in eight microhabitats in southern Lake Victoria, Africa. Striped portions of bars indicate proportion of species that are restricted to the given habitat. Total bar length indicates proportion of species that use the given habitat as one of their major habitats. The Nile perch (pictured, right), introduced in 1954, exterminated many species of cichlids, and patterns of extinction were habitat specific. Habitat shifts have subsequently occurred in many cichlid species, apparently as a means of avoiding predation.

a massive episode of extinction during the Nile perch's population explosion, what some conservation biologists have called "the largest mass extinction of contemporary vertebrates" in recent history (Seehausen et al. 1997). Two hundred endemic species of cichlids disappeared (Seehausen et al. 1997). Many of these species, as well as many other species that survived, were not randomly distributed in Lake Victoria, but were concentrated in particular habitats (fig. 9.9). The extinctions also were not randomly distributed, but rather habitat specific. Most extinctions occurred in offshore and sublittoral zones. Many species that survived the introduction of the Nile perch have made major shifts in habitat use. For example, in the pre-Nile perch era, *Haplochromis tanaos* and *H. plagiodon* were restricted to littoral sand bottom habitats on the east side

of Lake Victoria's Mwanza Gulf (Witte el al. 1992). In the 1990s, after the Nile perch had reached large population levels, these species were found in littoral and sublittoral mud bottom habitats on the west side of Mwanza Gulf (Seehausen et al. 1997). Such habitat shifts are consistent with a general pattern of habitat selection documented in terrestrial and aquatic species—namely, that in the presence of a predator, individuals shift from optimal foraging habitat to optimal cover habitat, or to any habitat where the predator is not present (Rosenzweig 1991). The population survives, but growth and reproduction are often reduced. Managers must consider that if nonindigenous predators enter a system, habitat management and conservation strategies may have to be fundamentally altered to preserve biodiversity.

MANAGEMENT OF FRESHWATER HABITATS FOR CONSERVATION

Managing Chemical and Physical Inputs to Aquatic Systems

Preserving and restoring the conservation value of aquatic systems can be accomplished only by active management. In North America, the leading threats to freshwater fauna are increased sediment loads and nutrient inputs from agriculture, interference from exotic species, altered hydrologic regimes associated with dams, and acidification. In particular, problems such as sedimentation, eutrophication, and acidification are *input-oriented* problems, and their best solution lies in input regulation.

Managing Sedimentation and Eutrophication

The sources of sedimentation and eutrophication are soil and fertilizer inputs, respectively, from surrounding lands, especially agricultural lands, and urban waste. Both are usually nonpoint pollution problems, aggravated through high levels of erosion associated with modern agricultural methods. Thus, the best management to address both problems would be sociopolitical in nature, occurring through laws and policies that (1) reduce the use of fertilizers, particularly on highly erodable lands and on lands near watercourses; (2) require removal of fertilizers, especially phosphorus, nitrate, and nitrite from urban sources, before allowing urban discharge to proceed downstream; and (3) reduce erosion on agricultural lands through increased vegetative cover bordering streams and through cultivation methods less destructive of soil structure. However, managers of specific aquatic systems, such as individual lakes and streams, lack power and jurisdiction to implement such sweeping changes over entire regions and drainage basins. The systems they are responsible to conserve may be degraded by inputs from detrimental land-use practices around them that they cannot directly control. In such cases, managers must use site-specific approaches within their jurisdiction. They must stop such inputs from entering the system even as they reach it, or they must remove or neutralize such inputs after they have entered.

The most immediate and direct ways to stop such inputs into an aquatic system, such as a lake or wetland, are (1) to install filters and other devices at the proximate source of input, such as the inflow stream, that remove the sediment and fertilizer when they arrive and (2) to surround shorelines and banks with vegetation that can remove high levels of phosphorus and nitrate/nitrite from runoff. The installation of filters and other devices can be expensive and the planting and management of appropriate vegetation both costly and labor intensive, but, when properly employed, both techniques can work.

Such practices may dramatically lower the amount of sediment and fertilizer entering an aquatic habitat, but reductions of fertilizer input will not necessarily restore the damage done by previous nutrient loading. What does one do with the phosphorus and other nutrients that have already entered and remain in the system? Remedies for this problem are dredging, chemical manipulation, and biomanipulation.

Dredging is the most direct approach. In this method, sediment from a eutrophied lake, pond, or wetland is physically scraped off the bottom using large, earth-moving machines. Sediment may then be placed in an artificially constructed basin where the phosphorus is removed by physical or chemical means. Thus purified, the sediment may then be returned to the original system. Although admirably direct, dredging is expensive, labor intensive, and disruptive to existing populations and communities, especially the benthos. Dredging may require temporarily draining the system, and the method is seldom suitable or effective in large, deep lakes. During the dredging operation itself, the aquatic habitat may not be suitable for other uses by humans.

Some chemical methods can convert phosphorus into chemical states that prevent it from entering or interacting in the system. One of the most well known is the so-called Riplox method ((Brönmark and Hansson 1998). In this method, the sediment surface is first oxidized, causing the phosphorus that is present to precipitate in metal complexes. Then calcium nitrate ($Ca(NO_3)_2$) and iron chloride ($FeCl_3$) are added to the sediment, increasing the levels of oxygen and iron concentrations present. The pH of the system, which would tend to decline at this point, is stabilized through the addition of calcium hydroxide ($Ca(OH)_2$). At a suitable pH, denitrifying bacteria in the sediment will transfer the nitrate in the added calcium nitrate to nitrogen gas (N_2), releasing it to the atmosphere. If these reactions proceed as planned, a chemical "lid" is placed over the surface of the sediment that prevents the release of phosphorus from the sediment into the water.

The third method, biomanipulation, attacks the eutrophication problem by manipulating populations of living creatures in the system. First, the densities of zooplanktivorous fish (generally the cyprinids) are reduced, either by adding piscivorous (fish-eating) species or by removing the cyprinids directly by trawling with gill nets or by poisoning. Theoretically, if the number of zooplanktivorous fish are reduced, zooplankton populations will grow and the grazing rate on algae and phytoplankton will increase. As a result, algal blooms will decrease and water clarity will improve. Biomanipulation has worked best where at least 80% of the zooplanktivorous fish are removed, and its success appears not to be due to the reasons originally believed. Rather, removal of the fish seems to lead to an increase in the levels of submerged macrophytic plants and periphytic algae at the sediment surface (recall the destructive effect of carp on aquatic vegetation). These plants in turn absorb large amounts of nutrients that are then no longer available for phytoplankton. Further, the plants oxidize the surface of the sediment, reducing the absorption of phosphorus into the water. Removal of fish reduces bottom disturbance by benthic-feeding fish, excretion of nutrients by fish, and phosphorus released into the water from the dead and decomposing fish.

Interestingly, lake systems can, with respect to phosphorus, exist in **alternative stable states**, in which, at similar nutrient levels, they may be dominated by submerged macrophytes in clear water or by high densities of phytoplankton and associated turbid water. The transition from one state to the other is not gradual but rapid. The theory can be illustrated visually by the "marble in a cup" model developed by Sheffer (1990)

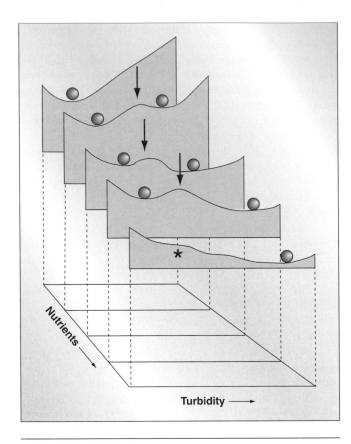

Figure 9.10

The "marble in a cup" model of alternative stable states of lakes relative to different levels of phosphorus inputs. Position of the marble(s) indicates the position(s) of one or more potentially stable states of the system. Stability of the system is achieved through a combination of biomanipulation and control of phosphorus inputs, but not one or the other exclusively. Typically, turbidity increases as nutrient levels increase. With small increases in nutrient levels, turbidity may not change unless a disturbance (represented by arrows) occurs. However, at a certain point (*) nutrient levels are too high for the water to remain in a clear state.

Adapted from Sheffer (1990).

(fig. 9.10) Under high levels of nutrient enrichment, the lake can exist only in a turbid state. As phosphate levels decline, alternate stable states are possible, depending on which way the system (marble) is pushed. If, for example, macrophytes and periphytic algae can be well established at intermediate nutrient levels, they can take in excess amounts of phosphorus (luxury uptake) that limits availability of phosphorus for phytoplankton and prevents their populations from increasing (and the clarity of the water from decreasing). The lower the level of nutrients in the water, the more stable the clear state becomes. If the model has conceived the system correctly, it demonstrates that the system's condition is a function not only of nutrient inputs, or even fish populations, but also of the state of populations of macrophytes and periphytic algae. Further, the system's future state is, in part, dependent on its present state, especially on how well established such populations of macrophytes and algae are and how much additional phosphorus they can absorb.

Although this model is conceptual and theoretical, it does have some empirical support in studies of Swedish lakes that exhibit such alternative stable states (Blindow, Hargeby, and Andersson 1997).

Managing Freshwater Systems Through Riparian Zones

Riparian vegetation is the plant community adjacent to a body of water, such as a lake or stream. Riparian zones, aside from their potential importance as corridors that link populations in different areas, profoundly affect the quality of freshwater ecosystems because they can modify, dilute, or concentrate substances from terrestrial environments in the drainage basin. Thus, for good or ill, riparian zones are the link between an aquatic system and its terrestrial context. For streams and rivers, riparian zones as limited as 10 to 30 m in width can substantially moderate temperatures, stabilize banks, and provide essential material inputs to biotic communities. Riparian vegetation of similar widths (9 to 45 m) can substantially reduce inputs of sediments from the surrounding landscape (Osborne and Kovacic 1993). Finally, riparian vegetation, deliberately arranged as buffer strips along streams, lakes, or wetlands, can substantially reduce inputs of nutrients such as phosphorous and nitrogen from a surrounding and heavily fertilized agricultural landscape.

The quality of riparian vegetation is often especially critical to egg, larval, fry, and juvenile stages of fish because they have more narrow environmental tolerances than adults. For example, removal of riparian vegetation in the South Umpqua River of Oregon has been a contributing factor to declines in this river's chinook salmon (*Oncorhynchus tshawytscha*) population. Removal of riparian vegetation, primarily for logging and road construction, has contributed to increased erosion and subsequent siltation that covers gravel substrates needed for egg-laying habitat, with associated decreases in oxygen concentration and light penetration. Destruction of riparian vegetation increases evaporation from the stream, leading to reduced summer stream flows. In spring, runoff increases during peak flows, washing out deposits of gravel and debris from streambeds that are essential elements of salmon habitat. The most adverse effect of eliminating riparian vegetation is that summer water temperatures in some sections of the South Umpqua have risen above lethal levels for salmon (26°C) in recent years (Ratner, Lande, and Roper 1997). According to Ratner, Lande, and Roper, who conducted a population viability analysis (chapter 7) on this population of salmon, "if habitat degeneration continues at the historical rate . . . the population has a 100% probability of going extinct within 100 years" (Ratner, Lande, and Roper 1997). Ratner and her colleagues advocate closing roads along the river and its tributary streams and beginning a process of active riparian vegetation restoration as essential steps to maintain the South Umpqua chinook salmon.

In an experiment comparing grass and forested buffers, Osborne and Kovacic (1993) determined that both reduced nitrate and phosphorous concentrations in surface water and in shallow groundwater by up to 90%. On an annual basis, the forested buffers reduced nitrate concentrations more (range 40 to 100%) than the grass buffers (range 10 to 60%), but were less effective than grass buffers at reducing phosphorus. Over time, both kinds of buffers "leaked" the nutrients they trapped, but such losses could be reduced by periodically harvesting the vegetation in the strips (Osborne and Kovacic 1993).

Managing Acidification

The most common direct method of restoring lakes suffering from acidification is a technique called **liming**, the direct addition of lime (calcium carbonate, $CaCO_3$). Properly applied, liming restores the pH to a neutral or alkaline state, and normally leads to an increase in species diversity in the lake as well as an increase in the abundance of most species. However, liming does nothing to alter the existing input of acidic substances into the lake. If these inputs remain unaltered, the benefits of liming will be lost and the process will have to be repeated.

Managing Wetlands

Vegetative buffer strips adjacent to wetlands, even if relatively monotypic and composed of common, inexpensive grass species, remove nutrients, including nitrates and phosphates, from runoff and permit fewer nutrients to enter the wetland (Rickerl, Janssen, and Woodland 2000). An interesting and sometimes unexpected outcome of planting buffering vegetation is that it may actually increase the diversity of the plant community around the aquatic system. In South Dakota, three species—smooth bromegrass (*Bromus inermis*), orchardgrass (*Dactylis glomerata*), and alfalfa (*Medicago sativa*)—were planted as buffer species in experimental plots around wetlands. After establishment, the buffered communities had 29 additional plant species not found in the wetland itself or in uplands around unbuffered wetlands (Rickerl, Janssen, and Woodland 2000).

Coordinated management of lake-wetland complexes can produce more effective results for conservation than managing each system separately. Managers can reduce the inputs of phosphorus and other nutrients into a lake by maintaining or creating wetlands around it. Wetland vegetation and associated wetland systems absorb far greater quantities of nutrients, especially phosphorus and nitrates, from the lake's drainage basin than can terrestrial vegetation. Wetlands can remove up to 79% of total nitrogen, 82% of nitrates, 81% of total phosphorus, and 92% of sediment in drainage water (Chescheir, Skaggs, and Gilliam 1992).

Wetlands, as noted earlier, often have disproportionately high levels of species richness, compared with terrestrial habitats of similar area. Wetlands often demonstrate species-area relationships similar to those documented in island flora and fauna (chapters 4 and 5). As the size of a wetland increases, so does its species richness. Thus, the conservation value of a wetland increases with size (Findlay and Houlahan 1997). Because wetlands are often radically different than their surrounding landscape, successful management of wetland species may require management of landscape-level processes that extend far from the wetland's borders. For example, in southeastern Ontario (Canada) Findlay and Houlahan (1997) determined that wetland species richness in plants, herptiles (amphibians and reptiles), and birds was negatively correlated with the density of paved roads within 2 km of the wetland edge. Further, species richness

of plants, herptiles, and mammals was positively correlated with the proportion of forest cover within the same distance of the wetland. Thus, a manager may be able to do as much to enhance biodiversity in a wetland by managing land-use processes as by managing the wetland itself.

Managing Biological Inputs to Aquatic Systems—Dealing with Invasive Species

Invasive species, both plant and animal, pose unique and particular problems for managers of aquatic systems. Some basic principles of managing invasive species have already been covered in chapter 7 and will not be repeated here. However, some management approaches are unique to aquatic species, especially plants.

Ashton and Mitchell (1989) note two basic strategies for the control of nonindigenous species: protection and intervention. Protectionist approaches are applicable in a variety of contexts and have been explored in chapter 7. Interventionist approaches are more species- and site-specific. There are six types of interventionist (control) techniques that can be especially successful against invasive aquatic plants. These are (1) manual removal; (2) mechanical control (using machines to mow, uproot, shred, or dredge out established plants); (3) chemical control (herbicides); (4) biological control (introduction of a specific parasite or predator to decimate the invader); (5) environmental manipulation (especially water-level manipulation); and (6) the direct use of the invasive species for some economic benefit (i.e., harvest) (Ashton and Mitchell 1989). Despite the daunting prospect of trying to eradicate an established invasive species, some such programs have succeeded. In successful control programs, the infestation was attacked early when the invasive plant was low in numbers and small in extent. Also, control efforts were successful when the invasive species was confined to one location. Under these conditions, all of the above techniques have been used effectively. Even the notorious carp (in this case, the grass carp, *Ctenopharyngodon idella*) has been put to good use as an agent of biological control to eradicate nonindigenous submerged plants (Ashton and Mitchell 1989).

Where invasive species are well established, some attempts at control, and even eradication, have been successful, but the range of effective techniques is more limited. Manual and mechanical removal are not practical when invasions become widespread, however, chemical and biological controls may still be effective. For example, an invasion of water hyacinth in Lake Hartbeespoort in The Republic of South Africa was eradicated with large-scale use of herbicides (fig. 9.11a). An infestation of the water fern *Salvinia molesta* was eradicated from Lake Moondarra in Australia through the introduction of another nonnative species, the Brazilian beetle or Salvinia weevil (*Cyrtobagous salviniae*) (fig. 9.11b). In the case of the beetle, environmental conditions also played an important role, with drought reducing populations of *Salvinia molesta* to low levels just prior to the beetle's introduction.

The dangers of biological control, especially of introducing a nonnative biological control agent like the Brazilian beetle to an African lake, are many, and have been reviewed in

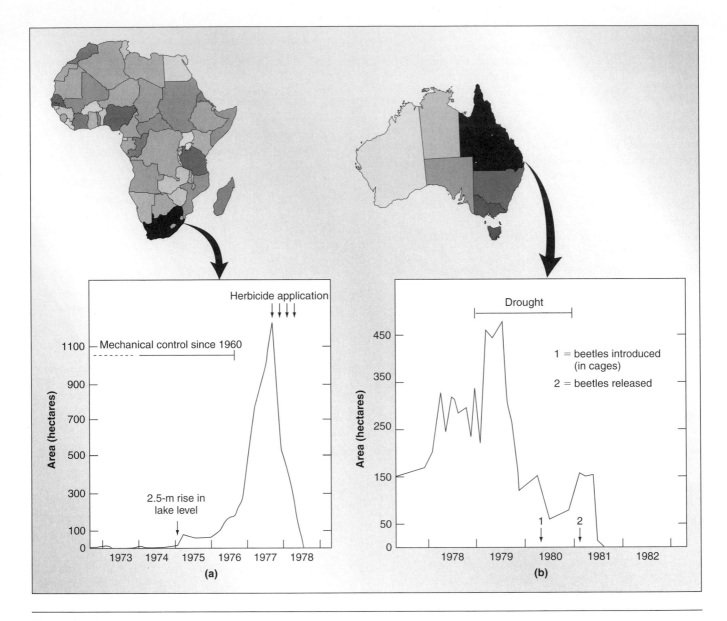

Figure 9.11

(a) Changes in the area covered by water hyacinth (*Eichornia crassipes*) on Lake Hartbeesport, South Africa, before and after herbicide application. (b) Changes in the area covered by water fern (*Salvinia molesta*) on Lake Moondarra, Australia, before and after introduction of the beetle *Cyrtobagous salviniae*.

Adapted from Ashton and Mitchell (1989).

chapter 7. Risks in biological control can, however, be reduced where a native species can be used as the control agent. For example, to control the previously mentioned invasive aquatic weed, Eurasian watermilfoil, Sheldon and Creed (1995) evaluated the effects of a native North American aquatic weevil, *Euhrychiopsis lecontei*. In a carefully controlled experiment, Sheldon and Creed compared the growth of Eurasian watermilfoil, as well as 10 native aquatic species, in enclosures with and without weevils (fig. 9.12). They found 50% less Eurasian milfoil in enclosures with weevils than in those without weevils, but weevils had no significant effect on any native species.

POINTS OF ENGAGEMENT—QUESTION 1

What elements of Sheldon and Creed's study eliminate or reduce risks often associated with biological control, especially biological control using a nonnative control agent? Does their study contain protocols that could be applied more generally to biological control of invasive species?

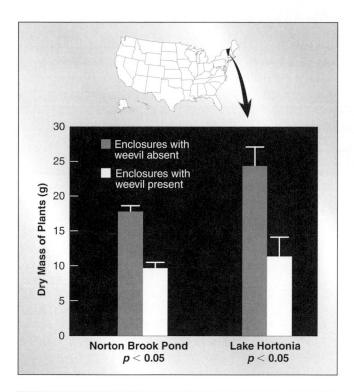

Figure 9.12

The effect of feeding by the native North American aquatic weevil *Euhrychiopsis lecontei* on Eurasian watermilfoil (*Myriophyllum spicatum*) in two Vermont lakes. Watermilfoil biomass is significantly lower where the weevil is present.

After Sheldon and Creed (1995).

Nonindigenous animal species are much more difficult to control, much less eradicate. Decades of chemical treatments, such as rotenone poisoning, or environmental manipulation (water drawdowns or complete drainage) to eradicate carp have usually had only short-term effects, if that, and have proved most effective at eliminating native species. The carp easily reestablished themselves in most cases, but the more desirable native species often did not. Mussels such as the Chinese clam (*Potamocorbula amurensis*), zebra mussel, and other invasive invertebrates have proved impossible to eliminate from aquatic environments once established, making preventionist approaches all the more important to maintaining the health of aquatic systems.

Legislation and Management for Freshwater Environments

The Wild and Scenic Rivers Act

In the United States, the most significant legislation protecting streams is the U.S. Wild and Scenic Rivers Act of 1968. Under this act, a stream or section thereof designated as a wild or scenic river is protected from any action by any federal agency that would adversely affect its water quality. Congress may also, by

a special act, designate a section of a river as a National River, such as Missouri's Current River, and extend similar protection (Benke 1990). However, by 1990, less than 2% of U.S. streams (less than 100,000 km out of an estimated 5.2 million km) had been deemed of sufficient quality to merit such federal protection.

The Clean Water Act and Indices of Biotic Integrity

In the United States, the Water Pollution Control Act Amendments of 1972, outgrowths of the earlier Clean Water Act, adopted a visionary, biologically oriented approach to the assessment of national waters. The amendments directed the Environmental Protection Agency to "restore and maintain the physical, chemical, and biological integrity of the nation's waters" and to enhance "all forms of natural aquatic life" (Meybeck and Helmer 1989). Unfortunately, the law's vision of protecting the integrity and diversity of entire systems was lost in the convenience of a reductionistic approach that favored chemical standards and an emphasis on point source pollution (Karr 1991). The former was easier to apply and the latter was easier to clean up. Unfortunately, it is possible for fresh waters to meet standards relating to physical and chemical contaminants and still not be capable of sustaining functional ecosystems. Such systems, being dependent on processes, do not support diverse biological communities if interactions among organisms, and interactions between organisms and the surrounding physical environment, are not properly functioning. Many environmental impacts that degrade ecosystems are simply too diverse and too complex to be detected and understood by chemical assays.

To address this deficiency, a growing emphasis among conservationists has been the use of various indices of biotic integrity (IBI) as alternative, ecologically based measurements of water quality, particularly in streams. Although IBIs vary in detail, most follow similar basic principles and procedures. A particular taxon, say fish, is rated and scored in different attributes (table 9.1). The sum of the scores is then used to provide a summary value for the IBI that is associated with an "integrity class" ranking for the site that provides a summary index of community characteristics (table 9.2) (Karr 1991). Most such indices use three groups of attributes (technically, "metrics") to make their assessment. These are species richness and composition, trophic composition, and fish abundance and condition (Karr 1991). Expectations for the values of individual metrics are based on those found in an undisturbed, but otherwise similar, stream. Many conservationists advocate the use of the IBI over more traditional, chemically based measurements of individual elements or compounds (including pollutants and toxins) in streams because the IBI (1) reflects and focuses upon distinct attributes of biological systems, not simply chemical properties; (2) measures the sampled stream against a minimally disturbed system, thereby establishing a clear baseline and biological expectation; and (3) requires the incorporation of professional ecological judgment in evaluating the stream's condition, not simply a check for compliance in terms of specified elements, compounds, or toxins (Karr 1991). IBIs are strongly associated with independently derived measures of overall watershed

Table 9.1   **Metrics Used to Determine an Index of Biotic Integrity (IBI) for Fish Communities**   Ratings of 5, 3, or 1 are assigned to each measurement according to whether its value approximates, deviates somewhat from, or deviates strongly from the value of the same measurement at a comparable but relatively undisturbed site. Adapted from Karr 1991. Originally developed for midwest U.S.

METRICS	RATING OF METRIC		
	5	3	1
Species Richness and Composition			
1. Total number of native fish species			
2. Number and identity of darter species (benthic species)			
3. Number and identity of sunfish species (water-column species)	Expectations for metrics 1–5 vary with stream size and region.		
4. Number and identity of sucker species (long-lived species)			
5. Number and identity of intolerant species			
6. Percentage of individuals identified as green sunfish (tolerant species)	<5	5–20	>20
Trophic Composition			
7. Percentage of individuals as omnivores	<20	20–45	>45
8. Percentage of individuals as insectivorous cyprinids	>45	20–45	<20
9. Percentage of individuals as piscivores (top carnivores)	>5	1–5	<1
Fish Abundance and Condition			
10. Number of individuals in sample	Expectations for metric 10 vary with stream size and other factors.		
11. Percentage of individuals as hybrids (exotics, or simply lithophils)	0	0–1	>1
12. Percentage of individuals with disease, tumors, fin damage, and skeletal anomalies	<2	2–5	>5

condition (Steedman 1988), and are inexpensive, simple, and sensitive to environmental change. If continued and increasing use of IBIs can recapture the original ecosystem emphasis of the U.S. Clean Water Act and other water pollution laws by focusing attention on the ecological condition of freshwater systems, they may provide greater incentive and more valuable insight for managers to restore such systems to their original functions and properties, rather than merely meeting chemically prescribed standards.

Table 9.2 An Example of Total Index of Biotic Integrity (IBI) Scores, Integrity Classes, and Associated Class Attributes for Fish Communities

TOTAL IBI SCORE*	INTEGRITY CLASS OF SITE	ATTRIBUTES
58–60	Excellent	Comparable to the best situations without human disturbance; contains all regionally expected species for the habitat and stream size, including the most intolerant forms, with a full array of age (size) classes; balanced trophic structure.
48–52	Good	Species richness somewhat below expectation, especially because of the loss of the most intolerant forms; some species are present with less than optimal abundances or size distributions; trophic structure shows some signs of stress.
40–44	Fair	Signs of deterioration include loss of intolerant forms, fewer species, highly skewed trophic structure (e.g., increasing frequency of omnivores and green sunfish or other tolerant species); older age classes of top predators may be rare.
28–34	Poor	Dominated by omnivores, tolerant forms, and habitat generalists; few top carnivores; growth rates and condition factors commonly depressed; hybrids and diseased fish often present.
12–22	Very Poor	Few fish present, mostly introduced or tolerant forms; hybrids common; disease, parasites, fin damage, and other anomalies are regularly observed.
†	No Fish	Repeated sampling finds no fish.

*The score is the sum of the 12 metric ratings. Sites with values between classes are assigned to the appropriate integrity class following careful consideration of individual criteria/metrics by informed biologists.
†No score can be calculated where no fish were found.

After Karr 1991

International and National Legislation for Wetlands

As noted in chapter 2, wetlands were one of the first cases in which international legislation, specifically the Ramsar Convention, focused on the protection of an ecosystem instead of a species. Recall that the Ramsar Convention obligated its signers to conduct land-use planning for wetlands and wetland preservation, to identify and designate at least one wetland as a "wetland of international importance," and to establish wetland nature reserves (Koester 1989). Canada's federal policy on wetland conservation provides one of the best national examples of implementing the ideals of Ramsar. The Canadian policy is a comprehensive federal plan that articulates strategies for sustainable use and management of the nation's wetlands. It aims to provide for the maintenance of overall wetland function on a national level; enhance and rehabilitate degraded wetlands; recognize wetland functions in planning, management, and economic decision making in all federal programs; secure and protect wetlands of national importance; use wetlands in a sustainable manner; and allow no net loss of wetlands on federal lands and waters (Rubec 1994). Although no policy is ever perfectly trans-

lated into practice, the Canadian wetlands policy has experienced remarkable success, primarily through its nonregulatory approach. Each Canadian province, following directives of federal policy, has developed its own public review and consultation process for wetlands conservation (Rubec 1994). Federal wetland directives led to the publication of a standardized manual, the *Wetlands Evaluation Guide* (Bond et al. 1992). With an estimated endowment of nearly one-quarter of the world's remaining wetlands, Canada's leadership in wetlands conservation policy is not only commendable but strategic.

Although Canada has provided a commendable example of integrating international wetlands conservation with national and provincial policies, other nations also have developed extensive wetlands conservation legislation. Wetlands conservationist Michael Williams, a native of the United Kingdom, considers the best example of national wetlands legislation to be that of the United States, which he asserts is "the most elaborate and complex legislation in place for the longest time" (Williams 1990). A number of U.S. legislative acts address wetlands conservation, and most lead to increasing preservation and restoration of

Table 9.3 Federal Legislation Affecting the Conservation of Wetlands in the United States

PROGRAM OR ACT	PRIMARY IMPLEMENTING AGENCY	EFFECT ON WETLANDS
Discouraging or Preventing Wetland Conversions		
Regulation		
Section 404 of the Federal Water Pollution Control Act (1972) amended as the Clean Water Act (1977)	U.S. Army Corps of Engineers, Department of Defense	Regulates activities that involve disposal of dredged or fill material
Acquisition		
Migratory Bird Hunting and Conservation Stamps (1934)	U.S. Fish and Wildlife Service (FWS)	Acquires or purchases easements with revenue from fees paid by hunters for Duck Stamps
Federal Aid to Wildlife Restoration Act	FWS	Provides grants to states for acquisition, restoration, and maintenance of wildlife areas
Wetlands Loan Act (1961)	FWS	Provides interest-free loans for acquisitions of and easements for wetlands
Land and Water Conservation Fund	Forest Service, Bureau of Land Management, FWS, National Park Service	Provides funds that can be used to aquire wetland wildlife areas
Water Bond Program (1970)	Agriculture Stabilization and Conservation Service, U.S. Department of Agriculture (USDA)	Leases wetlands and adjacent upland habitat from farmers for waterfowl habitat over 10-year period
U.S. Tax Code	Internal Revenue Service (IRS)	Provides tax deductions for donors of wetlands
Other General Policies or Programs		
Executive Order 11988 Floodplain Management (1977)	All agencies	Minimizes wetland loss and degradation from federal activities
Executive Order 11990 Protection of Wetlands (1977)	All agencies	Minimizes impacts on wetlands from federal activities
Coastal Zone Management Act (1972)	Office of Coastal Management	Provides funding (up to 80%) for state wetland protection initiatives associated with estuaries and other coastal zone habitats
Food Security Act (1985)	USDA	Withholds subsidies for agricultural improvements involving wetland conversion
Encouraging Wetlands Conversion		
U.S. Tax Code	IRS	Encourages farmers to drain and clear wetlands by providing tax deductions and credits for all types of general development activities
Payment-in-kind program	USDA	Indirectly encourages farmers to place previously unfarmed areas, including wetlands, into production

Adapted from Williams 1990.

wetlands (table 9.3). For example, the 1985 Food Security Act contained a provision designed to arrest the process of draining wetlands on private agricultural lands before the last of these wetlands were lost. This provision, popularly known as "Swampbuster," denies most U.S. Department of Agriculture benefits to farmers who drain wetlands on their land. Swampbuster creates an eligibility requirement for farmers to receive commodity price supports, disaster payments, Farmers' Home Administration loans, and other benefits. Amended in 1990 as the Food, Agriculture, Conservation and Trade Act, the Swampbuster provision was supported in this amendment by the creation of the federal Wetland Reserve Program (WRP), which provides for payment of subsidies to farmers who remove croplands from production in former wetland areas and reestablish the land as wetlands. To enroll in WRP, the landowner's plan must include drainage alterations and the establishment of marsh plants on the enrolled site. The WRP was begun as a pilot program with a limited budget, and because the program's budget is small, so is its enrollment. In 1992, only 20,200 ha (20%) were enrolled in nine states out of 100,800 ha eligible for enrollment. Nevertheless, if even a fraction of all U.S. croplands enrolled in the program, the WRP would become the largest wetlands restoration program in U.S. history (Lant, Kraft, and Gillman 1995).

Despite an extensive network of wetlands conservation legislation, supported by national policies and executive orders (e.g., Executive Order 11990 of 1977) that make wetlands conservation a matter of national priority, wetlands loss in the United States continues, in part because (1) there is a lack of agency coordination in wetland conservation; (2) most legislation does not regulate private activity on private lands, which remain the greatest single source of wetland losses; and (3) some U.S. legislation still encourages, directly or indirectly, the draining of wetlands. For example, the U.S. tax code provides tax deductions and credits for farmers for many types of development activities, including draining wetlands (Williams 1990).

Setting Priorities for Conservation in Freshwater Habitats

The World Wildlife Fund–United States (WWF–U.S.) recently made a priority assessment of North American lakes and streams by region using two criteria: biological distinctiveness and conservation status of watersheds within a region (Abell et al. 2000). In ranking biological distinctiveness, WWF–U.S. gave priority to those regions that contained one or more systems that made important contributions to biodiversity at four different levels (globally outstanding, continentally outstanding, bioregionally outstanding, or nationally important). In ranking conservation status, regions were ranked as critical (intact habitat reduced to small, isolated patches with low probability of persistence over the next decade without immediate action); endangered (intact habitat of isolated patches of varying length with low to medium probability of persistence over the next 10 to 15 years without immediate or continuing protection or restoration); vulnerable (intact habitat remains in both large and small blocks, persistence is likely over next 10 to 20 years if the area

receives adequate protection and restoration); relatively stable (disturbance and alteration in certain areas, but function linkages among habitats still largely stable, surrounding landscape practices do not impair aquatic habitat or could be easily modified to reduce impacts); and relatively intact.

Categories I–V were assigned on the basis of integration of these two criteria, with I being the most critical and V being the least critical (fig. 9.13). Following a triage philosophy of conservation, the greatest need for protection was assigned to globally outstanding areas in endangered and vulnerable status. Critical areas were considered too degraded and at risk to have high hopes of saving, and stable or intact systems were considered not to require immediate action. Among systems in the endangered and vulnerable categories, conservation priority declines as the importance of the system decreases in scope.

The WWF–U.S. prioritization system is far from perfect, but it is extremely useful at two levels. As a specific prioritization of aquatic conservation needs, the assessment uses objective criteria to identify key areas in need of immediate protection. As a method of conservation assessment, the ranking system can be adapted to other regions of the world or to smaller scales while preserving its intended purpose: focusing conservation efforts in areas that will reward the efforts with the greatest contribution to biodiversity. For example, conservation biologists working to manage or establish a system of local preserves may have no aquatic systems that are globally or continentally outstanding, but they may have systems that are outstanding at smaller scales, such as state or local levels. The need for such assessment, followed by appropriate management, is critical. Although the North American assessment found that Arctic lakes and rivers were, for the most part, intact and stable, there were no large temperate lakes or rivers that could be so described. The majority of temperate lakes and rivers were classified as endangered or critical (Abell et al. 2000).

To set management and conservation priorities, managers must understand the causes of habitat loss and their effects on aquatic diversity. Many aquatic and wetland species show dramatic shifts in distribution over relatively short time spans. Managers must determine if such changes represent the effects of habitat loss or environmental change or are simply random events. Making an accurate determination is critical to making an appropriate management response. But managers cannot make these determinations without systematic assessment and decision-making processes.

One approach to making such assessments is the use of **rule-based models** that evaluate possible mechanisms of distributional changes in species. Skelly and Meir (1997) used a rule-based approach to evaluate possible causes of changes in distributions of 14 species of amphibians across a landscape of 32 ponds in Michigan. Specifically, they attempted to explain changes using three different models: (1) an *isolation model* that assumed that changes in distribution were driven by distances between ponds (i.e., by dispersal abilities of the amphibians); (2) a *succession model* that assumed that distribution was determined by changes in vegetation in and around the ponds; and (3) a *null model* that assumed that changes were random events. Their basic data set was simple; namely, presence-absence data on 14 amphibian species based on annual surveys from 1967 to

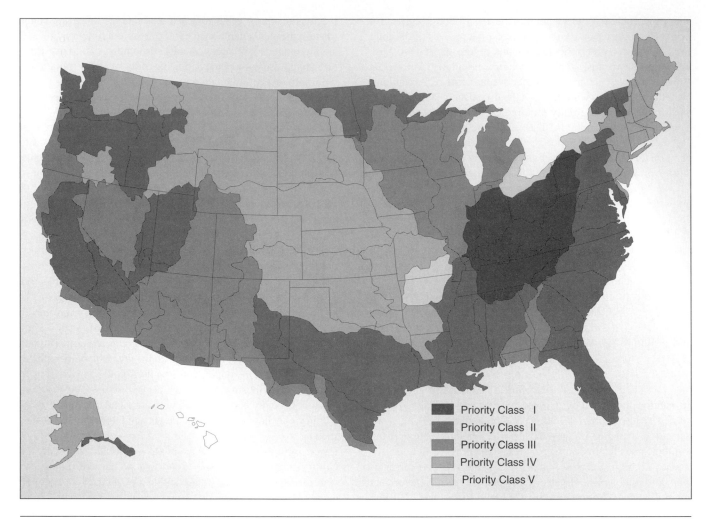

Priority Class   I
Priority Class   II
Priority Class   III
Priority Class   IV
Priority Class   V

Figure 9.13

Priority categories for conservation of freshwater ecoregions in the United States. Prioritization is based on combined ranking of biological value (i.e., relative contribution to global biodiversity) and current management/conservation status. Conservation efforts will vary within the same priority category because of differences among watersheds in habitat, beta diversity, and resilience.

After Abell et al. (2000).

1974 and from 1988 to 1992. The underlying hypothesis of each model was used to divide the ponds into three classes based on (1) the distance from the pond to the nearest population of each species (isolation model), (2) the vegetational characteristics in and around the pond (succession model), or (3) random assignment of each pond to one of three classes (random model). For individual species, the succession model made fewer mistakes in predicting the occurrence of three species at individual ponds and was better at predicting species richness of amphibans at ponds (Skelly and Meir 1997).

These results suggested that the presence of amphibians in this landscape of wetlands could be best managed by managing the vegetation characteristics of the ponds, not by changing the distribution of ponds. Skelly and Meir note that the ability to explain a pattern with a rule-based model is not the same as showing causation between a factor and its effect. To accomplish that, managers would have to manipulate vegetation in and around the ponds experimentally and monitor amphibian response. What the rule-based approach does provide is insight

about *which* experiments might be most useful to conduct. The authors conclude that "even relatively coarse information on presence and absence can be put to an . . . important use: as survey information accumulates it becomes a source of insight for managers interested in determining *why* species distributions are changing, not just *if* they are changing" (Skelly and Meir 1997). Rule-based models can be used in other contexts, but their application here shows how a manager, informed only by simple survey data, could use rule-based models to evaluate management actions and plan experiments to determine the causes of changes in species presence and distribution.

MARINE HABITATS AND BIODIVERSITY

It is beyond our scope to consider the multitude of marine habitats that contribute to earth's total biodiversity. However, we briefly review three that make disproportionately large contributions and face specific and significant threats from human activity. These are communities associated with coral reefs, bottom sediments, and hydrothermal vents.

Coral Reefs

Coral reefs have been called the tropical rain forests of the oceans. Worldwide, over 600 species of coral contribute to this remarkable habitat, and individual reefs may harbor up to 400 species of coral, 1,500 species of fishes, 4,000 species of mollusks, and 400 species of sponges (Hinrichsen 1997). Although the bulk of any coral reef is nonliving matter, the surface layer of living creatures is composed mostly of coral polyps. Relatives of jellyfish and anemones, the polyps have column-shaped bodies topped with stinging tentacles. These creatures secrete calcium carbonate as a metabolic product, and from such secretions fashion cup-shaped structures that serve as their homes and that they attach to one another. Over many years and generations of coral, these calcium carbonate secretions build a coral reef, each new generation enlarging the reef by building on the bodies of their ancestors.

Coral reefs are centers of marine biodiversity because they combine the elements of structure, nutrients, water quality, and light to create a favorable and productive environment for living things. Physically, the body of the reef provides a substrate and point of attachment for many species, especially more sedentary taxa such as crustaceans and mollusks. Even among more active species, the physical characteristics of the reef provide cavities for shelter and breeding. Upon this structure, high densities of prey species attract proportionally high densities of predators.

The coral polyps that build the reef constantly secrete calcium, an essential nutrient for photosynthetic organisms such as phytoplankton. Additional inputs of calcium come from the ongoing erosion and breakdown of dead corals that form the body of the reef. Because the reef forms in well-lit waters, light is available in combination with calcium and other nutrients, creating a favorable environment for photosynthesis to take place. Interacting with nutrient and light availability is a generally high water quality, produced in part by abundant populations of sponges on the reef's surface. Sponges, using the reef as a surface for support, circulate and cleanse the surrounding water through their own bodies, enhancing water quality, lowering turbidity, and allowing penetration of light to greater depths.

Benthic Communities

Of 29 nonsymbiont animal phyla known on earth, all but one have representatives in the ocean, and all of these have representatives in benthic communities. In fact, most of the diversity found in marine ecosystems consists of invertebrates that live in or on bottom sediments (Snelgrove 1999). We are only now beginning to appreciate the biodiversity of such communities. For example, 64% of polychaete (tubeworm) taxa identified in a recent deep-sea study were previously unknown to science (Grassle and Maciolek 1992). Given such ignorance,

it is not surprising that we do not know the exact number of marine benthic species, but estimates have ranged from a low of 500,000 (May 1992) to a high of more than 10 million (Lambshead 1993). What is it about benthic habitats that leads to such high biodiversity?

There is enormous variability in benthic habitats and their associated communities, but some general patterns hold worldwide. Benthic habitats in extreme environments, such as estuaries, eutrophied areas, and high-energy regions with low organic content, have lower diversity than sediments in aquatic habitats without these characteristics (Snelgrove 1999). In addition, the diversity in sediment grain size is directly correlated with the diversity of the benthic community, probably because a greater diversity of sediment sizes naturally provides a higher diversity in sizes of food particles (Whitlatch 1977). Finally, diversity in seagrass bed sediments is higher than in adjacent sediments associated with open areas (Peterson 1979).

In shallow water, the distribution of benthic organisms is determined primarily by abiotic habitat features, including temperature, salinity, depth (which affects both light and pressure), surface productivity, and sediment dynamics. Many species have specific tolerances to temperature, salinity, and pressure because these factors affect their osmotic balance and the functioning of cellular enzymes. Despite poor swimming abilities and the lack of a central nervous system, the planktonic larvae of many benthic invertebrates show some ability to select favorable benthic habitat (i.e., they display habitat preference) (Butman, Grassle, and Webb 1988; Snelgrove 1999). It is not clear what environmental cue the larvae are responding to, although substrate organic content has been suggested (Butman 1987). What is demonstrable is that where most larvae end up, in terms of sediment type, is both adaptive and nonrandom (by definition, preferential).

Communities Associated with Hydrothermal Vents

A habitat like none other on earth is found in association with marine hydrothermal vents in some of the deepest parts of the sea, along the fissures and edges of tectonic plates. These habitats and their associated communities are a relatively recent discovery, first reported in 1977. Most benthic communities, both freshwater and marine, that exist below euphotic (lighted) depths must depend upon the input of organic matter from outside (allochthonous matter). Marine benthic communities around these vents, however, are unique in that the foundation of their productivity is bacteria that convert heat energy from the vent into chemical energy, analogous to the way in which photosynthetic organisms convert light energy into chemical energy. This process, known as chemosynthesis, provides a radical alternative to our traditional view of community production, structure, and function. Although the process may be new to human understanding, the hydrothermal vent communities appear to be very old, and extremely stable. They furnish habitats for relict species from the mesozoic era, including many species of crinoids and the most primitive living sessile barnacles (fig. 9.14) (Barry and Dayton 1991).

Figure 9.14

A hydrothermal sea vent community. Such communities are not only characterized by high biodiversity, but contain many unique "relict" marine species not found in other habitats. This photo, from the Galapagos Trench in the Pacific Ocean, reveals tube worms, vent fish, and Galapagos trench crabs.

CONSERVATION CHALLENGES OF MARINE HABITATS

Although problems of marine habitat and species preservation vary locally and regionally, the major threats to marine environments are consistent throughout the world. Some are similar or identical to threats facing freshwater environments, whereas others are unique to the marine system. The most important global threats include exploitation of commercial species, direct destruction of marine habitats, indirect degradation of marine habitats from land-based sources including eutrophication, pollution (primarily from radioactive wastes, heavy metals, and petroleum products), the degradation of coastal zones (from erosion, development, and habitat destruction), (VanDeVeer 2000), and nonindigenous species (Ruiz et al. 1997).

In the 1940s and 1950s, the emerging science of fisheries management perceived fish stocks as renewable resources that could be managed for a maximum sustainable yield (MSY), whose value could be calculated precisely by using estimates based on catch per unit effort (Ricker 1958). All that was thought to be required for a sustainable fishery was a reproductive surplus. Today the concept of MSY has all but disappeared from fisheries, along with many of the fish stocks mismanaged under its assumptions. The United Nations Food and Agriculture Organization (FAO) estimates that almost 70 percent of the world's marine stocks are fully to heavily exploited, overexploited, or depleted and in need of urgent conservation and management (UNFAO 1992). In U.S. fisheries alone, 45% of all species are considered overharvested (Ruckelshaus and Hays 1998). As fish biologists have learned more about fish populations, they have found that most such populations (1) show widely ranging cycles of high and low abundance, (2) do not necessarily show a strong correlation between recruitment and number of adults present, and (3) do not necessarily show advance warning of impending population decline or crash from overexploitation (Hilborn, Walters, and Ludwig 1995). The decline may be sudden, and stocks may not recover

to harvestable levels in the short term even when given complete protection.

The effects of overexploitation on targeted commercial species are not surprising, but the effects on nontarget species can be equally devastating. The removal of prey species may severely reduce the populations of predator species, and not of fish only, but also of birds and mammals. The clearest examples of this effect have been seen in the decline of Peruvian seabirds following the decimation of the offshore anchovy fishery, and the decline of sea otter populations off the California coast following overfishing of abalones (Agardy 1997).

Some cases of this type have had legal as well as biological ramifications. For example, in 1998, a coalition of environmental organizations sued the North Pacific Fishery Management Council under the Endangered Species Act for failing to protect critical foraging habitat for the Stellar sea lion (*Eumetopias jubatus*) by allowing unregulated pollock (*Theragra chalcogramma*) fishing in the sea lion's main foraging areas (Stump 2000). Lack of food had previously been identified as a primary cause of the decline in sea lion populations, and pollock is a principal prey species of sea lions. The plaintiffs argued that it made no sense to allow unregulated fishing in critical foraging habitat, and violated the ESA's directive that required "reasonable and prudent alternative (RPA) measures" be taken to avoid inflicting "adverse modification" on the critical habitat of a species. A U.S. district court agreed and ordered the National Marine Fisheries Service to revise its regulations. Exactly what regulations will be imposed on the pollock industry is still being debated among the parties involved, but the debate has shifted from whether or not a problem existed to what must be done to solve it (Stump 2000).

Because the removal of a prey species can cause population declines in the predator, the removal of the predator can cause changes in prey populations, and those changes do not always lead to uniform or long-term increases (Goeden 1982). Overexploitation disrupts equilibria of many populations (Agardy 1997), and can make them more susceptible to declines associated with environmental and demographic stochasticity, such that stocks may continue to decline even after take is restricted or stopped altogether (Lauck et al. 1998). The take of nontargeted species in commercial fishing also continues to be a serious problem despite concern, attention, legislation, and supposedly improved technologies. In some fisheries, such as shrimp, the discarded biomass of by-catch exceeds the targeted catch worldwide (Agardy 1997). Species such as sea turtles, dolphins, sharks, rays, and benthic organisms continue to be affected as by-catch species.

Causes of Marine Habitat Degradation

The destruction of marine habitats can occur through a variety of means, most of which are associated with commercial fishing. One of the most obvious and deadly is the use of explosives, such as dynamite, to harvest coral reef species. A single blast can devastate hundreds or thousands of cubic meters of coral reef, destroying not only individual fish, but also the structure upon which the community depends. Destruction of physical reef structures instantly eliminates what may have taken hundreds or thousands of years for marine organisms to build (Agardy 1997).

Figure 9.15

A portion of the Atlantic Ocean bottom before (top) and after (bottom) being swept by a trawling net. Prior to trawling, a complex and diverse community was present in and on the sediments, but trawling obliterated the community.

After Auster (1998).

The structure of benthic communities is significantly altered by the use of bottom trawling nets. Auster (1998) provides pictures of a site on the bottom of the Gulf of Maine off the east coast of the United States before and after bottom trawling (fig. 9.15). The top three photographs (before) reveal a complex and diverse assemblage of creatures, including tubeworms, sponges, and many other forms of life. The bottom two photographs (after) show the same spot after a trawl net was dragged across it. The complexity of the habitat has been obliterated, along with all its residents.

Table 9.4   **A Classification of Fish Habitat Types on the Outer Continental Shelf of the Temperate Northwest Atlantic**

CATEGORY	DESCRIPTION	RATIONALE	COMPLEXITY SCORE*
1	Flat sand and mud	Areas with no vertical structure such as depressions, ripples, or epifauna.	1
2	Sand waves	Troughs provide shelter from current; previous observations indicate that species such as silver hake hold position on the downcurrent sides of sand waves and ambush drifting demersal zooplankton and shrimp.	2
3	Biogenic structures	Burrows, depressions, cerianthid anemones, hydroid patches; features that are created or used by mobile fauna for shelter.	3
4	Shell aggregates	Provide complex interstitial spaces for shelter; also provide a complex, high-contrast background that may confuse visual predators.	4
5	Pebble-cobble	Provide small interstitial spaces and may be equivalent in shelter value to shell aggregate, but less ephemeral than shell.	5
6	Pebble-cobble with sponge cover	Attached fauna such as sponges provide additional spatial complexity for a wider range of size classes of mobile organisms.	10
7	Partially buried or dispersed boulders	Partially buried boulders exhibit high vertical relief; dispersed boulders on cobble pavement provide simple crevices; the shelter value of this type of habitat may be lower or higher than previous types based on the size class and behavior of associated species.	12
8	Piled boulders	Provide deep interstitial spaces of variable sizes.	15

*Habitat complexity scores do not increase at a constant rate, but reflect cumulative effects of structural components added at each succeeding level. After Auster 1998.

This vivid visual example of marine habitat destruction can be understood more generally through a conceptual model of the effects of fishing gear upon different marine habitats, such as might be found on a continental shelf. Consider eight different categories, ranging, at the simplest level, from flat sand or mud to the most complex, piled boulders (table 9.4). Auster (1998) assigned a "numerical complexity score" to each habitat category. Note that, as habitats become more complex, scores do not increase linearly. For example, category 6, pebble-cobble with sponge cover, receives 5 (not 1) additional points because it contains elements of all previous categories plus dense emergent epifauna. Category 7 receives 10 points for containing all the elements of category 6 plus 2 points for shallow boulder crevices and current refuges. Finally, category 8 receives an additional 3 points for its addition of deep crevices

(Auster 1998). The effect of intensive fishing activity, primarily trawls and dredges, is to reduce habitat complexity by smoothing bedforms (habitat categories 1 and 2), removing epifauna (categories 3, 4, and 6) and removing or dispersing physical structures (categories 5, 7 and 8). Such a model predicts that the effect of fishing activity on habitat complexity is nonlinear (fig. 9.16). The more complex the original habitat, the greater the loss of complexity that results (Auster 1998).

Marine habitats are even more degraded from land-based sources. This indirect, but extensive degradation has multiple causative agents (table 9.5). Many of these, such as eutrophication, sedimentation, and thermal pollution, are proximity based relative to the source of the pollution, and thus have their greatest effects on coastal and estuarine environments. But others, such as radioactive wastes and persistent toxins, such as PCBs,

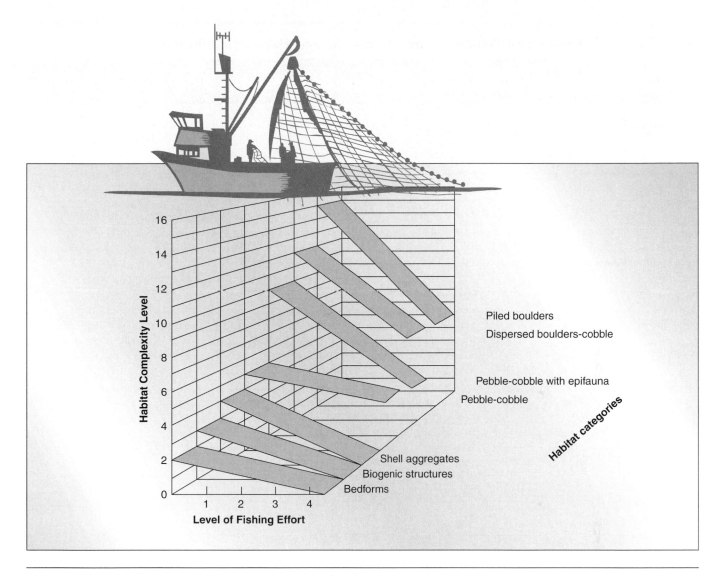

Figure 9.16

A conceptual model of the effects of fishing gear on seafloor habitat. Note that increases in fishing effort produce disproportionately greater reductions in habitat complexity in heterogeneous habitats than in simpler habitats.

Adapted from Auster (1998).

DDT, and similar or derivative compounds, travel long distances in ocean currents, or may be deposited far out to sea through atmospheric circulation patterns. Similarly, some kinds of military wastes, such as radioactive material or chemical weapons, may be deliberately transported long distances from shore before being deposited. These pollutants can cause habitat destruction and devastate populations thousands of miles from their source.

As in freshwater and terrestrial habitats, nonindigenous species pose a significant threat to the stability of marine communities and the habitats that support them. Historically, most invasions were by so-called fouling organisms that attached themselves to the hulls of ships (Ruiz et al. 1997). Today, as metal hulls have replaced wooden ones and the speed of ocean vessels has increased, these types of invaders have actually declined in importance, but four other means of invasion remain.

These include (1) intentional releases of aquaculture, commercial or sport fishery, or bait species; (2) the connection of waterways through canals; (3) the release of species associated with the pet industry with other types of management practices; and (4) the release of organisms in the ballast water of ships. Of these, the last has often been the most destructive to native communities and habitats, perhaps because it is the least intentional yet introduces the largest volume of water into new areas.

The intentional releases of certain species have sometimes provided new and important sources of commercial and sport fishing or aquaculture. For example, the Pacific oyster (*Crassostrea gigas*) was transported from Japan to San Francisco Bay to establish an oyster fishery. Other planned introductions, however, have had unforeseen and sometimes devastating consequences. The connection of different marine environments by canals has allowed two-way invasions between

Table 9.5   Some Land-Based Pollutants That Degrade Marine Habitats and Ecosystems

POLLUTANT	EFFECTS ON MARINE BIOTA
Herbicides	• Have serious effects at low concentrations. • May destroy or damage zooxanthellae in coral, free-living phytoplankton, algal, or seagrass communities. Basic food chain processes are destroyed or damaged.
Pesticides	• May selectively damage zooplankton or benthic communities (planktonic larvae are particularly vulnerable) and cause immediate or delayed death of vulnerable species. • May accumulate in animal tissues and affect physiological processes such as growth, reproduction, and metabolism.
Antifouling Paints and Agents	• May selectively damage elements of zooplankton or benthic communities. • Are prevalent in harbors, near shipping lanes, and in enclosed, poorly mixed areas with heavy recreational boat use.
Sediments and Turbidity	• May exceed or smother the clearing capacity of benthic animals, particularly filter feeders. • Reduce light penetration, likely to alter vertical distribution of plants and animals in shallow communities such as coral reefs. • May absorb and transport other pollutants.
Petroleum Hydrocarbons	• May cause local necrosis if organisms are briefly exposed, whereas long-term exposure eventually causes death. • Are detrimental to reproduction and dispersion. • Water-soluble hydrocarbons cause mucus production, abnormal feeding, changes in a wide range of physiological functions, and, with longer exposure, death. • Residual hydrocarbons may lead settling larvae to avoid affected areas, and thus block recolonization and repair.
Sewage and Detergent—Phosphates	• Have effects at very low levels. • Inhibit a wide range of physiological processes and increase vulnerability of affected biota to a range of natural and human-induced impacts. • Inhibit calcification (e.g., in corals and coralline algae).
Sewage and Fertilizers—Nitrogen	• Distort competition and predator/prey interactions in biological communities (i.e., coral reefs, which are characterized by low levels of natural nitrogen) because of increased primary production in phytoplankton and benthic algae. • Increased sedimentation because of increased detritus from planktonic communities. • Increased nutrient level in benthos from sedimentary organic material. • Favor the growth of some filter or detritus feeders (e.g., sponges and holothurians).
High- or low-salinity water—freshwater runoff, effluents (low-salinity water floats on top of water column; high-salinity water sinks, prior to mixing and dispersion)	• Species highly tolerant of the changed regime may alter biological communities, particularly in shallow, poorly mixed, or enclosed waters. • May affect settlement and physiology of shallow benthic and reef organisms. • May cause physiological stress.
High or low water temperature from industrial plant heating or cooling	• May cause physiological stress. • May affect settlement and physiology of shallow benthic and reef organisms.

Continued

Table 9.5   *Continued*

POLLUTANT	EFFECTS ON MARINE BIOTA
Heavy Metals (e.g., Mercury and Cadmium)	• Accumulation has severe effects on filter feeders and species higher in the food chain. • May interfere with physiological processes such as the deposition of calcium in skeletal tissue. • May cause physiological stress.
Surfactants and Dispersants	• Most are toxic to marine biota. • Have synergistic effects when mixed with hydrocarbons: mixtures can be more toxic than individual components. • Can interfere with a wide range of physiological processes (e.g., photosynthesis).
Chlorine	• At low levels, inhibits external fertilization in some invertebrates (e.g., sea urchins). • Can be lethal to individual species.

Adapted from Kenchington 1990.

established communities in different areas, sometimes from radically different environments. Today the Mediterranean Sea has over 240 exotic species, and 75% are attributed to migration through the Suez Canal, primarily from the Red Sea (Ruiz et al. 1997).

The most extensive and often-used mechanism of invading species is through the ballast water of ships (Carlton 1985). One ship can carry more than 150,000 metric tons of ballast water for trim and stability, which it may dump in an estuary at the end of a voyage. In estuaries associated with major port systems, the amount of water dumped from foreign oceans can be staggering. For example, the port system of the Chesapeake Bay has been estimated to receive over 10,000,000 metric tons annually, and U.S. and Australian ports may receive over 79,000,000 metric tons each year. This amounts to more than 9 million liters of water per hour! At this rate of input, it is not surprising that estuaries appear to receive more exotic invaders than open oceans. For example, 212 nonindigenous species are known from San Francisco Bay, but fewer than 10 have been found along its adjoining outer coast (Ruiz et al. 1997).

Although invasions in terrestrial and freshwater systems are notorious for their devastating results on native species, marine communities appear to be more resistant to their effects. The U.S. Fish and Wildlife Service considers exotic species to be a significant cause of the decline of 160 native threatened or endangered species, but few of these are marine. In fact, there are relatively few recent extinctions of marine and estuarine species, and these extinctions did not appear to be caused by exotic species. Nevertheless, some exotic species have devasted native populations, marine environments, and commercial fisheries. The recent invasion of San Francisco Bay by the Chinese clam has altered marine communities in ways similar to the effects of the zebra mussel on freshwater systems. Chinese clams have become so numerically dominant, achieving densities of over 10,000 individuals/m^2, that they have replaced other benthic organisms, cleared plankton from overlying water, and eliminated seasonal plankton blooms (Snelgrove 1999). The invasion of the green crab (*Carcinus maenas*) along the northeastern U.S. coast has significantly reduced clam and mussel fisheries. A larger and more voracious predator than native U.S. crabs, the green crab can devastate local populations of oysters, clams, and other shellfish. The American comb jelly (*Mnemiopsis leidyi*) has unexpectedly contributed to a collapse of commercial fisheries in the Black and Azov Seas in Europe because it competes more effectively for the same food source (copepods) as native commercial fish (Ruiz et al. 1997).

Marine invasions are not as well studied or understood as those that occur in terrestrial habitats or in fresh water, so it is difficult to identify general trends or effects common to most invaders. There is some evidence that invading species decrease the abundance and evenness of remaining native species, decrease variation among communities (reduction in beta diversity, chapter 4), and alter gene flow within and among communities (Macdonald et al. 1989; Drake 1991; Ruiz et al. 1997). Overall, marine environments around island areas and estuarine environments appear to be more susceptible to invasion than do communities in open oceans.

The array and variety of threats to marine habitats means that there is no single strategy that can address all problems at once. However, one emerging strategy designed to address multiple threats is the concept of the marine protected area (MPA).

Marine Reserves: Management Context, Goals, and Strategies

All parks and reserves face the problem of defining appropriate biological boundaries that ensure the persistence of what the park is established to preserve. However, this problem is greater in aquatic environments, particularly in marine environments.

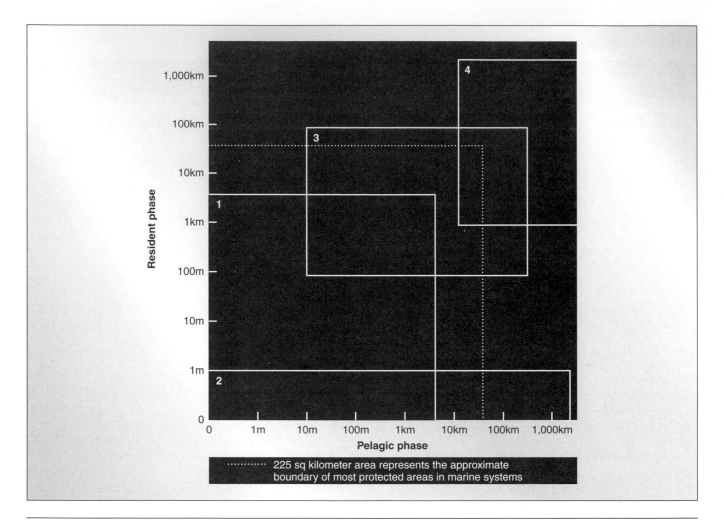

Figure 9.17

Four categories of life cycles characteristic of marine creatures, with respect to spatial scale. Box 1 represents species with no planktonic (drifting) phase and fixed or restricted movement in their adult phase. Box 2 represents species in which one phase is fixed and the other is planktonic or pelagic (open ocean). Box 3 represents species in which adults have large, defined territories but larvae are planktonic. Box 4 represents species in which phases of life are all planktonic or all pelagic.

After Kenchington (1990).

Land preserves are essentially two-dimensional, defined by their length and width on the earth's surface. Air may be a medium for flight and some passive dispersion, and contains essential elements and compounds for respiration and photosynthesis, but it is also relatively homogeneous. In contrast, marine reserves are three-dimentional, and their third dimension—the water column—is much more dynamic and critical to the marine community than is air in a terrestrial environment. In addition to plant and animal communities on the ocean floor, the water column itself contains communities of its own, perpetually drifting or swimming in and through it. Spores, eggs, and young of even the most sedentary species must use the water column for reproduction, dispersal, and development (Kenchington 1990). At most times and places, most photosynthesis, respiration, and transport of matter and energy take place within this water column, not on the seabed. Thus, in prescribing the

boundaries of a marine reserve, one must consider carefully the properties of the water mass within such boundaries, because on these properties all life within the mass depends.

The water mass has enormous effects on issues of reserve scale. During early phases of development, most marine species have far greater dispersal distances than terrestrial species. Some remain highly mobile throughout life, whereas others become essentially immobile as adults. Kenchington (1990) identifies four basic life-history categories of marine creatures relevant to the question of spatial scale (fig. 9.17): (1) species with fixed or restricted movement in their adult phase, but no planktonic (drifting) phase (box 1); (2) species in which one phase is fixed and the other is planktonic or pelagic (box 2); (3) species in which adults have large, defined territories, but larvae are planktonic (box 3); and (4) species in which all phases of life are either planktonic or pelagic (box 4).

Ecosystem-Level Protection: Australia's Great Barrier Reef Marine Park

One of the best examples of a large marine reserve is Australia's Great Barrier Reef Marine Park (GBRMP), one of the world's premier protected areas, and part of the Biosphere Reserve and World Heritage Site programs. The Great Barrier Reef itself is a vast complex of some 2,900 individual reefs and 250 cays (low islands or reefs made of sand or coral) stretching along the continental shelf of northeast Australia from just south of the Tropic of Capricorn to the Torres Strait. The system possesses 71 genera of coral alone. The Great Barrier Reef was relatively inaccessible to humans until the 1960s. The GBRMP that attempts to preserve it is in many ways unique among marine preserves. The preserve was not established to stop or solve an existing problem or degradation of the reef, but was actually established in anticipation of future problems. In 1967, a private Australian firm filed an application for permission to take coral limestone from a part of the reef for use in the production of agricultural lime. The Wildlife Preservation Society of Australia perceived this application as setting a precedent for dangerous and destructive processes that could eventually destroy the reef. With other conservation groups joining the lead of The Wildlife Preservation Society, public outcry led to the refusal of the permit application by the provincial government (Queensland). Further controversies over offshore oil drilling in the reef area and outbreaks of the crown-of-thorns starfish (*Acanthaster planci* which destroyed reef corals) led to legislation that established the GBRMP (Kenchington 1990).

Today the GBRMP is a vast multiple-use area managed by establishing different zones within the park for different uses, through which it has successfully accommodated a variety of user groups (Agardy 1999) (fig. 9.18). However, even in this exemplary park there are serious problems. For all its size and jurisdictional power, the GBRMP authority that manages the park has no control over land-based inputs that pose significant threats to its coral reefs, commercial fish stocks, and endangered species. Its jurisdiction stops at the shoreline, and it cannot stop influxes of land-based sediments and chemical pollutants that pour into its aquatic system (Agardy 1999).

Tourist-Recreation Marine Reserves: The Bonaire Marine Park

Marine reserves are not the exclusive domain of large nations, nor are they established exclusively to protect fisheries or commercial stocks. The tiny Netherlands Antilles off the northern coast of Venezuela established the Bonaire Marine Park (BMP) around the island of Bonaire in the early 1980s. The Bonaire Marine Park is neither a vast, multiple-use area like the GBRMP nor a strictly no-take, closed marine reserve for scientific research and conservation. It belongs to a unique category that could be called "tourist-recreation reserves."

The BMP was established primarily to preserve the beauty of local marine resources for the enjoyment of snorkel and scuba divers, a mainstay of the island's tourist-driven economy (Dixon, Scura, and van't Hof 1995). Nearly 17,000 scuba

---

Marine conservation legislation and marine reserves are designed to meet three goals simultaneously: (1) protect marine and coastal biodiversity, (2) ensure that marine productivity is not undermined by uncontrolled exploitation, and (3) focus efforts for restoration of vital areas that may be presently degraded but have potential to support healthy marine ecosystems in the future (Agardy 1997). To these ends, marine reserves have been established worldwide with a variety of names, jurisdictions, and specific purposes. Within these reserves, areas closed to all types of marine fishing and harvesting are often designated as "harvest refugia" or "no-take zones." These are generally designed to protect a particular commercial stock or group of stocks from overexploitation. At large scales, "biosphere reserves," administered by the United Nations Educational, Scientific, and Cultural Organization (UNESCO), are usually divided into three zones. "Core" reserves are areas with little or no harvesting or other activities, "buffer" areas are those where limited harvest and other activities are permitted (Agardy 1999), and "transition" areas are zones that are least protected, and are administered with regulations most like those outside the reserve (Sobel 1993). At small scales, more limited reserves may be established to achieve a more limited set of conservation objectives, or even only one.

Efforts to establish marine reserves have varied in effectiveness according to region and country. There are 135 legally protected marine and coastal areas in the Greater Caribbean Basin alone (Dixon, Scura, and van't Hof 1995); France has 5 fully operational reserves; Spain has designated 21; and Italy has established 16. The United States has 12 designated marine reserves, administered under the National Marine Sanctuary Program (NMSP) and officially known as National Oceanic and Atmospheric Administration (NOAA) Marine Sanctuaries. The U.S. program has been criticized because its reserves are considered too small (less than 1% of U.S. territorial waters) and unprotected (less than 0.1% are no-take areas) (Agardy 1999). But the United States has established large marine reserves off the Florida Keys and the central California coast, and the NMSP has demonstrated a strong commitment to reserve persistence and effective protection. Within this variety of management goals, strategies, and national efforts, we can examine some specific case histories of individual marine reserves to better understand their role in conservation.

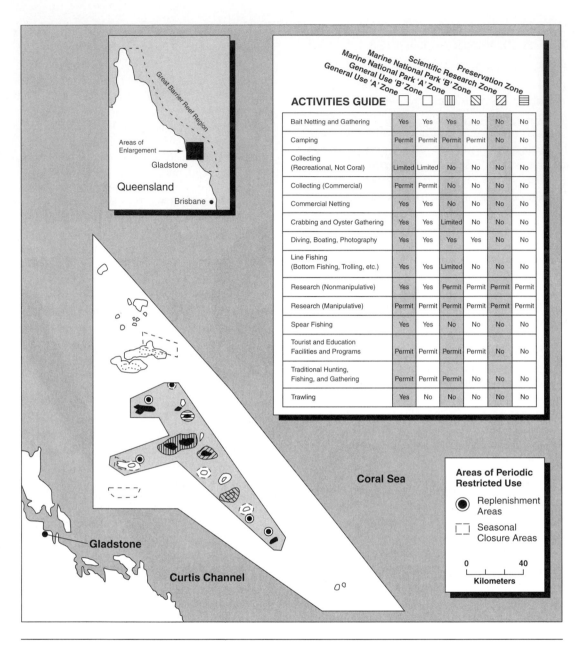

Figure 9.18

Management zones in the Capricornia section of the Great Barrier Reef Marine Park, Australia.

After World Resources Institute et al. (1992).

divers visited Bonaire in 1991, and the number has been increasing at the rate of 9 to 10% annually. To accommodate divers within the marine park, Bonaire established a "snorkel trail," as well as a series of freestanding platforms throughout the reef area. Studies of the park show that tourism and conservation are interactive joint products of the marine park, but that use levels by tourists cannot increase indefinitely, even in a relatively "nonconsumptive" activity like diving. A "threshold" level exists in the park for diving pressure on reefs. Underwater areas around platforms that receive 4,000 to 6,000 dives per year begin to show signs of stress and wear, and both coral cover and species diversity begin to decline at this point. How-

ever, the distribution of funds from diving creates an environment that produces pressure to increase the number of divers. For most divers, diving at Bonaire is part of a prepaid travel package arranged with agents in the United States and Europe. As part of the arrangement, the diver receives "vouchers" that cover most expenses such as lodging, transportation, and food. Divers who come under these conditions often spend very little additional money. The locals are reimbursed for a portion of the vouchers by sending them back to the U.S. or European agents, but only after large commissions are deducted. As a result, income to locals from diver visitation may be marginal, and the economic benefit of each additional diver is important to residents. However, if increasing stress leads to a loss of world-class diving experiences at Bonaire, fewer visitors will come and total income will decline. Assuming that diving will continue at least at its present rate (the local economy has few other sources of income beyond subsistence agriculture and fishing), current suggestions to maintain the quality of the marine park include better distribution of divers, better diver education and training in "diver etiquette," and better regulation of underwater activities. However, as Dixon, Scura, and van't Hof (1995:142) noted, "These management measures do not *increase* the tolerance of marine systems to stress, rather they help to distribute the burden more evenly across the ecosystem. Such measures require both money and legal authority." Greater legal authority is possible, but local citizens have been reluctant to grant the park more regulatory authority than it already has. Locals have been especially reluctant to yield greater authority to regulate diving operators and cruise boats in the park, practices through which many local citizens make their living. Increasing revenues from divers may be an easier matter, as previous amounts charged to divers (U.S. $10) are only about half what diver surveys indicate as their average willingness to pay ($20) (Dixon, Scura, and van't Hof 1995).

The Bonaire Marine Park illustrates the dilemma of conflicting values that was addressed conceptually in chapter 3 (Values and Ethics in Conservation). If the real value of the marine resources is viewed as economic rather than intrinsic, then the resources themselves may be degraded even as economic revenues rise from fees for seeing and photographing these resources. It is possible that such degradation might have no adverse economic effects because divers would gradually become accustomed to the decreasing quality of the diving experience. Some marine conservationists have advocated that tourism and recreation should become the primary uses of the marine environment, the basis for appreciation and enjoyment of marine environments, and the foundation of long-term social and economic benefits for the local, national, and global community (Kenchington 1990). The experience of the Bonaire Marine Park shows that this optimism is premature, or perhaps misplaced altogether. Tourism can have a destructive effect on marine populations and habitats, and recreational use that is not well planned will lead to degradation of valued resources, conflicts between conservation values and economic interests, and little benefit to individuals in the local economy. In contrast, properly planned ecotourism can move beyond conflict, and even coexistence between conservation and economics, to a symbiotic relationship in which local citizens take responsibility and ownership of the resource and its values, marketing opportunities to enjoy the resource in profitable but nondestructive ways (Kenchington 1990). But in order for this to happen, practical management steps must be taken: (1) the use of the resource, even if nonconsumptive, must be restricted to a level that the resource can sustainably support; (2) the users must be optimally dispersed to avoid concentrations of use that could be destructive to the resource; and (3) where possible, the resource sites must be "hardened" by facilities and supporting structures that allow the sites to bear the level of use allowed without degradation.

Tourism and recreation will grow in importance as uses, values, and generators of economic wealth in human communities associated with marine environments. However, tourism and recreation do not exhaust the potential uses of marine resources or meet the needs of human populations for food. Ultimately, marine reserves may have an important role to play in commercial fishing as well as in tourism and research.

Marine Protected Areas and Commercial Fisheries

In 1982, most nations of the world adopted the conventions established at the United Nations Third Conference on the Law of the Sea (UNCLOS III). The most radical change in international law that emerged from this convention was the extension of national jurisdiction over territorial waters from the historic 12-nautical-mile standard to 200 nautical miles, a move estimated to place 90% of marine fishery resources within the jurisdiction of individual nations (Lauck et al. 1998). These enlarged areas of national jurisdiction, or exclusive econonic zones (EEZs) (Kaitala and Munro 1995) were seen as the saviors of international marine fishing. With this change in international maritime law, it was optimistically believed that the oceans' commercial fishing stocks would avoid becoming an example of Hardin's "tragedy of the commons" (Hardin 1968). More conservative harvesting policies of individual nations, backed by the power of international law, would lead to wise use based on national interest and local ownership of the fishery resource, resulting in long-term sustainability of marine fishery harvest.

These happy, hopeful visions have yet to come true. Despite the extension of territorial limits to 200 nautical miles and more exclusive use of fisheries stocks by individual nations, commercial fisheries have collapsed all over the world. One of the saddest and most dramatic failures of UNCLOS III occurred in what was historically one of the world's largest and most dependable commercial fisheries, the cod fishery of the Grand Banks of Newfoundland. In the early 1900s, the northern cod fishery was producing annual harvests of approximately 250,000 tons (Ruckelshaus and Hays 1998), but by the 1980s declining catches provoked the Canadian government to enforce drastic cuts. The stock still failed to recover, so the Canadian government instituted a 2-year moratorium on cod, but cod continued to decline even after the moratorium was in place. By 1996, the moratorium had become permanent (Lauck et al. 1998).

Lauck and colleagues (1998) argue that the answer to the problem of sustainable commercial fisheries may be the marine reserve. Far from being simply a means to enhance tourism or to preserve unique ecosystems or rare species, they assert that marine reserves should become the foundation of a new form of fisheries management that is based on a radical change of perspective. Namely, they argue we should abandon the concept that every available commercial fish stock should be exploited optimally and replace it with the strategy of "bet hedging." In other words, fisheries science should assume that high levels of uncertainty are a permanent and persistent dimension of estimating the size of fish populations and their future trends. If high uncertainty is taken as a given, then the optimal strategy is not to attempt to harvest a population wherever it occurs, but to harvest *some* of the populations at the predicted (but uncertain)

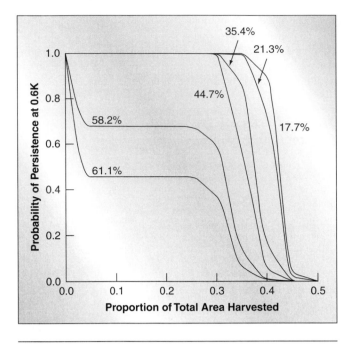

Figure 9.19

The probability of a fish stock remaining above 60% of its carrying capacity for 20 years depends on the fraction of area available for harvesting. In this simulation model, when more than 30% of the stock's total area is available for harvesting (i.e., outside the marine reserve), the probability of maintaining a population size that is > 0.6 K (K = carrying capacity) drops rapidly. Each line represents a different value for coefficient of variation associated with the average harvest (CV is defined as the standard deviation of the harvest fraction/mean of the harvest fraction).

After Lauck et al. (1998).

optimal level and leave a large portion unharvested as a protection against unforeseen declines in the harvested stock.

Lauck and colleagues articulate their ideas quantitatively in a model whose goal is to retain a fish population at more than 60% of its carrying capacity for at least 40 years (fig. 9.19). Through a series of equations that permit estimation of the proportion of the population available for harvest outside a closed area, they model the probability that the population could persist for the specified periods and levels. They assumed that half the available population outside the reserve was captured every year, but with coefficients of variation (the measure of uncertainty about the mean) assigned at six levels from 18 to 61%, and they varied the fraction of the total area available for harvesting (i.e., the size of the marine reserve) (Lauck et al. 1998).

The results were dramatic. Even with a moderate amount of variation in the catch ($CV < 50\%$), the probability of the population persisting for 40 years dropped drastically when the amount of exploitable area became greater than 30%. If the catch percentage was more variable, the probability of the population's persistence was less than one (not certain) even if only 5% of the area was available for harvest. The probability of successfully protecting the stock of fish increased in the model if the harvest was reduced to lower levels, and at lower levels, more of the total area could be made available to fishing.

Two conclusions emerged from the model. First, "a reserve can simultaneously lead to stock protection and a higher level of catch," and "it is possible to maximize catch while protecting the stock" (Lauck et al. 1998). That is, Lauck and colleagues conclude that marine protected areas provide the "simplest and best approach to implementing the precautionary principle and achieving sustainability in marine fisheries" (Lauck et al. 1998).

Empirical data from marine reserves support their value in restoring fish populations. Russ and Alcala (1996) compared density and biomass of large predatory fish at two small marine reserves in the Philippines with two similar control sites. They found that the longer the reserve was protected from fishing, the greater the increase in density and biomass of large predatory fish (fig. 9.20). But they also noted that unregulated fishing within the reserves, even for a short time, eliminated gains in biomass and density that had taken years to achieve. Russ and Alcala (1996) conclude that "funding in support of marine reserves as fisheries management tools must be long-term, and . . . management measures used to implement and maintain marine reserves must be robust in the long term, i.e., on scales of decades."

POINTS OF ENGAGEMENT—QUESTION 3

Examine figure 9.21. Points above the diagonal line indicate species that increased in abundance after establishment of the reserve. Points below indicate species that were less abundant after reserve establishment. In your opinion, do these data support the assertion that marine reserves lead to increased populations of fish and invertebrates? Why or why not?

Not all marine areas will be placed in marine reserves. Most marine populations and their associated habitats will continue to be exploitable in open seas or in unprotected territorial waters. An alternative strategy to relieve the twin pressures of exploitation and degradation on these systems is that of mariculture, the intensive commercial cultivation of certain species in limited areas.

Mariculture—The Case History of the Giant Clam

Some forms of mariculture, such as the pearl industry and oyster farming, have been practiced successfully for centuries. Other forms are relatively recent developments. However, given an ever-accelerating human demand for marine creatures as food and for other products such as jewelry or decoration, it is certain that maricultural techniques will increase in size, scope, and diversity in the next decade. Like intensive agriculture in terrestrial landscapes, mariculture concentrates disturbance of the environment; increases, intensifies, and concentrates pollution; and reduces systems to the lowest possible levels of species diversity and ecological complexity, effectively eliminating most ecosystem services. Like intensive terrestrial

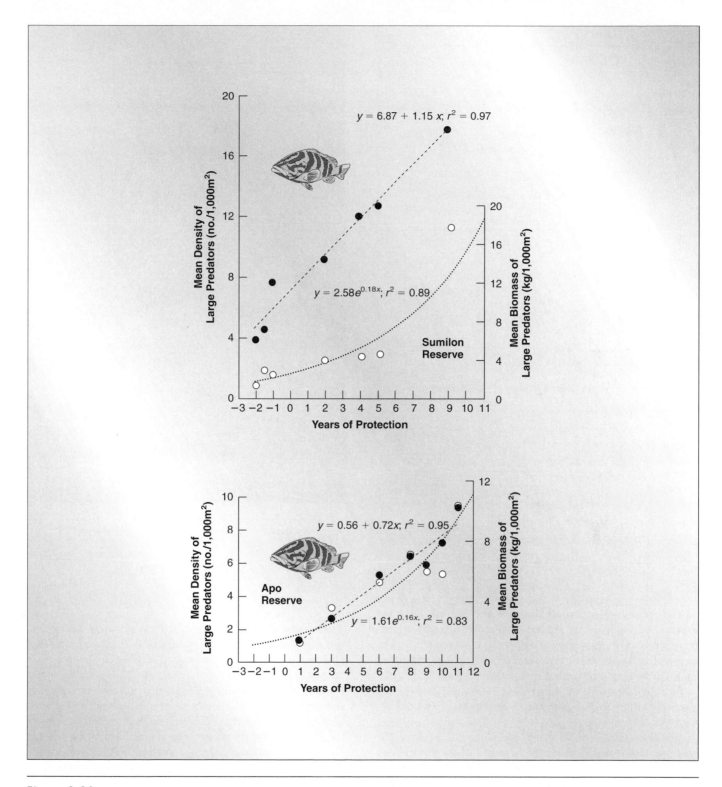

Figure 9.20

Changes in density (solid circles, dashed lines) and biomass (open circles, dotted lines) of large predatory fish at two small marine reserves in the Philippines, compared with two similar control sites. The longer the reserve was protected from fishing, the greater the increase in density and biomass of large predatory species.

Adapted from Russ and Alcala (1996).

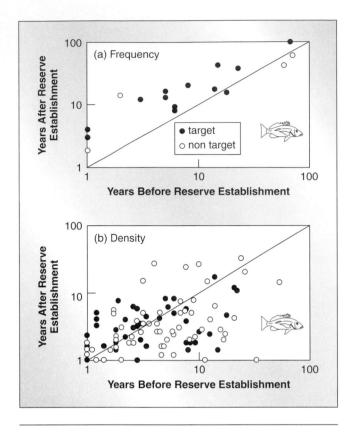

Figure 9.21

(a) Frequency of occurrence and (b) density of commercially targeted (solid circles) and nontargeted (open circles) of fish species before and after the establishment of marine reserves. Symbols above the diagonal lines indicate species that were more frequent or had higher densities after reserve establishment.

Data compiled and figures adapted from Ruckelshaus and Hays (1998).

Figure 9.22

The giant clam (*Tridacna gigas*), an endangered species that has responded favorably to intensive mariculture.

agriculture, however, mariculture can also provide large per area, per effort yields of food and other products from the creatures subjected to its management. Because mariculture can be so effective and efficient, it can reduce the need to disturb or exploit natural systems and their populations, which may not be resilient to disturbance or exploitation even at very low levels. The case history of the giant clam illustrates the potential advantages of mariculture.

Giant clams (*Tridacna* spp. and *Hippopus* spp.) include nine species of marine clams that live in shallow tropical and subtropical waters, often on coral reefs, in the Indo-Pacific, primarily in the Indo-Malay region (fig. 9.22) (Lucas 1997). Only one species, *Tridacna gigas,* could truly be called "giant," having a maximum shell length of 137 cm and a mass of about 500 kg (Lucas 1994). Other species range from 15 to 50 cm in length and average about 15 kg in weight. Nevertheless, one adult of even the smallest species would amply fill the average dinner plate, and a high demand for giant clams as food leads many to wind up there. Giant clams are limited to shallow waters because they live in a symbiotic relationship with microalgae known as zooxanthellae. The zooxanthellae, which are

photosynthetic, use the clam's mantle as a point of attachment, and from this substrate transfer some of the organic products of photosynthesis to their clam hosts. The clam fulfills its part of the symbiosis not only by acting as the substrate for the algae, but also by providing inorganic nutrients to the zooxanthellae and exposing them to sunlight in the shallow waters. The relationship can be considered essential for both organisms because the clam obtains many nutrients from these algae (Lucas 1997).

Because of their large size, their high value as food, and their accessibility in shallow waters, giant clams have been heavily exploited. This has led to a ban on international trade in clam products, bans on fishing for giant clams in marine reserves, limits on collection effort and harvestable size of clams outside of reserves, aggregation of remaining populations to facilitate reproduction, and replenishment of wild stocks with cultured clams. It is these "cultured clams" that deserve a more detailed examination.

After fertilization, the planktonic clam eggs are dispersed passively by ocean currents. Upon hatching, the clams develop into a free-swimming trochophores (fig. 9.23), which, in turn,

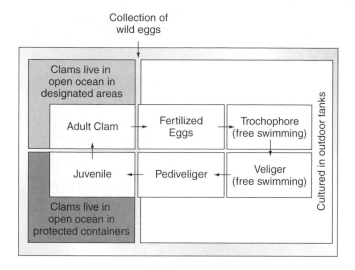

Collection of wild eggs

Figure 9.23
The life cycle of the giant clam (*Tridacna gigas*).

develop into small, filter-feeding, bivalved larvae called veligers. After increasing in size and developing a prominent foot, the veligers settle out of the current onto a reef surface, where they will grow and metamorphose. Although the clam may move slightly after settling, its initial location is likely to be its habitat for life.

Clam mariculture makes use of this life-history strategy by collecting eggs from wild clams or, more often, using eggs of existing domestic stock, and maintaining the hatching, larval, and juvenile stages in outdoor tanks. Juveniles at larger stages are moved into protective containers in the ocean, and larger individuals are later cultured without protection in the open sea. Amazingly, the mariculture of giant clams has no deleterious environmental effects. Even the feces produced by clams at high densities are so packed with algae (recall the clam's symbiotic relationship with the zooxanthellae) that they are rapidly and readily consumed by plankton-feeding fishes that take up residence, in abundance, around the clam colonies.

The mariculture of giant clams offers an environmentally friendly way to gain valuable resources from a fragile environment, the coral reef, while at the same time providing the means to supplement wild populations of clams with individuals raised in captivity. However, even this apparent success story cannot be accepted uncritically. Like sea turtle farming (chapter 3) (Ehrenfeld 1992), the mariculture of giant clams has drawbacks, some of which are the fruits of its own success. If effective, the increased supply of giant clams from mariculture can fuel increased demand for giant clams as food and ornaments, and encourage mariculture operators to remove additional quantities of eggs and adult clams from wild populations. In a climate of higher demand, pressure will increase to take clams directly from wild stocks. Consumers would not know the difference, and wild clams could be harvested with only a fraction of the time and effort needed to raise clams by mariculture. Poaching would become attractive, especially to individuals in the local culture who possess the skill to collect giant clams on their own. This last objection has been addressed, at least in

part, through the development of a village-based, clam farming program in the Solomon Islands by the International Centre for Living Aquatic Resources Management Coastal Aquaculture Centre (Lucas 1997). Local villagers own and work in all stages of the mariculture enterprise, receiving the profits and sharing the risks directly. However, not every area where clam mariculture is practiced can expect to gain this degree of local ownership. In those cases, the potential and stimulus for poaching giant clams could be high.

Multiple and Conflicting Jurisdictions over Marine Resources

Lamenting the current state of marine environmental law and policy, W. M. von Zharen of the Texas Institute of Oceanography wrote, "the present management of the marine ecosystem is based on a series of regimes that are directed at the various parts rather than the whole and that are, as such, ineffectual" (von Zharen 1999). Marine conservationist Elliot Norse agreed, noting that a successful marine conservation strategy must be "cross-sectoral, embracing all categories of marine ecosystems and species, all types of human use, and all sources of threats" (Norse 1993:281). As we have already explored in our discussion of law and policy (chapter 2), international and national laws, and their respective interests, are often at odds in the conservation of marine resources and habitat. National jurisdictions over territorial waters, for example, do not always coincide with the distributions and movements of commercial fish populations, leaving these stocks vulnerable to depletion by international harvesting. Inputs of pollution from one country may be moved, through various currents and discharges, into the territorial waters of another country, where they degrade that nation's marine resources. And the discharge of ballast water from ships of distant countries into estuaries, bays, and coastal waters of another may transfer nonindigenous species that destroy local stocks of native marine creatures. Solutions to these problems are not possible without international cooperation and mutually enforced international conservation law.

The primary documents that serve as sources for an international conservation strategy are *Agenda 21* (United Nations 1992), the *Global Biodiversity Strategy* (World Resources Institute et al. 1992), and *Caring for the Earth: Strategy for Sustainable Living* (IUCN et al. 1991). Although these documents differ in details that are beyond the scope of this discussion, they agree that international strategies should include efforts to reduce population growth and the consumption and wasteful use of marine resources; the development of an open, nondiscriminatory, equitable, and environmentally sound international, multilateral trading system; and ratification of major UN documents establishing regional and global laws, policies, protocols, and organizations for marine ecosystem management, especially ratification of the third United Nations Convention on the Law of the Sea (UNCLOS III) (Norse 1993).

One attempt to develop more consistent patterns of international cooperation has been the work of the International Organization for Standardization (ISO), which developed out of the 1992 Earth Summit meetings (von Zharen 1999). The ISO has played a leading role in developing international and regional environmental management standards (EMSs) that attempt to establish consistent, internationally accepted protocols for managing resource use and pollution in marine environments. Core principles of the ISO include a commitment to environmental management as an organizational priority; identification of appropriate legislative and regulatory requirements; identification of the environmental aspects of an organization's activities, products and services; development of management processes for achieving objectives and targets; appropriate financial and human resources to achieve targets; assignment of clear procedures for accountability; establishment of a maintenance review and audit process; and development and maintenance of communication with interested parties (von Zharen 1999).

Management actions will differ in local context, but global strategies for protecting marine ecosystems endorsed by the World Resources Institute, the International Union for the Conservation of Nature, the United Nations Environmental Programme, and other international conservation organizations focus on three things.

1. Establish a commission on ecosystem restoration to provide technical guidance and help secure funding for nations seeking to restore the sustainability of their coastal and fresh waters.

2. Map, using GIS technology, all macroscopic structure-forming species including coral, oyster and worm reefs, kelp and seagrass beds, and mangrove forests that provide essential habitat for other species.

3. Develop a marine biogeographic scheme based on patterns of species endemism that can be used to establish a global system of marine protected and special management areas and use this scheme to establish a global network of marine parks (Norse 1993).

Synthesis

We know too little about aquatic habitats, especially the marine habitats that cover 71% of the earth's surface and fill more than 90% of the planet's livable volume. Yet we make extensive withdrawals of natural resources from these poorly understood systems. The more important problem is not that we know too little, but that we may know too little, too late.

Problems of degradation and destruction of aquatic habitats result from both unmanaged inputs and unconstrained exploitation. To restore aquatic habitats, we must control what we put in and reduce what we take out. In the next decade, the successful conservation of aquatic habitat will require (1) control of inputs to aquatic systems through management of land-use practices surrounding these systems; (2) the establishment of an extensive, well-defined, and properly enforced aquatic reserve system, consisting of designated lakes, rivers, and marine areas that preserve high levels of the global biodiversity of aquatic communities; (3) more aggressive, persistent, and comprehensive research efforts to understand the workings of aquatic systems, unfamiliar worlds in which we do not live and which, without great effort, we cannot even observe; (4) reduction and restriction of our use of aquatic resources; and (5) international cooperation, jurisdiction, and ownership of the problems of marine environments.

Learning Online

Visit our webpage at www.mhhe.com/conservation for case studies, animations, practice quiz questions, and additional readings to help you understand the material in this chapter. You'll also find active links to the following topics:

Extinction and Marine Habitats
Water Use and Management
Water Pollution
Red Tides
Coral Bleaching

Groundwater
Resources from the Sea
Fisheries
Impact of Humans on the Sea:
 Harvesting

Impact of Humans on the Sea:
 Pollution
Fisheries and Conservation Issues
 Concerning Teleosts

Managing Commercial Fisheries Through Marine Reserves—A Directed Discussion

Reading assignment: Lauck, T., C. Clark, M. Mangel, and G. Munro. 1998. Implementing the precautionary principle in fisheries management through marine reserves. *Ecological Applications* 8:S72–S78.

Questions

1. Compare and contrast the authors' concept of a marine reserve with more traditional views of the role of marine reserves in commercial fisheries.

2. In the appendix of this paper, the authors note a critical assumption of their model: "The reserve boundaries are set for harvesting but the stock moves smoothly across the boundary and fills the entire fishing ground."

Identify a biological theory that supports this assumption. What alternative assumption is possible? Is there any support for the alternative in theory or specific knowledge of the natural history of harvested species?

3. Critics of Lauck et al. could argue that reducing the level of harvest achieves the same result as establishing a marine reserve. Why might these different strategies produce different outcomes? Which strategy is best at reducing risk to the fishery? Why?

4. Lauck et al. note an important cost of their strategy—enforcing the boundaries and protective regulations of the marine reserve. If you were the manager of a new reserve, how would you propose to pay for such costs? How would you persuade both conservationists and commercial fishers to support the plan with sufficient enthusiasm to pay some of the costs?

Literature Cited

Abell, R. A., D. M. Olson, E. Dinerstein, P. T. Hurley, J. T. Diggs, W. Eichbaum, S. Walters, W. Wettengel, T. Allnutt, C. J. Loucks, and P. Hedao. 2000. *Freshwater ecoregions of North America: a conservation assessment.* Washington, D.C.: Island Press.

Agardy, T. 1997. *Marine protected areas and ocean conservation.* Austin, Tex.: Landes and Academic Press.

Agardy, T. 1999. Creating havens for marine life. *Issues in Science and Technology* 16:37–44.

Ashton, P. S., and D. S. Mitchell. 1989. Aquatic plants: patterns and mode of invasion, attributes of invading species and assessment of control programs. In *Biological invasions: a global perspective,* eds. J. A. Drake, H. A. Mooney, F. diCastri, R. H. Groves, F. J. Kruger, M. Rejmánek, and M. Williamson, 111–54. Chichester, England: Wiley.

Auster, P. J. 1998. A conceptual model of the impacts of fishing gear on the integrity of fish habitats. *Conservation Biology* 12:1198–1203.

Barbier, E. B. 1995. Tropical wetland values and environmental functions. In *Biodiversity conservation: problems and policies,* eds. C. A. Perrings, K. G. Mäler, C. Folke, C. S. Holling, and B.-O. Jansson, 147–69. Dordrecht, The Netherlands: Kluwer Academic Publishers.

Barry, J. P., and P. K. Dayton. 1991. Physical heterogeneity and the organization of marine communities. In *Ecological heterogeneity,* Ecological Studies 70, eds. J. Kolasa and S. T. A. Pickett, 270–320. New York: Springer-Verlag.

Benke, A. C. 1990. A perspective on America's vanishing streams. *Journal of the North American Benthological Society* 9:77–88.

Blindow, I., A. Hargeby, and G. Andersson. 1997. Alternative stable states in shallow lakes—what causes a shift? In *The structuring of submerged macrophytes in lakes,* eds. E. Jeppesen, M. Søndergaard, and K. Christoffersen, 353–68. Berlin: Springer.

Bond, W. K., K. W. Cox, T. Heberlein, E. W. Manning, D. R. Witty, and D. A. Young. 1992. Wetland evaluation guide: final report of the Wetlands Are Not Wastelands Project. Sustaining wetlands issues paper, no. 1. Ottawa, Ont.: North American Wetlands Conservation Council.

Brix, H. 1994. Constructed wetlands for municipal waste treatment in Europe. In *Global wetlands: old world and new,* ed. W. J. Mitsch, 325–33. Amsterdam: Elsevier.

Broecker, W. S. 1991. The great ocean conveyor. *Oceanography* 4:79–89.

Brönmark, C., and L. A. Hansson. 1998. *The biology of lakes and ponds.* Oxford, England: Oxford University Press.

Butman, C. A. 1987. Larval settlement of soft-sediment invertebrates: the spatial scales of pattern explained by active habitat selection and the emerging role of hydrodynamical processes. *Oceanography and Marine Biology: An Annual Review* 25:113–65.

Butman, C. A., J. P. Grassle, and C. M. Webb. 1988. Substrate choices made by marine larvae settling in still water and in a flume flow. *Nature* 333:771–73.

Carlton, J. T. 1985. Transoceanic and interoceanic dispersal of coastal marine organisms: the biology of ballast water. *Oceanography and Marine Biology: An Annual Review* 23:313–74.

Chescheir, G. M., R. W. Skaggs, and J. W. Gilliam. 1992. Evaluation of wetland buffer areas for treatment of pumped agricultural drainage water. *Transactions of the American Society of Agricultural Engineers* 35:175–82.

Cowardin, L. M., V. Carter, F. C. Golet, and E. T. LaRoe. 1979. *Classification of wetlands and deepwater habitats of the United States.* Washington, D.C.: U.S. Government Printing Office.

Crivelli, A. J. 1983. The destruction of aquatic vegetation by carp. *Hydrobiologia* 106:37–41.

Dixon, J. A., L. F. Scura, and T. van't Hof. 1995. Ecology and microeconomics as "joint products": the Bonaire Marine Park in the Caribbean. In *Biodiversity conservation: problems and policies,* eds. C. A. Perrings, K.-G. Mäler, C. Folke, C. S. Holling, and B.-O. Jansson, 127–45. Dordrecht, The Netherlands: Kluwer Academic Publishers.

Drake, J. A. 1991. Community assembly mechanics and the structure of an experimental species ensemble. *American Naturalist* 137:1–26.

Ehrenfeld, D. 1992. The business of conservation. *Conservation Biology* 6:1–3.

Farrell, D. A. 1998. Restoring stream corridors. *Agricultural Research* 46:2.

Findlay, C. S., and J. Houlahan. 1997. Anthropogenic correlates of species richness in southeastern Ontario wetlands. *Conservation Biology* 11:1000–9.

Goeden, G. B. 1982. Intensive fishing and a "keystone" predator species: ingredients for community instability. *Biological Conservation* 22:273–81.

Gosselink, J. G., and E. Maltby. 1990. Wetland losses and gains. In *Wetlands: a threatened landscape,* ed. M. Williams, 296–322. Oxford, England: Basil Blackwell.

Grassle, J. F., and N. J. Maciolek. 1992. Deep-sea species richness: regional and local diversity estimates from quantitative bottom samples. *American Naturalist* 13:313–41.

Hardin, G. 1968. The tragedy of the commons. *Science* 162:1243–48.

Hayes, T. 1998. Conservation of native freshwater mussels: an overview. *Endangered Species Update* 15(6):108–10.

Hilborn, R., C. J. Walters, and D. Ludwig. 1995. Sustainable exploitation of renewable resources. *Annual Review of Ecology and Systematics* 26:45–67.

Hinrichsen, D. 1997. Coral reefs in crisis. *BioScience* 47:554–58.

International Union for the Conservation of Nature (IUCN), United Nations Environmental Programme, and World Wide Fund for Nature. 1991. *Caring for the earth: a strategy for sustainable living.* Gland, Switzerland: IUCN.

Johnson, P. D., and R. S. Butler. 1999. Conserving a treasury of diversity. *Endangered Species Bulletin* 24(4):16–17.

Johnson, L. E., and J. T. Carlton. 1996. Post-establishment spread in large-scale invasions: dispersal mechanisms of the zebra mussel *Dreissena polymorpha. Ecology* 77:1686–90.

Kadlec, R. H. 1994. Wetlands for water polishing: free water surface wetlands. In *Global wetlands: old world and new,* ed. W. J. Mitsch, 335–49. Amsterdam: Elsevier.

Kaitala, V., and G. Munro. 1995. The management of trans-boundary resources and property rights systems: the case of fisheries. In *Property rights and the environment,* eds. S. Hanna and M. Munasinghe, 69–84. Washington, D.C.: Beijer International Institute of Ecological Economics and the World Bank.

Karr, J. R. 1991. Biological integrity: a long-neglected aspect of water resource management. *Ecological Applications* 1:66–84.

Kenchington, R. A. 1990. *Managing marine environments.* New York: Taylor and Francis.

Koester, V. 1989. *The Ramsar Convention on the Conservation of Wetlands: a legal analysis of the adoption and implementation of the convention in Denmark.* Gland, Switzerland: IUCN.

Lambshead, P. J. D. 1993. Recent developments in marine biodiversity research. *Oceanus* 19:5–24.

Lant, C. L., S. E. Kraft, and K. R. Gillman. 1995. The 1990 farm bill and water quality in corn belt watersheds: conserving remaining wetlands and restoring farmed wetlands. *Journal of Soil and Water Conservation* 50:201–5.

Lauck, T., C. W. Clark, M. Mangel, and G. R. Munro. 1998. Implementing the precautionary principle in fisheries management through marine reserves. *Ecological Applications* 8:S72–S78.

Lucas, J. S. 1994. The biology, exploitation, and mariculture of giant clams (Tridacnidae). *Reviews in Fisheries Science* 2:181–223.

Lucas, J. S. 1997. Giant clams: mariculture for sustainable exploitation. In *Conservation and the use of wildlife resources,* ed. M. Bolton, 77–95. London, England: Chapman and Hall.

Macdonald, I. A. W., L. L. Loope, M. B. Usher, and O. Hamann. 1989. Wildlife conservation and the invasion of nature reserves by introduced species: a global perspective. In *Biological invasions: a global perspective,* eds. J. A. Drake et al., 215–55. Chichester, England: Wiley.

Maltby, E., and R. E. Turner. 1983. Wetlands of the world. *Geography Magazine* 55:12–17.

Matthews, E., and I. Fung. 1987. Methane emission from natural wetlands: global distribution, area, and environmental characteristics of sources. *Global Biogeochemical Cycles* 1:61–86.

May, R. 1992. Bottoms up for the oceans. *Nature* 357:278–79.

Meybeck, M., and R. Helmer. 1989. The quality of rivers: from pristine stage to global pollution. *Paleogeography, Paleoclimatology, and Paleoecology* 75:283–309.

Mitsch, W. J., R. H. Mitsch, and R. E. Turner. 1994. Wetlands of the old and new worlds: ecology and management. In *Global wetlands: old world and new,* ed. W. J. Mitsch, 3–56. Amsterdam: Elsevier.

Neverman, D., and W. A. Wurtsbaugh. 1994. The thermoregulatory function of diel vertical migration for a juvenile fish, *Cottus extensus. Oecologia* 98:247–56.

Norse, E. A. 1993. *Global marine biodiversity: a strategy for building conservation into decision making.* Washington, D.C.: Island Press.

Osborne, L. L., and D. A. Kovacic. 1993. Riparian vegetated buffer strips in water-quality restoration and stream management. *Freshwater Biology* 29:243–58.

Peterson, C. H. 1979. Predation, competitive exclusion, and diversity in soft-sediment benthic communities of estuaries and lagoons. In *Ecological processes in coastal and marine systems,* ed. R. J. Livingston, 223–64. New York: Plenum.

Ratner, S., R. Lande, and B. B. Roper. 1997. Population viability analysis of spring Chinook salmon in the South Umpqua River, Oregon. *Conservation Biology* 11:879–89.

Ricker, W. E. 1958. Maximum sustained yields from fluctuating environments and mixed stocks. *Journal of the Fishery Research Board of Canada,* Bulletin no. 191.

Rickerl, D. H., L. L. Janssen, and R. Woodland. 2000. Buffered wetlands in agricultural landscapes in the prairie pothole region: environmental, agronomic, and economic evaluations. *Journal of Soil and Water Conservation* 55:220–25.

Robel, R. J. 1961. The effect of carp populations on the production of waterfowl food plants on a western waterfowl marsh. *Transactions of the North American Wildlife and Natural Resources Conference* 26:147–59.

Rosenzweig, M. L. 1991. Habitat selection and population interactions: the search for mechanism. *American Naturalist* 137:S5–S28.

Rubec, C. D. A. 1994. Canada's federal policy on wetland conservation: a global model. In *Global wetlands: old world and new,* ed. W. J. Mitsch, 909–17. Amsterdam: Elsevier.

Ruckelshaus, M. H., and C. G. Hays. 1998. Conservation and management of species in the sea. In *Conservation biology: for the coming decade,* 2d edition, eds. P. L. Fiedler and P. M. Kareiva, 112–56. New York: Chapman and Hall.

Ruiz, G. M., J. T. Carlton, E. D. Grosholz, and A. H. Hines. 1997. Global invasions of marine and estuarine habitats by non-indigenous species: mechanisms, extent, and consequences. *American Zoologist* 37:621–32.

Russ, G. R., and A. C. Alcala. 1996. Marine reserves: rates and patterns of recovery and decline of large predatory fish. *Ecological Applications* 6:947–61.

Seehausen, O., F. Witte, E. F. Katunzi, J. Smits, and N. Bouton. 1997. Patterns of remnant cichlid fauna in southern Lake Victoria. *Conservation Biology* 11:890–904.

Sheffer, M. 1990. Multiplicity of stable states in freshwater systems. *Hydrobiologia* 200/201:475–86.

Sheldon, S. P., and R. P. Creed Jr. 1995. Use of a native insect as a biological control for an introduced weed. *Ecological Applications* 5:1122–32.

Sioli, H. 1975. Tropical rivers as expressions of their terrestrial environments. In *Tropical ecological systems,* eds. F. B. Golley and E. Medina, 275–88. New York: Springer-Verlag.

Skelly, D. S., and E. Meir. 1997. Rule-based models for evaluating mechanisms of distributional change. *Conservation Biology* 11:531–38.

Snelgrove, P. V. R. 1999. Getting to the bottom of marine biodiversity: sedimentary habitats. *BioScience* 49:129–38.

Sobel, J. 1993. Conserving biological diversity through marine protected areas: a global challenge. *Oceanus* 36:19–26.

Steedman, R. J. 1988. Modification and assessment of an index of biotic integrity to quantify stream quality in southern Ontario. *Canadian Journal of Fisheries and Aquatic Sciences* 45:492–501.

Stump. K. 2000. Trouble in the North Pacific: sea lions vs. fisheries. *Earth Island Journal* 15(2):11.

United Nations. 1992. Adoption of agreements on environment and development: Agenda 21, United Nations Conference on Environment and Development. United Nations Document A/CONS.151/26. New York: United Nations.

United Nations Food and Agriculture Organization. 1992. *Marine fisheries and the law of the sea: a decade of change.* UN Food and Agriculture Organization Fisheries circular no. 583. Rome: UN Food and Agriculture Organization.

VanDeVeer, S. D. 2000. Protecting Europe's seas. *Environment* 42:10–26.

von Zharen, W. M. 1999. An ecopolicy perspective for sustaining living marine species. *Ocean Development and International Law* 30:1–41.

Whitlatch, R. B. 1977. Seasonal change in the community structure of the macrobenthos inhabiting the intertidal sand and mud flats of Barnstable Harbor, Massachusetts. *Biological Bulletin* 152:275–94.

Williams, M. 1990. Protection and retrospection. In *Wetlands: a threatened landscape,* ed. M. Williams, 323–53. Oxford, England: Basil Blackwell.

Witte, F., T. Goldschmidt, J. Wanink, M. van Oijen, K. Goudswaard, E. Witte-Maas, and N. Bouton. 1992. The destruction of an endemic species flock: quantitative data on the decline of the haplo-chromine cichlids of Lake Victoria. *Environmental Biology of Fishes* 34:1–28.

World Resources Institute, International Union for the Conservation of Nature, and United Nations Environmental Programme. 1992. *Global biodiversity strategy.* Washington, D.C.: World Resources Institute.

Only if we can comprehend and envision the entity we are trying to shape as a dynamic whole can we have any hope of dealing with it creatively.

—J. T. Lyle, 1985

Ecosystem Management

In this chapter you will learn about:

1 what ecosystem management is.

2 how and why the concept of ecosystem management developed.

3 the scientific basis of ecosystem management.

4 the prerequisites for ecosystem management.

5 the obstacles to ecosystem management and their potential solutions.

THE CONCEPT OF ECOSYSTEM MANAGEMENT

One definition of *ecosystem* is *all the organisms in a given area interacting with the physical environment so that a flow of energy leads to trophic structure, biotic diversity, and material cycles* (Odum 1971). Put simply, ecosystems are energy- and nutrient-processing systems with physical structures and functions that circulate matter and distribute energy (Berry et al. 1998).

Although the ecosystem concept dates to the early twentieth century, the idea of *managing* ecosystems is relatively new. Of all modern efforts in conservation, none has proven more elusive in definition or more controversial in implementation than "ecosystem management." Conservationist Michael Bean wrote of ecosystem management, "rarely has a concept gone so directly from obscurity to meaninglessness without any intervening period of coherence" (Bean 1997:23). Less cynically but not more optimistically, Berry and associates (1998) wrote, "No single operational definition of ecosystem management exists, although its basic principles are understood."

In the United States, 18 federal agencies have adopted or are considering adoption of programs based on ecosystem management (Congressional Research Service 1994; Christensen et al. 1996; Haeuber 1996; Haeuber and Franklin 1996). Philosophical support for their efforts comes from a wealth of definitions of

ecosystem management, many written by the agencies themselves (table 10.1). However, despite the intensity of effort represented by these definitions, policy analyst Richard Haeuber (1996) described ecosystem management as "a loose collection of agency specific concept papers, policy guidance documents, and potential—or only partially implemented—administrative changes." None of the current definitions is entirely satisfactory or unequivocal. One of the best efforts is that of Noss and Cooperrider (1994:360–361), who define ecosystem management as *any land management system that seeks to protect viable populations of all native species, perpetuates natural disturbance regimes on the regional scale, adopts a planning time line of centuries, and allows human use at levels that do not result in long-term ecological degradation.*

Although some professionals view ecosystem management as a new paradigm, others have asserted that it is instead a vacuous phrase "desperately seeking a paradigm" (Lackey 1998). Views on the concept of ecosystem management are diverse. Various authors have asserted that the ecosystem management concept has been around since the 1930s (Haeuber 1996), that it is a dressed-up version of the U.S. Forest Service's old "multiple use" management (Czech and Krausman 1997), that it is the same as "public lands management because public lands are ecosystems" (Czech 1995), that it is what managers have been doing all along (Irland 1994; More 1996; Berry et al. 1998), that it is the same as conservation because it has the same goal and therefore should be renamed "ecosystem conserva-

Table 10.1 **Some Definitions of Ecosystem Management from U.S. Federal Agencies**

AGENCY	DEFINITION
Department of Agriculture	The integration of ecological principles and social factors to manage ecosystems to safeguard ecological sustainability, biodiversity, and productivity.
Department of Commerce, National Oceanic and Atmospheric Administration	Activities that seek to restore and maintain the health, integrity, and functional values of natural ecosystems that are the cornerstone of productive, sustainable economies.
Department of Defense	The identification of target areas, including Department of Defense lands, and the implementation of a "holistic approach" instead of a "species-by-species approach" in order to enhance biodiversity.
Department of Energy	A consensual process based on the best available science that specifically includes human interactions and management and uses natural instead of political boundaries in order to restore and enhance environmental quality.
Department of the Interior: Bureau of Land Management	The integration of ecological, economic, and social principles to manage biological and physical systems in a manner safeguarding the long-term ecological sustainability, natural diversity, and productivity of the landscape.
Fish and Wildlife Service	Protection or restoration of the function, structure, and species composition of an ecosystem, recognizing that all components are interrelated.
National Park Service	A philosophical approach that respects all living things and seeks to sustain natural processes and the dignity of all species and to ensure that common interests flourish.
U.S. Geological Survey	Ecosystem management to emphasize natural boundaries, such as watersheds, biological communities, and physiographic provinces, and bases management decisions on an integrated scientific understanding of the entire ecosystem.
Environmental Protection Agency	To maintain overall ecological integrity of the environment while ensuring that ecosystem outputs meet human needs on a sustainable level.
National Science Foundation	An integrative approach to the maintenance of land and water resources as functional habitat for an array of organisms and the provision of goods and services to society.

Compiled from U.S. Congressional Research Service 1994.

tion" (Czech 1995), that it is a conspiracy to reduce the extractive use of natural resources and expel private citizens from public lands (Christensen et al. 1996), that it is not a new paradigm at all (Slocombe 1993; Taylor 1993; Czech 1995), and that Aldo Leopold thought of it first (Czech 1995; Knight 1995; Grumbine 1998).

Such diverse views make for entertaining reading, but they do not resolve the questions of what ecosystem management is or what its actual or potential role in conservation ought to be. Conservation goals are not achieved by arguing about words and names, and science does not create concept by fiat, but by function. But words do matter when they embody dis-

tinctive and objective concepts. If we follow a classical authority such as Aristotle, a good definition expresses the essence or nature of an entity and not merely its accidental properties (Abelson 1967; More 1996). An ideal definition would be one that includes "all instances and only instances" of the category we define and specifies both the essence of ecosystem management and its boundaries so that when we apply it, we will be able to determine if something is or is not ecosystem management (More 1996).

"Ecosystem management" is a concept and practice that will take an ever-increasing role in conservation biology and management. As we grow in our appreciation of the amount

Table 10.2   **Fundamental Differences Between Resource Management and Ecosystem Management Paradigms**

	RESOURCE MANAGEMENT	ECOSYSTEM MANAGEMENT
Entity of Value	Resource	Ecosystem
Application of Value	Beneficial use	Continuing function
Management Unit	Species or abiotic factor	Landscape elements
Time Scale	Relatively short	Relatively long
Management Jurisdiction	Single government agency	Multiple government agencies and private landowners
Management Decision Making	Single government agency	Multiple government agencies and private stakeholders
Management Goals	Production and use of resource commodities	Productivity and sustainability of ecosystem functions and processes

of area populations need to remain viable and persistent, as we include more sectors of human society in conservation effort, and as we recognize the necessity of cooperation among management agencies, conservation organizations, and public and private sectors of human activity, an ecosystem management approach will become essential. But it will also be ineffective if no one understands what ecosystem management really means. To make further progress, we must distinguish ecosystem management from other, superficially similar, alternatives.

Development of the Ecosystem Management Paradigm

From the 1960s, managers of public lands, as well as academics in applied sciences like wildlife management, range management, fisheries, and forestry commonly spoke and wrote about "ecosystem concepts in management" (Major 1969; Van Dyne 1969; Wagner 1969; Wagner 1977). The term *ecosystem management* had become common by the late 1970s (Czech and Krausman 1997). However, when one examines original contexts, authors from this period almost always used the phrase to describe either the management of populations as commodities within ecosystems or the manipulation of processes, structures, and functions of ecosystems to produce desired levels of animal populations or plant biomass (Major 1969; Wagner 1969; Wagner 1977).

If this is all that "ecosystem management" means, then the concept would certainly not meet the criteria for a scientific paradigm, and it would not represent a genuine "paradigm shift" to any new concepts or ideas. Ultimately, paradigms come to incorporate and express the values, theories, methodologies,

and tools that a professional community prescribes and believes in to achieve a desired condition (Kuhn 1970; Czech 1995). The modern concept of ecosystem management is very different from earlier ideas, and the increasing adoption of what is called "ecosystem management" does represent a genuine transfer of popular, scientific, and professional loyalty from one group of ideas and values to another. In the case of ecosystem management, this transfer has been from a more traditional view of what could be called "resource management" to the modern concept of "ecosystem management." The critical elements of ecosystem management are most clearly understood when they are directly compared with parallel elements in the more traditional resource management paradigm (table 10.2).

In the United States, major federal agencies have always had jurisdiction *over* ecosystems, but they have, until recently, never managed their jurisdictions *as* ecosystems. In the resource management paradigm, the entity of value is a particular "resource," either an individual species or an abiotic component of the system such as water, soil, or minerals. The resource is seen as a commodity and its value is "use." Units of management are the species or abiotic factors and the sites on which they occur. In a resource management approach, single species are managed on a site-specific basis, usually through direct intervention. Time scales are relatively short and management decisions occur in individual agencies. The long-term goal is optimal, renewable, and sustained production of multiple natural resources as commodities, both economic and noneconomic, for multiple uses. Within this larger aim, management objectives are set and determined by demand for commodities that the system can supply.

The resource management paradigm increasingly has had difficulties in dealing with contemporary conservation problems, creating anomalies that helped prepare a climate of acceptance for an alternative approach to management. For example, as the number of threatened and endangered species has grown, species preservation on a case-by-case basis has become too expensive and ineffective (Franklin 1993; Sparks 1995). Population viability analysis (PVA) has proven to have serious deficiencies, including high requirements for data and human effort and difficulty in validating results (Franklin 1993; Poiani et al. 2000). A species-oriented approach to habitat management—the construction of habitat-suitability models (HSMs)—also is limited by ignorance, particularly with regard to aquatic ecosystems, where there may be many species for which insufficient information exists to construct an HSM (Sparks 1995). Further, optimizing habitats for a few species may create suboptimal habitat for many others. Finally, many of the "smaller" taxonomic categories such as fungi, protozoans, and bacteria have gone unnoticed in their declines toward extinction because single-species approaches have invested most heavily in large, high-profile vertebrates (Franklin 1993). The resource management paradigm has also been eroded by the growing realization that its command and control structures, which reside in individual agencies, cannot cope with modern environmental problems, such as air and water pollution, toxic waste disposal, acid precipitation, soil erosion, and stream sedimentation, because the sources and scope of the problems exceed the agencies' jurisdictional powers. Further, the resource management paradigm has not been able to cope successfully with what could be called the "nationalization" of environmental values. In the past, local ecosystems were of primarily local interest. The resource management paradigm often produced outcomes with local benefits, especially in providing jobs and incomes. More recently, strong, geographically diverse national constituencies have arisen that pursue ecosystem preservation for aesthetic and recreational values, and the resource management paradigm has not been able to successfully address these constituencies or their concerns, and so has often lost broad-based support.

Although the resource management paradigm has maintained high levels of support among commodity-oriented constituencies such as timber, livestock, fisheries, hunting, and mining interests, its legal foundations have eroded with the passage of laws that contain a stronger ecosystem emphasis and a shift in public values away from viewing ecosystems as sources of commodities (Marsden 1991) (chapter 2). The inability of the resource management paradigm to deal with increasing numbers of endangered species, changing legal mandates, and shifting public values and uses of ecosystems created an opening for an alternative way of viewing and managing public and private lands. The alternative paradigm is ecosystem management.

Characteristics of the Ecosystem Management Paradigm

Entity of Value and Sustainability

Ecosystem management places value in the ecosystem rather than the resources it contains. The ecosystem itself is perceived as an object worthy of respect and admiration, valued for its beauty, complexity, history, and cultural significance. Ecosystem management values long-term delivery of ecosystem services, the stability and persistence of ecosystem compo-

nents, and the long-term stability of transfers of matter and energy within the system. Management goals regarding how much can be extracted from or used in the system are set by its capacity to deliver the desired goods and services, not by the demand for those good and services. Although goods and services may be outputs of the ecosystem, the ecosystem manager's ultimate goal is sustainability of the system, not the delivery of resources. Because of this shift in the entity of value from resource commodities to ecological systems, the ecosystem management paradigm is capable of dealing with changing biological and sociopolitical structures that frustrate the resource management paradigm.

Management Mechanisms and Decisions

The mechanisms of ecosystem management are primarily the identification and guidance of landscape-scale processes. Management time scales are relatively long, and areas of management are too large to fit within the jurisdiction of a single agency. In fact, jurisdiction usually is not exclusively within government control at all, but also includes significant private ownership. Thus, management decisions must incorporate decision-making strategies that involve all agencies with jurisdiction over lands or processes in the ecosystem, private landowners within or adjacent to the system, and nonresidents who use the system on a seasonal basis or who have specific and vested interests in the condition of the system.

With these distinguishing features in mind, a "classical" definition of ecosystem management (i.e., one that distinguishes ecosystem management from other types of land and resource management) might sound something like *a pattern of prescribed, goal-oriented environmental manipulation that (1) treats a specified ecological system as the fundamental unit to be managed; (2) has a desired outcome of assuring the persistence of historical components, structure, function, products, and services of the system within biological and historical ranges and rates of change over long time periods; (3) uses naturally occurring, landscape-scale processes as the primary means of achieving management objectives; and (4) determines management objectives through cooperative decision making by individuals and groups who reside in, administer, or have vested interests in the state of the ecosystem.* In this context, Grumbine's 10 themes of ecosystem management (Grumbine 1994), the Ecological Society of America's eight primary characteristics of ecosystem management (Christensen et al. 1996), and More's (1996) five dimensions of ecosystem management can be seen as parallel expressions of similar values (table 10.3).

In all its contexts and definitions, ecosystem management has consistently included and stressed three foundational premises: (1) the ecosystem, rather than individual organisms, populations, species, or habitats, is considered the appropriate management unit; (2) emphasis is placed on the development and use of adaptive management models, which treat the ecosystem as the subject of study and research, and management activities as experimental and uncertain; and (3) those with vested interests in the persistence, health, and services of the ecosystem (stakeholders) should participate in management decisions. In ecosystem management, actions should follow

Table 10.3   Parallel Descriptions of Ecosystem Management

THE ECOLOGICAL SOCIETY OF AMERICA'S EIGHT PRIMARY CHARACTERISTICS OF ECOSYSTEM MANAGEMENT (CHRISTENSEN ET AL. 1996)		GRUMBINE'S 10 DOMINANT THEMES OF ECOSYSTEM MANAGEMENT (GRUMBINE 1994)		MORE'S 5 DIMENSIONS OF ECOSYSTEM MANAGEMENT (MORE 1996)	
Sustainability	Managers aim to create improvements that are not ephemeral.	Hierarchical Context	Managers connect all levels (genes, species, populations, ecosystems, and landscapes).	Recognition of Ecosystem Health	Managers focus on integrity, functions, protection, critical habitats, habitat relationships, and restoration of the ecosystem.
Goals	Managers set goals that specify future processes and outcomes necessary for sustainability.	Ecological Boundaries	Managers are concerned with the ecological boundaries that often cross administrative or political boundaries.	Maintenance and Enhancement of Biodiversity	Managers maintain or restore native/primeval species and care for old-growth and older forest stands.
Sound Ecological Models and Understanding	Managers organize research at all levels of ecological organization.	Interagency Cooperation	Managers interact with the legal mandates and management goals of other agencies to whom they are responsible for a component of the ecosystem (i.e., federal and state officials and private land owners).		
Complexity and Connectedness	Managers recognize that biological diversity and structural complexity strengthen ecosystems against disturbance and supply the genetic resources necessary to adapt to long-term change.	Ecological Integrity	Managers maintain or restore native species, populations, and ecosystems.	Emphasis on Sustainability	Managers holistically consider the long-term effects of their plans.
Wide Temporal and Spatial Scale	Ecosystem processes operate over a wide range of temporal and spatial scales and their behavior at any given location is greatly affected by surrounding systems. Thus, there is no single appropriate scale or time frame for management.			Wide Temporal and Spatial Scale	Managers avoid fragmentation, protect waterways, and focus on landscape-scale trends and conditions.
Humans as Ecosystem Components	Managers value the active role of humans in achieving sustainable management goals.	Humans Embedded in Nature	Managers consider the impact of humans upon the ecosystem and also the impact the ecosystem has on humans.	Legitimacy of Human Dimensions	Managers recognize the need for human communities to utilize some ecosystem resources.

Continued

Table 10.3 *Continued*

THE ECOLOGICAL SOCIETY OF AMERICA'S EIGHT PRIMARY CHARACTERISTICS OF ECOSYSTEM MANAGEMENT (CHRISTENSEN ET AL. 1996)	GRUMBINE'S 10 DOMINANT THEMES (GRUMBINE 1994)	
The Dynamic Character of Ecosystems	**Monitoring**	Managers establish patterns of regular monitoring of ecosystem components and processes.
Managers avoid attempting to halt the evolution of ecosystems.		
Adaptability and Accountability	**Adaptive Management**	Managers are flexible and adapt to uncertainty. Management is viewed as a learning process during which the results of continual experiments are utilized to establish management.
Managers realize that paradigms and current knowledge are not infallible. Approaches to management are viewed as experiments whose results are carefully examined.		
	Organizational Change	Management agencies may need to change their structure and the way they operate (e.g., forming an interagency committee, changing professional norms, or altering power relationships).
	Data Collection	Managers research the ecosystem and collect ecosystem data.
	Values	Managers keep in mind that human values play a dominant role in ecosystem management goals.

Compiled from Christensen et al. 1996, Grumbine 1994, and More 1996.

careful experimental design, include environmental controls (unmanipulated sites or subjects), and be carefully monitored over time. If the experimental design is sound, the results of the management action should be relatively unambiguous, but must still be interpreted stochastically (within a range of outcomes with differing probabilities), rather than as a deterministic outcome generated by simple cause-and-effect relationships.

Legislative Mandate for Ecosystem Management

There is no existing environmental statute in the United States that provides an unequivocal basis for ecosystem management (Haeuber 1996). This does not mean that there is no legal authority for taking an ecosystem management approach, but such an approach has, to this point, been based on implied directives found in other environmental legislation.

The Wilderness Act (1964) established a national wilderness system "where man himself is a visitor who does not remain," and the Wild and Scenic Rivers Act (1968) established a similar aquatic system where most forms of human impacts were limited or prohibited altogether. By implication, the Wilderness Act and the Wild and Scenic Rivers Act acknowledged the value of intact ecosystems and assumed that, by isolating such systems from human impacts, the systems would be protected and preserved (Christensen et al. 1996). In 1969, the National Environmental Policy Act (NEPA) established environmental quality as a national priority. Although NEPA was a type of "impact assessment" legislation, the policies and agency regulations that evolved in the course of its implementation encouraged new and more active forms of public participation in the decisions of federal land management agencies. Such a participation-driven process created a climate that required the development of the "stakeholder" concept, because individuals and groups with legitimate vested interests in management decisions had to be identified. Cooperative decision making by diverse agencies and private citizens began to be seen as the appropriate goal, and even an ideal, of public land management. Such an ideal is a foundational principle of ecosystem management.

The Endangered Species Act (ESA, 1973) focused primarily on individual species and subspecies, but it contained explicit language that required protection of the habitats and ecosystems of endangered or threatened species to ensure their recovery. Thus, the ESA highlighted the connectedness of species and the interrelationships between species and their surrounding landscapes, and indirectly encouraged agencies to coordinate and adopt principles of ecosystem management.

The Clean Water Act (CWA, 1977) established a goal of restoring and maintaining the physical, chemical, and biological integrity of the nation's waters. Its wording gives strong support for an ecosystem management approach. For many years, the U.S. Environmental Protection Agency (EPA) continued to follow the precedents of earlier versions of the Clean Water Act that put the emphasis on enforcing standards of water purity and testing the efficacy of pollution abatement programs (Sparks 1995). Recently, the EPA has begun a gradual but significant shift toward using the Clean Water Act as a means of ecosystem protection. In 1990, the EPA's Science Advisory Board advised the agency to "communicate to the general public a clear message that it considers ecological risks to be just as serious as human health and welfare risks, because of the inherent value of ecological systems and their link to human health" and it encouraged the EPA to take action on the protection of aquatic ecosystems (Norton and Davis 1997). From these recommendations, the EPA formed a working group, the Habitat Cluster, to formulate a strategy for problems facing the nation's ecosystems. The cluster identified six areas as critical for ecosystem protection:

1. improving the use of existing regulatory authorities,
2. improving the focus of the EPA's nonregulatory programs for ecosystem protection,
3. improving the habitat science base,
4. providing better habitat information management,
5. forming effective public and private partnerships, and
6. using a risk-based approach for setting priorities and making decisions.

This change in emphasis led to not only new regulations and policies within the EPA, but also to the improved use of existing programs and protocols, such as the Clean Water Act's watershed-oriented programs and geographic initiatives to protect aquatic ecosystems and reverse trends in their degradation. By 1994, EPA Administrator Carol Browner stated unequivocally, "We need ecosystem protection," and began systematic changes within the EPA to make its methods more responsive to that need (Norton and Davis 1997). The new approach, called *community-based environmental protection,* is intended to improve the effectiveness of EPA actions by making its national-scale programs more relevant and responsive to people and ecosystems at local scales. The Clean Water Act has many limitations, including relatively limited power in controlling nonpoint pollution, harvest, nonindigenous species, wetland losses, and many types of habitat alteration. Nevertheless, the CWA does offer some important mechanisms for ecosystem protection, such as water quality standards, that were used recently in the initial phases of restoring the Florida Everglades and in restoring salmon fisheries in the Pacific Northwest (Norton and Davis 1997).

The National Forest Management Act (NFMA) of 1976 was drafted primarily to constrain where and under what conditions timber harvesting could take place in national forests, but it has become one of the most important pieces of federal legislation for ecosystem management. The NFMA contained requirements for environmental protection of national forests, and, under its requirements, the Forest Service began to move away from its traditional emphasis on timber production as its most important priority. In 1982, the NFMA was revised with the intention of weakening its environmental provisions, but the revisions actually had the effect of strengthening them. One of the most important changes instructed the Forest Service to provide for "diversity of plant and animal communities based on the suitability and capability of the specific land area in order to meet overall multiple-use objectives . . . (and) to preserve the diversity of tree species similar to that existing in the region." Although not explicitly calling for ecosystem management, the NFMA all but demanded an ecosystem management approach. The NFMA directs the Forest Service that "habitat shall be managed to maintain viable populations of existing native and desired nonnative vertebrate species in the planning area." Such habitat "must be well distributed so that those individuals can interact with others in the planning area" (i.e., individuals should be distributed so as to avoid negative effects of low densities, or Allee effects, chapter 7). This directive is difficult to accomplish without managing at landscape scales, and so has been interpreted as an implied mandate for ecosystem management.

The California Desert Protection Act of 1994 was perhaps the first U.S. legislation to provide protection for an ecosystem. Passage of this act occurred through a confluence of concerns over endangered species such as the desert tortoise (*Gopherus agassizii*), ecosystem degradation from the use of off-road vehicles, and strong regional public support. Although noteworthy in recognizing ecosystems as entities entitled to legal protection, the act is highly regional in nature and provides no national mandate for ecosystem protection.

A single act of legislation did not give birth to the modern concept of ecosystem management; instead, a combination of legislative mandates converged upon this concept and its inherent values. We explore these values further to better understand the concept and implementation of ecosystem management.

Values in Ecosystem Management

The fundamental value of ecosystem management is that, rightly practiced, it preserves the essential functions of the ecosystem and the products and services that such functions deliver. The products and services of ecosystems can be considered together as support functions, production functions, regulation functions, and information functions (table 10.4) (Vos and Zonneveld 1993). Ecosystems manifest these functions in numerous ways, many of which are specific to individual ecosystems; however, some important functions and services are common to many ecosystems.

Climate and Hydrology

Large ecosystems regulate global climate through removal of atmospheric carbon, release of oxygen, and transfer of water (Christensen et al. 1996). The alteration of large ecosystems has been shown to have significant effects on local climatic factors such as temperature and rainfall.

Ecosystems store and purify water and reduce detrimental downstream effects associated with erosion and flooding. This function is particularly important for forest, wetland, riparian, and river ecosystems. For example, floodplains and wetlands associated with river ecosystems reduce the extent of downstream flooding and the height of the floodwaters; thus, constriction of the floodplain decreases the ability of the system to absorb water inputs. Ironically, intensive flood management techniques such as higher levees and flood protection devices exacerbate floods by preventing water from moving to and being stored in the floodplain and thus increase the height of the flood (Leopold 1994; Sparks 1995). The service of water storage and flood prevention by intact ecosystems is so important that many have advocated making watersheds the fundamental unit of ecosystem management (Norton and Davis 1997; Rama Mohan Rao et al. 1999).

Table 10.4 Four Categories of Ecosystem Functions

FUNCTION	DEFINITION
Support	Offers a foundation for human constructions and transport
Production	Provides mineral materials and gives the opportunity to produce agricultural products
Regulation	Guarantees a sustainable environment based on the resilience of ecological systems
Information	Offers information regarding the various aesthetic, scientific, and historical interests of humankind

After Vos and Zonneveld 1993.

Biological Productivity and Components of Biodiversity

Functioning ecosystems produce large amounts of plant and animal biomass. Some of this biomass can be used as sources of food, structural materials, and medicine. In addition, ecosystems produce and store large quantities of genetic resources that provide new strains of crop plants or enhance existing ones. Whereas the paradigm of resource management focuses efforts on biotic elements that can be treated as commodities because of their instrumental or noninstrumental value, ecosystem management focuses on preserving systems and the processes that provide resources to sustain those elements.

Even if ecosystem processes are preserved, that does not mean that species-specific management efforts will never be needed. Although ecosystem processes are a *necessary* objective for conservation, they are not, by themselves, a *sufficient* objective for conservation. For example, in Hawaii, most ecosystem processes have been maintained, but 90% of native vertebrate species and 10% of native plant species have disappeared (Nott and Pimm 1997). In this case and others like it, species remain appropriate targets for conservation, but no species can be maintained without suitable, functioning ecosystem processes.

Establishment of Scientific Baselines

Ecosystem management requires relatively undisturbed areas as controls if it is to be both experimental and adaptive (Christensen 1997). Baseline data on ecosystem processes and components are obtained through monitoring and manipulation of functioning and intact ecosystems. For example, major advances in understanding the interactive effects of fire and grazing on plant communities have been achieved through studies of the Serengeti-Mara ecosystem in Africa (Sinclair and Arcese 1995). Landscape-scale appreciation of the effects of fire on plant and animal communities has been gained from studies in the Greater Yellowstone ecosystem since the fires of 1988 (Keiter and Boyce 1991; Greenlee 1996). And significant insights into the effects of commercial fishing on fish populations and coral reefs have been gained from baseline studies of protected versus unprotected marine ecosystems in coastal African waters (McClanahan and Obura 1996) (chapter 9).

Aesthetic Experience

Aldo Leopold expressed the qualities of pristine ecosystems precisely in terms of aesthetic experience, defining a wilderness, for example, as an area that could absorb a two-week backpacking trip (Leopold 1966). Intact ecosystems also have aesthetic value through historical association. Today people seek experiences in intact, functioning ecosystems to experience recreation, beauty, physical challenge, opportunity to practice primitive skills, and spiritual renewal.

Commodity Production

Although its fundamental management unit is the ecosystem, ecosystem management can include the production of commodities. Functioning, well-managed ecosystems can be expected to produce sustained outputs of timber, water, minerals, game, fish, range, and recreational opportunity. Because ecosystem management demands consideration of human dimensions, human residents within ecosystems are tied to the commodities that the system produces and need to profit from such production in order to remain resident. However, recall that the distinction in ecosystem management, compared with resource management, is that yields are determined by the system's capacity, not by external demand, and that levels of commodity production are set at what is socially optimal, in light of noncommodity values, not what is economically optimal (Ficklin, Dunn, and Dwyer 1996).

This does not mean that economic optimality can be ignored. Noncommodity value considerations also constrain the type of management and harvest systems that are used to extract commodities. As resource economists Ficklin, Dunn, and Dwyer (1996) speaking of conflicts between spotted owl management and timber harvesting in the U.S. Pacific Northwest, point out, "since the economy of the Pacific Northwest is dependent on its forests, not everyone can afford *not* to produce timber products." In this example, timber cutting creates a cost to parties outside the timber industry, and not cutting timber creates a cost to the industry. Noncommodity value considerations also constrain the type of management and harvest systems that are used to extract commodities. The production and sale of commodities from ecosystems provides a benefit to those who extract and sell the commodity, such as timber, and they bear the cost of extracting and marketing the commodity they sell. This is called an *internal* cost, for it is borne by the buyers and sellers of the commodities directly (internally). But commodity extraction and production from ecosystems also carries *external* costs that are borne by other humans or by the system itself and its nonhuman components, such as loss of ecosystem structure and function, degradation of ecosystem processes and habitats, and

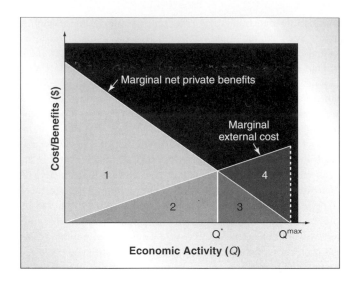

Figure 10.1

The relationship between marginal net private benefits and marginal external cost. The optimal level of economic activity occurs at point Q^*. Here production is reduced from the level that maximizes profit (Q^{max}) but reduces external costs associated with production. Area 1 is the net benefit to society at Q^*, 1 + 2 is the total profit from the economic activity, 3 is the amount of profit lost as economic activity is reduced from Q^{max} to Q^*, and 3 + 4 is the reduction in external costs associated with the same reduction in economic activity.

After Pearce and Turner (1990).

endangerment or extinction of species. For example, in the case of the spotted owl (*Strix occidentalis caurina*), we see the nature of external and internal costs for one particular system (temperate old-growth forests) and its residents. Initially the timber industry and the Forest Service disregarded the needs, and the value, of an endangered species and a rare ecosystem, valuing instead the (internal) economic benefits to timber producers. But U.S. courts enforced the value of the species and the ecosystem, saying, in effect, that the external costs of logging old-growth forests were too high, and put a stop to the commodity production of timber from old-growth forests.

Determining an optimal level of commodity production or other economic activity can be understood as a relationship between net private benefits and external costs (fig. 10.1). A socially optimal level of economic activity is reached when marginal net private benefits are balanced by marginal external costs. It is relatively easy to estimate private benefits from market prices, but it is more difficult to estimate external costs by methods such as contingent valuation (chapter 3). The value of successfully estimating these costs and benefits could be very great, and has important implications for ecosystem management. As Ficklin, Dunn, and Dwyer (1996) noted, "If the relationship between marginal costs and marginal benefits could be determined for an entire ecosystem, then it would be possible to manage ecosystem resources for the socially optimal allocation of resources to various uses." That estimate can be made if we know the costs and benefits of the commodities, the external costs of extracting and marketing the commodities, and the effect of different management practices (to enhance commodity production) on the external costs. Recognizing this as a legitimate means of evaluating commodity production in ecosystems also has important implications. It means that neither the total protection of the ecosystem (no commodity production) nor total harvest of all available commodities (maximum commodity production) represents a social optimum. Thus, ecosystem management is neither "ecosystem preservation" nor "multiple-use management." Effective ecosystem management must accurately estimate costs and benefits, and do so on an ecosystem-wide basis, to maintain the sustainability of the ecosystem and the permanent residence of humans within it.

THE SCIENTIFIC BASIS OF ECOSYSTEM MANAGEMENT

Ecosystem management is challenging because it attempts to manage a constantly changing entity. As Berry and colleagues (1998) have noted, "There exists a large gap between the operational needs of those responsible for ecosystems management and the knowledge required to meet those needs." To understand how ecosystems change and how such change might be managed, an ecosystem manager must determine (1) the physical boundaries of the system to be studied and the fundamental landscape units within the system; (2) how to construct meaningful ecological models of the system; (3) how to collect and monitor relevant data at scales appropriate to the system and its model(s); (4) how to identify, measure, and manage the most important processes affecting the transfer of matter and energy in the system; (5) how ecological processes interact with landscape processes and scales; and (6) how to design and conduct small-scale experimental manipulations within the system to test the predictions of the model(s) and determine the system's most likely responses to natural disturbances and management practices.

The Problem of Definition—What Is an Ecosystem?

Just as a general definition of *ecosystem* can be elusive, the particular limits of an individual ecosystem to be managed can be hard to define, even when there are no jurisdictional constraints and when the best available scientific information is employed. Ecosystems have notoriously "leaky" boundaries (Boyce 1998); they are open systems, not closed to gains or losses in terms of organisms, matter, and energy. Different agencies and individuals approach the problem of ecosystem definition in a variety of ways, depending on ecosystem characteristics and management objectives.

Traditional political delineations are the easiest way to delimit boundaries of an ecosystem, and may have the attraction of placing the management area under a single administrative jurisdiction, or at least under a group of related jurisdictions. Unfortunately, this method of delineating an ecosystem is almost never meaningful because ecological processes rarely match the borders of management jurisdictions. Examples of the shortcomings of politically based ecosystem designations

are as numerous as they are depressing. One of the most tragic is the case of the Everglades ecosystem of South Florida. The combined holdings of the Everglades National Park, the Big Cypress National Preserve, Biscayne National Park, three water conservation areas, and numerous smaller state and private reserves protect about 67% of the land area of the original ecosystem. But, although the majority of the Everglades land area was protected, its water flows were not. The construction of 2,200 km of canals and levees, over 40 pumps and spillways, and the impoundment of neighboring Lake Okeechobee permanently altered the amount and timing of water through the ecosystem, causing long-term, severe degradation (Lockwood and Fenn 2000). A long-embattled, but finally approved restoration plan will require a multibillion dollar effort over a period of more than 30 years (chapter 11).

A number of scientists, managers, and management agencies have begun to define ecosystems as collections of watersheds within a defined area or that empty into a common source (D'Erchia 1997; Norton and Davis 1997; Rama Mohan Rao et al. 1999; Richarson and Gatti 1999). For example, managers in India divided the country into 20 land resource regions and 186 land resource areas based on an integrative assessment of differences in soil, rainfall, forest cover, land-use practices, water resources, and elevations. This system was refined into a scheme of 17 soil conservation regions that were classified by several categories including climate, rainfall, mean annual temperature, elevation, watershed boundaries and land use (table 10.5) (Rama Mohan Rao et al. 1999). In Wisconsin, managers used watersheds as fundamental units in an ecosystem management approach to restoring drained agricultural wetlands (Richardson and Gatti 1999). Using a geographic information system (GIS), managers first combined satellite imagery with information on Wisconsin wetlands from state and federal wetland inventories to produce an integrated, digitized database that was used to identify wetlands and account for changes in wetland status over the previous 12 years. Wetlands were then ranked according to their rates of soil loss, and sediment delivery of all wetlands within a watershed was then summed. Using these criteria, highest priority for restoration and management was given to wetlands at the highest elevations (upper ends of watersheds). The restoration of upper-elevation wetlands had the greatest effect on reducing water velocities to all downstream sites, thus reducing erosion and sediment delivery to downstream areas (Richardson and Gatti 1999).

The Problem of Information—What Data Should Be Collected and Interpreted for Ecosystem Management?

Ecosystem managers face difficult choices about which information to collect, how to use it, and how to interpret it. There is an abundance of general-purpose data on ecosystem management, but information that meaningfully relates such data to management decisions is scarce. Recognizing the value of the latter kind of information, some have gone so far as to demand that ecosystems should not be managed without establishing "quantitative and measurable standards of ecosystem structure and functions" (Wagner and Kay 1993:268). This expectation is not realistic.

Table 10.5   Characteristics of Soil Conservation Regions in India for Watershed-Based Management*

REGION	CLIMATE	RAINFALL (cm)	MEAN TEMP (°C)	GROWING PERIOD (degree days)
1. Glacier	Cold arid	0–100	< 20	0–90
2. Karewas	Cold semiarid (dry) to humid and perhumid	60–250	20–22.5	90–300
3. Shiwalik	Semiarid to humid and perhumid	20–150	20–27.5	120–300
4. Indogangetic Plain	Subhumid dry to subhumid moist	100–150	22.5–27.5	180–210
5. Arid	Typic/arid to hyperarid	5–50	25–27.5	0–90
6. West Alluvial Plain	Semiarid dry to semiarid moist	50–150	25+	90–150
7. Beehar	Semiarid dry to moist	40–150	22.5–27.5	120–150
8. Southern Malwa	Semiarid moist to subhumid dry	75–150	22.5–27.5	120–180
9. Plateau	Semiarid dry to semiarid moist	50–275	20–27.5	90–150
10. Chalka	Arid (typic) to semiarid moist	50–250	20–27.5	60–150
11. Western Ghat	Subhumid to perhumid	100–250	25.0–27.5	240–270
12. Central Eastern Upland	Subhumid dry to subhumid moist	75–150	22.5–27.5	150–180
13. Eastern Ghat	Semiarid dry to subhumid dry	100–150	25.0–27.5	120–210
14. Diara	Subhumid dry to subhumid moist	100–150	25.5–27.5	150–210
15. Sunderban and Eastern Valley	Subhumid dry to perhumid	100–150	20–27.5	210–300
16. North Eastern Hill	Humid to perhumid	150–250	20–22.5	270–300
17. Island	Humid to perhumid	160–300	20–28	240–300

*Watershed boundaries and land use not included in this table.
After Rama Mohan Rao et al. 1999.

Fluctuations in ecosystem structure, processes, and component populations are great and often unpredictable (Boyce 1998). Ecosystems are large units encompassing multiple jurisdictions and categories of land ownership. The types of data that are gathered, stored, and desired by different agencies and stakeholder groups are diverse and usually uncoordinated. For example, the EPA commissioned an independent review group of scientists in 1995 to assess the current status of ecosystem man-agement research in the Pacific Northwest. After a comprehensive review of published and unpublished literature, interviews with and extensive surveys of scientists, managers, interest groups, and national policy makers involved in the area, and in-depth conversations with focus groups of all of the above, the scientists came to a depressing and shocking conclusion: *no main source of ecosystem management research information exists in the Pacific Northwest region* (Berry et al. 1998). Thus, ecosystem managers are caught between the extremes of those demanding an impossibly high standard of ecosystem knowledge as a basis for ecosystem management and the reality of interagency uncoordination that does not even know where the information is.

Although managers may not have the luxury of always being able to define "quantitative and measurable standards," neither are they doomed to ignorance because of poor interagency coordinating efforts in ecosystem management. Managers and their agencies can take the initiative in making intelligent and informed decisions by consistently collecting and monitoring appropriate ecosystem data at appropriate scales. From such information, they can determine the range of fluctuations of ecosystem processes and components in different systems, and whether existing data in their system are within those ranges. Several important methodologies and types of data—both old and new—exist to help achieve ecosystem management goals.

Biological data have been used with some success to evaluate the heath and quality of some ecosystems. The Ohio Environmental Protection Agency has developed a monitoring and assessment program for its streams and rivers that uses both biological and chemical data to assess water quality standards. Repeated biological surveys provide baseline criteria to judge water quality and overall ecosystem health. Indices for "normal" or "healthy" ecosystems were derived from sampling at 300 minimally impacted sites. Fish and macroinvertebrate data were used to establish attainable, baseline expectations for different habitats and ecoregions across the state. Using chemical data alone, 9% of sites failed to meet water quality standards. With the addition of biological data, 44% failed (Norton and Davis 1997). Based on these results, the Ohio EPA has affirmed that biological components are key indicators of ecosystem health and continues to use them in defining water quality standards.

One of the most objective measures of ecosystem health is the presence of ecological stress. Because ecosystems are successional in nature and often show signs of stress only over relatively long periods, long-term data are needed to reliably evaluate ecosystem condition. There are essentially four kinds of long-term data available for ecosystems: (1) regularly collected data, (2) archived data from previous studies, (3) data from long-term natural repositories, and (4) data from preserved areas (not subjected to disturbance) within the ecosystem.

Regularly Collected Data

Regularly collected data, sometimes incorrectly called "continuously collected data," are obtained from surveys or samples that measure the same variables, at the same locations, at regular intervals. For example, in specific regions and ecosystems, vegetation data are often collected at annual or otherwise regular intervals along permanently established transects. The historic "Parker transects" of Yellowstone National Park have provided estimates of range condition in the park since 1954, although some measurements were begun as early as the 1930s (Coughenour, Singer, and Reardon 1994). Measurements taken inside and outside long-established exclosures (wire pens that exclude ungulates) have helped determine the long-term effects of grazing and browsing of ungulates on plant communities.

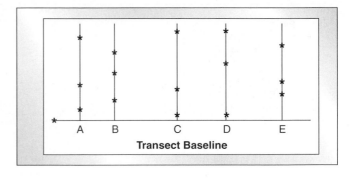

Transect Baseline

Figure 10.2

A schematic diagram of one way to establish a baseline with randomly selected sampling transects and plots.

Similarly, McInnes and associates used exclosures established on Isle Royale National Park in Michigan between 1948 and 1950 as controls to evaluate the effects of moose browsing on forests (McInnes et al. 1992).

Where long-established transects or other sampling units do not exist in an ecosystem, managers are well advised to establish them using standard survey/census methods and to begin regular sampling and monitoring. A random or systematic arrangement of sampling points can be chosen by using a landscape grid to ensure that (1) samples are chosen randomly or systematically, (2) samples are representative of the entire area, and (3) adequate numbers of sites are sampled (fig. 10.2).

Although grid sampling can provide large amounts of data, it has many problems. Any particular grid assumes that the sampled environment is relatively homogeneous at the sampling scale. Where the environment is obviously heterogeneous, multiple grids may be necessary, one for each kind or category of landscape or ecosystem encountered. As grids or sampled points within grids multiply, sampling becomes increasingly labor intensive and costly. In remote ecosystems, some randomly selected sampling points may be difficult to reach without disproportionate expenditures of time and effort. There will be tension between the number of variables sampled at each grid point and the number of grid points that can be sampled. Sampling more variables provides more site-specific information, but the increase in time investment per point reduces the sample size and reliability of observed differences. Because of this tension, some ecologists advocate giving more attention to selection of variables to sample than to maximization of the number of grid points (Loehle 1991). For example, variables that are not independent of other variables can be eliminated, and those variables that exert the greatest influence on the system can be given greater weight.

The historical practice of collecting information from such sampling points directly through field observation invariably led to underinformed management because limits of costs and human resources were reached quickly in such time- and labor-intensive methods. Today long-term, regularly collected data acquisition and analysis are facilitated by remote sensing techniques and geographic information systems.

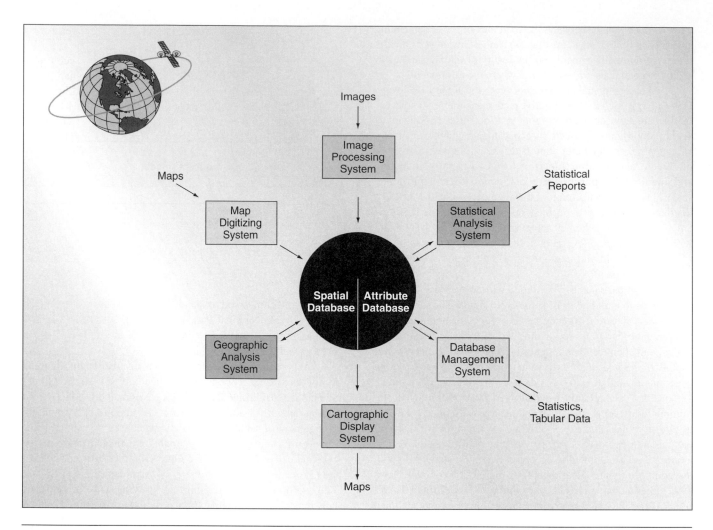

Figure 10.3

The conceptual organization of a GIS in terms of components, relationships, and outputs.

After Eastman (1995).

In a very real sense, ecosystem management was not possible prior to the development of remote sensing, whereby an array of satellites orbiting the earth provide information on a multitude of environmental variables. The primary source of satellite imagery used for analysis of biological data at geographic scales are the Landsat TM satellites, designed for high-resolution photography and arranged in orbits so as to systematically photograph all parts of the earth's surface at regular intervals. When such photographs are visually enhanced and their spectral images digitized, they yield vast amounts of information, especially when integrated through the technologies of GIS.

A GIS is a computer-assisted system designed for the acquisition, storage, manipulation, analysis, and display of geographic data. It consists of multiple pieces of integrated software that, together, can solve complex problems in spatial analysis (fig. 10.3). Additionally, a GIS can process complex images, such as aerial or satellite photographs, to yield more information than could be acquired by human visual perception alone (fig. 10.4). Among the most common and important uses of GIS are (1) creating data management systems for geographic information that allow users to enter "attribute data" (e.g., elevation, soil

type, vegetative cover, land use, and other variables) into files that can then be manipulated to display such data in new ways that are geographically or spatially sensistive; (2) determining and displaying the union and intersection of different geographic and biological variables through virtual, computer-generated overlay maps; (3) organizing and displaying the distribution of geographic, climatic, or biological patterns at large scales (e.g., regional, continental, or global) (fig. 10.4); (4) converting raw remote sensing data to digital information that can be redisplayed as an interpretive map using prescribed classification procedures; (5) performing statistical analyses of geographic information; and (6) providing "decision support systems" that use preprogrammed decision rules and criteria to assist managers in making decisions about land use or resource allocation, visually represented by "multicriteria suitability maps" that indicate the best management practices on particular land areas.

To show unions and intersections of spatial variables, GIS software can construct maps of the same area for different variables (e.g., soils, vegetation, and elevation), and then create a single virtual overlay map that shows the union and intersection of different variables with one another. Using such a map,

Landscape	%
Forest	83.9
Geysers	15.5
Rock	0.6

Figure 10.4

A satellite image of Yellowstone National Park, taken on July 22, 1988.

http://observe.ivv.nasa.gov/nasa/exhibits/eyes_sky/jellystone1.html

potential habitats for species whose environmental tolerances require a union of multiple variable states (e.g., vegetation, elevation, and moisture) can then be identified.

Satellite images or aerial photographs can be used to identify vegetation or land-use types on the basis of the spectra they emit. From this information, an appropriately constructed GIS program can determine the proportion and quantity of areas associated with different land-use practices or habitat types in an area, a first step toward an inventory of the condition of an ecosystem.

For example, Shinneman, Watson, and Martin (2000) recently used a GIS to determine the relative level of ecosystem protection in the southern Rocky Mountains of the United States. They first converted existing vegetation maps of the region into GIS formats, then used GIS to reclassify the vegetation into 13 types of regional ecosystems. They created a separate map showing the boundaries of "land stewardship categories" that represented different levels of protection for each land area and closely followed existing GAP (Gap Analysis Program) (chapter 8) categories in that region. In addition to the land stewardship categories, the existing GAP also integrated the occurrence of rare and endangered species. Recall that GAP is a nationally instituted effort to use GIS to identify "gaps" in biodiversity protection by integrating maps of natural vegetation with distributions of species (D'Erchia 1997). By creating an overlay map that showed the union and intersection of the land stewardship categories and ecosystem types, integrated with rare and endangered species associated with such ecosystems, Shinneman, Watson, and Martin (2000) were able to use GIS to determine the area of each ecosystem associated with each stewardship category. Their findings were disturbing: only about one-third (31%) of protected areas were below 10,000 ft in elevation and only 3 of 13 regional ecosystem types had more than 10% of their area within protected lands. Lower-elevation sites had far lower levels of protection, but contained the vast majority of rare and endangered species. Compared with protected lands, unprotected roadless areas contained nearly two-thirds (62%) of their area below 10,000 ft. If such roadless areas were protected, 11 of 13 ma-

jor ecosystems would have at least 10% of their area within protected status.

A GIS can also use data from the long-term ecological research (LTER) program to detect historical changes in ecosystems. For example, D'Erchia (1997) used a GIS to compare land-cover changes along the Upper Mississippi River at LTER sites associated with impoundments constructed on the river from 1891 to 1989. Specifically, the GIS approach used in this effort permitted an analysis of the effects of levee placement on habitat, floodwater levels, and sedimentation rates in this ecosystem over an extended period of time.

POINTS OF ENGAGEMENT—QUESTION 1

Ecosystem management developed partly in response to the increasing complexity of environmental problems and to shifting public attitudes toward the value of ecosystems. How was its development and acceptance also facilitated by the development of remote sensing and GIS technologies that were occurring at about the same time? How, in your view, did technical opportunity, management need, and public sentiment interact to facilitate development of the ecosystem management concept?

Archived Data and Historical "Experiments"

Another important long-term comparison and evaluation is the examination of archived data available from early studies of some ecosystems. In the United States and many other countries, areas where all or part of the ecosystem has a long history of public ownership and agency jurisdiction often have a long history of ecological investigations. Many such investigations take the form of unpublished reports, unprocessed data in agency files, personal journals, or historical photos of the ecosystem at various sites. Although such archived data may vary in quality and almost always require extra effort in analysis and interpretation, such information can be extremely valuable. For example, Chadde and Kay (1991) compared historical and contemporary photos from the same sites to document the decline of tall willow communities in Yellowstone National Park from the late 1800s to the present (fig. 10.5).

Data from Long-Term Natural Repositories

Not all archives are found in file cabinets, field journals, or old photograph albums. An ecosystem keeps its own long-term records that can provide outstanding documentation of past events, if one knows where to look. Lake and bog sediments, for example, provide long-term records of biological change in ecosystems compiled over hundreds or thousands of years. Such sediments are repositories of pollen grains that are dispersed annually by plants. Pollen grains are resistant to decay and distinctive by species, and their proportional abundance in sediments, appropriately corrected for differences in pollen production by different species, gives an index of the proportional abundance of plant species around the lake or bog. Properly extracted, a core of sediment provides a "profile" of the

Figure 10.5

Yancey's Hole on the northern range of Yellowstone National Park, Wyoming. In 1915 (left photo), tall willow (*Salix* spp.) communities are prominent. At the same spot in 1987 (right photo), willows were absent.

1915 (left) photo courtesy of Charles C Thomas, Publisher, Springfield, Illinois. Right photo courtesy of Charles E. Kay.

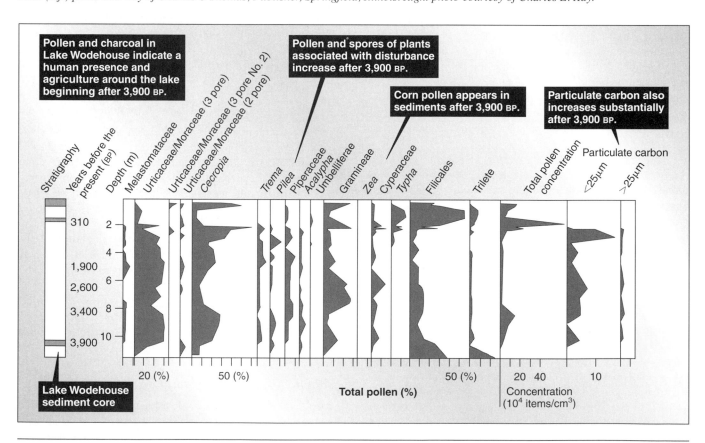

Figure 10.6

Pollen and particulate carbon in the sediments of Lake Wodehouse, Panama, dating from 3,900 years before the present (BP).

Data from Bush and Colinvaux (1994).

abundance of different species of pollen deposited in the lake over time. As the abundance of pollen of different species changes, one can make inferences about changes in the ecosystem, and estimate the time period associated with such changes if one can accurately estimate the rate of sediment deposition.

Notice, for example, how the relative abundance of different pollen types changed in the sediments of a Panama lake over time (fig. 10.6).

Sediments also keep careful records of material that precipitates from the surrounding water and air. Two such precipitates are ash and charcoal, substances usually produced by fires. From the abundance and depth of these materials in lake and bog sediments, one can infer the relative frequency of fire in the surrounding ecosystem over an extended period of time.

An indirect repository of recorded ecological events also may be found in distribution patterns of long-lived vegetation. For example, Heinselman (1973) estimated the timing and distribution of major fires in the Boundary Waters Canoe Area of the United States and Canada between 1727 and 1868 by measuring the distribution of old-growth, even-age stands of different ages established during that period.

Data from Undisturbed Areas

Nature preserves or other specially protected areas within an ecosystem serve as controls or "before impact" sites when evaluating ecosystem status and change; they are what some ecologists call "time-control substitutes" (Loehle 1991). Ecosystem structure and function in these sites can be compared with otherwise analogous disturbed sites within the ecosystem to determine how the system has changed over time in response to the disturbances that affect it. Such comparisons can be helpful, but require careful interpretation and many qualifiers. Regional sources of disturbance, such as air pollution, water pollution, or invasive species, may affect all sites regardless of their "protected" status. Multiple types and levels of disturbances may affect unprotected sites, so differences between them and protected areas are unlikely to be traceable to a single cause. Finally, protected, undisturbed sites are often smaller than surrounding unprotected areas. As discussed in chapter 8, ecological structure and function in small preserves may be affected, often adversely, by reduced size and edge effects.

Using Natural Processes in Ecosystem Management

Attempts at ecosystem management often fail because they protect land but not the ecological processes that determine the characteristics of its ecosystems. We will examine four processes that shape ecosystems, that are potentially amenable to management, and that can substantially affect management strategies and actions. These are fire, water flow, herbivory, and predation. This list is not exhaustive, but neither is it arbitrary. Informed managers know that there are scores of processes that affect ecosystems, but they cannot control most of these. Instead, they must make strategic and practical choices about which processes can be directly manipulated to achieve management ends, or that so affect other management actions that the consideration of the process must be a permanent component of every management policy. Thus our list is short, but chosen to meet these practical criteria.

Fire

The ecological effects of fire vary, but in most ecosystems fire increases (1) habitat heterogeneity and amount of edge; (2) plant and animal diversity; (3) nutrient uptake by plants, especially graminoids and forbs; (4) loss of nutrients from soil; (5) rates of erosion and surface runoff; (6) rates of stream flow; (7) grazing and browsing of burned areas by ungulates; and

(8) establishment of early successional species (Knight and Wallace 1989; Van Dyke, DiBenedetto, and Thomas 1991; Leach and Givnish 1996; Van Dyke, DeBoer, and Van Beek 1996). All but the first of these effects are usually of short duration, generally lasting only 1 to 3 years. In contrast, the increase in habitat heterogeneity after a fire may be a long-term effect, although its magnitude varies with the scale and intensity of fire. Small, dispersed fires tend to increase habitat heterogeneity, whereas large, contiguous fires make landscapes more homogeneous, particularly if the fire is hot and spreads rapidly.

Habitat fragmentation and fire suppression are often correlated, particularly in grassland and prairie habitats, making ecosystem management approaches more problematic, but also more essential. For example, in 1986–1987, Leach and Givnish (1996) recensused 54 Wisconsin prairie remnants that were originally surveyed in the 1940s and 1950s. Habitat fragmentation alone predicted plant species loss, but Leach and Givnish made three additional predictions that would be true if species loss was related to fire suppression as well as fragmentation. First, short species with leaves on or close to the ground would be lost at a more rapid rate than taller species because shorter species would have an advantage on sparsely vegetated sites (such as might be the case after burning), but would be shaded by taller competitors on less-disturbed, more densely vegetated sites. Second, plants associated with nitrogen-fixing bacterial symbionts would be lost at a more rapid rate because these species would have an advantage on open, sunny, nutrient-poor sites (created by recent burning, because fire volatizes nitrogen). Third, plants with small seeds would be lost at a more rapid rate because their recruitment is favored on frequently disturbed, less-vegetated microsites, such as are created by fire, rather than on less-disturbed, more heavily vegetated sites. Each prediction proved correct, and differences were especially large between nitrogen-fixers and nonfixers (table 10.6) and between small-seeded plants and large-seeded plants (table 10.7) (Leach and Givnish 1996). Thus, the absence of fire, combined with fragmentation, appears to have severe effects on system biodiversity.

Ecosystem managers can use fire to maintain or enhance biodiversity, increase habitat heterogeneity, increase plant nutrient uptake, create conditions attractive to particular species or other desired outcomes on a site-specific basis, or permit naturally ignited fires to burn over wider areas to create such effects through the ecosystem. However, even if fire is allowed to burn without suppression throughout the system, it is unlikely to produce a stable pattern of landscape structure or to replicate historical landscape patterns (Baker 1989). Although site-specific application of fire can be beneficial, even essential, in fragmented habitats, the regional or ecosystem-wide use of prescribed fire is unlikely in the foreseeable future because of the lack of appropriate technology, insufficient financial and human resources in any one agency to control large-scale fires, and strong public and political opposition to large-scale fires as a management technique. Even the choice to allow fires to burn in relatively uninhabited ecosystems is problematic because large fires may spread to inhabited areas, result in loss of human life and property, and they cannot be effectively controlled once they reach certain critical sizes. Although fire remains one of the most powerful and long-standing forces that have shaped

Table 10.6 Local Extinction and Recruitment in Plant Species in Wisconsin Prairies With and Without Nitrogen-Fixing Symbioses, Based on Total Numbers of Species Occurrences Across 54 Sites in the 1940s and 1950s Versus the Late 1980s

GUILD	ORIGINAL SPECIES TOTAL	PERSISTING (*P*)	RECRUITMENT (*R*)	ABSOLUTE EXTINCTION (%)	*R/P* RECRUITMENT (%)	NET EXTINCTION (%)
N-Fixers	278	171	41	38.5	24.0	23.7[*†]
Nonfixers	2,913	1,926	830	33.9	43.1	5.4[*]

[*]Net loss differs from zero (*P* < 0.001; McNemar change test).
[†]Net loss among N-fixers is higher than among nonfixers (*P* < 0.03, χ^2 = 5.20 for 1 df).
After Leach and Givnish 1996.

Table 10.7 Local Extinction and Recruitment as a Function of Seed Mass, Based on Total Numbers of Plant Species Occurrences Across 54 Sites in the 1940s and 1950s Versus the Late 1980s

SEED MASS (mg)	ORIGINAL SPECIES TOTAL	PERSISTING (*P*)	RECRUITMENT (*R*)	ABSOLUTE EXTINCTION (%)	*R/P* RECRUITMENT (%)	NET EXTINCTION (%)
< 0.1	136	54	43	60.3	79.6	28.7[*]
< 1	1,087	665	302	38.8	45.4	11.0[*]
< 10	1,316	947	290	28.0	30.6	6.0[*]
< 20	290	187	94	35.5	50.3	3.1
> 20	238	176	66	26.1	37.5	−1.7

[*]Net loss differs from zero (*P* < 0.01; McNemar change test).
After Leach and Givnish 1996.

ecosystems historically, its application as a management tool in ecosystems will likely remain limited to individual, relatively small sites.

Water Flow

Today most wetland and riparian systems are managed on a site-specific basis (Fredrickson 1997), but the processes that control them—in particular, water flow—extend over vast areas. Whereas most lake systems remain relatively constant in water volume and flow, streams experience seasonal flow variations that significantly affect the stream ecosystem and associated terrestrial ecosystems (chapter 9), particularly those within its floodplain. Variation in rate of water flow is often one of the most important controlling variables of ecosystem structure and function. Such changes in water flow promote exchanges of nu-

trients and organisms among habitats and enhance system productivity. A flood pulse may also provide a dimension of "seasonality" to environments that are otherwise unseasonal, such as tropical rain forests, making them more productive and diverse (Sparks 1995). Systems that experience flow variations include some of the most species-rich places on earth, such as the Amazon rain forest, the papyrus marshes of the Nile, the swamps of the Okavango River in Botswana, and the shallow wetlands and lakes of the Gran Pantanal of the Paraguay River in South America. These areas also support important commercial fisheries (Welcomme 1985).

Changes in water flows may have landscape-level effects, often increasing habitat heterogeneity. Natural flooding over a historical floodplain also increases habitat heterogeneity during floods because small differences in topography in the floodplain, such as small depressions, will hold water longer and at greater depths, providing habitat for waterfowl, amphibians, and other terrestrial species that use ephemerally flooded areas. The recently developed technology of "laser leveling" (establishing a single slope across an entire field to provide irrigation to crops from floodwaters) eliminates such small depressions and associated habitat heterogeneity (Fredrickson 1997), with predictable declines in species diversity.

Flooding lowers total biomass production over the short term, but usually does not have long-term effects. In the southeastern United States, large, infrequent floods regulate the development of longleaf pine forests through differential mortality via complex interactions of forest landscape position, associated landforms, and sizes of individual trees. Infrequent but large-scale flooding can shift a population of trees from an uneven-aged stand to an even-aged one because the immediate postflood period synchronizes germination of new individuals in the flooded area. Floods also move downed trees, limbs, and brush into stream channels, increasing channel structure, surface area, and roughness, promoting sediment retention, and increasing the stability of the channel surface. Such woody biomass in streams increases invertebrate activity, provides cover for fishes, and increases the habitat diversity of the stream channel (Michener et al. 1998).

Most stream species are adapted to flow variations, and decline in abundance if such flows are reduced or eliminated (Sparks 1995). Such species often use variations in flow as cues for seasonal breeding and feeding activity because optimal conditions for both are different and do not usually occur at the same time. If variations in flow rates cease, breeding and feeding activities may be curtailed or aborted. For example, in New Zealand, the endangered black stilt (*Himantopus novaezealandiae*) nests on gravel bars that are formed immediately after mountain snowmelt leads to spring floods. Construction of dams to hold or divert water for agriculture in this area changes flooding schedules and submerges stilt nests (Boyce and Payne 1997). Similarly, the reduction in sandbar development in the Missouri River in the United States has reduced nesting habitat for the endangered piping plover (*Charadrius melodus*) and the least tern (*Sterna antillarum*). In addition, most species of freshwater mussels depend on fishes to transport their larvae up- or downstream; thus, stream impoundments that limit movement of fish may isolate and endanger mussel populations.

Floods cause extensive erosion of stream channels and export large amounts of nutrients (Michener et al. 1998), and these effects are beneficial, even essential, for some species. The wood turtle (*Clemmys insculpta*), for example, endangered in the U.S. states of Iowa, Minnesota, and Wisconsin, depends on bare, sandy, eroded banks along stream channels as sites to dig nest burrows (Boyce and Payne 1997).

During flooding, many fish use backwater areas that are normally dry for spawning or as refuge from predators. In the long run, floodplain systems tend to conserve biodiversity during periods of harsh climatic change, as evidenced by the fact that the Mississippi River and its tributaries retain one-third of the 600 fish species of North America and most of the 297 species of freshwater mussels (Sparks 1995). As streams, particularly larger rivers, form new floodplains, they also create new habitat. In the southern United States, the Atchafalaya River, a branch of the Mississippi, is building new delta floodplains into the Gulf of Mexico and increasing the size of the largest remaining river overflow swamp in the United States (Sparks 1995).

In terms of human safety, property, and economics, floods are negative events. Thus, traditional management responses have been to control or eliminate flooding through impoundments that reduce fluctuations in flow; to construct levees and other physical structures that reduce the size of the floodplain; and, on larger streams, to physically remove downed timber and other woody debris from stream channels to increase the ease, speed, and safety of navigation. In a context of ecosystem management, all of these are legitimate concerns. However, if human interests are the exclusive concerns of managers, it is likely that flood-adapted components of the system will decline because flood pulses are the primary source of energy in many freshwater aquatic systems to facilitate habitat structure, nutrient exchange, and organism movement. Moreover, humans receive indirect benefits from floods. Worldwide, fish yield per acre is greater in rivers with flood pulses and floodplains than in associated impoundments where flooding is reduced or absent (Sparks 1995). Flooding also renews floodplain soil by depositing sediment and nutrient loads that would otherwise be lost through erosion, depleted by crops, or sequestered in the soil.

If flow is restricted to a narrower channel such that the traditional floodplain cannot receive and store the floodwaters, flood heights and damages will increase at other locations. It is noteworthy that many dams on major rivers were built to aid navigation by maintaining river depth at a sufficiently high level during periods of lowest flow, not to stop floods (Sparks 1995). Dams may be removed to restore natural flows and increase productivity. For example, a hydroelectric dam was destroyed in Maine to restore a natural migration of Atlantic salmon (Souers 2000), and the destruction of several dams in the Pacific Northwest is now being considered (Joseph 1998).

Herbivory

Herbivores, especially large ones, often exert profound controlling influences on ecosystem components, structure, and function. In many cases, ecosystem management is impossible without herbivore management. The manager who fails to manage the ecosystem's herbivores often finds that the herbivores will manage the ecosystem themselves, and not always to the manager's intended ends. Herbivores affect ecosystem processes primarily through regulation of habitat, regulation of energy flow, regulation of plant nutrient cycling, and effects on plant nutrition. Both browsers (herbivores feeding on woody vegetation) and grazers (herbivores that feed on herbaceous vegetation) achieve these ends, although not always through the same means.

Browsers often prevent the vertical development of sapling vegetation of their preferred food species, contributing to spreading, shrubby growth forms that keep the affected plant biomass within their reach (McNaughton, Ruess, and Seagle 1988). In a northern boreal forest, moose prevent saplings of preferred species from growing into the tree canopy, resulting in a forest with fewer canopy trees and a well-developed

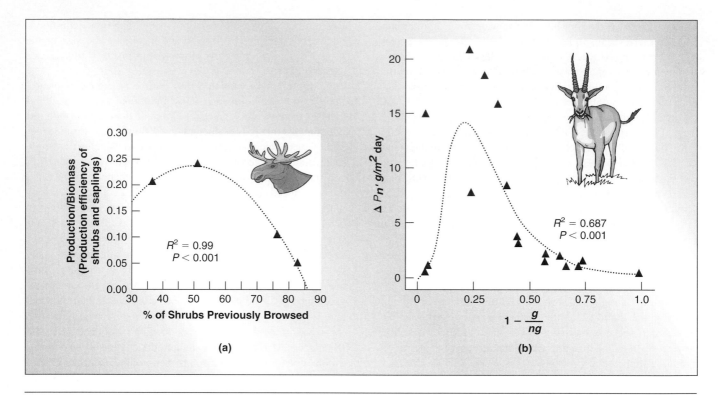

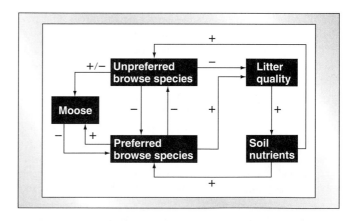

Figure 10.7

The relationship of (a) browsing and (b) grazing to production efficiency. In (a), light-to-moderate browsing by moose stimulates higher ratios of plant production to biomass in shrubs. In (b), grazing produces a similar response function, with greatest increases in productivity (g/m²/day) associated with intermediate levels of grazing intensity (1 − g/ng, where g is the biomass in grazed areas unprotected by fencing and ng is the biomass in a permanent exclosure in which no grazing occurs).

After (a) McInnes et al. (1992) and (b) McNaughton (1979).

understory of shrubs and herbs, all within reach of the moose (McInnes et al. 1992). In fact, light-to-moderate browsing leads to increased production efficiencies (higher rates of production per unit biomass) in shrubs and saplings that are browsed (fig. 10.7a). Through browsing, moose also reduce the quantity and quality of litter and soil nutrients, driving a complex set of ecological interactions among browse, litter quality, and soil nutrients (McInnes et al. 1992) (fig. 10.8). Similar effects are seen in mixed deciduous-coniferous forests, where moose typically browse preferentially on deciduous hardwoods. This pattern of feeding not only changes forest composition, but, more generally, reduces nitrogen mineralization, nitrogen inputs, and overall primary productivity of the forest because browsing reduces the quantity and quality of litter returned to the soil (Pastor et al. 1993). Moose, in conjunction with snowshoe hare (*Lepus americanus*), also can reduce fine root production in plants as a result of their herbivory on aerial biomass (Ruess, Hendrick, and Bryant 1998).

In coastal marsh ecosystems in Louisiana, herbivores, especially nutria (*Myocastor coypus*) and wild boar (*Sus scrofa*), reduced above-ground biomass from 460 g/m² (ungrazed plots) to 112 g/m² (grazed plots). In addition to these large removals of vegetation, herbivores also reduced soil-building processes by lowering below-ground production and expansion of the plant root zones. These effects could lead to a destruction of wetland habitat where sediment accumulation is low and soil building

Figure 10.8

Relationships between levels of moose browsing, preferred and unpreferred species of browse, litter quality, and soil nutrients. Arrows indicate direction of interaction. Plus signs indicate positive effects; minus signs indicate negative effects. Thus, a series of ecological interactions is set in motion by browsing.

After McInnes et al. (1992).

processes proceed slowly (Ford and Grace 1998). Such an effect would be exacerbated by the rising sea levels currently being experienced worldwide.

Table 10.8   Beneficial Effects of Herbivory on Plant Growth and Metabolism

EFFECTS OF HERBIVORY ON PLANTS

1. Photosynthetic rates increase in the remaining tissue.
2. Older tissues, functioning at levels below maximum photosynthetic level, are removed.
3. The active photosynthetic period of residual tissue is prolonged as the rate of leaf senescence is reduced.
4. Substrates are circulated through the plant.
5. Removal of overshadowing tissue intensifies light on potentially more active underlying tissues.
6. Increased leaf growth and tillering result from the division and elongation of cells; the activation of remaining meristems increase due to the plant's hormonal response; and growth is also promoted by chemicals in ruminant saliva.
7. Transpiration surface is reduced. Consequently, soil moisture conservation increases.
8. Nutrients are recycled from dung and urine.

Adapted from McNaughton 1979.

In Africa, mixed feeders that combine browsing and grazing, such as elephants (*Loxodonta africana*), regulate the abundance of woody browse and underlying grasses, often by pushing over trees or stripping their bark (McNaughton, Ruess, and Seagle 1988). These activities reduce the abundance of woody vegetation and create openings that grasses subsequently invade. High densities of elk (a North American mixed feeder) can suppress height and survivorship of trembling aspen (*Populus tremuloides*), willow (*Salix* spp.), and a variety of conifer species through browsing (Chadde and Kay 1991; Kay and Wagner 1994; Romme et al. 1995). Similarly, white-tailed deer (*Odocoileus virginianus*) can reduce the survivorship of some species of long-lived forbs, like trillium (*Trillium* spp.), where most of the leaf area and reproductive structures can be removed in one bite. Where deer densities are high and forest habitats are fragmented, there is experimental evidence that deer could extirpate trillium populations in individual fragments and inhibit efforts to restore such populations (Augustine and Frelich 1998).

Grazing species affect a variety of components of ecosystem structure and function. In some systems, grazers can remove up to 40% of standing plant biomass, significantly lowering ecosystem production. In other systems, grazers may initiate changes in plant morphology and physiology that lead to higher levels of plant and ecosystem productivity. For example, in grazed African systems, dominant grasses are dwarfed, low-growing forms with short internodes, whereas in ungrazed systems in the same area, dominant grasses are tall-growing species (McNaughton, Ruess, and Seagle 1988). When grazed, many plants respond by increasing biomass concentration in their tissues (the ratio of mass to volume, often measured in milligrams of plant biomass per cm^3), creating more "biomass per bite" for herbivores. Gregarious herbivores exploit this response by actively creating "grazing lawns," intensely grazed areas where the herbivores' own grazing activity keeps plant heights low and biomass concentration high. Although this reduces total plant biomass density in the grazing lawn compared with ungrazed areas, it increases foraging efficiency because of the increases in biomass concentration (McNaughton 1984). Grazing also tends to increase photosynthetic rates in plants, in-

crease rates of nutrient allocation to growing plant tissues, increase growth rates in plants, and produce other effects that often benefit herbivores (table 10.8). Although there is an optimum level of grazing, above which plants are damaged and their productivity is reduced, light-to-moderate grazing by native herbivores tends to produce positive responses in growth rates, metabolic efficiency, and nutrient concentrations (McNaughton 1979). In fact, in an experimental analysis of grazed systems in the Serengeti-Mara regions of Tanzania and Kenya, McNaughton (1979) determined that over two-thirds ($r^2 =$ 0.69) of variations in plant productivity could be explained by grazing intensity alone (fig. 10.7b), producing a response function similar to that produced by the effects of browsing (fig. 10.7a).

Ecosystem managers must consider not only the effects of native herbivores, but also domestic livestock. Cattle and sheep, for example, produce very different ecosystem effects. When grazing by livestock is light or absent, grasses and sedges in upland Ponderosa pine (*Pinus ponderosa*) and mixed conifer forests of the Rocky Mountains outcompete tree seedlings for space and nutrients. Tree density remains low and individual trees are large and widely spaced. When grazing in such systems is more intense, cattle and sheep reduce biomass and density of understory grasses and sedges, leading to increased survivorship and densities of trees, which some investigators assert leads to more frequent and severe fires (Belsky and Blumenthal 1997) (fig. 10.9). As grazers, livestock also reduce the cover of herbaceous plants and litter, increase soil disturbance and compaction, reduce water infiltration rates to soil, and increase rates of soil erosion (Belsky and Blumenthal 1997).

Predation

Although herbivores are predators on plants, we will use the term *predation* in this discussion in its more traditional sense (one animal killing another animal for food), and we will refer to the animals that do so as predators or carnivores. Of all the factors in our short list, predation is perhaps the most variable in its effects on ecosystems and the most difficult to predict and control, but among the most important considerations in

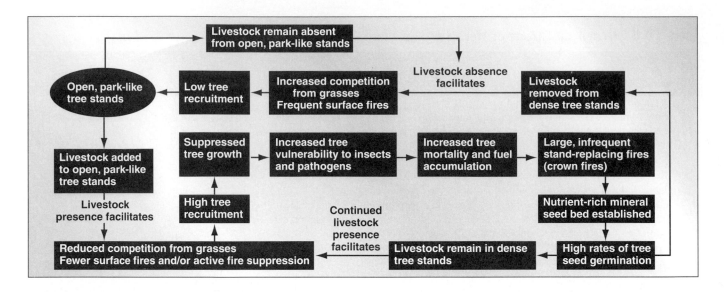

Figure 10.9

A conceptual model of the effects of livestock grazing on stand dynamics of western, interior coniferous forests of the United States. In the absence of livestock grazing, competition from grasses and sedges reduces survivorship of tree seedlings. Grasses and sedges also provide fuel for frequent, low-intensity surface fires that also kill tree seedlings. Both effects lead to low tree recruitment and low-density, open, park-like forests. Livestock grazing removes grasses and sedges, leading to increased seedling recruitment and high-density forests that can carry more intense stand-replacing canopy fires.

Adapted from Belsky and Blumenthal (1997).

ecosystem management. The effects of predators vary widely in ecological context. In some systems, predators can generate increases in biodiversity (Paine 1966, 1969) (chapter 4), but this effect is not universal. Under most conditions, large vertebrate carnivores, such as wolves, mountain lions, and bears (*Ursus* spp.), regulate their own numbers through social interaction and behavior (Seidensticker et al. 1973; Beecham 1983; Fuller 1989). Carnivores commonly can make functional (dietary shifts) and numerical (changes in density) responses to changes in prey abundance, but under relatively few conditions have large carnivores been shown to regulate prey populations (Bergerud and Ballard 1988; McLaren and Peterson 1994). Even when regulation does occur, it is often short lived or only during harsh conditions when prey are especially vulnerable. Large predators are suspected to control densities of smaller predators (mesofauna) (Soulé et al. 1988), an effect with important implications in ecosystems, but evidence for such control is still in the form of anecdote, observation, and correlation, rather than experimental results.

Despite ambiguity about the importance of predators in ecosystems, predation, especially by large vertebrate carnivores, is essential to consider in ecosystem management. Large vertebrate carnivores have the largest home ranges of any group of terrestrial vertebrates. Thus, managers have always believed that, if large carnivores were featured species of protection in ecosystem management, their large area needs would guarantee the survival of other species with smaller area needs that lived under the carnivore's "umbrella." Although intuitively appealing, this concept has never been carefully tested. One first approximation is offered by Noss and associates (1996), who compared the percent of species protected in different taxa under two different grizzly bear protection plans in the state of Idaho (Shaffer 1992; U.S. Fish and Wildlife Service 1993). Amphibian, bird, and mammal species distributions are relatively well protected under one or both plans, but reptile species distributions find little protection under either (table 10.9).

Large predators are also driving forces in ecosystem management because their large dispersal distances compel managers to consider the connectivity of regional ecosystems. Specifically, does there exist, or can there be made to exist, habitat corridors between adjacent ecosystems that large predators will use in dispersal and through which they can move with high rates of survivorship? The efficacy of corridors is still desperately in need of experimental study (chapter 8), but observational and anecdotal evidence suggests that wide-ranging predators can use corridors, even those of apparent low habitat quality, to move from one regional ecosystem to another (Noss et al. 1996). The question of survivorship in corridors is not a matter of habitat characteristics and spatial considerations only, but also one of public attitude and education. Favorable public attitudes toward predators tend to increase their densities, range, and persistence. Unfavorable public attitudes lead to their extermination. The presence of large predators brings national attention to the ecosystems in which they live, with attendant powerful national constituencies that act as the predators' advocates and can effectively exert powerful political and social pressure on management decisions. However, no amount of advocacy for predators will change the fact that large predators are dangerous. They routinely kill livestock, pets, and, in some cases, human beings. The presence of predators requires

Table 10.9   The Number of Terrestrial Vertebrate Species with 10% of Their Predicted Statewide Distribution Protected by a Recovery Zone of a U.S. Fish and Wildlife Service (1993) Grizzly Bear Recovery Plan and an Alternative Recovery Plan (Shaffer 1992) in Idaho, U.S.A.*

CLASS	STATE TOTAL SPECIES	USFWS ZONES (%)	ALTERNATIVE ZONES (%)
Amphibians	8	4 (50)	5 (63)
Reptiles	13	0 (0)	2 (15)
Birds	126	66 (52)	100 (79)
Mammals	68	32 (47)	51 (75)
Total	215	102 (47)	158 (73)

*Ubiquitous and peripheral species not included.

ecosystem managers to devise multiple management strategies for the control, or in some cases, eradication of predators, in different ecological and human contexts. For example, "zonation management" is one proposed method of managing wolves at an ecosystem level that would protect wolves in "core" areas but subject them to increasingly higher levels of control in areas with higher densities of livestock and humans (Mech 1995) (fig. 10.10).

Thus, although managers may not always, or even often, be able to use predation as a controllable ecosystem process, the presence of large predators forces managers to expand management from local to regional interests, and to include political, sociological, and educational dimensions that encompass many stakeholder groups.

Implementing Ecosystem Management

Adaptive Management Approaches

Ecosystem management is the management of uncertainty, in terms of (1) unknowable responses and complexities of ever-changing systems, (2) lack of human understanding of the processes and components that determine ecological condition, and (3) unreliabililty of human-collected data and the human biases and limits inherent in its interpretation (Christensen et al. 1996). None of these uncertainties can ever be eliminated, although the effects of the second and third forms can be reduced. Lack of understanding can be lessened through carefully designed research, strategically targeted to ask the most appropriate and urgent questions about ecosystem behavior. Biases and human limitations can be constrained with a dedication to high levels of professionalism in the analysis and interpretation of data, careful peer review of research efforts, and humility about one's own favorite hypotheses and theories. Conservation biologists cannot remove uncertainty, but they can manage it (Christensen et al. 1996). Their most important tool in this ef-

fort is the practice of adaptive management, which has been defined as *a process that combines democratic principles, scientific analysis, education, and institutional learning to manage resources sustainably in an environment of uncertainty* (Lee 1993). Adaptive management requires both experimental manipulation of the system and willingness to change research priorities according to critical management needs. It likewise requires ongoing interaction and communication with public and private stakeholders to disseminate research results in meaningful ways and to learn what the stakeholders consider meaningful research. Lee (1993) states that adaptive management requires (1) commitment to improving the outcomes of management over biological time scales; (2) awareness of the experimental nature of management; (3) the willingness to accept the risk of perceived failures; (4) a common understanding, with stakeholders, of the goals, strategies, and uncertainties of management; and (5) a mandate for action from the stakeholders or other authorities.

In adaptive management, management goals and strategies are hypotheses to be tested by experiments. Experimental results must be compared with predictions made beforehand. Thus, experimental design and manipulation are critical components of such management. Because ecosystems are such large units, their direct manipulation is challenging. Because ecosystem structure and function are affected by many variables, experimental design is difficult. Despite these problems, experimental manipulation can be performed in an ecosystem context. But precisely because experimental manipulation is necessary, natural and undisturbed areas are needed as experimental controls.

McClanahan and Obura (1996) performed an ecosystem-scale "experiment" on African coral reef systems, using Kenyan marine reserves as controls and comparing them with adjacent marine areas open to collecting and commercial fishing. The combined species richness of coral, snails, and fishes was higher in the protected areas, and the differences in species richness between protected and unprotected areas

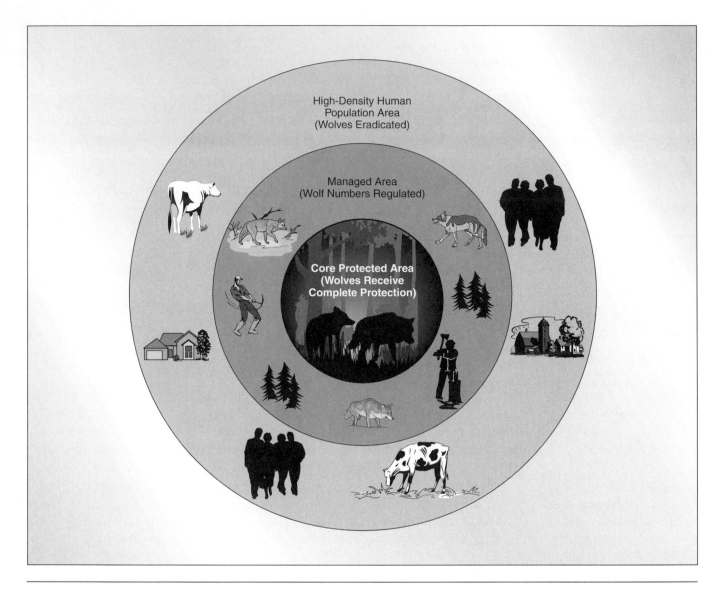

Figure 10.10

"Zonation management" for wolves or other large, mobile predators. In a core protected area with low human densities and minimal human impacts, wolves receive complete protection. In a surrounding area (management area), wolf numbers are regulated and individual wolves that kill livestock or pets are destroyed. In surrounding areas of high human population densities and impacts, wolves are killed if they enter the area.

Based on a concept described by Mech (1995).

increased with the size of the sampled area (fig. 10.11). Such a study cannot be considered truly experimental because it lacks baseline information (namely, the species richness of the reserve areas before they became reserves) and additional research is needed to more precisely determine the specific causes of decline for particular species. However, such efforts have merit as first attempts to document the differences that ecosystem protection may make in this context, and to begin to identify fishing and collection methods that are most harmful to the diversity of the system.

Adaptive management of ecosystems must begin by articulating clear goals that are sustainable and practical (Christensen et al. 1996). It must reconcile management actions to appropriate spatial scales and ecosystem processes, and to

differing objectives of various stakeholders. And it must reconcile temporal scales in management, such as short-term political and economic concerns, with long-term ecosystem patterns and processes. Ecosystem management processes must demonstrate long-term planning, long-term commitment, and an ability to make short-term decisions. Institutions of management must be adaptable and able to manage uncertainty and risk.

Ecosystem Modeling

Ecological modeling has been described as "indispensable and always wrong" (Lee 1993). Simple compartmental models serve to organize information and express connections and re-

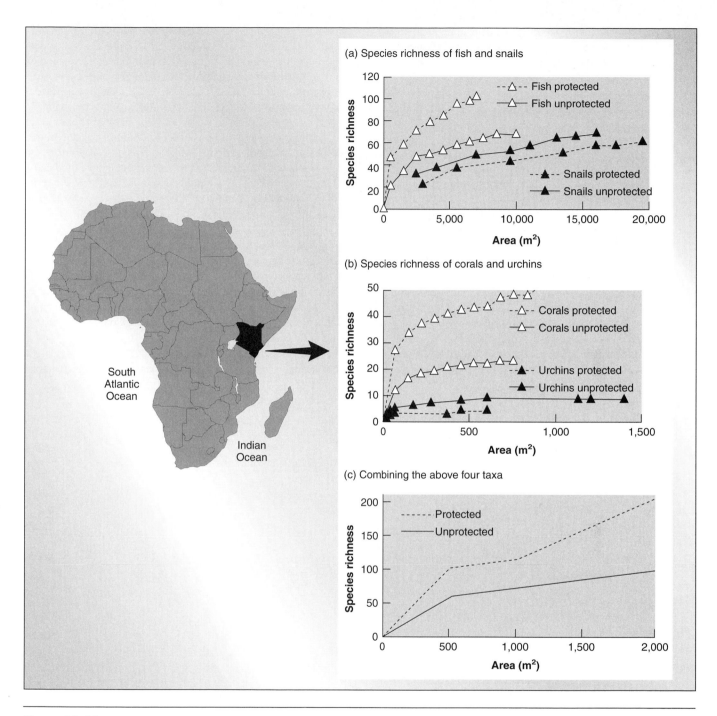

Figure 10.11

Species richness in Kenyan marine reserves and adjacent areas open to collecting and commercial fishing.

After McClanahan and Obura (1996).

lationships. Complex simulation models involve the mathematical estimation of many state and transition variables through time and across landscapes (Christensen et al. 1996). Models specifically designed for use in ecosystem management typically serve at least one of the following functions.

1. *Prediction models* estimate the outcome of individual actions or decisions using mathematical or conceptual relationships.

2. *Research coordination models* include factors that each research effort studies and identify the relations between the research efforts.
3. *Policy evaluation models* explore different management scenarios through simulation by evaluating an array of different management actions.
4. *Institutional memory models* inform future scientists and managers of the workings of the system as their predecessors understood it.

5. *Management training models* train decision makers in anticipating possible responses of the systems they manage and the actions they may take.

6. *Optimization models* determine the "best" action to achieve a particular objective.

7. *Hypothesis-testing models* provide statements of a scientific hypothesis and means of comparing the hypothesis with other models and data (Hillborn 1995).

Models are constructed by determining issues the model should address, appropriate indicators of system performance, management actions that may be considered, spatial and temporal measurement scales, desired resolution, model components, and appropriate flow of information among components (Hillborn 1995). Although models differ in detail and purpose, ecosystem managers must define the values of five types of entities to build a working model. *Stocks* refer to amounts of levels of a variable of interest that the model counts or monitors. *Sources* and *sinks* are entities from which the stock originates (sources) or into which the stock is absorbed (sinks). *Flows,* generally expressed as equations, determine rates of movement of stocks to and from sources and sinks or, in some cases, from one stock to another (e.g., the conversion of plant biomass into animal biomass). *Parameters* or *converters* are values of variables used to determine rates of flow. *Connectors* show the path through which material is transferred from one stock to another, or to and from sources and sinks. If a manager can accurately estimate the initial value of pertinent stocks, determine the conceptual framework (often, equations) that regulate the change in stocks or the rate at which stocks move or are converted to other ecosystem components, the values of parameters in this framework, and the paths through which the stocks are transferred, then, components of interest in the ecosystem can be modeled.

Take, as a highly simplified example, a system comprising two stocks, populations of seaweed and their primary herbivores, limpets (a gastropod mollusk). An estimation of initial levels of these two stocks, knowledge of values of six parameters (reproduction rates of limpets and seaweeds, the effect of limpet feeding on seaweed, limpet death rates, and density-dependent constraints on limpet and seaweed population growth), and an understanding of connections between parameter values and rates of change in stocks (expressed in the equations) are sufficient to track changes in the two populations (fig. 10.12) (Brennan et al. 1970).

Models that attempt to address all components of an ecosystem are much more complex than the previous example.

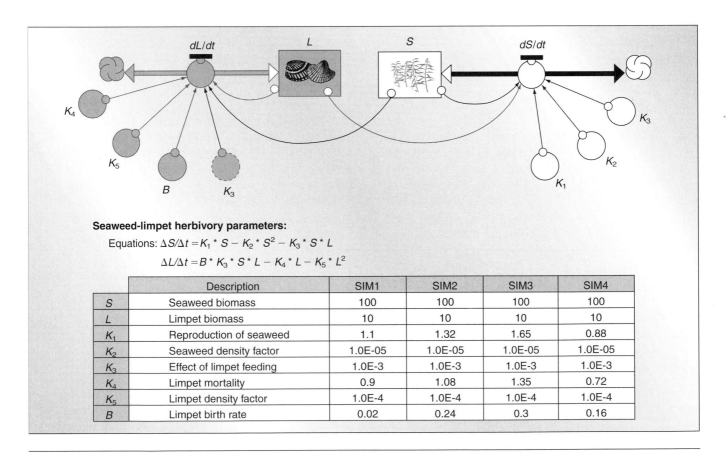

Seaweed-limpet herbivory parameters:

Equations: $\Delta S/\Delta t = K_1 * S - K_2 * S^2 - K_3 * S * L$

$\Delta L/\Delta t = B * K_3 * S * L - K_4 * L - K_5 * L^2$

	Description	SIM1	SIM2	SIM3	SIM4
S	Seaweed biomass	100	100	100	100
L	Limpet biomass	10	10	10	10
K_1	Reproduction of seaweed	1.1	1.32	1.65	0.88
K_2	Seaweed density factor	1.0E-05	1.0E-05	1.0E-05	1.0E-05
K_3	Effect of limpet feeding	1.0E-3	1.0E-3	1.0E-3	1.0E-3
K_4	Limpet mortality	0.9	1.08	1.35	0.72
K_5	Limpet density factor	1.0E-4	1.0E-4	1.0E-4	1.0E-4
B	Limpet birth rate	0.02	0.24	0.3	0.16

Figure 10.12

Conceptual illustration, equations, parameter values, and simulation results (SIM1–SIM4) of a model of limpet herbivory on seaweed. Boxes (L and S) respresent populations (stocks) of limpets and seaweed in the system. K_1–K_5 and B are parameters (constants) affecting rates of flow (dL/dt and dS/dt) between stocks and ecological sinks (clouds).

Equations and parameters derived from Brennan et al. (1970). Illustration designed by J. D. Schmeling. Used by permission.

Table 10.10   **Attributes of Landscape Structure Based on U.S. Geological Survey Land Use and Land Cover Digital Data**

LANDSCAPE ATTRIBUTE		DESCRIPTION
Composition		Proportion of the landscape in:
	F	Forest.
	W	Wetlands.
	A	Agriculture (cropland and pasture).
	U	Urban land.
Diversity/Edge	H'	Shannon diversity index. $$H^1 = \sum_{i=1}^{n} P_i \ln P_i \,,$$ where P_i is the proportion of land type i, and n is the number of land types in the landscape.
	D	Land type dominance, the tendency for one or a few land types to make up a majority of the landscape. Calculated as: $$D = \ln n + \sum_{i=1}^{n} P_i \ln P_i \quad.$$ Dominance is scaled to vary over 0–1, where $D \to 1$ reflects landscapes dominated by one or a few land types.
	C	Land type contagion, the extent to which land types are aggregated in contiguous patches. $$C = n \ln n + \sum_{i=1}^{n} \sum_{j=1}^{n} P_{ij} \ln P_{ij} \,,$$ where P_{ij} is the probability that land type i is adjacent to land type j. Contagion is scaled to vary over 0–1, where $C \to 1$ indicates a highly clumped pattern (O'Neill et al. 1988).
	E_T	Total edge among all land types.
	$E_{F/A}$	Edge between forest and agricultural land.
	$E_{F/U}$	Edge between forest and urban land.

Continued

However, models that make predictions about single effects or states may be much simpler. For example, in the Serengeti, Wolanski and colleagues (1999) conducted extensive field studies and hypothesized that water quality and quantity were the dominant forces driving ecological events. Background studies clearly established that the migration of ungulates was related to the level of rainfall and river flows, but these variables still did not predict animal movements with precision. Further investigations revealed that water quality, specifically salinity, was the most accurate predictor of ungulate movements. Where the water is fresh, wildlife remain. Where it becomes saline, they leave. Variation in salinity by decade also was the most

Table 10.10 *Continued*

LANDSCAPE ATTRIBUTE		DESCRIPTION
Patch Characteristics	F_s	Average size of forest patches.
	$F_\#$	Number of forest patches.
	A_s	Average size of agriculture patches.
	$A_\#$	Number of agriculture patches.
	U_s	Average size of urban patches.
	$U_\#$	Number of urban patches.
Patch Configuration	$D_{P/A}$	Fractal dimension based on the perimeter-area method (Krummel et al. 1987). Measures the complexity of forest patch shape.
	D_G	Fractal dimension based on the grid method (Milne 1991). Measures the dispersion of forest patches.

Adapted from Flather and Sauer 1996.

important controlling factor in determining the discontinuity between grasslands and wooded savannas. Modelers, using the value of average salinity from a single lake, were able to predict the movement of animals from grasslands to woodlands within 1 week when salinity values rose above a predetermined threshold (Wolanski et al. 1999).

Modeling and monitoring are intimately connected because modeling identifies the critical ecosystem variables that are the best indices of changes in status or function of an ecosystem, and so make the monitoring effort more efficient and cost-effective. By knowing what to sample and monitor, we eliminate the waste of time and money that could be spent monitoring variables that are *not* sensitive to ecosystem processes.

Managing Landscape Processes

Because ecosystem management must encompass large spatial scales, it depends heavily on the developing science of landscape ecology to understand how to link landscapes and ecological processes. The fundamental unit of a landscape is a patch, and landscapes in an ecosystem are composed of discrete, bounded patches that differ from one another in biotic and abiotic structure or composition (Pickett and Cadenasso 1995). One or more cover types form a matrix in which other, different patches appear. The spatial configuration of these patches is called a *landscape pattern* (Opdam et al. 1995).

Landscape ecology assumes that spatial heterogeneity in a landscape will affect biological processes, that the observed effects will show recognizable patterns (Bowers and Matter

1997), and that these patterns will not be restricted to a single spatial or temporal scale (Risser 1987). Therefore, a fundamental goal of landscape ecology is to determine how landscapes are structured, and then to determine how landscape structure affects the abundance, distribution, and demographic performance of organisms.

Measures of landscape structure are composition, diversity/edge, patch characteristics, and patch configuration. Composition refers to the proportion and amount of the landscape in different habitat or land-use types. Measures of diversity or edge may be measures of biological diversity at alpha, beta, or gamma levels, measures of land-type dominance, or measures of "land-type contagion," which assesses the extent to which land types are arranged in contiguous patches. "Edge" can be measured as the length of the boundaries among all habitat types or between specific, paired combinations of habitat types. Patch characteristics may be expressed as the size and number of patches of different habitat types, and patch configuration can be presented as the relationship between the amount of area in a patch and the length of its perimeter. Flather and Sauer (1996) offer an example of land cover attributes in these four categories that can be determined from maps, aerial photographs, or satellite imagery (table 10.10).

If such measures can be calibrated at an ecosystem scale with a GIS or from aerial or satellite photographs, and then linked to spatially specific, ecosystem-wide biological survey data, it is possible to observe the effects of landscapes on species distribution (Flather and Sauer 1996). As discussed in chapter 7 (The Conservation of Populations), different arrangements of landscape structure create different patterns of population demography and dispersal. A key question is, "How do populations in different landscapes vary in performance?" (Bowers and Matter 1997). But determining correlations between landscapes and performance is not enough. Correlation is not causation. Rather, observed correlations between landscape structure and ecosystem properties should be used to generate hypotheses about the processes that might create such correlations, and managers should then perform small-scale, but well-designed experiments to test particular hypotheses about these relationships. This is the correct approach to meet a goal of ecosystem management—determining how ecological processes are linked to landscape patterns. Because this relationship exists, a fundamental problem for ecosystem management is how natural processes such as fire, water flows, and herbivores will affect patch structure at landscape scales and how the structure will change over time. Landscape ecology is a critical foundation of ecosystem management because it can demonstrate how processes operating at different spatial scales integrate to shape the structure and function of an ecosystem (Pickett and Cadenasso 1995).

Identifying Functional Ecosystems as Management Units for Conservation

Poiani and associates (2000) argue that an ecosystem can be considered "functional" if it (1) possesses the historic composition and structure of the ecosystem and its species within a natural range of variability; (2) has dominant environmental regimes controlled by natural processes; (3) is of sufficient size to possess at least one minimum dynamic area (50 times the size of the average disturbance patch); and (4) is connected to other essential landscape elements, among which species are free to move. But even if an ecosystem meets these criteria, how do we decide if it merits a major conservation effort?

Many conservation biologists would argue that this question is ultimately answered by an examination of endangerment. Is the ecosystem facing an immediate threat of irreversible alteration, or are its component species in immediate danger of extinction? If we take this view, then a reasonable approach is to evaluate the ecosystem through geographic and environmental attributes shared by endangered species. For example, Flather, Knowles, and Kendall (1998) ranked counties in the United States, adjusted for size differences, by the number of threatened and endangered species in each, separated species groups by taxonomic units, and then identified areas where endangered species were concentrated (hot spots) as the top 5% for each group. A land resource classification system developed by the U.S. Department of Agriculture arranged counties that supported many endangered species into regions of similar climate, physiography, soil, vegetation, and land use. Each species was assigned an endangerment factor or factors that represented the main reasons for its decline, and then a "factor diversity" index was calculated for each region that represented the complexity of endangerment factors for that taxonomic group in that region.

The larger the index, the greater the complexity of endangerment factors and the more complex the recovery strategy required (Flather, Knowles, and Kendall 1998). Such a method could be adapted to any region or ecosystem to provide an index of the distribution of threatened and endangered species and the complexity of the causes of their endangered status, and would help identify where the risks of extinction are concentrated. At the ecosystem level, habitat patches could serve as more natural, ecological, and fundamental units than counties for the initial assessment of spatial distribution. This approach focuses on factors that affect many species throughout the system and serves to prioritize management efforts because it identifies which species cover multiple areas and which are unique to particular areas, as well as identifying areas affected by a few or many endangerment factors. Widely distributed species are likely to respond more strongly to management actions that change endangerment factors throughout the entire ecosystem than are species confined to a more limited portion of the system. Areas affected by only a few endangerment factors may be easier to protect with simpler, more easily implemented management strategies than those affected by many factors. This method can also be used to identify ecosystems or parts of ecosystems that are subject to greater endangerment stress (Flather, Knowles, and Kendall 1998).

Obstacles to Implementing Ecosystem Management

As noted earlier, although 18 federal agencies in the United States offer definitions of ecosystem management, attempts to implement it often have not been successful. Federal agencies have political, land-use, and agency-specific limits that often bear little relation to the functional boundaries of real ecosystems. Aside from the problem of diverse and confused definitions of ecosystem management, we must examine several other obstacles that must be removed for ecosystem management to become a concept that contributes to conservation.

The Need for a Unified Vision and Values for Ecosystems and Ecosystem Management

Ecosystem management must incorporate patterns of sustainability and the human dimension. To achieve these goals, the various stakeholders need unifying values and purposes to participate constructively in managing ecosystems. Working toward unifying values and goals does not mean that stakeholders will not continue to hold personal, even self-centered, positions about the best use of the ecosystem. Such stakeholder-specific interests must be heard, understood, and addressed. But the effort to move toward a unifying vision and set of values remains essential if ecosystem management is to progress beyond a spirited discussion of stakeholder interests to coordinated and constructive action that can be supported by multiple stakeholders. Without such values and purposes, we continue in the present reality of polarized and fragmented groups of stakeholders with different, often conflicting, values and visions of the ecosystem, leading to separate agendas that foster distrust and conflict. Conservation biologists simply cannot accept this condition if conservation is to succeed at ecosystem levels. They must work to change it.

The Need for Unified Sources of Information and Analysis

To cooperate effectively in ecosystem management, agencies need a common clearinghouse of information and analysis regarding ecosystem processes and their responses to management systems. No such clearinghouse exists, and there are currently no plans to create one. Existing information is dispersed among scientific literature, proceedings of professional conferences and symposia, and intraagency technical reports. The information varies in quality, reliability, focus, format, and accessibility, and there are no uniformly accepted standards for data collection among agencies.

The Need for Interagency Cooperation in Decision Making and Public Trust

Ecosystem management requires the creation of permanent committees, boards, or working groups in which all agencies with jurisdiction or interest in the ecosystem are represented. Currently, most attempts at ecosystem management simply create ad hoc groups for individual ecosystem management projects. To be effective, ecosystem management groups must be permanent, have their own budgets, and possess the authority to make meaningful decisions about ecosystem management policies. Interagency groups must move the concept of ecosystem management from the *discussion agenda* (ideas discussed or defined by agencies) to the *decision agenda* (ideas that are the subject of decisive agency action). To make such movement possible, the interagency group must be able to define ecosystem management in terms of the problems to be solved rather than the concepts to be defined. If the interagency team can do this successfully, it translates the concept of ecosystem management into a pattern of guidelines for action in all represented agencies. Some practical examples of this concept are beginning to emerge in the recent and increasing formation of so-called "watershed councils." Such councils consist of interagency bodies of public and private stakeholders that ecosystem management issues within individual watersheds. For example, the Pacific Rivers Council funded a recent comprehensive study of salmon populations in the U.S. Pacific Northwest that produced methods and guidelines for prioritizing these populations for conservation (Allendorf et at. 1997).

It is not enough for such interagency groups to agree among themselves. The public must be involved in the decision-making process because, in democratic societies, people do not support what they do not create. If local citizens do not help to shape ecosystem policy, they will not support the policies created by outside interests, and their lack of support will doom the policy to failure. Further, unless information is communicated effectively to the public, more information does not lead to better decisions or to increased public support for those decisions. Ecosystem managers must first determine what constitutes a sense of place for local residents and how that sense is expressed in their activities, social relationships, and use of the ecosystem. Once that is understood, managers must use language that communicates shared meanings with local residents and their sense of place if they want local support and cooperation. Thus, communicating goals for ecosystem management cannot be only about scientific or management issues. Management goals also must address issues of self-interest to local residents and to the social relations and context in which they live (Cantrill 1998).

The Need for a Mechanism of Research Policy Translation

If ecosystem management is to have a basis in science and a foundation of professional credibility, it must have the means to smoothly translate reliable research findings into informed policy. This condition requires established and ongoing channels of communication and high levels of trust among researchers, managers, and lawmakers. At present, no such mechanism exists. As a result, no legislation has been passed or is being seriously considered that creates a mandate for ecosystem management (Haeuber 1996).

Some scientists and conservationists have called for something like a "national ecosystems act," "native ecosystems act," or "endangered ecosystems act" that would provide such a mandate in management coupled with substantive protection for endangered ecosystems (Noss and Scott 1997). Under such legislation, endangered ecosystems would receive protection analogous to the protection that species receive under the Endangered Species Act. The protected ecosystems would be prohibited from "takings" that would degrade them in any way and become the subjects of "recovery plans" designed to restore historic structure and function. Such an act also would seek to ensure representation of viable examples of all types of native ecosystems in the United States, and would provide guidelines for inventory, research, and monitoring of these systems.

No such legislation has been passed, or even proposed by those within the federal legislative system, and none is likely to be given serious consideration in the foreseeable future (Haeuber 1996). Instead, the present situation is a discontinuity between research and policy, such that policy is typically uninformed by research, and research and policy interests have low levels of mutual trust. A legislative mandate could help create federal agency agendas driven by long-term perspectives in management, budget, policy, and research. At present, agency agendas are driven by short-term perspectives of political considerations and perpetuation of existing influence and power.

POINTS OF ENGAGEMENT—QUESTION 2

Consider the definition of ecosystem management offered earlier. Can the concepts expressed in this definition serve as unifying ideas for diverse agencies in developing a common paradigm and value system of ecosystem management? If so, through what mechanisms? If not, what else must be clarified?

There are cases in which unique combinations of circumstances force multiple agencies, with private citizens, to abandon traditional, vested interest approaches and adopt an ecosystem management approach to a conservation problem that cannot be solved by other methods. The northern spotted owl *(Strix occidentalis caurina)* offers a case history that reveals why an ecosystem approach may become necessary to solve problems that initially appear to be issues of single-species management.

WHY ECOSYSTEM MANAGEMENT MATTERS IN SPECIES MANAGEMENT: THE CASE OF THE NORTHERN SPOTTED OWL

As recently as the late 1960s, little was known about the uncommon and rarely seen northern spotted owl (fig. 10.13), which inhabits the U.S. Pacific Northwest. In the summer of 1967, Eric Forsman, an undergraduate student at Oregon State University, learned that he could usually elicit the owl's response if it was present by imitating its call (Meslow 1993). When Forsman and a fellow undergraduate, Richard Reynolds, began to use this technique to search for spotted owls in Oregon, they discovered that the birds could regularly be found in old-growth forests, but not usually in other habitats. Forsman and Reynolds brought their data to the attention of professor Howard Wight of Oregon State University. Wight viewed the data with interest, and by 1972, Forsman had begun graduate research on the spotted owl under Wight's direction (Meslow 1993).

Forsman's studies of the spotted owl revealed a pattern of habitat use and population distribution that was directly in conflict with management policies of the U.S. Forest Service (Forsman, Meslow, and Wight 1984). The owls were almost exclusively restricted to old-growth forests, which forest managers viewed as areas of low productivity (individual trees were no longer adding significant annual biomass), with many individuals near the end of their life span (overmature stands). From an economic perspective, the rational management policy was to cut old-growth timber; however, timber management policy on public lands operates within a context of environmental law, and two laws enacted during the studies of the spotted owl had enormous effects.

In 1973, the U.S. Fish and Wildlife Service included the spotted owl in its "Red Book," an early version of the official national list of endangered species. After the owl's Red Book listing, an interagency group, the Oregon Endangered Species Task Force, recommended that management agencies retain 300 acres of old-growth forest around every spotted owl nest site (Caldwell, Wilkinson, and Shannon 1994). This recommendation was rejected by both the Forest Service and the U.S. Bureau of Land Management (BLM) because both agencies wanted a statewide population management goal established for spotted owls before implementing site-specific management practices (Meslow 1993). When the ESA became law later in 1973, the northern spotted owl was not listed as an endangered species, and thus received no legal protection or benefit from the act.

In 1976, the newly passed National Forest Management Act (NFMA) directed the Forest Service to "maintain viable populations of existing native and desired non-native verte-

Figure 10.13

The northern spotted owl (*Strix occidentalis caurina*), a species that can be effectively preserved only with an ecosystem management approach to its obligate habitat, old-growth conifer forests.

brate species" in national forests (Wilcove 1993). In other words, the NFMA and its attendant policies told the Forest Service that it was not allowed to create any more endangered species, nor was it allowed to destroy portions of a species' range or habitat (Meslow 1993). Although the concepts and techniques of population viability analysis were still in their infancy, it was clear that the initial recommendation of 300 acres for each pair of owls was inadequate because the protection of such small areas would not protect enough individuals to sustain the population (Wilcove 1993).

The Oregon Endangered Species Task Force then recommended a goal of maintaining 400 pairs of spotted owls on public lands in Oregon. The plan called for protecting habitat in ways that would provide for clusters of three to six pairs of spotted owls, although single-site management was still permitted. Core areas for clustered pairs were to be no more than 1.6 km apart and each pair was to have a core area of 300 acres.

Conservation biologist David Wilcove called these rules "management by minima—choosing the smallest number of owls and the smallest amount of old-growth per pair of owls in a weak effort to fulfill legal obligations" (Wilcove 1993). Subsequent studies on radio-tracked owls (Forsman 1980; Forsman and Meslow 1985) demonstrated that individual owls needed at least 1,000 acres, not 300 acres, of old-growth forest for permanent territories. In light of this data, the spotted owl management plan was revised and recommendations were changed to 1,000 acres of old-growth forest for each owl pair within 1.5 miles of its nest site. However, the recommendations were rejected by the BLM and only partially followed by the Forest Service (Meslow 1993).

Supported with an empirical basis of estimating the home range of the owls, a population viability analysis by Russell Lande concluded that the spotted owl population was declining, and that the population could not be conserved unless significant portions of the landscape remained in old-growth forests. A patchwork of small, old-growth reserves would not be adequate (Wilcove 1993). In subsequent analyses with additional data, Lande concluded that the population was stable under current conditions, but refined his model to estimate the probability of population persistence at differing levels of habitat loss (Lande 1988). On the basis of this analysis, Lande determined that the spotted owl population could not persist with less than 20% of the landscape in old-growth forests. In contrast, the U.S. Forest Service proposed management guidelines that would have conserved only 6% of the landscape in old-growth forests (Wilcove 1993). A coalition of environmental groups, led by the Seattle Audubon Society, sued the Forest Service for failing to adopt a credible conservation strategy in compliance with NEPA, ESA, and NFMA, eventually gaining an injunction on 135 timber sales in spotted owl habitat (Caldwell, Wilkinson, and Shannon 1994).

In the midst of the controversies created by the Lande analyses, in 1987 the U.S. Fish and Wildlife Service (FWS) was petitioned to list the spotted owl as an endangered species under the ESA. The FWS determined that the listing was unwarranted, but in 1988 a coalition of conservation groups filed an appeal in federal court, which ruled that the FWS decision was not scientifically based and directed it to readdress the issue. In the same year, the Oregon Endangered Species Task Force proposed new management recommendations for the northern spotted owl that, for the first time, addressed its entire range through Washington, Oregon, and northern California. The recommendations were not followed by any of the land management agencies.

Meanwhile, in 1989 the successful litigation effort of the Seattle Audubon Society led Forest Service Chief Dale Robertson to appoint an Interagency Scientific Committee (ISC) to "develop a scientifically credible conservation strategy for the northern spotted owl" (Meslow 1993). The ISC was headed by wildlife biologist Jack Thomas and composed of representatives from the Forest Service, Fish and Wildlife Service, Bureau of Land Management, and National Park Service. The ISC recommended a strategy, subsequently known as the Thomas Report (Thomas et al. 1990), that called for a system of habitat conservation areas (HCAs) on public forest land in Washington, Oregon, and California. In Washington and Oregon, each HCA would accommodate 20 pairs of owls, spaced at 19-km inter-

vals. In California, where old-growth forests are more fragmented, HCAs were to accommodate at least 10 pairs of owls and be no more than 10 km apart (Harrison, Stahl, and Doak 1992). No timber harvesting would be allowed in the HCAs (Caldwell, Wilkinson, and Shannon 1994), and the "50-11-40" rule provided for dispersal of juvenile owls from one HCA to another by requiring a certain amount of landscape timber coverage (50%) of acceptable size (at least 11 in. diameter) and associated canopy closure (40%) in each quarter township in areas lying between adjacent HCAs (Harrison, Stahl, and Doak 1992; Franklin 1993; Meslow 1993; Wilcove 1993). Nearly 8 million acres were protected under the plan, with 30% of the landscape preserved in old-growth forests. The Thomas strategy also allowed about 500,000 acres of old-growth forest outside of HCAs to be cut (Harrison, Stahl, and Doak 1992). Concurrent with these recommendations, the FWS reexamined the status of the spotted owl in 1989 and proposed listing it as a threatened species (Wilcove 1993). By 1990, the owl was officially listed.

The Thomas strategy was the subject of more lawsuits against the Forest Service in 1991 that prevented it from selling timber under the new plan. Eventually, U.S. District Judge William Dwyer ruled that the plan carried significant risks to the owl and that the Forest Service had failed to consider the needs of other species inhabiting old-growth forests (Harrison, Stahl, and Doak 1992).

New management recommendations call for increasing levels of input from community-based groups representing all stakeholder groups affected by the decision, and require increased consideration of economic and human community considerations (Raphael and Marcot 1994). It is not possible to write a final, or even pleasant, ending to this story or to foretell that the owls will avoid extinction and live happily ever after in their old-growth forests. But despite its controversies and limitations, the Thomas Report was a watershed event in the practical development of ecosystem management. It irrevocably shifted management emphasis from site-specific considerations to a focus on functional ecosystems, and its recommendations have been incorporated into numerous subsequent studies of old-growth forests and their management (Wilcove 1993).

The story of the northern spotted owl illustrates the conceptual evolution of ecosystem management and enables us to track the paradigm shift from resource management to ecosystem management in a particular case. Note the progression in the management strategies of the spotted owl.

Individual pairs

↓

Small isolated clusters of pairs

↓

Small clusters of pairs arranged in a habitat network
(spatial structure)

↓

Large clusters of pairs connected by landscape habitat corridors
(spatial structure with connectivity)

Just as the management of populations changed over time in the case of the spotted owl, so did the management of its surrounding environment. Conceptually, that progression looks like this:

Site-specific protection of small habitat units

↓

Protection of cluster habitat units

↓

Protection of a network of habitat around clusters

↓

Protection of an ecosystem (old-growth forests) as a specific percentage of landscape

This story also illustrates the scientific, social, legal, and political elements necessary for ecosystem management to succeed:

Scientific research establishes conservation concern

↓

Concern is connected to agency management practices

↓

Agency management practices are affected by legislation, public input, and judicial review

↓

Judicial review forces interagency cooperation to achieve compliance

↓

Interagency cooperation increases inclusiveness in decision making and requirements for empirical data and scientific reliability

↓

Increased inclusiveness and reliability strengthens support for management proposals

Many issues and controversies regarding ecosystem management in general, and the spotted owl in particular, remain unresolved. The price for ecosystem management has been high in economic, social, and political currencies and its outcomes remain uncertain. Although the controversy of the spotted owl is a dramatic example of the development of ecosystem management from a theoretical concept to a practical management strategy, the strength of opposition generated against the spotted owl also led to significant revisions—some would say weakening—of the Endangered Species Act (chapter 2) and to lasting, more well-organized social opposition to the application of ecosystem management in other contexts.

POINTS OF ENGAGEMENT—QUESTION 3

What elements of the spotted owl's demographics and habitat preferences created anomalies that a resource management approach was unable to solve? What features of an ecosystem management approach were better suited to address these same problems?

Synthesis

Ecosystem management, like management at any biological level, should be undertaken with care, caution, and humility. It is sobering to remember the findings of ecologist C. S. Holling, who determined, after an examination of 23 managed ecosystems, that it was management activities that led to the collapse of the systems (Holling 1995). Ecosystem management must have a scientific basis that can identify what kinds of questions to ask, what data to collect and how to collect them, how to model the system to be managed, and how to create adaptive management mechanisms that are responsive to changing ecosystem needs and human concerns.

The concept of ecosystem management represents a genuine paradigm shift in our conception of management. Ecosystem management has, indeed, become necessary, inescapable, and urgent because of the failure of the resource management paradigm to address current conservation problems, the causes and sources of regional environmental degradation, and the nationalization of environmental values. Species management approaches too quickly exhaust limited human and financial resources as the number of threatened and endangered species continues to rise, and they overlook species that are small, difficult to classify, or not appealing to public sentiment or consciousness.

Ecosystem management has yet to prove technically feasible or politically acceptable in most contexts. To make progress toward meaningful ecosystem management, conservationists must develop a common, accepted, and operational definition of the concept, devise practical ways to implement it in varying contexts, and create political support and legislative mandates to translate ecosystem management concepts into enforced policy directives. Progress toward ecosystem management will necessitate conflict and clarification of values. Individual agency jurisdictions must be replaced with permanent working groups or boards with independent budgets, regional authority, and legal mandates. Without these, ecosystem management will remain a compelling theoretical concept, but a frustrating and unfulfilled practice.

Progress toward ecosystem management is most impeded by the lack of a transagency definition of what ecosystem management is and by a coherent vision of exactly what ecosystem management would look like if it were actually working. Articulating that definition and developing that vision are essential prerequisites to moving toward functional ecosystem management. The alternatives—default visions of species management, habitat management, site management, or commodity (resource) management—are unsustainable, but we cannot break away from these perspectives until conservation biologists have made ecosystem management a more credible and desirable alternative. Defining ecosystem management and forming a compelling vision of its functions will be one of the great conservation challenges in the coming decade.

Making Ecosystem Management Work—A Directed Discussion

Reading assignment: Berry, J., G. D. Brewer, J. C. Gordon, and D. R. Patton. 1998. Closing the gap between ecosystem management and ecosystem research. *Policy Sciences* 31:55–80.

Questions

1. Berry et al. report that interagency attempts to implement ecosystem management were marked by frustration, fragmentation, and polarization. What factors may have led to these results in interagency ecosystem management efforts? What could be done to reduce or eliminate these factors?
2. Traditionally, the needs and wants of ecosystem managers and stakeholders have directed much of the research of conservation biologists. Do you think conservation biologists should have a greater role in determining what questions to ask, or should they focus on providing answers to questions managers ask?
3. Berry et al. note that the Walla Walla group was an exception to negative tendencies of interagency efforts to pursue ecosystem management. What elements made this group successful in implementing ecosystem management?
4. Berry et al. propose the creation of a coordinating board of ecosystem management research, with appropriate funding, legislative mandate, and support structures. Do you think this approach would remove current obstacles to ecosystem management? If so, why? If not, what is wrong with the proposal, and what alternative approach might be more successful in making ecosystem management work?

Learning Online

Visit our webpage at www.mhhe.com/conservation for case studies, animations, practice quiz questions, and additional readings to help you understand the material in this chapter. You'll also find active links to the following topics:

Field Methods for Studies of
    Ecosystems
Species Abundance, Diversity, and
    Complexity

Food Webs
Nutrient Cycling
Succession and Stability
Landscape Ecology

Geographic Ecology
Global Ecology

Literature Cited

Abelson, R. 1967. Definition. In *The encyclopedia of philosophy,* Vol. 2, ed. P. Edwards, 314–24. New York: MacMillan and Free Press.

Allendorf, F. W., O. Bayles, D. L. Bottom, K. P. Currens, C. A. Frissell, D. Hankin, J. A. Lichatowich, W. Nehlsen, P. C. Trotter, and T. H. Williams. 1997. Prioritizing Pacific salmon stocks for conservation. *Conservation Biology* 11:140–52.

Augustine, D. J., and L. E. Frelich. 1998. Effects of white-tailed deer on populations of an understory forb in fragmented deciduous forests. *Conservation Biology* 12:995–1004.

Baker, W. L. 1989. Landscape ecology and nature reserve design in the Boundary Waters Canoe Area, Minnesota. *Ecology* 70:23–35.

Bean, M. J. 1997. A policy perspective on biodiversity protection and ecosystem management. In *The ecological basis of conservation: heterogeneity, ecosystems, and diversity,* eds. S. T. A. Pickett, R. S. Ostfeld, M. Shachak, and G. E. Likens, 23–28. New York: Chapman and Hall.

Beecham, J. J. 1983. Population characteristics of black bears in west central Idaho. *The Journal of Wildlife Management* 47:405–12.

Belsky, A. J., and D. M. Blumenthal. 1997. Effects of livestock grazing on stand dynamics and soils in upland forests of the interior West. *Conservation Biology* 11:315–27.

Bergerud, A. T., and W. B. Ballard. 1988. Wolf predation on caribou: the Nelchina herd case history: a different interpretation. *The Journal of Wildlife Management* 52:344–57.

Berry, J., G. D. Brewer, J. C. Gordon, and D. R. Patton. 1998. Closing the gap between ecosystem management and ecosystem research. *Policy Sciences* 31:55–80.

Bowers, M. A., and S. F. Matter. 1997. Landscape ecology of mammals: relationships between density and patch size. *Journal of Mammalogy* 78:999–1013.

Boyce, M. S. 1998. Ecological-process management and ungulates: Yellowstone's conservation paradigm. *Wildlife Society Bulletin* 26:391–98.

Boyce, M. S., and N. F. Payne. 1997. Applied disequilibriums: riparian habitat management for wildlife. In *Ecosystem management: applications for sustainable forest and wildlife resources,* eds. M. S. Boyce and A. Haney, 133–46. New Haven, Conn.: Yale University Press.

Brennan, R. D., C. T. de Wit, W. A. Williams, and E. V. Quattrin. 1970. The use of digital simulation language for ecological modeling. *Oecologia* 4:113–32.

Bush, M. B., and P. A. Colinvaux. 1994. Tropical forest disturbance: paleoecological records from Darien, Panama. *Ecology* 75:1761–68.

Caldwell, L. K., C. F. Wilkinson, and M. A. Shannon. 1994. Making ecosystem policy: three decades of change. *Journal of Forestry* 92(4):7–10.

Cantrill, J. G. 1998. The environmental self and a sense of place: communication foundations for regional ecosystem management. *Journal of Applied Communications Research* 26:301–18.

Chadde, S. W., and C. E. Kay. 1991. Tall-willow communities on Yellowstone's northern range: a test of the "natural regulation" paradigm. In *The Greater Yellowstone ecosystem: redefining America's wilderness heritage,* eds. R. B. Keiter and M. S. Boyce, 231–62. New Haven, Conn.: Yale University Press.

Christensen, N. L. 1997. Implementing ecosystem management: where do we go from here? In *Ecosystem management: applications for sustainable forest and wildlife resources,* eds. M. S. Boyce and A. Haney, 325–44. New Haven, Conn.: Yale University Press.

Christensen, N. L., A. M. Bartuska, J. H. Brown, S. Carpenter, C. D'Antonio, R. Francis, J. F. Franklin, J. A. MacMahon, R. F. Noss, D. J. Parsons, C. H. Peterson, M. G. Turner, and R. G. Woodmansee. 1996. The report of the Ecological Society of America committee on the scientific basis for ecosystem management. *Ecological Applications* 6:665–91.

Congressional Research Service. 1994. *Ecosystem management: federal agency activities.* Washington, D.C.: Congressional Research Service, Library of Congress.

Coughenour, M. B., F. J. Singer, and J. Reardon. 1994. The Parker transects revisited: long-term herbaceous vegetation trends on Yellowstone's northern winter range. In *Plants and their environments: proceedings of the first biennial scientific conference on the Greater Yellowstone Ecosystem,* ed. D. G. Despain, 73–95. Technical Report NPS/NRYELL/NRTR-

93XX. Denver, Colo.: U.S. Department of the Interior, National Park Service, Natural Resources Publication Office.

Czech, B. 1995. Ecosystem management is no paradigm shift; let's try conservation. *Journal of Forestry* 93:17–23.

Czech, B., and P. R. Krausman. 1997. Implications of an ecosystem management literature review. *Wildlife Society Bulletin* 25:667–75.

D'Erchia, F. 1997. Geographic information systems and remote sensing applications for ecosystem management. In *Ecosystem management: applications for sustainable forest and wildlife resources,* eds. M. S. Boyce and A. Haney, 201–25. New Haven, Conn.: Yale University Press.

Eastman, J. R. 1995. *IDRISI for Windows users guide version 1.0.* Worcester, Mass.: Clark University.

Ficklin, R. L., E. G. Dunn, and J. P. Dwyer. 1996. Ecosystem management on public lands: an application of optimal externality to timber production. *Journal of Environmental Management* 46:395–402.

Flather, C. H., and J. R. Sauer. 1996. Using landscape ecology to test hypotheses about large-scale abundance patterns in migratory birds. *Ecology* 77:28–35.

Flather, C. H., M. S. Knowles, and I. A. Kendall. 1998. Threatened and endangered species geography. *BioScience* 48:365–76.

Ford, M. A., and J. B. Grace. 1998. Effects of vertebrate herbivores on soil processes, plant biomass, litter accumulation and soil elevation changes in a coastal marsh. *Journal of Ecology* 86:974–982.

Forsman, E. D. 1980. Habitat utilization by spotted owls in the west-central Cascades of Oregon. Ph.D. dissertation. Corvallis: Oregon State University.

Forsman, E. D., and E. C. Meslow. 1985. Old-growth forest retention for spotted owls, how much do they need? In *Ecology and management of the spotted owl in the Pacific Northwest,* eds. R. J. Gutierrez and A. B. Carey, 58–59. General Technical Report PNW-185. Washington, D.C.: U.S. Forest Service.

Forsman, E. D., E. C. Meslow, and H. M. Wight. 1984. Distribution and biology of the spotted owl in Oregon. *Wildlife Monographs* 87:1–64.

Franklin, J. F. 1993. Preserving biodiversity: species, ecosystems, or landscapes? *Ecological Applications* 3:202–5.

Fredrickson, L. H. 1997. Managing forested wetlands. In *Ecosystem management: applications for sustainable forest and wildlife resources,* eds. M. S. Boyce and A. Haney, 147–77. New Haven, Conn.: Yale University Press.

Fuller, T. K. 1989. *Population dynamics of wolves in north-central Minnesota. Wildlife Monographs* 105.

Greenlee, J. M. 1996. The ecological implications of fire in Greater Yellowstone. *Proceedings of the second biennial conference on the Greater Yellowstone Ecosystem.* Fairfield, Wash.: International Association of Wildland Fire.

Grumbine, R. E. 1994. What is ecosystem management? *Conservation Biology* 8:27–38.

Grumbine, R. E. 1998. Seeds of ecosystem management in Leopold's *A Sand County Almanac. Wildlife Society Bulletin* 26:757–60.

Haeuber, R. 1996. Setting the environmental policy agenda: the case of ecosystem management. *Natural Resources Journal* 36:1–27.

Haeuber, R., and J. Franklin. 1996. Forum: perspectives on ecosystem management. *Ecological Applications* 6:692–93.

Harrison, S., A. Stahl, and D. Doak. 1992. Spatial models and spotted owls: exploring some biological issues behind recent events. In *The landscape perspective: readings from Conservation Biology,* ed. D. Ehrenfeld, 177–80. Cambridge, Mass.: Blackwell Science and The Society for Conservation Biology.

Heinselman, M. L. 1973. Fire in the virgin forests of the Boundary Waters Canoe Area, Minnesota. *Quaternary Research* 3:329–82.

Hillborn, R. 1995. A model to evaluate alternative management strategies for the Seregenti-Mara Ecosystem. In *Serengeti II: dynamics, management, and conservation of an ecosystem,* eds. A. R. E. Sinclair and P. Arcese, 617–37. Chicago: University of Chicago Press.

Holling, C. S. 1995. What barriers? What bridges? In *Barriers and bridges to renewal of ecosystems and institutions,* eds. L. H. Gunderson, C. S. Holling, and S. S. Light, 3–25. New York: Columbia University Press.

Irland, L. 1994. Getting from here to there: implementing ecosystem management on the ground. *Journal of Forestry* 92(8):12–17.

Joseph, P. 1998. The battle of the dams. *Smithsonian* 29(8):48–54.

Kay, C. E., and F. H. Wagner. 1994. Historical condition of woody vegetation on Yellowstone's Northern Range: a critical evaluation of the "natural regulation" paradigm. In *Plants and their environments: proceedings of the first biennial scientific conference on the Greater Yellowstone Ecosystem,* ed. D. G. Despain, 151–69. Technical Report NPS/NRYELL/NRTR-93XX. Denver, Colo.: U.S. Department of the Interior, National Park Service, Natural Resources Publication Office.

Keiter, R. B., and M. S. Boyce. 1991. *The Greater Yellowstone ecosystem: redefining America's wilderness heritage.* New Haven, Conn.: Yale University Press.

Knight, D. H., and L. L. Wallace. 1989. The Yellowstone fires: issues in landscape ecology. *BioScience* 39:700–706.

Knight, R. L. 1995. Ecosystem management and Aldo Leopold. *Rangelands* 17(6):182–83.

Krummel, J. R., R. H. Gardner, G. Sugihara, R. V. O'Neill, and P. R. Coleman. 1987. Landscape pattern in a disturbed environment. *Oikos* 48:321–24.

Kuhn, T. S. 1970. *The structure of scientific revolutions.* 2d ed., enlarged. Chicago: University of Chicago Press.

Lackey, R. T. 1998. Ecosystem management: desperately seeking a paradigm. *Journal of Soil and Water Conservation* 53:92–94.

Lande, R. 1988. Demographic models of the northern spotted owl (*Strix occidentalis caurina*). *Oecologia* 75:601–7.

Leach, M. K., and T. J. Givnish. 1996. Ecological determinants in species loss in remnant prairies. *Science* 273:1555–58.

Lee, K. N. 1993. *Compass and gyroscope: integrating science and politics for the environment.* Washington, D.C.: Island Press.

Leopold, A. 1966. A *Sand County Almanac with essays on conservation from Round River.* New York: Sierra Club/Ballantine.

Leopold, L. B. 1994. Flood hydrology and the floodplain. In *Water resources update: coping with the flood: the next phase,* Issue no. 94–95, eds. G. F. White and M. F. Myers, 11–14. Carbondale, Ill.: University Council on Water Resources.

Lockwood, J. L., and K. H. Fenn. 2000. The recovery of the Cape Sable seaside sparrow through restoration of the Everglades ecosystem. *Endangered Species Update* 17(1):10–14.

Loehle, C. 1991. Managing and monitoring ecosystems in the face of heterogeneity. In *Ecological heterogeneity,* Ecological Studies 86, eds. J. Kolasa and S. T. A. Pickett, 144–59. New York: Springer-Verlag.

Lyle, J. T. 1985. *Design for human ecosystems: landscape, land use, and natural resources.* New York: Van Nostrand Reinhold.

Major, J. 1969. Historical development of the ecosystem concept. In *The ecosystem concept in natural resource management,* ed. G. M. Van Dyne, 9–22. New York: Academic Press.

Marsden, E. 1991. Will the Bush administration choose reform? *High Country News* 7 (October):13.

McClanahan, T. R., and D. O. Obura. 1996. Coral reefs and nearshore fisheries. In *East African ecosystems and their conservation,* eds. T. R. McClanahan and T. P. Young, 67–99. New York: Oxford University Press.

McInnes, P. F., R. J. Naiman, J. Pastor, and Y. Cohen. 1992. Effects of moose browsing on vegetation and litter of the boreal forest, Isle Royale, Michigan. *Ecology* 73:2059–75.

McLaren, B. E., and R. O. Peterson. 1994. Wolves, moose, and tree rings on Isle Royale. *Science* 266:1555–58.

McNaughton, S. J. 1979. Grazing as an optimization process: grass-ungulate relationships in the Serengeti. *American Naturalist* 113:691–703.

McNaughton, S. J. 1984. Grazing lawns: animals in herds, plant form, and coevolution. *American Naturalist* 124:863–86.

McNaughton, S. J., R. W. Ruess, and S. W. Seagle. 1988. Large mammals and process dynamics in African ecosystems. *BioScience* 38:794–801.

Mech, L. D. 1995. The challenge and opportunity of recovering wolf populations. *Conservation Biology* 9:270–78.

Meslow, E. C. 1993. Spotted owl protection: unintentional evolution toward ecosystem management. *Endangered Species Update* 10(3–4):34–38.

Michener, W. K., E. R. Blood, J. B. Box, C. A. Couch, S. W. Golladay, D. J. Hippe, R. J. Mitchell, and B. J. Palik. 1998. Tropical storm flooding of a coastal plain landscape. *BioScience* 48:696–705.

Milne, B. T. 1991. Lessons from applying fractal models to landscape patterns. In *Quantitative methods in landscape ecology,* eds. M. G. Turner and R. H. Gardner, 199–235. New York: Springer-Verlag.

More, T. A. 1996. Forestry's fuzzy concepts: an examination of ecosystem management. *Journal of Forestry* 94(8):19–23.

Norton, D. J., and D. G. Davis. 1997. Policies for protecting aquatic diversity. In *Ecosystem management: applications for sustainable forest and wildlife resources,* eds. M. S. Boyce and A. Haney, 276–300. New Haven, Conn.: Yale University Press.

Noss, R. F., and A. Cooperrider. 1994. *Saving nature's legacy: protecting and restoring biodiversity.* Washington, D.C.: Defenders of Wildlife and Island Press.

Noss, R. F., H. B. Quigley, M. G. Hornocker, T. Merrill, and P. C. Paquet. 1996. Conservation biology and carnivore conservation in the Rocky Mountains. *Conservation Biology* 10:949–63.

Noss, R. F., and J. M. Scott. 1997. Ecosystem protection and restoration: the core of ecosystem management. In *Ecosystem management: applications for sustainable forest and wildlife resources,* eds. M. S. Boyce and A. Haney, 239–64. New Haven, Conn.: Yale University Press.

Nott, M. P., and S. L. Pimm. 1997. The evaluation of biodiversity as a target for conservation. In *The ecological basis of conservation: heterogeneity, ecosystems, and diversity,* eds. S. T. A. Pickett, R. S. Ostfeld, M. Shachak, and G. E. Likens, 125–35. New York: Chapman and Hall.

Odum, E. P. 1971. *Fundamentals of ecology.* 3d ed. Philadelphia: Saunders.

O'Neill, R. V., J. R. Krummel, R. H. Gardner, G. Sugihara, B. Jackson, D. L. DeAngelis, B. T. Milne, M. G. Turner, B. Zygmunt, S. W. Christensen, V. H. Dale, and R. L. Graham. 1988. Indices of landscape pattern. *Landscape Ecology* 1:153–62.

Opdam, P., R. Poppen, R. Reijnen, and A. Schotman. 1995. The landscape ecological approach in bird conservation: integrating the metapopulation concept into spatial planning. *Ibis* 137:S139–S146.

Paine, R. T. 1966. Food web complexity and species diversity. *American Naturalist* 100:65–75.

Paine, R. T. 1969. The *Pisaster-Tegula* interaction: prey patches, predator preference, and intertidal community structure. *Ecology* 50:950–61.

Pastor, J., B. Dewey, R. J. Naiman, P. F. McInnes, and Y. Cohen. 1993. Moose browsing and soil fertility in the boreal forests of Isle Royale National Park. *Ecology* 74:467–80.

Pearce, D. W., and R. K. Turner. 1990. *Economics of natural resources and the environment.* Baltimore, Md.: Johns Hopkins University Press.

Pickett, S. T. A., and M. L. Cadenasso. 1995. Landscape ecology: spatial heterogeneity in ecological systems. *Science* 269:331–34.

Poiani, K. A., B. D. Richter, M. G. Anderson, and H. E. Richter. 2000. Biodiversity conservation at multiple scales: functional sites, landscapes, and networks. *BioScience* 50:133–46.

Rama Mohan Rao, M. S., V. N. Sharda, S. C. Mohan, S. S. Shrimali, G. Sastry, P. Narain, and I. P. Abrol. 1999. Soil conservation regions for erosion control and sustained land productivity in India. *Journal of Soil and Water Conservation* 54:402–9.

Raphael, M. G., and B. G. Marcot. 1994. Key questions and issues: species and system viability. *Journal of Forestry* 92(4):45–47.

Richardson, M. S., and R. C. Gatti. 1999. Prioritizing wetland restoration activity within a Wisconsin watershed using GIS modeling. *Journal of Soil and Water Conservation* 54:537–42.

Risser, P. G. 1987. Landscape ecology: state of the art. In *Landscape heterogeneity and disturbance,* ed. M. G. Turner, 3–14. New York: Springer-Verlag.

Romme, W. H., M. G. Turner, L. K. Wallace, and J. S. Walker. 1995. Aspen, elk, and fire in northern Yellowstone National Park. *Ecology* 76:2097–2106.

Ruess, R. W., R. L. Hendrick, and J. P. Bryant. 1998. Regulation of fine root dynamics by mammalian browsers in early successional Alaskan taiga forests. *Ecology* 79:2706–20.

Seidensticker, J. C., M. G. Hornocker, W. V. Wiles, and J. P. Messick. 1973. Mountain lion social organization in the Idaho primitive area. *Wildlife Monographs* 35:1–60.

Shaffer, M. L. 1992. *Keeping the grizzly bear in the American West: a strategy for real recovery.* Washington, D.C.: Wilderness Society.

Shinneman, D. J., J. Watson, W. W. Martin. 2000. The state of the southern Rockies ecoregion: a look at species imperilment, ecosystem protection, and a conservation opportunity. *Endangered Species Update* 17(1):2–9.

Sinclair, A. R. E., and P. Arcese. 1995. *Serengeti II: dynamics, management, and conservation of an ecosystem.* Chicago: University of Chicago Press.

Slocombe, D. S. 1993. Implementing ecosystem-based management. *BioScience* 43:612–23.

Souers, A. 2000. Salvation for the dammed. *Sierra* 85(2):20, 22.

Soulé, M. E., D. T. Bolger, A. C. Alberts, J. Wright, M. Sorice, and P. Hill. 1988. Reconstructed dynamics of rapid extinctions of chaparral-requiring birds in urban habitat islands. *Conservation Biology* 2:75–90.

Sparks, R. E. 1995. Need for ecosystem management of large rivers and their floodplains. *BioScience* 45:168–82.

Taylor, S. 1993. Practical ecosystem management for plants and animals. *Endangered Species Update* 10(3–4):26–29.

Thomas, J. W., E. D. Forsman, J. B. Lint, E. C. Meslow, D. R. Noon, and J. Verner. 1990. *A conservation strategy for the northern spotted owl.* Portland, Ore.: U.S. Forest Service, Bureau of Land Management, U.S. Fish and Wildlife Service, and National Park Service.

U.S. Fish and Wildlife Service. 1993. *Grizzly bear recovery plan.* Missoula, Mont.: U.S. Fish and Wildlife Service.

Van Dyke, F., M. J. DeBoer, and G. M. Van Beek. 1996. Winter range plant production and elk use following prescribed burning. In *Proceedings of the conference on the ecological implications of fire in Greater Yellowstone,* eds. D. G. Despain, P. Schullery, and J. M. Greenlee, 193–200. Fairfield, Wash.: International Association of Wildland Fire.

Van Dyke, F. G., J. P. DiBenedetto, and S. C. Thomas. 1991. Vegetation and elk response to prescribed burning in south-central Montana. In *The Greater Yellowstone ecosystem: redefining America's wilderness heritage,* eds. R. B. Keiter and M. S. Boyce, 163–79. New Haven, Conn.: Yale University Press.

Van Dyne, G. M., ed. 1969. *The ecosystem concept in natural resource management.* New York: Academic Press.

Vos, C. C., and J. I. S. Zonneveld. 1993. Patterns and processes in a landscape under stress: the study area. In *Landscape ecology of a stressed environment,* eds. C. C. Vos and P. Opdam, 1–27. London, England: Chapman and Hall.

Wagner, F. H. 1969. Ecosystem concepts in fish and game management. In *The ecosystem concept in natural resource management,* ed. G. M. Van Dyne, 259–307. New York: Academic Press.

Wagner, F. H. 1977. Species vs. ecosystem management: concepts and practices. *Transactions of the North American Wildlife and Natural Resources Conference* 42:14–24.

Wagner, F. H., and C. E. Kay. 1993. "Natural" or "healthy" ecosystems": are U.S. parks providing them? In *Humans as components of ecosystems,* eds. M. J. McDonnell and S. T. A. Pickett, 157–270. New York: Springer-Verlag.

Welcomme, R. L. 1985. River fisheries. Food and Agriculture Organization Fisheries Technical Paper 262. Rome: Food and Agriculture Organization of the United Nations.

Wilcove, D. S. 1993. Turning conservation goals into tangible results: the case of the spotted owl and old-growth forests. In

Large-scale ecology and conservation biology: the 35th symposium of British Ecological Society with the Society for Conservation Biology, eds. P. J. Edwards, R. M. May, and N. R. Webb, 313–29. Oxford, England: Blackwell Scientific Publications.

Wolanski, E., E. Gereta, M. Borner, and S. Mduma. 1999. Water, migration, and the Serengeti ecosystem. *American Scientist* 87:526–33.

PART THREE

Applications

All understanding seeks application. The final chapters of this text transfer conservation understanding to conservation application at three levels. Chapter 11 examines the application of conservation to the restoration of populations, habitats, and ecosystems. Chapter 12 explicates the application of conservation to broader patterns of human behavior in economics and development. Chapter 13 describes a different kind of application altogether. Specifically, it is about the application of transference, the transference of knowledge about conservation into a career commitment to conservation.

Many applications of conservation knowledge and theory have already been made to specific problems in earlier chapters. These final chapters are placed in a different category because the applications are more systemic. Although not every conservation effort is necessarily an effort at restoration, many, perhaps most, are. Thus, chapter 11 offers a systematic approach to using effective management techniques to achieve a variety of conservation goals. In doing so, I make the assumption that an important dimension of being human is to have an interactive and beneficent relation to the land and non-human creatures around us, that the highest and best interaction of humans and nature is not a relationship of mutual exclusion, but an interaction of mutual benefit in which human actions are actually directed to increase biodiversity, enhance populations and their habitats, and improve ecosystem function.

Conservation biology is not merely the collection of information about natural scarcity and diversity, it is the application of such information to achieve certain social outcomes that are considered desirable. Because conservation is inescapably normative, it must address underlying economic patterns and motives for resource use, the application of market mechanisms and incentives to achieve conservation goals, and the consideration of alternative economic systems and processes that might be more effective at achieving desirable long-term states of environmental quality. These issues reflect the content of chapter 12.

Chapter 13 addresses you, the student, directly to explain how to make a career commitment to the field of conservation biology and how to make that commitment effective in its personal and professional outcomes. Specifically, this chapter explores how to make the transition from being a student in a course to being a citizen of a discipline and how to progress from the status of student to colleague. The final section of this chapter examines four trends now operative in conservation biology, and offers suggestions and strategies on how you might prepare for their future effects.

CHAPTER 11

Restoration Ecology

The process of ecological restoration provides an ideal basis for the development of a modern system of rituals for negotiating our relationship with the rest of nature.

—Frederick Turner, 1987

In this chapter you will learn about:

1 the definition and development of restoration ecology.

2 procedures and protocols for implementing restoration ecology.

3 techniques for monitoring and managing restored ecosystems.

4 obstacles, problems, and rewards of restoration efforts in real-world examples.

DEFINITION AND DEVELOPMENT

The Concept of Restoration

In 1935 a small group of Civilian Conservation Corps workers began replanting tallgrass prairie on a portion of an abandoned farm near Madison, Wisconsin, that had recently been acquired by the University of Wisconsin to serve as an arboretum. The project was directed by one of the university's professors, Aldo Leopold. The same year, Leopold began the restoration of a tract of abandoned farmland about an hour's drive north of the university. Reflecting on the meaning of this work and place years later, Leopold wrote, "On this sand farm in Wisconsin, first worn out and then abandoned by our bigger-and-better society, we try to rebuild, with shovel and axe, what we are losing elsewhere" (Leopold 1966: xviii). Arguably, these two events, connected in time, space, and one man's life, mark the beginnings of modern restoration ecology.

Webster's *Seventh New Collegiate Dictionary* defines **restoration** as "bringing back to a former position or condition," or "a representation or reconstruction of the original form." In an ecological context, *restoration* has been defined as "returning the land to its former use and condition" (Bradshaw and Chadwick 1980). The United States Bureau of Land Management, the nation's largest federal landholder, defines ecological restoration

as a process designed "to repair ecological systems by restoring essential structure and function that promote long-term stability" (Van Haveren et al. 1997). A more explicit definition was provided by Joel Brown (1994), who described restoration ecology as "that applied branch of evolutionary ecology that attempts to understand and implement the reintroduction of species, the recreation of habitats and species assemblages, and the revitalization of threatened organisms and communities."

Related concepts of *revegetation, rehabilitation,* and *reclamation* imply preliminary or partial steps toward restoration. **Revegetation** is a term normally used to describe the process in which plants recolonize a disturbed site from which a previous disturbance event has removed all or part of the former vegetation. Revegetation does not necessarily imply that the *original* vegetation has been reestablished, only that some sort of vegetation now occupies the site. **Rehabilitation** describes visual improvements of disturbed land (Bradshaw and Chadwick 1980), whereby the site is aesthetically enhanced to "look like" its former self. Rehabilitation does not necessarily include reestablishment of the land's former components, or their interactions or functions. **Reclamation**, a term with a long history in land management, refers to the preparation and enhancement of degraded land to fulfill its former use or a new use. Reclamation can be considered a weak form of restoration, in that the *functions* of the land—but not necessarily its *components* and *structure*—are restored. For example, the Federal Surface Mining

307

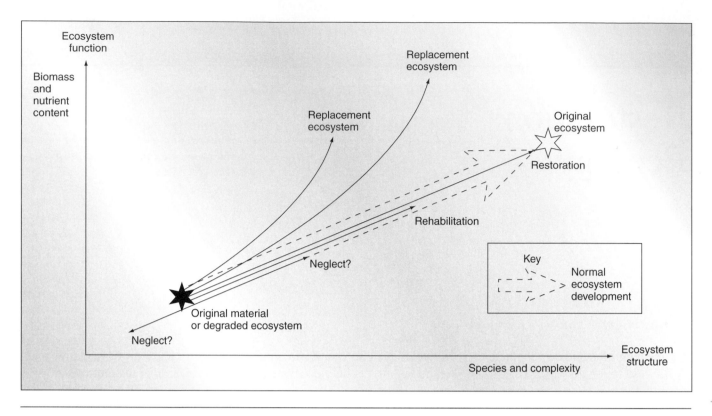

Figure 11.1

The relationship between ecosystem structure and ecosystem function in restoration. In natural selection, there is an increase in both ecosystem function and structure as demonstrated by the "normal ecosystem development" arrow. When ecosystems are degraded, both dimensions are reduced. If degraded ecosystems are neglected, they may slowly recover by natural processes or degrade further by erosion or landslip. When attempts are made at returning ecosystems to their original form, "restoration" describes complete success, whereas "rehabilitation" is only partial restoration. These efforts may also result in "replacement," in which an ecosystem different from the original is produced.

After Bradshaw (1987).

Control and Reclamation Act of 1977 requires that land disturbed by surface (strip) mining must be returned to its original contour, have its topsoil replaced, be protected from adverse hydrological effects (such as excessive erosion), and be planted with vegetation "similar to" the vegetation prior to mining.

Ecological restoration differs from revegetation, rehabilitation, and reclamation in three fundamental ways. First, restoration seeks to reestablish not only the function of the site, but also its components, structure, and complexity. Whereas "reclamation" may be complete when a land unit or system has an appropriate predisturbance appearance and fulfills all or most of its previous functions, such as water storage, soil conservation, or commodity production, "restoration" is not accomplished until the system possesses these traits *and* regains previous levels of diversity, complexity, sustainability, biological structure, and species composition (fig. 11.1). As the British restoration ecologist A. D. Bradshaw put it, "The objective . . . cannot therefore be just to establish vegetation. It must be to establish a complete ecosystem which, because we do not want to have to keep propping it up, should be self-sustaining" (Bradshaw 1984).

Second, restoration is not based on the exclusion of human influence from natural systems, but rather relies on intentional, purposeful, and constructive human activities. The science of restoration ecology takes into account that interactions and

interdependencies between ecological systems and humans can even be creative, making a system function better than it would otherwise. For example, prescribed fire is used in restoring degraded prairies, and Jordan (1994) has argued that "the need of the prairie for fire dramatizes its dependence on *us,* and so liberates us from our position as naturalists or observers of the community into a role of real citizenship. The burning of prairies is more than a process or a technology, it is an expressive act—and what it expresses is our membership in the land community." Restoration is not preservation. It takes the view that the most appropriate response to nature is not protection from human influence, but the constructive use of human intervention.

Third, restoration does not merely attempt to imitate what a system was, but also aims to replicate what a system did, and thus create a self-generating, sustainable, and persistent system (Turner 1987). A restored ecosystem is capable of sustaining itself, can resist invasions by new species, is as productive as the original, retains nutrients as well as the original, and has biotic interactions similar to the original (Ewel 1987).

A fourth quality of an ideal restoration is that it should be *aesthetically pleasing.* Whereas aesthetics depend on human perception, and therefore are not essential attributes of ecosystem function, restorationists can add value to their efforts if, when possible, they can create systems and landscapes of high

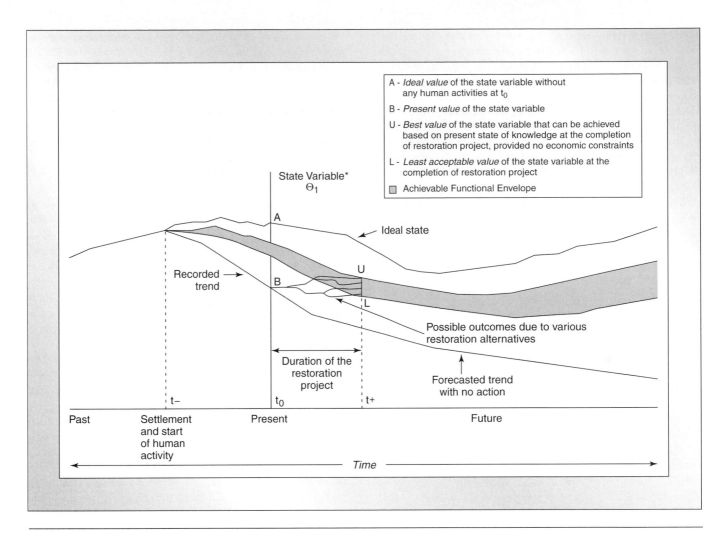

Figure 11.2

A conceptual model of restoration as movement from a present (degraded) state toward an ideal state of an ecological variable or condition.

From the Committee on Restoration of Aquatic Ecosystems (1992).

aesthetic quality. Ecologist Raymond Dasmann asserts that there are four primary dimensions to an aesthetically pleasing environment: order, diversity, health, and function (Dasmann 1966). *Order* draws attention to the fact that natural vegetation communities are typically dominated by a relatively small number of species, and these species form a unifying background for the site and its appearance in the landscape. Attention to *diversity* in restoration requires addition of numerous, rare components. These components often not only contribute to the system's function, but also add elements of contrast in color, texture, and other visual elements in the site or landscape. *Health* and *function* are related concepts, imperfectly but objectively indexed in the capacity of species to reproduce and perpetuate themselves (health) and to individually and collectively provide various ecological services, including soil generation and maintenance, control of erosion and flooding, micro- and macroclimate regulation, and regulation of water flow (function).

A conceptual model of restoration depicts the movement of a system from its present state toward an objectively defined ideal state (fig. 11.2). It may not be possible to attain the ideal

state because of constraints of knowledge, time, and money, but restoration ecologists can identify a range of acceptable states or an "achievable functional envelope" (Committee on Restoration of Aquatic Ecosystems 1992), most or all of which can also be designed with positive aesthetic qualities.

The Legal Foundations of Restoration: U.S. Examples

The legal foundation and mandate for restoration are recent in origin but accelerating in influence and development. U.S. environmental legislation illustrates the overall pattern of such development and provides examples of particular statutes that define the legal understanding and incentives for restoration efforts. In the United States, one of the earliest relevant statutes was the Federal Aid in Wildlife Restoration Act (Pittman-Robertson Act) of 1937. The Federal Aid in Wildlife Restoration Act levied a 10% tax on the sale of sporting arms and ammunition (later increased to 11%), the proceeds of which are pooled and then redistributed to individual states that contribute funds at a 1:3 ratio (one state dollar for every three federal dollars). Monies are used for research and management efforts that benefit wildlife, including efforts to enhance, acquire, or restore wildlife habitat. The success of this act spawned a later, parallel statute for fish, the Federal Aid in Fish Restoration Act (Dingell-Johnson Act) of 1950. In the Federal Aid in Fish Restoration Act, a 10% excise tax on certain kinds of fishing tackle provides the basis of funding, pooled and redistributed to the states by similar formulae and ratios as the Federal Aid in Wildlife Restoration Act (Bolen and Robinson 1995).

Other restoration-oriented legislation in the United States followed. The Federal Water Pollution Control Act of 1948 and its subsequent amendments provided funding and legal authority for the restoration of freshwater lakes and streams. The National Environmental Policy Act also provided legal incentive for restoration resulting from actions on federal lands. In the 1970s the Coastal Zone Management Act of 1972 and the Outer Continental Shelf Lands Act Amendments of 1978 provided legal authority and support for coastal restoration. The Endangered Species Act of 1973, although not focusing on restoration of sites, provided legal authority for restoration of populations and, if needed, restoration of habitat upon which those populations depend. The Federal Surface Mining Control and Reclamation Act of 1977 created regulations for coal mine reclamation and established minimum reclamation performance standards that included restoration to approximate original contour, segregation and replacement of topsoil, establishment of vegetation comparable to premining conditions, and protection from adverse hydrologic effects. Under the terms of the act, operators were required to take responsibility for successful revegetation for 5 years after seeding. In areas where annual precipitation is less than 66 cm, this requirement is increased to 10 years. Restoration costs are funded by a coal tax (Bradshaw and Chadwick 1980).

Finally, one of the most recent, and most famous pieces of restoration legislation was the Comprehensive Environmental Response, Compensation, and Liability Act of 1980, subsequently amended by the "Superfund" Amendments and Reauthorization Act of 1986 (Trefts 1990). This act, often referred to simply as "Superfund," gave the U.S. Environmental Protection Agency (EPA) authority to, among other things, require cleanup and restoration of toxic industrial waste sites. It is a popular misconception in the United States that the "Superfund" established under the act pays the costs of cleanup. Actually, individual polluters are responsible for the cost of cleanup and restoration of the site, regardless of costs. The monies in Superfund are used primarily to clean up sites in which no responsible party can be identified, to pay the EPA's cost of overseeing the cleanup to ensure compliance, or to pay the costs of litigation if the polluter or the polluter's insurers contest the finding of the EPA in court.

Types of Restoration and Criteria for Evaluation

Many restoration ecologists argue that the *process* of restoration ecology matters as much as the *product* of the restoration. From this perspective, restoration can be an organizing principle for ecological research. Jordan, Gilpin, and Aber (1987) assert,

"The acid test of our understanding is not whether we can take ecosystems to bits on pieces of paper, but whether we can put them together and make them work." Many restoration efforts have passed this test, including some conducted on formerly derelict lands that today make important contributions to conservation. For example, 17 current wildlife refuges or sites of scientific importance in the United Kingdom were once abandoned waste sites, such as lead mines, gravel pits, clay pits, chalk pits, and peat cuttings (Bradshaw and Chadwick 1980).

Most restoration efforts could be assigned to one of two categories of approach. *Compositional approaches* to restoration make community structure and components the organizing principles of the restoration effort (Howell 1986). The goal is to replace or restore species on the site that match a historic or "target" community as closely as possible, in order to construct (restore) a "snapshot" of a site's ecological past, rather than a moving picture. Although compositional approaches do not ignore the importance of species interactions and ecological function, they tend to assume that function follows composition, not vice versa. Compositional approaches tend to place greater value on populations and communities than on ecological functions and services of the system. Thus, compositional approaches are often used in efforts that feature the restoration of an endangered, threatened, or "favorite" species or group of species to an area.

In contrast, *functional approaches* emphasize the restoration of functions or services that the system provides, not the individual species responsible for those services (Howell 1986). Under a functional approach, it matters less which species are in the system than that the species provide the intended function or service. For example, in a forest restoration, a functional approach might use any tree species of appropriate height characteristics and growth rates if it provided the percent of canopy closure desired. Functional approaches, ironically, are often used precisely to benefit key species or groups of species. For example, "cover species" of plants may be sown on a degraded site, such as a degraded prairie, to provide concealment and protection from predators for upland game birds such as the ring-necked pheasant. But the plant species sown, for example cheatgrass (*Bromus tectorum*) may not be a native prairie species.

Perhaps the best example of the difference between compositional and functional approaches can be seen in how each approach would view the historic loss of the American chestnut (*Castanea dentata*) from forests in the eastern United States. The chestnut was historically the dominant tree of mature eastern forests in both numbers and biomass. Practically exterminated in the wild by a fungal disease, the chestnut blight, in the nineteenth century, its loss radically altered the community composition and structure of the eastern forest. But, based on available information and on simulation studies that have attempted to reconstruct ecosystem processes in past forest systems, ecosystem function in these forests was substantially unaffected (Allen and Hoekstra 1992). From the standpoint of a compositional approach, the loss of the chestnut was a catastrophe. From the standpoint of a functional approach, it was a nonevent. Restorationists must determine in advance what they are intending to restore, and whether it is ecological function or ecological composition that is the entity of value in the restoration effort.

The dichotomy of compositional and functional approaches should not be pressed too hard, because the boundaries between the two approaches are often blurred in individual restoration efforts. However, explicit and intentional recognition of the differences in these approaches is helpful to ensure that the goals of the restoration effort match the method that is being used to pursue them.

Regardless of whether a predominantly compositional or functional approach to restoration is used, some common objective criteria exist to determine the relative success of restoration efforts. These criteria include *sustainability* (the restored system can perpetuate itself), *resistance to invasion* (the restored system can resist significant invasions by nonindigenous species), *productivity* (the restored system is as efficient at processing energy and producing biomass as the original), *nutrient retention* (the restored community retains nutrients as well as the original), *functional relation* (the restored community has biotic interactions and relations among its components similar to the original, even if the components themselves are different from the original), and *genetic appropriateness* (populations in the restored community are, if possible, derived from genetic stocks similar to those of historic populations, adapted to local environmental conditions, and not a threat to the genetic integrity of other, neighboring populations in which gene flow is occurring or may occur in the future) (Ewel 1987).

These are ambitious criteria. By their standards, ecological restoration is achievable, but it also can be difficult, frustrating, and slow. We often know little about the historical ecosystems we attempt to restore. Key species may have been exterminated from the site or system, physical conditions may have changed radically, and cost and time may render any attempt at restoration prohibitive (Van Haveren et al. 1997). Before examining specific techniques and case histories of ecological restoration, we examine a general model to describe the processes of habitat degradation as a result of human activity, and the inadequacy of preservation as a conservation strategy.

A GENERALIZED MODEL OF HABITAT DEGRADATION

Throughout their history, humans have converted habitats from one form to another for their own use and convenience. In some instances, conversion was simply an exchange of one type of habitat for another. However, most modern conversions usually result in habitat degradation, or the loss of functional attributes and services that the system could provide in an unimpaired state, with a loss of diversity and structural complexity (i.e., loss of species and their spatial arrangements). General patterns of habitat conversion are well documented for many parts of the world (fig. 11.3) (Dobson, Bradshaw, and Baker 1997). For example, a common pattern, particularly in the tropics, is the conversion of forest to agricultural land, and then to degraded land. Each year, 1 to 4% of tropical forests are converted to other land uses, usually agriculture, and the resulting type of agricultural land typically is useful for only 3 to 5 years and is then abandoned. Ecological succession on such abandoned lands is slow

and, without nearby sources of indigenous forest species, the sites may never return to their previous levels of species diversity and complexity.

Dobson, Bradshaw, and Baker (1997) offer a simple mathematical model to describe patterns and rates of habitat degradation using four coupled differential equations. The equations assume that an original area of forest (F) is first converted to agricultural land (A), which after a period of time ($1/a$, where a is the rate at which agricultural land is converted to unused land) becomes unused land (U). The unused land recovers through succession or restoration to forested land after a given time interval ($1/s$, where s is the rate of succession to forested land). Unused land can also be restored to agriculturally viable land after a time interval of $1/b$, where b is the rate of recovery to a productive condition for agriculture. The rate of habitat conversion is assumed to be a function of the number of humans (P) using the land at any given time. The change in the amount of forest land over time (t) is equal to rate of return of unused land to forest land minus the change in human population density, times the amount of forest land in the area, or mathematically,

$$\frac{dF}{dt} = sU - dPF.$$

The second equation describes the change in the amount of agricultural land over time based on the change in the human population density, times the amount of forest land in the area plus the rate at which unused land is converted to agricultural land, and the rate at which current agricultural land is degraded:

$$\frac{dA}{dt} = dPF + bU - aA.$$

Third, the change in the amount of unused land over time is equal to the change in agricultural land minus the combined rate at which unused land is converted to forest and agriculture:

$$\frac{dU}{dt} = aA - (b + s)U.$$

Fourth, the change in the number of humans over time is equal to the numerical increase in the human population (rP) times the ratio of land available for agriculture in excess of subsistence needs ($A-hP$) to total agricultural land:

$$\frac{dP}{dt} = rP\frac{A - hP}{A}.$$

Equilibrium expressions for each of the principal variables can be derived from these equations, permitting predictions of how long it will take, under different assumptions of human growth rates, succession rates, and conversion rates, for forest land to be converted to agriculture, or degraded to unused land (fig. 11.4). As the length of time that land can be used for agriculture increases, the percentage of forest that remains in the landscape decreases. When land can be used only for a short time for agriculture and human populations are low and grow slowly, much of the landscape remains forested, a condition we see today where indigenous people still practice nomadic, "slash and burn" agriculture (Dobson, Bradshaw, and Baker 1997). Although this particular model uses forests as the habitat being converted, it can be applied equally well to conversion of

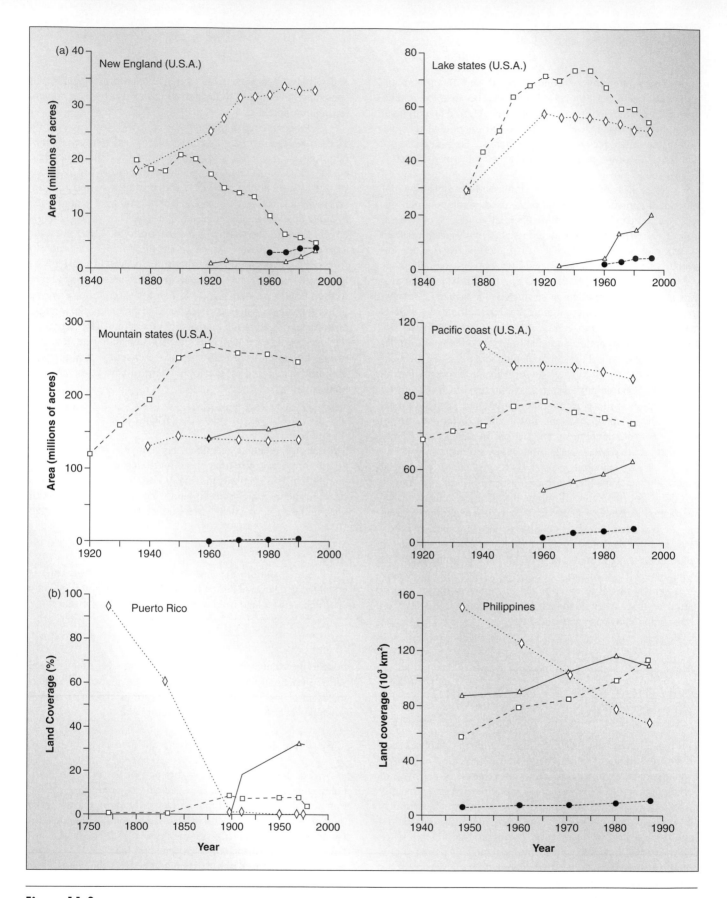

Figure 11.3

Patterns of habitat conversion in different regions of the world. Squares indicate farm area, diamonds are forest area, circles are urban areas, and triangles are miscellaneous forms of degraded land.

After Dobson, Bradshaw, and Baker (1997).

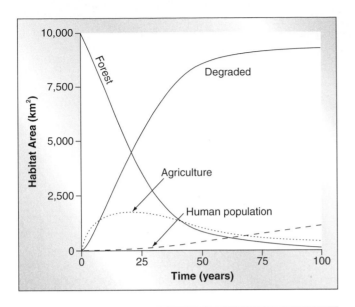

Figure 11.4

Patterns of land-use change toward equilibrium levels in conversions among forest, agriculture, and degraded lands. The amount of degraded land increases as forest land decreases.

After Dobson, Bradshaw, and Baker (1997).

prairies to agricultural land, or to industrial conversions, in which one industry is succeeded by another and derelict land is immediately reused. In all cases, the generalized, and pessimistic, prediction of the model is that complex habitats are rapidly converted to degraded sites. Thus, if this model accurately reflects patterns of land conversions (and historical data suggest that it does), then preservation alone will not succeed in protecting habitats. To be successful, conservation biology must include effective strategies for restoring previously degraded lands. Dobson, Bradshaw, and Baker (1997) wrote of their model's implications that "In a world where development should be curtailed yet . . . cannot be stopped, there is intense psychological and practical importance to growing new, or restored, ecosystems. Indeed, restoration can now be considered a critical element in managing the world's environment. It provides a powerful way of reducing the length of time for which habitat remains in the degraded and unused category of the simple model described above."

PROTOCOLS, PROCEDURES, AND EXAMPLES OF EFFECTIVE ECOLOGICAL RESTORATION

Disturbances of various types are natural processes in all ecosystems (chapters 8 and 10). However, some types of disturbances, particularly those caused by humans, have negative consequences for ecosystem function and lead directly to site degradation. An important goal in restoration ecology is to determine

the nature and consequences of past disturbances (identify the problem) and then determine how these consequences can be corrected (determine a strategy for solution).

The target criteria for restoration of a given site may be historical/compositional (to restore a site or system to its former community of species) or functional (to restore a site or system for a specific purpose). An assessment of the site should include structural and functional attributes of the system and comparative historical data of the same variables before degradation. Alternatively, the assessment may use variables from a "target system" that approximates a desired future state of the degraded system.

Some examples of physical and chemical soil characteristics that may be measured in a preliminary assessment of a degraded site are given in table 11.1. In addition to these individual features of soil, the ecological systems that a site supports also possess *emergent properties*, characteristics that cannot be predicted from individual ecological components such as soil, but instead appear or emerge in the totality of the system. These include resilience, a measure of the system's ability to recover from disturbance; persistence, the system's ability to undergo successional process through time without excessive human intervention; and verisimilitude, a measure of how similar the restored ecosystem is to its historical or targeted standard of comparison. An assessment of a site to be restored should evaluate emergent properties before restoration is implemented.

Restoring Terrestrial Ecosystems

Steps toward restoring terrestrial ecosystems include:

1. stabilizing land surfaces by stopping excessive erosion;
2. controlling inputs of pollution;
3. achieving initial improvements in the visual appearance of the site;
4. restoring productivity;
5. restoring diversity;
6. restoring to the extent possible, the species composition; and
7. restoring the functional properties that characterized the site prior to its degradation.

These steps do not always occur in this order, nor does each one necessarily occur on every site. Patterns of restoration are typically site and system specific, and vary with the type of system being restored. To make the above principles more specific, we examine them in the context of particular systems.

Soil Restoration

Soil and its associated nutrients are the foundation of a functioning terrestrial ecosystem. Soil may be lost because of a single catastrophic event or disturbance (e.g., the removal of soil by mining activities) or by ongoing processes (high rates of erosion). In the first case, restoration is achieved by directly replacing the same or a similar kind of soil on the site. In the second case, the rate of erosion must be changed by (1) altering the topography of the site by changing, adding, or removing landscape features; (2) increasing the capacity of the soil on the

Table 11.1 Physical and Chemical Characteristics of Derelict Land Materials. Symbols (see footnotes) indicate level and range of deficiency or excess

| MATERIALS | PHYSICAL VARIABLES | | | | CHEMICAL VARIABLES | | | | |
	TEXTURE AND STRUCTURE	STABILITY	WATER SUPPLY	SURFACE TEMPERATURE	MACRO-NUTRIENTS	MICRO-NUTRIENTS	pH	TOXIC MATERIALS	SALINITY
Acid rocks	△△△[a]	✓[b]	△△	✓	△△	✓	△	✓	✓
Bauxite mining	△△/✓[c]	✓	✓	✓	△△	✓	✓	✓	✓
Calcareous rocks	△△△	✓	△△	✓	△△△	✓	▲[d]	✓	✓
Coastal sands	△△/✓	△△△/✓	△/✓	✓	△△△	✓	✓	✓	✓/▲
Fly ash	△△/✓	✓	✓	✓	△△△	✓	✓/▲▲▲	✓/▲	✓/▲▲
Gold wastes	△△△	△△△	△	✓	△△△	✓	△△△	✓	✓
Heavy metal wastes	△△△	△△△/✓	△△/✓	✓	△△△	✓	△△△/▲	✓/▲▲▲	✓/▲▲▲
Iron ore mining	△△△/✓	△△/✓	△/✓	✓	△△	✓	✓	✓	✓
Land from sea	△△	✓	✓	✓	△△	✓	✓/▲	✓	✓/▲▲
Oil shale	△△	△△△/✓	△△	✓/▲▲	△△△	✓	△△/△	✓	✓/▲▲▲
Roadsides	△△△/✓	△△△	△△/✓	△/✓	△△	✓	△/✓	✓	✓/▲▲
Sand and gravel	△/✓	✓	✓	✓	△/✓	✓	△/✓	✓	✓
Strip mining	△△△/✓	△△△/✓	△△/✓	✓/▲▲▲	△△△/✓	✓	△△△/✓	✓	✓/▲▲
Urban wastes	△△△/✓	✓	✓	✓	△△	✓	△/✓	✓/▲▲	✓

[a] Multiple large open triangles indicate increasing levels of deficiency (△ = slight, △△ = moderate, △△△ = severe).
[b] A check mark indicates adequate conditions.
[c] Slashes indicate ranges of conditions from low extreme (left side of slash) to high extreme (right side of slash). Modified from Bradshaw and Chadwick 1980 and Bradshaw 1983.
[d] Solid triangles indicate increasing levels of excess in the same pattern.
Modified from Chadwick 1980 and Bradshaw 1983.

Figure 11.5

A "water bar" on a trail in the Absaroka-Beartooth Wilderness Area in Montana (U.S.A.). Water bars divert sheet flows of precipitation off the trail, reducing erosion and preventing the formation of rills and gullies.

site to receive and hold water, thus reducing rill and sheet flow over the soil surface; (3) installing structures that divert water from the site or alter its rate and direction of movement to reduce its erosive effects; or (4) changing the amount and type of vegetation on the site to increase soil stability and decrease rates and amounts of water movement.

Where restoration is restricted to relatively small sites, altering topography and landscape features is feasible. Steep slopes can be converted to shallower slopes with earth-moving equipment. Basins or depressions can be created to collect water and release it slowly over time. Alternatively, increasing the amount of slope causes water to drain more rapidly from the site, creating a drier environment. Physical structures, even very simple ones, can, when strategically placed, significantly change patterns of surface water flow. For example, the National Park Service and U.S. Forest Service install "water bars" across trails to prevent excessive erosion that could cause rutting and gullying (fig. 11.5). The water bar, usually a log, post, or timber approximately the width of the trail, is laid perpendicular to the trail in a well-fitting rut or groove that holds the bar in place, but is shallow enough to allow the bar to protrude slightly above the trail's surface. Water flowing down the trail is stopped (dammed) by the bar and flows off the trail on the downhill side. Placing the bars at short, regular intervals where the trail ascends a steep

slope prevents surface runoff from concentrating and gaining speed along the trail and limits the amount of sediment lost from the trail to the surrounding landscape. At a larger scale, terraces can be constructed on steep slopes, particularly those plowed for row crop production, that intercept downward-moving sheets of water from uphill slopes and prevent soil carried in the water from moving further downslope.

Planting grass, shrubs, or trees increases soil stability for several reasons. Plants and plant litter on the soil surface intercept precipitation before it strikes the ground directly, thus reducing its force and erosive impact. Plants absorb some water directly and release it later, and more slowly, through transpiration and evaporation, or bind it chemically to compounds in their own cells. Finally, plant roots bind soil below the surface, reducing the ability of water to move soil downslope.

Even where soil remains in place, nutrients may be lost because of disturbance or ongoing degradation. The reason restoration is often such a slow process is because degraded lands are typically nutrient poor. However, although soil degradation, toxicity, and nutrient deficiency are among the most basic problems of restoration, they are also problems for which there exist some of the simplest and most direct solutions (table 11.2).

In initial restoration efforts on degraded terrestrial landscapes, nitrogen is often the key limiting nutrient. Although abundant as a diatomic gaseous molecule (N_2) in the atmosphere, atmospheric nitrogen is not directly usable by plants, which must receive nitrogen in the form of ammonia (NH_3), nitrate (NO_3), or nitrite (NO_2^-). Where these compounds have been lost, plants themselves can be used to change nutrient availability in soil. Many species of plants, especially legumes (family Fabaceae), live in symbiotic association with nitrogen-fixing bacteria, particularly members of the genus *Rhizobium*, which exist in colonies on the plant's roots. *Rhizobium* facilitate a chemical reaction in which nitrogen gas (N_2) is split and atoms of hydrogen are then attached, producing ammonium (NH_4^+)(fig. 11.6). Through such reactions, many legumes can cause the accumulation of nitrogen in soils, many at rates of 50 to 150 kg of nitrogen/ha/year, and a few at rates of over 1,000 kg of nitrogen/ha/year (Bradshaw 1984). Two examples are clover (*Trifolium* spp.) and lespedeza (*Lespedeza* spp.). Both have been used successfully in the restoration of pasture lands on coal wastes in the United Kingdom and Australia (Dobson, Bradshaw, and Baker 1997). Some trees, such as *Casuarina* spp. and *Acacia* spp., concentrate nitrogen in soils through similar mechanisms, and have been used to restore forests on metal and coal mine wastes in India (Dobson, Bradshaw, and Baker 1997). Thus, if degraded soil is properly inoculated with *Rhizobium* and appropriate legumes are then planted, other species, otherwise unable to colonize the site because of nitrogen deficiencies, can rapidly become established if other factors are properly adjusted. The reactions described above do not normally work well in acidic soils, so the pH of the soil must be 5 or higher. This can be accomplished by an application of calcium carbonate ($CaCO_3$). Legumes are highly sensitive to phosphate deficiency, so phosphorous must often be applied on degraded sites. However, the application must be made after the legume seeds have germinated, because the seeds of legumes are sensitive to phosphorus and do not germinate well in high concentrations of it.

Table 11.2   Common Soil Problems Associated with Degraded Land and Potential Short- and Long-Term Treatments

CATEGORY OF PROBLEM	TYPE OF PROBLEM	SPECIFIC CONDITION	SHORT-TERM TREATMENT	LONG-TERM TREATMENT
Nutritional	Macronutrients	Insufficient nitrogen	Add fertilizer.	Revegetate with legume.
		Others	Add fertilizer and lime.	Add fertilizer and lime.
	Micronutrients	Various	Add fertilizer.	Various
Physical	Structure	Too compact	Rip or scarify.	Revegetate.
		Too open	Compact or cover with fine material.	Revegetate.
	Stability	Unstable	Add stabilizer/mulch.	Regrade or revegetate.
	Moisture	Too wet	Drain.	Drain.
		Too dry	Add organic mulch.	Revegetate.
Toxic	pH	Too high	Add pyritic waste or organic matter.	Allow weathering.
		Too low	Add lime.	Add lime.
	Heavy metals	Too high	Add organic mulch or metal-tolerant cultivar.	Add inert covering or metal-tolerant cultivar, and revegetate with tolerant species or cultivar.
	Salinity	Too high	Allow weathering or irrigate.	Revegetate with tolerant species or cultivar.

Adapted from Bradshaw 1983.

Many sites throughout the world have been degraded not by loss of soil or soil nutrients, but by the dumping of toxic wastes or by activities, such as mining, that lead to the accumulation of toxins. The most common types of toxins that degrade terrestrial sites are heavy metals and certain kinds of organic chemicals used in industrial production, such as toluene, xylene, benzene, and other complex hydrocarbons. Some kinds of toxins, particularly volatile organic compounds (VOCs) can be eliminated over time by techniques as simple as repeatedly plowing up the contaminated soil to bring the VOCs into contact with atmospheric gases. Upon contact, the VOCs, true to their name, combine with gases and leave the soil (volitalization). Other types of compounds are not so easily removed. One of the most effective, and increasingly widespread techniques for removing more chemically inert toxins is **bioremediation**, the use of living organisms to remove toxic chemicals from the soil. Many kinds of bacteria, once established in soil, will chemically alter VOCs and other kinds of complex hydrocarbons, rendering them innocuous. In **phytoremediation**, plants are used to remove toxic chemicals (phytoextraction), especially metallic ions and compounds. Phytoextraction involves physically removing large chemical deposits or pieces of metal from the soil, then planting the contaminated site with a species that can accumulate the remaining metals and related metallic compounds in its tissues. Many plants that are native to metal-

liferous soils can tolerate exceptionally high tissue levels of zinc, nickel, cadmium, lead, copper, and cobalt. The plants are harvested, then safely disposed of, taking with them the toxic metals and metal compounds from the soil. If a phytoremediation process is iterated over several years, many plant species can reduce metal concentrations to very low, nonharmful levels. In some cases, this result can be achieved in a single growing season. A natural community can be reestablished, and the site no longer poses a risk to resident organisms or to human visitors.

One variant of phytoremediation is **phytostabilization**, in which plants release chemicals into the soil that render toxins inert, stabilizing or immobilizing them so that they cannot react further, enter living organisms, or move through the soil to different locations. On a Superfund site in Kansas, researchers used poplar trees (*Populus* sp.) to immobilize high concentrations of lead and zinc and revegetate a site that had been unvegetated for nearly 80 years (Dobson, Bradshaw, and Baker 1997).

Grassland Restoration

Grasslands, including plains, prairies, and savannas throughout the world, are among the most endangered ecosystems on every continent. Causes of degradation include urban, residential, and agricultural development, overgrazing, desertification, and fire suppression. Many techniques exist for restoring degraded

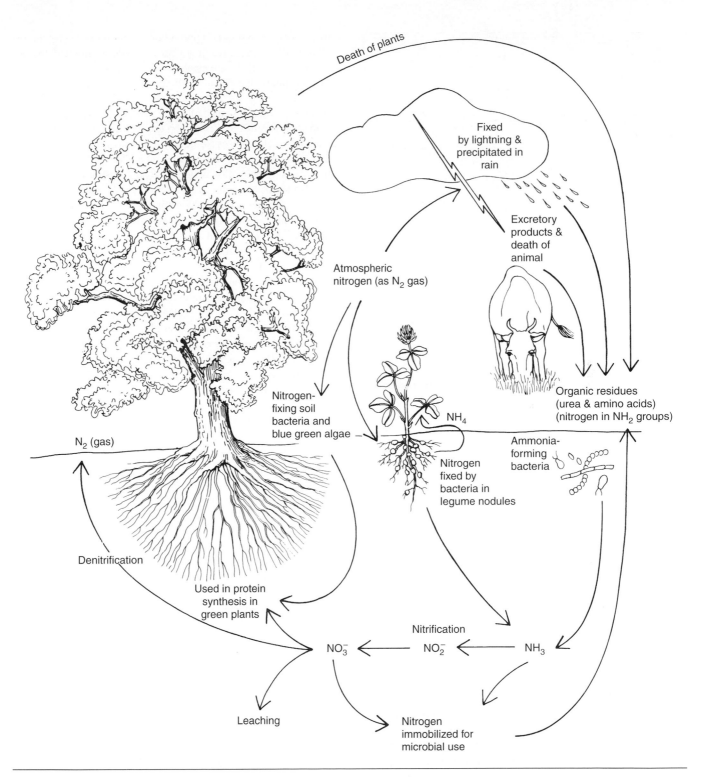

Figure 11.6
The nitrogen cycle in a terrestrial ecosystem. Note that atmospheric nitrogen is converted to a form usable by plants (NH_4^+) through the actions of bacteria that exist in symbiotic relations with certain plant species. These reactions are essential for restoring levels of nitrogen-based plant nutrients on a degraded site.

grasslands, but one of the most widely applied, and most consistently effective, is prescribed burning.

Historically, grassland ecosystems experienced widespread, recurrent fires. In North American prairies, early spring fires remove dead vegetation (duff) and allow sunlight to strike

the soil directly, increasing soil temperatures. Native species, still belowground in early spring, germinate sooner in warmer soil, and their growing season is lengthened (Pauly 1997).

Fire also rejuvenates grasslands by increasing soil nutrient availability. Fire increases microbial activity in soil, causing

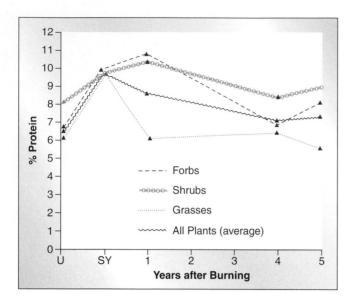

Figure 11.7

Changes in plant protein levels in forbs, grasses, and shrubs following a fire in a sagebrush community in south-central Montana. Protein levels increase because nitrogen-based nutrients are more available after a fire due to increased soil microbial activity and from nitrogen-based compounds in ash on the soil surface. U = unburned sites, SY = same year (late summer estimate) following early spring burn.

After Van Dyke, DiBenedetto, and Thomas (1991).

greater mobility of nutrients and increasing their uptake by plants, and the ash left behind from burned plants is high in nutrients, particularly nitrates. Nutrients in the ash are easily leached into the upper layers of the soil, where they can be readily absorbed by germinating plants. Increases in nitrates and other nitrogen-containing compounds often result in dramatic short-term increases in plant protein levels, making plants that regenerate on a burned area particularly nutritious for herbivores (fig. 11.7). Fire also retards the encroachment of woody vegetation that otherwise would shade or outcompete indigenous grasses and forbs. Woody vegetation may be killed outright, and fire favors germination of grasses and forbs because it leaves behind warm, bare ground where they are better competitors than trees and shrubs. In the United States, managed tall- and midgrass prairies, typically dominated by "warm season" grasses that germinate best in warm soil, are normally burned every 3 to 4 years to allow for sufficient accumulation of fuel that will carry the fire and generate sufficient temperatures for optimum germination of preferred species and complete removal of dead vegetation. However, where warm season plant communities face continuing threats of invasion from "cool season" grasses and Eurasian weeds, burning is often conducted annually (Pauly 1997).

The effectiveness of prescribed fire for prairie management is matched only by its potential danger to life and property. Therefore, proper protocols for prescribed, controlled fires in prairies are essential. Fire management techniques vary regionally in consideration of differences in vegetation and cli-

mate, but certain standard procedures are appropriate for almost all prescribed fires. Prescribed burns should be initiated only after proper firebreaks are established around the site. These may include landscape features already present, such as streams, plowed (nonvegetated) fields, roads or trails, or (in early spring burns) snowbanks (especially effective for fires on a dry south slope that ascends to a ridge top to meet a snow-covered north slope). If such features are not present, they must be created, either by digging a fire line to expose bare soil or by burning narrow strips of vegetation under close supervision to remove fuel in advance of the main fire. All firebreaks, whether natural or created, are often widened by setting "backfires" (fires that burn into the prevailing wind) in advance of setting the main prescribed fire (Thompson 1992). The appropriate window of weather conditions for safely igniting prescribed fires in grasslands is usually narrow, and typically requires wind velocities below 15 mph (enough to carry the fire forward but not enough to cause it to jump the firebreaks), humidity above 40%, and temperatures below 70°F. Because these conditions rarely occur in summer in temperate areas, burning is best done in spring or fall. As noted above, spring fires provide advantages for the germination of warm season prairie grasses because they remove vegetation from the previous season that impedes germination and create warmer soil temperatures that enhance it. Spring fires also have the advantage of making nutrients, especially nitrogen, maximally available to plants when it is most needed—in the early stages of germination and growth. However, early spring fires often destroy the nests of some birds, especially early ground nesters such as the ring-necked pheasant, northern bobwhite (*Colinus virginianus*), and gray partridge (*Perdix perdix*). To protect wildlife, managers can resort to fall burns or, in spring burns, set aside a portion of the grassland to be left unburned.

Next to burning, the management techniques that have the greatest effects on grasslands are grazing and mowing. Natural grassland ecosystems throughout the world are well adapted to grazing by their own native communities of herbivores. Grazing can increase productivity of grasses and enhance ecosystem function in grasslands through a variety of mechanisms (previously discussed in chapter 10) (McNaughton 1979). It is not always possible or practical to maintain population densities of grazing animals in native grasslands that will achieve desired outcomes, but mowing can sometimes be used as an alternative management practice. Because it produces some physical effects on plants that are similar to grazing, mowing also can generate some of the same responses from plants, including increases in vigor, abundance, and primary production. Mowing also can partially mimic the effects of grazers in changing the vertical and horizontal structure of the grassland community, and, like prescribed burning, can prevent or retard the encroachment and invasion of woody plants. Mowing stimulates aboveground grass production, limits growth of nonindigenous species, and improves germination of native species, especially forbs, by increasing the amount of light reaching the soil.

Although burning, grazing, and mowing can be effective methods for rejuvenating and subsequently managing degraded grasslands, these techniques are of limited value in establishing new grasslands. To establish grasslands on sites where native vegetation has been lost, managers must select the species and plant them intentionally. Some prairie and grassland species germinate well with simple hand broadcasting methods, but others require soil coverage to ensure germination, and thus more sophisticated planting techniques. Many native grassland and prairie species require special treatments (e.g., extended exposure to cold temperatures) to ensure successful germination, and finding appropriate genetic stocks closely adapted to local environmental conditions can be a major limitation. Once planted, native restored grasslands require frequent and regular monitoring, evaluation, and adjustment of management strategies for eventual success. The establishment of initial grassland communities normally takes 3 to 5 years on newly restored sites, and may take longer under some conditions. Thus, managers must be patient and persevering if they intend to succeed.

Some of these management principles can be seen on a small scale in one effort in northwest Iowa, a region once dominated by tallgrass prairie, but today intensively cultivated to row crops. On 13 acres of such former cropland, my students and I established a native tallgrass prairie in 2000. With the help of staff from the local office of the Natural Resource Conservation Service (NRCS) and funding from the NRCS's Conservation Reserve Program and Pheasants Forever (a nongovernmental conservation organization dedicated to preserving habitat for ring-necked pheasants) we purchased and seeded 26 species of native forbs and 6 species of native grasses on the site using a no-till grain driller in mid-May. To eliminate native and exotic weed species, we monitored the site throughout the summer and spot sprayed with an herbicide when necessary, particularly to eliminate infestations of thistles (*Cirsium arvense*). This was done before most of the planted species emerged to avoid any herbicide-caused mortality. To further aid the growth of established species, we mowed the site at regular intervals three times during the summer, a technique that increased the vigor of native grasses but helped reduce infestations of quackgrass (*Agropyron repens*), a nonnative species. Of the 32 species planted, 20 (63%) germinated in the first year, including five of

Figure 11.8

A portion of a restored tallgrass prairie established on former row crop (corn and soybean) agriculture land in northwest Iowa. Northwest Iowa was dominated by tallgrass prairie vegetation prior to settlement.

the six native grass species (fig. 11.8). More species of forbs were added in the second year, and germination success of the original species pool increased to 81% with the germination of six additional species.

Two years is a very short time in a long-term prairie restoration effort, and the levels of biodiversity in this restoration are still well below those of established native prairies. Most restoration efforts proceed slowly. For example, in a study of restoration efforts on European grasslands, Bekker and associates (1997) found that restoration on sites that had a history of intensive agricultural use and poor nutrient conditions typically took more than 20 years. If species had disappeared from the site, there was low recruitment from the soil seed bank. Thus, Bekker and associates (1997) concluded that the best sites for potential restoration were those which had the shortest history of agricultural use, and the maintenance of existing species-rich sites should receive top conservation priority.

Forest Restoration

Successful forest restoration begins with appropriate site selection. Forest restoration is appropriate, and normally will only succeed, in previously forested areas. Because of the enormous alteration of landscapes and vegetation by human activity, present vegetation may be no guide to the past. However, in humid, temperate regions, the soil order most strongly associated with forest vegetation are **alfisols.** Alfisols have a shallow layer of humus, well-developed horizons, and an accumulation of clay in the subsoil (B horizon). Insoluble oxides of iron and aluminum are often present. From these elements (Al and Fe) the alfisols derive their name. The presence of alfisols is often indicative of past forest communities on a site, even if such forests are no longer present.

Forest restoration requires a determination of whether the goal is compositional (e.g., to recreate a historic forest or to produce a forest that can provide a sustainable supply of certain species of trees) or functional (e.g., to protect the quality of a watershed, reduce soil erosion, or create an aesthetically pleasing landscape). Once the overall goal is clear, the manager will have clear criteria with which to select species of trees, subcanopy, and understory plants appropriate to the restoration objective.

Regardless of what objective is chosen, critical attention must be paid to the eventual characteristics of the restored forest canopy. Forests create microclimates markedly different from nonforest habitats, primarily through their own canopy development. The shade created by a forest canopy reduces ambient and soil temperatures, changes the amount and types of light reaching the surface, lowers rates of subcanopy evapotranspiration and evaporation of soil moisture, and increases relative humidity within the forest. Characteristic forest flora and fauna depend on these conditions for survival and reproduction. Thus, successful restoration of the forest canopy is arguably the single most important aspect of successful forest restoration. There are five general approaches to reestablishing a forest canopy:

1. Plant the species intended to form the canopy at desired densities; mulch or spray to control weeds in the immediate vicinity of the trees; and encourage invasion by mid- and understory species as shade develops.

2. Plant the canopy species at lower than desired densities, then add more trees and understory species in subsequent years. This creates an uneven-aged stand with greater structural diversity.

3. Plant the canopy species at greater-than-desired densities, then thin later or allow self-thinning to take place. Add understory species after thinning. This approach gives greater control of the site and greater resistance against invasive species.

4. Plant short-lived, fast-growing trees to "capture" the site, then underplant slow-growing, shade-tolerant trees of the desired canopy species. This approach accelerates natural selection.

5. Allow woody plants to invade without interference from nearby sites. Selectively remove undesirable species later. This approach takes the longest—potentially several human lifetimes (Howell 1986; Thompson 1992).

Although canopy management is often the critical element in temperate forest restoration, other types of forests must be managed differently. One such category is that of riparian forest communities that develop in association with riparian areas and river floodplains. Their high biodiversity offers high value in conservation, but they are among the world's most endangered forest habitats and present unique challenges in management.

Riparian Forest Communities

Riparian forest communities present special problems to restoration efforts because they are typically dependent on flood-pulse interactions with a local river or stream. As noted in previous discussions of ecosystem management (chapter 10), most major rivers in the world have been altered to reduce flooding and seasonal flow variations in favor of constant flow rates that are more beneficial to commercial shipping, human settlement and recreation, and agriculture. When historically variable flows and seasonal flooding are reduced or eliminated in a riparian ecosystem, associated riparian forests are significantly altered, become more susceptible to invasion by nonnative species, and may be eliminated altogether. Unlike upland forest restoration efforts that can be successful with site-specific approaches, any attempt to restore riparian forest communities must adopt an ecosystem management approach that focuses on restoring historic variations in flow rates and flood pulses.

Some of the most successful and best studied efforts in riparian forest restoration have been conducted in the extensive cottonwood (*Populus deltoides* ssp.)-willow (*Salix* spp.) forests along the middle Rio Grande River in central New Mexico. This forest, known locally by its Hispanic name of *bosque,* is the most extensive cottonwood-willow forest remaining in the southwestern United States. But flood control and flow regulation management have eliminated flooding and the forest is rapidly senescing (Molles et al. 1998). New stands of cottonwood are not being established, and the stabilized river flow has favored invasion by nonnative trees, including salt cedar (*Tamarix ramosissima*) and Russian olive (*Elaeagnus angustifolia*).

Manuel Molles, Jr., and his colleagues and students at the University of New Mexico have used two different ap-

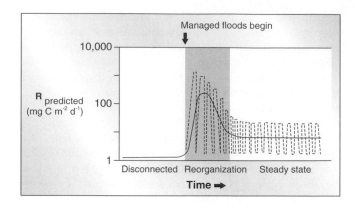

Figure 11.9

The reorganization model for riparian forest restoration based on analysis of experimentally and naturally flooded cottonwood forests along the middle Rio Grande River in New Mexico. Restoration involves three phases: the disconnected phase, prior to reestablishment of flooding; the reorganization phase, induced by reestablishing a seasonal flood pulse; and the steady-state phase in which rates of ecological processes (in this case, respiration) achieve relatively constant levels. *R* is the predicted response in forest floor respiration in average annual response (solid line) and expected variation (dotted line).

After Molles et al. (1998).

proaches to restoring the bosque. The first approach is to uproot salt cedar and then establish cottonwood forests directly by planting pole-sized cottonwoods or through wet soils management that favors germination of cottonwood seeds. The second approach uses managed flooding to reconnect now disconnected riparian forests to the Rio Grande River at seasonal intervals. To evaluate the second approach more analytically, Molles and his colleagues created experimentally flooded sites and associated dry and flooded control sites. They discovered that, with flooding, cottonwood forests could not proceed directly from their disconnected condition to steady-state, flood-regulated cottonwood forests. Rather, they passed through a "reorganization" phase (similar to that described by Bormann and Likens [1979] in the "shifting mosaic" model of upland forest development) before establishing a self-generating cottonwood forest with fairly constant rates of respiration (fig. 11.9). During the reorganization phase, large amounts of surface organic matter, accumulated in the absence of flooding, created high biological oxygen demand during flooding that rapidly rendered floodwaters anoxic. However, repeated flooding progressively dampens the level of respiration as decomposition of forest floor material proceeds. At the naturally flooded site, dissolved oxygen was never depleted from floodwaters but remained at 4 mg/L or higher throughout the flooding period (Molles et al. 1998).

These experimental efforts suggest that restoration of riparian forest systems may be more complex than terrestrial forest systems, take longer to achieve, and require landscape- or ecosystem-level approaches rather than site-specific techniques. Different ecological processes appeared to move toward restoration at different rates. Thus, the overall system of a disconnected forest may require at least a decade or more of repeated annual flooding before it begins to approach a historic riparian cottonwood forest in all its ecological functions, processes, and self-generating capacities (Molles et al. 1998).

Restoring Aquatic Ecosystems

Restoration of aquatic ecosystems involves: (1) reestablishing historical or targeted hydrologic characteristics, especially flooding and flow regimes; (2) reducing excessive delivery of sediments and chemical contaminants and removing their accumulations where appropriate; and (3) revegetating shoreline, riparian, and wetland areas with native species and removing nonindigenous species (Committee on Restoration of Aquatic Ecosystems 1992).

Wetland Restoration

As noted in chapter 9, wetland restoration often must address the problems of eutrophication and acidification, which are *input-oriented* problems best solved by input regulation. The causes of eutrophication in freshwater wetlands are usually phosphate inputs from surrounding agricultural and urban areas. Ideally, management would be based on laws and policies that reduce or prohibit the use of phosphates in fertilizer and domestic use, require removal of phosphorus from urban sources before allowing discharge to proceed downstream, and reduce erosion on agricultural lands through increased vegetative cover bordering streams and cultivation methods less destructive of soil structure. However, wetland managers lack power and jurisdiction to implement such sweeping changes over entire regions and drainage basins, and they are faced with detrimental land-use practices around them that they cannot directly control. In such cases, managers must prevent phosphorus from entering the system or remove or neutralize inputs.

The most immediate and direct ways to stop inputs of phosphorus into a wetland system are to install filters at the source of input to remove phosphorus when it arrives, or to surround shorelines and banks with vegetation that can extract phosphorus from runoff. Vegetative buffer strips adjacent to wetlands, even if relatively monotypic and composed of common, inexpensive grass species, remove nutrients, including nitrates and phosphates, from runoff (Rickerl, Janssen, and Woodland 2000) (chapter 9). Buffer plantings can be done on managed lands that surround a lake, wetland, or on adjacent nonpreserve lands through the use of incentives for landowners. In the United States, private landowners can be paid cash subsidies under either the Conservation Reserve Program or the Wetland Reserve Program for buffers planted on former wetlands converted to crop production, partially offsetting revenue lost from the reduction of cropland acreage (chapter 9).

Selection of wetland restoration sites should be based on historical or functional criteria, with the highest priority given to restoring wetlands in areas that are underrepresented in the drainage basin or restoring types of wetlands that are now underrepresented (Galatowitsch and van der Valk 1994). Historic wetland sites can usually be identified by using a soils map to locate **hydric soils**, which are "saturated, flooded or ponded long enough during the growing season to develop anerobic conditions that favor the growth and regeneration of hydrophytic vegetation" (USDA 1987). Extended saturation in soils is often indicated by gray and olive colors (so-called gley soils) and by accumulations of organic matter at the soil surface or by development of organic soil horizons (Thompson 1992). Such indicator soils include aquasols, histosols, and fluvaquents, soils with substantial organic (peat) composition, or mineral soils with gray subsoil.

Although hydric soils may indicate the site of a past wetland, their presence is not sufficient to determine whether a wetland restoration effort is called for. If most of the wetlands in the area have been drained, the groundwater will be much further below the surface than it was when the wetland was present. Without other sources of water or changes in hydrology, a restored wetland probably would not be able to retain water throughout the growing season. In some cases, such as the intensively farmed agricultural region of the southern prairie pothole region of the United States, restoring wetland hydrology can be as simple as excavating or plugging drainage tiles on the periphery of the wetland basin (Thompson 1992). Precisely because hydrology is so important to maintaining a restored wetland, priority should be given to the restoration of wetland complexes and entire basins rather than individual, isolated wetlands or parts of a basin. As a general rule, restored wetlands should have at least 2.5 acres of watershed for each basin, with a minimum watershed area of 4 acres for temporary and seasonal wetlands, and at least 4 acres of watershed for each acre of basin with a minimum of 17 acres of watershed. If one of the objectives of restoration is to affect flood flows, at least 0.5% of the watershed should be covered by a wetland. Priority should be given to wetlands drained within the last 20 years because they have the best potential for natural revegetation without excessive management effort. If the wetland is being restored for wildlife habitat, one should determine what target group of animals will benefit. Finally, one should determine how much of the basin can be restored without creating conflicts with neighboring landowners.

By restoring wetlands, managers may hope to achieve (1) improved groundwater recharge and discharge; (2) flood control through increased flood storage and flood desynchronization; (3) habitat preservation for plants and animals, especially endangered or threatened species; (4) prevention of erosion and shoreline anchorage; (5) improved water quality by absorbing sediment and nutrients from surrounding agricultural lands and preventing their movement to downstream sources; and (6) increased opportunities for outdoor recreation and education (Thompson 1992).

As noted earlier (chapter 9), wetland habitats are disproportionately endowed with populations of both economically important and endangered species; however, restoration of a wetland does not ensure that threatened, endangered, or economically important species will appear there. "If you flood it, they will come," is not always true with respect to wetlands restoration. For example, natural Iowa wetlands typically contain between 20 and 32 invertebrate species, but restored Iowa wetlands usually contain only 4 to 16 species (Galatowitsch and van der Valk 1994). Moreover, one cannot assume that a natural seed bank still remains on all sites that historically held wetlands. Studies of U.S. prairie wetlands have shown that, if a site designated for restoration has been dry for more than 20 years, it is unlikely to revegetate naturally from on-site seed deposits (Galatowitsch and van der Valk 1994), and it is likely to have an impoverished plant community even if restored 5 years after draining (Weinhold and van der Valk 1989). Restored wetlands on sites that have been without sufficient water more than 20 years usually contain agricultural weeds and a few wind-dispersed wetland species. Such sites do acquire some wetland species, but their abundance and composition is strongly influenced by the distance of the restored sites from other wetlands, the species present in neighboring wetlands, and the species-specific dispersal distances of wetland plants. In degraded environments, where only scattered remnants of historic wetlands remain, *Typha* spp., which is wind pollinated and a good disperser, often dominates remaining wetlands although it was relatively rare historically (Galatowitsch and van der Valk 1994). Recreating historic plant communities on wetlands that were drained more than 20 years ago may require planting desired species, especially if the restored wetland is too far from other wetlands for those species to colonize.

Wetlands can, under the right conditions, trap large quantities of sediment and prevent it from being deposited downstream. In some areas, wetland vegetation is used in tertiary treatment for sewage effluents (Thompson 1992) (fig. 11.10); however, wetland plant communities can perform such functions only if their main source of water is surface runoff from surrounding agricultural lands. Wetlands whose hydrology is determined primarily by groundwater or direct precipitation input will have less effect on the amount of sediment and nutrients that reach downstream locations, and, consequently, on downstream water quality. Additionally, there is a limit to the amount of sediment and nutrients wetlands can absorb without becoming degraded themselves; therefore, restored wetlands should not receive drainage from farmsteads, feedlots, corrals, sewage lines, or septic fields (Galatowitsch and van der Valk 1994).

In designing a wetland restoration program, dikes and water control structures must be properly sized and constructed to control the location and extent of flooding. After implementation of the design, regular visits to the restored sites are needed to monitor water depth gauges in a deep portion of the basin, construct annual lists of plant species, compare such species to those found in the historic or target wetland, census or sample target species, and identify problems with vegetation, structure, or function that need correction.

Ideally, restoration will include not only the wetland, but also the surrounding area. Specific management practices will vary among wetlands, but some general principles that apply to most wetland restorations are:

1. Burning the area to be restored prior to flooding to increase germination of native wetland species, followed by regular spring burning to reduce cattail invasions.
2. *Not* stocking the restored wetland with fish.
3. Placing screens in inlets and outlets of wetlands, if present, to prevent entry of nonindigenous species, such as carp.
4. Eradicating pest or nonindigenous species, such as purple loosestrife.
5. Making special management efforts to preserve rare or threatened plant and animal species.
6. Encouraging soil conservation practices on lands surrounding the wetland basin to reduce sediment input.
7. Encouraging reduced pesticide and fertilizer use on surrounding lands to avoid eutrophication and accumulation of toxins (Galatowitsch and van der Valk 1994).

Lake Restoration

To reduce inputs of phosphorus and other nutrients into a lake, restoration may involve maintaining or creating adjacent wetlands, which can remove up to 79% of total nitrogen, 82% of nitrates, 81% of total phosphorus, and 92% of sediment in drainage water (Chescheir, Skaggs, and Gilliam 1992).

In many cases, the key to successful lake restoration may be management practices in adjacent terrestrial systems. For example, in Maryland's Chesapeake Bay, nitrate inputs, primarily from groundwater, have contributed to declining water quality and to lower harvests of commercial species. Long-term efforts to reduce erosion from surrounding croplands reduced surface nitrate inputs, but did little to reduce overall nitrate levels. Recently, phytoaccumulation has been used to address the problem more successfully. Winter cereal grains such as rye (*Secale cereale*) were planted on croplands immediately after harvest of corn (*Zea mays*), removing most of the available soil nitrate and concentrating it in their own tissues (fig. 11.11). Groundwater nitrate levels fell from 10 to 20 mg/L to less than 5 mg/L (Staver and Brinsfield 1998). One key to the effectiveness of this technique appeared to be planting the cover crop as soon as possible; plantings delayed by as little as 30 days were much less effective than plantings done immediately after harvest. This technique would have to be practiced for many years before it could significantly affect water quality in Chesapeake Bay, but it holds promise for reducing nitrate inputs from agricultural lands without taking those lands out of agricultural production.

Figure 11.10

"Constructed" wetlands can be established to treat domestic wastewater through phytoaccumulation.

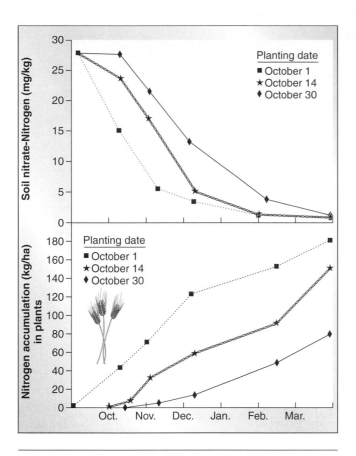

Figure 11.11

Levels of nitrate in soil and plant tissues of rye (*Secale cereale*) planted immediately after harvest of a corn crop in Maryland, monitored in 1998 and 1999.

After Staver and Brinsfield (1998).

Because water in lakes is replaced very slowly, incoming biota and material are not easily removed, and abiotic material that continues to enter a lake will accumulate there. But lakes also present unique opportunities for restoration precisely because the source and type of inputs can be identified relatively easily, and, if inputs can be controlled, the lake ecosystem will change. Because they receive inputs from throughout the surrounding drainage basin, lake ecosystems are strongly affected by land-use practices, and by physical and chemical processes characteristic of the surrounding landscape. Heavy applications of fertilizer, while increasing crop yields, also produce high nutrient loads in nearby lakes. This problem can occur in small lakes or ponds, as well as in large lakes that receive inputs from throughout a region. A good example of such degradation, and evidence of effective—if incomplete—rehabilitation, is Lake Erie, one of the North American Great Lakes.

Of the five Great Lakes in North America, Lake Erie is the shallowest, southernmost, and warmest, and its shores are surrounded by the highest overall densities of humans and the greatest amount of industrial and agricultural activity. It receives discharges from Lake Superior, Lake Michigan, and Lake Huron via the Detroit River (fig. 11.12). As a result,

Lake Erie is particularly vulnerable to environmental degradation. Water quality had begun to decline severely in Lake Erie by the 1950s, evidenced by massive algal blooms, the disappearance of many benthic species, oxygen depletion in bottom waters, population eruptions of introduced forage fish such as the alewife (*Alosa pseudoharengus*), collapse of commercial fisheries of walleye (*Stizostedion vitreum vitreum*) and other species (Boyce et al. 1987; Makarewicz and Bertram 1991), and extermination of the endemic blue pike (*Stizostedion vitreum glaucum*) (Werner 1980) (fig. 11.13).

After researchers identified heavy phosphorus loading as the principal cause of Lake Erie's degradation, the U.S. and Canadian governments began a program in 1972 to improve municipal waste facilities and bring major municipal discharges into compliance with a 1.0 mg/L phosphorus limitation on effluent. This effort was helped by the introduction of nonindigenous species of Asiatic clams into the lake, whose high rates of filtration during feeding also contributed to clearing large quantities of phosphate from the water (chapter 9). By 1985, phosphorus loading from municipal discharges in both countries had been reduced by 84% (Makarewicz and Bertram 1991). Concurrent with phosphorus reductions, walleye fishing was prohibited in U.S. waters, and salmonids (primarily king and coho salmon) were introduced.

Since the implementation of the phosphorus control program, phytoplankton biomass has decreased throughout the lake. *Aphanizomenon flos-aquae,* a nuisance blue-green alga, declined 89% from 1970 to 1985, and another eutrophic indicator species of diatom (*Fragilaria capucina*) declined by 94%. A major recovery of walleye was evident by the mid-1980s, with concurrent declines in alewife, spottail shiner (*Notropis hudsonius*), and emerald shiner (*N. atherinoides*) in the western and central portions of the lake (Makarewicz and Bertram 1991). Dissolved oxygen levels in bottom waters have risen throughout the lake, and oxygen-depletion rates have fallen.

Although it is unlikely to restore Lake Erie's historic assemblages of biota or water chemistry, this long-term restoration effort, by focusing on reducing phosphorus and protecting commercial fishing stocks, has produced notable improvements. Commercial fish populations have risen, increased oxygen in bottom waters has permitted reestablishment of more diverse benthic communities, and water quality has improved dramatically because of reductions in eutrophic algae and bacteria.

River and Stream Restoration

River and stream ecosystems are intimately tied to terrestrial systems and the associated physical, chemical, and biological processes that occur throughout their drainage basins. Thus, complete "restoration" of major river ecosystems is a daunting, if not impossible task, in part because of interactions between the main channel and adjacent, low-velocity floodplain habitats that develop during overbank flooding (chapter 10). Throughout the world, such areas typically contain high densities of humans, and are often economically important, especially in agricultural production. Thus, in the strict sense, the restoration of large river ecosystems would be impossible without great economic loss, human hardship, and massive population displacement. Gore

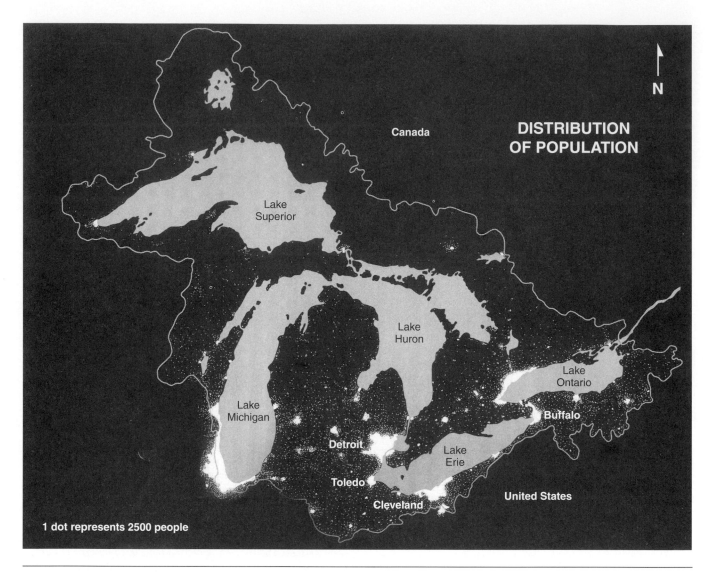

Figure 11.12

The North American Great Lakes. This map shows the distribution of population in the Great Lakes region; each light dot represents 2,500 people. Note the location of Lake Erie, which receives discharge from four major U.S. cities (Buffalo, Cleveland, Detroit, and Toledo), each of which appears as a patch of light on the map. These population centers represent significant sources of phosphorus pollution, resulting in eutrophication.

and Shields (1995) suggested that "rehabilitation" or "the recovery of some . . . ecological functions and values," is a more realistic goal for rivers. Criteria used to evaluate the success of rehabilitation are functional rather than historical, and the targets are particular values of key system variables, or "abiotic endpoints" (Milner 1994).

Although major river systems throughout the world are degraded by pollution, overharvesting of biotic resources, and destruction of supporting riparian and floodplain ecosystems, the most fundamental disruption of most major river ecosystems is the alteration and regulation of flow. Dams are an obvious source of flow alteration, although others exist. Today 60% of the world's total stream flow is regulated, and almost no major river remains undammed. In addition to major dams, smaller structures such as dikes and levees often prevent movement of water from the main channel into historic flood-

plain areas. Such structures do not necessarily prevent flooding, but do disrupt the regularity and periodicity of floodplain ecology (chapter 10). A less obvious but more pervasive effect on stream flow is the process of channelization, the straightening, deepening, and altering of river and stream channels by human engineering that transforms meandering, multiple-channel rivers with slow water movement and large floodplains into straight, single-channel systems of high water velocity and little or no floodplains (Gore and Shields 1995). Such changes alter the habitats associated with the stream channel as well as their biological and chemical properties, and convert associated aquatic habitats of riparian and floodplain areas into terrestrial habitats.

In addition to the general requirements of restoration described for other systems and the need to restore historic flow regimes, river rehabilitation requires:

Figure 11.13

The extinct blue pike (*Stizostedion vitreum glaucum*), a subspecies of walleye that was the basis of an important commercial fishery in Lake Erie in the first half of the twentieth century.

1. replacement of riparian and floodplain vegetation;
2. development of structures to restore geomorphic diversity needed to support biological diversity;
3. action to relieve floodplain isolation;
4. remediation and control of water-quality degradation, especially control of pollution inputs; and
5. restoration of native populations of plants and animals, including restocking of depleted populations of commercial importance.

Techniques for river rehabilitation include manipulation of the river channel, backwater treatments, alterations of the riparian zone and floodplain, and regulation of flow.

Many techniques exist for manipulating the river channel itself. Some procedures directly manipulate substrate quality and distribution. For rehabilitation of specific sites, especially small rivers and streams, new gravel or boulders can be poured directly into the channel. Such substrates create additional habitat and new patterns of eddies and currents favorable to many organisms. However, in large rivers, this approach is neither cost-effective nor humanly practical. In small streams, excavating a deep hole in the channel will reduce water velocity and preserve gravel and rock substrates downstream. The slower-moving water possesses less energy and releases much of its transported sediment. Periodically, sediment is excavated from the hole to prevent it from filling and allow the process to continue. Over time, large quantities of sediment can be removed from stream channels degraded by erosion from surrounding lands.

Clean, high-velocity riffle areas of gravel and stone may be created in small streams by appropriate placement of structures like deflectors or log weirs (fig. 11.14) that deflect or accelerate flows over a defined area. The deflector or weir also creates, behind it, a low-flow pool area that provides relief from current

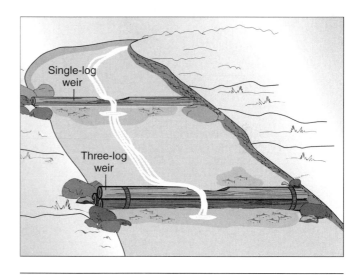

Figure 11.14

A log weir placed in a stream to obstruct flow.

and protective cover from predators (Gore and Shields 1995). In large rivers, these structures are impractical because they are impediments and hazards to navigation; instead, spur dikes (fig. 11.15) are often placed just upstream of shoals of gravel and cobble to accelerate the flow of water. By keeping substrates free of sediment, spur dikes benefit organisms that require such sediment-free substrate for foraging, resting, or spawning.

On a larger scale, it is sometimes possible to manipulate the river channel directly by creating or recreating meanders in the river's path. One such meander restoration project is underway in the Kissimmee River in Florida. Before restoration was begun, a small-scale demonstration project diverted flow

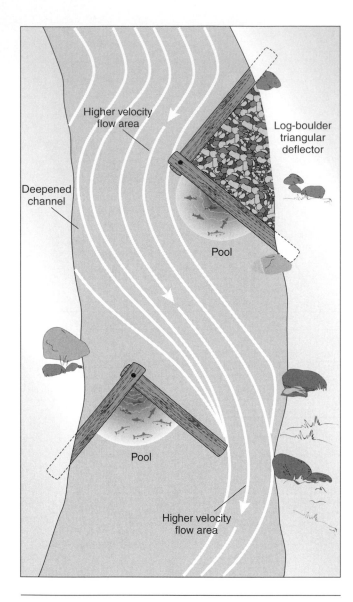

Figure 11.15
Spur dikes placed in a stream to divert the flow of water. Spur dikes are often placed just upstream of shoals of gravel and cobble to create riffle habitat and to keep gravel and cobble free of sediment.

through former meanders and floodplains. The newly flooded meanders and floodplain were rapidly invaded by hydrophytic plants; current-seeking insects increased in the restored sections and midge populations declined; use by wading birds increased; and there was a 10-fold increase in fish densities (Gore and Shields 1995).

Other techniques of river restoration focus on backwater habitats, riparian zones, and floodplains. In some cases, backwater areas that have become filled with sediment can be restored by dredging, and backwaters isolated from the river by channelization can be reconnected through diversion of the main channel's flow (Patin and Hempfling 1991). Riparian zones are degraded by erosion, human development, or bank stabilization structures that are, ironically, designed to reduce erosion (Gore and Shields 1995). Recently developed technolo-

gies use large vertical vanes placed on the channel bed to divert flows away from erodable shoreline areas, reducing disturbance to riparian zones. Where these structures are impractical or unaffordable, woody debris or stones are used. In some areas, floodplains have been rejuvenated by the use of levees positioned inland from the main channel (so-called setback levees). Setback levees permit periodic flooding of the portion of the floodplain nearest the main channel while protecting human residences and agricultural activities farther from the channel (Welcomme 1989). Two simpler, but equally effective, methods of partially restoring floodplain hydrology are to turn off the pumps that remove water from the floodplain during floods (Sparks et al. 1990) and to excavate breaches in levees at selected points (Kern 1992).

Perhaps the most direct method of restoring the functions of river ecosystems is through restoration of flow. The majority of the world's large rivers are regulated to create relatively constant rates of flow that are beneficial to transport, agriculture, and human residence and development along the river and its floodplains. In contrast, most river species are adapted to variable and seasonal flows, and many species cannot persist without such fluctuations. In some cases, existing dams and other structures can be used to create seasonal variations in flow. Managers of the Columbia River Basin's dams in the Pacific Northwest have proposed a seasonal shift in flow rates, to be achieved by selling hydroelectric power to other regions in spring and summer, thereby increasing flow and benefiting spring salmon runs. In winter, the period of highest demand for hydroelectric power, power companies would purchase electricity from other regions. This approach has been estimated to cost significantly less than additional fish restoration efforts and structures to take fish around dams, but it is controversial and has generated strong opposition from politicians and local landowners (Gillis 1995). In other cases, more extreme measures are being considered, including the destruction and removal of dams (chapter 10).

POINTS OF ENGAGEMENT—QUESTION 1

Suppose that salmon populations could be restored to targeted levels by either of two strategies: supplementing natural populations with hatchery-raised fish or increasing rates of reproduction and survivorship in natural populations through restoration of river ecosystems. What is the difference in the two strategies in their primary entity of value, and what is the difference in types of overall costs and benefits?

GENETIC AND POPULATION-LEVEL CONSIDERATIONS IN RESTORATION ECOLOGY

Much current literature and conservation effort focuses on broad goals of compositional or functional restoration of sites, communities, landscapes, and ecosystems. Such efforts aim to ensure that sufficient habitat within functioning ecosystems will be available for species persistence. But the question then becomes, "Which species?" Just as habitats, landscapes, and ecosystems have been destroyed, so have the populations they contained. Surviving adjacent or highly mobile populations may become reestablished naturally on a restored site once its fundamental structure and function are present, but more remote and less mobile populations will not. As in other dimensions of restoration, their return must be an intentional human act. Further, such populations must have particular genetic characteristics to survive and reproduce. The practice of bringing in "whatever we can get," although a strategy with many historic examples, is not only poor science, it may be illegal, particularly where endangered species are involved. Further, the established populations must not only be *adapted* to the site, they must be *adaptable* to its future changes. The first quality requires an intelligent assessment of the match between the genetic characteristics of the introduced population and the characteristics of the site, and the second requires a sufficient level of genetic diversity in that population. Thus, we now examine genetic and population dimensions of the restoration process.

Genetic Match and Genetic Diversity in Reintroduced Populations

When a site is degraded to the point that entirely new populations must be brought in to restore it, the first question is, "How do you match genotypes and environments?" The most direct method is to directly compare genetic material between individuals that remain on the site or were once present on the site and material from populations considered candidates to recolonize the site. The closest matches to persistent or historic populations are likely to be local ecotypes of plants and adjacent populations of animals.

Direct DNA analysis is not always possible or practical, and, where habitat degradation is regional, no local populations may exist that can serve as colonists. Faced with this problem on a wider geographic scale of restoration, the U.S. Forest Service uses "seed zone boundaries" as a basis for selecting the best sources for plants to be used in restoring degraded or devegetated sites. Seed zone boundaries are "drawn on the basis of natural diversity and structure, but topographic, climatic, and edaphic divisions are often used when genetic information is incomplete or lacking" (Knapp and Dyer 1998). A more general approach is to use regional climatic data and soil, selecting populations that grow in areas with similar moisture and temperature regimes and soil type compared to the target site. If multiple populations of the same species must be used to have sufficient numbers of seeds for the sites, it is essential that these be populations capable of outcrossing. Even if this is the case, outbreeding depression can result from crosses among members of multiple populations. However, relatively few studies have shown outbreeding depression to be a problem in plants (Waser and Price 1994), and the problem does not seem to be a threat to the overall success of well-planned restoration efforts (Knapp and Dyer 1998). The problems of inbreeding and introgression, previously discussed in chapter 6, are probably more significant concerns in a restoration effort, especially when at-

tempting to restore populations of rare species. Thus, managers should make every effort, particularly in plant populations, to prevent rare species from hybridizing with more common, genetically compatible species.

In plants, the level of genetic diversity achieved and maintained in a population may be as dependent on the types and species of pollinators as on the genetic composition of the population. Thus, restoration efforts that aim to restore genetic diversity in plant populations must provide habitat not only for the plant, but for the appropriate pollinators. When the habitat needs of the pollinator are different from the plant, this may mean that additional types of habitat must be included in the restoration effort and that these additions must be within the effective activity range of the pollinator. Similarly, some plants that depend on animals for dispersal will not persist if the animal dispersers are not present in sufficient numbers. Here, too, managers may have to add areas that contain habitat that such animals require to ensure their role in maintaining the plant population and its genetic diversity (Knapp and Dyer 1998). For example, the attempt to protect the endangered salt marsh bird's-beak (*Cordylanthus maritimus maritimus*), a hemiparasitic annual plant, through the restoration of wetlands associated with Sweetwater Marsh National Wildlife Refuge in San Diego Bay, California, required not only wetland habitat for the plant but also adjoining upland habitat for its primary pollinators, Halictine bees (Zedler 1998).

Trend Analysis: A General Approach to Population Restoration

Restoring individual populations of plants or animals normally addresses one of two conditions. One is the restoration of threatened or endangered populations. The other is the restoration of more common species on sites from which they have been exterminated or to sites where they were not historically present but are expected to do well. In the first case, the goal is usually to preserve existing populations *in situ* and, through various management and protective measures, create an environment in which the population increases to levels at which it will no longer be in imminent danger of extinction. These types of restoration efforts typically follow a six-part process:

1. Inventory: a geographically based assessment of rare taxa that documents their existence within mapped political units.
2. Survey: an ecologically based assessment of populations in the field that identifies their habitat(s) and endangerment factors.
3. Habitat protection: an application of land-use restrictions that can be applied, negotiated, or that generate the least political resistance to benefit the endangered population(s).
4. Management.
5. Monitoring.
6. Recovery.

The simplest form of monitoring involves collecting regular census or survey data to determine the status of the endangered population. Although this gives an index of the population's status, it cannot and does not tell us the prospects

Table 11.3 Demographic Parameters That Serve as General Indicators of Population Stability in Analysis of Trends in Endangered Plant Populations

PARAMETER	LIFE FORM	POPULATION IS STABLE IF:
Survivorship	Annual	Mortality inflection point on survivorship curve (Type I) follows onset of seed production.
	Perennial	The number of individuals in a new cohort equals or exceeds the number of established individuals after inflection point on survivorship curve (Type III).
Seed bank	All	Density of viable seeds in soil prior to season of germination far exceeds the average density of established individuals.
	Annuals and herbaceous perennials	Year-to-year changes in density of viable seed are not correlated with changes in the density of established, reproductive individuals.
Seed production	All	Seed production per individual of an endangered taxon equals or exceeds that of a nonendangered relative with similar life form.
Age structure	Perennial	Number of established, reproductive individuals is less than the number of established juveniles and/or the number of recruited seedlings.
Frequency of establishment	Annual	Frequency of establishment is less than the half-life of seeds in the seed bank.
	Perennial	Frequency of establishment is less than the half-life of established, reproductive plants.

Based on Harper 1977 as summarized by Pavlik 1994.

for the population's long-term persistence or the causes of its decline or continuing low levels. Both are essential for successful restoration and recovery.

Demographic monitoring is the most effective method for determining both the probability of persistence and the causes of decline or chronically low population levels. Unlike census or survey data, demographic monitoring uses some method of following the fates of individuals in the population over time and makes repeated on-site measurements to do so. Two monitoring tools are most important: **trend analysis** and **factor resolution** (Pavlik 1994).

Trend analysis calculates one or more specific demographic variables in one or more populations and, from such calculations, determines if the population is growing, stable, or declining. One of the most common measures in trend analysis is λ, the population's finite rate of increase, which can be calculated from the annual birth and death rates of the population, determined by monitoring the fates of specific individuals. We have already seen the use of lambda to assess trends in endangered species populations like the cheetah (chapter 7). A lambda of 1 indicates a stable population, more than 1 an increasing population, and less than 1 a declining population. Population viabilty analysis (PVA) (chapter 7) is an integrated form of trend analysis in which multiple demographic variables are used to estimate the probability of a population of a specific size persisting for a specified time period.

Trend analysis alone cannot identify the exact cause of the population's low numbers or decline, but it can lead to intelligent guesses about what kinds of factors to investigate. Some general, nonmathematical indicators of population stability can be used with trend analysis to make an intelligent assessment as to whether an introduced plant population is stable and likely to persist at its new site (table 11.3). A plant population with high rates of annual adult survivorship but chronically low rates of seed germination would suggest that environmental variables affecting seeds, such as soil moisture, seed predation, or soil temperature, may be more important to the population than factors affecting adult plants, such as ambient temperatures, herbivory, or competition. For example, trend analysis of populations of *Amsinckia grandiflora*, an endangered annual forb (family Boraginaceae) native to dry grasslands of California, indicated that competition with other plant species was the primary limiting factor in population recovery (Pavlik 1994). Thus, a suggested management strategy was to treat sites where *Amsinckia grandiflora* occurred in ways that would remove more common, competing species. The two most commonly used treatments were herbicides (which killed surrounding grasses) and prescribed burning. Figure 11.16 shows an example of an experimental design, derived from such trend analysis, that was used to determine which treatment had the greatest effect on populations, and table 11.4 gives the results of the experiment (Pavlik, Nickrent, and Howard 1993). We see

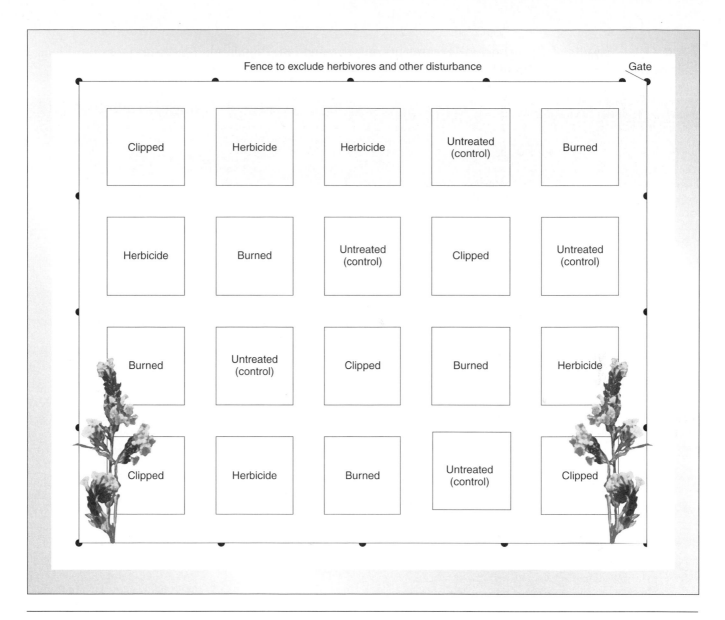

Fence to exclude herbivores and other disturbance   Gate

Clipped	Herbicide	Herbicide	Untreated (control)	Burned
Herbicide	Burned	Untreated (control)	Clipped	Untreated (control)
Burned	Untreated (control)	Clipped	Burned	Herbicide
Clipped	Herbicide	Burned	Untreated (control)	Clipped

Figure 11.16

Experimental design used to create a new population of the endangered plant *Amsinckia grandiflora* within its historic range. Each of the 20 plots was either a treatment (burned, hand clipped, or herbicide [fusilade] treated) or a control in order to measure the effects of competition from nonnative grasses.

After Pavlik, Nickrent, and Howard (1993).

from this example how trend analysis can guide the second phase of demographic monitoring, *factor resolution*. In factor resolution, experimental tests of one or more suspected factors limiting population growth are conducted in the field. The results of these tests may determine which factor or factors are actively limiting population growth. This analytical approach allows managers to determine which variables to manipulate to produce the best chance for population recovery.

As noted above, the second type of problem in population restoration is to create new populations on a site where they do not exist. Here the manager reverses the strategy. He or she begins with factor resolution and ends with trend analysis. Experimental tests are conducted in the field to determine the

most important factors limiting the growth of the founding population. Their results are then used to determine the trajectory of population growth and what, if anything, ought to be done to enhance it.

Although some restored populations remain chronically small and require the use of trend analysis and factor resolution to determine the means to increase population sizes, there are notable exceptions to this pattern. In some cases, chronically small populations have increased in number and distribution, not through direct introduction of populations but through an array of indirect factors such as legal protection, changing public attitudes, and altered regional habitat distribution. In these cases, restored or recovering populations may

Table 11.4   Results of Experimental Treatments on Germination, Population Size, Survivorship, Plant Size, and Nutlet (Fruit) Production in the Endangered Plant *Amsinckia grandifolia*

TREATMENT	GERMINATION	POPULATION SIZE (REPRODUCTIVE PLANTS/PLOT)	SURVIVORSHIP (% OF GERMINATION) TO REPRODUCTION	MEAN MAXIMUM PLANT SIZE (cm)	NUTLET (FRUIT) PRODUCTION (# / PLANT)
Control	55.4 ± 5.2 [a]*	38.6 ± 15.8[a]	42.7 ± 16.5[a]	26.0 ± 3.1[a]	15.1 ± 10.1[a]
Burn	55.4 ± 9.9[a]	67.2 ± 19.8[a]	75.3 ± 11.6[b]	33.7 ± 5.3[b]	29.1 ± 14.4[a]
Clip	54.1 ± 4.8[a]	57.8 ± 16.5[a]	63.1 ± 12.0[a]	23.1 ± 3.7[a]	6.6 ± 5.6[a]
Herbicide	54.0 ± 8.1[a]	56.4 ± 15.6[a]	64.4 ± 10.8[a]	40.5 ± 4.1[b]	53.5 ± 16.5[b]

*Values (mean + SD) in a column followed by the same letter are not statistically different ($P < 0.05$, ANOVA). Data from Pavlik, Nickrent, and Howard 1993.

increase markedly to the point that control of the population must be considered. One such example is the restoration of the gray wolf in the Great Lakes states of Minnesota, Wisconsin, and Michigan.

POINTS OF ENGAGEMENT—QUESTION 2

Consider the data shown in table 11.4. All three management treatments are designed to achieve the same result—increased populations of *Amsinckia grandiflora* through elimination of competing species—but achieve it through different means. Based on the results displayed here, which treatment or combination of treatments would you use to achieve the objective of the largest population of *Amsinckia grandiflora*, and why?

The Gray Wolf: A Case History of Natural Population Restoration

The gray wolf, once widely distributed throughout North America, was exterminated throughout most of the United States by the early 1900s. One exception was a remnant population persisting in northeast Minnesota. With the passage of the Endangered Species Act of 1973, the wolf received federal protection as a listed threatened species in Minnesota and as an endangered species in the rest of the conterminous United States. Protected by the act, the Minnesota population nearly tripled between 1973 and 1990, increasing from approximately 700 to 2,000 individuals (Fuller et al. 1992; Mladenoff et al. 1997).

By the late 1970s, biologists in the region noted that wolves were expanding their range to the neighboring regions of northern Wisconsin and Upper Michigan (Mladenoff et al. 1997). Wolf protection coincided with two developments that also aided population growth. First, the general public became more favorable to the presence of wolves, leading to less frequent illegal shootings. Second, changes in forest habitat in the upper Great Lakes region, especially the creation of young, intensively managed forests with many openings, led to increases in populations of the wolf's primary prey, the white-tailed deer. This triumvirate of legal protection, public support, and abundant prey created favorable conditions for wolf population growth, leading to exponential increases in populations in all three states (fig. 11.17).

Expanding wolf populations in the upper Great Lakes region have forced rethinking of traditional wisdom about wolf ecology. Once considered the very symbol of wilderness, wolves in this region are now moving southward into areas with higher human population densities and more altered habitats. Wolf researchers Mladenoff and colleagues write, "Recent research and monitoring of wolf behavior . . . and recent wolf population growth have shown that wolves are not the wilderness species they were once assumed to be. . . . If wolves are not killed, and ungulate prey are adequate, they can apparently occupy semi-wild lands formerly thought to be unsuitable. Our work and that of others suggests that current population growth will continue and that dispersal ability and adaptability of wolves will allow them to colonize increasingly developed areas" (Mladenoff et al. 1997).

The recovery of the wolf in the upper Great Lakes area is an incubator of two developing and related dilemmas in restoration efforts. First, if wolf populations continue to grow at present rates, negative interactions between wolves and humans, particularly involving livestock and pets, are certain to increase. Wolf populations will have to be reduced. As L. David Mech, one of North America's leading authorities on the wolf, points out, the wolf could live in far more places than it does today if

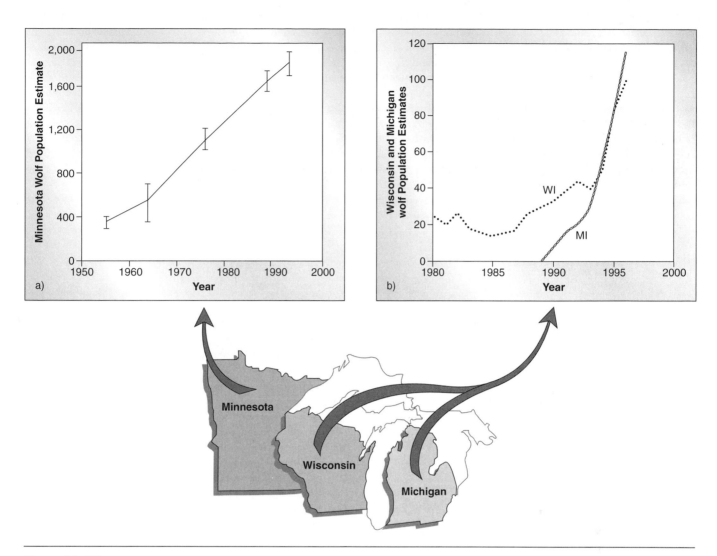

Figure 11.17

Estimated recent wolf population growth in (a) Minnesota, and (b) Wisconsin and Michigan.

After Mladenoff et al. (1997).

control (killing) could be carried out by the public (private landowners protecting their own livestock) instead of exclusively by government officials (Mech 1995). Indeed, a larger segment of the public would find the wolf an acceptable species where they actually live if the public had some role in wolf control. The tragic irony is that "some people revere wolves so much that, rather than having wolves face control, these people would rather not restore wolves to areas where they would have to be controlled" (Mech 1995).

Second, the increasing deer densities needed to support wolf populations in the upper Midwest would reduce the biodiversity of the understory forest community and reduce recruitment of some tree species (Mladenoff and Stearns 1993; Alverson, Waller, and Solheim 1995). The Forest Service has been sued by botanists on this issue (Mlot 1992) and is increasingly inclined, in cooperation with state agencies, to institute management programs that will reduce deer populations. If deer densities decline, so will wolf productivity and pup survival (Fuller et al. 1992). As a result, wolf numbers and ranges can be expected to contract.

The success of wolf restoration in the U.S. upper Great Lakes illustrates two important principles inherent in restoration efforts that often receive inadequate consideration. First, the creation of a legal, social, and ecological climate favorable to a species may have more effect on its recovery than direct manipulation of the population. Second, managers would be prudent to determine, in advance, an acceptable maximum population and range for restored and recovering populations before they reach levels that produce negative interactions with humans, habitat degradation, loss of biodiversity, or steep declines in populations that could pose new threats to population persistence.

It is precisely the potential success of restoration that creates an intriguing paradox: if we truly desire restored ecosystems and their species to be a part of everyday life, we must be prepared to alter our own attitudes and actions. Traditionally, conservationists have affirmed this sentiment, always assuming that it meant that nonconservationists would have to alter their "incorrect" opinions and behavior to allow for the presence of other species. But restoration is a dangerously reciprocal process, and conservationists, as well as nonconservationists, must accept some limits on nonhuman species if they would see those species become a more permanent part of a landscape in which humans reside. If we cannot suffer the presence of wild creatures where we actually live, then our parks are functionally no more than prisons in which the wardens must shoot any of the larger inmates if they attempt to escape. This is, in fact, the policy that is practiced in many parks in many places. Such so-called conservation is nothing but anthropocentrism of the most selfish form. In contrast, a comprehensive view of restoration seeks ways in which functioning ecosystems may ultimately be restored even in the context of permanent human residence and work. This more comprehensive view is addressed in an approach that incorporates human presence, activity, and cultural attitudes into the entire restoration process, and is known as "biocultural restoration."

BIOCULTURAL RESTORATION

Reclamation and restoration are not merely scientific efforts to rehabilitate nonhuman elements of an impoverished and degraded landscape. To succeed, ecological restoration must incorporate cultural values and land ownership. In many cases, restoration efforts begin with a legal mandate imposed by a governmental authority on local residents. Such a pattern, although sometimes a necessary first step, does not foster an environment conducive to long-term restoration. If local residents and nonresidents who use the site do not take ownership of the restoration effort, it is likely to fail. Restoration efforts must intentionally move local residents from a reactive posture that denies responsibility for the current state of a site to a proactive posture that not only accepts, but also anticipates responsibility for the site and does even more than is required in its restoration (Clarkson 1995) (table 11.5). This level of involvement is needed not only to ensure the success of the restoration from an ecological standpoint, but also to reconnect local residents and their culture with the restored site as a place in which they find

valuable products, services, knowledge, and experience. We can illustrate these principles with examples of restoration as a "value-added" dimension of private property exchange in a developed setting and an example of tropical forest restoration in a developing country that intentionally includes local culture in the restoration effort.

The California counties of Contra Costa, Santa Cruz, and Solano include residential property valued among the highest in the United States. In these counties, resource economists Carol Streiner and John Loomis estimated the economic value of various measures of stream restoration to private property values using a "hedonic property model" (Freeman 1993) that treats property value (P) as a function of its structural characteristics (S), neighborhood (N), and environmental quality (Q), expressed in the identity

$$P_i = f(S_i, N_i, Q_i),$$

where the subscript i refers to each value for an individual (ith) property (Streiner and Loomis 1996). Because attributes of property value are typically grouped, Streiner and Loomis created "restoration packages" that could be applied to different properties (table 11.6). Restoration package A included improving fish habitat and the acquisition of additional land along the stream by the California Department of Water Resources Urban Stream Restoration Program for a streamside education trail. Restoration package B featured properties where streams were restored in ways that reduced flood damage, cleaned, revegetated, and stabilized the stream bank, cleared obstructions from the stream channel, and added aesthetic elements such as check walls, rock or stone walls, or wood plank walls along the stream.

Streiner and Loomis determined that individual elements of package A added $15,000 to $19,000 in property value. In package B, only stabilization and reduced flood damage added value, but these increased worth by up to $7,800. A joint model incorporating dimensions of both packages added over $19,000 in average value to individual properties (table 11.6) (Streiner and Loomis 1996). This analysis demonstrated that not every type of restoration added significantly to private property value, but many did. Restorations

Table 11.5 Categories of Participant Attitudes and Strategies in Restoration Efforts

PARTICIPANT ATTITUDE	PARTICIPANT STRATEGY	PARTICIPANT PERFORMANCE
Reactive	Denies responsibility	Does less than required
Defensive	Admits responsibility but resists it	Does the least that is required
Accommodative	Accepts responsibility	Does all that is required
Proactive	Anticipates responsibility	Does more than is required

For restoration to be completely successful and enduring, managers must ultimately move all participants from reactive to proactive categories through increasing participant levels of involvement and ownership in the restoration effort.
Modified from Clarkson 1995.

Table 11.6   Values of Alternative "Restoration Packages" Associated with Stream Restoration Efforts Adjoining Private Residential Property in Contra Costa, Santa Cruz, and Solano Counties, California

RESTORATION MEASURE	ABSOLUTE VALUE OF RESTORATION	VALUE OF RESTORATION RELATIVE TO PROPERTY VALUE (%)
Restoration Package A		
Fish habitat improvement	$15,571	11
Land acquisition for education trail	$19,123	13
Education trail established	$17,560	12
Restoration Package B		
Streambank stabilization	$4,488	3
Reduced flood damage	$7,804	5
Joint Model		
Education trail established with streambank stabilization	$19,078	13

Adapted from Streiner and Loomis 1996.

that added value included both efforts that restored inherent functions of the stream (e.g., improved fish habitat) and those that more directly benefited the property owner (e.g., reduced flood damage).

POINTS OF ENGAGEMENT—QUESTION 3

The above example describes a situation where stream restoration was necessary because of degradation caused by private residential development. Imagine yourself as a private developer of residential property that included riparian habitat. What development strategy would you employ to conserve the values described above so that they added value to the initial sale of the property rather than being lost in the construction process?

Although this example demonstrates that restoration can add value to private property, the fixed boundaries of such property and the limited influence of individual owners make it unlikely that the entire system will be restored. In this example, private property owners received the benefit, while a government agency bore the cost. Thus, the owners remain primarily concerned about their own property, not ecosystem structure or function. The property's increased value, although beneficial, does not necessarily make the owners better stream stewards. More effective and pervasive outcomes in restoration result when comprehensive cultural values are linked to social benefits at larger scales, particularly where local residents, not government bureaucracies, become the primary agents of restoration. This kind of effort is exemplified in the work of tropical ecologist Daniel Janzen in the area of forest restoration.

Janzen has argued that conservation efforts focusing on the few remaining intact patches of habitat exclude more than 90% of local residents from direct ecosystem benefits. "Restoration," said Janzen (1988), "is most needed where people live." Indeed, Janzen has found that a successful approach to tropical restoration includes a restoration of local culture to appreciate and value what is conserved. To achieve this, managers must gain explicit and public agreement on management goals. The next step, says Janzen, is to "Choose an appropriate site, obtain it, and hire some of the former users as live-in managers. . . . The challenge is to turn the farmer's skills at biomanipulation to work for the conservation of biodiversity" (Janzen 1988). In addition, the restoration effort must be supported by diverse and imaginative education programs taught within the site being restored, and local people must be intellectually involved in the restoration and management process.

Practical considerations and problems are paramount concerns in cultural dimensions of restoration. Guanacaste National Park is one example of restoration efforts operating in a unique way (Allen 1988) (fig. 11.18). Rather than attempting to establish a new national park from remaining fragments of undisturbed tropical forests (which were rare), the Costa Rican government proposed a reverse process of creating a park from a combination of degraded forest land and existing forest fragments. Whereas national parks traditionally are efforts to protect pristine areas from further degradation before restoration becomes necessary, the process of creating Guanacaste required restoration to make the park viable. In helping to establish Guanacaste National Park in Costa Rica, Janzen had to convince 14 farm families to stop clearing forests on their farms until their land was purchased by the park. The farmers insisted on receiving cash, so Janzen arranged for all transactions to take place at a local bank on the same day, and each farmer would deposit the money in a personal savings account. He did this to protect the farmers from robbery and to protect himself and the park from bad public relations. The purchases were transacted successfully, the farmers' money was protected, and the park was established.

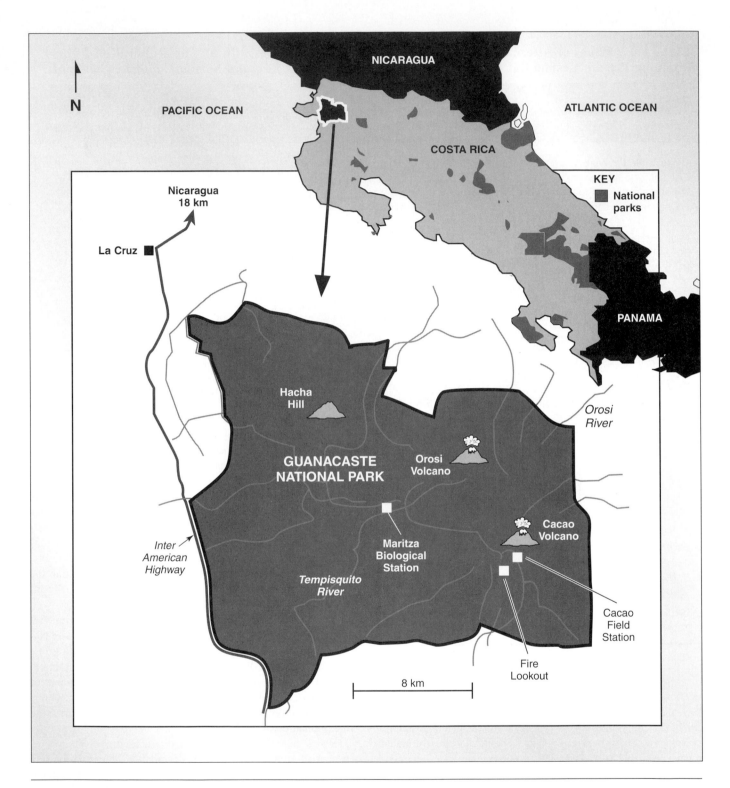

Figure 11.18

Guanacaste National Park, Costa Rica, was formed not through preservation of pristine habitats, but through restoration of land formerly used for agriculture and forestry.

Figure 11.19
Tropical ecologist Daniel Janzen works with local residents in management and education efforts in Guanacaste National Park, Costa Rica, an example of biocultural restoration. In Guanacaste, local residents take responsibility for management, policy, and environmental education within the park.

Today Guanacaste is supported by an intensive environmental education program, run by local residents for all Costa Ricans, especially school children. A typical day for a teacher at Guanacaste begins with the arrival of grade-school children at the park at 7:30 A.M. Hikes, lectures, field activities, and discussions fill a day that usually ends with the students' departure at about 4 P.M. (Allen 1988). The Guanacaste program has inspired schools to develop curricula in environmental education that make use of the park and its natural heritage. Local projects and internships for local students include seed collection and planting, species inventory, trail maintenance, and cutting and burning fire lanes, all of which can lead directly to professional employment as naturalists within the park. Local residents also provide room and board to visiting scientists who come to the park for research in tropical ecology, and help make arrangements for professional and scien-

Figure 11.20
The Everglades of South Florida, an enormous wetland complex now targeted for a massive restoration effort.

tific conferences that take place at the park. Local landowners graze livestock in the park at regulated levels, and the livestock play an important role in dispersing seeds of tropical trees. Janzen believes that "the most practical outcome is that this program will begin to generate an ongoing populace that understands biology. In 20 to 40 years, these children will be running the park, the neighboring towns, the irrigation systems, the political systems. When someone comes along with a decision to be made about conservation, resource management, or anything else, you want that person to understand the biological processes that are behind that decision because he or she knew about them [since] grade school" (fig. 11.19) (Allen 1988).

A CASE HISTORY OF ECOLOGICAL RESTORATION: THE FLORIDA EVERGLADES

One of the most ambitious restoration projects in the history of conservation began in south Florida in 1992. This was the effort to restore and achieve ecological sustainability of the regional ecosystem known as the Everglades.

The Everglades is a complex mosaic of riverine wetland communities that once meandered over most of South Florida, linked by freshwater flow and dependent on interannual variability (fig. 11.20). It is essentially a wide, shallow, slow-moving river (the Kissimmee) of sawgrass (*Cladium jamaicense*) that travels 158 km in a series of oxbows and meanders (Brumback 1990). Slight rises on the east and west coasts of Florida create a shallow valley with excellent water-holding capacities in the southern third of the state. Historically, water from Lake Okeechobee to the north would overflow its banks during the rainy season and move slowly southward through this "valley," eventually reaching Florida Bay. This "river of grass," as the Everglades has been called, varied from 50 to 60 miles in width but was only 1 to 2 feet deep. A unique wetland and one of the world's largest wetland ecosystems, the Everglades once covered nearly 4 million hectares.

The original (historic) ecosystem owed its unique characteristics to a hydrologic regime with dynamic storage capacity and sheetflow of low-nutrient water across the landscape that eventually entered the Florida and Biscayne Bays. The channel of the Kissimmee was so diverse and braided in its original form that, as late as the early 1900s, passengers on steamships moving along the river in opposite directions would see one another early in the morning, but then spend a full day traveling the sinuous and braided channel of the river before finally meeting late in the afternoon (Culotta 1995). The same hydrology that created this unique, shallow sheetflow of the Kissimmee over the landscape organized the ecosystem's primary and secondary production, established salinity gradients, and concentrated production into a network of dry season refugia used by fish and invertebrates. Because of the system's ability to store large quantities of water for extended periods, precipitation received in the wet season was released slowly into the system and maintained its productivity into the dry season. Minor differences in topography created habitat heterogeneity through distinctive hydroperiods at different sites, and enhanced biodiversity in the ecosystem. Interannual variation in water flows interacted with another dominant ecological process in the landscape—fire. Annual variations in lengths of wet and dry seasons affected the number and extent of fires, which further contributed to habitat heterogeneity.

Beginning in the 1880s, a series of flood control projects, including dikes, channelization, and drainage were initiated to drain land and provide fresh water for agriculture and urban development and, after severe hurricanes in the 1920s and 1940s, to provide flood control. The majority of these efforts were undertaken by the Army Corps of Engineers. Their dams and levees broke the Everglades into small, more easily drained plots. With these, the Corps built a network of canals that swept water out of the center of south Florida, east to the Atlantic and west to the Gulf of Mexico. Lake Okeechobee was diked to serve as a regulated freshwater reservoir. Such efforts reduced the variability of annual water flows and lowered the overall amount of water passing through the system. The most devastating ecological effects were achieved in the 1960s when the Corps completed its masterstroke, channeling the formerly 158-km watercourse into a straight 77.2-km white-walled canal. This canal was known to the Corps as C-38, but came to be known among conservationists as "The Wicked Ditch" (Brumback 1990). Thus, by the late 1960s, the Corps had accomplished its objectives of controlling floods, creating land for agriculture, and providing fresh water for crops and people. It also succeeded in draining the Everglades.

Today the water that historically spread slowly over a 60-mile expanse is more swiftly and narrowly channeled through 2,200 km of canals and levees, stored in parks called "water conservation areas," and partitioned by 143 water-control structures (Brumback 1990; Culotta 1995). These changes have reduced the overall size of the system and its heterogeneity, leading to the system's decline to a degraded remnant of its historic condition. Water, once the most important characteristic of the Everglades

system, has now become its most important limiting factor. In abnormally wet years, the canals cannot drain the water from the land fast enough and water must be released from Lake Okeechobee to protect people and property from potential hurricane flooding. The excess water is released into water conservation areas, which then turn into vast shallow lakes, with disastrous effects on wildlife. When the water conservation areas are full and urban and agricultural areas are threatened, water is released into Everglades National Park. During dry years, water is diverted from water conservation areas to agricultural and urban areas and never reaches the park. The outcome of these water control structures and management practices is that Everglades National Park is now wetter than it should be in wet years and drier than it should be in dry years (Brumback 1990).

It is not only the quantity of water that is critical to the Everglades, but the quality as well. The water of the Everglades was historically nutrient poor; its limestone substrate absorbed most phosphorus entering the system from natural sources. Today, however, water coming into the Everglades from sugarcane fields and other croplands in the surrounding Everglades agricultural area (EAA) contains high levels of phosphorus and other nutrients. Algal blooms and fish kills occurred in Lake Okeechobee in the late 1980s, leading many to predict its demise (Brumback 1990). Controversies continue as to what the appropriate level of incoming phosphorus should be, with some environmentalists arguing for levels as low as 10 ppb, and farmers predictably resisting such low levels because that would make it difficult to fertilize their crops and produce a worthwhile profit.

The system is also threatened with invasion by nonindigenous species, such as melaleuca (*Melaleuca* spp.), a tree native to Australia but flourishing in South Florida. Melaleuca does well on wetland sites and forms dense, homogeneous stands that can eliminate native vegetation and, because of the tree's high water demands, dry up the site. The large, tropical marine toad (*Bufo marinus*) escaped from pet owners and established populations in suburban Miami. The marine toad's large size, voracious appetite, and generalist diet make it a serious threat to native amphibians as both a competitor and a predator, and to many species of invertebrates indigenous to the Everglades. But perhaps the most serious potential invader is the Asian swamp eel (*Monopterus albus*). Now established in Miami canals by escapees from hobbyists' aquariums, the Asian swamp eel can reach 3 feet in length, cross extensive land areas, and has a voracious appetite for fish, frogs, crayfish, and other invertebrates. It is literally only a floodgate away from entering the Everglades, where it could wreak havoc on native populations. Florida game officials are considering trying to block the eel by electrifying the gates, a technique that has worked in the Pacific Northwest, preventing sea lampreys (*Petromyzon marinus*) from entering the Columbia River. It may already be too late. In December 2000, U.S. Geological Service biologists discovered a population of Asian swamp eels at the eastern border of Everglades National Park near Homestead, Florida. The swamp eel's invasion of the Everglades may now be a certainty whose effects managers must now prepare for.

Given these changes, the Everglades is now considered an endangered ecosystem. In 1994, the U.S. Army Corps of En-

Table 11.7   Six Alternative Plans Developed During the 1990s to Restore Everglades National Park and the Associated Everglades Ecosystem

ALTERNATIVE	CHARACTERISTICS	COST
1	No changes in water delivery structures. Flows managed through new operational schedule using existing structures to balance ecosystem and human needs.	$5 million
2	Widen canals through Everglades agricultural area to increase volume of water flowing from Lake Okeechobee to the Everglades to the south. Build new levee to protect Miccosukee Indian Reservation from flooding. Create water preserve area east of the Everglades to store additional water, replenish Biscayne Aquifer, and provide short hydroperiod marshes. U.S. Army Corps of Engineers removes 50 miles of existing water barriers. Government purchases or takes 35,000–78,000 acres of agricultural land.	$1.2–1.5 billion
3	Same as 2 but adds goal of reducing ecological fragmentation created by South Florida's water management system. Levees and canals removed from three water conservation areas (WCAs) north of Everglades National Park, allowing water to move through WCAs as sheet flow, similar to hydrology of presettlement Everglades. Removes 71 to 117 miles of canals and levees and 35,000 acres of agricultural land.	$1.27–1.35 billion
4	Same as 2, but instead of widened canals, constructs flow ways 1 to 3 miles wide through Everglades agricultural area creating greater sheet flow and more storage capacity. Would not address fragmentation issues associated with WCAs. Removes 58 miles of water barriers and 54,000–85,000 acres from agricultural production.	$1.3–1.55 billion
5	Combines restored connectivity of WCAs of Everglades National Park (plan 3) with flow ways (plan 4). Removes 105 miles of canals and levees and 55,000 acres of agricultural land.	$1.7 billion
6	Creates a single flow way 7 to 13 miles wide through WCAs to reconnect them and provide for maximum water flow and storage. All levees within Everglades agricultural area removed. WCAs enlarged to form a continuous buffer between the Everglades and developed portions of South Florida. Removes 117 miles of canals and levees and 167,000 acres of agricultural land.	$2 billion

gineers released a preliminary study that outlined six possible alternatives for altering the hydrology of the Everglades, but without endorsement or consensus as to which was the best option (table 11. 7). Five of the six alternatives call for one or more of three actions: (1) drawing more water from Lake Okeechobee and finding more natural ways (than the present canal system) to convey it south to the Everglades; (2) creating more water storage capacity in the system, allowing water to be released more slowly over a larger part of the year; or (3) reconnecting the sections of the natural system that have been isolated from one another by the present system of levees, dikes, and canals. The most expensive and costly of the options would create a flow way 7 to 13 miles wide through existing water conservation areas, reconnecting them as a wetland complex and providing for maximum flow and storage. All levees within the environmentally affected area would be lowered to ground level. Water preserve areas would be expanded to form a con-

tinuous buffer around the Everglades, standing between them and the developed areas of South Florida. One hundred seventeen miles of canals and levees would be removed and 167,000 acres would be taken out of agricultural production at a cost of approximately $2 billion.

An Interagency Ecosystem Management Task Force was formed to develop a strategy for ecosystem restoration. The team used an approach called *scenario consequence analysis* to develop a hypothetical set of conditions that are internally consistent and scientifically defensible and that specify factors needed to evaluate effects. These conditions or scenarios provide a means to assess the implications of management actions on the ecosystem and society and to compare relative costs and benefits of alternative management strategies. The one sustainable scenario is as follows. More flood control water is released from Lake Okeechobee into the Everglades system, the natural dynamic storage capacity of the upper end of the Everglades basin is reinstated, and more natural patterns of sheet flows, hydroperiods, and hydropatterns are established. This scenario requires a larger core protected area and a buffer area that would be a contiguous system of interconnected marshes, detention reservoirs, seepage barriers, and water treatment areas. This scenario also retains sustainable agriculture around the area, protecting surrounding lands from conversion to nonsustainable agriculture or urban development, both of which would degrade the system.

Smaller, less costly projects are already underway. One is to divert more water into Shark Slough, an ecologically important wetland complex in the northeastern part of the Everglades. A second is to create a buffer strip between wetlands and drained crop fields along the Everglades' eastern border. A third, perhaps the most significant, is the restoration of the Kissimmee River's twists and turns by the same Army Corps of Engineers that straightened it. The initial effort to restore the river's historic meandering course will cost an estimated $370 million, but this is miniscule compared to the larger, regional restoration effort that will be conducted over the next 15 to 20 years. It is this latter effort that is intended to completely replumb the Everglades, including 14,000 km^2 of wetlands and engineered waterways (Culotta 1995).

Resolution of the competing alternatives ultimately came on December 11, 2000, when President Bill Clinton signed the Everglades Restoration Act (technically, Section 528 of the Water Development Act of 1996). More costly and ambitious than any of the earlier alternatives, the act authorizes a $7.8 billion, 30-year effort to restore historic sheet flows of water to the Everglades. The U.S. Department of State has called the plan the "world's largest environmental restoration" and "history's largest environmental restoration effort." It includes a provision to develop 181,250 acres of above-ground reservoirs with a total capacity of 1.54 million acre-feet, 300 wells to store and retrieve water from underground aquifers, and 35,600 acres of wetlands to filter polluted runoff. The act authorizes the removal of 240 miles of levees and canals and provides funding to reuse 220 million gallons of wastewater per day to reduce water demand. This generation will witness, in this effort, whether the human species has both the technology and the will to complete the restoration of an entire regional ecosystem.

Synthesis

The world and its creatures have sustained too much damage to persist through protective isolation. Restoration, not preservation, is the best hope for achieving the goals of conservation biology. And the human experience will continue to be impoverished by the absence of living creatures and functioning ecosystems if we see ourselves only as the cause of their demise, and our separation from them the only cure. We will reclaim our full knowledge, our cultural heritage, and our proper role only through active participation in restoring the world around us.

Historical benchmarks provide good guides for restoration, but they are not the only appropriate targets. Restoration is not simply turning back the clock on natural processes because living systems constantly change, and those creatures that live within such systems adapt to and depend on change in order to survive. Moreover, no such historical view may be available to guide us in many cases; historical data on most ecosystems are sketchy at best, absent at worst. To restore ecosystems effectively, we must combine respect for a system's past with the ability to imagine, create, and invent its future, often aiming toward functional goals rather than historical measures.

Restoration is essential to conservation. There are simply not enough pristine areas left for all the remaining individuals of the world's other species to survive solely in areas that humans choose to "preserve." Restoration changes the equation of human use and degradation from an endless process of habitat loss and degradation to, potentially, an equilibrium in which humans give back some of what they have received, and habitats are created for populations to grow, not merely preserved for populations to persist.

Restoration rightly expands, as well as dignifies, the human ecological role. As Frederick Turner said, "Potentially, at least, human civilization can be the restorer, propagator, and even creator of natural diversity, as well as its protector and preserver" (Turner 1994). Restoration is the process that actualizes humanity's best relationship to nature—that of necessary contributors to the health and beauty of the world around us—and requires that we reflectively and intelligently participate in making what is damaged work, and making what is degraded whole. That humans are often the cause of ecological damage and degradation does not diminish the value of present and future restoration. In this we find the true meaning and ultimate expression of conservation.

Conflicts and Controversies of Biocultural Restoration—A Directed Discussion

Reading assignment: Janzen, D. H. 1988. Tropical ecological and biocultural restoration. *Science* 239:243–244.

Questions

1. Janzen argues that restoration ecologists working in the tropics must use the "living biotic debris" of fragmented and degraded tropical habitats for restoration and that "restoration is most needed where people live." What is Janzen's rationale for these assertions? How is this argument different from (or even contradictory to) the concept of restoration espoused by many federal agencies and private conservation organizations in North America?

2. Janzen proposes that restoration should include management of the restored ecosystem by local residents, explicit and public agreement on management goals, and managed use of cattle grazing as a restoration mechanism. What constituencies would support these approaches to restoration in North America? Which groups would most likely be in conflict over any proposal to use such mechanisms in North American habitats?

3. Janzen views the educational use of restored tropical habitats and ecosystems as necessary for their persistence. What avenues does he suggest or imply for linking educational value to restoration management? Are such linkages clearly defined or fully used in North American restoration efforts? If so, which mechanisms are used? If not, what mechanisms should be initiated to make the connection stronger?

4. What normative values does Janzen state or assume in his advocacy for tropical restoration? Is he justified in assuming these values, or should he explicitly defend them? Is Janzen's example—of merging conservation science with activism and becoming involved in sociopolitical processes—one that conservationists should follow or reject?

Learning Online

Visit our webpage at www.mhhe.com/conservation for case studies, animations, practice quiz questions, and additional readings to help you understand the material in this chapter. You'll also find active links to the following topics:

Restoration Ecology   Geographic Ecology
Landscape Ecology   Global Ecology

Literature Cited

Allen, T. F. H., and T. W. Hoekstra. 1992. *Toward a unified ecology.* New York: Columbia University Press.

Allen, W. H. 1988. Biocultural restoration of a tropical forest. *BioScience* 38:156–61.

Alverson, W. S., D. M. Waller, and S. L. Solheim. 1995. Forests too deer: edge effects in northern Wisconsin. In *Readings from* Conservation Biology: *the landscape perspective,* ed. D. Ehrenfeld, 119–29. Cambridge, Mass.: Blackwell Science.

Bekker, R. M., G. L. Verweij, R. E. N. Smith, R. Reine, J. P. Bakker, and S. Schneider. 1997. Soil seed banks in European grasslands: does land use affect regeneration perspectives? *Journal of Applied Ecology* 34:1293–1310.

Bolen, E. G., and W. L. Robinson. 1995. *Wildlife ecology and management.* 3d ed. Englewood Cliffs, N.J.: Prentice Hall.

Bormann, F. H., and G. E. Likens. 1979. Catastrophic disturbance and the steady state in northern hardwood forests. *American Scientist* 67:660–69.

Boyce, F. M., M. N. Charlton, D. Rathke, C. H. Mortimer, and J. Bennett. 1987. Lake Erie research: recent results, remaining gaps. *Journal of Great Lakes Research* 13:826–40.

Bradshaw, A. D. 1983. The reconstruction of ecosystems. *Journal of Applied Ecology* 20:1–17.

Bradshaw, A. D. 1984. Ecological principles and land reclamation practice. *Landscape Planning* 11:35–48.

Bradshaw, A. D. 1987. The reclamation of derelict land and the ecology of ecosystems. In *Restoration ecology: a synthetic approach to ecological research,* eds. W. R. Jordan III, M. E. Gilpin, and J. D. Aber, 53–74. New York: Cambridge University Press.

Bradshaw, A. D., and M. J. Chadwick. 1980. The restoration of land: the ecology and reclamation of derelict and degraded land. *Studies in ecology*. Vol. 6. Berkeley: University of California Press.

Brown, J. S. 1994. Restoration ecology: living with the prime directive. In *Restoration of endangered species: conceptual issues, planning, and implementation*, eds. M. L. Bowles and C. J. Whelan, 355–80. Cambridge, England: Cambridge University Press.

Brumback, B. C. 1990. Restoring Florida's Everglades: a strategic planning approach. In *Environmental restoration: science and strategies for restoring the earth*, ed. J. J. Berger, 352–61. Washington, D.C.: Island Press.

Chescheir, G. M., R. W. Skaggs, and J. W. Gilliam. 1992. Evaluation of wetland buffer areas for treatment of pumped agricultural drainage water. *Transactions of the American Society of Agricultural Engineers* 35:175–82.

Clarkson, M. B. E. 1995. A stakeholder framework for analyzing and evaluating corporate social performance. *Academy of Management Review* 20:92–117.

Committee on Restoration of Aquatic Ecosystems: Science, Technology, and Public Policy. 1992. *Restoration of aquatic ecosystems: science, technology, and public policy*. Washington, D.C.: National Academy Press.

Culotta, E. 1995. Bringing back the Everglades. *Science* 268:1688–90.

Dasmann, R. 1966. *Aesthetics of the natural environment*. Washington, D.C.: Conservation Foundation.

Dobson, A. P., A. D. Bradshaw, and A. J. M. Baker. 1997. Hopes for the future: restoration ecology and conservation biology. *Science* 277:515–21.

Ewel, J. J. 1987. Restoration is the ultimate test of ecological theory. In *Restoration ecology: a synthetic approach to ecological research*, eds. W. R. Jordan III, M. E. Gilpin, and J. D. Aber, 31–33. Cambridge, England: Cambridge University Press.

Freeman, A. M., III. 1993. In *The measurement of environmental and resource values*, 367–420. Baltimore, Md.: Johns Hopkins University Press.

Fuller, T. K., W. E. Berg, G. L. Radde, M. S. Lenarz, and G. B. Joselyn. 1992. A history and current estimate of wolf distribution and numbers in Minnesota. *Wildlife Society Bulletin* 20:42–55.

Galatowitsch, S. M., and A. G. van der Valk. 1994. *Restoring prairie wetlands: an ecological approach*. Ames: Iowa State University Press.

Gillis, A. M. 1995. What's at stake in the Pacific Northwest salmon debate? *BioScience* 45:125–28.

Gore, J. A., and F. D. Shields, Jr. 1995. Can large rivers be restored? *BioScience* 45:142–52.

Harper, J. L. 1977. *The population biology of plants*. London, England: Academic Press.

Howell, E. A. 1986. Woodland restoration: an overview. *Restoration and Management Notes* 4:13–17.

Janzen, D. H. 1988. Tropical ecological and biocultural restoration. *Science* 239:243–44.

Jordan, W. R., III. 1994. "Sunflower forest": ecological restoration as the basis for a new environmental paradigm. In *Beyond preservation: restoring and inventing landscapes*, eds. A. D. Baldwin, J. De Luce, and C. Pletsch, 17–34. Minneapolis: University of Minnesota Press.

Jordan, W. R., III, M. E. Gilpin, and J. D. Aber. 1987. Restoration ecology: ecological restoration as a technique for basic research. In *Restoration ecology: a synthetic approach to ecological research*. eds. W. R. Jordan III, M. E. Gilpin, and J. D. Aber, 3–21. Cambridge, England: Cambridge University Press.

Kern, K. 1992. Restoration of lowland rivers: the German experience. In *Lowland floodplain rivers*, eds. P. A. Carling and G. E. Betts, 279–97. Chichester, England: Wiley.

Knapp, E. E., and A. R. Dyer. 1998. When do genetic considerations require special approaches to ecological restoration? In *Conservation biology: for the coming decade*, 2d ed., eds. P. L. Fielder and P. M. Kareiva, 345–63. New York: Chapman and Hall.

Leopold, A. 1966. *A Sand County almanac*. New York: Sierra Club/Ballantine Books.

Makarewicz, J. C., and P. Bertram. 1991. Evidence for the restoration of the Lake Erie ecosystem. *BioScience* 41:216–23.

McNaughton, S. J. 1979. Grazing as an optimization process: grass-ungulate relationships in the Serengeti. *American Naturalist* 113:691–703.

Mech, L. D. 1995. The challenge and opportunity of recovering wolf populations. *Conservation Biology* 9:270–78.

Milner, A. M. 1994. System recovery. In *The rivers handbook*, Vol. 2, eds. P. Calow and G. E. Petts, 76–97. London, England: Blackwell Scientific.

Mladenoff, D. J., R. G. Haight, T. A. Sickley, and A. P. Wydeven. 1997. Causes and implications of species restoration in altered ecosystems. *BioScience* 47:21–31.

Mladenoff, D. J., and F. Stearns. 1993. Eastern hemlock regeneration and deer browsing in the northern Great Lakes region: a re-examination and model simulation. *Conservation Biology* 7:889–900.

Mlot, C. 1992. Botanists sue Forest Service to preserve biodiversity. *Science* 257:1618–19.

Molles, M. C., C. S. Crawford, L. M. Ellis, H. M. Valett, and C. N. Dahm. 1998. Managed flooding for riparian ecosystem restoration. *BioScience* 48:749–56.

Patin, J. W. P., and T. E. Hempfling. 1991. Environmental management of the Mississippi. *Military Engineer* 83:9–11.

Pauly, W. R. 1997. Conducting burns. In *The tallgrass restoration handbook: for prairies, savannas, and woodlands*, eds. S. Packard and C. F. Mutel, 223–43. Washington, D.C.: Island Press.

Pavlik, B. M. 1994. Demographic monitoring and the recovery of endangered plants. In *Restoration of endangered species: conceptual issues, planning, and implementation*, eds. M. L. Bowles and C. J. Whelan, 322–53. Cambridge, England: Cambridge University Press.

Pavlik, B. M., D. L. Nickrent, and A. M. Howard. 1993. The recovery of an endangered plant. I. Creating a new population of *Amsinckia grandiflora*. *Conservation Biology* 7:510–26.

Rickerl, D. H., L. L. Janssen, and R. Woodland. 2000. Buffered wetlands in agricultural landscapes in the prairie pothole region: environmental, agronomic, and economic evaluations. *Journal of Soil and Water Conservation* 55: 220–25.

Sparks, R. E., P. B. Bayley, S. L. Kohler, and L. L. Osborne. 1990. Disturbance and recovery of large floodplain rivers. *Environmental Management* 14:699–709.

Staver, K. W., and R. B. Brinsfield. 1998. Using cereal grain winter cover crops to reduce groundwater nitrate contamination in the mid-Atlantic coastal plain. *Journal of Soil and Water Conservation* 53:230–40.

Streiner, C. F., and J. B. Loomis. 1996. Estimating the benefits of urban stream restoration using the hedonic price method. *Rivers* 5:267–78.

Thompson, J. R. 1992. *Prairies, forests, and wetlands: the restoration of natural landscapes in Iowa.* Iowa City: University of Iowa Press.

Trefts, D. C. 1990. State coastal zone resource restoration and the common law public trust doctrine. In *Environmental restoration: science strategies for restoring the earth,* ed. J. J. Berger, 335–46. Washington, D.C.: Island Press.

Turner, F. 1987. The self-effacing art: restoration as imitation of nature. In *Restoration ecology: a synthetic approach to ecological research,* eds. W. R. Jordan, III, M. E. Gilpin, and J. D. Aber, 47–52. Cambridge, England: Cambridge University Press.

Turner, F. 1994. The invented landscape. In *Beyond preservation: restoring and inventing landscapes,* eds. A. D. Baldwin, Jr., J. De Luce, and C. Pletsch, 35–66. Minneapolis: University of Minnesota Press.

USDA National Technical Committee for Hydric Soils. 1987. *Hydric soils of the United States.* Washington, D.C.: USDA Soil Conservation Service.

Van Dyke, F. G., J. P. DiBenedetto, and S. C. Thomas. 1991. Vegetation and elk response to prescribed burning in south-central Montana. In *The Greater Yellowstone ecosystem: redefining America's wilderness heritage,* eds. R. B. Keiter and M. S. Boyce, 163–79. New Haven, Conn.: Yale University Press.

Van Haveren, B. P., J. E. Williams, M. L. Pattison, and J. R. Haugh. 1997. Restoring the ecological integrity of public lands. *Journal of Soil and Water Conservation* 52:226–31.

Waser, N. M., and M. V. Price. 1994. Crossing-distance effects in *Delphinium nelsonii:* outbreeding and inbreeding depression in progeny fitness. *Evolution* 48:842–52.

Weinhold, C. E., and A. G. van der Valk. 1989. The impact of duration of drainage on the seed banks of northern prairie wetlands. *Canadian Journal of Botany* 67:1878–84.

Welcomme, R. L. 1989. *River fisheries.* Food and Agriculture Organization Fisheries Technical Paper 262. Rome: Food and Agriculture Organization of the United Nations.

Werner, R. 1980. *Freshwater fishes of New York State.* Syracuse, New York: Syracuse University Press.

Zedler, J. 1998. Replacing endangered species habitat: the acid test of wetland ecology. In *Conservation biology: for the coming decade,* 2d ed., eds. P. L. Fielder and P. M. Kareiva, 364–79. New York: Chapman and Hall.

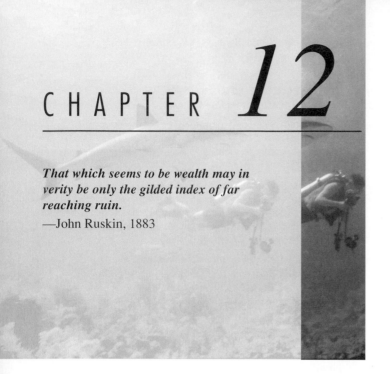

CHAPTER *12*

That which seems to be wealth may in verity be only the gilded index of far reaching ruin.

—John Ruskin, 1883

Conservation, Economics, and Sustainable Development

In this chapter you will learn about:

1 the role of economics in conservation theory and practice.

2 the characteristics of, and differences between, traditional (neoclassical) economic theory

and alternative views associated with environmental economics.

3 economic processes that can contribute to environmental protection and conservation.

4 definitions, characteristics, strategies, and applications of sustainable development.

The human population of the earth, now in excess of 6 billion, takes approximately 40% of the planet's net primary productivity for its own use. Despite this already enormous consumption of the earth's ecological output, more than 1 billion humans suffer, by clear and objective criteria, from malnutrition, poverty, lack of safe drinking water, adequate shelter and warmth, and basic health services. Thus, even if the human population were to be frozen at its present number, demands on the earth's space and resources would increase. But the human population is not frozen; it is growing by more than 80,000 each day, and stability is not expected until it reaches 10 to 12 billion.

Directly or indirectly, humans necessarily appropriate resources at the expense of other living creatures; indeed, the resources taken often *are* the living creatures. Yet the conservation of genetic diversity, populations, habitats, landscapes, and ecosystems will require that humans find ways to use living resources that do not degrade or destroy their capacity for renewal. Conservation biologists must explicate the relationships between human consumption and the persistence of other living creatures, and identify ways of changing patterns of human population growth and resource consumption to achieve conservation goals.

One of the most basic and generalizable expressions of the relationships among economics and the environment is the so-called Ehrlich Identity, formalized by biologists Paul and Anne Ehrlich (Ehrlich and Ehrlich 1990) as

$$I = P \times A \times T,$$

where I is environmental impact, P is population, A is affluence (a measure of consumption), and T is technology (an index of efficiency of resource use and pollution abatement). The identity is of little value mathematically (for one thing, it is very difficult to express I in meaningful units or find common units for all the variables), but it is of value conceptually. Basically, the Ehrlich Identity asserts that environmental impact is not simply a function of human population density, but also of per capita consumption (A) and efficiency of resource use (T). Furthermore, the relationship between impact and other variables is complex and often nonlinear. For example, extreme poverty (very low value of A) often results in great environmental damage because of the direct and destructive manner in which impoverished peoples obtain resources. Understanding this relationship requires knowledge of how human populations grow.

HUMAN POPULATION GROWTH—A RECENT HISTORY

For most of the last 10,000 years, humans obtained resources through cooperative efforts in hunter-gatherer societies, in which energy used to obtain resources was provided by human or animal labor. Human efforts relied on *flows* of energy that, directly or indirectly, were tied to solar input. Beginning in the seventeenth century, industrial technology began to use previously unexploited *stocks* of energy, namely fossil fuels such as coal, oil, and natural gas. Although the rate at which solar energy can be used has historically been dependent on the rate at which it is received, the rate at which a stock of energy is used depends on the rate of consumption and the technology for converting the stock into useful work. Thus, it is possible to use stocks of fossil fuels at much higher rates than they are produced. The development of technology for converting such stocks into energy allowed human effort to become more efficient and more productive, and resulted in an increase in the human standard of living and faster population growth.

By 1800 the human population had reached 1 billion. By 1930 it had doubled, growing at a little over 0.5% per year. The third billion was added by 1958, and a growth rate of nearly 2% per year resulted in a population of 4 billion by 1975 (and a doubling time of only 45 years). Although overall growth rates then began to slow, the large size of the population led to faster recruitment of additional billions. The 5 billionth human was born in 1987 (Erickson 1995), and the 6 billionth only 11 years later (fig. 12.1).

Since 1600, human populations in industrialized nations have progressed through a "demographic transition" characterized by two distinct phases. At the beginning of the Industrial Revolution, world human population was characterized by a near-zero growth rate, average life expectancy of less than 30 years, and a combination of high birth and death rates. In the first phase of the demographic transition, death rates fell and life expectancy increased. As a consequence,

population density rose. In the second phase of the demographic transition, birth rates also fell to levels similar to death rates. Most industrialized nations have passed through this transition; individuals in their populations typically have life spans of over 70 years, and the population collectively has similarly low birth and death rates (Piel 1995). In the transitions of individual nations, increasing life span is the apparent trigger of restraints on fertility. Demographer Gerard Piel wrote, "People who can look forward to the full biologically permitted human lifetime permit themselves to be future dwellers. Assured of the survival of their first infant(s), they can plan their families, . . . they see that the fewer, the more—for each" (Piel 1995).

World population is still increasing, but at a decreasing rate. The rate of human population growth peaked at around 2% per year from 1950 to 1970 and has been in decline since. Nevertheless, most reliable population estimates predict stabilization of the human population at around 10 billion individuals, or even 12 billion. Projections of human population growth vary, and so do predictions about the level of affluence humans will enjoy at a given population density.

In complex systems, there are four observed, recurring patterns of growth (fig. 12.2). In the exponential model, the population grows at an ever increasing rate, with no environmental limit on its growth (fig. 12.2a). The population, in the words of economist Herman Daly, "has no environment" (Daly 1991). Instead, population growth is determined by the difference between birth rates and death rates (r, the intrinsic rate of increase) and the previous size of the population (N_{t-1}, see equation in chapter 7). Exponential growth has been observed in many populations, but only temporarily. No population has even been observed to exhibit exponential growth indefinitely.

In the logistic growth model (fig. 12.2b), the population increases at an ever decreasing rate as it smoothly approaches an upper asymptote. In this pattern of growth, density-dependent factors "signal" the population, or individuals in it, when they are approaching environmental limits, and those signals are translated into behaviors that precisely control the population's growth trajectory.

A variation on logistic growth, sometimes referred to as "delayed logistic" (fig. 12.2c), includes definite environmental limits, but environmental signals are unclear, lag behind demographic events, or are insufficiently precise to "brake" the population's growth rate. As a result, the population exceeds its "limit" and the excess individuals degrade the system and lower its capacity to support the population. The population crashes to a very low level or, in extreme cases, declines to extinction.

The fourth pattern of growth also incorporates an absolute environmental limit combined with imperfect signals and imprecise control mechanisms, leading the population to overshoot the limit (fig. 12.2d). However, in this case, both the system and the population are able to recover, leading to a pattern of "stable oscillation" around an average population density associated with the environmental limit.

All four patterns of population growth occur in animal populations in nature. The pertinent question, however, is, "Which pattern will the human population follow?" It is

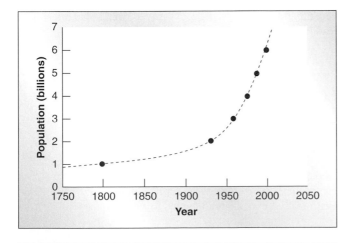

Figure 12.1

Human population growth since 1800.

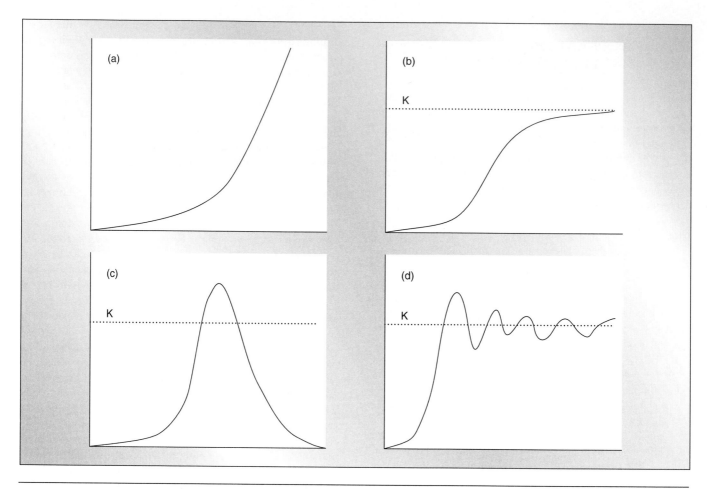

Figure 12.2

Four patterns of growth observed in natural populations: (a) exponential growth (no environmental limit); (b) instantaneous logistic growth (smoothly approaches environmental carrying capacity K); (c) time-specific logistic growth without resilience (overshoots environmental carrying capacity and then, as a result of environmental degradation and demographic factors, declines to extinction); and (d) time-specific logistic growth with resilience (overshoots environmental carrying capacity but then recovers because of resilience of the population and the environment; may experience regular and predictable oscillations above and below a mean population level).

reasonable to exclude the possibility of sustained exponential growth, although some technological optimists argue that, theoretically, there is no environmental limit to human population density on earth. However, even if advances in technology were to make human population growth theoretically unlimited, empirical data demonstrate that the human population growth rate has already slowed and is likely to slow even more. That is, even if humans could maintain exponential growth indefinitely, it appears that they are not, in fact, doing so.

Thus, the more likely alternatives are different types of logistic growth. Extreme optimists favor a pattern of instantaneous logistic growth (usually with an environmental limit still far above present or projected population levels), whereas extreme pessimists predict a doomsday scenario based on the third category, discrete (time-delayed) logistic growth. Although the "back halves" of the three logistic curves have different shapes (representing different outcomes), they all display identical "front half" curves. That is, all the possible outcomes look the same in the beginning. Under these conditions, economist J. R. Ehrenfeld astutely remarked, "foretelling the future by looking

at the past is virtually impossible" (Ehrenfeld 2000). Instead, we will examine the traditional view of economics (neoclassical economics) and the assumptions derived from it that affect how resources are used by and distributed among humans.

ECONOMICS AND THE ENVIRONMENT

Development of Neoclassical Economics

Economics, the study of how people allocate scarce resources among competing ends to satisfy unlimited human wants, began in the eighteenth century as a branch of moral philosophy. Historically, economists attempted to answer questions such as:

1. What should be produced?
2. How many goods should be produced?
3. In what manner should goods be produced?
4. How should such goods be allocated?

As a subdiscipline of moral philosophy, economics was originally concerned not merely with what *could* be done regarding the production, acquisition, and distribution of goods, but also, perhaps more importantly, with what *should* be done. Classical economists asked questions not merely about how wealth might be created and acquired, but also about what purpose wealth serves. For example, economist J. C. L. Simonde de Sismondi (1773–1842) questioned the view of growth in economic activity as an end in itself. To illustrate his point, de Sismondi told the story of the Gandalin, a man who gave temporary lodging to a sorcerer. Gandalin noticed that every morning the sorcerer said a few words to a broom handle and turned it into a water carrier. The broom handle would bring him as many pails of water as he desired. Intrigued, Gandalin hid behind the door one morning and heard the magic words of command, but he could not hear what the sorcerer said to undo the command. Unconcerned by the imperfection of his knowledge, Gandalin waited for the sorcerer to go out that day, and then pronounced the magic words over the broom handle. The broom handle went to river and fetched a bucket of water. Then it repeated the task until the water tank was overflowing and water was beginning to flood the room. The broom handle ignored every command Gandalin shouted at it to stop. In desperation, Gandalin took an axe to the broom handle and chopped it into pieces, whereupon every piece rose up and began to repeat the task that the broom handle had previously been doing alone. The entire house would have been flooded if the sorcerer had not returned and revoked the spell.

Although this story, popularized as "The Sorcerer's Apprentice," is now associated with a Mickey Mouse cartoon, de Sismondi was attempting to make a serious point. "Water," he wrote, "just as much as work, just as much as capital, is necessary for life. But one is able to have too much, even of the best things of life" (Smith 1993). To de Sismondi, Gandalin epitomized the industrialist of his day, and the broom handle and its fragments represented the processes of mass production. In de Sismondi's view, industrialists knew how to make more things, but they did not know how or why to stop. So the mass production of things went on and on until the world became "flooded" with products beyond anyone's need or good.

The classical English economist John Ruskin stated the purpose of economics in a similar way. "The real science of political economy," wrote Ruskin, "which has yet to be distinguished from the bastard science, as medicine from witchcraft, and astronomy from astrology, is that which teaches nations to desire and labor for the things that lead to life, and which teaches them to scorn and destroy the things that lead to destruction" (Ruskin 1883).

To Ruskin, wealth was more than the measurement of one's possessions; it included the capability to use them in an appropriate manner. Ruskin wrote, "'Having' is not an absolute, but a graduated, power; and consists not only in the quantity or nature of the thing possessed, but also (and in a greater degree) in its suitableness to the person possessing it and in his vital power to use it." Ruskin noted further that, "A nation which desires true wealth, desires it moderately, and can therefore distribute it with kindness, and possess it with pleasure; but one which desires false wealth, desires it immoderately, and can neither dispense it with justice, nor enjoy it in peace."

Other philosophers and economists, including John Stuart Mill, argued that perpetual growth in material well-being was neither possible nor desirable. Following Aristotle's concepts of causality, classical economists addressed the issues of (1) material cause (What is the source or resource base of economic production, and are the best materials being used?), (2) efficient cause (What or who directs and assembles the resources into a functioning economic system, and does this assembly lead to efficient use of resources?), (3) formal cause (What is the economic plan or blueprint that underlies economic direction and organization, and is it a good plan?) and (4) final cause (What is the purpose of the economic system, and what need is met by its existence?). Especially because of their interest in the question of final cause, classical economists were suspicious of the prospect of unlimited economic and material growth. However, by the late nineteenth century, the modern neoclassical emphasis in economics had begun to replace the classical one. Unlike classical economists, who believed that the value of a good depended on the labor expended to produce it, neoclassical economists believed that value was driven by scarcity, and thus price represented an interaction of supply and demand. The emphasis of neoclassical economics became the study of relationships between small changes in prices and quantities as economic activity was extended by small amounts, leading to predictive models of aggregate economic behavior (marginal analysis). Neoclassical economists consciously tried to model their theoretical systems on physics, even going so far as to speak of individuals as "economic molecules" (Prugh et al. 1999). In terms of causality, neoclassical economists largely ignored the question of material cause (because material resources were seen as unlimited and perfectly substitutable) and final cause (the purpose of the system was to maximize individual human welfare, which was arbitrated by individual preferences, not collective or social good) and merged the efficient and formal causes into a single question (Smith 1993). Although classical economists believed that economic growth should and would eventually stop, neoclassical economists saw economic growth as a perpetual, desirable, and necessary condition of economic activity.

Various perceptions of the environment in neoclassical economics are illustrated in figures 12.3 and 12.4. In figure 12.3, the economy is all there is. Inputs of raw material either come from outside the system (or, literally, from nowhere) or are considered a sector of the economy (the "extractive" sector, such as mining, forestry, or fisheries). Products and wastes are exported out of the system (to nowhere). In an alternative but similar neoclassical view of the environment in figure 12.4, the environment is more explicitly recognized, but it still exists only as a subset of the economy. Here, the environment provides natural capital (resources) that serve as factors of production, and resources provided and supplied by the environment are valuable but not indispensable. Natural capital (resources from the environment) and human capital (human-created artifacts, labor, or intelligence used in production) are essentially perfect substitutes. Therefore, whatever is scarce in, or altogether lost from the environment, can be replaced by human technology, labor, intelligence, or means of production. Because such substitution is infinite, there are no real limits to economic growth. Human technology overcomes scarcity through the substitution of human capital for natural capital and ever-higher efficiencies of use. This continued growth raises all

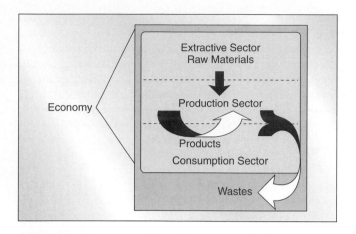

Figure 12.3

Neoclassic theory views the economy as the total system with such sectors as extraction, production, and consumption. Natural resources (raw materials) are production inputs and wastes are production outputs. The environment is not considered a larger system that supports the economy; instead, inputs from the environment are considered a sector of the economy. Wastes generated in production are transported out of the economy.

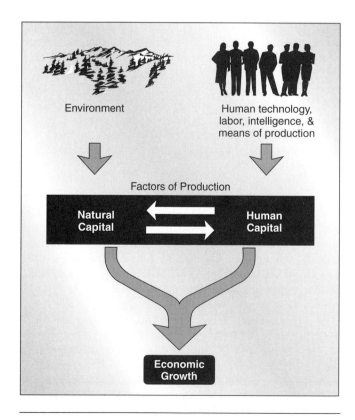

Figure 12.4

Incorporation of an environmental subsystem into neoclassical economic theory. Natural capital and human capital are substitutable. Natural resources serve as factors of production; if any particular natural resource is exhausted, human ingenuity provides a substitute.

standards of living and solves, or will eventually solve, problems associated with poverty without the necessity for redistribution of current wealth. Because technology allows increasingly more efficient use of natural resources, the problem of environmental degradation also will eventually cease to exist. Thus, human welfare is maximized when each individual pursues his or her selfish and individual interests in market transactions (Prugh et al. 1999).

Neoclassical economics eventually replaced classical economics in part because of its success in predicting aggregate economic and market behavior through quantitative mathematical models. Neoclassical analysis also provided an objective means of determining the valuation of consumer goods and predicting appropriate levels of supply and efficient means of production and distribution. Further, neoclassical economics was compatible with the increasingly secular spirit of both culture and academia. By relegating questions of value to the sovereignty of individual welfare and personal economic preference, it conveniently avoided, or at least minimized, confronting issues of social responsibility or collective good in economic behavior. And finally, neoclassical economics promoted an optimistic view of continual economic growth and continually increasing material affluence. However, problems arose in areas that neoclassical economic theory had failed to take into account.

Problems of Neoclassical Economics

Externalities

The economist A. C. Pigou defined an **externality** as a phenomenon that is external to markets and hence does not affect how markets operate, when in fact it should (Costanza et al. 1997). For example, waste and pollution—end products of many economic processes—cost money to clean up and reduce the level of environmental services from systems that they degrade, such as the availability of clean air or clean water. Other externalities can include increased human health costs because of increased incidence of disease associated with polluted environments, or damage to property and physical structures. However, these costs may not be included in the price of the good whose production created the pollution and waste. Instead, they are external costs or *externalities* shared by all who live in and use the polluted system(s). Traditional neoclassical economics considers such systems and their services free of charge because their services, such as waste disposal and assimilation, have no market. And because the systems and their services cannot be appropriated by any one person (e.g., no one person can appropriate the use of the atmosphere for breathing), the systems are not covered by the traditional rights that protect private property. Thus, neoclassical economics hides the true cost of production through externalities.

In fact, environmental externalities from even a single activity, such as power generation, can be both extensive and expensive. For example, the European Commission identified eight kinds of external costs associated with power generation from coal, oil, and gas that would add from 1.7 to 4.1 cents per kilowatt-hour if they were included in the market price (table 12.1) (European Commission 1995).

Table 12.1 **Externalities Associated with the Production of Power Using Coal, Oil, and Gas as Calculated by the European Commission's ExternE Project**

EXTERNALITY	COAL	OIL	GAS
Agriculture	Sulfur, acidification, and ozone: crop and soil impacts	Sulfur, acidification and ozone: crop and soil impacts	N/A
Amenity	Noise: operational road and rail traffic impacts	N/A	Noise: operational impacts
Forests	Sulfur, acidification, and ozone damage	Sulfur, acidification, and ozone damage	N/A
Global Warming	CO_2, CH_4, and N_2O damage	CO_2, CH_4, and N_2O damage	CO_2, CH_4, and N_2O damage
Marine	Acidification impacts	Accidents with oil tankers	Fishery: extraction impacts
Materials	Sulfur and acidification damage on surfaces	Sulfur and acidification damage on surfaces	Sulfur and acidification damage on surfaces
Occupational Health	Diseases from mining and accidents during mining, transport, construction, and dismantling	Accidents: death and injury impacts	Accidents: death and injury impacts
Public Health	Particulate matter, ozone, and accidents: mortality, morbidity, and transport impacts	Particulate matter and ozone: mortality, morbidity, and transport impacts	Particulate matter: mortality, morbidity, and transport impacts
Total Estimate (U.S. Cents/kWh)	2.8–4.1	2.7–2.9	1.7

European Commission 1995. Adapted from Söderholm and Sundquist 2000.

Modern neoclassical economic theory does consider the value of externalities important for assisting market processes and for making socially efficient choices. Its preferred mechanism of correction is through taxes and subsidies that accurately reflect external costs and create a system of economic rewards and punishments that will lead a profit-maximizing firm to select a mix of goods and production technologies that will best satisfy environmental and economic goals (Söderholm and Sundquist 2000).

Substitution

Neoclassical economics defines *capital* as the means of production and distinguishes between *manufactured capital* (physical artifacts created by humans to produce goods) and *natural capital* (natural objects or resources that can be transformed into consumer or producer goods). In traditional neoclassical economics, manufactured capital and natural capital are substitutes; a unit of one can be exchanged for a unit of the other. In this view of substitutability, production cannot be limited by diminishing levels of natural resources because human capital can substitute for them. What is the problem? Consider the relationship between trees and saws. In order to produce lumber for a large number of consumers, one needs trees (natural capital) and saws (manufactured capital) to cut the trees down. If one has a large supply of trees but only one saw, it might be wise to invest in the production of more saws and cut a larger number of trees. But if one reaches a point where one has nothing but saws (exclusively manufactured capital) and no trees (an absence of natural capital), production cannot continue; more saws cannot be substituted for fewer trees. No matter how many loggers have saws or how efficiently they cut, lumber cannot be produced without natural capital. Neoclassical economists would argue that metal or plastic could be substituted for wood, but these materials are not the same as lumber. And even if they prove to be perfect substitutes, they also are ultimately resource based.

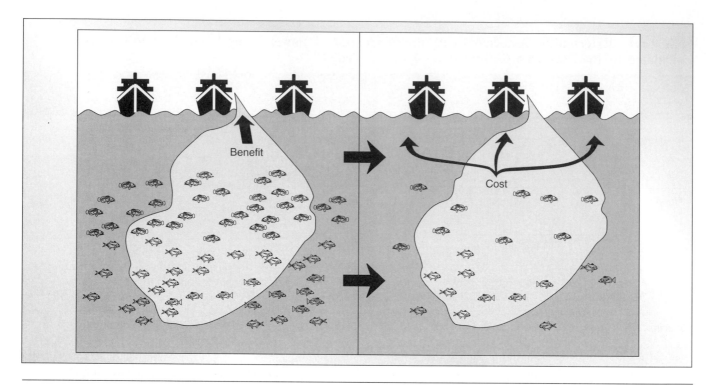

Figure 12.5

A visual example of Garrett Hardin's "tragedy of the commons" (Hardin 1968), in which a common-access resource (a fishery) is inevitably degraded by the "rational" economic behavior of each individual. In this case, it is always in the fisher's best interest to take all the fish that can be caught, as the individual receives all the benefit while the cost (depletion of the fishery) is split among all fishers. With all fishers acting according to this principle of maximizing personal benefits and minimizing personal costs, the fish population will be depleted.

Common-Access Resources

Neoclassical economics considers common-access resources, such as air, water, and marine fisheries, to be free of charge. Such resources are neither protected by traditional property rights nor exchanged in markets where price would be set by supply and demand. Removal of stocks (e.g., fish from the ocean) may lower the long-term sustainability of the open-access system to produce more resources, thus degrading the system. The profits that come from selling the stocks (such as the fish) accrue solely to the seller who removed them; however, all users share the cost of the long-term decline in the system. This is the logic behind Garrett Hardin's famous essay, "The Tragedy of the Commons," in which Hardin (1968) tells of a group of herders who keep animals in a common pasture. For many years, natural limits on both the herders and their livestock keep numbers in check, but gradually the number of herders begins to grow. As this growth occurs, herders are faced with a choice: they can add one more animal of their own to the pasture and reap the reward, but the addition will degrade the pasture's ability to support the aggregate population of livestock. Although there are both benefits and costs to adding one more animal, benefits and costs are not equal. If the herder adds an animal, the reward belongs to the herder alone, but the cost of degrading the pasture is shared by all the herders. Thus, the herder who adds the animal pays only a portion of the cost. Logically, then, it will always be in the herder's best (and selfish) interest to choose adding another animal, but such logical behavior eventually leads to the destruction of the pasture.

We can illustrate the concept visually by changing our example to another common-access resource. Imagine a fishery in the ocean that is open to all fishers who can reach it. Each fisher is rationally motivated to capture all the fish he or she can, because the benefit belongs to the fisher alone, while the cost (long-term depletion of the stock) is shared among all the fishers (fig. 12.5). We can understand the situation more analytically with the aid of figure 12.6, a graphical relation of fishing effort to revenues and costs. Note that maximum net benefit (greatest difference between revenue and cost) occurs at point E_1. But fishing effort will not stop here. Rather, excessive fishing effort will occur as a result of rational individual economic behavior because, as long as the fishery is profitable, current fishers continue to expand their fishing effort and new fishers continue to enter the fishing ground to point E_2, beyond which point cost exceeds revenue and fishing should cease. However, fish stocks are much more likely to be depleted at E_2 than at E_1 and the resource may suffer irreparable degradation and collapse. Recall from chapter 9 some unfortunate historical cases, such as the Grand Banks cod fishery (Lauck et al. 1998; Ruckelshaus and Hays 1998), in which this is precisely what happened.

Traditional neoclassical economics assumes that individual choice, constrained by market behavior, inevitably leads to the greatest good, but this is not the case in the tragedy of the commons. The tragedy can be averted by the intervention of a regulatory authority (government) or through a social process in which herders begin to work cooperatively for the common

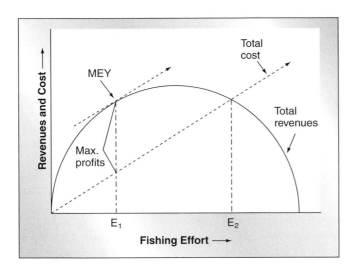

Figure 12.6

The relationship between fishing effort, revenues, and costs in an open-access fishery. Excessive fishing effort occurs because current fishers expand effort beyond the point of maximum profitability (MEY) and new fishers continue to enter the fishery. Every fisher continues to make a profit through increased fishing effort to point E_2, the point where total revenue and total costs are equal. Beyond this point, costs exceed revenues and fishing becomes economically irrational, but the stock already may have been irreparably depleted by increasing effort from E_1 to E_2.

Developed from concepts formulated by Gordon (1954). Figure adapted from Costanza et al. (1997).

good, but there is no solution intrinsic to neoclassical economics theory itself; thus, the theory faces significant problems in dealing with the allocation and sustainability of open-access resources.

General Market Failure

A key element in the success and eventual ascendancy of neoclassical economics was its concept of markets. In neoclassical understanding, markets organize economic activity by using the price of goods and services to communicate the wants and limits of a diverse society, and to coordinate economic decisions in an efficient manner. Markets decentralize the process of decision making and exchange and thus are attractive to democratic societies that place a high value on personal freedom and individual choice. When markets operate properly, no central planner is needed to allocate resources, and private decisions are expected to lead collectively to optimal social outcomes. But markets can, and routinely do, fail to allocate resources efficiently, often because of the inability of supporting social institutions to establish well-defined property rights for biological services. For example, market prices may not communicate society's desires and values accurately for natural resources, endangered species, or rare ecosystems. Market prices may understate the range of services provided by species or ecosystems; alternatively, markets for these entities may not exist, and thus send no signal to the marketplace about the social value of the asset. Economists Shogren and Haywood (1998) wrote,

"When individual decisions impose costs on or generate benefits for other individuals who are not fully compensated for losses or who do not fully pay for gains, a wedge is driven between what individuals want unilaterally and what society wants as a collective." Traditional markets also do a poor job of valuing *nonrival* and *nonexclusive* goods and services, such as those produced by ecosystems. **Nonrival goods** and services are those in which the use of the good or service by one person does not reduce or restrict its use by another (e.g., breathing oxygen from the atmosphere). **Nonexclusive goods** and services are those for which it would be extremely difficult and costly to exclude anyone from receiving the benefits (e.g., protection from ultraviolet radiation by the ozone layer). Traditional market operations, as envisioned by neoclassical economic theory, thus face serious problems in valuing environmental goods and services. These problems in valuation often lead to market failures in which goods and services are not properly valued, resources are not allocated efficiently, and the collective good of society is not achieved.

Ethical Problems: Determining Ultimate Means and Ends

In neoclassical economics, markets provide the mechanisms for determining how to do things efficiently, but not for answering the question of what one should do. Neoclassical economics assumes that, although individual human wants can be satisfied, the aggregate of human wants is insatiable; therefore, economic growth will be perpetually driven by consumer demand. Is there some point of material wealth beyond which further accumulation is not helpful, and may even be harmful, to the highest and best ends of the individual and society? That question is not asked by neoclassical economists, who tend to ignore the Aristotelian question of *causa finalis* (What is the purpose that causes the action?). The goal of neoclassical economics is the maximization of human welfare, hence its other common name, "welfare economics." Thoughtful people routinely ask, "Is that all there is, or are there alternative ends, even ultimate ends that transcend personal welfare and individual preference?"

Neoclassical economics emphasizes how skillful we can be at doing something and our plan for doing it; in such a view, "doing" becomes its own justification. Neoclassical economics rarely explores the questions, "Doing what, for what?" or "Doing what, for what and with what?" (Smith 1993). Because of its emphasis on personal preference, autonomous choice, and individual wants, neoclassical economics offers no incentive to make choices based on any entity but the individual. It neither considers nor encourages community-based decisions for collective welfare, nor does it address values that may lie outside individual choice.

Thus, although neoclassical economics has demonstrated great usefulness in explaining and predicting market-based economic behavior, it ignores many realities that exist outside markets, especially the reality of the physical environment. Unless these failings are addressed, economic activity will continue to be adversarial to environmental protection generally and to particular conservation goals specifically. To address this problem, we will first consider potential solutions offered by neoclassical economists (what we might call "paradigm-repair solutions") and then consider more radical alternatives (alternative paradigm solutions) to the conflicts between economic activity and environmental protection.

MARKET-BASED SOLUTIONS TO ECONOMIC-CONSERVATION CONFLICTS

Managing "Hummingbird" Economies

Some economists would argue that all conflicts between economic activity and conservation goals are fundamentally based on a failure to clearly define and properly allocate property rights. Economist Bruce Yandle describes the problem with an analogy of hummingbirds at a backyard nectar feeder. A dominant bird vigilantly guards the feeder and attacks other birds when they attempt to use the nectar, but when the dominant bird is engaged in one dogfight, another bird swoops in and gets some of the nectar. The dominant bird's effort fails because it has no way to completely secure the feeder and stake a claim to it. Resources are depleted because of insecure and ill-defined property rights (Yandle 1997).

But Yandle argues that people are not hummingbirds, and they may successfully develop "hummingbird economies" based on common-access resources without recourse to government control. Drawing directly on Hardin's tragedy of the commons

motif, Yandle describes a nontragic ending to Hardin's story of the herders (whom Yandle calls "shepherds," making their livestock "sheep") and their common pasture. Rather than simply wrecking the pasture through unregulated grazing, the shepherds form a "shepherd's club" in which they share information and act cooperatively. Yandle states that "they may learn that each additional sheep reduces the weight gained for the collective flock. This marginal product, which is the change in total weight gained by the flock with the addition of one more sheep, declines faster than the average weight gained per sheep" (Yandle 1997:14). Yandle notes the shepherds will learn that "flock production would improve if fewer sheep were placed on the pasture. Pastures are depletable resources, but they are also sustainable. With appropriate safeguards, a shepherd community can engage in pasture rotation and maintain an economically efficient level of sustained use" (Yandle 1997:15).

Like all good economists, Yandle proves his point with a graph (fig. 12.7). Sheep enter the pasture as long as average weight gain is at least as high as the next best opportunity (any alternative pasture). Herd expansion stops when the marginal gain becomes negative. Using this figure, Yandle notes that "the club would allow sheep in the pasture until opportunity cost was equal to the marginal product of the pasture. . . . Operating at that point, the shepherd club would produce the largest amount of weight gain possible: each member could conceivably be wealthier than before, depending on the rule for output sharing" (Yandle 1997:15).

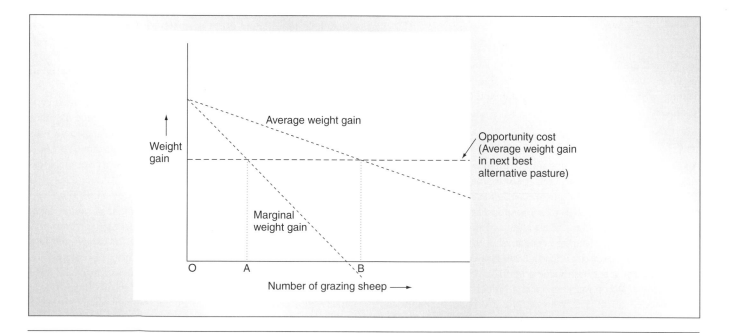

Figure 12.7

Hardin's (1968) tragedy of the commons expressed as a problem of marginal gains. The number of sheep entering the pasture (x axis) should be increased as long as average weight gain (y axis) is at least as good as the next best opportunity (any alternative pasture, represented by the line labeled "opportunity cost"). Herd expansion stops when the marginal gain becomes negative. Logically, a community of shepherds would add sheep in the pasture until opportunity cost was equal to the marginal product of the pasture, line segment OA on the x axis. Operating in this region, the shepherds would produce the largest amount of weight gain possible. If sheep are added past point B, opportunity cost (average weight gained in next best pasture) exceeds average weight gain. At this point, shepherds should place additional sheep on the alternative pasture.

After Yandle (1997).

Economist Donald Leal formalizes the conditions necessary for the potential tragedy of the commons to have the happy ending described by Yandle:

1. The boundaries of resource use are clearly defined.
2. Group decisions are based on established rules that determine how the group parcels out the value of the resource.
3. The rules are linked to time- and place-specific knowledge of resource constraints so that resulting decisions are economically efficient.
4. Some resources of the community must be devoted to monitoring the state of the resource and the behavior of the users to prevent "cheating," to provide rewards to those who keep the rules, and to apply sanctions against those who break them.
5. Where conflicting demands are likely to arise among group members, there must be mechanisms in place for bargaining.
6. The rules must not be changed by higher levels of government (Leal 1998).

Averting the Tragedy of the Commons: Historical and Contemporary Examples

Sockeye Salmon and Native Alaskans

Leal's (1998) six conditions can be seen clearly in one historical example involving the allocation of fishery resources among native Alaskans. Although there are five species of salmon (*Oncorhynchus* spp.) that commonly spawn in Alaskan rivers, the sockeye salmon *(O. nerka)* has long been the most prized by natives. Sockeyes arrive earliest and remain longest in their spawning streams, have the lowest historical variation in annual return numbers, the highest nutritional value and, in the opinion of some, the best taste (Leal 1998). Thus, the Tlingit and Haida tribes, two groups of coastal native Alaskan peoples, placed a high value on the sockeye. Coupled with their values described above, sockeye salmon also were scarce, but definable as property resources because sockeyes migrate only in stream systems that include a freshwater lake (Leal's condition 1). The combination of scarcity and definable resource boundaries enabled the Tlingit and Haida to make rules limiting access to particular streams to an individual clan or house group (condition 2). Thus, the number of individuals consuming the resource was appropriate to the sustainable level of production of the resource and was based on the size of the stream and the number of individuals required to set traps and weirs across the stream's mouth. Management decisions (how many sockeye to take) were linked to time- and place-specific knowledge of resource constraints (condition 3) by placing them in the hands of the *yitsati*, the eldest clan male. The yitsati could use his power to enforce the regulations and to punish those who violated them (condition 4), but enforcement costs were low because fishing rights could not be transferred outside the clan, and thus there were relatively few conflicts among users and little need for bargaining mechanisms to resolve differences (condition 5). The system and the sockeye salmon resource collapsed when other cultures, first Russia and then the United States, ignored the established system and allowed anyone to place traps and weirs at the mouths of rivers (violation of condition 6).

The Tlingit and Haida system of preserving the commons is one among a number of documented historical examples of sustainable use of a common-access resource through normative social values (Ostrom 1990). Ostrom, commenting on the success and similarities of these societies in resource management, noted that all behaved as if the systems they managed had intrinsic value, not merely a value in terms of commodity output. Ostrom asserts that the social norms reflect "valuations that individuals place on actions or strategies in and of themselves, not as they are connected to immediate consequences. When an individual has strongly internalized a norm related to keeping promises, for example, the individual suffers shame and guilt when a personal promise is broken. If the norm is shared with others, the individual is also subject to considerable social censure for taking an action considered to be wrong by others" (Ostrom 1990:35; quoted in Leal 1998:284). This perception is in sharp contrast to that of Hardin (1968) who claimed that, although such social conscience could prevent the tragedy of the commons, conscience was really only a kind of manipulative communal trick employed to delude others into acting against their own best interests. Hardin summed up his argument with a quote from the philosopher Neitzsche: "A bad conscience is a kind of illness" (Hardin 1968). It is, ironically, just this sort of "illness" that has been an essential element in promoting and preserving the ecological health of a variety of complex resource use systems worldwide, while the contrasting "health" of a purely self- and profit-centered system of values has led to their demise. Thus, although such systems had clearly defined property rights and some market-driven incentives, their ultimate values were not market based. A similarly essential role for socially normative values, coupled with private property and market-based incentives, also has emerged in many contemporary examples of conservation.

Resource Management by Market Incentive, Local Control, and Social Norms

Conflicts between conservation efforts to preserve many African wildlife species, such as elephants (*Loxodonta africana*), and local landowners, whose crops, homes, and lives are threatened by such wildlife led to the development of the Communal Area Management Program for Indigenous Resources (CAMPFIRE) program in Zimbabwe, a radical departure from traditional command-and-control, government-run programs to conserve African wildlife. In CAMPFIRE, management authority is delegated to district administrators of communal lands. Working with these district administrators, CAMPFIRE assists in establishing local citizen cooperatives that have territorial rights (i.e., well-defined property rights) over wildlife resources. The majority of revenue generated from wildlife must be returned to the communities that bear the costs of living with the wildlife. CAMPFIRE has generated considerable revenue for local communities using this approach, primarily from safari hunting and ecotourism (Kreuter and Simmons 1995).

A similar approach, in a very different setting, is found in the world's largest commercial cod fishery, the Lofoten fishery of Norway. Remarkably, there have never been any quota regulations in this fishery, nor any special licensing system (Leal 1998). Rather, the government of Norway has, since 1897, given legally delegated management authority to 15 "control districts," each of which has a well-defined territory and broad powers to develop and enforce district-specific fishing regulations. Enforcement is carried out by inspectors elected from within the fishing community (note the conformity here to Leal's condition 4) for each type of fishing gear. Judgments against violators are rendered by local magistrates. All fishers must register with and obey the rules of the district in which they fish.

Private Property, Market Incentives, and Conservation—Current Conflicts and Proposed Solutions

Just as some neoclassical economists see poorly defined property rights as the fundamental cause of all conservation problems, they also see abuse of property rights as the fundamental grievance that has fueled the anticonservation "property rights rebellion," especially against the Endangered Species Act (ESA) and wetland conservation legislation, particularly Section 404 of the Clean Water Act, which has been interpreted in U.S. courts as giving federal agencies authority to prevent destruction of wetlands by private landowners (Fitzsimmons 1999).

Both of these acts, and others, give the U.S. federal government, in particular federal agencies such as the Fish and Wildlife Service, the Army Corps of Engineers, and the Environmental Protection Agency, broad powers to prevent private landowners from any kind of activity or development on privately owned lands that could harm an endangered species or degrade or destroy wetlands. Many economists and other advocates of private property rights see such restrictions as effectively reducing the value of the owner's land without providing any compensation for the landowner, and thus violating the Fifth Amendment to the U.S. Constitution, "nor shall private property be taken for public use without just compensation." The argument is, if the destruction of wetlands and endangered species represents a loss to the general welfare of the U.S. public (which is exactly the claim that conservationists make), then their preservation represents a public benefit. A benefit has a value, and the private landowner who provides such public benefit by protecting the endangered species or not developing the wetland is entitled to just compensation from the public (i.e., the government) (Ceplo 1995).

Many proponents of such views argue that they are not against conservation. Rather, they claim to be in favor of better, more economically efficient conservation through greater use of market forces, abetted by government-supported incentives, that make conservation marketable and profitable. Specific suggestions are that the government:

1. Provide just compensation to private landowners for maintaining endangered species, wetlands, or other entities of conservation value on their land at their expense.

2. Create a market for some conservation entities, such as wetlands, through government issuance of tradeable certificates, such as Wetland Protection Certificates, that the landowner would receive for every land unit (e.g., acre or hectare) of wetland preserved. Certificate holders would be allowed to exchange the permits with other landowners or for governmental permission to develop other wetlands (the ratio to be determined according to wetland type and value). Thus, development could not be denied to landowners who offset the development with the preservation of other wetlands of equal or greater value.

3. Buy conservation easements, as private conservation organizations do, that constrain land use to those activities that will benefit, or cause no harm to, endangered species, wetlands, or other entities of conservation value.

4. Provide matching funds to conservation organizations, such as The Nature Conservancy or Ducks Unlimited, who buy land or easements from private landowners to achieve conservation goals.

5. Provide tax benefits to landowners who provide habitat for endangered species, preserve or create wetlands, or perform other services that further define conservation goals.

6. Permit the sale of threatened or endangered species under circumstances that allow the landowner to profit from maintaining a viable population of the endangered species on the landowner's property (Fitzsimmons 1999).

The views and suggestions outlined here can operate within traditional free market structures, although they may require new roles for both government and private landowners in order to be effective. However, many economists argue that more radical alternatives are necessary to achieve conservation goals. We now consider some of the foundational principles that define these alternative economic views, collectively referred to as "environmental economics."

POINTS OF ENGAGEMENT—QUESTION 1

Are all economic and land-use conflicts in conservation conflicts of property rights, or are they also (or alternatively) conflicts about what should be valued? Historically, why did unrestricted free enterprise and market incentives lead to degradation and, in some cases, permanent loss of resources, even when property rights were clearly defined?

ENVIRONMENTAL ECONOMICS

In his famous "An Essay on the Principle of Population," the English cleric-turned-economist Thomas Malthus (fig. 12.8) wrote, "Population, when unchecked, increases in a geometric ratio. Subsistence increases only in an arithmetic ratio. A slight acquaintance with numbers will shew the immensity of the first power in comparison with the second. . . . In two centuries and a quarter, the population would be to the means of subsistence

Figure 12.8

Thomas Malthus, an English cleric and economist, whose work, *An Essay on the Principle of Population* proposed that human populations are limited by environmental constraints and resource scarcity.

as 512 to 10." Malthus predicted that populations would grow beyond their means of subsistence, and it is only "misery and vice" (which included, for Malthus, disease, famine, war, and pestilence) that keep numbers in check (Piel 1995).

Charles Darwin credited Malthus with inspiring his insights into the theory of natural selection. Yet despite his mathematical logic, Malthus's conclusions were generally disregarded because they appeared to be refuted by human experience. Human ingenuity and technology have, in fact, shown far more than arithmetic increase. World human population is now in its third doubling since Malthus published his "Essay," and has been sustained by more than five doublings of "the means of subsistence" (world per capita GDP). Even since 1950, industrial technology (overriding any arithmetic constraint) has twice doubled the output of material goods (Piel 1995). With this record of achievement and growth in both population and affluence, neoclassical economists could safely ignore Malthus's ideas and treat his essay as an historical footnote.

Not all economists, however, were prepared to dismiss Malthus's views on growth and its limits. One of the most influential modern economists to address the problem of environmental constraints was Nicolas Georgescu-Roegen. In his classic work, "The Entropy Law and the Economic Problem," Georgescu-Roegen argued that, "What goes into the economic process represents *valuable natural resources* and what is thrown out of it is *valueless waste*" (Georgescu-Roegen 1993:76, emphasis his). "Matter-energy," Georgescu-Roegen continued, "enters the economic process in a state of *low entropy* and comes out of it in a state of *high entropy*." Recall that entropy is a measure of the amount of disorder or unusable energy in a system; as entropy increases, the amount of energy available in the system for work decreases. To illustrate his point, Georgescu-Roegen asserted that, "a piece of coal can only be used once. And, in fact, the entropy law is the reason

why an engine (even a biological organism) ultimately wears out and must be replaced by a *new* one, which means an additional tapping of environmental low entropy" (Georgescu-Roegen 1993:80).

To better illustrate the relationship between economic processes and entropy, Georgescu-Roegen distinguished between *stocks* of free energy of the mineral deposits within the earth and *flows* of solar radiation intercepted by the earth. Whereas humans have almost complete command of the energy stocks, for all practical purposes, we have no control over the flow of solar radiation. To Georgescu-Roegen, the implications of this understanding of stocks and flows were profound: "There is an important asymmetry," he noted, "between our two sources of low entropy. The solar source is stock abundant, but flow limited. The terrestrial source is stock limited, but flow abundant (temporarily). Peasant societies lived off the solar flow; industrial societies have come to depend on enormous supplements from the unsustainable terrestrial stocks." Thus, to Georgescu-Roegen, the principal question was not, "How many people can the Earth support?" but rather, "How long can a population of any given size be maintained?" With a long-term view in mind, Georgescu-Roegen wrote, "Every time we produce a Cadillac, we irrevocably destroy an amount of low entropy that could be used for producing a plow or a spade. In other words, every time we produce a Cadillac, we do it at the cost of decreasing the number of human lives in the future" (Georgescu-Roegen 1993:85).

Georgescu-Roegen compared the earth's resources to an hourglass full of sand (fig. 12.9), in which the sand in the upper part of the hourglass represented the store of low-entropy resources. Its rate of movement into the bottom of the hourglass was its "flow," controlled by rates of solar inputs. As the amount of low-entropy resources diminished, the amount of high entropy waste (in the bottom of the hourglass) increased and accumulated. Some of the upper sand coalesced into clumps and might move through the neck of the hourglass all at once, analogous to fossil fuels that might accumulate large quantities of solar energy and then can be tapped at higher rates of flow. But regardless of the rate of flow, the sand in the upper half is destined to run out. And, unlike a real hourglass, this one cannot be turned over!

Geoergescu-Roegen's view of the economic process leads to radically different conclusions than those of traditional neoclassical economics. First, capital and resources are not substitutes, but complements. Agents of transformation cannot create the materials they transform or the materials out of which the agents themselves are made. Second, Georgescu-Roegen questioned how long economic growth, fueled by energy from stocks of nonrenewable resources, could last. Analytically, Georgescu-Roegen (1993:84–85) expressed it this way:

> Let S *denote the present stock of terrestrial low entropy and let* r *be some average annual amount of depletion. If we abstract (as we can safely do here) from the slow degradation of* S, *the theoretical maximum number of years until the complete exhaustion of that stock is* S/r. *This is also the number of years until the* industrial *phase in the evolution of mankind will forcibly come to its end. Given the*

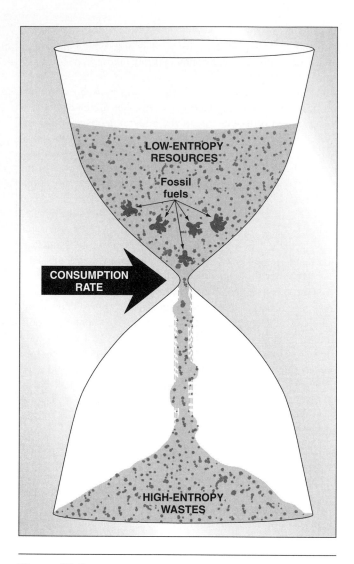

LOW-ENTROPY RESOURCES

Fossil fuels

CONSUMPTION RATE

HIGH-ENTROPY WASTES

Figure 12.9

The "hourglass analogy" of economist Nicolas Georgescu-Roegen illustrates the relationship between entropy and economics. The sand in the upper part of the hourglass represents earth's low-entropy resources. As humans consume these resources, high-entropy wastes are produced. Regardless of the consumption rate, the sand in the upper half is destined to run out.

fantastic disproportion between S and the flow of solar energy that reaches the globe annually, it is beyond question that, even with a very parsimonious use of S, the industrial phase of man's evolution will end long before the sun will cease to shine.

Characteristics of Environmental Economics

The views of Georgescu-Roegen and others led to new ways of thinking about interactions between environmental and economic processes and to the growth of environmental economics as a distinct economic paradigm. Unlike traditional growth economics, environmental economics asserts that manufactured capital cannot, in the long run, substitute for

natural capital in providing raw materials and energy, stock (nonrenewable) resources, flow (renewable) resources, a sink for wastes, and key life support systems, including water, air, climate regulation, food, and biodiversity. Thus, environmental economics saw the human economy not as a self-sufficient system that could draw material from or dump material into the environment without restraint, but as an environmentally dependent subsystem of human activity that would cease to function without environmental goods and services (fig. 12.10). In the past, the illusion of independence was created because the human economy was small relative to the biosphere, and sources of raw materials and sinks for wastes were relatively large (fig. 12.10a). But as the human economy has grown, the source and sink regions of the biosphere have diminished because of use and degradation, and so have their capacities to provide resources and absorb waste (fig. 12.10b). Thus, the human economy must increasingly make environmental constraints a more explicit consideration in producing goods and services and disposing of the waste that such production creates (Costanza et al. 1997).

Just as the environment provides economically valuable functions, it also imposes economic constraints in the form of:

1. Increasing capital costs of obtaining raw materials and energy when depletion occurs.
2. Increasing inputs required to produce each output from the same capital because of diminishing economic returns (e.g., more fertilizer and pesticides may be needed for each yield unit in agriculture).
3. Increasing demand for more effective and expensive pollution prevention and cleanup (expressed as higher input costs or higher government or household expenditures).

Although the alternatives to growth economics are diverse, common threads among environmental economists are that economic activity should be practiced on a sustainable scale, use methods and practices of fair distribution of economic goods and services, and provide for efficient allocation of resources. In addition, systems of environmental economics require:

1. A redefinition of "growth" and a differentiation among types of growth.
2. An explicit determination and measurement of environmental constraints on economic growth.
3. Definition of the functions of the environment and natural resources in economic systems.
4. The creation and organization of markets for environmental goods and services, combined with methodologies for valuing natural capital.
5. Alternative measures of human well-being.

Types of Growth

Environmental economics distinguishes three types of economic "growth": (1) growth of biophysical throughput, (2) growth in production or income, and (3) growth of human welfare. "Throughput" refers to flows of matter and energy from the first stage of production through consumption. Growth in throughput is probably the most environmentally harmful, because it is characterized by high rates of consump-

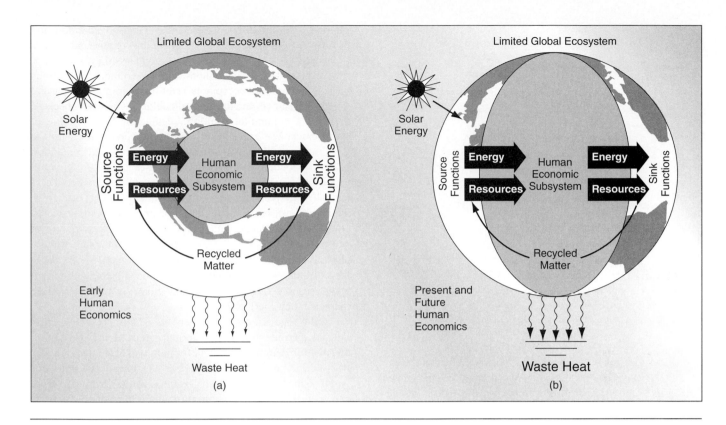

Figure 12.10

An alternative view of the relationship between the economy and the environment, as understood within the framework of environmental economics. The physical environment provides raw materials and energy, stock (nonrenewable) resources, flow (renewable) resources, a sink for wastes, and key life support systems including water, air, climate regulation, food, and biodiversity. However, it is limited globally. (a) In the past, the human economy was small relative to the biosphere, and sources of raw materials and sinks for wastes were relatively large. (b) As the human economy has grown, the source and sink regions of the biosphere have diminished because of use and degradation, and so have their capacities to provide resources and absorb waste. Thus, the human economy must increasingly make environmental constraints a more explicit consideration in producing goods and services and disposing of the waste that such production creates.

Adapted from Goodland, Daly, and El Serafy (1992) and Costanza et al. (1997).

tion of matter and energy, relatively low efficiency of energy use, low durability and rapid replacement of material artifacts, and increasing rates of waste production.

Growth in production or income has the potential to be more environmentally benign. If increased production is characterized by more durable goods generated with less energy per effort, it may put less stress on environmental systems in the long run. "Income" is technically defined as *the flow of service through a period of time that is yielded by capital* (Daly 1991), but increases in income do not necessarily represent more material consumption and accumulation. Rather, they represent increases in "services" delivered and satisfaction with those services. Thus, it is theoretically possible for income to increase while environmental degradation decreases.

Growth in human welfare is at once the most important and most difficult to measure of the growth indices. Traditional neoclassical economics has used per capita GNP or per capita GDP as its usual index, but environmental economists question whether either statistic measures human welfare. One alternative measure to per capita GDP is the Index of Sustainable Economic Welfare (ISEW), which integrates:

1. Income distribution (difference between the richest one-fifth of the population and the remaining lower four-fifths).
2. Net capital growth (measured as total net capital growth by adding increases to manufactured capital and subtracting the amount required to maintain the same per capita level).
3. Natural resource depletion and environmental damage (measured as the depreciation of natural capital by subtracting an estimate of the income lost to future generations through the depletion of exhaustible resources and adding estimates of environmental damage).
4. Unpaid household labor.

The ISEW also adds the value of expenditures on good streets and highways, public health, and education, and it subtracts defensive expenditures on health and education (e.g., trauma care and remedial reading programs), spending on national advertising (aimed at increasing demand), and the costs of urbanization, commuting, and auto accidents. The ISEW is expressed as a dollar value that can be compared with per capita GNP, the

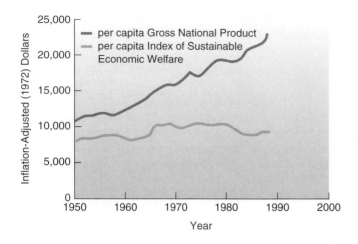

Figure 12.11

Changes in the U.S. Gross National Product (GNP) and Index of Sustainable Economic Welfare (ISEW) since 1950. Although the GNP has increased, the ISEW has failed to grow.

traditional measure of economic well-being. Whereas the per capita GNP and GDP have continually increased in the United States and other industrialized countries, the ISEW has decreased in recent years (fig. 12.11).

Methods for Valuing Environmental Goods and Services

A deficiency of neoclassical economics has been its failure to define markets for environmental good and services, but markets remain powerful and effective tools for determining value. Therefore, one of the challenges of environmental economics is to define and create markets for environmental entities that did not previously exist. Properly developed, environmental markets stimulate market efficiency and improve overall economic efficiency. Environmental markets are most effective when they operate under one or more of the following three principles:

1. *Polluter pays principle*—Economic strategies force polluters (rather than society) to pay for the pollution they create and prevent them from externalizing pollution costs.
2. *Precautionary principle*—If the environmental outcome of an economic activity is uncertain, one should err on the side of caution and place the burden of proof on the potential polluter to demonstrate that economic activity will not do irreversible harm.
3. *Polluter pays precautionary principle*—If uncertainty exists regarding the environmental effect of a proposed economic activity, make the polluter pay in advance for the potential costs of environmental remediation and restoration, with the investment to be returned if no pollution occurs.

Government Regulation

Government regulation is not, in the strict sense, a market mechanism for valuing the environment. Rather, it is an attempt to correct market failures when markets do not value the envi-

ronment appropriately or efficiently. Through coercion and mandatory regulations, the government may require individuals or businesses to meet environmental standards they would otherwise disregard as uneconomical. Government regulations provide a minimum acceptable level of environmental protection for consumers and the general public, but precisely because they are a nonmarket solution, regulations have significant weaknesses. For example, the burden of proof lies with the government to demonstrate that an environmental regulation has been violated, and proving such cases can be difficult and expensive. Thus, regulation encourages a "cops and robbers" mentality between regulators and potential polluters, often producing at best, reluctantly minimal compliance and, at worst, covert noncompliance. Enforcing regulations effectively is expensive. Enforcing regulations ineffectively, although it costs less money, generates high levels of noncompliance, and defeats the purpose of the regulation. Additionally, regulation typically ensures that environmental concerns remain outside market and business culture.

But regulation also has benefits. The impetus of regulation has spurred innovation in pollution-reduction technologies, effectively creating a "market" for pollution abatement devices where none previously existed. For example, regulations imposed on the U.S. auto industry by the Clean Air Act led to rapid and radical developments in automobile technology, such as catylitic converters, unleaded gasoline, more fuel-efficient engines, hybrid autos, and, eventually, electric and hydrogen-powered cars.

Regulation has a strong, and mostly favorable, historical precedent. Most of the early, landmark environmental legislation in the United States, and later in other countries, initially took the form of mandatory regulations imposed by the government. The National Environmental Policy Act, the Clean Water Act, the Clean Air Act, and the Endangered Species Act all derive their primary authority from the ability to regulate individual and corporate behavior according to objective, preset standards, and to impose fines or imprisonment for failure to comply. Interestingly, however, these acts have been successful mostly because of their requirements for information, rather than their threats of penalties. All require full disclosure of facts regarding individuals or businesses engaged in activities covered by legislation, so that the public, the press, professionals, the courts, and legislatures can evaluate them. In a society with a free press and a well-informed public, public access to information, and the public consequences of it, has probably done more to create compliance with environmental laws than the threat of penalty (Quinn and Quinn 2000).

To avoid some of the problems associated with mandated programs and regulations, governments may adopt and promote regulation through voluntary programs that provide information, technical assistance, and cash and material subsidies to encourage compliance. Voluntary programs reduce or eliminate the costs of enforcement and tend to create more favorable relations between government regulators and private enterprise. The major disadvantage of voluntary programs is that they may require heavier tax burdens to operate and thus result in *deadweight social losses* or distortions in economic efficiency and market function caused by the diversion of earned income to the government through taxes. For example, increased taxes on income usually lead to reduced demand for consumer goods and reduced production of such goods. In this case, the reduction in demand is not real, but is rather a "deadweight loss" imposed on economic activity by the government through taxation. Nevertheless, voluntary programs are economically more efficient than mandatory ones if (and only if) the tax revenues needed to support the voluntary program are low *and* the costs of government services relative to the private cost of the same services are low, *or* if the voluntary program costs less than the mandatory program, or both (Wu and Babcock 1999). In general, voluntary programs are most cost-efficient when the number of individuals or businesses involved is large relative to the total population, and the government services provide nonrival public goods such as information or technical assistance.

Voluntary and mandatory programs to address the same environmental concern are not necessarily mutually exclusive. Complementary voluntary programs and enforced mandatory programs can achieve high compliance and efficient economic results. In such a carrot-and-stick approach, the government provides subsidies, information, technical assistance, and material capital to those who comply, while punishing the noncompliant with fines, confiscation of property or capital, or even imprisonment.

Taxation

The first action by governments to curb pollution and protect the environment usually takes the form of mandatory regulations, followed soon after by voluntary programs. Taxation is normally the third step, although it sometimes is introduced concurrently with voluntary programs. Taxation imposed on undesirable activities, a more marketlike strategy for environmental protection, follows the "polluter pays principle" and can be one of the best ways to correct market failure, especially when used to force a producer of pollution to assume the burden of otherwise external costs. Taxation is an especially appropriate and effective method for controlling pollutants that are widely dispersed because the government is the institution best equipped to address the large areas and numbers of people affected. For example, if the government taxes a power plant on a per unit basis for the hydrogen sulfide emissions produced in generating electricity, it accomplishes two things at once. First, the government is effectively reimbursing itself for the social costs of air pollution, including increased costs of health care (because of respiratory diseases caused or aggravated by the pollution), increased costs of property damage to the government's own public buildings, structures, and lands (from the effects of acid rain), and increased costs of preserving species, habitats, and ecosystems that may decline because of pollution. Second, the government creates an incentive for the pollution producer to reduce pollution on its own, because every unit reduction in pollution lowers cost and increases the margin of profit. Two indirect benefits often result from the second effect. First, pollution control becomes part of the intrinsic "organizational culture." Second, if the pollutants are material in nature and are removed by the producer prior to emission, taxation may open up new markets for the pollutants to be used in beneficial ways. For example, "scrubbers" in smokestacks remove pollutants created by coal burning and thereby accumulate a "sludge." Although harmful as an air pollutant, the sludge is high in sulfur and can, if appropriately applied, be used as a fertilizer to supply an important plant nutrient. The Tennessee Valley Authority, one of the nation's largest producers of electrical power, now makes from 6 to 10 million dollars annually by selling the sludge gathered from its scrubbers. The Indianapolis Power and Light Company is even planning to adjust its operating conditions to produce higher-quality sludge (Hoffman 2000).

Environmental Property Rights

One approach to creation of property rights for the environment is to sell "pollution rights" on a per unit basis. Like taxation, provision of environmental property rights is a manifestation of the "polluter pays principle" through which the government sells "rights to pollute" on a per unit basis rather than taxing each unit produced. Typically, the government establishes some absolute standard for operation (e.g., a maximum number of pollutants allowed or a minimum level of environmental quality), outside of which producers are not allowed to operate. Within the boundaries of the standard, any producer must pay, in advance, for each unit of pollution emitted. In effect, this method tacitly asserts that an ecological system, such as the atmosphere, has value through the services it delivers. To degrade ecological services through pollution represents a social cost, for which the polluter must pay. Before purchasing pollution rights, however, the polluter receives a reward, in the form of a cost reduction, if he or she successfully reduces pollution and therefore is able to operate with a reduced outlay of expense for pollution rights. An additional incentive is the transferability of pollution rights; that is, if a polluter does not "use up" all the pollution rights (because of increased efficiency and cleaner production processes), they can be sold to another polluter for their market value. Cleaner producers gain an economic advantage, and "dirty" polluters must pay more up front or buy more rights to pollute from their competitors.

Tradable permits force regulators to determine an absolute maximum acceptable level of aggregate pollution or depletion that is ecologically sustainable. Once that limit has been determined, regulators must distribute the rights to pollute in some fair manner so that the market can attain efficient allocation of permits through trading. Pollution rights can operate in a free market, but only after ecological and political boundaries have been clearly established (Daly 1999).

Insurance Against Environmental Damage

Governments and private citizens can require persons or businesses that contract with them for goods and services to provide proof of insurance against environmental loss or degradation. This approach is based on the "precautionary principle" and works very much like ordinary insurance approaches, except that it is applied to the environment. For example, such insurance might stipulate that if the logging practices of a timber company cause sedimentation above specified levels in surrounding streams, the cost of rehabilitation will be paid by the insurance company. As in the case of car insurance, the premium paid by the individual or business to the insurance company will depend on their environmental record. Persons and businesses with records of environmental abuse would pay high premiums because they represent high risk to the insurance company, whereas policyholders with records of environmental protection would represent low risks and pay correspondingly low premiums.

Although environmental insurance is still not widely used, its applications are increasing. In addition to traditional forms of insurance against environmental damage, many larger banks and other large lending institutions are increasingly careful to inspect the environmental record of a loan applicant especially for loans on projects with potential environmental effects. Applicants with poor environmental records increasingly are considered bad investment risks (Costanza et al. 1997).

A variation on traditional insurance policies against environmental damage is a practice called *environmental insurance bonding,* or alternatively, "flexible assurance bonding" (Costanza et al. 1997), which is based on what might be called the "polluter pays precautionary principle." An individual performing work that has the potential for environmental harm puts up, in advance, a bond equal in value to the cost of repairing such harm, should it occur. The party for whom the work is done places the bond in an interest-bearing account where it remains until the work is completed. If the party performing the work keeps environmental damage and costs within previously specified limits (i.e., performs as well or better than expected in terms of environmental damage), the value of the bond, plus some of the interest, is returned. If not, the money is used to repair the damage to the environment, if possible.

Another expression of an insurance-based approach to environmental protection is manifested in habitat conservation plans (HCPs), previously discussed in chapter 2. Recall that in an HCP, a landowner agrees not to use some land commercially for the protection of an endangered species using the property. In return, the landowner receives the equivalent of a long-term contract (the "no-surprises" agreement) that guarantees (insures) against any additional future restrictions and regulations by the government. The typical length of the agreement is 100 years (Quinn and Quinn 2000).

Note an important shift of focus illustrated by these various forms of environmental insurance. Whereas government regulations place the burden of proof on the public and the government (i.e., the polluter is presumed innocent unless a violation is documented), environmental insurance and environmental insurance bonding place the burden of proof on potential polluters. Now such potential polluters must make a financial pledge asserting that they will *not* pollute, and forfeit the value of their pledge if they fail to live up to prescribed agreements.

Empowering Stakeholder Interests

Another economic strategy that government can adopt to protect the environment is to empower the interests of private stakeholders against polluters. For example, government agencies and their officials may use their resources to arrange meetings among stakeholders with diverse environmental interests, and may, in some cases, serve as mediators or arbitrators of those interests. The government may serve as a clearinghouse of information for stakeholders, or provide legal counsel and advice, or funding to obtain it, to private stakeholders when they are opposed by larger corporate stakeholders.

The strategy of empowering stakeholder interests can accomplish things by itself, but it is often most effective when it operates under the shadow of government regulation. That is, the government may encourage and empower private stakeholders to reach agreements about environmental protection on their own, but simultaneously inform the stakeholders that, if they are unable to reach an agreement among themselves, the government will step in and impose an agreement on them. This threat of government intervention is often a powerful incentive for cooperation and negotiation among otherwise adversarial interests. For example, in 1995, an agreement was reached among private stakeholders for a 30-year effort to restore water quality and fisheries in San Francisco Bay. This agreement was signed into effect by the parties only hours before a deadline set by the U.S. Environmental Protection Agency, after which the EPA would have imposed its own water-quality plan under the authority of the Clean Water Act.

The above strategies offer general approaches for valuing environmental goods and services, but we still have not offered an example of a method that could be used to address a particular issue in a specific context. In the following case history, we examine how one method of valuation, travel cost valuation, can answer a particular question pertinent to conservation: How do we estimate the value of admission to a national park?

A Case History of Place and Method—Using Travel Cost Analysis to Estimate the Value of a Costa Rican National Park

Travel Cost Method

The travel cost method (TCM) is an economic method of determining environmental values that belongs to a larger family of behavior-based, analytical economic techniques called *revealed preference methods.* TCM works on a simple but reasonable assumption: the more valuable an environmental resource or amenity, the farther people are willing to travel to experience it, the more they are willing to spend per trip, and the more trips they are willing to make. Such variables can be integrated, at least in part, through a TCM demand curve such as that shown in figure 12.12. Here the cost per trip (a reflection of distance, *y* axis) is related to the number of trips a person makes to the site that contains the amenity or resource (*x* axis), such as a national park or a hunting area. Note that as the cost of the trip rises, the number of trips decreases. TCM also can be effective if it is used in a "before" and "after" approach to a site, such as a comparison of visitation before and after an environmental improvement (e.g., creating a lake for fishing).

In neoclassical economics, the supply and demand curve relates price to quantity demanded; in a TCM curve, travel costs are the analog of price and number of trips the analog of quantity demanded. If a person is willing to make five trips, each costing $20, then the person's willingness to pay is 5 × $20 or $100. This willingness to pay is also referred to as *consumer surplus.* Aggregate consumer surplus is calculated as the area under the TCM demand curve (Loomis 2000). If we want to compare two different areas, and we suspect that one area is of higher environmental quality, we could calculate the TCM demand curves for both areas. If our hypothesis is correct, the

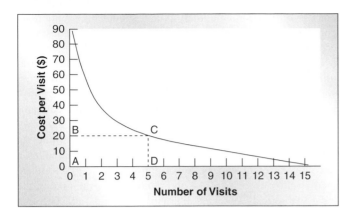

Figure 12.12

A travel cost method (TCM) demand curve that estimates the value of an environmental amenity, such as a national park. Unlike a neoclassical supply and demand curve, the TCM demand curve makes travel costs (y axis) serve as the analog of price, and number of trips to the park (x axis) the analog of demand. Willingness to pay for the environmental amenity or service is represented by area ABCD and its value is the product of travel costs and number of trips. In this example, if it costs $20 for a trip to the park, five trips will be taken, so $100 represents the value of the park to the individual.

TCM demand curve from the better environmental area should lie to the right of the other.

In addition to its usefulness in estimating values of consumer surplus, values of environmental improvements, and relative values of different areas, TCM also can be useful to managers in setting prices for entry fees to parks and reserves. The Costa Rican National Park Service learned, through TCM analysis, that a common entrance fee was not the most economically efficient method of raising revenue because tourist demand differed among parks (fig. 12.13) (Chase et al. 1998). Not only did the demand curves show different *y* intercepts and slopes, but they also had different shapes. The Costa Rican National Park Service could potentially use such data in four alternative ways, depending upon their objective:

Objective One: Maximize revenue generated from entrance fees and estimate total maximum revenue from the three parks.

This objective can be achieved by determining the maximum product of *x* (number of visitor days) and *y* (entrance fee). For example, if the managers charge a $15 daily fee at Manuel Antonio Park, they can expect approximately 4,000 visitor days, or a revenue of $60,000. If, on the other hand, they charge an entrance fee of $22.50, visitor days are reduced to approximately 3,000, but revenue increases to $67,500.

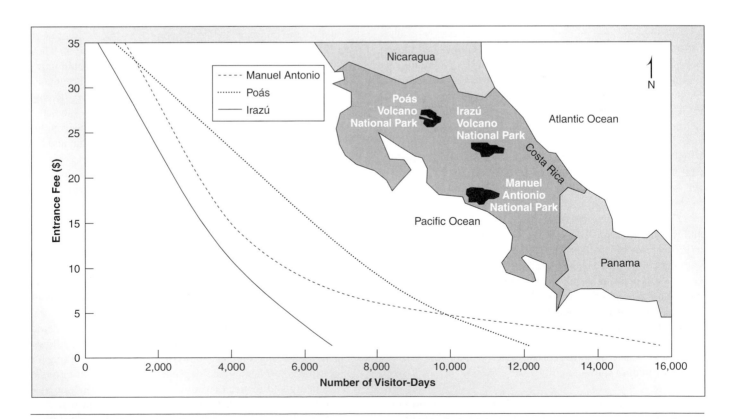

Figure 12.13

Demand curves associated with three Costa Rican national parks that relate entrance fees to visitor days. Such demand curves, properly estimated, can be used to answer questions about various economic and ecological strategies for the parks, including how much to charge for admission to achieve an optimal level of visitation.

After Chase et al. (1998).

Objective Two: Minimize financial cost to the parks.

Perhaps the National Park Service has a limited budget and insists that parks be self-supporting through the collection of daily fees. If we assume that costs increase with increasing numbers of visitors, then it should be possible to determine a "supply curve" that relates visitor days to park costs. Where park costs per day intersect the demand curve of the daily entrance fee, supply equals demand. If the entrance fee is set below this point, visitor days increase as costs increase, but revenue decreases, creating a deficit. If the fee is set at a higher level, revenues may exceed cost, but the park is "underused" in terms of services that could be provided. Of course, park costs may be a constant; that is, perhaps it costs just as much to manage the park whether any visitors come or not. If this is the case, then the "supply" curve is simply a horizontal line. But it is still valuable to determine the point at which it intersects the demand curve of the entrance fee–visitor day relationship because that point represents the *minimum* entrance fee that must be charged to recover costs of operation.

Objective Three: Minimize environmental cost to the parks.

As noted in chapter 9 and the case history of the Bonaire Marine Reserve, increased visitation almost certainly will increase environmental degradation to a park. If the Park Service can determine a maximum acceptable threshold of such damage, beyond which further harm would degrade or destroy the park's value and purpose, they can theoretically determine the threshold number of visitor days allowable, and set revenues to create this level of demand. Using this strategy, price is actually used as a tool to control or limit demand, and through such control to limit degradation to the environment. By limiting degradation, park managers may enhance persistence and diversity of populations that can continue to reside within park boundaries.

Objective Four: Ensure that Costa Rican nationals are not "priced out" of their own parks by wealthier foreign tourists.

Here the demand curve can be used to address an issue of access and social equity. An entrance fee of $17 per day may increase revenues compared with one of $5, and European or North American tourists may be willing to pay it. Average-income Costa Ricans however, may not be able to visit their own park! If the Park Service considers that such a condition represents an injustice to its own citizens, it would choose a fee appropriate to average national income. Using its demand curve, it would then be able to predict expected visitor days and make its management plans accordingly. In fact, many countries do take such income discrepancies into account, and accordingly charge a lower entrance fee for their own citizens than for foreign visitors.

Optimizing outcomes for these four objectives illustrates the usefulness of the demand curve to achieve an economically efficient solution. This example also illustrates the inadequacy of economic data by themselves to achieve anything "good" unless we first determine which "good" we wish to achieve. Thus, moral choice—the deliberate and predetermined objective to reach the highest and best outcome—is an essential element in our use of an economic analysis and should be intentionally considered and scrutinized when we consider the relationship of economic activity and conservation.

The previous methodologies assume the validity of traditional assumptions of neoclassical economics. However, some environmental economists believe that these assumptions are fundamentally flawed, and therefore such techniques are of little or no value in achieving conservation goals or establishing a correct long-term relation between economic activity and the environment. Rather, what is needed is a radically different way of looking at economic systems and their relation to the physical and biological world. One of the most theoretically explicit alternative systems, and one that speaks most directly to the issues of conservation biology, is the paradigm of steady-state economics.

STEADY-STATE ECONOMICS

Foundational Concepts

The eighteenth-century moral philosopher and classical economist John Stuart Mill was perhaps the first advocate of an economic steady state, rather than a constant state of growth. Mill wrote that "a stationary condition of capital and population implies no stationary state of human improvement," and that, in a stationary (nongrowth) state, people would be more likely and better able to improve "the art of living . . . when minds ceased to be engrossed by the art of getting on" (quoted in Daly 1996:3). Mill envisioned a state in which people would enjoy the fruits of their savings once they had attained an acceptable standard of living. Mill did not believe that a continual increase in material prosperity was possible or "natural," in the sense that nonliving systems tended toward equilibrium; however, his ideas were forgotten, and growth became both the goal and the paradigm of modern neoclassical economics.

In the late 1960s, the economist Herman E. Daly put forward a systematic economic theory of the steady state. Daly (1991:17) defined a **steady-state economy** as one *with constant stocks of people and artifacts, maintained at some desired, sufficient levels by low rates of maintenance "throughput" (i.e., by the lowest feasible flows of matter and energy from the first stage of production [depletion of low-entropy materials from the environment] to the last stage of consumption [pollution of the environment with high entropy wastes and exotic materials]).*

Three variables (what Daly calls "magnitudes") determine the condition of a steady-state economy: stocks, flows, and services. *Stocks* represent physical levels of capital (people and artifacts). *Flows* are rate-dependent movements of *throughput* (energy and matter that move through the economic system). *Services* are defined as "want satisfaction" provided by the stocks, with satisfaction being measured as a "psychic magnitude," a level or sense of well-being (fig. 12.14). The value of both the stocks and the throughput is determined by the value attributed to the satisfaction of wants that the stocks provide. The purpose of managing levels of stocks is to achieve what Daly calls *satisficing*, or choosing some level of stocks "that is

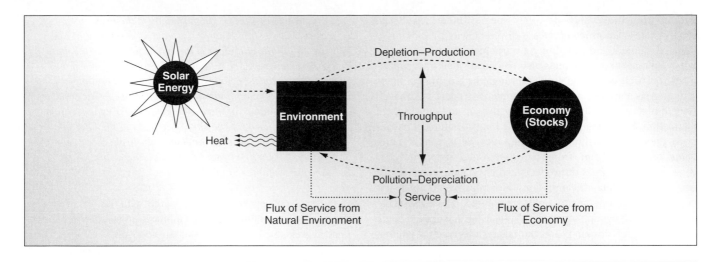

Figure 12.14

A schematic depiction of key variables and relations in a steady-state economy. Stocks of people and artifacts (goods) are maintained at some constant, desired, and sufficient level through low rates of maintenance "throughput" (i.e., by the lowest feasible flows of matter and energy from the first stage of production [depletion of low-entropy materials from the environment] to the last stage of consumption [pollution of the environment with high-entropy wastes and exotic materials]).

Adapted from Daly (1977).

sufficient for a good life and sustainable for a long future" (Daly 1991). Whereas traditional growth economics attempts to maximize throughput, steady-state economics seeks to minimize it. Whereas growth economics seeks to increase consumption (and, of necessity, depletion) of stocks, steady-state economics seeks to maintain a constant level of stocks at high quality and high productivity. Within the constraints of these constant stocks, service is to be maximized. Daly (1991:16) summarized these concepts this way:

> *In steady-state economics, two things do not grow (two magnitudes are held constant): the population of human bodies and the population of artifacts (stock of physical wealth). The cultural, genetic inheritance, knowledge, goodness, ethical codes and so forth embodied in human beings are not held constant. Likewise, the embodied technology, the design, and the product mix of the aggregate total stock of artifacts are not held constant. Nor is the current distribution of artifacts among the population taken as constant. If we use growth to mean quantitative change and development to mean qualitative change, then we can say that a steady-state economy develops but does not grow.*

These ideas are related more systematically in what Daly calls the "key identity" of steady-state economics:

$$\frac{\text{Service}}{\text{Throughput}} = \frac{\text{Service}}{\text{Stock}} \times \frac{\text{Stock}}{\text{Throughput}}$$

Although steady-state economics does not aspire to "growth," which is defined in this context as increases in throughput, it does aspire to "development," or increases in services. In the above identity, development means increasing the value of the second and third ratios. Increasing the second ratio means that, for every unit of stock used, a greater quantity of service is rendered. Thus, steady-state economics focuses on increasingly

efficient use of stock and increasing the quality of the stock itself. Raising the third ratio means decreasing the level of throughput associated with the production or use of each unit of stock. In other words, steady-state economics seeks to increase efficiency of production and productivity of human capital. Thus, in defining development as an increase in services (satisfaction), Daly (1991:36) noted, "The steady-state economy would force an end to pure growth but would not curtail, and would in fact, stimulate, development."

In the steady-state paradigm, emphasis and attention are shifted from throughput to stocks, specifically to the quality of stocks and the distribution of stock ownership. Unlike growth economics, which advocates further growth as the answer to poverty, steady-state economics advocates redistribution of resources as the answer to poverty.

The Steady-State Economy and the Environment

A steady-state economy sees economic activity as "an open subsystem of a finite and nongrowing ecosystem (the environment)" (Daly 1993). The economy operates by importing low-entropy matter-energy (raw materials) and exporting high-entropy matter-energy (waste). Thus, as a subsystem of the larger ecosystem, the economy must itself at some point also become nongrowing.

Recall that, within traditional economic thinking, the environment was either invisible (fig. 12.3) or a subsystem of the human economy (fig. 12.4). Steady-state economics, like environmental economics generally, explicitly incorporates the environment and views the economy as a subsystem of it (fig. 12.14). The optimal scale of the economy decreases as the degree of complementarity between natural and manufactured capital increases, as our desire for the direct experience of nature increases, and as our estimate of both the intrinsic and instrumental value of other species increases. Thus, steady-state economics creates a different way of looking at the environment and of valuing environmental entities.

Steady-state economics also recognizes that the environment constricts economic growth through the biophysical limits of finiteness, entropy, and ecological interdependence. Although these limits can and do act alone, they also create limits through interaction. Recycling pushes back the limits of finitude, but entropy prevents everything from being recyclable, just as it prevents the same item from being recycled indefinitely. If environmental sources of products and environmental sinks for wastes were infinite, the limitations of entropy would not be important, but environmental sinks are limited. The atmosphere can be polluted beyond breathability; fresh water can contain too much waste for drinking; and the oceans can reach a point at which they can no longer assimilate or absorb certain kinds of toxins. The fact that both environmental sources and sinks are finite, combined with the reality of entropy, means that "the ordered structures of the economic subsystem are maintained at the expense of creating a more-than-offsetting amount of disorder in the rest of the system" (Daly 1996:33).

Ethical Implications of Steady-State Economics

In traditional neoclassical economics, ethical decisions about what to buy, and how much of it, rest entirely with the individual. This ethical sovereignty of persons is not shared by steady-state economics. Although steady-state economics recognizes the reality and importance of individual choice, it also recognizes that:

1. The ecosystem is the ultimate source of all products and services.
2. Individuals exist in communities, and community values must, in some cases, supersede individual values if the community is to persist.
3. Ultimate ends, that is, a person's greatest and highest good, may not be achieved, and may even be harmed, by a constant fixation with and accumulation of material to supply intermediate ends. Therefore the intermediate ends of material wants, although important and real, are superseded by ultimate ends to develop virtue and character that distinguish the highest and best condition of being human (Daly 1996).

As noted earlier, classical economists addressed these sorts of questions naturally and without apology. Thus, they distinguished between the states of "wealth" and "illth" (Smith 1993). Wealth was, in a very literal sense, the possession of things that made one well, or whole. In contrast, "illth" was a state of possessing things that, literally, made one ill or did one harm. In this view, wealth was not simply measured as the sum of one's possessions, but as one's capacity to enjoy and use them. Thus, to a man who knew how to play a flute, the possession of a flute was a measure of wealth. But to the man who could not, possessing the flute was worthless. Further, some possessions, or an excess of possessions, was not considered to be merely without worth, but to be positively harmful. Economic historian G. A. Smith clarified the problem using a medical analogy.

The practice of medicine may require the prescription of an addictive stimulant for the sake of good health. The amount of the stimulant is finite and limited by the end. When, however, one takes a stimulant for its own sake, the desire for it becomes infinite since it is no longer limited by a final goal but is an end in itself. The same is true of the output of the economic process which, rather than being used for the sake of achieving the final goal of life, tends to become the final goal itself. Since output is then not limited by any final goal, the desire for it becomes infinite (Smith 1993).

The concept of illth disappeared with the rise of modern neoclassical economics and its emphasis on personal preference and aggregate growth. The loss of this concept was indicative of a greater loss of any sense of moral purpose to economic enterprise, or the moral outcomes of the unrestricted pursuit of wealth. Modern economists, like other modern scientists, are embarrassed when pressed on questions of whether a state of constantly increasing wealth is really good for a person's character or additive to personal happiness. But these questions must be asked because their answers strike at the heart of whether "conservation" of natural resources is possible, or even desirable.

Neoclassical economics assumes that consumer preferences are fixed and that the goal of economics is to satisfy them. All sovereignty of choice, market and moral, rests with the individual as a consumer. However, by the end of the twentieth century, some economists had begun to argue that communities were essential to define the social good, adapt the social order, and manage environmental systems (Costanza et al. 1997). The alternative offered by environmental economics generally, and by steady-state economics particularly, is to first come to a social consensus about what is good, right, fair, and just. This is the concept of "community sovereignty" (Costanza et al. 1997), the alternative to consumer sovereignty.

Steady-state economics raises an important question that growth economics does not address. Namely, is there some point beyond which further accumulation of physical artifacts is useless or even harmful in achieving our best and highest purposes? In other words, are the growth of material wealth and the satisfaction of wants ultimate ends or intermediate ends? As Herman Daly put it, "Could it be that one of our wants is to be free from the tyranny of infinite wants?" (Daly 1991:21).

Growth economics posits that, in the aggregate, physical wants, those satisfied by the accumulation of wealth, are insatiable. Steady-state economics proposes that physical wants can be satisfied. If physical wants cannot be satisfied, then ecosystems and their products and services must continue to produce more (if renewable) or be further depleted (if nonrenewable) and conservation efforts must work with whatever is not needed to satisfy human desire. But if physical wants can be satisfied, then humans can direct their efforts toward something other than satisfying their own wants, use of resources can be set at fixed and sustainable levels, and conservation efforts can be pursued as part of a larger good, the vision of a sustainable planet that provides for a diversity of life within an array of complex ecosystems.

Steady-state economics provides not only an alternative economic paradigm, but renders the service of raising ethical issues that speak to our core beliefs and assumptions about our economic choices, issues that are ignored in traditional neoclassical economics. The outcome of an ecological view of economics, of which steady-state economics is one possibility, is a state of what has been called **sustainable development.**

SUSTAINABLE DEVELOPMENT

Definitions and Characteristics

Sustainable development has been described as a term that "everyone likes, but nobody is sure of what it means" (Daly 1996:1). According to the most common and widely used definition of sustainable development (contained in the 1987 Bruntland Commission Report), development is sustainable if it "meets the needs of the present without compromising the ability of future generations to meet their own needs." Other definitions include, "living within environmental constraints of absorptive and regenerative capacities" (Costanza et al. 1997:173), promoting "values that encourage consumption standards that are within the bounds of the ecologically possible and to which all can reasonably aspire" (Stefanovic 2000:5), and "practices that simultaneously create economic vitality, environmental stewardship, and social equity" (Audirac 1997, quoted in Weinberg 2000).

Vague definitions may be politically advantageous if they facilitate the achievement of consensus. However, if definitions do not mature in clarity, they become not the basis of agreement but the breeding grounds for disagreement. As Herman Daly put it, "Acceptance of a largely undefined term sets the stage for a situation where whoever can pin his or her definition to the term will automatically win a large political battle for influence over our future" (Daly 1996:2). As recently as 2000, economist A. J. Hoffman lamented that "no practical definition of sustainable development yet exists" (Hoffman 2000). Further, if the goals of sustainable development are no clearer than its definition, there is little hope that the goals can be met. Thus, creating an operational definition of sustainable development is vital.

Recall the distinction made in steady-state economics between "growth" and "development": growth refers to a constantly increasing rate of flow of matter and energy through the economic system, whereas development refers to an increase in services or satisfaction received. Thus, development can be considered realistically sustainable because dimensions of efficiency, quality, durability, service, and satisfaction can be constantly improved. "Growth," on the other hand, cannot be considered sustainable because physical resources and environmental sinks are finite and energy is constantly degraded. Thus, growth is defined by increasing use of matter and energy and increasing accumulation of physical artifacts. Development, on the other hand, is associated with increases in efficiency of energy use, increased quality and durability of capital and consumer goods, and increased qual-

ity of services rendered with increased capacity to enjoy and appreciate the services.

Some economists distinguish between "weak" and "strong" sustainable development. "Weak" sustainable development treats natural and manufactured capital as substitutes, and therefore assumes that sustainable growth is possible because, through substitution, resources are not actually limiting. "Strong" sustainable development treats natural and manufactured capital as complements, and assumes that the correct economic strategy is to enhance the limiting factor. If a renewable resource is limiting, a strong sustainable strategy assumes that the proper response is to lower the production and use of manufactured capital and make investments to increase the stock of natural capital. In a contemporary context, sustainable development emphasizes economic prosperity, environmental quality, and social equity, the so-called triple bottom line of sustainable development (Hoffman 2000).

Resource economist Lester Brown noted eight characteristics of a sustainable economy that would encourage sustainable development:

1. Human births and deaths are in balance.
2. Soil erosion does not exceed the natural rate of new soil formation.
3. Tree cutting does not exceed tree planting.
4. The fish catch does not exceed the sustainable yield of fisheries.
5. The number of cattle on a range does not exceed the range's carrying capacity.
6. Water pumping does not exceed aquifer recharge.
7. Carbon emissions and carbon fixation are in balance.
8. The number of plant and animal species lost does not exceed the rate at which new species evolve (Brown 1996).

Why is it so difficult to achieve these conditions? More specifically, what patterns of social behavior prevent the achievement of goals that seem so reasonable and good? To understand the problem more clearly, we examine the phenomenon of the "social trap."

Sustainable Development and Social Traps

One of the principal barriers to sustainable development is the existence of so-called social traps that reinforce unsustainable behaviors. By definition, a social trap is said to occur when local, individual incentives lead to behaviors that are not consistent with the overall goals or good of the system (Costanza et al. 1997). Smoking cigarettes is a social trap that effectively ensnares people by temporarily separating benefits (increased social acceptance or status, pleasure, and stimulation) and costs (addiction to nicotine, heart and lung disease, increased health costs). In conservation, social traps are often related to system sustainability because practices that yield the largest short-term benefits are often those that can most degrade the system. The ultimate environmental social trap is Hardin's (1968) "Tragedy of the Commons," in which rational economic behavior by individuals (adding an additional animal to the pasture) leads inexorably to social

Figure 12.15

A symbolic representation of a social trap (represented by a road) and four methods of avoiding it. In alternative (a), travelers are warned (educated) about the dangers of the wrong road before they begin to follow it. In (b), travelers take out an insurance policy in advance in case of a wreck on the ill-advised road. In (c), a superordinate authority, such as the government, forbids the use of the road by closing it. Finally, in (d), the trap is turned into a trade-off in which travelers choosing the high-risk road pay a toll that reflects the potential cost of using the road.

After Costanza (1987).

ruin (the destruction and loss of the pasture for everyone). As economist Robert Costanza expressed it, "We go through life making decisions about which path to take based largely on 'road signs,' the short-run, local reinforcements that we perceive most directly" (Costanza 1987).

There are four primary ways to avoid social traps that prevent or delay the achievement of sustainability, and each one involves altering the "road signs" in some way (fig. 12.15). One of the most effective is education, which changes the road sign to include a warning about the long-term outcome of the unsustainable behavior. If farmers know that current rates of local soil erosion will lead to significantly lower yields within 20 years *and* are educated about the means to reduce soil erosion in the present, they are more likely to take such steps now to avoid the long-term costs. However, education itself has a cost. It often takes many years of educational input before there is evidence of any output in changed behavior. Education is expensive, and its cost must be borne by the government or the individual. Education requires a high degree of long-term commitment to receive the promised benefits. Thus, although education is an excellent long-term solution, it is not an approach that can offer immediate relief to urgent environmental crises.

A second approach to avoiding social traps is obtaining insurance. This approach is analogous to taking out an insurance policy before going down the road, just in case the road proves so bad that you wreck the car! For example, a national forest might set aside a portion of the forest as off-limits to oil drilling, even if experimental studies show that certain methods of oil drilling do not cause significant negative environmental effects. If the study proves misleading, the entire system will not be affected by incorrect information. As noted earlier, insurance can discourage risky, unsustainable behavior

because high-risk behavior typically carries a higher premium, but having insurance can also create a false sense of security. Some things simply cannot be recompensed if lost, no matter how large the payoff. No amount of money can restore the benefits of radiation protection provided by the ozone layer if it is destroyed, and no fund, no matter how large, can buy back an endangered species that is driven to extinction. Insurance is effective when the system is actually *insurable* against loss, and *replaceable* by what the insurance provides. But insurance offers no protection against unsustainable behavior that destroys an irreplaceable system or resource.

Superordinate authority is a third means of avoiding social traps. An authority, such as the government, may have sufficient perspective, resources, and expertise to require behavior that is detrimental to some individuals, but will achieve the common, social good. In our road sign analogy, invoking superordinate authority is equivalent to placing a "road closed" sign on a path that you simply refuse to let people take. Examples include some provisions of the Endangered Species Act that prevent landowners from modifying their lands in ways harmful to endangered species on their property. Such restraint is not in the landowner's best interest, but government authority overrules personal interest to protect a perceived interest of the community or nation at large. Similarly, religious traditions or institutions, shared family or community values, or cultural norms can act as superordinate authorities. If the authority is government, there will be considerable public expense to enforcing the rules (hiring the police force). Other authorities, such as religious traditions, incur no such expense and may enjoy more support, but their force is reduced as communities become more diverse.

A fourth approach to avoiding social traps is to convert the trap into a trade-off. This means that individuals must accept, ahead of time, that choosing an unsustainable behavior carries a cost, and the cost must be paid in advance. This approach is equivalent to placing an additional sign on the road to unsustainable behavior that says, "Stop ahead. Pay toll." A common manifestation of the trade-off approach is to tax unsustainable behaviors, and the trade-off is most clear when the tax revenues are spent to undo the damage caused by the behavior. Taxes on pollution are an example.

POINTS OF ENGAGEMENT—QUESTION 2

In many industrialized nations, farmers cultivating land immediately adjacent to urban centers may be offered large sums of money for their land by real estate developers anticipating growth in suburban populations. If the farmland is sold, the individual short-term outcome is a large profit to the farmer realized in a lump sum payment far greater than the net worth of the crops produced. Long-term, aggregate outcomes usually include increased distances between agricultural production and urban centers, increasing suburban sprawl, and net loss of individual family farms in the community or region. Does this scenario represent a social trap or a rational economic decision? If a social trap, which strategy or combination of strategies described above would be the best way to prevent undesirable consequences?

These four methods of avoiding social traps are not mutually exclusive, and none of them is perfect. They are often most effective when used in combination, and when the approach is designed to stop a particular kind of unsustainable behavior. Some social traps can be avoided simply by changing the system of objectives and rewards, the "rules of the game." Imagine sitting down with two friends at a table, in the middle of which are 15 poker chips that represent resources. You are told that the way to "win" the game is to have the most chips when the game is over. On every turn, each player may take one, two, or three chips on his or her turn. The chips are renewable, and the number left at the end of a turn will be doubled before the next turn starts. Under these rules, a player will "lose" if he or she does not take 3 chips. Acting rationally to achieve this objective, the players remove 9 chips, leaving 6. The 6 are doubled to 12. In round two, the players repeat their behavior. Now there are 3 chips, and these are doubled to 6. When round three starts, the first two players again take 3. The third player gets none and "loses," and the game ends in a tie. Each of the other two players finishes with 9 chips and there are no chips left.

But suppose the rules were changed so that the outcome was to maximize the total number of chips accumulated by all three players. If each player takes two chips on each turn instead of three, the number of poker chips continues to increase, and the players can acquire poker chips indefinitely. Analogously, if the goals of resource use are changed from rewarding a single "winner" who maximizes short-term gain at the expense of others, to rewarding groups that can accumu-

late the largest total yields over time, behavior toward the resource also changes. A social trap is avoided and a resource provides a long-term sustainable yield.

The Problem of Policy: What Government Strategies Encourage Sustainable Development?

Brown (1996) suggests three critical policies needed to promote sustainable development: elimination of subsidies for unsustainable activities, institution of a global carbon tax, and replacement of income taxes with environmental taxes.

The first issue is one of the most pervasive, and bizarre, social obstacles to sustainable development. Governments worldwide have multiple programs that provide cash payments or other forms of remuneration for activities that cannot be sustained in the long term. Price supports in agriculture, for example, require direct cash payments to farmers when market prices fall below profitable levels. In the United States, the recently passed Freedom to Farm Act was designed to eliminate such subsidies, but it has been replaced by major federal loans and subsidies in every year in which large harvests resulted in falling prices for corn and soybeans. Soil erosion rates, pesticide and fertilizer hazards, declines in native populations of plants and animals, and continued degradation of groundwater and surface water in the most intensively farmed areas of the midwestern United States indicate that traditional, row-crop corn and soybean farming is not a sustainable practice, ecologically or economically. Market forces would reduce corn and soybean production because of their unprofitability under current conditions, leading to alternative, more sustainable land uses. Government loans and subsidies absorb farmers' losses and encourage unsustainable farming practices.

A global carbon tax is at least as controversial a proposal as eliminating subsidies for unsustainable behavior, but perhaps even more necessary. Carbon dioxide emissions represent perhaps the single greatest threat to global climate stability. A tax on carbon emissions would internalize the cost of such emissions, which are currently external to the production process, but quite real in the form of increased worldwide temperatures and their unprofitable effects on crop yields, increased incidence of violent and unstable weather events, and rising sea levels. As industries sought to lower production costs, one of their first priorities, under a carbon tax, would be to invest in technologies that reduce carbon emissions and increase efficiency of energy use, slowing the negative effects associated with global climate change. As noted earlier, such a tax would make reducing carbon emissions part of "corporate culture" by making the reward for reduced emissions intrinsic to a company's profit and loss, or to a nation's GNP.

Income generation and savings are beneficial economic activities, yet governments discourage both through taxation. An alternative would be to create a tax structure based on environmentally detrimental activities. The more an individual or business participated in these activities or purchased products associated with them, the greater would be their tax burden. For individuals, environmental taxes might take the form of higher sales taxes on less-fuel-efficient cars, taxes on nonrecyclable or nonbiodegradable products, or taxes on home heating or cooling practices that produced high levels of pollution. For business and industry, there might be an expansion of existing taxes and fees on pollution and waste on a per unit basis.

Industrial Patterns of Sustainable Development

Economist J.R. Ehrenfeld proposes that, if industrial processes are to become sustainable, they ought to be patterned after the organization of natural systems: "Three collective features of stable ecosystems seem very important; connectedness, community, and cooperation. Other characteristics such as tightly closed material loops and thermodynamically efficient energy flows offer important themes for technological and institutional design" (Ehrenfeld 2000). Connectedness, community, and cooperation run counter to the modern emphasis on individualism, but are essential foundations of sustainability. As in the case of the players in the poker chip game, individualistic goals degrade systems; connected relationships, communal goals, and cooperative behavior sustain them. Just as detritivores close material loops in ecosystems, Ehrenfeld advocates greater emphasis on and more respect for the "scavenging" functions in human society. "Scavengers, those at the bottom of the food chain that take the wastes and turn it into something useful are as important as the predators at the top, but not in human society. Scavengers have historically lived in a social niche outside the mainstream of society. They are the untouchables" (Ehrenfeld 2000). In stable ecosystems, materials are circulated through a web of interconnections with scavengers at the bottom of the food web turning wastes into food. Industrial processes that involve closed-loop, or nearly closed-loop, material processing have the greatest potential for sustainability because each and every waste product is made to reenter the production process at some point, or the waste is processed into a product needed in some other production process. If the latter, the producer finds a suitable market for the waste, which becomes a factor of production in another process.

Stable ecosystems are often those that operate with the least degree of dissipation of energy (and materials) thermodynamically possible. As biological systems approach minimum entropy production, their degree of organization and interconnection increases. Similar patterns in industrial production result in similar outcomes, and more sustainable production for the long term.

POINTS OF ENGAGEMENT—QUESTION 3

Suppose that all manufacturers of consumer products were required to charge a deposit to their customers on the retail packaging materials in which the product was presented, and also were required to accept the packaging material back from consumers and to pay back their deposit. What effect would this have on (1) the price of consumer goods, (2) the type and amount of packaging materials, and (3) the production process?

Synthesis

No conservation effort can ultimately succeed without intimate connection to value. And no conservation value can endure with vitality unless it finds expression in economic behavior. Some conservationists, eager to see endangered species, vital habitats, and rare ecosystems stand toe-to-toe with industrial output, residential real estate development, and intensive agriculture, have developed or employed a variety of creative measures to document the dollar values of their concerns, whereas others, equally creative and passionate, have laid elaborate plans through which humanity can continue to take more but, through its increased ingenuity, degrade the environment less.

Both these approaches, although well intentioned, have got the question backward. The first fails because it does not ask whether current systems of individual-preference, market-driven valuations can ever rightly determine what is good for many, or how people will ever become better than their own self-centered appetites if those appetites are all that determine their economic behavior. The second errs because it sees human activity as an endless process of acquisition and degradation, progressively made more efficient to acquire what people want with less environmental harm, but never directing anyone to want the right things.

Conservation biologists, if they are to help construct a future in which conservation can succeed, must offer a different set of assumptions about economic behavior and ask a different array of questions. Specifically, what are our rightful obligations to other species, habitats, and ecosystems and how can we design economic systems that recognize, and even reward meeting them? How can we better choose what we shall value, instead of taking our immediate appetites, wants, and desires as givens that must be satisfied, regardless of environmental cost? And finally, how can we restructure the human enterprise of growing and gathering our food and making and using our goods so that such activities not only cease to degrade the world, but make human activity and human presence agents of ecological restoration?

It has been said that most people want progress as long as they don't have to change very much to get it. We can see, in individual communities and isolated efforts, that it is possible to make economic activity the reflection of value rather than the determinant of it, and it is possible to make human activity a restorative ecological force rather than an agent of ecological degradation and destruction. These conditions, although rare, are possible, and represent true progress. Today, private economic incentives can aid conservation because social values have been changed, and that change has itself been shaped by laws and policies of the government that set certain environmental and conservation values, such as endangered species, wetlands, clean air and clean water, above and beyond market forces. These laws and policies were themselves expressions of changing social values that, in democratic societies, lawmakers were forced to address because the public demanded it. The call by some to make all conservation efforts entirely market driven is a mistake at best and a misrepresentation of environmental history at worst. Markets and property rights can be harnessed to achieve conservation goals efficiently when they are made to serve socially normative conservation values enforced by law and policy. But markets and property rights cannot intrinsically generate conservation value, and their historic failure to do so is an inarguable witness of the human experience. Unguided by any end but the force of collective individual desire, human want and appetite inevitably overwhelm all other concerns. Conservation biologists, with environmental economists, must offer a careful and well-designed integration of conservation as an expression of human economic behavior that is guided toward conservation goals established outside the economic process itself. And in doing so, conservation biologists must work to make conservation a normal pattern of economic behavior, not simply a series of heroic (but ultimately futile) efforts to save things that no one ever really valued.

Distinctions of Sustainability and Why They Matter—A Directed Discussion

Reading assignment: Goodland, R., and H. Daly. 1996. Environmental sustainability: universal and non-negotiable. *Ecological Applications* 6:1002–1017.

Questions

1. Differentiate among environmental sustainability, economic sustainability, and social sustainability. Why is it important to make the distinction? Suggest ways for conservation biologists to clarify these concepts for scientists and the public. How might conservation efforts benefit from clarification of these three types of sustainability?

2. Of the four types of environmental sustainability described in the article (weak, intermediate, strong, absurdly strong), which is most likely to achieve conservation goals? Which can best achieve the conditions of the input-output rules (defined on page 1008)? What is the difference between environmental sustainability and sustained yield?

3. Goodland and Daly explicitly or implicitly advocate many policies that have socio-political outcomes, including reductions in material resource consumption, technology transfers from rich to poor countries, and redistribution of wealth. Should conservationists accept the prevailing economic system as a "given" and define conservation goals within the limits of the prevailing system? Alternatively, should conservation biologists work to make existing economic systems and practices more sensitive to conservation goals? Justify your choice.

4. Are all four of the World Bank's strategies for environmental sustainability likely to achieve conservation goals? Does any seem counterproductive? Why? What new policies, or changes in existing policies, might create a more explicit link between achieving environmental sustainability and preserving biodiversity?

Learning Online

Visit our webpage at www.mhhe.com/conservation for case studies, animations, practice quiz questions, and additional readings to help you understand the material in this chapter. You'll also find active links to the following topics:

Food and Agriculture
Sustainable Agriculture

The Informed Consumer
Human Population Growth

Literature Cited

Audirac, I. 1997. *Rural sustainable development in America.* New York: Wiley.

Brown, L. R. 1996. We can build a sustainable economy. *Futurist* 30:8–12.

Ceplo, K. J. 1995. Land-rights conflicts in the regulation of wetlands. In *Land rights: the 1990s property rights rebellion,* ed. B. Yandle, 103–49. Lanham, Md.: Rowman and Littlefield.

Chase, L. C., D. R. Lee, W. D. Schultze, and D. J. Anderson. 1998. Ecotourism demand and differential pricing of national park access in Costa Rica. *Land Economics* 74:466–82.

Costanza, R. 1987. Social traps and environmental policy. *BioScience* 37:407–12.

Costanza, R., J. Cumberland, H. Daly, R. Goodland, and R. Norgaard. 1997. *An introduction to environmental economics.* Boca Raton, Fla.: St. Lucie Press.

Daly, H. E. 1977. *Steady-state economics.* San Francisco: Freeman.

Daly, H. E. 1991. *Steady-state economics,* 2d ed. Washington, D.C.: Island Press.

Daly, H. E. 1993. The steady-state economy: toward a political economy of biophysical equilibrium and moral growth. In *Valuing the earth: economics, ecology, ethics,* eds. H. E. Daly and K. N. Townsend, 325–63. Cambridge, Mass.: MIT Press.

Daly, H. E. 1996. *Beyond growth: the economics of sustainable development.* Boston, Mass.: Beacon Press.

Daly, H. E. 1999. *Environmental economics and the ecology of economics: essays in criticism.* Cheltenham, England: Edward Elgar.

Ehrenfeld. J. R. 2000. Industrial ecology: paradigm shift or normal science? *American Behavioral Scientist* 44:229–44.

Ehrlich, P., and A. E. Ehrlich. 1990. *The population explosion.* New York: Simon and Schuster.

Erickson, J. 1995. *The human volcano: population growth as a geologic force.* New York: Facts on File.

European Commission. 1995. *ExternE: externalities of energy.* Vol. I. Summary. Luxembourg: European Union.

Fitzsimmons, A. K. 1999. *Defending illusions: federal protection of ecosystems.* Lanham, Md.: Rowman and Littlefield.

Georgescu-Roegen, N. 1993. The entropy law and the economic process. In *Valuing the earth: economics, ecology, ethics,* eds. H. E. Daly and K. N. Townsend, 75–88. Cambridge, Mass.: MIT Press.

Goodland, R., H. E. Daly, and S. El Serafy. 1992. *Population, technology, and lifestyle.* Washington, D.C.: Island Press.

Gordon, H. S. 1954. The economic theory of a common property resource. *Journal of Political Economy* 62:124–42.

Hardin, G. 1968. The tragedy of the commons. *Science* 162:1243–48.

Hoffman, A. J. 2000. Integrating environmental and social issues into corporate practice. *Environment* 42:22–33.

Kreuter, U. P., and R. T. Simmons. 1995. Who owns the elephants? The political economy of saving the African elephant. In *Wildlife in the marketplace,* eds. T. L. Anderson and P. J. Hill, 147–65. Lanham, Md.: Rowman and Littlefield.

Lauck, T., C. W. Clark, M. Mangel, and G. R. Munro. 1998. Implementing the precautionary principle in fisheries management through marine reserves. *Ecological Applications* 8:S72–S78.

Leal, D. R. 1998. Cooperating on the commons: case studies in community fisheries. In *Who owns the environment?* eds. P. J. Hill and R. E. Meiners, 283–313. Lanham, Md.: Rowman and Littlefield.

Loomis, J. B. 2000. Can environmental economic valuation techniques aid ecological economics and wildlife conservation? *Wildlife Society Bulletin* 28:52–60.

Ostrom, E. 1990. *Governing the commons: the evolution of institutions for collective action.* New York: Cambridge University Press.

Quinn, J. B., and J. F. Quinn. 2000. Forging environmental markets. *Issues in Science and Technology* 46:45–52.

Piel, G. 1995. Worldwide development or population explosion: our choice. *Challenge* (July/August):13–22.

Prugh, T., R. Costanza, J. H. Cumberland, H. E. Daly, R. Goodland, and R. B. Norgaard. 1999. *Natural capital and human economic survival,* 2d ed. Boca Raton, Fla.: Lewis Publishers.

Ruckelshaus, M. H., and C. G. Hays. 1998. Conservation and management of species in the sea. In *Conservation biology: for the coming decade,* 2d ed., eds. P. L. Fiedler and P. M. Kareiva, 112–56. New York: Chapman and Hall.

Ruskin, J. 1883. *Unto this last: four essays on the principles of political economy.* New York: Wiley.

Shogren, J. F., and P. H. Haywood. 1998. Biological effectiveness and economic impacts. In *Private property and the Endangered Species Act: saving habitats, protecting homes,* ed. J. F. Shogren, 48–69. Austin: University of Texas Press.

Smith, G. A. 1993. The purpose of wealth: a historical perspective. In *Valuing the earth: economics, ecology, ethics,* eds. H. E. Daly and K. N. Townsend, 183–209. Cambridge, Mass.: MIT Press.

Söderholm, P., and T. Sundqvist. 2000. Ethical limitations of social cost pricing: an application to power generation externalities. *Journal of Economic Issues* 34:453–62.

Stefanovic, I. L. 2000. *Safeguarding our common future: rethinking sustainable development.* Albany: State University of New York Press.

Weinberg, A. S. 2000. Sustainable economic development in rural America. *Annals of the American Academy of Political and Social Science* 570:173–85.

Wu, J. J., and B. A. Babcock. 1999. The relative efficiency of voluntary versus mandatory environmental regulations. *Journal of Environmental Economics and Management* 38:158–75.

Yandle, B. 1997. *Common sense and the common law for the environment: creating wealth in hummingbird economies.* Lanham, Md.: Rowman and Littlefield.

CHAPTER 13

Professional Effectiveness and Future Directions in Conservation Biology

In this chapter you will learn about:

1 the value of a personal mission statement in conservation biology and the principles for writing one.

2 elements in educational experiences and professional

relationships that lay the foundation for opportunity and service in conservation biology.

3 objective criteria for selecting educational programs, mentors, and jobs.

4 emerging trends in conservation biology, their potential significance, and their relationship to your educational experience.

PEOPLE AS AGENTS OF CONSERVATION—THE THINGS TEXTBOOKS NEVER TELL YOU

When a celebrated alumna of a prestigious university was asked what she thought of her undergraduate training, she replied, "It was all very well done. Quite comprehensive. They taught me everything but how to get a job."

This lament is not unique to any particular field. In an effort to avoid appearing "vocational" or "prescriptive," colleges and universities often sidestep—at least in classes and textbooks—the issue of how people cease being students and become effective professionals in a particular field. Implicit in this silence is the assumption that ideas in textbooks will equip students to function at the forefront of their disciplines. The truth is, acquiring information about conservation is not the same as doing conservation. Textbooks do not perform conservation; people do. And the people who accomplish the most are those who become conservationists. Conservation as a career merits attention alongside conservation biology as an academic pursuit.

What follows is an unconventional chapter on the problem of moving from knowledge *about* conservation biology to effective involvement *in* conservation biology. Following that

examination, I consider some emerging trends in conservation biology that students and professionals may prepare to encounter in the future.

CONSERVATION BIOLOGY AS VOCATION: BEGINNING AN EFFECTIVE CAREER

Articulating Your Personal Mission and Purpose in Conservation

Yogi Berra, the colorful ex-baseball player, once remarked, "If you don't know where you're going, you'll probably end up someplace." Unfortunately, "someplace" may not be a satisfactory destination. Having a clear understanding of one's own mission and purpose is an essential first step toward the kind of self-mastery that produces meaningful success and professional effectiveness. Clarifying a personal mission makes one's motivations clear and explicit, enabling internal restatement of motivations or explanation to others at any time and under any circumstances. Clarity of mission produces perseverance through discouraging circumstances and events. The act of writing a personal mission statement, or statement of purpose, helps define not only one's mission, but also one's sense of *self*—an identity that exists independently of performance or external evaluation. This kind of self-knowledge provides the confidence to try new, unorthodox ways of accomplishing meaningful goals, the resilience to cope with setbacks and loss, and the freedom to fail and learn from one's failure.

An effective, personal understanding of purpose should be simple, clear, and memorable. Many students confuse a personal mission statement with the answer to the question, "What do I want to do with my life?" A personal mission statement, however, does not tell you what to do. It points you in a productive direction to discover what you might be capable of doing, and why it might be meaningful to do it. A simple, general statement of purpose for a person considering a major in conservation biology or a related field might be, "I want to determine what problems conservation biology tries to solve and if there is an opportunity for me to contribute to the solutions." Other possibilities include, "I want to determine what the field of conservation biology will require of me in terms of knowledge, skills, and attitudes," or even, "I want to determine the defining characteristics and ideas that distinguish conservation biology from other disciplines in biology and determine if my interests are compatible with this discipline's distinct qualities."

Although many students enter college with fundamental questions about what vocation to choose, and how to choose one, others think they already know, and move past the question of "What shall I do with my life?" to "How can I be most effective in the vocation I have chosen?" So, many students who know that they want to be conservation biologists may want to frame their personal mission statements in more specific ways. To do so, one must identify the essential qualities of one's role in conservation biology *without which a career in the field would not be satisfying.* For example, one student's motivations might lead to the following statement of purpose: "I want to work for an organization that deals directly with the management and conservation of endangered species." Such a statement is probably broad enough to allow the individual to consider both governmental and nongovernmental organizations at national and international levels, but specific enough to limit and discipline the inquiries and preparations for such a career. Without that specificity, students can become overwhelmed and confused by the variety of possible preparations they could undertake, or the variety of organizations that are connected to the work of conservation.

Pursuing Your Mission Through Education: Defining Your Core Curriculum

The number of colleges and universities throughout the world that offer degrees in conservation biology is still relatively small. However, practicing conservation biologists earned their degrees in a variety of disciplines, including general biology, ecology, environmental biology, wildlife management, fisheries management, botany, zoology, forestry, range management, and many others. The name of the degree program one chooses is important, but less so than the courses taken to earn the degree. Attempts to define ideal undergraduate preparation for a career in conservation biology result in irresolvable debate; however, most conservation educators would identify five components as essential: (1) examination of basic biological processes and entities at cellular/genetic, organismal, population, and ecosystem levels; (2) training in mathematical analysis, interpretation, and presentation of complex sets of quantitative information, and practice in designing experiments to generate quantitative information; (3) study of physical and chemical processes that govern basic biological processes and shape the external environment; (4) use of technologies widely applied in conservation; and (5) consideration of the social, political, and cultural forces that shape the practice of conservation in the context of human society.

A curriculum that encompasses all of these components would be commendably ambitious, but unrealistic if it attempts to cover all areas in depth. The current and correct emphasis on interdisciplinary education should not obscure the need for intensive training in at least one keystone discipline. What the field of conservation biology needs are not students who know a little about a lot of things, but students who have depth in a core discipline and sufficient breadth to converse with and contribute to other disciplines to achieve conservation goals.

It is difficult to accomplish all these objectives during a 4-year undergraduate experience. Most conservation biologists working in the field today would admit to being "undereducated" (and some would confess that they were completely uneducated) in at least some of the categories of knowledge and skill they use daily. Yet they function effectively as conservation professionals because they acquired the missing components through graduate education, on-the-job experience, or by working as one member of a team of individuals with diverse expertise, each focusing on his or her own strength. Students should not panic if, upon inspecting the curriculum, they discover that some areas are not covered as thoroughly as others; however, students should not be complacent. They should strive to take courses, either as requirements or as electives, that will develop their knowledge and competence in all the above areas and simultaneously build a foundation of well-developed interpersonal skills.

Making the Transition from Student to Colleague: The Nature and Necessity of Interpersonal Skills

The lecture format is an efficient way of communicating large amounts of information quickly to a large audience of students. Unfortunately, lecturing rewards and reinforces all the wrong behaviors for success in the student's eventual working environment, a community of fellow professionals in which two-way communication is critical. As a student, you will not be able to avoid lectures. You may have heard many of them in this conservation biology course. But though the knowledge base of conservation biology is rooted in the natural sciences, human interaction is the primary process of achieving the mission of conservation. Because success in this mission is tied directly to interpersonal skills, every student must, at some point, recognize the need for these skills and begin to practice them. The transformation from "recipient of information" to "contributing citizen of a professional community" does not occur by accident, but by intention and design. Through a series of steps, students develop the skills essential for the shift from student to practitioner.

Recognize the Obstacles to Professional Acceptance

Explicit hurdles for undergraduate students are passing courses and gaining a degree. For graduate students, the explicit hurdle is preparation of a thesis. For the new employee in the agency, it is meeting goals and targets. But lying beneath the surface of all three cultures is a similar implicit hurdle—to attain the status of a colleague among one's associates.

Attaining the status of a colleague begins with understanding that you must manage an array of relationships, not just an array of courses, projects, or data. The first step toward learning how to be a colleague is the intentional initiation of relationships with your associates. For students, this means engaging professors in conversations that are not always oriented around the need to clarify or complain about a particular assignment, but rather are discussions of issues of mutual interest and concern. In the same way, students should initiate peer group discussions about professional and scientific issues and form permanent discussion and reading groups to make this a regular practice. In the environment of a conservation organization or agency, it means starting the same kinds of conversations with other workers, including superiors. Initiate relationships with others by venturing outside your own peer group, educational status, or employment level and by addressing issues with them that are about their needs, not yours. These words should not be interpreted as encouraging insincere concern or false friendship. Make your interest in and concern for others genuine. The interest you demonstrate in others and in their work lays the foundation for the second step, in which you strive to become a solution to others' problems and an asset in their efforts.

The demands of conservation biology in academic and professional settings make time a scarce resource. As you begin to understand what your professors or fellow workers are doing, you will also begin to appreciate the difficulty of the problems they face. When you see the opportunity, offer yourself and your skills as a part of the solution to one of their problems. For instance, assist with the collection of data in the field or with transporting equipment to a field site. Help with the preparations for a class or lab exercise. From this foundation, undergraduate students should begin to seek out opportunities to work for faculty as teaching aids or research assistants. Graduate students should look for opportunities for collaborative research, presentation, and publication with faculty. And new employees in an agency should seek cooperative efforts among others that pursue common goals important to the agency's mission.

Not everyone will want to treat you like a colleague. Some professors will always want to dominate students whom they can control, not develop younger colleagues with whom they could one day collaborate. Work with those who treat you with respect and who demonstrate an interest in helping you grow from student or servant to colleague and associate.

Establish Credibility Independent of Academic Courses

In academic cultures, students tend to see their professional preparation in terms of fall and spring semesters, and their summers as "vacations," or ways to earn money at lucrative but nonconservation jobs to pay the high cost of college. But summers offer the best opportunity for students to begin to gain the experience and credentials that will facilitate their transformation to citizens of a professional culture.

All types of conservation agencies and organizations hire seasonal workers, and many professors hire additional field assistants to help in ongoing research. Some conservation organizations, such as The Nature Conservancy, have special summer programs specifically designed to employ students, whereas others, like the Student Conservation Association, focus exclusively on placing students in conservation positions. Many undergraduates have three summers between enrollment and graduation. If each summer is used wisely, and complemented with experiences gained during the academic year, students can build an impressive array of accomplishments and credentials that mark them as distinctive applicants for jobs and graduate research. In fact, students can distinguish themselves on the basis of what they do professionally much more effectively than by their grade point averages. Consider two case histories.

Case History 1—Sarah Bowdish. Sarah Bowdish (fig. 13.1) grew up in rural northeastern Iowa, and went to college at a small liberal arts institution in another part of the same state. During her freshman year, she developed an interest in conservation, but her college had no major in conservation biology. Sarah chose a closely related major, biology/environmental science, to pursue her goals. By her sophomore year, she had learned, through building relationships with her professors, of ongoing research on the effects of burning and mowing on native prairie plant communities and their associated communities of birds. With strong recommendations and a good academic record, Sarah applied for and gained a position as a research assistant on such a study at the DeSoto National Wildlife Refuge in Iowa. The work was difficult and demanding. She was in the field daily before sunrise to start bird surveys on different experimental treatments. When the bird counts were finished by midmorning, Sarah began intensive plant sampling that often kept her in the prairies until sunset.

Sarah's dedication, diligence, and growing knowledge of the prairies at DeSoto did not go unnoticed by the Fish and Wildlife Service staff. Increasingly, they consulted Sarah about plant and bird responses to the management treatments, and began to make use of her knowledge of specific sites. Sarah was asked to continue as an assistant for a second summer. During that year, she coauthored the progress reports, wrote grant proposals with her professor, and learned the nuances of working with and gaining grant support from agency administrators. By the end of her second summer of research, the quality of her work and management recommendations had thoroughly impressed the DeSoto staff. Staff administrators arranged for Sarah to speak to members of the Society for Range Management (SRM) at their annual meeting in nearby Omaha, Nebraska.

Figure 13.1
Sarah Bowdish, graduate research assistant in the Department of Botany and Microbiology, University of Oklahoma. As an undergraduate, Sarah's involvement in tallgrass prairie research and management, combined with an interest in environmental and conservation policy, created a record of accomplishment and a network of contacts that prepared her for graduate research.

Figure 13.2
Nathan De Jager, habitat restoration specialist with the Natural Resource Conservation Service. Nathan's summer experiences coordinating the Conservation Reserve Program in Sioux County, Iowa, helped him learn techniques of prairie and wetland restoration, as well as how to write research and management grant proposals to pay for local restoration efforts.

Sarah's presentation to the SRM provided her with many new professional contacts, as well as many compliments from the society's leaders and members.

Sarah then submitted her work for presentation at the annual meeting of The Wildlife Society, to be held that year in Austin, Texas. After review, Sarah's abstract was accepted. Her presentation was not only an important contribution to the meeting, but provided another opportunity to establish professional contacts. During the academic year, Sarah attended two student conferences in Washington, D.C., on issues of environmental law and policy. She used her interpersonal skills to become a leader among student caucuses, and an effective and persuasive advocate for conservation with her own senators and members of Congress.

By the end of Sarah's junior year, her contributions to conservation management and policy began to receive wider recognition. She won one of 75 Morris K. Udall Scholarships in Environmental Policy; this $5,000 national award is given to undergraduate students who show promise of being able to make exceptional contributions to environmental policy issues.

In 2000, Sarah accepted a graduate assistantship at the University of Oklahoma to study the effects of global warming on the species composition and phenology of prairie plant communities. Along with fieldwork in the state, she traveled to the

Czech Republic for additional, more specialized studies in multivariate analysis of ecological data. Her interests in politics and her strong interpersonal skills were soon having a pronounced effect on her colleagues at Oklahoma. Within a year she had been elected a senator to the Graduate Student Senate of the University and elected president of the University of Oklahoma Botanical Society. After completing her master's degree, Sarah plans to enter law school to study and practice conservation law and provide expertise in conservation policy.

Case History 2—Nathan De Jager. Nathan De Jager (fig. 13.2) enrolled in the same liberal arts college as Sarah at about the same time. Although a gifted athlete in football and baseball, he had little to recommend him as a student or future scholar. Nathan had grown up on a farm only a few miles from the college, and his interest in the land around him and its care led him to choose a major in biology and environmental science. Through personal contacts, Nathan's major professor learned that the local office of the Natural Resources Conservation Service (NRCS) wanted to hire a technician for the summer to help with the county's Conservation Reserve Program (CRP), a federal program that subsidizes farmers to take selected lands of high conservation value out of production. Although his academic credentials were not yet impressive, Nathan gained a chance to interview for the position. His knowledge of farming and of the local landscape and local farmers impressed the NRCS staff and Nathan was hired. He proved so skillful at persuading local farmers to convert farmland to native prairie and natural riparian habitat that after only a few weeks, Nathan was placed in charge of the CRP program for the entire county. Over the course of the next two summers, Nathan enrolled nearly 400 additional acres in the conservation reserve program, assisted in the restoration of three wetland areas, and helped design mitigation measures for the construction of two new livestock facilities to reduce their pollution impacts. In addition to allowing Nathan to continue to administer the CRP program, his supervisors placed him on two additional agency working groups on water pollution and riparian habitat restoration.

In the course of his work, Nathan discovered that his own college owned a farm within his jurisdiction that qualified for the Conservation Reserve Program, and that it included an important local stream. He convinced the college administration to alter its traditional farming practices, convert a portion of the farm to native prairie and riparian vegetation, and then raised $7,000 in external grants from the Iowa Department of Agriculture and Land Stewardship and Pheasants Forever, a private conservation organization, to pay for the restoration. Under Nathan's supervision, the restoration made excellent progress in its first 2 years (some of which is described in chapter 11). After graduation, Nathan accepted a full-time job with the Natural Resources Conservation Service in Boone, Iowa, where he continued to work with habitat restoration through the conservation reserve program. In this position, Nathan has been directly responsible for the restoration of over 1300 acres of wetland and associated upland habitat. Following this experience, Nathan began his graduate education at the University of Minnesota–Duluth in August 2002. His field work will take him to Alaska for studies of the effects of moose browsing on native plant communities.

Common Threads in Different Cases—Patterns of Successful Transition from Conservation Students to Conservation Professionals

Sarah and Nathan have different stories, hopes, and ambitions, but both made a successful transition from classroom students to productive young colleagues. Both have made contributions to conservation entirely on their own merits by following four practices that any student of conservation can emulate.

1. As undergraduates, Sarah and Nathan used their summers to become involved in conservation practice.
2. Sarah and Nathan used interpersonal skills and technical knowledge to build trust, establish cooperative relationships, and persuade both scientists and the public to change practices and behaviors in ways that were of benefit to conservation goals.
3. Sarah and Nathan articulated clear personal conservation missions and pursued them through personal initiative. For example, Nathan wanted to change agricultural prac-

tices in his local landscape to reduce soil erosion and preserve native prairie.
4. By making constructive contributions of value to their colleagues, Sarah and Nathan gained an increasingly wide network of contacts that expanded their influence, added to their credentials, and increased their effectiveness as professional conservationists.

Sarah and Nathan did not have the advantage of an explicit curriculum in conservation biology, and they did not enjoy the benefits of highly specialized courses or extensive institutional resources that are typical of large state universities. Both eventually produced good, but not exceptional, academic records. But neither was "lucky." The achievements that Sarah and Nathan eventually gained were inevitable consequences of consistent, professional behavior, rightly applied to well-chosen ends through intelligent means.

POINTS OF ENGAGEMENT—QUESTION 1

What opportunities do you have in your current educational environment that can create the same kinds of outcomes that Sarah and Nathan achieved? What actions should you begin to take to transform these opportunities into tangible accomplishments and credentials?

Building a Professional Network of Contacts and References

People who want to work in any professional field, including conservation biology, need to face some sobering statistics. In the United States, it is estimated that approximately 80% of all new positions are filled without ever being advertised (Hart 1996). Likewise, it has been estimated that the national average percent success of an individual applicant in getting an advertised position is less than 7%. But when applicants have direct contact or secondary contact (e.g., one of the applicant's references knows the evaluator) with the evaluator, the likelihood of success is estimated at 86% (Hart 1996). These statistics highlight the importance of *networking*. Why do networks make such a difference in success, and how does one use them effectively?

Every human being has a fundamental need for lasting and satisfying personal and professional relationships. People tend to associate with others who are like them and share their fundamental values and goals. Through regular and repeated actions with one another, cooperation grows, and will increase even more among groups that share a common functional objective. Through regular interactions, and the consequent trust people establish, comes a foundation for permanent organizational success which benefits all individuals in the group.

Paradoxically, educational programs in most colleges that claim to prepare their students for professional success often fail to prepare them for these relational realities. The typical college curriculum in conservation biology or related subjects rewards (through grades) only student demonstrations of knowledge and technical ability. This reward system tells students that if they increase in knowledge and ability, they will gain more and more reward. Further, it leads students to believe that they will always be evaluated fairly relative to other individuals. Such a belief system can survive only in the confines of an academic classroom.

In vocational settings, knowledge and technical ability are the primary determinants of success only in relatively low-level positions with fairly fixed tasks and deterministic job descriptions involving management of data and information. Higher-level positions require increasing skill, not only in managing data, but in managing relationships. Management and networking expert W. E. Baker tells the story of a mythical manager, Bill, who suffered from the kind of occupational myopia "that restricted his field of vision to the technical part of his job. He didn't know that his ability to get the job hinged on his success in cultivating, maintaining, and mobilizing a vast array of relationships. He didn't realize that his success depended so much on people he didn't yet know. Because he didn't build relationships and get hooked into the network, Bill wasn't able to discover critical information, influence key decision makers, negotiate successfully, or implement his strategy. He didn't see the world as a network of relationships. . . . Oh, yes. You're probably wondering what happened to Bill. Well, he was fired" (Baker 1994:4).

Although Baker is using a scenario from the world of corporate business, interpersonal skills take on additional importance in conservation biology because social and political outcomes are necessary for conservation to occur. Thus, effective interactions with other individuals of diverse backgrounds and training are essential for success. Cannon, Dietz, and Dietz (1996) note, "The human interaction processes critical to the work of conservation biology include sharing information, explaining ideas and values, listening to others, communicating a clear understanding of the opinions and feelings of others, and working together to solve problems, resolve disputes, and carry out action plans."

Despite the essential and urgent need for these kinds of skills, academic programs have been slow to make them an explicit part of the curriculum. Cannon, Dietz, and Dietz (1996) surveyed 298 graduate programs in conservation biology and closely related fields as well as 702 public and private organizations that employed conservation biologists. A majority of respondents in both groups identified seven key areas in which training was needed: (1) written and oral communication, (2) explaining science and values of biodiversity to the lay public, (3) group decision making, (4) interpersonal skills, (5) group planning, (6) leadership, and (7) advocacy. Despite the perceived need for high levels of training in all of these areas, few academic institutions and even fewer conservation organizations offered courses in human interaction skills. Sixty-four percent of graduate faculty respondents and 78% of organization respondents considered these areas to be as important or more important than scientific knowledge and technical skills (fig. 13.3). Yet there is a gap between the perceived need and desire for such skills and their actual offerings in conservation curricula in educational and professional settings (tables 13.1 and 13.2). If this gap is not closed, academic course work in conservation will in no

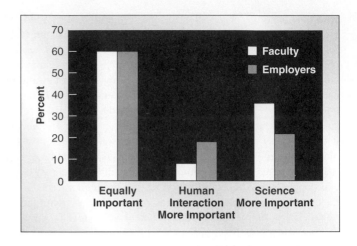

Figure 13.3

Employer and faculty comparisons of the relative importance of science knowledge and skills compared with human interaction skills.

Data from Cannon, Dietz, and Dietz (1996).

way equip students for success and effectiveness in real conservation efforts.

Such relational skills are essential in building effective professional networks, and networks are increasingly essential for professional success. As organizations grow in size and applicants grow in number, evaluators face ever more difficult decisions in selecting applicants for positions. More and more evaluators rely on their personal relationships with other professionals, whom they do know, to make decisions about the selection of applicants, whom they do not know.

How does one form such contacts, and how does one judge whether or not a potential contact is a good one? Studies of human interactions reveal that persons who work together toward common goals tend to form stable, positive, and mutually supportive relationships, even when personality differences are extreme. Thus, a first step in forming networks is to return to your personal mission, use it to identify goals of primary importance, find other individuals (in person or electronically) who share and are working toward these goals, and join them in mutually effective efforts. Joining and working in an organization in which members share a common mission is an effective beginning. To be successful, this will involve, at minimum, a commitment to attend the organization's meetings, participate in its discussions, hold membership in a committee or office, and contribute to work that defines the organization's purposes. In an electronic context, discussion groups abound that are defined by common interests and concerns for specific issues or problems. Conservation biology is no exception. By joining such discussion groups and participating constructively in the discussion, relationships are formed that may mature into effective contacts.

Although participation in organizations and discussion groups is helpful in building an effective network, the most committed long-term relationships are likely to develop in more focused efforts, and often in employer-employee or mentor-student relationships. There are many foundations for such relationships. They may begin through mutual interest in the subject matter of a course, continue through common

Table 13.1   **Faculty Ratings of Training Needs and Current Course Offerings and Requirements at Academic Institutions** [*]

SKILL CATEGORY	PERCEIVED NEED FOR TRAINING (% OF RESPONDENTS)				COURSES OFFERED (% OF INSTITUTIONS)		COURSES REQUIRED (% OF INSTITUTIONS)	
	HIGH (3)	MEDIUM (2)	LOW (1)	AVERAGE RATING	YES	NO	YES	NO
Written and oral communication	96.4	3.6	0.0	2.96	75.0	25.0	46.3	53.7
Explaining science and values of biodiversity to lay public	76.8	23.2	0.0	2.77	27.4	72.6	6.1	93.9
Group decision-making	68.7	26.5	4.8	2.64	38.1	61.9	10.8	89.2
Interpersonal relationships	66.3	28.9	4.8	2.61	27.4	72.6	8.6	91.4
Group planning	64.2	30.9	4.9	2.59	33.3	66.7	8.5	91.5
Leadership	53.1	43.2	3.7	2.49	31.0	69.0	8.6	91.4
Advocacy	51.9	40.7	7.4	2.44	19.0	81.0	3.7	96.3
Negotiating or dispute resolution	45.1	45.1	9.8	2.35	26.2	73.8	2.5	97.5
Knowledge of more than one language and culture	32.5	55.4	12.0	2.20	61.9	38.1	14.6	85.4
Interactive economic valuation	26.8	54.9	18.3	2.09	45.2	54.8	13.4	86.6
Interactive program evaluation	19.5	63.4	17.1	2.02	31.0	69.0	6.1	93.9

[*]Categories are listed in rank order based on the average rating received.
After Cannon, Dietz, and Dietz 1996.

efforts in the discipline outside the course, and mature through the student's desire for guidance from the faculty member and the faculty member's desire to help the student. The circle of primary contacts will expand for students who actively pursue research and vocational experiences, especially experiences that lead them off their own campus to other colleges or outside agencies. The greater the diversity of experiences and organizations, the greater the number of primary contacts established.

Primary contacts can become helpful as "network partners" if they possess certain traits. A good network partner is a person who respects you, likes and understands you, and is involved and influential in an area related to your objective. Individuals who lack the third trait may be good and valued friends, but will be of little help in gaining employment or graduate education opportunities. Individuals who possess the third quality but lack the first two traits are potentially powerful but usually unhelpful. The presence of all three traits in one individual is what defines a person who will be most helpful in forming effective networks that lead you to discover opportunities in employment and education, and that can convince other people, whom you do not know, that you are worth their investment and support.

Conservation as a Social Process: Involvement in Professional Societies

Like other humans, scientists are social. They form communities of common purpose, not only to achieve their objectives, but also to support and encourage one another emotionally to continue to strive toward those purposes. As Michael Soulé wrote about the origins of conservation biology, "just as the idea of a 'star wars' defense shield had to await the development of nuclear missiles, so the idea of conservation biology had to await a serious threat to biological diversity. . . . Conservation biology began when a critical mass of people agreed that

Table 13.2   Employer Ratings of Training Needs and Current Course Offerings and Requirements at Conservation Organizations[*]

SKILL CATEGORY	PERCEIVED NEED FOR TRAINING (% OF RESPONDENTS)				COURSES OFFERED (% OF INSTITUTIONS)		COURSES REQUIRED (% OF INSTITUTIONS)	
	HIGH (3)	MEDIUM (2)	LOW (1)	AVERAGE RATING	YES	NO	YES	NO
Written and oral communication	90.4	8.8	0.8	2.90	20.0	80.0	27.5	72.5
Explaining science and values of biodiversity to lay public	75.2	24.0	0.8	2.74	11.8	88.2	17.6	82.4
Interpersonal relationships	72.8	24.0	3.2	2.70	16.4	83.6	9.9	90.1
Leadership	65.6	32.8	1.6	2.64	19.1	80.9	9.9	90.1
Group decision-making	62.4	35.2	2.4	2.60	16.4	83.6	9.9	90.1
Group planning	55.6	41.9	2.4	2.53	14.5	85.5	11.0	89.0
Advocacy	56.9	32.5	10.6	2.46	10.9	89.1	14.3	85.7
Negotiating or dispute resolution	45.6	44.0	10.4	2.35	13.6	86.4	5.5	94.5
Interactive economic valuation	29.4	50.0	20.6	2.09	1.8	98.2	8.8	91.2
Interactive program evaluation	22.8	54.5	22.8	2.00	7.3	92.7	8.8	91.2
Knowledge of more than one language and culture	16.4	45.1	38.5	1.78	1.8	98.2	4.4	95.6

[*]Categories are listed in rank order based on the average rating received.
After Cannon, Dietz, and Dietz 1996.

they were conservation biologists. There is something very social and very human about this realization" (Soulé 1985).

Once a professional society is formed, it soon has its own journals, conferences, bylaws, membership requirements, and certification standards. Although such societies are formed to advance a common mission, they are also formed to provide personal and professional benefits to their members. There are many scientific societies engaged in various aspects of conservation, but the most identifiable and intentional of them is the Society for Conservation Biology (SCB). What does belonging to and participating in this society do for you and your development as a conservation biologist?

The SCB publishes a journal, *Conservation Biology,* which all members receive. The journal serves not only to inform members of ongoing research in the discipline, but also, through its editorial policies that determine what will and will not be published in the journal, functionally defines the issues of conservation biology. Through its editorials and commen-

taries, it also serves to form, and inform, community-based views on conservation issues and values. This is not to suggest that a society's journal can dictate what its members think. Anyone who has ever attended a faculty meeting or a professional conference knows that there is no environment more likely to produce disagreement than a room full of scholars. But despite appearances to the contrary, professional societies and their journals do function synergistically to shape a collective view of what a discipline is. In this way they provide members with professional identity, a function that is of great value to each person in further refining his or her own personal mission statement.

The SCB holds international, national, and regional meetings at which its members can present the results of their research, meet one another, form associations of common goals and interest, and recognize and affirm significant accomplishments of individual members and the society. Such meetings also produce official organizational statements on important issues that allow the society to speak with a unified voice to the general public or to the political process.

Finally, the SCB maintains and makes available resources on jobs and graduate programs in conservation. The more members that are gainfully employed in conservation, and the more satisfied they are with the value of their work, the more likely they are to remain active and productive members of the society. If such satisfaction erodes and is replaced by professional discontent, the society suffers.

It is valuable to visit the SCB website at http://conservationbiology.org to see how a collection of scientists with common mission and purpose present themselves socially to one another and to the world. Compare this presentation with other organizations that have related missions, such as The Wildlife Society or the Ecological Society of America, and note both the similarities and differences. If you wish to make your contributions to conservation maximally effective, you must express them in a social context of other professionals in some way and at some point. Without this dimension of professional life, even the best efforts are never fully effective. For some individuals, social contact is natural and easy. For others, it is uncomfortable and difficult. Persons who are shy or retiring by nature may find it easier to relate to others in activities focused around a common task or purpose, whereas naturally gregarious individuals may prefer purely social or recreational settings. Choose social interactions and contexts that you find most appropriate to your own temperament and interests, but do not neglect this dimension of professional life.

Building Research and Vocational Experience

Earning good grades demonstrates that one has learned how to function successfully as a student. But graduate positions and jobs will require you to function much more independently in initiating programs, research efforts, and even organizational change with minimal direction. Performance in research and vocational positions during undergraduate years will often be used as a determining factor in measuring whether an individual can handle these tasks.

Throughout the world, the best undergraduate programs in all sciences require undergraduates to participate in at least one intense research experience. This experience may begin as early as the sophomore year, and often culminates in a senior project or thesis. Outstanding programs in conservation biology are no exception. Whether research experience is an explicit requirement or not, undergraduates should actively pursue it, beginning in their freshman year. Many opportunities for research experiences exist—research-intensive courses or research with institutional faculty and graduate students, faculty at other institutions, and independent agencies or research organizations.

Where available, students should give serious efforts to research-intensive courses built into the undergraduate program of study. Many programs include and require a course with a title such as Senior Thesis or Independent Research Effort. Student who enroll in such courses should invest diligently to make the research effort of the highest quality, as original an investigation as possible with a final product that is as near to publishable journal standards as possible.

If there is no such course in the curriculum, there may be advanced courses that require a literature review paper in a particular subject. Students should view these assignments as opportunities to accomplish two things. First, they represent opportunities to produce a permanent document that can be copied, read, and judged by future evaluators. Although a literature review paper in a particular course or a senior research thesis may never actually be published, if its quality demonstrates that the student can do publishable work, it can make a favorable and distinctive impression that grades alone cannot. Second, the preparation of a research paper or thesis helps the student clarify his or her own interests and express these interests in terms of a knowledgeable discussion of recent developments in the field. Thus, when evaluators ask for a statement of research interests, the student who has prepared a senior thesis or professional literature review can speak with greater power and precision than a student who can express ideas only in terms of vague preferences.

Even if you do not encounter the opportunity to prepare such a work as a course or curricular requirement, it is wise to take on such a project personally. For example, you might consult with a professor to design a literature review that the professor would find useful for his or her own research, and offer to provide the literature review. This approach creates an accountability structure, provides ongoing input from a professional perspective, and can help build a long-term relationship of trust and respect.

Faculty and graduate students at colleges and universities invariably are conducting individual research efforts. In some cases, they may advertise for assistants. Wise undergraduates should respond enthusiastically, even if the positions are unpaid. However, many such positions are not advertised, but will be offered to students who have performed well in classes and fostered positive relationships with faculty and graduate assistants. Some positions may be initiated by students themselves by offering their services as volunteers to assist in ongoing research efforts. Persons who prove themselves reliable in these roles invariably soon find themselves with greater responsibilities and opportunities because they progressively gain the trust of those with whom they work.

A student's opportunities are not limited to the local campus. Faculty who obtain grants at other colleges and universities may advertise widely for assistants. Such ads typically appear as posters or notices placed on the departmental bulletin board, in the newsletters of professional organizations, and as notices on websites. Some types of grants, such as those administered by the National Science Foundation (NSF), may have required procedures and protocols for publication and be posted at various types of Internet "clearinghouses." In some cases, such grants may be specifically targeted for undergraduates and may be funded for long-term research, permitting applications to be made year after year.

Research and vocational training also may come through direct employment with a government conservation agency or nongovernmental conservation organization. In the United States, agencies such as the National Park Service, the U.S. Forest Service, Bureau of Land Management, U.S. Fish and Wildlife Service, Natural Resources Conservation Service, and others hire thousands of undergraduates each summer for agency work. In many cases, the agencies have established specific programs for the professional development and training of undergraduates, often with the explicit intention of grooming successful participants for future employment with the agency. Many state and local government agencies have similar programs, although typically the number of positions available is smaller. In the United States, many private and public conservation organizations, such as Mount Desert Island National Laboratory, Savannah River Ecology Laboratory, and Konza Prairie, to name a few, also offer extensive programs of summer research experiences designed for undergraduates. These opportunities are often prominently displayed on the department bulletin board. Too few students pursue them.

Evaluation Criteria for Employment and Graduate School—Grades, Recommendations, and the Graduate Record Exam

Two important evaluation criteria used by graduate schools and potential employers are a student's academic record and letters of recommendation. From the academic record, the evaluator learns about a student's preparation by examining which courses have been taken. From the grades in these courses, the evaluator sees indices of the student's motivation, knowledge, and ability. Thus, good grades, combined with wise course selection, are critical to favorable evaluation. From letters of recommendation, evaluators learn if the student has formed significant relationships with professors and other professionals and how the student is perceived in these relationships. Additionally, if the letter is well written, the evaluator also gains a context of what the student's achievements mean and what the student had to do to attain them. Carefully crafted letters of recommendation do not merely convey information. They are deliberate and persuasive arguments that tell the evaluator exactly *why* the student should be admitted or hired.

Important as grades and recommendations are, both have limitations. Standards for grades vary among institutions, and high grades are more common, less distinctive, and less valuable than they once were. Recommendations, although important, are viewed with greater skepticism today because some references exaggerate, write letters too quickly, make misleading statements, or even lie. Unfortunate recent examples of students suing professors over poor recommendations have tarnished the influence of letters of recommendation even further. Although students now typically sign a waiver giving up the right to see letters of reference, the spectre of lawsuits or other forms of personal reprisal makes many references reluctant to put negative comments in writing. This often leads to letters of recommendation that contain entirely positive, but mostly meaningless, statements. To avoid this problem, students should determine, to the best of their ability, those professors and supervisors who are genuinely comfortable and enthusiastic about writing a letter of recommendation on their behalf. It is entirely appropriate to ask a potential reference, "Could you write a positive recommendation about me?" Students should be prepared to accept the answer "no" as graciously as they would accept the answer "yes."

Despite the best of effort and investigation to avoid these and other potential problems, there is no denying that letters of recommendation today no longer command the same level of influence, respect, and trust they once did. Similarly, favorable employment evaluations in conservation research and management, while viewed as an asset, are also treated with a certain skepticism. Thus, contemporary evaluators are faced with a dilemma. Grade reports, recommendations, and work evaluations are all valuable but subjective. They may be biased and distorted, and not easily comparable among applicants. Evaluators, particularly reviewers for applicants to graduate schools, have only one standardized measure that can be used as both an absolute and comparative standard—the Graduate Record Examination, or GRE.

The GRE is a standardized, multiple-choice test equally weighted in three areas of ability: analytical, quantitative, and verbal. The analytical section tests the respondent's ability to think logically and draw conclusions or inferences from given information. The quantitative section tests mathematical and numerical skills and knowledge. The verbal section tests the extent of the applicant's vocabulary—skills in understanding the meaning of words and ability to interpret written expression of ideas. Answering every question correctly in any section receives a score of 800.

Perfect scores are rare, and not even the best graduate programs demand perfection. What constitutes an acceptable, or at least admissible, score for an applicant to a graduate program or a job varies. Some graduate programs have a defined minimum score, below which applicants will not be considered, regardless of the strength of credentials in other areas. Although there is no national standard, a common minimum is either a composite score of 1,800 or a minimum of at least 600 in each section. Any student aspiring to graduate work in conservation biology must take the GRE. The most strategic time is typically in the spring of the junior year, the following summer, or, at the very latest, early in the fall of the senior year. If the scores are sufficiently high on the first try, the student has this important credential in hand and can proceed with preparing other application materials. If the scores are not satisfactory, there is still time to retake the test before most application deadlines pass. Although the GRE is promoted as an examination that tests breadth of educational experience, and therefore is not easily "studied for," most individuals can raise their scores on a second attempt, some significantly. Performance on the GRE usually improves with intelligent preparation, study, and practice. Various GRE test preparation services and aids exist, ranging from free online preparation services to more sophisticated and intensive efforts via agencies that charge for their services.

Graduate Studies in Conservation Biology

Choosing a Program

In 2001, fifty-six colleges and universities offered programs in conservation biology in the United States. These programs varied slightly by name, by emphasis (graduate or undergraduate), by specific curriculum, and by degrees offered. Applicants should ask specific questions about each program they consider:

1. *Is its curriculum accredited by appropriate professional organizations?* For example, persons with a particular interest in forest conservation biology can determine if their program is accredited by the Society of American Foresters, which maintains a list of schools with accredited curricula. Many other professional organizations with interests in conservation may accredit or in some way evaluate curricula.

2. *Do the faculty have a successful record of publication and grantsmanship?* Because a student graduating with an M.S. or Ph.D. in conservation biology or a related field will be judged not only on the degree itself, but also on the publications that result from his or her graduate work, it is wise to choose a program in which graduate theses and dissertations normally lead to one or more publications authored or coauthored by the student. In a quality program, this pattern is the rule, not the exception. In most programs, you can inspect the credentials of faculty members online, determining not only their interests, but their productivity and success.

3. *Do graduate students in the program receive adequate financial support and other compensation?* Annual stipends vary by institution, but today are normally ranging from $9,000 to $15,000 for master's students and from $15,000 to $24,000 for Ph.D. candidates in conservation-related programs. Most waive or reduce tuition, and many provide health insurance. Although salary should not be the only factor in choosing a program, it is generally unwise to commit to a graduate program that does not fund its graduate students. A lack of financial support means that the program has made little meaningful commitment to or investment in its students' success.

4. *Do graduates of the program enjoy a good record of employment or placement in programs of more advanced graduate study?* Statistics on employment of graduates can usually be obtained, with a little perseverance and digging, from departmental or university records. A poor record of placement suggests that the program, and/or the work of its students, are not being perceived as credible by outside reviewers. A strong record not only indicates the opposite, but suggests that an incoming graduate student enjoys a preexisting network of supportive contacts who are successful products of the program.

Other program-specific factors are important, but sometimes more subjective and difficult to measure. Many things can be learned only by on-site visits and interviews. These are essential to making a good decision. When meeting with faculty and students of the program in person, visiting applicants should try to determine how well the faculty and graduate students work with one another. Do they seem to enjoy one another, affirm one another's efforts and accomplishments, and work cooperatively in joint research and institutional efforts? An absence of these qualities or, worse, evidence of personal hostilities among faculty or graduate students, identifiable "camps" or "followings," or negative comments about fellow faculty or students are symptoms of an unhealthy personal and professional environment that should be avoided.

Choosing a Project, Graduate Professor, and Mentor

Although the research project, graduate professor, and mentor are technically three different things, they are invariably related, and may all be tied to the same person. An ideal project is one that combines several traits. First, is the research of genuine interest to the student? Does the student want to have her or his name and reputation associated with this problem in the professional community, and perhaps, in the eye of the public? Although research interests will grow and expand in any healthy and developing career, it is likely that a student's name will be associated with his or her first publications for a very long time. Students should consider carefully if a proposed project is one that they respond to with enthusiasm, apathy, or dread. Poor fit between student interest and research effort leads to low levels of motivation, and that, in turn, leads invariably to poor research and low levels of subsequent professional success. Second, is the project of significant interest to the broader professional and scientific community? That is, does it address questions of foundational interest to the discipline, especially questions that, if answered, may provide general illumination on current theoretical or management predictions? Third, does the research address significant current issues or problems in conservation? Fourth, is the research doable within existing constraints of time, expertise, and funding? Doing good research is always challenging, but the probability of success should be significantly greater than zero. The practicality and feasibility of research will usually be evident if the student makes a careful analysis of proposed experimental designs, past successes (or failures), faculty expertise in the proposed effort, and background in technical and logistic support for the project within the department and the university. Although some elements of a new project are always uncertain, projects that are overly doubtful are usually the product of poor planning and poor underlying support structures.

Although it is fashionable for academics and professionals to talk more about mentoring today than in the past, good mentors remain rare. The word *mentor* is derived from the Greek. It was not originally a word, but a name. When the Greek hero Odysseus departed for the war against Troy, he entrusted his friend Mentor with the education of his son, Telemachus. Out of the ideal of this noble relationship and responsibility has grown the *definition* of what a mentor ought to be: "a trusted counselor or guide, a tutor or coach." Mentors are distinguished from teachers because they are not merely interested in imparting knowledge and skills, but are actively interested in the total welfare and growth of the student or younger colleague. A true mentor possesses the wisdom to discern what is good for a student and has the power to bring it about. Thus, an ideal mentor is a person who takes a sincere and selfless interest in a student's welfare, sees his or her potential for growth as a person and a professional, understands what he or she needs to achieve such growth, and has sufficient personal and professional influence to arrange the resources and opportunities necessary to see that such growth takes place. Mentors are also distinguished from teachers by their active efforts to train their students to make wise choices and avoid pitfalls or traps that would slow or stop their development. Students should be alert to the opportunity to develop

Table 13.3   Contrasting Motivations, Goals, and Constraints of Conservation Managers and Academic Research Biologists

JOB COMPONENT	CONSERVATION MANAGER	ACADEMIC RESEARCHER
Motivation	Questions driven by need to answer specific problem, eye toward application.	Questions driven by theory, basic science.
Goal(s)	Provide data to manager to guide management; derive guidelines for action.	Publish in high-quality journals; compete for research funding.
Service	Explicit responsibilities to agency; realistic goals.	State public and idealistic goals.
Time frame/work schedule	Work quickly to obtain data; long planning range of agency budget process.	Conform to class, academic schedules; projects chosen to fit thesis schedule.
Staffing	Cost-effective workers.	Train students in modern techniques, find them jobs; recruit and support students.
Financial considerations	Need to accomplish as much as cost-effectively possible.	Grant funds important, especially indirect cost recovery for some institutions.

After Huenneke 1995.

relationships with true mentors and pursue them actively. There is no other single influence that can affect their future success as much.

Choosing a Vocational Setting

Should I Take This Job?

If you are a student now, it is vital to remember that graduate school is only one path to the goal of effective conservation work. And for all but those who become college professors, academic life is something that will not be the last stop in a career journey. Thus, one must consider what kind of vocational setting is appropriate to one's goals. In any vocation, the work environment is an important determinant of personal satisfaction and professional productivity. We sometimes have the luxury of choosing from among several possible work environments. Traditional occupational elements such as location, salary, and benefits should receive their due consideration. But most people find that these are not the primary satisfying factors in work. For an emerging conservation biologist, other aspects are more likely to exert a greater influence over personal satisfaction in work. Here are four additional criteria to consider.

1. *Is there strong correspondence between the organization's mission and my own personal mission as a conservation biologist?* All effective organizations place definite limits on goals and priorities. Their effectiveness is, in large measure, a result of such limitations. By focusing on se-

lected missions and targets for which they are uniquely suited, they achieve success. An important initial consideration, as a potential employee, is to look for alignment and correspondence between your mission and the organization's mission, between your interests and their interests, and between your abilities and their needs. The closer the match in mission, interest, and need, the more satisfying the work is likely to be, and the more you will be valued and esteemed by the organization. You must also consider the varying reward systems associated with different professional cultures and determine, in advance, what you want to be rewarded for and what you are good at producing that will gain rewards. For example, the reward systems of academic and nonacademic careers in conservation biology are very different (table 13.3). Academic researchers are rewarded for publication and grantsmanship, whereas conservation managers in nonacademic settings, such as government, are rewarded for providing data to guide specific management actions and policies and doing it quickly. These different incentive and reward structures create very different professional cultures and working environments.

2. *Does the organization reach the same audience that I want to reach?* As in the previously discussed issues of scope in conferences and journals, so conservation agencies and organizations operate at international, national, regional and local levels. Does the sphere of influence in which you want to operate match the organization's sphere of influence? If the answer is yes, then work is

likely to be highly satisfying. If the answer is no, it is likely to be extremely frustrating. A biologist who wants to address national issues of conservation is likely to be miserable working for an organization that deals only with local or regional concerns.

3. *Would I be working with people I can respect and trust, and with whom I share common interests?* It is often uncertain, given limited contacts with potential co-workers, exactly how relationships will develop in an organizational environment. A measure of optimism is appropriate. Blind faith is not. If potential co-workers display obvious behaviors, attitudes, or ideas that you cannot accept, practice patterns of work and activity that lack professional or personal integrity, or display a philosophy of work or management to which you are opposed, then decline the position. Never, under any circumstance, accept a position with individuals whom you know beforehand you do not like and respect, regardless of their professional prestige.

4. *Is my job something that I can do effectively and with satisfaction?* Even in the best organizations and among the best co-workers, jobs vary. Is the proposed job description one that offers an opportunity for you to display your strengths, grow in your competencies, and contribute to the organization's goals in ways meaningful to the organization and to you? An effective and satisfying position should provide a clearly defined organizational role, and should involve the individual in decision making that is appropriate to that role and relevant to his or her expertise.

After You Take the Job

How Can I Excel in My Job?

In more menial jobs, tasks are specific, deterministic, and easily evaluated. Retention is based on meeting minimum standards of performance and productivity. In more meaningful work, such as a career in conservation biology, both the job and its evaluation are more fluid and require creativity and imagination. Meeting only minimal expectations within the limits of written job descriptions is unlikely to lead to satisfaction or advancement, and may even result in dismissal. Effective workers follow certain principles in approaching their assigned tasks and in making strategic decisions about which organizational tasks to undertake when they have the freedom to choose tasks themselves.

1. *Make work output responsive to the needs of others.* Every occupational task and output, whether in the form of written reports, oral presentations, management decisions, or data analysis interfaces with others in the organization who might use such output in their own tasks. Before undertaking a specific task, determine: (a) Who will receive the output of my work? (b) What questions should this output answer to be most helpful to them in accomplishing their own goals and meeting their own needs? (c) In what format should this output be presented so that it can be adapted with minimum effort to other uses and contexts? Employees who consistently produce output with their co-workers' needs in mind,

and who frame their work in ways that make it easy for others to use and integrate their efforts in other ways, are employees who become more influential in organizational life and who are most appreciated and respected by others.

2. *Determine the indispensable needs of your organization or work group, and make yourself the person who meets these needs through your work output.* Although all organizations manifest a diversity of work output, there are certain core objectives in work that must be accomplished for the organization to function. These core objectives are usually not included in any specific job description or posted on the agency bulletin board, but they are quite real nonetheless, and can be determined by a discerning employee. Given a measure of freedom in tasks and priorities, a strategically thinking employee will choose those tasks that make him- or herself indispensable to an organization. This is done by determining what these sorts of indispensable needs are, which of them one can accomplish, and then fulfilling these indispensable needs in a relentlessly consistent manner.

Quality of Relationships

Current research on organizational productivity reveals that the most effective and productive organizations are those that manage through relationships or, as it is sometimes expressed, those that manage networks rather than tasks. Effective leadership does not attempt to control others but to inspire others to share common goals and take ownership of organizational tasks that will accomplish those goals. Such an approach requires a high quality of relationships in a work environment. Quality of relationships can be maintained by consistently and intentionally pursuing certain principles in relationships with fellow workers.

1. *Seek to build relationships in a work environment on the basis of common goals, shared tasks, and shared credit for accomplishment.* To the extent possible, expand your own work tasks and objectives to include the work of others. When this strategy is followed, it produces not only better work with wider applications, but also a work environment in which people share in the fruits of accomplished tasks.

2. *Take a genuine interest in the welfare of others and build trust by understanding their needs and being an active part of meeting them.* Organizational relationships without trust, no matter how efficiently designed, simply do not work. Trust is established by deliberately being attuned to both the professional and personal needs of fellow workers and, whenever possible, being a resource that helps such needs to be met.

3. *Whenever possible, involve others in decision making.* People do not support what they do not create. Each person who has a stake in creating an organizational objective has a stake in achieving it. Whenever possible, deliberately seek out and create opportunities for others to share in decision making in work that affects them, and to which they can make a constructive contribution.

The education you pursue today will not match your profession as you find it tomorrow. Students who are preparing for work in conservation now will not enter the field as fully established professionals for years to come. During their gestation as undergraduate students, graduate assistants, and agency interns, many things will change. How should students prepare themselves now for the professional environments and conservation problems they may confront in the next 5 years?

I do not know the answer to this question, but I know it is the right question to ask, because education is an expression of planning for the future, a future we cannot yet clearly see. Therefore, in the final section of this final chapter, we will consider four emerging trends in conservation biology and their implications for conservation education. Because change is life's only constant, successful living requires foresight. However, fortune-telling has never been a particularly steady or reputable line of work. And the vocation carries multiple risks. Vague predictions are worthless. Specific predictions can be wrong. The predictions I offer are short term and uncertain, but specific and suggested by already developing trends. I cannot imagine every possibility, but here are some trends for which one might prepare.

EMERGING TRENDS IN CONSERVATION BIOLOGY—THE NECESSITY AND RISK OF PREDICTING THE FUTURE

Trend One: An Increasing Confluence of the Small-Population Paradigm and the Declining-Population Paradigm in Addressing Species Management

According to population ecologist Graeme Caughley, conservation biology has only two paradigms: the small-population paradigm and the declining-population paradigm. The small-population paradigm seeks to understand "the effect of smallness on the persistence of a population" and the declining-population paradigm "deals with the cause of smallness and its cure" (Caughley 1994).

Conservation biology emerged as a distinct discipline through an increasingly sophisticated development of the small-population paradigm, particularly in genetics. The development of the small-population paradigm was explicated in the concept of the "extinction vortex," "that compound snare of feedback loops by which inbreeding depression, demographic stochasticity and genetic drift might combine to render a population smaller" (Caughley 1994 chapter 5). The strength of the small-population paradigm was its ability to make generalizations and predictions about small populations that were applicable regardless of species. But the small-population paradigm was not always helpful or predictive in real-world management of rare species.

The declining-population paradigm has traditionally been the realm of the applied sciences, and its applications are evident in management. It seeks to understand what causes populations to decline and how such causes might be removed. Unfortunately, such investigations almost always have operated on a case-by-case basis. Few generalizations have emerged about causes of decline that could be applied to all populations or that could make specific predictions for multiple contexts. Although this paradigm could not provide a theoretical underpinning for conservation biology, it did provide the means for most of conservation biology's specific successes because it assumed that any population decline had a tangible and specific cause that could be identified and removed.

Future efforts in conservation biology must move toward an increasing synthesis of these approaches. The managed recovery of the Lord Howe Island woodhen (chapter 7) is an example of such a synthesis, whereby on-site captive breeding (small-population paradigm) was coupled with astute diagnosis and treatment of proximate causes of decline (declining-population paradigm) to produce a successful recovery. The previously cited work by Pavlik (1993, 1994) to restore the endangered plant *Amsinckia grandiflora* by using an experimental approach and trend analysis (Bowles and Whelan 1994) is another example of a successful merger of the small-population paradigm and the declining-population paradigm. Here, investigators considered both the factors that keep populations small and those that cause their decline, in order to facilitate the recovery of a single species.

As regional and ecosystem management approaches become pervasive, the theoretical problems associated with small populations will gain increased attention by managers who do not have the resources to address the recovery of every small population on a case-by-case basis. This trend will be coupled with an increasing tendency to treat management recommendations as testable hypotheses and management actions as experiments. Recall that this approach is called **adaptive management** (chapter 10), which is *a process that combines democratic principles, scientific analysis, education, and institutional learning to manage resources sustainably in an environment of uncertainty* (Lee 1993). Adaptive management requires experimental manipulation of the system and willingness to change research priorities according to critical management needs. In adaptive management, experimental results must be compared with predictions made in advance. Thus, experimental design and manipulation become critical to management success.

An implication of this trend is that recovery plans for declining populations must increasingly adopt an experimental approach, and the agent of decline must be determined via elimination of alternative explanations through rigorous experimental tests. Because hypotheses are derived from theoretical principles, experiments to determine the agent of population decline must examine the theoretical basis of the small-population paradigm. The increasing scrutiny we can expect in the management of endangered species and the growing use of adaptive management approaches will result in a greater need, and more intentional research, to elaborate general theories of decline in populations and to draw upon existing theories of small populations to design sound experiments.

POINTS OF ENGAGEMENT—QUESTION 2

If the above prediction proves correct, will remaining distinctions between the small-population paradigm and the declining-population paradigm eventually be lost? What do *you* predict will be the effect on conservation biology if the small-population paradigm is more widely applied to specific cases and the declining-population paradigm develops more theory about general and predictable causes of population decline?

Trend Two: A Growing Need to Assess and Manage Multiple Human Interests, Values, and Relationships at Landscape and Ecosystem Scales to Achieve Meaningful Conservation Goals

As noted in chapter 10, movement toward ecosystem management in conservation biology has, to date, been focused on refining conceptual issues and definitions. Although an operational and more widely accepted definition of ecosystem management is essential to progress in conservation, we are already seeing an emerging trend toward the application of ecosystem management as a functional conservation necessity, even though its definition is not yet clear. In a comprehensive review of ecosystem management, Yaffee (1996) identified 77 U.S. projects that have used ecosystem management approaches. All grew out of practical necessities associated with the goals of particular conservation efforts, not because the parties involved were initially attracted to or had knowledge of ecosystem management concepts. In fact, most developed because initial approaches using traditional management concepts failed. For example, Ohio's Big Darby Creek is a relatively small (132 km) stream inhabited by 40 species of mussels and 86 species of fish. The challenge of preserving the water quality and aquatic habitat associated with Big Darby was exacerbated by the fact that very little of Big Darby's stream course flowed through protected lands or nature reserves. Conservationists were forced to build a broad coalition of support to achieve adequate protection and management. The outcome was the Big Darby Creek Partnership, a conservation coalition that eventually included stakeholders from more than 40 private and public organizations. Although many management actions are site- and stakeholder-specific, the partnership has succeeded in establishing landscape-level management policies and practices. Some of these include prescribed burning on both public and private lands to rejuvenate native vegetation, as well as a voluntary "grass banking" program in which ranchers receive grass for livestock in exchange for donating permanent conservation easements on land adjacent to the stream (Yaffee 1996).

The expanded application of ecosystem management approaches appears likely to continue in the short term, not through government‐mandates and federally prescribed ecosystem management "councils," but through the personal initiatives of private and public groups. Increasingly, the success of ecosystem management will depend on the initiative of managers who possess high levels of interpersonal skills and know how to use them. As a consequence, students of conservation biology must learn how to use such skills as precisely as they use technical skills like GIS and radio-telemetry. Students interested in building coalitions for conservation will make the acquisition of interpersonal skills to facilitate cooperative behavior, group decision making, and effective interpersonal communication part of their academic curriculum and vocational training. Further, faculty and other curriculum planners in conservation biology must be more deliberate in creating curricular structures that enforce thinking across disciplines, working cooperatively and effectively in teams, and developing effectiveness in a wide range of communication skills (Touval and Dietz 1994). But if students find that their curriculum contains no such requirements, they must gain these skills through their own initiative, selecting appropriate elective courses and gaining summer professional experiences that require them to practice such skills.

Human presence in and use of ecosystems are likely to become even more important factors in ecosystem management in the future. The "nature under glass" mentality that has pervaded North American conservation during the past century will have to be discarded in many cases for ecosystem management to work. And the traditional curricular structures developed in disciplines such as wildlife management, forestry, and range management will be inadequate to deal with the changing environment of ecosystem management. In India and other developing countries, for example, Saberwal and Kothari (1996) advocate radical revision, if not abandonment, of the traditional North American conservation model in favor of an approach based on the social sciences and humanities. Their call for revision is not based on a faddish desire to seem more interdisciplinary, but on fundamental differences in the interactions between people and parks in developing countries and North America.

"Unlike the west, the great majority of national parks and sanctuaries in developing nations also serve to meet the daily subsistence requirements of millions of rural poor. In India, for example, three million people were living inside protected areas in 1996, with millions more living in areas immediately adjacent to them and dependent on their resources for their subsistence. . . . The ensuing conflicts between humans and wildlife and between local populations and state institutions over access to and use of these wildlife resources has increasingly taken center stage in conservation debates in many parts of the third world" (Saberwal and Kothari 1996).

Because parks and reserves in most parts of the world must continue to provide resources for the subsistence of human populations, future conservation managers in these contexts must be able to make analytical assessments of sociological and economic conditions, and then combine their knowledge with sensitivity to cultural traditions and practices within and around nature reserves and parks. Students who have

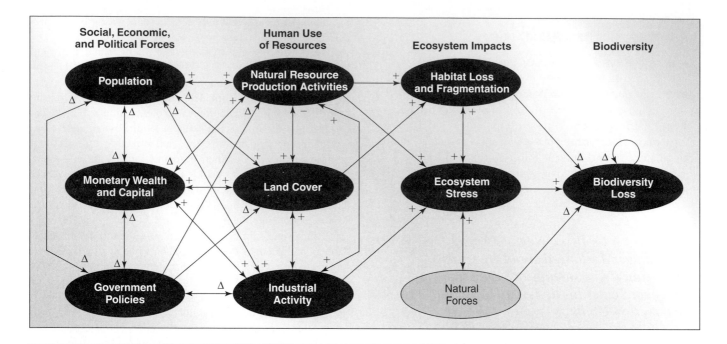

Figure 13.4

A preliminary predictive model of human effects on biodiversity loss. Arrows represent direction of influence of variables. Signs depict how an increase in activity in one variable will hypothetically affect activity in the second variable (increasing [+], decreasing [−], or causing an unspecified change [Δ]). The gray oval (Natural Forces) represents an essential part of the system that is not a human impact. Feedback loops between biodiversity and other model components are not illustrated. The model illustrates the importance of socioeconomic and political factors on ecosystem impacts and biodiversity loss.

After Forester and Machlis (1996).

incorporated courses in the social sciences and humanities into their conservation biology curriculum may possess an advantage over classically trained natural scientists in successfully implementing conservation strategies under these conditions.

Conservation efforts in the United States and Canada also must incorporate human factors if conservation efforts in ecosystem and restoration management are to expand to larger regional scales. In particular, ongoing efforts to restore populations of large carnivores such as the timber wolf and grizzly bear will involve large land units, diverse interests and stakeholders, and varied private and public land jurisdictions so that regional success will be impossible without cooperative planning and decision making. Conservationists who emerge as leaders in this developing sociopolitical environment will be those who are best able to manage relationships and lead groups of diverse interests, not those who have the greatest scientific or technical expertise.

Michael Soulé, the founder of modern conservation biology, demonstrated characteristic prescience in envisioning the social and political role of a conservation biologist. Soulé anticipated the vocation of "consulting ecologist," "whose contributions to economic and social well being would be on a par with other tactically focused professionals such as engineers and architects" and who could produce "pragmatic solutions to disputes between conservationists and those in the labor- and market-oriented sections of the economy" (Coleman, Mattice, and Brocksen 1996). Soulé worked to actualize this concept by creating the Center for Biodiversity Analysis and

Management (CBAM) at the University of California at Santa Cruz. Comprised of interdisciplinary faculty, CBAM conducts research designed to expand the theory and practice of "socially sound, politically viable, and scientifically robust approaches to regional conservation" (Coleman, Mattice, and Brocksen 1996).

Technical and theoretical tools to support this more interdisciplinary approach to conservation biology and management are now being developed and put to use. Forester and Machlis (1996), for example, constructed an interdisciplinary model of biodiversity loss that attempts to evaluate the effect of human factors on the decline of biodiversity. Their model (1) integrates social, economic, and political forces that affect rates of human use of resources; (2) predicts the effects of these forces on various aspects of the ecosystem; and (3) makes predictions about biodiversity loss (fig. 13.4). Students who can understand, use, and construct similar models will be well equipped to achieve conservation goals at ecosystem and landscape levels in the face of human need and resource use.

Managing human relationships and interests to achieve conservation goals at landscape and ecosystem levels means that traditional "command and control" approaches to conservation by government agencies will be complemented with emerging coalitions of private and public organizations that pursue common conservation goals on the basis of shared values and mutually beneficial interactions, not mutual fears of imposed fines and punishments. The development of this trend is tied to a third trend emerging in the short-term future of conservation biology.

Trend Three: The Need for More Coordinated Relationships Between Government Conservation Regulations and Market-Driven Conservation Incentives

Students of conservation history learn quickly that market forces, left to themselves, do not achieve conservation. The decimation of global fisheries, the whaling industry, and the upper Great Lakes forests of the United States, among many examples, illustrate the potential of market-driven free enterprise to exhaust resources. Modern environmental regulation, beginning in the United States in the 1960s, provided definitive protection for diminishing resources and endangered species, acted forcefully to stop unreasonable levels of pollution, and established conservation as a national priority. What has been less noticed, but not less effectual, is the growth of a national foundation of citizen-based support for conservation values. These values have been spread globally through the statutory cloning of U.S. conservation legislation in other countries. Citizen-based support is not monolithic, but is racked with controversy from within and attacked by well-organized opponents from without. This citizen base does not support conservation values to the full extent that most practicing conservation biologists would like. But public support is genuine, wide, and deep, and forms an effective cultural environment in which conservation can progress.

Private enterprise and free markets cannot achieve conservation goals through market processes, but they can be useful servants in reaching such goals if guided by community-based values and government constraints. Although recent conservation history has shown the power of government to impose conservation as a national priority which it advanced by legal prohibition against environmental destruction, even conservationists sympathetic to the government's ends must admit the inefficiency of some of its means.

Chapter 12 provided a brief survey of some market-driven structures and mechanisms that can be used to serve conservation goals, and can do so more efficiently than government-imposed regulations. Today there is a growing trend toward increased use of market-based incentives to achieve conservation goals. One example from Montana illustrates how this trend expresses itself and how conservation biologists must be prepared to deal with it.

Figure 13.5

Paddle fish (*Polyodon spathula*) roe provides the basis for a growing caviar industry in Glendive, Montana.

A Case History of Government-Free Market Coordination in Conservation—Paddlefish Caviar from Glendive, Montana

In eastern Montana, the town of Glendive is home to approximately 6,000 residents. Its traditional economic base was agriculture, particularly cattle ranching. Located along the lower reaches of the Yellowstone River, Glendive is also famous for its paddlefish (*Polyodon spathula*) (fig. 13.5), which spawn over gravel bars in swifter sections of this part of the Yellowstone. Paddlefish are highly valued as food, and even one can provide a lot of it. Typical adult paddlefish may be 5 to 7 feet long (including their paddlelike snout) and weigh 60 to 120 pounds. Each year, the paddlefish season (15 May to 30 June) attracts about 3,000 anglers to Glendive. Because paddlefish feed mostly on plankton and other microscopic organisms, they cannot be caught with conventional bait and lures, but must be snagged. Fishers use large treble hooks on weighted, heavy lines, heaved into the river with surf rods 8 to 12 feet long.

Paddlefish roe (eggs) are used as caviar, comparable in quality to the more famous caviar of sturgeons (family Acipenseridae). In the southern United States, paddlefish populations in the Mississippi River and its larger tributaries have been overexploited, resulting in the closure of many state fisheries in this region (Anderson and Leal 1997). Local entrepreneurs in Glendive, however, recognized that the fishers harvesting paddlefish from the Yellowstone and the nearby Fort Peck Reservoir and Lake Sakakawea, had no interest in the roe. In fact, fishers had historically dumped several tons of roe on the banks of the Yellowstone near Glendive each spring, attracting large concentrations of flies and rats, and creating a public nuisance and health hazard. Glendive business planners conceived a plan to open a caviar-processing plant and market the caviar internationally. Although the U.S. Fish and Wildlife Service was reluctant to approve a project that might increase demand for paddlefish products, approval in the Yellowstone rested with state officials of the Montana Department of Fish, Wildlife, and Parks. Biologists in the department determined that local populations of paddlefish appeared stable and secure. The paddlefish habitat in the river, reservoir, and lake also was reasonably well protected and not faced with any serious threats of pollution. The plan did not increase mortality to the population or decrease recruitment because the roe would come only from fish harvested by private anglers. No additional paddlefish would be taken exclusively for roe. However, to ensure the profit motives did not lead to overharvest, state officials stipulated that fishers had to donate their roe, not receive payment for it. To make donations more attractive, the Glendive community agreed to employ individuals who would clean an angler's paddlefish without charge. In return for the cleaning service, the cleaner would keep the roe and give it to the caviar factory. Finally, the state stipulated that, in addition to a regular fishing licenses, paddlefish anglers would be required to purchase special paddlefish tags that must be affixed to each fish caught (Anderson and Leal 1997). In order not to favor the affluent, the cost of the paddlefish tag is low (in 2002, $2.50 for Montana residents and $7.50 for fishers from out of state), but the number of tags that can be sold is fixed. In this way, state officials were able to control, with near certainty, the maximum number of paddlefish that could be harvested in a season, while making a food resource and recreational opportunity accessible to all citizens. Finally, the Glendive business community agreed to devote half of the net revenues from paddlefish caviar to the state of Montana for paddlefish research and management (Anderson and Leal 1997). On 3 March 1989, Montana Governor Stan Stevens signed a bill that made it legal for the Glendive caviar factory to sell paddlefish eggs, effectively providing an exception to a state law that prohibited the sale of wild game products. In its first 7 years of operation, the caviar project in Glendive grossed over $1 million in revenue.

In our examination of values and ethics in conservation (chapter 3), we noted the dangers of tying conservation to market-driven demand, particularly in the example of the green turtle. We would be wise to remember the words of conservation biologist David Ehrenfeld (1992) regarding this example, that "the power of global demand erodes all safeguards . . . the commercial ranching of green turtles inevitably brings us around again on the downward spiral—a little closer to the extinction of the remaining populations. By no stretch of the imagination is this conservation." But there are differences between this case history of paddlefish caviar and sea turtle ranching. These differences can be isolated in six elements of the Montana case history that have general applicability to a successful interaction between governmental authority and free market systems.

Element One: The resource to be used is derived from stable populations whose habitats are protected and managed for the population's benefit. The state, not the market, made biological determinations of the status of the population and its habitat. The health and sustainability of both were mandated by a state regulatory authority as prerequisites for any use of the paddlefish as a human resource.

Element Two: The persistence of the resource in perpetuity is mandated, regardless of market comparisons of present versus future values or considerations of opportunity costs. The market was not consulted about whether paddlefish habitat could be used more profitably for other purposes, or whether the current value of paddlefish at higher rates of exploitation had more value than the future or option values of paddlefish populations in future generations. The persistence of paddlefish in Montana was taken as given and nonnegotiable, set above and beyond the reach of the market by the state's statutory authority, implicit in both its laws and in regulatory policies that enforced them.

Element Three: The rate of exploitation of the resource is determined by biological criteria, not economic criteria. Optimal exploitation rates of paddlefish were not determined according to the criteria of profit maximization, supply and demand, or human welfare. The exploitation rate was determined by the biological productivity of the local paddlefish population as determined by biologists.

Element Four: The harvest is administered by the state in a manner that ensures that maximum sustainable yield is not exceeded, and in a way that removes profit incentives to violate yield restrictions. By requiring every harvested paddlefish to be tagged and having state game wardens arrest violators, the state enforced the conditions described above without regard to market values. By making the tags relatively cheap despite their scarcity, the state made the opportunity to harvest the resource independent of market forces that would have favored the affluent. By requiring that roe be donated rather than sold by fishers, the state removed profit incentives to harvest more paddlefish than had been allowed through tag sales.

Element Five: Free enterprise is allowed to create a market for a natural resource, to allocate the distribution of the resource according to supply and demand, and to permit the private sector to receive an economic incentive. With restrictions in place that prevented the market from determining the *amount* of the resource that could be harvested, the market was permitted to determine the *value* of the resource that was harvested. A portion of the profits was received by those who marketed the resource, giving them incentive to continue, and placing a specific economic value on the resource itself.

Element Six: Profits are tied to resource sustainability by mandating that a portion of the profits be reinvested in the productivity of the resource, not in expanding the production or harvest capacities of the market. Historically, profits from commodity resources have been used to increase the capacity of exploiters to take more resources at faster rates. This pattern increased short-term profits, but decreased or destroyed long-term resource sustainability. In this case history, however, profits are used to improve the productivity of the resource. More profits can come from larger harvests, but only if the investments lead to larger populations that produce larger yields. Thus, an important positive feedback system is established that couples economic profit and biological productivity.

Transferable Principles from the Glendive Experiment

The harvesting program for paddlefish flesh and roe in Glendive illustrates two principles that conservation biologists are inclined to forget: (1) not all forms of resource use lead to resource depletion, and (2) not all forms of resource use lead to habitat degradation. When these two conditions can be met, conservationists can work with the private sector to create markets for resources that assign specific economic values to benefits derived from such resources. The overall lesson from Glendive, and other examples like it, is straightforward when coupled with lessons from conservation history. The market is a bad master for conservation values, but it can be made a useful servant to achieve conservation ends, if its mechanisms are channeled to achieve profit according to predetermined, community-based conservation values enforced by government regulation. Administered wisely, the coupling of conservation values and market incentives can produce and efficiently distribute benefits that build broad-based, community-level support for conservation more efficiently and more effectively than legislative mandate alone. Such efforts can help move the concept of sustainability beyond the academic and professional culture of conservation biology into the culture of private business and economics. In this case, the people of Glendive came to understand that a healthy, stable population of paddlefish is an index of their community's well-being. Decline in the paddlefish population would be viewed by them as a symptom of distress. The status of the paddlefish population is quantifiable, and the level of harvest is scientifically defensible. Thus, the Glendive paddlefish industry is truly sustainable, and rests on a well-defined cultural and economic foundation.

Students of conservation biology should note this trend of coupling conservation goals with private economic benefit, and make themselves competent in economic concepts and analysis in order to recognize opportunities for such couplings when they are present. Practitioners trained in this manner will be able to build increasingly stable foundations for conservation by fusing public regulation with private incentive.

POINTS OF ENGAGEMENT—QUESTION 4

What conservation efforts can you name or envision that could be more effective if the conservation goal for a given population was coupled, under appropriate regulation, to a market-driven incentive, and how could such a coupling be created without threatening the population?

The final trend we will examine is the most uncertain, but lies at the heart of what conservation biology is and what it will become. The use of market-based mechanisms as an aid to paddlefish conservation assumes that paddlefish ought to be conserved. The question for the future is, "Can conservation biology successfully articulate the basis of this assumption, not only for paddlefish, but for any species?"

Trend Four: Increased Tension and Debate in Defining Conservation Biology as Information Driven Versus Mission Driven

In a recent letter, a correspondent argued that Michael Soulé's concept of conservation biology as a mission-driven, value-laden discipline was misguided and harmful, that it put scientists on the slippery slope of advocacy when they ought to stay focused on providing objective scientific information. My correspondent's views are not an isolated perspective. They have become increasingly common in the professional culture of conservation biology, both implicitly and explicitly.

Brussard, Murphy, and Tracy (1994) argued that activism by conservation biologists should occur "outside of our professional society" and that conservation biologists and conservation activists should, as groups, maintain "a strategic and measurable distance" (Brussard, Murphy, and Tracy 1994). Conservation biologist Dennis Murphy (1990) defined conservation biology as "the application of classical scientific methodology to the conservation of biological diversity." Murphy asserts, "Conservation biology only exists because biological information is needed to guide policy decision-making." He concludes that "the practice of conservation biology ends where science ends and where advocacy begins." Thus, conservation advocacy, specifically the act of working toward a normative outcome or condition, is not, by this definition, conservation biology. Rather, the role of conservation biology is to provide scientific knowledge to resolve technical questions associated with the formation of conservation policy, but no more (Murphy 1990). Murphy's views are supported implicitly in recent patterns of publication. For example, in addressing why biodiversity ought to be maintained, Walker (1992) tells the readership of *Conservation Biology*, "I exclude here the moral and ethical arguments, not because they are unimportant but because they are nonscientific." Despite Walker's protest, the moral and ethical arguments, with their accompanying normative values, also would seem to be inappropriate to discuss in the text of the article. This pattern of thought has increasingly become the norm in featured articles in the journal.

Overtly normative papers addressing issues of value were common as feature articles in *Conservation Biology* in its early issues of the 1980s. In the 1990s, however, writings addressing normative values, ethics, and mission have increasingly been placed in editorials, essays, and letters, but not in articles (Barry and Oelschlaeger 1996). This pattern is typical of many scientific movements and professional societies. For example, early issues of *The Journal of Wildlife Management* in the 1930s also showed a greater willingness to address normative values and professional mission (e.g. Errington and Hamerstrom 1937).

Some conservation biologists see these trends as a normal and healthy process of professional maturity in the discipline, a pattern that prevents it from becoming a "pseudoscience" (Murphy 1990). Others see such trends as expressions of intellectual error and moral capitulation. Caughley (1994), for example, asserts that "the saving of a species from extinction has always been a paramount responsibility within the field of biology." But to assert that science has a "responsibility" to save species is a normative, value-laden statement of advocacy that makes no sense unless one appeals to a standard of what "ought" to be the correct application of scientific knowledge. The birth of conservation biology is rooted in such normative assumptions that formed the basis of its identify and mission (chapter 1), and the rapid growth of conservation biology has been a testament to their appeal. In attempting to define the discipline in its early years, Michael Soulé wrote, "conservation biology is a crisis discipline grounded in the recognition that humans are causing the death of life—the extinction of species and the disruption of evolution" and that conservation biology is a response by "those scientists who feel compelled to devote themselves to the rescue effort" (Soulé 1991).

The proposition that conservation biologists can retain value neutrality by simply providing information to managers and policy makers is suspect on four counts. First, conservation biologists must make a priori decisions about what information to present and how to present it. This is not a question of unscrupulous manipulation, but of legitimate scientific perspective, and such perspective is determined, in part, by what the biologist understands to be the normative uses and values of the information. Second, managers and policy makers often interpret objective information, such as model results, as offering normative diagnoses and prescriptions (such as why a population is declining and what to do about it) instead of seeing model results as objective descriptions of "what would happen if" certain conditions prevail. Clouded by this misconception, policy makers often present normative recommendations to the public as if such recommendations were objective and inescapable conclusions dictated by the model, not by their own normative values. For example, managers and policy makers often use ecological and population models to legitimize policy decisions rather than inform the public (or, in some cases, themselves) about the consequences of possible decisions they could make. In the words of modeler J. B. Robinson, "By cloaking a policy decision in the ostensibly neutral aura of scientific forecasting, policy makers can deflect attention from the normative nature of that decision" (Robinson 1992). As noted earlier (chapter 4), Porter and Underwood (1999) assert that management decisions are inherently normative, and that conservation managers are constantly engaged in a process of trying to both understand and value ecological process, precisely because they must choose the point at which they intervene in the process. They must then communicate their reasons to diverse public interests. "Whether we define ecological integrity in terms of species or processes, we must inevitably make a decision as to where in the . . . sequence we choose to intervene. That choice represents a value judgment" (Porter and Underwood 1999). Management decisions are therefore inescapably expressions of normative values. Thus, conservation biologists should ask themselves if they would be better off to consider and evaluate normative values in their presentation of data and recommendations or to let managers and policy makers use their results (and their own credibility as scientists) to make normative policy decisions that are disguised as value-neutral statements dictated by model results or other kinds of scientific facts.

This second problem leads naturally to a third. If conservation biologists refuse to consider normative values associated with their research, they may find themselves confined to addressing trivial questions instead of issues of real significance. For example, Caughley (1994) asserts that the primary query answered by population viability analysis—namely, how long a population will persist—is a trivial question. Having a specific answer does not make the question any more significant. The significant questions are, "What is putting the population's persistence in jeopardy?" and "What can we do about it?" But to answer the latter question implies a normative end, that the population *ought to persist,* and its solution requires not merely provision of information but changes in human behavior, as well as specific social and political outcomes that change conditions causing endangerment. Working toward these ends sounds suspiciously like activism.

The fourth problem associated with a failure to evaluate and articulate normative values associated with preserving biodiversity was reviewed previously in chapter 3 by legal ethicist Lawrence Tribe: If conservation biologists do not articulate the preservation of biodiversity as a normative value in conservation, then managers and policy makers are most likely to frame the value of species, habitat, and ecosystems entirely in terms of human utility. Recall Tribe's (1974) own description of the problem: "Policy analysts typically operate within a social, political, and intellectual tradition that regards the satisfaction of individual human wants as the only defensible measure of the good, a tradition that perceives the only legitimate task of reason to be that of consistently identifying and then serving individual appetite, preference, or desire." Tribe (1974) argues that when a conservation biologist feels compelled to frame the debate in this way, he is "helping to legitimate a system of discourse which so structures human thought and feeling as to erode, over the long run, the very sense of obligation which provided the initial impetus for his own protective efforts." Thus, to remove discussion of normative values from conservation is not an intellectually neutral decision. It is, rather, an intellectual commitment with enormous implications. "By treating human need and desire as the ultimate frame of reference, and by assuming that human goals and ends must be taken as externally given, . . . rather than generated by reason, environmental policy makes a value judgment of enormous significance. And once that judgment has been made, any claim for the continued existence of threatened wilderness areas or endangered species must rest on the identification of human wants and needs which would be jeopardized by a disputed development" (Tribe 1974). Only by continuing to articulate and explain the rationale for the preservation of biodiversity as a normative value in conservation do conservation biologists prevent the public debate from becoming an analysis of cost/benefit ratios, and from becoming a discussion framed entirely in terms of human needs and benefits.

Barry and Oelschlaeger (1996) express the dangers of what they perceive as a false "maturity" of conservation biology that separates normative values from scientific effort. They write, "to deserve its title conservation biology must be ethically overt—that is it must affirm its mission to be the protection of habitat and the preservation of biodiversity. Otherwise its name is as linguistically dishonest as the claim that 'war is peace' (Orwell 1949)." Specifically, without an evaluative component, there is no "conservation" in conservation biology. This lack of evaluation is, in fact, perfectly consistent with Murphy's (1990) earlier quoted definition of the field: "the application of classical scientific methodology to the conservation of biological diversity." Barry and Oelschlaeger (1996) note the dangers implicit in such a definition, arguing that "conservation biology is not just applied biology but rather hinges on an explicit evaluative judgment: Biodiversity is good and should be preserved. Apart from such a value judgment, one must wonder why there would be any reason to invest effort in conservation biology."

The choice of sides in this debate is not merely a matter of personal taste, but a determination of what conservation biology is, what issues its practitioners can and should address, and what constitutes "success" in conservation. If conservation biology should only supply correct information to policy makers and predict the outcomes of management actions, it bears no responsibility for those outcomes and it may legitimately define its success in terms of "grants received, awards won, dissertations completed, and papers published" (Barry and Oelschlaeger 1996). Conservation biology should, in this view, be understood and defined as a subdiscipline of ecology that provides management application without prescription. If, on the other hand, its purpose is to preserve biodiversity, then its success is defined by its ability to make persuasive management recommendations that save species from extinction. The grants, awards, dissertations, and papers may be means to this end, or accidental correlations with it, but they are not ends in themselves.

If conservation biology is to be defined as information driven, then its research is to be dedicated to needs defined by management and policy and to the consequences of management decisions. It need seek no normative outcomes. Further, it need not be overly concerned with interdisciplinary study because specialization will remain the most productive and efficient path to generate the greatest quantity and precision of information. On the other hand, if conservation biology is value driven, it must not only pursue research defined by management need and by the consequences of management decisions, but it must also engage the process of management itself, offering *recommendations* as well as *information*. In the latter case, interdisciplinary approaches become essential because conservation biologists would need to speak conversantly with the public and with other disciplines, to consider and take responsibility for societal outcomes, and to evaluate rigorously an array of such outcomes against different, sometimes conflicting, normative standards.

Conservationists rightly want their science to be taken seriously and regarded as truth, not propaganda. Many believe that, paradoxically, only a clear distinction between scientific effort and conservation activism can advance the goals of conservation activism because recommendations for conservation must be based on rigorous science (Brussard, Murphy, and Tracy 1994). The debate described here will not weaken, but intensify as its issues become more precisely defined. The outcome of this debate has profound implications for the design of conservation education. Whatever the choice, it must be an intentional one, not a slow, haphazard slip into one state or another. Barry and Oelschlaeger (1996) summarize the value of consciously examining the contextual and social values of scientific practice: "the practice of conservation biology is strengthened rather than weakened by examination of the underlying values upon which a consensus has been reached within the scientific community. The alternative . . . is that evaluative judgments remain hidden, outside the context of open discussion."

Synthesis

The trends apparent in conservation biology form a series of emerging and fascinating paradoxes. To save species, conservation biologists must become increasingly rigorous in experimental design to analyze correctly the causes of endangerment and decline. But, in so doing, conservation biologists must become increasingly sophisticated in their understanding and analyses of conservation values in order to determine what problems to solve and how to apply the solutions toward appropriate management objectives. The second paradox is that the success and growth of conservation biology as a discipline, manifested in the proliferation of graduate programs, the stature and respect of its journals, and the increase in its membership and funding, are forces that can alter the unique characteristics that inspired such success and can change the discipline's mission, goals, and engagement with social outcomes. The third paradox is that, even as social, political, and economic conditions are devel-

oping to create more pervasive, widely supported conservation efforts that reach beyond the professional community of scientists and managers, conservation biologists are debating whether to reduce their commitments to the normative values that have created this climate of support, and adopt a more familiar, traditional stance of "scientific objectivity" that defines their role solely as providers of information, not as informed advocates for correct conservation action.

Perhaps the most interesting paradox of all is that those engaged in studying conservation biology today must make choices in the present on the basis of assessments of an uncertain future. This last paradox is not new or unique. In every discipline, students are trained to deal with a professional world that may have changed radically by the time they enter it. Science requires a long apprenticeship. Students cannot afford to take the "catalog knows best" approach when choosing the courses that define their undergraduate education. The requirements for their major may be well designed, or not. Their curriculum may define almost everything, leaving little room for elective choice. Or it may define almost nothing, leaving the choices to the student, along with the responsibility for their outcome.

This chapter has taken the risk of offering some prescriptions and predictions because conservation is performed by conservationists, not by words in a textbook that magically assemble themselves into correct conservation actions. Predictions are necessary because the future must be considered, along with the past, to act wisely in the present. I encourage all students of conservation biology to consider what the future may hold for themselves and their discipline. We make the future every day by every daily choice, and our choices reflect our commitments to what we truly value. Personally, it is my belief that if conservation biologists disguise, or worse, grow hostile toward or ashamed of their field's historic normative commitment to preserve biodiversity, they will in time forget it. Then they will have their reward—a morally eviscerated, uninspired discipline of biological application indistinguishable from its predecessors.

Reflection upon a profession's normative values is an inherent, healthy element of scientific objectivity. It advises us of our own mission and perspective, so that we may articulate it to others rather than presume they should support it because we say so. Such reflection takes appropriate account of the social nature of scientific knowledge. If reflection on value and mission is lacking, then conservation biology will become "intellectually and functionally sterile and incapable of averting an anthropogenic mass extinction" (Barry and Oelschlaeger 1996).

In practice, conservation biology is inescapably normative. Otherwise, it ceases to be conservation. Fortunately, most conservation biologists practice and articulate conservation as if that were the case, whatever philosophical position they claim to defend. Thus, no less than 42 of the field's most eminent professionals agreed that "the goal of conservation should be to secure present and future options *by maintaining biological diversity at genetic, species, population and ecosystem levels; as a general rule neither the resource nor other components of the ecosystem should be perturbed beyond natural boundaries of variation*" (Mangel et al. 1996, emphasis mine). This appears to be a blatantly normative statement about a desired outcome, not simply an aspiration for a better understanding of why species decline and disappear. It is my hope that the coming generation of conservation biologists will clearly articulate and affirm the historic values and mission that birthed their discipline. Their decisions on such matters will shape the professional environment in which they will have to live and work. In defining what we value, we define what we shall be, professionally and personally.

What Kinds of Research Should Conservation Biologists Do?—A Directed Discussion

Reading assignment: Underwood, A. J. 1995. Ecological research and (and research into) environmental management. *Environmental Applications* 5:232–47.

Questions

1. Underwood asserts that managers too often set the research agenda by telling researchers what questions to answer, generating responses that Underwood describes as "confused and uncertain." Based on this course or your own experience, can you cite examples in which this seems to be true? Would conservation goals be more effectively achieved if researchers told managers what questions should be answered?

2. Underwood describes a second category of research in which researchers make specific tests of the results of management decisions. Drawing on this book or your own knowledge and experience, identify a management decision that could be subjected to an experimental test. Formulate a research hypothesis and experimental design. What effect would consistent use of this approach have on managers and management decisions in conservation?

3. Underwood advocates increased research into how management decisions are made, but notes that "this is not normally a scientific study." What assumptions do conservationists make by choosing not to systematically study the management decision process? Are these assumptions valid? What obstacles and objections might confront conservation biologists engaging in this type of research?

4. If Underwood is correct about the importance of communities of scientists concurrently pursuing all the categories of research he describes, how would this affect the design of an "ideal" curriculum in conservation biology? What changes would you make in your present curriculum that would better equip you to investigate any or all of these research categories?

Learning Online

Visit our webpage at www.mhhe.com/conservation for case studies, animations, practice quiz questions, and additional readings to help you understand the material in this chapter. You'll also find active links to the following topics:

Careers in Science
Environmental and Ecological
    Organization Sites
Individual Contributions to
    Environmental Issues
Environmental Organizations
Miscellaneous Environmental
    Resources

Introductory Materials and
    Governmental Sites
Utility and Organizational Sites
General Environmental Sites
General Ecology Sites
Ecosystem Management
Conservation and Management of
    Habitats and Species

Environmental Policy, Law and
    Planning
Environmental Ethics
Biodiversity

Literature Cited

Anderson, T. L., and D. R. Leal. 1997. *Enviro-capitalists: doing good while doing well*. Lanham, Md.: Rowman and Littlefield.

Baker, W. E. 1994. *Networking smart*. New York: McGraw Hill.

Barry, D., and M. Oelschlaeger. 1996. A science for survival: values and conservation biology. *Conservation Biology* 10:905–11.

Bowles, M. L., and C. J. Whelan, eds. 1994. *Restoration of endangered species: conceptual issues, planning, and implementation*. Cambridge, England: Cambridge University Press.

Brussard, P. F., D. D. Murphy, and C. R. Tracy. 1994. Cattle and conservation biology—another view. *Conservation Biology* 8:919–21.

Cannon, J. R., J. M. Dietz, and L. A. Dietz. 1996. Training conservation biologists in human interaction skills. *Conservation Biology* 10:1277–82.

Caughley, G. 1994. Directions in conservation biology. *Journal of Animal Ecology* 63:215–44.

Coleman, W. G., J. Mattice, and R. W. Brocksen. 1996. Soulé's conservation biology as a foundation for econometric ecosystem management. *Conservation Biology* 10:1494–99.

Ehrenfeld, D. 1992. The business of conservation. *Conservation Biology* 6:1–3.

Errington, P. L., and F. N. Hamerstrom. 1937. The evaluation of nesting losses in juvenile mortality of the ring-necked pheasant. *The Journal of Wildlife Management* 1:3–20.

Forester, D. J., and G. E. Machlis. 1996. Modeling human factors that affect the loss of biodiversity. *Conservation Biology* 10:1253–63.

Hart, R. 1996. *Effective networking for professional success*. London, England: Kogan Page.

Huenneke, L. F. 1995. Involving academic scientists in conservation research: perspectives of a plant ecologist. *Ecological Applications* 5:209–14.

Lee, K. N. 1993. *Compass and gyroscope: integrating science and politics for the environment*. Washington, D.C.: Island Press.

Mangel, M., and 41 co-authors. 1996. Principles for the conservation of wild living resources. *Ecological Applications* 6:338–62.

Murphy, D. D. 1990. Conservation biology and the scientific method. *Conservation Biology* 4:203–4.

Orwell, G. 1949. *1984*. New York: New American Library.

Pavlik, B. M. 1994. Demographic monitoring and the recovery of endangered plants. In *Restoration of endangered species: conceptual issues, planning, and implementation*, eds. M. L. Bowles and C. J. Whelan, 322–53. Cambridge, England: Cambridge University Press.

Pavlik, B. M., D. L. Nickrent, and A. M. Howard. 1993. The recovery of an endangered plant. I. Creating a new population of *Amsinckia grandiflora*. *Conservation Biology* 7:510–26.

Porter, W. F., and H. B. Underwood. 1999. Of elephants and blind men: deer management in the U.S. national parks. *Ecological Applications* 9:3–9.

Robinson, J. B. 1992. Of maps and territories: the use and abuse of socio-economic modeling in support of decision-making. *Technological Forecasting and Social Change* 42:147–64.

Saberwal, V. K., and A. Kothari. 1996. The human dimension in conservation biology curricula in developing countries. *Conservation Biology* 10:1328–31.

Soulé, M. E. 1985. What is conservation biology? *BioScience* 35:727–34.

Soulé, M. E. 1991. The "two point five society." *Conservation Biology* 5:255.

Touval, J. L., and J. M. Dietz. 1994. The problem of teaching conservation problem solving. *Conservation Biology* 8:902–4.

Tribe, L. H. 1974. Ways not to think about plastic trees. *Yale Law Journal* 83:1315–48.

Walker, B. H. 1992. Biodiversity and ecological redundancy. *Conservation Biology* 6:18–23.

Wigner, E. P. 1992. The recollections of Eugene P. Wigner as told to Andrew Szanton. New York: Plenum Press.

Yaffee, S. L. 1996. Ecosystem management in practice: the importance of human institutions. *Ecological Applications* 6:724–27.

GLOSSARY

A

acidification   The process through which the pH of surface fresh waters, especially lakes, declines because of inputs of acidic precipitation in the form of rain, snow, or fog.

age structure   The proportion of individuals in a population at each age, or in each age category.

alfisols   Soils with a shallow layer of humus, well-developed horizons, and an accumulation of clay in the subsoil.

Allee effect   The decrease in population growth rate in a small population, occurring when individuals do not often encounter potential mates.

allozyme electrophoresis   A molecular technique to assay genetic variation by separating enzyme proteins through their movement in a chemical medium (gel) to oppositely charged poles in an electric field.

alpha diversity   The diversity of species within a community (i.e., species richness).

alternative stable states   In lake systems, the potential for different conditions to prevail at similar nutrient levels, with rapid transitions occurring between states (e.g., from abundant submerged macrophytes in clear water, to dense phytoplankton and turbid water).

B

bequest value   The value of knowing that something is preserved for future generations.

beta diversity   A measure of the rate of change in species composition of communities across a landscape.

biodiversity   The entire array of earth's biological variety, contained in genes, populations, communities, and ecosystems.

bioremediation   The use of living organisms to remove toxic chemicals from soil or water.

C

capital   In neoclassical economics, the means of production.

cells   Discrete vertical zones in which water circulates in lakes as a result of wind activity, important in transporting matter and energy through the water column.

climax   The final, stable stage of ecological succession.

community   All populations occupying a given area at a particular time.

compression hypothesis

compression hypothesis   In the Equilibrium Theory of Island Biogeography, the concept that the addition of species to an island results in smaller niches occupied by each species.

connectedness   The presence of physical linkages between landscape elements.

connectivity   A parameter of landscape function that measures the processes by which subpopulations of organisms are interconnected into a functional demographic unit, achieved only if organisms actually move between connected units.

contingent valuation   Assigning economic values of nonmarket goods through analytical methods such as "willingness to pay" and "willingness to accept compensation."

corridor   A linear pathway that connects habitat patches and allows organisms to move among them.

D

declining-population paradigm   A body of concepts focusing on the deterministic processes responsible for population decline and how to mitigate or reverse threats to population persistence.

demographic stochasticity   Random fluctuations in birth and death rates, emigration and immigration, or sex ratio and age structure of a population.

deterministic factors   Factors that affect a population in a constant relation to the population's size.

E

ecological restoration   Repairing degraded ecosystems by reestablishing their original structures and functions.

ecological succession   A pattern of continuous, directional, nonseasonal change of plant populations on a site over time.

ecosystem   A biotic community interacting with its physical environment.

edge effects   A suite of processes and factors associated with edge environments, which become more pronounced when habitat is fragmented and the relative amount of core habitat decreases.

effective population size   The size of an "ideal" (randomly mating) population that would undergo the same amount of genetic drift as a particular real population.

endemism   The restriction of a species to a particular area or region, resulting in vulnerability to extinction through local changes in climate or land use.

393

endogenous disturbance   A disturbance generated from within an ecosystem, such as gap-phase replacement in a forest.

environmental stochasticity   Fluctuations in the probability of birth and death in a population because of temporal variation in habitat parameters, climatic variation, competitors, parasites, predators, and diseases.

epistasis   The interference with or suppression of the effect of a gene by a different gene, often manifested in the interactive effects of co-adapted genes.

Equilibrium Theory of Island Biogeography   Theory that relates the equilibrium number of species on an island to the size of the island and its distance from the mainland, also applied by conservation biologists to the design of nature reserves.

eutrophication   The process in which the release of nutrients, particularly phosphorus, into streams, lakes, or estuaries triggers a chain of events resulting in oxygen depletion, turbidity, and radical alteration of the biological community.

evolutionary species concept   The concept that a species is one lineage evolving separately from other lineages.

existence value   The value of knowing that something exists.

exogenous disturbance   A disturbance caused by forces outside an ecosystem (e.g., fire, flood, or climate change).

exponential population growth   A model in which population size increases at an ever-increasing rate, and only the population's size (N) and intrinsic rate of increase (r) determine the change in numbers of individuals.

externality   A cost that is not included in the price of a good whose production had negative effects, such as pollution or waste, but is instead shared by all who live in and use the system.

extinction   The termination of a population's existence, almost always preceded by population decline.

F

factor resolution   A population monitoring tool in which experiments are conducted to determine which factors actively limit population growth.

fecundity   The number of gametes produced per female per unit time.

filtering   In the Equilibrium Theory of Island Biogeography, a reduction in the number of species, genera, or higher taxonomic categories that are either unable to disperse from the mainland to an island or unable to find suitable habitat on the island.

fixation of harmful alleles   In a small population, an extreme condition in which all individuals possess only the harmful allele at a particular locus.

functional analogs   Species that play the same role in a community (members of the same functional type).

functional types   The various ecological roles (niches) in a community, also called "guilds" among animals and "life-forms" among plants.

G

gamma diversity   The rate of change of species composition with respect to distance in a landscape.

gap-phase replacement   The process through which gaps in the forest canopy and on the forest floor are created and subsequently filled with new species.

genetic drift   Random fluctuations in gene frequencies that occur as a result of nonrepresentative combinations of gametes during mating, especially in small populations.

genetic species concept   The separation of species according to genetic differences, such as restriction fragment length polymorphisms and amino acid sequence similarity.

genetic stochasticity   Fluctuations in demographic parameters, especially of small populations, through increased rates of inbreeding, genetic drift, and accumulation of unfavorable mutations.

H

habitat   The place, or type of place, in which a species can persist.

habitat fragmentation   The disruption of patterns or processes associated with habitats, including subdivision of large blocks of habitat into smaller, isolated blocks.

habitat generalists   Species that can exploit a variety of habitats in a given geographic range and, thus, are relatively invulnerable to extinction through habitat loss or land-use changes.

habitat heterogeneity   Differences in habitats, at a variety of spatial scales, which may be natural (rich internal structure of differing habitat patches) or artificial (fragmented habitats resulting from human activity).

habitat isolation   The separation of blocks of habitat from other blocks of similar habitat, a result of fragmentation.

habitat loss   Destruction, by human activities, of habitat for a particular species.

habitat specialists   Species that are typically highly successful in only one or a few types of habitat and, thus, are vulnerable to extinction through loss of the preferred habitat.

haplotypes   Mitochondrial DNA (mtDNA) groups that can be used in genetic techniques to determine rates of gene flow among populations.

hybridization   Mating between individuals of different species.

hydric soils   Soils that remain wet enough to support hydrophytic vegetation.

I

inbreeding   The mating of individuals with close relatives, with whom they may share many genes.

inbreeding coefficient   The probability that two alleles at the same locus in an individual are identical by descent.

inbreeding depression   A sequence of events initiated by matings between closely related individuals, especially in small populations of normally outbreeding species, whereby heterozygosity and fecundity are reduced and mortality is increased through expression of deleterious, recessive alleles.

indicator species   A species whose conservation status is assumed to reflect the status of other species with which it shares the community.

intermediate disturbance hypothesis   The prediction that biodiversity will be highest when disturbances are intermediate in intensity and frequency.

introgression   The long-term acquisition and incorporation of genetic material from one species into the genome of another species, especially when individuals of a rare species hybridize with those of a closely related, but more numerous (often, nonindigenous) species.

K

keystone species   A species, often a top predator, with strong effects on community or ecosystem processes and biodiversity.

L

landscape   An area that contains discrete, distinct habitat patches.

landscape pattern   The spatial arrangement of patches in a landscape.

langmuir rotations   Circulation patterns in lakes, established by prevailing winds and extending vertically throughout the water column.

lentic environment   A standing-water environment, such as a lake.

lethal equivalent   Deleterious, homozygous recessive gene combinations, resulting from inbreeding, which cause direct mortality in some individuals.

life table   A tabulation of age-specific rates of birth, mortality, survivorship, fecundity, and other population parameters, used to identify the traits of populations that determine patterns of growth over time.

liming   The direct addition of calcium carbonate to restore acidified lakes.

linear habitat   Long, narrow strips of habitat that provide resources and, potentially, pathways for movement.

logistic population growth   A model of population growth in which population size increases at a decreasing rate as it smoothly approaches an upper asymptote, set by environmental limits.

lotic environment   A flowing-water environment, such as a stream.

M

metapopulation   A population that exists as spatially disjunct subunits at different densities in habitat patches of varying carrying capacity.

metapopulation theory   A conceptual model to describe collections of subpopulations of a species in a given area, each occupying a suitable patch of habitat in a landscape of otherwise unsuitable habitat.

microsatellites   A type of satellite DNA that consists of short tandem repeats 2–4 nucleotides long, and whose variability is useful in determining pedigrees of individuals.

minimum viable population (MVP)   The minimum number of individuals required for a population to persist for a specified length of time, at a specified probability level.

minisatellites   A type of satellite DNA that consists of sequences up to 100 base pairs long, and whose variability forms the basis for DNA fingerprinting.

mitigation   In restoration ecology, lessening the effects of human disturbances.

multiple-use module   An approach to habitat conservation in which a fully protected core area is surrounded by concentric zones of natural areas used in progressively more intense fashion for recreation and commodity production.

mutational meltdown   The cyclic accumulation of deleterious mutations, declines in fitness, and declines in population size, which can lead to extinction of small populations.

N

natural catastrophes   Extreme forms of normal environmental variation (e.g., flash flooding or severe and prolonged drought) that have the potential to eliminate all individuals in a small population.

no surface occupancy   A method of mitigation, in which mining and drilling operations extract resources from beneath high-quality habitat, while drilling on a remote, low-quality habitat.

nonexclusive goods and services   Goods and services for which it would be extremely difficult and costly to exclude anyone from receiving the benefits (e.g., protection from ultraviolet radiation by the ozone layer).

nonindigenous species   Species introduced beyond their native ranges.

nonrival goods and services   Goods and services whose use by one person does not reduce or restrict use by others (e.g., breathing oxygen from the atmosphere).

O

occupancy model   Another name for the Levins' metapopulation model, based on its assumption that patches are either occupied (at carrying capacity) or unoccupied (no individuals in the patch).

oligonucleotide   A short piece of DNA used in the polymerase chain reaction.

optimal niche gestalt   An approach to habitat management, based on the idea that identifiable structural features of an environment allow a species to thrive, rather than merely persist.

option value   The value of a resource's expected future use, or what a person would be willing to pay to guarantee that the resource would be available for future use.

outbreeding depression   The decline in fitness that may occur when individuals from normally inbreeding populations breed with individuals from other populations of the same species, breaking up uniquely coadapted genetic combinations.

P

panmictic index   A measure of inbreeding as a deviation from the heterozygosity frequency expected under random mating.

patch   The fundamental unit of a landscape, containing only one type of habitat.

patchiness   A quality of habitat arrangement, manifested as contrasting, discrete states of physical or biotic phenomena.

pedigree inbreeding   Inbreeding by descent, or the measure of an individual's ancestry shared in its maternal and paternal lines.

phenotype   The attributes of individuals, determined by genotype, environment, and their interaction.

phytoremediation   The use of plants to remove toxic chemicals, especially metallic ions and compounds, from soil.

phytostabilization   The process in which plants release chemicals into the soil and render toxins inert.

polymerase chain reaction (PCR)   A genetic technique that uses DNA polymerase to repeatedly copy (amplify) a short region of a DNA molecule for various types of analysis, such as direct sequence of the PCR products to determine genotypes of individual animals.

pool   Prominent stream habitat feature found above and below riffles, which forms where stream gradient decreases or volume of the stream channel increases, and biomass accumulates.

population   A group of individuals of the same species that is spatially, genetically, or demographically discontinuous with other groups.

population bottleneck   A drastic, temporary reduction in population size through catastrophe or dispersal of individuals to a new area, resulting in loss of genetic variation.

population viability analysis (PVA)   The use of analytical or simulation models to make precise estimates of the likelihood of species persistence within a defined time period at a given level of probability.

Q

quasi-option value   The value of preserving options, given an expectation of growth in knowledge that might lead to a future, but as-yet undiscovered or unrealized use of the resource.

R

random amplified polymorphic DNA (RAPD) analysis   A genetic technique that requires only a small amount of material from a living creature (e.g., discarded hair, feathers, antlers, or eggshells) and uses one random oligonucleotide primer to generate essentially unlimited numbers of loci for analysis.

reaction-diffusion model   A type of species invasion model in which populations travel as a wave of a given velocity (V) determined by the population's intrinsic rate of increase (r) and rate of movement (D).

reclamation   The preparation and enhancement of degraded land to fulfill its former use or a new use.

recruitment   The entry of young organisms into the adult population.

rehabilitation   The process of aesthetically enhancing degraded ecosystems.

restoration ecology   The science of restoring degraded habitats by identifying and mitigating processes that caused habitat degradation, establishing restoration goals and measures of success, and applying methods to achieve them.

restriction fragment length polymorphisms (RFLPs)   Variations in the length of restriction fragments produced from identical regions of the genome, which can be used to measure variation in nuclear DNA among individuals.

revegetation   The process in which plants (but not necessarily the original vegetation) recolonize a disturbed site from which a disturbance removed all or part of the vegetation.

riffle   Prominent stream habitat feature associated with turbulence caused by relatively rapid stream flow over an uneven bottom, dominated by periphyton and characterized by primary production.

rule-based model   A type of model used to evaluate possible mechanisms of distributional changes in species.

S

satellite DNA   Short, highly repetitive segments of DNA in an organism's genome with base sequences differing from those of other forms of DNA.

scientific paradigm   Within a scientific community, a way of thinking that determines what is studied and how it may be studied.

seiche   Recurring, rocking water movement pattern within a lake, generated by prevailing winds or by differences in density of water in different layers.

sex ratio   The ratio of males to females in a population.

sink   In a metapopulation, an area of low-quality habitat in which the population cannot replace itself without immigration.

small-population paradigm   A general theory of the characteristics of small populations, causes and effects of threats to small populations, and how those threats can be managed.

source   In a metapopulation, an area of high-quality habitat in which population surpluses are produced.

spatial scale   A measure of habitat patchiness that relates interpatch distance to a species' dispersal ability.

spatially explicit model   A metapopulation model that incorporates differing degrees of connectedness between population subunits and features "localized interactions."

spatially implicit model   A metapopulation model (e.g., Levins' model) in which habitat patches and local populations are discrete, but equally connected with one another.

spatially realistic model   A metapopulation model that incorporates the specific geometry (e.g., size, shape, and arrangement) of particular patches.

species-abundance curve   The graphical representation of a species-area relationship.

species-area relationship   The basic equation of island biogeography, $S = cA^z$, which describes the number of species on an island (S) as a constant power of the island's area (A), mediated by two constants, c and z.

species richness   The number of species present in a community.

steady-state economy   As defined by Herman Daly, an economy that develops (changes qualitatively) but does not grow in population or stocks of wealth.

stepping-stone islands   In the Equilibrium Theory of Island Biogeography, smaller islands that lie between a recipient island and its mainland colonizing source.

stochastic factors   Factors whose effects on a population vary randomly, but usually within a limited range.

stratified diffusion model   A type of species invasion model that incorporates long-distance dispersal and density-dependent rates of spread.

structure-based indicators   Indexes of changes in biodiversity calculated through assessment of changes in ecological structure, such as forest stand complexity or foliage height diversity.

T

temporal scale   The duration of a habitat relative to a species' generation time.

timing limitations   A method of mitigation, whereby seasonally occupied habitats are used for some human activities when animals are absent.

trend analysis   A population monitoring tool for calculating one or more specific demographic variables in a population and, from such analysis, determines whether the population is growing, stable, or declining.

typological species concept   The notion that species are distinguished by morphological characteristics that can be determined by gross observation.

U

umbrella species A species, or group of species (e.g., large mammals), of particular conservation or public interest, and whose protection will benefit many species in other taxa.

use value The economic value derived from the actual use of a resource.

V

veligers Free-swimming, filter-feeding larvae of mussels, such as the zebra mussel (*Dreissena polymorpha*).

visual minimization A method of mitigation that involves reducing the distance at which animals can see objects associated with human disturbance.

W

wetlands Ecosystems in which the water table is at or near the surface, or the land is covered by shallow water, and characterized by high species richness.

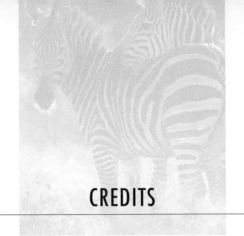

CREDITS

Chapter 1
Figure 1.1: Geoffrey Clements/ CORBIS; 1.2: Underwood and Underwood/CORBIS; 1.3: Photo by Rockwood Photo Co., 1903 Library of Congress, Prints and Photographs Division # cph 3q09872; 1.4 USDA Forest Service, Grey Towers, N. H. L., Milford, PA; 1.5: University of Wisconsin Archives; 1.6: Erich Hartmann/Magnum; 1.7a: ©Richard Baetson, courtesy of USFWS; 1.7b: Dave Menke; 1.7c: Courtesy Don Klosterman; Figure 1.8: From Andrewartha & Birch, THE DISTRIBUTION & ABUNDANCE OF ANIMALS. Copyright 1954. Reprinted with permission of The University of Chicago Press.

Chapter 2
Figure 2.1: ©C. C. Lockwood/Animals Animals; 2.2: ©Ann B. Swengel/ Visuals Unlimited; 2.4: ©BIOS (S. Dawson)/Peter Arnold. Figure 2.3: From Weiss, Edith Brown and Harold K. Jacobson, editors, ENGAGING COUNTRIES: STRENGTHENING COMPLIANCE WITH INTERNATIONAL ENVIRONMENTAL ACCORDS. Copyright 1998. Reprinted by permission of MIT Press.

Chapter 3
Figure 3.2: ©Bud Nielsen/Visuals Unlimited; 3.4: ©Frank Hanna/Visuals Unlimited; Table 3.3: From TOWARD UNITY AMONG ENVIRONMENTALISTS, First Edition by Bryan G. Norton. Copyright 1991. Reprinted by permission of Oxford University Press.

Chapter 4
Table 4.9: From N. Myers, ENVIRONMENTALIST 8(3). Copyright 1998. Reprinted with permission of Kluwer Academic Publishers; Figure 4.10: ©Breck P. Kent/Animals Animals; 4.11: ©Dani/Jeske/Earth Scenes; 4.14: ©Leo Keeler/Animals Animals. Figure 4.12: Figure 1, p. 144 from "Mapping Biodiversity Value Worldwide" by Williams et al. from PROCEEDINGS OF THE ROYAL SOCIETY OF LONDON B264. Copyright 1997. Reprinted with permission of The Royal Society.

Chapter 5
Figure 5.3: From CONSERVATION AND EVOLUTION by O. H. Frankel & M. E. Soule. Copyright 1981. Reprinted with the permission of Cambridge University Press; Figure 5.4: ©A. & M. Shah/Animals Animals; 5.18: Courtesy of Bob Paty; 5.20: ©Michael Gadomski/Earth Scenes; 5.21: ©Bill Kamin/Visuals Unlimited; 5.22: Eric and David Hosking/CORBIS; 5.24: ©John Sohlden/Visuals Unlimited; 5.25: ©Richard Hermann/Visuals Unlimited. Table 5.2: Data from BIOLOGICAL CONSERVATION, Vol. 29, Pages 27–46. Copyright 1984. Reprinted with permission from Elsevier Science. Figure 5.13: From ECOLOGY: CONCEPTS AND APPLICATIONS by M. Molles. Copyright 2002. Reprinted with permission of McGraw-Hill Education; Figure 5.19: From METAPOPULATIONS AND WILDLIFE CONSERVATION, edited by D. R. McCullough. Copyright 1996. Reprinted by permission of Island Press.

Chapter 6
Figure 6.1: Courtesy of Shawn Stewart; 6.3a: ©Robert M. Plowes; 6.3b: John Cancalosi/DRK Photo; 6.7: ©Jane McAlonan/Visuals Unlimited; 6.9: ©Zig Leszczynski/Visuals Unlimited. Figure 6.3: From AMERICAN JOURNAL OF BOTANY by Dolan et al. Copyright 1999. Reprinted with permission; Figure 6.4: Reprinted with permission from SCIENCE, Vol. 266, October 14, 1994 by Jimenez et al. Copyright 1994 American Association for the Advancement of Science.

Chapter 7
Figure 7.1: ©Darren Bennett/Animals Animals; 7.4: Courtesy ANSP, Steven Holt/Stockpix.com; 7.6: ©John Gerlach/ Visuals Unlimited; 7.9: Pam Gardner/ CORBIS; 7.10: ©Klaus Uhlenhut/ Animals Animals. Table 7.1: From ECOLOGY, Vol. 65, pp. 1617–1628 by P. W. Sherman and M. L. Morton. Copyright 1984. Reprinted with permission of the Ecological Society of America.

Chapter 8
Figure 8.1 Courtesy of Shawn Stewart; 8.4: ©David Welling/Earth Scenes; 8.5: ©Bill Beatty/Visuals Unlimited; 8.11: ©John Cancalosi/Peter Arnold. Figure 8.7: From "Relative Effects of Habitat Loss and Fragmentation on Population Extinction" by L. Fahrig in JOURNAL OF WILDLIFE MANAGEMENT, Vol. 61, pp. 603–610. Copyright 1997. Reprinted with permission; Figure 8.16: From "Response of Elk to Installation of Oil Wells" by F. Van Dyke and W. C. Klein. JOURNAL OF MAMMOLOGY, Vol. 77, pp. 1028–1041 Copyright 1995. Reprinted with permission of Alliance Communications Group.

Chapter 9
Figure 9.2: ©HBOI/Visuals Unlimited; 9.4: ©Al Grotell 1988; 9.14: WHOI/J. Edmond; 9.15: ©Paul J. Auster; 9.22: ©David B. Fleetham/Visuals Unlimited; Figure 9.10: From "Multiplicity of Stable States in Freshwater Systems" by M. Sheffer, HYDROBIOLOGIA, Vol. 200/201, pp. 475–486. Copyright 1990. Reprinted with kind permission from Kluwer Academic Publishers; Figure 9.18: From GLOBAL BIODIVERSITY STRATEGY by WRI, International Union for the Conservation of Nature. Copyright 1992 by The World Resources Institute. Reprinted with permission.

Chapter 10
Figure 10.4: Courtesy of USGS; 10.5a: From Vernon Bailey, Animal Life of Yellowstone National Park (1930) Courtesy of Charles C Thomas. Publisher, Springfield, IL; 10.5b: Courtesy of Charles E. Kay (photo no. 2895-25); 10.13 : © Kevin Schafer/CORBIS; Figure10.6: From Molles, ECOLOGY — CONCEPTS AND APPLICATIONS, Second Edition. Copyright 2002. Reprinted with permission of The McGraw-Hill Companies.

Chapter 11
Figure 11.2: Reprinted with permission and adapted from SCIENCE, TECHNOLOGY AND PUBLIC POLICY. Copyright 1992 by the National Academy of Sciences. Courtesy of the National Academy Press, Washington, DC; Figure 11.5: Courtesy of Patrick Pierson; 11.8: ©Jamie D. Schmeling; 11.10: Marty/Cordano/DRK Photo; 11.12: This graphic is used with permission of the Government of Canada and the United States Environmental Protection Agency, and is taken from their 1995 publication "The Great Lakes: An Environmental Atlas and Resource Book."; 11.13: Courtesy of David D. Hoekman; 11.19: Sal Dimarco, Jr./TimePix; 11.20: ©Brian Parker/Tom Stack & Associates; Figure 11.06: From ECONOMIC BOTANY: PLANTS IN OUR WORLD by Simpson & Ogorzaly. Copyright 2001. Reprinted with permission of The McGraw-Hill Companies; Figure 11.9: From BIOSCIENCE, Vol. 48 by Molles et al. Copyright 1998 by American Institute of Biological Sciences. Reproduced with permission of American Institute of Biological Sciences via Copyright Clearance Center.

Chapter 12
Figure 12.8 Bettmann/CORBIS.

Chapter 13
Figure 13.1: Courtesy of Northwestern College; 13.2: Courtesy of Northwestern College; 13.5: Courtesy of Glendive Chamber of Commerce & Agriculture, Glendive, MT.

INDEX